D1700169

The chemistry of
Organomagnesium Compounds

Patai Series: The Chemistry of Functional Groups

A series of advanced treatises founded by Professor Saul Patai and under the general editorship of Professor Zvi Rappoport

The **Patai Series** publishes comprehensive reviews on all aspects of specific functional groups. Each volume contains outstanding surveys on theoretical and computational aspects, NMR, MS, other spectroscopic methods and analytical chemistry, structural aspects, thermochemistry, photochemistry, synthetic approaches and strategies, synthetic uses and applications in chemical and pharmaceutical industries, biological, biochemical and environmental aspects.
To date, 118 volumes have been published in the series.

Recently Published Titles

The chemistry of the Cyclopropyl Group (Volume 2)
The chemistry of the Hydrazo, Azo and Azoxy Groups (Volume 2, 2 parts)
The chemistry of Double-Bonded Functional Groups (Volume 3, 2 parts)
The chemistry of Organophosphorus Compounds (Volume 4)
The chemistry of Halides, Pseudo-Halides and Azides (Volume 2, 2 parts)
The chemistry of the Amino, Nitro and Nitroso Groups (2 volumes, 2 parts)
The chemistry of Dienes and Polyenes (2 volumes)
The chemistry of Organic Derivatives of Gold and Silver
The chemistry of Organic Silicon Compounds (2 volumes, 4 parts)
The chemistry of Organic Germanium, Tin and Lead Compounds (Volume 2, 2 parts)
The chemistry of Phenols (2 parts)
The chemistry of Organolithium Compounds (2 volumes, 3 parts)
The chemistry of Cyclobutanes (2 parts)
The chemistry of Peroxides (Volume 2, 2 parts)
The chemistry of Organozinc Compounds (2 parts)
The chemistry of Anilines (2 parts)

Forthcoming Titles

The Chemistry of Hydroxylamines, Oximes and Hydroxamic Acids
The Chemistry of Metal Enolates

The Patai Series Online

Starting in 2003 the **Patai Series** is available in electronic format on Wiley InterScience. All new titles will be published as online books and a growing list of older titles will be added every year. It is the ultimate goal that all titles published in the **Patai Series** will be available in electronic format.
For more information see under **Online Books** on:

www.interscience.wiley.com

$R-Mg$

The chemistry of
Organomagnesium Compounds

Part 1

Edited by

ZVI RAPPOPORT

The Hebrew University, Jerusalem

and

ILAN MAREK

Technion-Israel Institute of Technology, Haifa

2008

John Wiley & Sons, Ltd

An Interscience® Publication

Copyright © 2008   John Wiley & Sons Ltd, The Atrium, Southern Gate, Chichester,
West Sussex PO19 8SQ, England

Telephone (+44) 1243 779777

Email (for orders and customer service enquiries): cs-books@wiley.co.uk
Visit our Home Page on www.wileyeurope.com or www.wiley.com

All Rights Reserved. No part of this publication may be reproduced, stored in a retrieval system or transmitted in any form or by any means, electronic, mechanical, photocopying, recording, scanning or otherwise, except under the terms of the Copyright, Designs and Patents Act 1988 or under the terms of a licence issued by the Copyright Licensing Agency Ltd, 90 Tottenham Court Road, London W1T 4LP, UK, without the permission in writing of the Publisher. Requests to the Publisher should be addressed to the Permissions Department, John Wiley & Sons Ltd, The Atrium, Southern Gate, Chichester, West Sussex PO19 8SQ, England, or emailed to permreq@wiley.co.uk, or faxed to (+44) 1243 770620.

Designations used by companies to distinguish their products are often claimed as trademarks. All brand names and product names used in this book are trade names, service marks, trademarks or registered trademarks of their respective owners. The Publisher is not associated with any product or vendor mentioned in this book. This publication is designed to provide accurate and authoritative information in regard to the subject matter covered. It is sold on the understanding that the Publisher is not engaged in rendering professional services. If professional advice or other expert assistance is required, the services of a competent professional should be sought.

Other Wiley Editorial Offices

John Wiley & Sons Inc., 111 River Street, Hoboken, NJ 07030, USA

Jossey-Bass, 989 Market Street, San Francisco, CA 94103-1741, USA

Wiley-VCH Verlag GmbH, Boschstr. 12, D-69469 Weinheim, Germany

John Wiley & Sons Australia Ltd, 42 McDougall Street, Milton, Queensland 4064, Australia

John Wiley & Sons (Asia) Pte Ltd, 2 Clementi Loop #02-01, Jin Xing Distripark, Singapore 129809

John Wiley & Sons Canada Ltd, 6045 Freemont Blvd, Mississauga, Ontario, L5R 4J3, Canada

Wiley also publishes its books in a variety of electronic formats. Some content that appears in print may not be available in electronic books.

British Library Cataloguing in Publication Data

A catalogue record for this book is available from the British Library

ISBN 978-0-470-05719-3

Typeset in 9/10pt Times by Laserwords Private Limited, Chennai, India
Printed and bound in Great Britain by Biddles Ltd, King's Lynn
This book is printed on acid-free paper responsibly manufactured from sustainable forestry in which at least two trees are planted for each one used for paper production.

Dedicated to the memory of

Yair Avni

Contributing authors

Jaap Boersma	Chemical Biology & Organic Chemistry, Faculty of Science, Utrecht University, Padualaan 8, 3584 CH Utrecht, The Netherlands
Katja Brade	Department Chemie und Biochemie, Ludwig-Maximilians-Universität München, Butenandtstr., 5-13, D-81377 München, Germany. Fax: +49-89-2180-77680
Gérard Cahiez	Laboratoire de Synthèse Organique Sélective et de Chimie Organométallique (SOSCO), UMR 8123 CNRS-ESCOM-UCP, 5 Mail Gay Lussac, Neuville s/Oise, F-95092 Cergy-Pontoise, France. Fax: +3-313-425-7383; e-mail: g.cahiez@escom.fr
Christophe Duplais	Laboratoire de Synthèse Organique Sélective et de Chimie Organométallique (SOSCO), UMR 8123 CNRS-ESCOM-UCP, 5 Mail Gay Lussac, Neuville s/Oise, F-95092 Cergy-Pontoise, France. Fax: +3-313-425-7383
Ben L. Feringa	Stratingh Institute for Chemistry, University of Groningen, Nijenborgh 4, 9747 AG, Groningen, The Netherlands. Fax: ++3-150-363-4296; e-mail: B.L.Feringa@rug.nl
Andrey Gavryushin	Department Chemie und Biochemie, Ludwig-Maximilians-Universität München, Butenandtstr., 5-13, D-81377 München, Germany. Fax: +49-89-2180-77680
Claude Grison	UMR CNRS-Université de Montpellier 2 5032, ENSCM, 8 rue de l'Ecole Normale, F-34296 Montpellier, France. Fax: +3-346-714-4342; e-mail: cgrison@univ-montp2.fr
Peter J. Heard	School of Science and Technology, North East Wales Institute, Mold Road, Wrexham LL112AW, UK; e-mail: p.heard@newi.ac.uk
Kenneth W. Henderson	Department of Chemistry and Biochemistry, 251 Nieuwland Science Hall, University of Notre Dame, Notre Dame, IN 46556, USA. Fax: +1-574-631-6652; e-mail: khenders@nd.edu
Katherine L. Hull	Department of Chemistry and Biochemistry, 251 Nieuwland Science Hall, University of Notre Dame, Notre Dame, IN 46556, USA. Fax: +1-574-631-6652

Torkil Holm	Department of Chemistry, Technical University of Denmark, Building 201, DK-2800, Lyngby, Denmark. Fax: +45-4-593-3968; e-mail: th@kemi.dtu.dk
Kenichiro Itami	Department of Chemistry and Research Center for Materials Science, Nagoya University, Chikusa-ku, Nagoya 464-8602, Japan. Fax: +8-152-788-6098; e-mail: itami@chem.nagoya-u.ac.jp
Johann T. B. H. Jastrzebski	Chemical Biology & Organic Chemistry, Faculty of Science, Utrecht University, Padualaan 8, 3584 CH Utrecht, The Netherlands. Fax: +3-130-252-3615; e-mail: j.t.b.h.jastrzebski@uu.nl
Jan S. Jaworski	Faculty of Chemistry, Warsaw University, 02-093 Warszawa, Poland. Fax: +4-822-822-5996; e-mail: jaworski@chem.uw.edu.pl
Gerard van Koten	Chemical Biology & Organic Chemistry, Faculty of Science, Utrecht University, Padualaan 8, 3584 CH Utrecht, The Netherlands. Fax: +3-130-252-3615; email: g.vankoten@uu.nl
Paul Knochel	Department Chemie und Biochemie, Ludwig-Maximilians-Universität München, Butenandtstr., 5-13, D-81377 München, Germany. Fax: +49-89-2180-77680; e-mail: paul.knochel@cup.uni-muenchen.de
Joel F. Liebman	Department of Chemistry and Biochemistry, University of Maryland, Baltimore County, 1000 Hilltop Circle, Baltimore, Maryland 21250, USA. Fax: +1-410-455-2608; e-mail: jliebman@umbc.edu
Fernando López	Departamento de Química Orgánica, Facultad de Química, Universidad de Santiago de Compostela, Avda. de las ciencias, s/n, 15782, Santiago de Compostela, Spain; e-mail: qofer@usc.es
Adriaan J. Minnaard	Stratingh Institute for Chemistry, University of Groningen, Nijenborgh 4, 9747 AG, Groningen, The Netherlands. Fax: ++3-150-363-4296; e-mail: A.J.Minnaard@rug.nl
Richard A. J. O'Hair	School of Chemistry, The University of Melbourne, Victoria 3010, Australia; Bio21, Molecular Science and Biotechnology Institute, The University of Melbourne, Victoria, 3010, Australia; ARC Centre of Excellence for Free Radical Chemistry and Biotechnology, Australia. Fax: +6-139-347-5180; e-mail: rohair@unimelb.edu.au
Koichiro Oshima	Department of Material Chemistry, Graduate School of Engineering, Kyoto University, Kyoto-daigaku Katsura, Nishikyo, Kyoto 615-8510, Japan. Fax: +8-175-383-2438; e-mail: oshima@orgrxn.mbox.media.kyoto-u.ac.jp
Mathias O. Senge	School of Chemistry, SFI Tetrapyrrole Laboratory, Trinity College Dublin, Dublin 2, Ireland. Fax: +3-531-896-8536; e-mail: sengem@tcd.ie.

Natalia N. Sergeeva	School of Chemistry, SFI Tetrapyrrole Laboratory, Trinity College Dublin, Dublin 2, Ireland. Fax: +3-531-896-8536
Tsuyoshi Satoh	Department of Chemistry, Faculty of Science, Tokyo University of Science; Ichigayafunagawara-machi 12, Shinjuku-ku, Tokyo 162-0826, Japan. Fax: 8-135-261-4631; e-mail: tsatoh@rs.kagu.tus.ac.jp
Suzanne W. Slayden	Department of Chemistry, George Mason University, 4400 University Drive, Fairfax, Virginia 22030, USA. Fax: +1-703-993-1055; e-mail: sslayden@gmu.edu
James Weston	Institut für Organische Chemie und Makromolekulare Chemie, Friedrich-Schiller-Universität, Humboldtstraße 10, D-07743 Jena, Germany. Fax: +49(0)-36-419-48212; e-mail: c9weje@uni-jena.de
Shinichi Yamabe	Department of Chemistry, Nara University of Education, Takabatake-cho, Nara, 630-8528, Japan. Fax: +81-742-27-9208; e-mail: yamabes@nara-edu.ac.jp
Shoko Yamazaki	Department of Chemistry, Nara University of Education, Takabatake-cho, Nara, 630–8528, Japan. Fax: +81-742-27-9289; e-mail: yamazaks@nara-edu.ac.jp
Hideki Yorimitsu	Department of Material Chemistry, Graduate School of Engineering, Kyoto University, Kyoto-daigaku Katsura, Nishikyo, Kyoto 615-8510, Japan. Fax: +81-75-383-2438; e-mail: yori@orgrxn.mbox.media.kyoto-u.ac.jp
Jun-Ichi Yoshida	Department of Synthetic Chemistry and Biological Chemistry, Graduate School of Engineering, Kyoto University, Nishikyo-ku, Kyoto 615-8510, Japan. Fax: +81-75-383-2727; e-mail: yoshida@sbchem.kyoto-u.ac.jp
Jacob Zabicky	Department of Chemical Engineering, Ben-Gurion University of the Negev, P. O. Box 653, Beer-Sheva 84105, Israel. Fax: +9-72-8647-2969; e-mail: zabicky@bgu.ac.il

Foreword

The present book, *The Chemistry of Organomagnesium Compounds*, is a continuation of the sub-group of volumes in 'The Chemistry of Functional Groups' series that deals with organometallic derivatives. Closely related to it are the two volumes, *The Chemistry of Organolithium Compounds* (Zvi Rappoport and Ilan Marek, Eds., 2003 and 2005) in three parts and the two parts of *The Chemistry of Organozinc Compounds* (Zvi Rappoport and Ilan Marek, Eds., 2006). Organomagnesium (or Grignard) reagents play a key role in organic chemistry. Although considered as one of the oldest organometallic reagents in synthesis, there have been a complete renaissance of the field in the last decade.

The two parts of the present volume contain 17 chapters written by experts from 11 countries. They include chapters dealing with structural chemistry, thermochemistry and NMR of organomagnesium compounds, formation of organomagnesium compounds in solvent-free environment, photochemistry of magnesium derivatives of porphyrins and phthalocyanines, and electrochemistry, analysis and biochemistry of organomagnesium derivatives. Special chapters are devoted to special families of compounds, such as magnesium enolates, ate-complexes, carbenoids and bonded-complexes with groups 15 and 16 compounds. Processes such as enantioselective copper-catalyzed 1,4-addition of organomagnesium halides, the iron-catalyzed reactions of Grignard reagents, and theoretical aspects of their addition to carbonyl compounds as well as carbomagnesiation reactions are covered in separate chapters. Both synthesis and reactivities of organomagnesium compounds are extensively discussed.

Unfortunately, the planned chapter on 'Theoretical Aspects of Organomagnesium Compounds' was not delivered. However, some theoretical aspects are covered in other chapters, especially Chapter 9. Another chapter on 'Mechanisms of Reactions of Organomagnesium Compounds' was not included after it was found that recent material on the topic was meager as compared with the coverage of the topic in Richey's book *Grignard Reagents, New Developments*, published in 2000. We gratefully acknowledge the contributions of all the authors of these chapters.

The literature coverage is mostly up to and sometimes including 2007.

We will be grateful to readers who draw our attention to any mistakes in the present volume or to omissions, and to new topics which deserve to be included in a future volume on organomagnesium compounds.

Jerusalem and Haifa	Zvi Rappoport
November 2007	Ilan Marek

The Chemistry of Functional Groups
Preface to the series

The series 'The Chemistry of Functional Groups' was originally planned to cover in each volume all aspects of the chemistry of one of the important functional groups in organic chemistry. The emphasis is laid on the preparation, properties and reactions of the functional group treated and on the effects which it exerts both in the immediate vicinity of the group in question and in the whole molecule.

A voluntary restriction on the treatment of the various functional groups in these volumes is that material included in easily and generally available secondary or tertiary sources, such as Chemical Reviews, Quarterly Reviews, Organic Reactions, various 'Advances' and 'Progress' series and in textbooks (i.e. in books which are usually found in the chemical libraries of most universities and research institutes), should not, as a rule, be repeated in detail, unless it is necessary for the balanced treatment of the topic. Therefore each of the authors is asked not to give an encyclopaedic coverage of his subject, but to concentrate on the most important recent developments and mainly on material that has not been adequately covered by reviews or other secondary sources by the time of writing of the chapter, and to address himself to a reader who is assumed to be at a fairly advanced postgraduate level.

It is realized that no plan can be devised for a volume that would give a complete coverage of the field with no overlap between chapters, while at the same time preserving the readability of the text. The Editors set themselves the goal of attaining reasonable coverage with moderate overlap, with a minimum of cross-references between the chapters. In this manner, sufficient freedom is given to the authors to produce readable quasi-monographic chapters.

The general plan of each volume includes the following main sections:

(a) An introductory chapter deals with the general and theoretical aspects of the group.

(b) Chapters discuss the characterization and characteristics of the functional groups, i.e. qualitative and quantitative methods of determination including chemical and physical methods, MS, UV, IR, NMR, ESR and PES—as well as activating and directive effects exerted by the group, and its basicity, acidity and complex-forming ability.

(c) One or more chapters deal with the formation of the functional group in question, either from other groups already present in the molecule or by introducing the new group directly or indirectly. This is usually followed by a description of the synthetic uses of the group, including its reactions, transformations and rearrangements.

(d) Additional chapters deal with special topics such as electrochemistry, photochemistry, radiation chemistry, thermochemistry, syntheses and uses of isotopically labeled compounds, as well as with biochemistry, pharmacology and toxicology. Whenever applicable, unique chapters relevant only to single functional groups are also included (e.g. 'Polyethers', 'Tetraaminoethylenes' or 'Siloxanes').

This plan entails that the breadth, depth and thought-provoking nature of each chapter will differ with the views and inclinations of the authors and the presentation will necessarily be somewhat uneven. Moreover, a serious problem is caused by authors who deliver their manuscript late or not at all. In order to overcome this problem at least to some extent, some volumes may be published without giving consideration to the originally planned logical order of the chapters.

Since the beginning of the Series in 1964, two main developments have occurred. The first of these is the publication of supplementary volumes which contain material relating to several kindred functional groups (Supplements A, B, C, D, E, F and S). The second ramification is the publication of a series of 'Updates', which contain in each volume selected and related chapters, reprinted in the original form in which they were published, together with an extensive updating of the subjects, if possible, by the authors of the original chapters. Unfortunately, the publication of the 'Updates' has been discontinued for economic reasons.

Advice or criticism regarding the plan and execution of this series will be welcomed by the Editors.

The publication of this series would never have been started, let alone continued, without the support of many persons in Israel and overseas, including colleagues, friends and family. The efficient and patient co-operation of staff-members of the Publisher also rendered us invaluable aid. Our sincere thanks are due to all of them.

The Hebrew University	SAUL PATAI
Jerusalem, Israel	ZVI RAPPOPORT

Sadly, Saul Patai who founded 'The Chemistry of Functional Groups' series died in 1998, just after we started to work on the 100th volume of the series. As a long-term collaborator and co-editor of many volumes of the series, I undertook the editorship and I plan to continue editing the series along the same lines that served for the preceding volumes. I hope that the continuing series will be a living memorial to its founder.

The Hebrew University	ZVI RAPPOPORT
Jerusalem, Israel	
May 2000	

Contents

1. Structural organomagnesium chemistry — 1
   Johann T. B. H. Jastrzebski, Jaap Boersma and Gerard van Koten

2. The thermochemistry of organomagnesium compounds — 101
   Joel F. Liebman, Torkil Holm and Suzanne W. Slayden

3. NMR of organomagnesium compounds — 131
   Peter J. Heard

4. Formation, chemistry and structure of organomagnesium species in solvent-free environments — 155
   Richard A. J. O'Hair

5. Photochemical transformations involving magnesium porphyrins and phthalocyanines — 189
   Natalia N. Sergeeva and Mathias O. Senge

6. Electrochemistry of organomagnesium compounds — 219
   Jan S. Jaworski

7. Analytical aspects of organomagnesium compounds — 265
   Jacob Zabicky

8. Biochemistry of magnesium — 315
   James Weston

9. Theoretical studies of the addition of RMgX to carbonyl compounds — 369
   Shinichi Yamabe and Shoko Yamazaki

10. Organomagnesium-group 15- and Organomagnesium-group 16-bonded complexes — 403
    Katherine L. Hull and Kenneth W. Henderson

11. Preparation and reactivity of magnesium enolates — 437
    Claude Grison

12. Functionalized organomagnesium compounds: Synthesis and reactivity — 511
    Paul Knochel, Andrey Gavryushin and Katja Brade

13.	Iron-Catalyzed Reactions of Grignard Reagents **Gérard Cahiez and Christophe Duplais**	595
14.	Carbomagnesiation reactions **Kenichiro Itami and Jun-ichi Yoshida**	631
15.	The chemistry of organomagnesium ate complexes **Hideki Yorimitsu and Koichiro Oshima**	681
16.	The chemistry of magnesium carbenoids **Tsuyoshi Satoh**	717
17.	Catalytic enantioselective conjugate addition and allylic alkylation reactions using Grignard reagents **Fernando López, Adriaan J. Minnaard and Ben L. Feringa**	771
Author index		803
Subject index		855

List of abbreviations used

Ac	acetyl (MeCO)
acac	acetylacetone
Ad	adamantyl
AIBN	azoisobutyronitrile
Alk	alkyl
All	allyl
An	anisyl
Ar	aryl
Bn	benzyl (PhCH$_2$)
Bu	butyl (C$_4$H$_9$)
Bz	benzoyl (C$_6$H$_5$CO)
c-	cyclo
CD	circular dichroism
CI	chemical ionization
CIDNP	chemically induced dynamic nuclear polarization
CNDO	complete neglect of differential overlap
Cp	η^5-cyclopentadienyl (C$_5$H$_5$)
Cp*	η^5-pentamethylcyclopentadienyl (C$_5$Me$_5$)
DABCO	1,4-diazabicyclo[2.2.2]octane
DBN	1,5-diazabicyclo[4.3.0]non-5-ene
DBU	1,8-diazabicyclo[5.4.0]undec-7-ene
DIBAH	diisobutylaluminium hydride
DME	1,2-dimethoxyethane
DMF	N,N-dimethylformamide
DMSO	dimethyl sulfoxide
E-	entgegen
ee	enantiomeric excess
EI	electron impact
ESCA	electron spectroscopy for chemical analysis
ESR	electron spin resonance
Et	ethyl (C$_2$H$_5$)
eV	electron volt

List of abbreviations used

Fc	ferrocenyl
FD	field desorption
FI	field ionization
FT	Fourier transform
Fu	furyl (OC_4H_3)
GLC	gas liquid chromatography
Hex	hexyl (C_6H_{13})
c-Hex	cyclohexyl (c-C_6H_{11})
HMPA	hexamethylphosphortriamide
HOMO	highest occupied molecular orbital
HPLC	high performance liquid chromatography
i-	iso
ICR	ion cyclotron resonance
Ip	ionization potential
IR	infrared
LAH	lithium aluminium hydride
LCAO	linear combination of atomic orbitals
LDA	lithium diisopropylamide
LUMO	lowest unoccupied molecular orbital
M	metal
M	parent molecule
MCPBA	m-chloroperbenzoic acid
Me	methyl (CH_3)
Mes	mesityl (2,4,6-$Me_3C_6H_2$)
MNDO	modified neglect of diatomic overlap
MS	mass spectrum
n-	normal
Naph	naphthyl
NBS	N-bromosuccinimide
NCS	N-chlorosuccinimide
NMR	nuclear magnetic resonance
Pen	pentyl (C_5H_{11})
Ph	phenyl
Pip	piperidyl ($C_5H_{10}N$)
ppm	parts per million
Pr	propyl (C_3H_7)
PTC	phase transfer catalysis or phase transfer conditions
Py	pyridine (C_5H_5N)
Pyr	pyridyl (C_5H_4N)

List of abbreviations used

R	any radical
RT	room temperature
s-	secondary
SET	single electron transfer
SOMO	singly occupied molecular orbital
t-	tertiary
TCNE	tetracyanoethylene
TFA	trifluoroacetic acid
TFE	2,2,2-trifluoroethanol
THF	tetrahydrofuran
Thi	thienyl (SC_4H_3)
TLC	thin layer chromatography
TMEDA	tetramethylethylene diamine
TMS	trimethylsilyl or tetramethylsilane
Tol	tolyl (MeC_6H_4)
Tos or Ts	tosyl (p-toluenesulphonyl)
Trityl	triphenylmethyl(Ph_3C)
Vi	vinyl
XRD	X-ray diffraction
Xyl	xylyl ($Me_2C_6H_3$)
Z-	zusammen

In addition, entries in the 'List of Radical Names' in *IUPAC Nomenclature of Organic Chemistry*, 1979 Edition, Pergamon Press, Oxford, 1979, p. 305–322, will also be used in their unabbreviated forms, both in the text and in formulae instead of explicitly drawn structures.

CHAPTER 1

Structural organomagnesium chemistry

JOHANN T. B. H. JASTRZEBSKI, JAAP BOERSMA and GERARD VAN KOTEN

Chemical Biology & Organic Chemistry, Faculty of Science Utrecht University, Padualaan 8, 3584 CH Utrecht, The Netherlands
Fax: +31-30-2523615;   e-mail: j.t.b.h.jastrzebski@uu.nl

I. INTRODUCTION	2
II. ORGANOMAGNESIATES	4
A. Introduction	4
B. Tetraorganomagnesiates $M_2[R_4Mg]$	5
C. Triorganomagnesiates $M[R_3Mg]$	12
D. Heteroleptic Organomagnesiates $M[R_2YMg]$ and $M[RY_2Mg]$	14
III. DIORGANOMAGNESIUM COMPOUNDS	23
A. Donor-base-free Diorganomagnesium Compounds	23
B. Diorganomagnesium Compounds Containing Multi-hapto-bonded Groups	25
C. Diorganomagnesium Compounds Containing Intramolecularly Coordinating Substituents	30
D. Donor–Acceptor Complexes of σ-Bonded Diorganomagnesium Compounds	36
E. Magnesium Anthracene Compounds	44
F. Donor–Acceptor Complexes of Diorganomagnesium Compounds with Multi-hapto-bonded Groups	47
IV. HETEROLEPTIC RMgY COMPOUNDS	54
A. Introduction	54
B. Monoorganomagnesium Cations	56
C. Monoorganomagnesium Compounds RMgY with Y = halogen (Grignard Reagents)	58
D. Monoorganomagnesium Compounds RMgY with Y = OR	69
E. Monoorganomagnesium Compounds RMgY with Y = NR_2	71
F. Monoorganomagnesium Compounds RMgY with Y = SR or PR_2	83
V. MIXED ORGANOMAGNESIUM TRANSITION-METAL COMPOUNDS	85

The chemistry of organomagnesium compounds
Edited by Z. Rappoport and I. Marek   © 2008 John Wiley & Sons, Ltd

I. INTRODUCTION

Although organomagnesium compounds are among the earliest reported organometallic compounds they were regarded as curiosities until 1900. At that time Victor Grignard, then a graduate student, worked in the laboratory of Professor Barbier at the University of Lyon in France. His task was to optimize conditions for what is now known as the Barbier reaction (equation 1)[1].

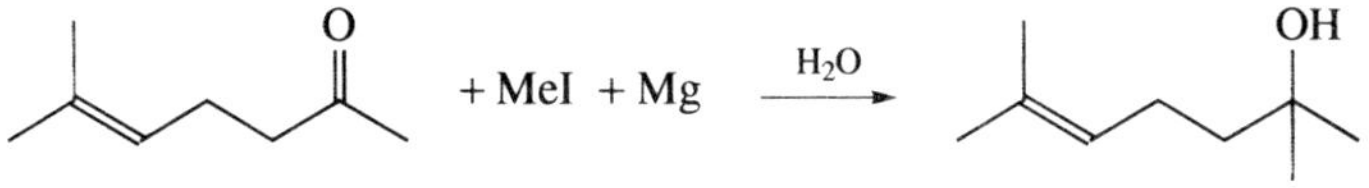

$$\text{(alkene-ketone)} + \text{MeI} + \text{Mg} \xrightarrow{H_2O} \text{(alkene-tertiary alcohol)} \quad (1)$$

Grignard proposed the intermediate in this reaction to be a RMgI species and concluded that yields might be improved by preparing this compound first and than adding it to the ketone. He found that alkyl halides indeed react readily with magnesium in diethyl ether as solvent to give compounds formulated as RMgX (equation 2). Addition of these reaction mixtures to a ketone or an aldehyde affords the corresponding alcohols in higher yields than when the Barbier procedure is used[2].

$$RX + Mg \xrightarrow{Et_2O} RMgX \quad (2)$$

Immediately, the synthetic potential of the Grignard reagents was recognized, resulting in an ever increasing number of investigations towards their preparation and application[3-5], and nowadays the Grignard reagent is one of the most powerful synthetic tools in chemistry. For Grignard's discovery and subsequent development of this finding, he was awarded the Nobel Prize in Chemistry in 1912.

Soon after its discovery an onium-type structure (1) for methylmagnesium iodide in ether was proposed by Baeyer and Villiger[6], while a somewhat different onium-type structure (2) was proposed by Grignard (Figure 1)[7].

Although it seemed that Standnikov had evidence to support Grignard's proposal[8], investigations by Thorp and Kamm demonstrated conclusively that Grignard reagents could not be represented by an onium type of structure[9]. A polar composition of the Grignard reagent R^- $(MgX)^+$ was proposed by Abegg[10] and he suggested the possibility of an equilibrium (equation 3), which nowadays is known as the Schlenk equilibrium.

$$2\ RMgX \rightleftharpoons R_2Mg + MgX_2 \quad (3)$$

With these proposals a debate started about the constitution of Grignard reagents in solution which lasted for about sixty years. This topic has been reviewed by Ashby[11].

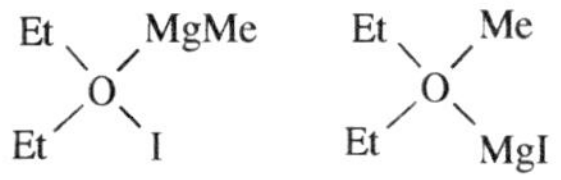

$$\underset{(1)}{\overset{Et}{\underset{Et}{>}}O\overset{MgMe}{\underset{I}{<}}} \qquad \underset{(2)}{\overset{Et}{\underset{Et}{>}}O\overset{Me}{\underset{MgI}{<}}}$$

FIGURE 1. The earliest proposed structures for methylmagnesium iodide in diethyl ether solution

Nowadays it has been well-established that the simple representation of Grignard reagents as RMgX, used in most organic text books, is far beyond the truth. Instead, in coordinating solvents like diethyl ether Grignard reagents exist as complicated mixtures of various aggregated species, in which the Schlenk equilibrium (equation 3) plays an important role[12]. The actual structures of the species present in solution depend on the nature of R, the nature of X, the properties of the coordinating solvent, concentration and temperature[3–5, 11, 13–15].

Modern techniques like X-ray absorption spectroscopy and large-angle X-ray scattering, which have been reviewed recently[16], have provided detailed information about the actual species present in solutions of organomagnesium compounds. Such studies are a prerequisite for a better understanding of the structure–activity relationships of organomagnesium compounds and in particular Grignard reagents, and the mechanisms involved in the reactions thereof[17, 18].

Elucidation of the structures of organomagnesium compounds in the solid state started in the early sixties of the previous century when modern X-ray crystallographic techniques became available. Single-crystal X-ray structure determinations of both the diethyl etherate of phenylmagnesium bromide and the diethyl etherate of ethylmagnesium bromide unambiguously showed that in the solid state these compounds exist as discrete monomers. In these structures the magnesium atom has a distorted tetrahedral coordination geometry as a result of the bonding of both the carbon atom and the bromine atom to magnesium and the coordination of two additional diethyl ether molecules to magnesium[19, 20]. Until then it was thought that Grignard reagents exist as asymmetric dimers in the solid state (Figure 2).

At the same time the structures in the solid state of Me_2Mg and Et_2Mg were determined by X-ray powder diffraction studies[21, 22]. Both compounds form polymeric chains as the result of the bridging of two methyl groups between two magnesium atoms, rendering the magnesium atoms almost perfectly tetrahedrally coordinated.

These early studies started a renaissance in the structural investigations of organomagnesium compounds in the solid state and nowadays hundreds of structures are known. In fact, in the January 2007 version of the CSD database[23] 423 structures containing at least one direct magnesium–carbon interaction have been found. The present chapter gives an overview of the structural investigations on organomagnesium compounds in the solid state, a topic that has been reviewed earlier by others[15, 24–28]. It should be noted that the structures of organomagnesium compounds obtained from X-ray crystallographic studies do not necessarily represent the structure as present in solution. Nowadays it is well known that organomagnesium compounds in solution are involved in complicated redistribution and aggregation equilibria. Such equilibria are driven by thermodynamics and therefore often the thermodynamic most stable species crystallize from such solutions. However, solubility properties and crystal packing effects also determine which particular organomagnesium compound crystallizes from solution.

According to its position in the Periodic Table of the Elements, magnesium is divalent and therefore should form organomagnesium compounds with two groups attached to it. However, because magnesium has only four electrons in its valence shell, this bonding

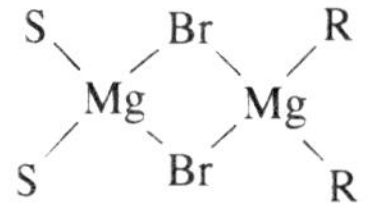

FIGURE 2. Proposed structure in the solid state for Grignard reagents before X-ray crystallography became available, S = Et_2O or THF

situation violates the octet rule. Therefore such a (linear) dicoordinate state, as e.g. found in simple diorganozinc compounds[29], is very rare. Organomagnesium compounds escape from such bonding situations by the additional coordination of donor molecules and/or by aggregation via bridging multi-center bonds or agostic interactions, resulting in most cases in a (distorted) tetrahedral coordination geometry at magnesium, which is the preferred one.

From a structural point of view, three classes of organomagnesium compounds can be distinguished, according to the number of carbon atoms directly bound to magnesium. These classes are: (i) ionic organomagnesium compounds in which the number of carbon atoms (three or four) bound to magnesium exceeds the valence number of magnesium, the so-called organomagnesiates, (ii) diorganomagnesium compounds and their coordination complexes and (iii) heteroleptic RMgX compounds in which X is an electronegative substituent like a halogen atom (Grignard Reagents) or a monoanionic group bound to magnesium via an electronegative atom like oxygen or nitrogen. Depending on the nature of X the latter class of compounds may be further divided into sub-classes. In the following sections the structures of these classes of compounds will be discussed.

II. ORGANOMAGNESIATES

A. Introduction

Alkali–metal ate compounds are among the first organometallic compounds reported. Already in 1858 the formation of a crystalline material formulated as 'Na[Et$_3$Zn]', obtained from the reaction of metallic sodium with Et$_2$Zn, was reported by Wanklyn[30]. It then took almost a century before the first organomagnesiate was reported. In 1951 Wittig and coworkers realized that organometallic compounds with anionic formulations, for which he coined the term 'ate', could be made[31]. In this paper the formation of Li[Ph$_3$Mg] and other 'ate'-type compounds from its homometallic components was described (equation 4).

$$Ph_2Mg + PhLi \longrightarrow Li[Ph_3Mg] \qquad (4)$$

The special and unique reactivities associated with this class of compounds were rapidly recognized. For example, the reaction of Li[Ph$_3$Mg] with benzalacetophenone yields mainly the 1,4-addition product while the same reaction with PhLi affords the 1,2-addition product. Wittig rationalized the chemistry of 'ate-complexes' in terms of anionic activation by which all of the ligands surrounding the metal were activated through an inductive effect[32]. In an early review of 'structures and reactions of organic ate-complexes' by Tochtermann this idea was emphasized[33].

When a diorganomagnesium compound and an alkali metal organic compound are mixed, an enhanced solubility of the resulting species in organic solvents is often observed, which is an indication of the formation of a mixed metal ate compound. This observation was reported by Coates and Heslop, who observed that Me$_2$Mg dissolves better in diethyl ether solutions that contain butyllithium than in the neat diethyl ether solvent. In this case the formation of a compound having the stoichiometry [Li(OEt$_2$)][Me$_2$BuMg] had been suggested[34]. A special feature of these 'ate' species with M[R$_3$Mg] and also M$_2$[R$_4$Mg] stoichiometry is that they exist in solution as an equilibrium mixture of various species with different metal-to-ligand molar ratios. For example, NMR studies of solutions containing a diorganomagnesium compound and an organolithium compound in various molar ratios established the presence in solution of at least three distinct different ate compounds in a rapid exchange equilibrium[35–37].

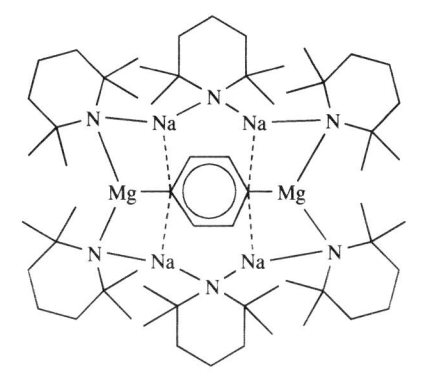

FIGURE 3. Example of an inverse crown containing a 1,4-phenylene dianion

The elucidation of the structures of organomagnesiates in the solid state started with the groundbraking X-ray crystallographic studies by Weiss on the structures of organoalkalimetal compounds including a series of organomagnesiate and organozincate compounds[38].

So far only organomagnesiates in which all the organic ligands are identical, i.e. the homoleptic organomagnesiates, have been considered. It should be noted, however, that this is not a prerequisite and organomagnesiates also exist having different organic groups. Another important class of organomagnesiates is that in which one or two of the monoanionic organic ligands are replaced by either a halide anion, or by an amido or alkoxide anion, the so-called heteroleptic organomagnesiates. During structural investigations of the latter type of compounds the concept of 'inverse crown' was discovered[39, 40]. These are aggregated compounds, usually built-up from magnesium amides or alkoxides and alkali metal amides or alkoxides, that have a very strong affinity to anionic species. Some of these are even capable of abstracting one or even two protons from an arene in a very regioselective way, forming heteroleptic organomagnesiates. Figure 3 shows an inverse crown containing bis-magnesiated benzene.

Examples of the application of organomagnesiates in organic synthesis are: (i) halogen–magnesium exchange reactions of (functionalized) aryl and alkenyl halides[41–43], (ii) the direct deprotonation of furans[44] and (iii) in highly selective addition reactions to ketones[45]. Another application of organomagnesiates is their use as a catalyst in the polymerization of butadiene to highly crystalline *trans*-1,4-polybutadiene[46].

B. Tetraorganomagnesiates $M_2[R_4Mg]$

Before discussing the structural features of tetraorganomagnesiates in the solid state it should be noted that structures in which the presence of separated anionic and cationic moieties can be distinguished are rare. In most cases such units are linked via electron-deficient bonds, i.e. two-electron three-center bridge-bonded carbon atoms between magnesium and the counter cation.

The first structure, unambiguously established by an X-ray crystal-structure determination, is Me_8Al_2Mg (**3**), that has the structural motif of four monoanionic carbon ligands bound to magnesium in a distorted tetrahedral coordination geometry (Figure 4)[47].

The molecular geometry of **3** comprises a central magnesium atom pairwise linked via four two-electron three-center bonded methyl groups to the two dimethylaluminium units. The four carbon-to-magnesium bonds (2.20, 2.22, 2.19 and 2.22 Å) are slightly elongated compared to the C–Mg distances observed in linear bis(neopentyl)magnesium

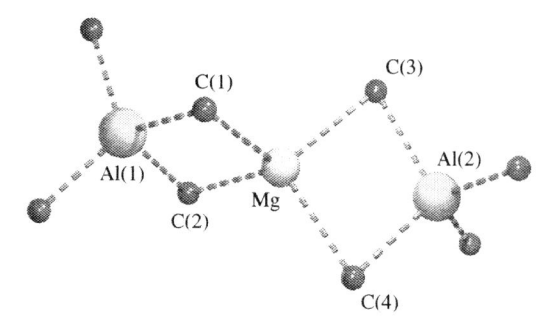

FIGURE 4. Molecular geometry of Me$_8$Al$_2$Mg (**3**) in the solid state

(2.13 Å) and [2,4,6-(*t*-Bu)$_3$C$_6$H$_2$]$_2$Mg (2.12 Å). Such an elongation is not unexpected for bridging methyl groups. The C(1)–Mg–C(2) and C(3)–Mg–C(4) bond angles (98.4 and 99.1°, respectively) are smaller than expected for the ideal tetrahedral value, but are compensated by larger values for the other C–Mg–C bond angles (average 115°). The acute Mg–C–Al bond angles of approximately 77° are in the range expected for bridging methyl groups. Arguably, this compound may be described as a true tetraorganomagnesiate comprising a central Me$_4$Mg^{2-} dianion linked to two Me$_2$Al$^+$ cations.

Crystalline [Li(TMEDA)$_2$]$_2$[Me$_4$Mg] (**4**) was obtained from the reaction of Me$_2$Mg, [MeLi]$_4$(TMEDA)$_2$ and TMEDA in diethyl ether as a solvent. Its X-ray crystal structure determination[48] revealed a molecular geometry (Figure 5) comprising a central Me$_4$Mg unit with average C–Mg distances of 2.260(8) Å. All C–Mg–C angles deviate less than 1° from the ideal tetrahedral value of 109.5°, pointing to an almost perfect tetrahedral coordination geometry around the magnesium atom. The four methyl groups interact pairwise with the lithium atoms of two Li(TMEDA) units. The relatively short C–Li distances range from 2.26(1) to 2.30(1) Å, values that are very close to the C–Mg distances, indicating that the methyl groups are symmetrically bridge-bonded between the magnesium and lithium atoms.

The solid-state structure of [Na(PMDTA)]$_2$[Ph$_4$Mg] (**5**)[49] shows a great similarity with that of **4**. Four aryl groups are bonded to the magnesium atom (C–Mg 2.29 Å, average) in an almost perfect tetrahedral arrangement (Figure 6). Two Na(PMDTA) units are linked to the central Ph$_4$Mg unit via bridge-bonding of two phenyl groups to each sodium atom,

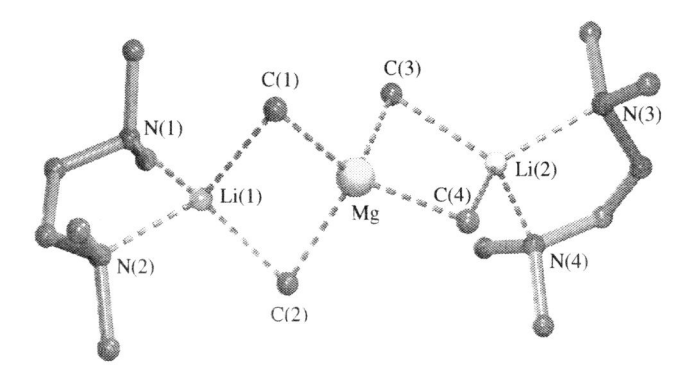

FIGURE 5. Molecular geometry of [Li(TMEDA)$_2$]$_2$[Me$_4$Mg] (**4**) in the solid state

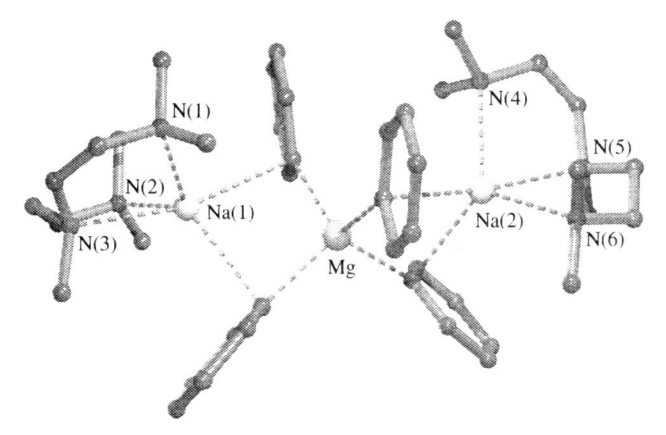

FIGURE 6. Molecular geometry of [Na(PMDTA)]$_2$[Ph$_4$Mg] (**5**) in the solid state

rendering these sodium atoms penta-coordinate. However, the bridging phenyl groups are less symmetrically bonded between magnesium and sodium than the bridging methyl groups between magnesium and lithium in **4**. This is shown by the longer C_{ipso}–Na bond lengths, ranging from 2.73 to 2.89 Å, compared to the C–Mg bond lengths of 2.29 Å. Furthermore, the C_{ipso}–Na vectors are orientated perpendicular to the planes of the aryl groups, pointing to a π-type interaction between C_{ipso} and sodium.

During the attempted preparation of an ethylmagnesiate from Et$_2$Mg and EtLi in a hexane/toluene solvent mixture it appeared that an unexpected metallation of toluene had occurred resulting in a compound with the formulation Li$_2$[Bn$_4$Mg]. According to a similar procedure, [Li$_2$(TMEDA)$_3$][Bn$_4$Mg] (**6**) was obtained from the reaction of Et$_2$Mg with EtLi in the presence of toluene and TMEDA (equation 5)[50].

$$\text{Et}_2\text{Mg} + \text{EtLi} \xrightarrow[\text{TMEDA}]{\text{toluene}} [\text{Li}_2(\text{TMEDA})_3][\text{Bn}_4\text{Mg}] \quad\quad (5)$$
$$(\mathbf{6})$$

An X-ray crystal-structure determination of **6** revealed a solid-state structure consisting of [Li(TMEDA)][Bn$_4$Mg] anionic and Li(TMEDA)$_2$ cationic units. The molecular geometry of the anion comprises a Bn$_4$Mg unit linked to the Li(TMEDA) moiety via two bridging benzyl groups (Figure 7). The C–Mg bond lengths of the bridge-bonding benzyl groups [C(3)–Mg 2.313(9) and C(4)–Mg 2.322(9) Å] are slightly elongated compared to those of the terminally bonded benzyl groups [C(1)–Mg 2.22(1) and C(2)–Mg 2.26(1) Å]. The observed C–Li bond lengths [C(3)–Li 2.27(1) and C(4)–Li 2.23(2) Å] point to a slightly asymmetric bridge-bonding of the benzyl groups between magnesium and lithium. The C(3)–Mg–C(4) bond angle of 104.7(4)° is smaller than the ideal tetrahedral value, but is compensated by a value of 111.1(5)° for the C(1)–Mg–C(2) bond angle. These deviations point to a slightly distorted tetrahedral coordination geometry at the magnesium atom. Finally, coordination saturation at the lithium atom is reached by a N,N'-chelate bonded TMEDA molecule.

The reaction of allylmagnesium chloride with methylaluminium dichloride affords, after workup of the reaction mixture and recrystallization from THF, a rather unexpected compound (**7**), which, according to its crystal-structure determination, appears to consist of [(allyl)$_2$Mg$_3$Cl$_3$(THF)$_6$]$^+$ cations and [(allyl)$_4$Mg]$^{2-}$ anions in a 2:1 molar ratio (Figure 8)[51].

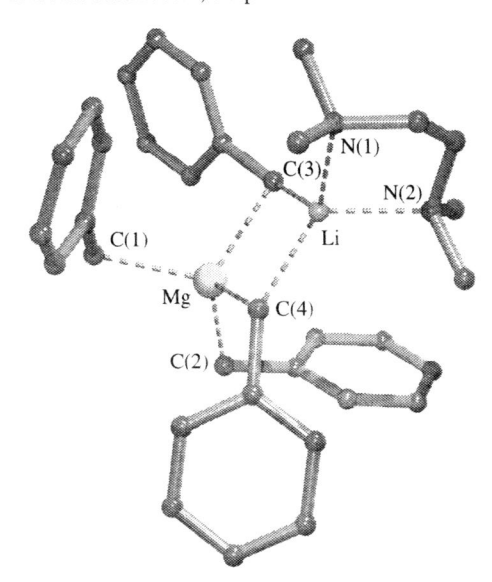

FIGURE 7. Molecular geometry of the [Li(TMEDA)][Bn₄Mg] anion of **6**

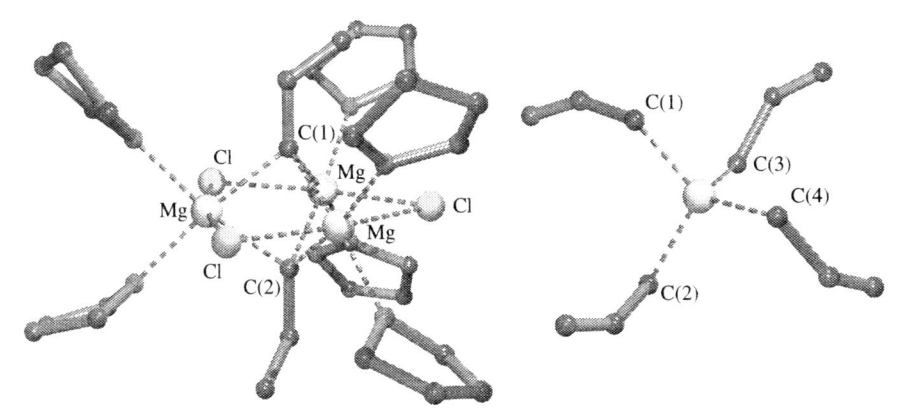

FIGURE 8. Cationic (left) and anionic (right) units of compound **7**

This compound is one of the few examples of tetraorganomagnesiates that contains isolated tetraorganomagnesium dianions in the crystal lattice. Due to the location of the magnesium atom at a special position (inversion center in space group Ibam) the four allyl groups in the anion are symmetry related. The C—Mg distances of 1.996(8) Å are relatively short. The C—Mg—C bond angles range from 108.4(6) to 110.6(6)°, indicating an almost perfect tetrahedral geometry around this magnesium atom.

The structure of the cationic part of **7** consists of a [Mg₃C₂] trigonal bipyramidal arrangement (Figure 8) with the magnesium atoms in the equatorial plane and the carbon atoms at the apical positions. The two allyl groups are μ_3-bonded (one above and one

1. Structural organomagnesium chemistry

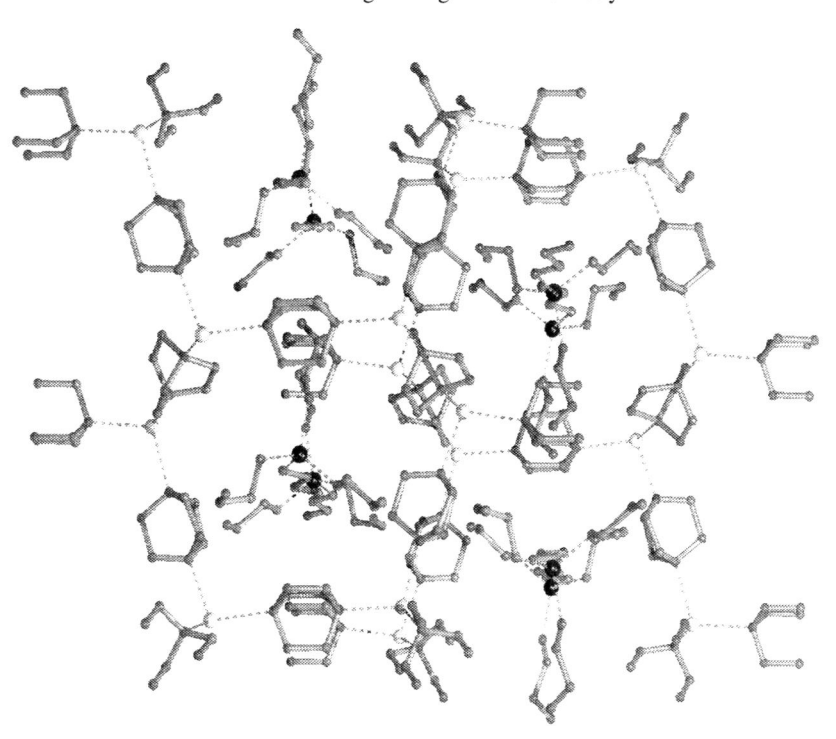

FIGURE 9. Part of the polymeric network of **8**. Disorder components and hydrogen atoms are omitted for clarity

below the trigonal plane) with their terminal carbon atoms to the three magnesium atoms. Each of the three chlorine atoms are bridge-bonded between two magnesium atoms in the equatorial plane.

The only other example of a tetraorganomagnesiate that contains isolated anions in its solid-state structure is $[Na_2(DABCO)_3(toluene)][Bu_4Mg]$ (**8**). Each DABCO and toluene molecule bridges two sodium atoms, forming a polycationic three-dimensional coordination network, in which isolated Bu_4Mg^{2-} anions are embedded (Figure 9)[52].

Like in **7** the Mg–C distances in the tetrabutylmagnesium dianion in **8** are relatively short [2.009(6), 2.010(7), 2.042(7) and 2.041(7) Å]. All C–Mg–C bond angles are close to 109°, indicating tetrahedral coordination geometry at the magnesium atoms.

Although the solid-state structure of $[Li(TMEDA)]_2[Ph_6Mg_2]$ (**9**) reveals that to each of the magnesium atoms four carbon atoms are bonded, this compound is best described as consisting of a central Ph_6Mg_2 dianion in which the magnesium atoms are linked via two symmetrically bridging phenyl groups [C(2)–Mg(1) 2.329(3) and C(2)–Mg(2) 2.286(3) Å] (Figure 10)[53]. Furthermore, to each of the magnesium atoms two phenyl groups are bridging between magnesium and lithium in an asymmetric way [C(1)–Mg(1) 2.186(3), C(1)–Li(1) 2.419(9) Å]. Coordination saturation at lithium is reached by a N,N'-chelate bonded TMEDA molecule.

During a study in which the cyclic tripod amine N,N',N''-trimethyltriazacyclononane (TAEN) was used as a solvent for Me_2Mg, a rather unexpected product was obtained

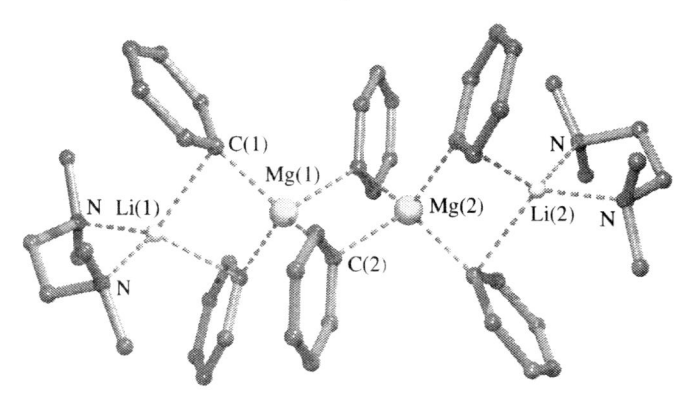

FIGURE 10. Molecular geometry of **9** in the solid state

which, according to its X-ray structural analysis, appeared to be $[\{Me_3Mg_2(TAEN)_2\}^+]_2$ $[Me_8Mg_3]^{2-}$ (**10**)[54]. This product is the result of a disproportionation reaction (equation 6).

$$7\ Me_2Mg + 4\ TAEN \longrightarrow [Me_3Mg_2(TAEN)_2]_2[Me_8Mg_3] \quad (\mathbf{10})$$

(6)

An X-ray crystal-structure determination of **10** revealed an asymmetric unit that contains two 'triple-decker' dimagnesium cations, a Me_8Mg_3 dianion and two benzene molecules (Figure 11). The two cations differ slightly with respect to bond distances and bond angles, but are chemically identical. In the cation the three methyl groups are symmetrically bridge-bonded (C–Mg average 2.354 Å) between the two magnesium atoms. To each magnesium atom a TAEN molecule is tridentate facially-coordinated with its

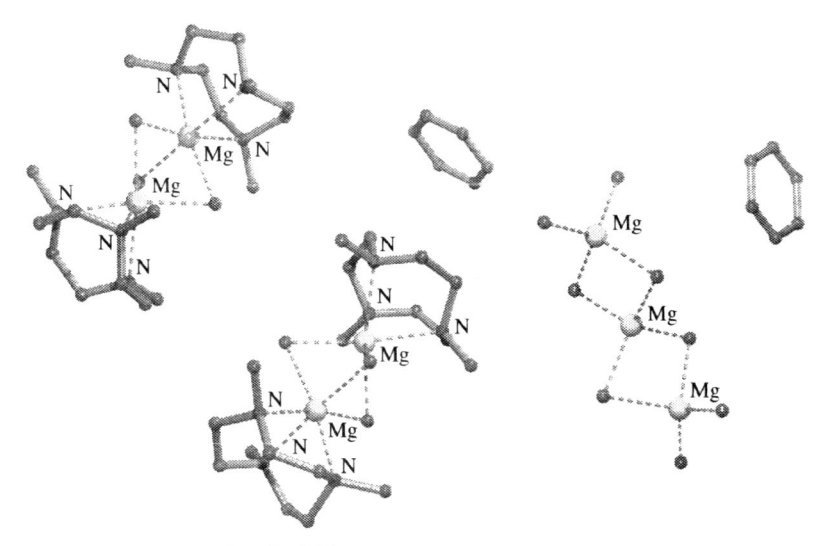

FIGURE 11. The asymmetric unit of **10**

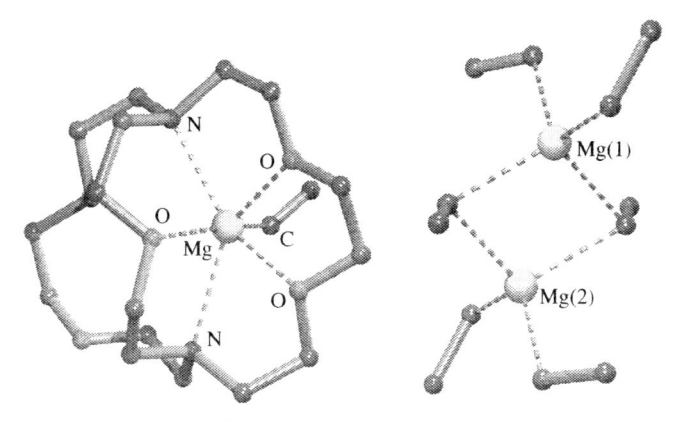

FIGURE 12. Molecular geometry of the cationic (left) and anionic (right) parts of **11**

three nitrogen atoms, resulting in a slightly distorted octahedral coordination geometry at each magnesium atom. The [Me$_8$Mg$_3$] dianion consists of a linear arrangement of three magnesium atoms with two symmetrically bridging methyl groups between each of the two magnesium atoms. To each of the two terminal magnesium atoms two further methyl groups are bonded, resulting in a tetrahedral coordination geometry at each of the magnesium atoms. As might be expected, the C—Mg bond distances of the terminal methyl groups (C—Mg average 2.161 Å) are slightly shorter than those of the bridging methyl groups (C—Mg average 2.294 Å).

Like TAEN also cryptands are capable of initiating a disproportionation reaction in diorganomagnesium compounds. From the reaction of Et$_2$Mg with 2,1,1-cryptand a crystalline product (**11**) was obtained. According to its X-ray crystal-structure determination **11** consists of isolated [EtMg(2,1,1-cryptand)]$^+$ cations and [Et$_6$Mg$_2$]$^{2-}$ anions in the crystal lattice (Figure 12)[55]. The dianion in fact is a dimer formed from two [Et$_3$Mg]$^-$ anions via two bridging ethyl groups between the two magnesium atoms. The two halves of the dimer are symmetry related via a crystallographic inversion center. The C—Mg bond distances, 2.336 Å for the bridging ethyl group and 2.223 Å for the terminal ethyl groups, are in the range as expected.

The cation contains a [EtMg]$^+$ moiety (C—Mg 2.157(9) Å) to which three oxygen atoms and two nitrogen atoms of the cryptand are coordinated.

It has been suggested that the formation of organomagnesiate anions from equilibria of dialkylmagnesium compounds with crown ethers, although in concentrations too low to be detectable by e.g. NMR spectroscopy, are responsible for the specific chemical behavior of such solutions (equation 7)[56].

$$3 R_2Mg + 15\text{-C-}5 \rightleftharpoons [RMg(15\text{-C-}5)][R_5Mg_2] \qquad (7)$$

A crystalline product with the formulation [MeMg(15-C-5)][Me$_5$Mg$_2$] (**12**) was obtained from a solution of Me$_2$Mg and 15-C-5, making use of special crystallization techniques. An X-ray crystal-structure determination revealed the presence of isolated [MeMg(15-C-5)]$^+$ cations. The anionic counter part consists of [Me$_5$Mg$_2$]$^-$ units in which two methyl groups are bridge-bonded between the magnesium atoms while one of the other methyl groups forms a bridge bond to a next [Me$_5$Mg$_2$]$^-$ unit, thus forming a polymeric chain (Figure 13)[57].

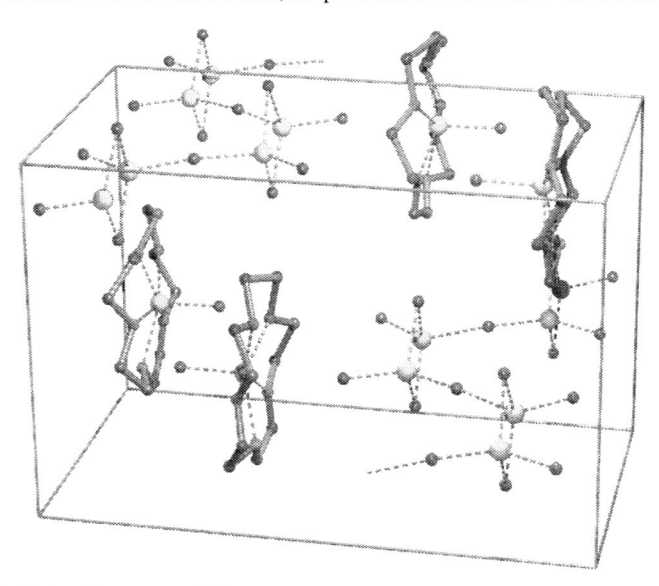

FIGURE 13. Unit cell contents of **12**

C. Triorganomagnesiates M[R₃Mg]

In contrast to their zinc congeners[29] only a very few compounds are known that contain a $[R_3Mg]^-$ structural unit in the solid state. The only compound having isolated anions and cations in its crystal lattice is [*neo*-PentMg(2,1,1-cryptand)][*neo*-Pent₃Mg] (**13**) obtained from a disproportionation reaction of *neo*-Pent₂Mg in the presence of 2,1,1-cryptand[55]. An X-ray crystal-structure determination (Figure 14) of **13** shows that the closest approach between the magnesium atom in the anion and a heteroatom in the cryptand is 5.71 Å, indicating the presence of isolated cations and anions in the crystal lattice. Although the C–Mg bond distances in the anionic [*neo*-Pent₃Mg]⁻ moiety vary slightly [C–Mg 2.125(12), 2.240(12) and 2.296(16) Å], the sum of the C–Mg–C bond angles around magnesium is 360° within experimental error, indicating a planar trigonal coordination

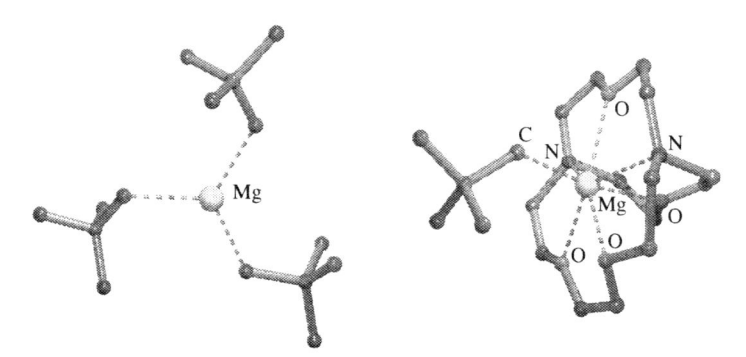

FIGURE 14. Molecular geometry of the anionic (left) and cationic (right) moieties of **13** in the solid state

geometry at the magnesium atom. In the [*neo*-PentMg(2,1,1-cryptand)]$^+$ cationic part all heteroatoms of the cryptand are involved in coordination to magnesium, resulting in an essentially pentagonal bipyramidal geometry at magnesium. Five of the heteroatoms of the cryptand and the magnesium atom lie approximately in a plane while one of the oxygen atoms of the cryptand and the bonding carbon atom of the neopentyl group occupy the apical sites.

A triarylmagnesiate [Li(THF)][{2,4,6-(*i*-Pr)$_3$C$_6$H$_2$}$_3$Mg] (**14**) was obtained from the stoichiometric reaction of the parent organometallic compounds. An X-ray crystal-structure determination of **14** revealed a molecular geometry in which the triarylmagnesiate and lithium are associated via bridging aryl groups (Figure 15)[58]. Both the magnesium atom and the lithium atom in **14** are three-coordinate, the magnesium atom as the result of the bonding of two bridging and one terminal carbon atom, and the lithium atom as the result of the bonding of two bridging carbon atoms and an oxygen atom of an additional coordinating THF molecule. The sum of the bond angles around both the magnesium atom and the lithium atom is close to 360°, indicating for both metals a trigonal planar coordination geometry. As expected, the Mg−C bond distances of the bridging carbon atoms [Mg−C(2) 2.249(4) and Mg−C(3) 2.206(4) Å] are somewhat longer than the terminal Mg−C bond [Mg−C(1) 2.147(4) Å]. The C−Li bond distances are relatively short [Li−C(2) 2.195(9) and Li−C(3) 2.251(9) Å] but not unexpected due to the bonding to a three-coordinate lithium atom. For the same reason also the bond distance of the coordinating oxygen atom of the THF molecule to magnesium is extremely short [Li−O 1.858(8) Å].

The mixed metal alkyl-amido base [BuNa(TMEDA)][TMP$_2$Mg] (TMP = 2,2,6,6-tetramethylpiperidine) is capable of deprotonating furan selectively at its 2-position in THF as a solvent. The product obtained is a complex tris-furylmagnesiate, with the empirical formula [Na$_2$(THF)$_3$][2-furyl$_6$Mg$_2$(TMEDA)] (**15**)[59]. An X-ray crystal-structure determination showed that in the solid state this compound exists as a coordination polymer of [Na$_2$(THF)$_3$][2-furyl$_6$Mg$_2$] units linked by bridging TMEDA molecules (Figure 16).

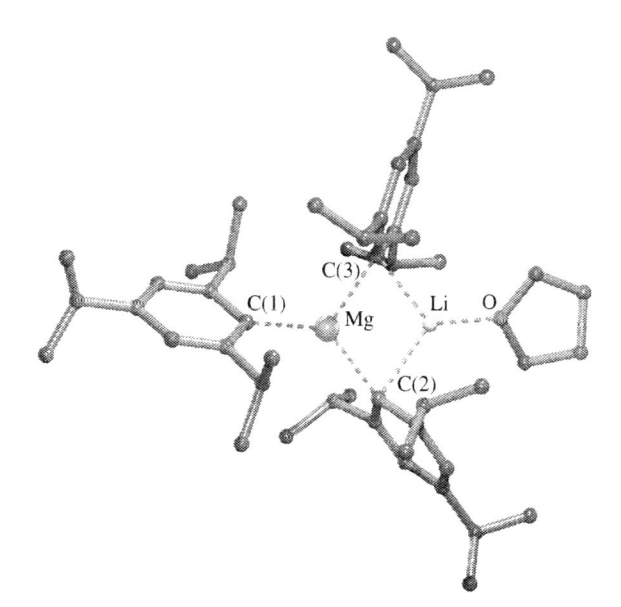

FIGURE 15. Molecular geometry of **14** in the solid state

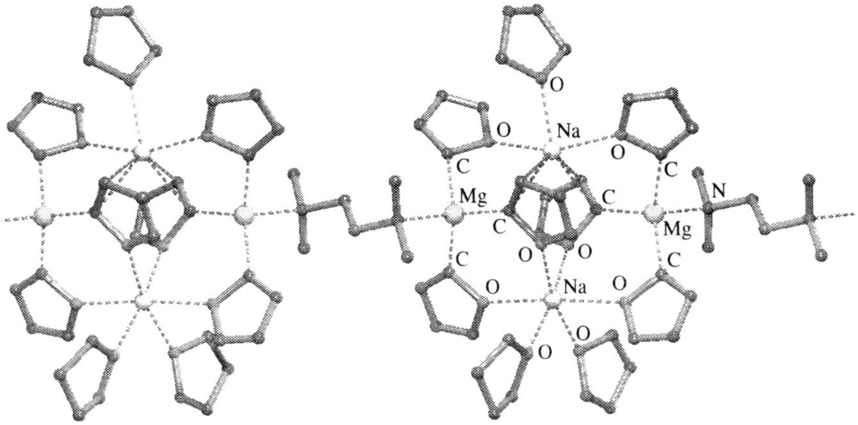

FIGURE 16. Part of the polymeric chain of inverse crown ether structure **15**

This compound represents an example of an inverse crown ether structure (Lewis acidic host–Lewis basic guest macrocyclic heterometallic alkoxides or amides)[45,46].

The repeating unit in **15** contains two [(2-furyl)$_3$Mg]$^-$ anionic moieties and two sodium cations assembled in a cyclic structure (Figure 16). Four of the furyl groups are pointing to the outside of this cycle and are coordinating pairwise to the sodium atoms with their furyl oxygen atoms. Two furyl groups are located inside the cycle and are coordinating with their oxygen atoms to the same sodium atom to which also two additional THF molecules are coordinated. These furyl groups also have a π-interaction with the other sodium atom to which one additional THF molecule is also coordinated.

Also aggregated magnesiates, containing acetylenic organic groups, [Li$_2$(TMEDA)$_2$][(PhC≡C)$_6$Mg$_2$][50] (**16**), [Na$_2$(TMEDA)$_2$][(t-BuC≡C)$_6$Mg$_2$][49] (**17**) and [Na$_2$(PMDTA)$_2$][(t-BuC≡C)$_6$Mg$_2$][49] (**18**), have been structurally characterized.

D. Heteroleptic Organomagnesiates M[R$_2$YMg] and M[RY$_2$Mg]

So far, only organomagnesiates have been considered consisting of an anionic moiety in which three or four carbon atoms are directly bound to the magnesium atom. However, also organomagnesiates exist in which the anionic moiety contains only one or two carbon atoms as well as one or two anions bound to the magnesium atom via a N- or O-heteroatom.

In particular, studies of the constitution of Grignard reagents in the solid state revealed that in addition to neutral organomagnesium species, also ionic structures exist that in fact are heteroleptic organomagnesiates. Three different types of species were observed in the solid state. The first are ionic [Mg$_2$(μ-Cl)$_3$(THF)$_6$][RMgCl$_2$(THF)] ones [R = t-Bu (**19**) and R = Ph (**20**)], which were obtained from THF solutions of t-BuMgCl and PhMgCl, respectively[60]. The second are neutral R$_2$Mg$_4$Cl$_6$(THF)$_6$ species [R = Me (**21**), R = t-Bu (**22**) and R = benzyl (**23**)], isolated from THF solutions of MeMgCl, t-BuMgCl and BnMgCl, respectively[60]. The last is ionic [Mg$_2$(μ-Cl)$_3$(THF)$_3$]$_2$[Ph$_4$Mg$_2$(μ-Cl)$_2$] (**24**), obtained from a THF solution of PhMgCl[60]. As the ratio of organic group to magnesium to chloride in these compounds is not 1:1:1, it is obvious that formation of these particular compounds can never be quantitative and that the remaining solutions must contain magnesium compounds having other stoichiometries.

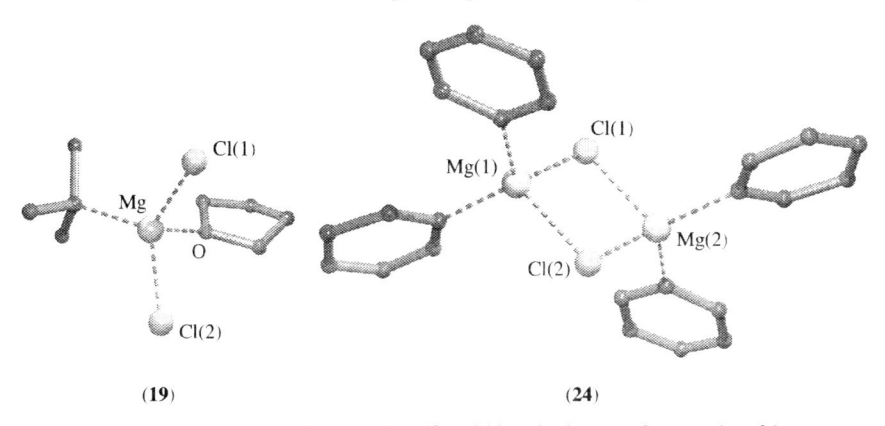

FIGURE 17. Monoanionic part of magnesiate **19** and bis-anionic part of magnesiate **24**

The magnesium atoms in the monoanionic moieties of **19** (Figure 17) and **20** have, as expected, a slightly distorted tetrahedral geometry. In **19** the Cl(1)–Mg–Cl(2) and C–Mg–O bond angles of 109.2(1)° and 110.9(1)°, respectively, are very close to the ideal tetrahedral value. The bis-anion [Ph$_4$Mg$_2$Cl$_2$]$^{2-}$ (Figure 17) in **24** may be regarded as being formed from the dimerization of two hypothetical [Ph$_2$MgCl]$^-$ moieties via chloride bridges. The almost equal Mg–Cl bond distances [Mg(1)–Cl(1) 2.432(1) and Mg(2)–Cl(1) 2.464(2) Å] indicate that the chlorides are symmetrically bridging. These bond distances are somewhat elongated compared to the terminal Mg–Cl bonds in **19** [both 2.232(2) Å], but this is not unexpected for bridging halogens. The ionic compounds, **19**, **20** and **24**, have in common that charge compensation is reached by the same cation, i.e. [Mg$_2$Cl$_3$(THF)$_6$]$^+$ in which the three chlorides are bridge-bonded between the two magnesium atoms while the three THF molecules provide to each of the magnesium atoms an octahedral coordination geometry.

The actual aggregate that crystallizes from a solution of a Grignard reagent is largely influenced by the nature of the solvent used. This became evident by the crystals obtained from a solution of MeMgBr in triglyme having stoichiometry [Mg$_2$Br$_2$(triglyme)$_2$] [Me$_2$MgBr$_4$] (**25**). The crystal structure determination of **25** revealed that the crystal lattice contains isolated [(μ-Me)$_2$Mg$_2$Br$_4$]$^{2-}$ magnesiate anions and [Mg$_2$(μ-Br)$_2$(triglyme)$_2$]$^+$ cations[61]. It is surprising that in the magnesiate anion of **25** the methyl groups rather than the bromide act as bridges between the two magnesium atoms. Usually, the softer halogen atoms have a stronger tendency to form bridges than the harder carbon atom.

Solutions of mixtures of alkali alkoxides and diorganomagnesium compounds have been studied in solution because of their relevance as initiators for styrene polymerization. From such solutions crystalline [Bu$_2$MgNaOBu-t(TMEDA)]$_2$ (**26**) and [Bu$_2$MgKOBu-t(TMEDA)]$_2$ (**27**) were isolated and structurally characterized[62]. They are aggregated species and may be regarded as heteroleptic organomagnesiates. Because **26** and **27** are isostructural, only the overall structural geometry (Figure 18) of **26** is discussed in more detail.

The central core of **26** is a flat four-membered O–Mg–O–Mg ring. One of the t-Bu groups is located above, and the other below this plane. To each of the oxygen atoms a sodium atom is bonded [Na–O 2.533(5) Å] while the four butyl groups are bridge-bonded between the sodium and magnesium atoms in a rather asymmetric way [C–Mg 2.190(6) and C–Na 2.852(7) Å]. Penta-coordination at each sodium atom is reached by the additional N,N'-chelate bonding of a TMEDA molecule.

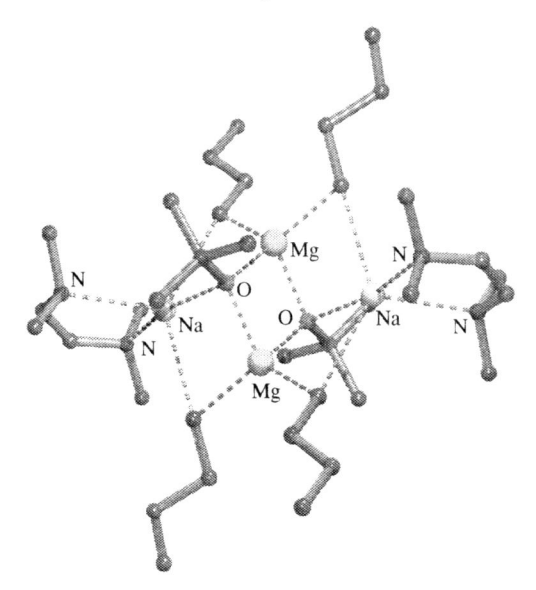

FIGURE 18. Molecular geometry of **26** in the solid state

(**27**)

FIGURE 19. Schematic structure of heteroleptic magnesiate **27**

Reaction of 2,2′-ethylidenebis(2,4-*tert*-butylphenol) (EDBP-H$_2$) with butyllithium and dibutylmagnesium in a 1:1:1 molar ratio in diethyl ether as a solvent affords the heteroleptic organomagnesiate [BuMgLi(EDBP)(OEt$_2$)]$_2$ (**27**) of which the structure is shown schematically (Figure 19)[63].

The butyl groups are slightly asymmetric bridge-bonded between lithium and magnesium [C–Li 2.263(7) and C–Mg 2.175(4) Å]. The bis-anionic EDBP ligands are O,O'-chelate bonded, with one of the oxygen atoms bridging between the two magnesium atoms, giving rise to a central O–Mg–O–Mg four-membered ring while the other oxygen atom is bridging between a magnesium atom and a lithium atom. To each of the lithium atoms an additional diethyl ether molecule is coordinated, affording a distorted trigonal coordination geometry at lithium. It is interesting to note that **27** is an efficient initiator for methyl methacrylate polymerization.

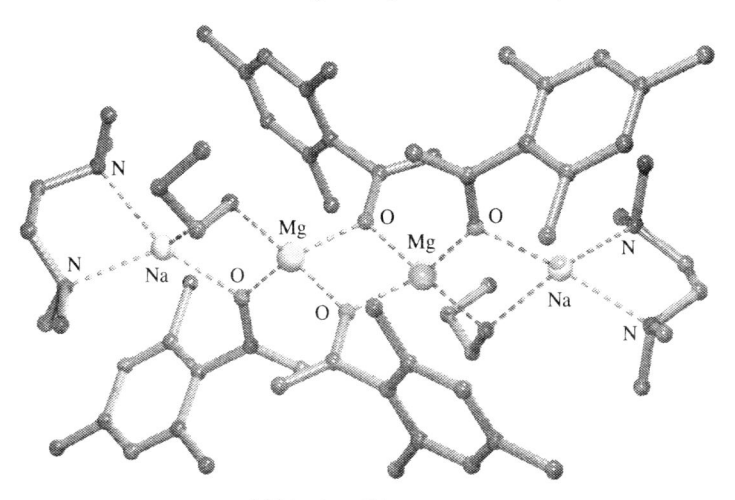

FIGURE 20. Molecular geometry of **28** in the solid state

Reaction of *in situ* prepared Na[Bu$_3$Mg] with 2,4,6-trimethylacetophenone and crystallization of the product from toluene in the presence of TMEDA afforded the heteroleptic organomagnesiate [BuMgNa{OC(=CH$_2$)Mes}$_2$(TMEDA)]$_2$ (**28**)[64]. It appeared that instead of deprotonation of the 2,4,6-trimethylacetophenone to give an enolate moiety, 1,2-addition had occurred. The X-ray crystal-structure determination of **28** (Figure 20) showed an almost linear Na ··· Mg ··· Mg ··· Na arrangement. The two magnesium atoms in this arrangement are linked by two bridging oxygen atoms of two enolate moieties while each of the sodium atoms is linked by one bridging carbon atom of the butyl group and one bridging oxygen atom of an enolate group. Coordination saturation at each of the sodium atoms is reached by the additional coordination of a TMEDA molecule.

During studies of the synthesis and structural characterization of mixed magnesium–lithium secondary amide aggregates, a heteroleptic organomagnesiate, [BuMgLi(N(SiMe$_3$)$_2$ (Py)] (**29**), crystallized as a by-product from a reaction mixture of *n*-BuLi, *sec*-Bu$_2$Mg and HN(SiMe$_3$)$_2$ in the presence of pyridine[65]. An X-ray crystal-structure determination of **29** revealed a monomolecular structure (Figure 21) in which two (Me$_3$Si)$_2$N amido groups are symmetrically bridge-bonded [N–Li 2.066(5) and N–Mg 2.090(3) Å] between magnesium and lithium while the butyl group is η^1-bonded to magnesium. An additional pyridine molecule is coordinated to the lithium atom. The sum of the bond angles around both lithium and magnesium is 360° within experimental error, indicating a trigonal planar coordination geometry around these metals. It should be noted that in the crystal lattice of **29** the η^1-bonded butyl group is chemically disordered with *n*-butyl and *sec*-butyl groups, indicating that prior to the amide-formation step scrambling of *n*-butyl and *sec*-butyl groups has occurred, most probably via a [(*sec*-Bu)$_2$(*n*-Bu)Mg]$^-$ magnesiate-type species.

The reaction of *t*-BuLi with [(Me$_3$Si)$_2$N]$_2$Mg in a 1:1 molar ratio in hydrocarbon solvents affords a crystalline product that appears to be the heteroleptic organomagnesiate [*t*-BuMgLi{N(SiMe$_3$)$_2$}$_2$] (**30**). Its X-ray crystal-structure determination revealed a structure (Figure 22) that shows similarities with that of **29**[66]. In **30** the two amide nitrogen atoms are symmetrically bridge-bonded between magnesium and lithium [N–Mg 2.079(1) and N–Li 2.055(2) Å] while the *t*-Bu group is σ-bonded to magnesium [C–Mg 2.174(1) Å]. The coordination unsaturation at lithium is released by an agostic interaction with the carbon atom [C–Li 2.563(3) Å] of a *t*-Bu group of a neighboring molecule,

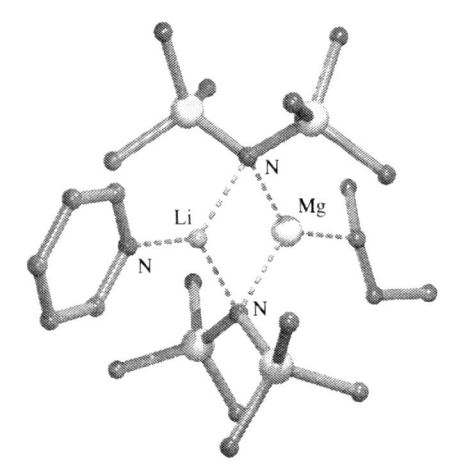

FIGURE 21. Molecular geometry of **29** in the solid state. (The minor *n*-butyl disorder component is omitted for clarity.)

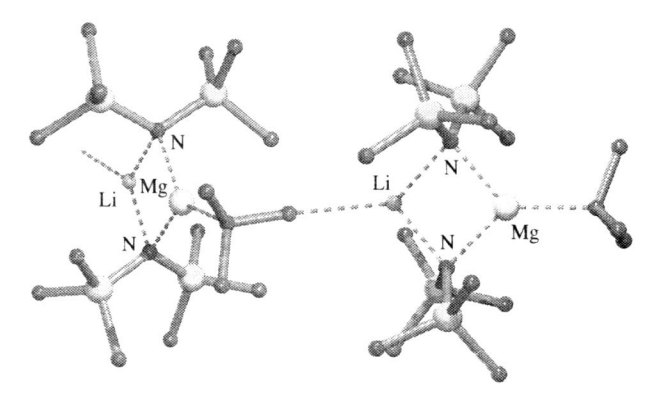

FIGURE 22. Two units of the polymeric structure of **30** in the solid state

resulting in polymeric chains (Figure 22). Like in **29** both the magnesium atom and the lithium atom have a trigonal planar coordination geometry.

The sodium analog of **30** was prepared from t-BuMgCl and two equivalents of [(Me$_3$Si)$_2$N]$_2$Na in diethyl ether as a solvent. The structure of the resulting heteroleptic organomagnesiate [t-BuMgNa{N(SiMe$_3$)$_2$}$_2$(OEt$_2$)] (**31**) shows great similarities with that of the repeating unit of **30**, but now with a diethyl ether molecule coordinated to the sodium atom instead of an agostic interaction, thus preventing the formation of polymeric chains.

It has been well-established that deprotonative metallation is one of the most widely studied and utilized tools in chemical synthesis. Selective di-metallation of arenes has been observed using mixed metal reagents. Reaction of a mixture of BuNa, Bu$_2$Mg and TMPH (2,2,6,6-tetramethylpiperidine) in 1:1:3 molar ratio in the presence of toluene or benzene affords the aggregated compounds [(MeC$_6$H$_3$)Mg$_2$Na$_4$(TMP)$_6$] (**32**) and [(C$_6$H$_4$)Mg$_2$Na$_4$(TMP)$_6$] (**33**), respectively, formed in a self-assembly process[67]. X-ray crystal-structure determinations of **32** and **33** (Figure 23) revealed macrocyclic structures with

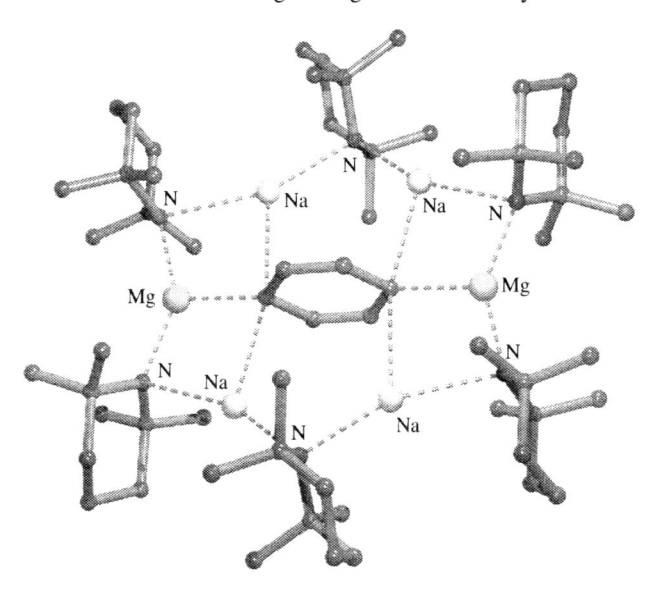

FIGURE 23. Molecular geometry of **33** in the solid state

six metal atoms (four sodium and two magnesium) alternating with six bridging amido nitrogen atoms in a twelve-membered ring, while a $[MeC_6H_3]^{2-}$ (in **32**) or $[C_6H_4]^{2-}$ (in **33**) dianion is located inside the ring. Each of the deprotonated carbon atoms has a relatively short C–Mg interaction [2.200(2) Å] and two longer C–Na bonds [2.691(2) and 2.682(2) Å]. Such structures represent examples of inverse crowns (Lewis acidic host–Lewis basic guest macrocyclic heterometallic amides)[45, 46]. It is noteworthy that in **32** the deprotonation of toluene is regioselective at its 2- and 5-position.

That small variations can have large influence on the ultimate structures of the aggregates formed during arene deprotonating and aggregation steps became evident when the same reaction under identical reaction conditions as for **33** was carried out using BuK instead of BuNa. Instead of the expected potassium analog of **33** an unprecedented twenty-four-membered $[(KNMgN)_6]^{6+}$ ring system was formed, which acts as a polymetallic host to which six mono-deprotonated arene anions are bonded. The X-ray crystal structures of $[(C_6H_5)Mg_6K_6(TMP)_{12}]$ (**34**) (Figure 24) containing six mono-deprotonated benzene molecules and of $[(MeC_6H_4)Mg_6K_6(TMP)_{12}]$ (**35**) containing six mono-deprotonated toluene molecules have been elucidated[68]. The twenty-four-membered ring is built up of six sequences of a potassium atom, a bridging amido nitrogen atom, a magnesium atom and again a bridging amido nitrogen atom. The deprotonated arene carbon atom is σ-bonded to magnesium [C–Mg 2.196(6) Å], while the other interatomic distances suggest that the hapticity of the aryl rings with respect to the potassium atoms is μ-η^3: η^2, i.e. three carbon atoms (one *ipso* and two *ortho*) on one face and two carbon atoms (one *ipso* and one *ortho*) on the opposing face.

When, under the reaction conditions outlined above for the preparation of **33**, ferrocene was added as the arene to be deprotonated, it appeared that 1,1′-di-deprotonation occurs resulting in an aggregated species $[\{(C_5H_4)_2Fe\}_3Mg_3Na_2(TMP)_2(TMPH)_2]$ (**36**)[69]. Unfortunately, its X-ray structure determination showed disordered moieties especially with respect to the coordinated TMPH molecules. Its lithium analog $[\{(C_5H_4)_2Fe\}_3Mg_3Li_2(TMP)_2(TMPH)_2]$ (**37**) was prepared in a similar way and, after treatment with pyridine (to

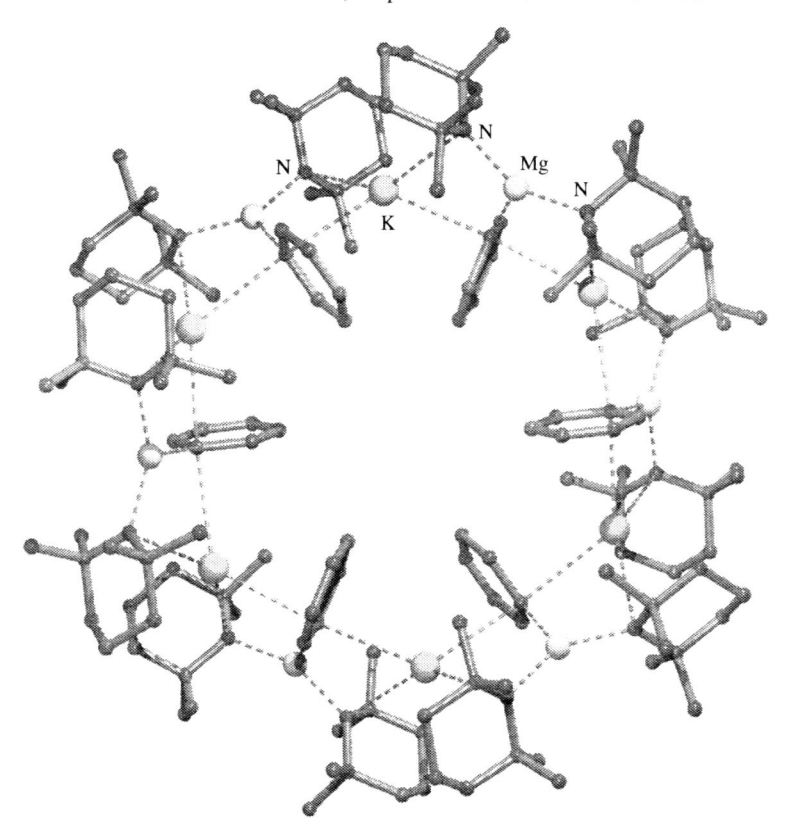

FIGURE 24. Molecular geometry of **34** in the solid state

effect substitution of the coordinated TMPH molecules by pyridine), crystalline [{(C$_5$H$_4$)$_2$Fe}$_3$Mg$_3$Li$_2$(TMP)$_2$(py)$_2$] (**38**) was obtained suitable for X-ray structural analysis. Its structure comprises a bent Li–Mg–Mg–Mg–Li arrangement to which two ferrocenyl dianions with one deprotonated carbon atom are bridge-bonded between lithium and magnesium, while the other deprotonated carbon atom is bridge-bonded between two magnesium atoms (Figure 25). The two deprotonated carbon atoms of the third ferrocenyl group are both bridge-bonded between two magnesium atoms. The two TMP groups are each bridge-bonded with their amide nitrogen atom between magnesium and lithium. Finally, a pyridine molecule is coordinated to each lithium atom, resulting in a trigonal planar coordination geometry of the lithium atoms.

Surprisingly, changing the secondary amine from TMPH to diisopropylamine results in the formation of entirely different structures. When three equivalents of diisopropylamine are added to *in situ* prepared Na[Bu$_3$Mg] and the resulting reaction mixture is used for deprotonating the metallocenes Cp$_2$Fe, Cp$_2$Ru or Cp$_2$Os, unprecedented inverse-crown architectures are obtained. For all three metallocenes isostructural architectures were obtained, consisting of a sixteen-membered [(NaNMgN)$_4$]$^{4+}$ host and a tetra-deprotonated metallocene guest[70]. For all three compounds [{(C$_5$H$_3$)$_2$Fe}Mg$_4$Na$_4$(*i*-Pr$_2$N)$_8$] (**39**), [{(C$_5$H$_3$)$_2$Ru}Mg$_4$Na$_4$ (*i*-Pr$_2$N)$_8$] (**40**) and [{(C$_5$H$_3$)$_2$Os}Mg$_4$Na$_4$ (*i*-Pr$_2$N)$_8$] (**41**) the structures were elucidated by X-ray crystallography. That of **40** is shown (Figure 26).

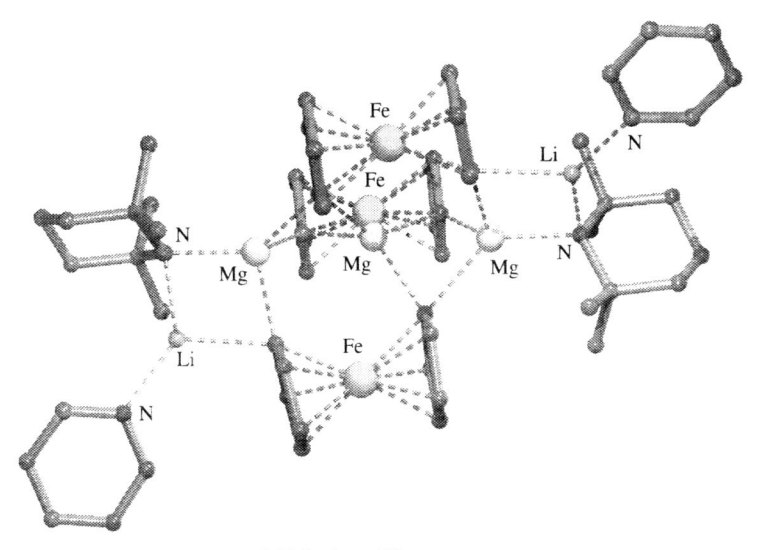

FIGURE 25. Molecular geometry of **38** in the solid state

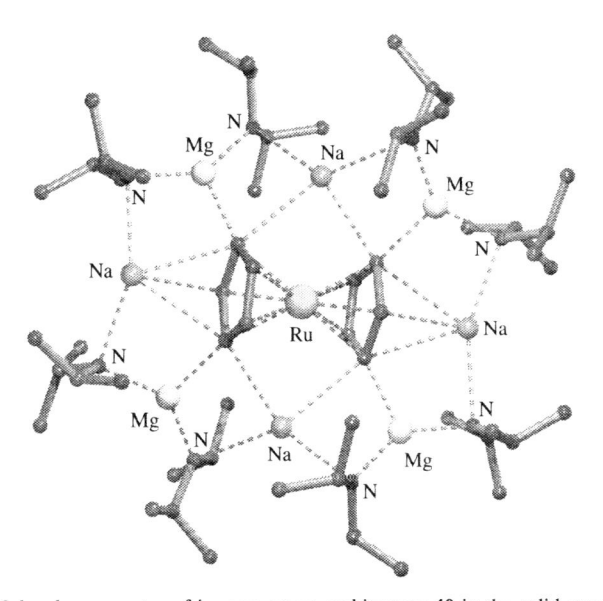

FIGURE 26. Molecular geometry of inverse crown architecture **40** in the solid state

The sixteen-membered ring consists of alternating magnesium and sodium atoms (four of each) with bridging amide nitrogen atoms between magnesium and sodium. The metallocene is selectively 1,3-1′,3′-tetra-deprotonated and each of the deprotonated carbon atoms forms a bridge-bond between magnesium and sodium while the 2- and 2′-carbon atoms have an additional interaction with a sodium atom (Figure 26).

In order to gain insight into the mechanism and species involved in the metallation of the arenes described above, the reaction steps prior to the metallation were studied in more detail[71]. The first step is the formation of Na[Bu$_3$Mg] from its parent organometallic compounds in a 1:1 molar ratio. The second step is the addition of three equivalents of the secondary amine, in this particular case three equivalents of TMPH. A detailed NMR spectroscopic study of this reaction mixture showed the presence of metal-bonded butyl groups and TMPH. From this observation it was concluded that not the anticipated Na[TMP$_3$Mg] but instead [BuMgNa(TMP)$_2$(TMPH)] had been formed. Most likely this latter compound is the actual intermediate that is active in the arene metallation step. Unfortunately, this compound was isolated as an oil and therefore its structural characterization by X-ray crystallography was impossible. However, addition of TMEDA afforded a crystalline compound with the formula [BuMgNa(TMP)$_2$(TMEDA)] (**42**). The X-ray crystal-structure determination of **42** reveals a structure with a central four-membered ring formed by a magnesium atom and a sodium atom with a butyl group bridge-bonded [C–Mg 2.200(2) and C–Na 2.669(2) Å] and an amido-nitrogen atom of one of the TMP groups bridge-bonded [N–Mg 2.079(1) and N–Na 2.452(1) Å] between these atoms (Figure 27). The other TMP group is σ-bonded [N–Mg 2.001(1) Å] to magnesium and the TMEDA molecule is chelate-bonded to sodium, giving this atom a tetrahedral coordination geometry.

Compound **42** is active in the deprotonation/metallation of arenes. When a solution of **42** is boiled under reflux in benzene, [PhMgNa(TMP)$_2$(TMEDA)] (**43**) is obtained which, according to its X-ray crystal-structure determination, is isostructural (bridging butyl group replaced by a bridging phenyl group) with **42**. In a similar way, using toluene, bis(benzene)chromium or bis(toluene)chromium, successful metallations to [3-(MeC$_6$H$_4$)MgNa(TMP)$_2$(TMEDA)] (**44**)[72], [{(C$_6$H$_6$)Cr(C$_6$H$_5$)}MgNa(TMP)$_2$(TMEDA)] (**45**)[73] and [{(MeC$_6$H$_5$)Cr(4-MeC$_6$H$_4$)}MgNa(TMP)$_2$(TMEDA)] (**46**)[74] were achieved. All three compounds were structurally characterized by X-ray crystal-structure determinations and are isostructural with **42** and **43**. It is noteworthy that toluene is selectively metallated at its 3-position, while the η^6-toluene group in bis(toluene)chromium is selectively metallated at its 4-position.

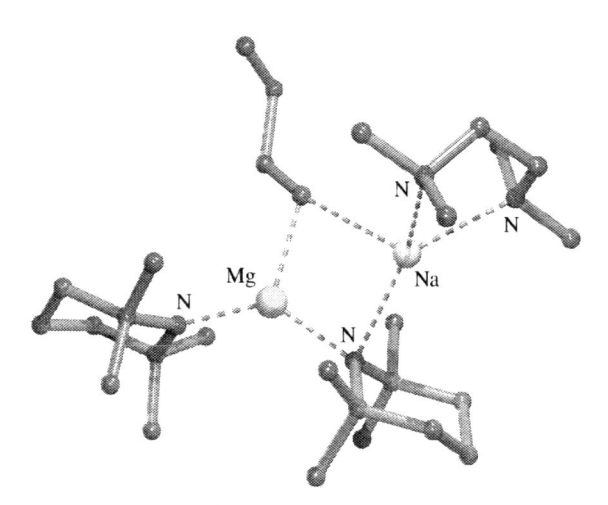

FIGURE 27. Molecular geometry of **42** in the solid state

III. DIORGANOMAGNESIUM COMPOUNDS
A. Donor-base-free Diorganomagnesium Compounds

As outlined before, diorganomagnesium compounds with two linear σ-bonded alkyl or aryl groups are very rare due to coordination unsaturation at magnesium. Usually, magnesium avoids such bonding situations by binding additional Lewis bases, by aggregating via three-center two-electron bonding or by forming agostic interactions. So far, the structures of only three diorganomagnesium compounds are known in which magnesium is two-coordinate.

Bis(neopentyl)magnesium (**47**) occurs in benzene solution as a trimer, for which both a linear structure **I** and a cyclic structure **II** (Figure 28) have been proposed[75].

Due to the high volatility of **47**, its structure in the gas phase could be determined by gas-phase electron diffraction[76]. This study showed that **47** exists as discrete monomers with a linear C—Mg—C arrangement with Mg—C bond distances of 2.126(6) Å in the gas phase.

The only two diorganomagnesium compounds with di-coordinated magnesium of which the structure in the solid state has been determined by X-ray crystallography are bis[(trimethylsilyl)methyl]magnesium (**48**)[77,78] and bis[(2,4,6-tri-*t*-butylphenyl]magnesium (**49**)[79]. Like observed in the gas phase for **47**, the magnesium atom of **48** in the solid state (Figure 29) has a perfect linear coordination geometry (C—Mg—C 180°). The observed Mg—C bond distance of 2.116(2) Å is also very close to the value observed for this bond in **47** in the gas phase.

In contrast to the linear structure of **47** and **48**, the X-ray crystal structure determination of **49** shows that in the solid state the di-coordinate magnesium atom has a bent structure [C-Mg-C 158.4(1)°]. This bending may be a consequence of the steric requirements of

FIGURE 28. Proposed structures for bis(neopentyl)magnesium **47** in solution

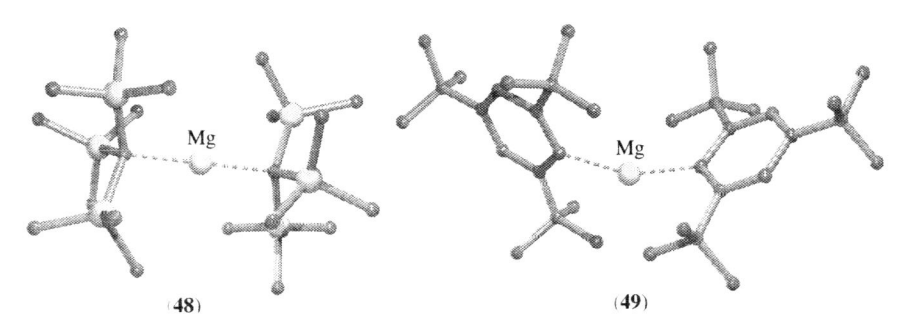

FIGURE 29. Molecular geometries of diorganomagnesium compounds **48** and **49** in the solid state

the bulky *t*-Bu groups, but also secondary (agostic) interactions of hydrogen atoms of the *ortho-tert*-butyl groups, two of which are in close proximity (2.28 Å) to the magnesium atom, might play a role. It was observed that **49** does not form donor complexes with Lewis bases like diethyl ether or THF, which is a striking difference with the 2,4,6-trimethyl- or 2,4,6-tri-isopropyl analogs of **49**[58].

When the bulky $(Me_3Si)_3C$ groups in compound **48** are replaced by less sterically demanding $(Me_3Si)_2CH$ groups, the structure of the resulting diorganomagnesium compound $[(Me_3Si)_2CH]_2Mg$ (**50**) in the solid state is entirely different. Its structure was determined both by X-ray and by neutron diffraction data and revealed a polymeric network of $[(Me_3Si)_2CH]_2Mg$ molecules linked via intermolecular agostic interactions with methyl groups of neighboring $[(Me_3Si)_2CH]_2Mg$ molecules (Figure 30)[80]. The intramolecular Mg−C distances are 2.117(4) and 2.105(4) Å, respectively, while the intermolecular (agostic) Mg−C interaction is 2.535 Å. This latter distance is considerably shorter than the sum of the Van der Waals radii (3.4 Å). Although the individual C−Mg−C bond angles deviate from 120°, the intramolecular C−Mg−C bond angle being 140.0(2)°, the sum of these bond angles is 360°, pointing to a distorted trigonal planar coordination geometry at magnesium.

When the steric congestion in **49** is slightly released, i.e. by replacement of the *t*-Bu groups by Et groups, again an entirely different structure for the corresponding diorganomagnesium compound in the solid state is observed. The X-ray crystal-structure determination of bis[2,6-diethylphenyl]magnesium (**51**) revealed a dimeric structure in which two of the four aryl groups are bridge-bonded [C−Mg 2.259(7) and 2.263(7) Å] between two magnesium atoms forming a central flat C−Mg−C−Mg four-membered ring (Figure 31)[81]. Furthermore, to each of the magnesium atoms an aryl group is terminal-bonded (C−Mg 2.121 Å) resulting in trigonal planar coordination at the magnesium atoms. The aryl groups are rotated out of the central C−Mg−C−Mg plane in a propeller-like fashion by angles in the range of 42.7 to 74.6°.

Bis-*tert*-butylmagnesium (**52**) also forms dimeric aggregates in the solid state. Its molecular geometry comprises two *t*-Bu groups each bridging between two magnesium atoms [C(1)−Mg(1) 2.3044(9), C(1)−Mg(2) 2.2978(8), C(2)−Mg(1) 2.3057(8) and

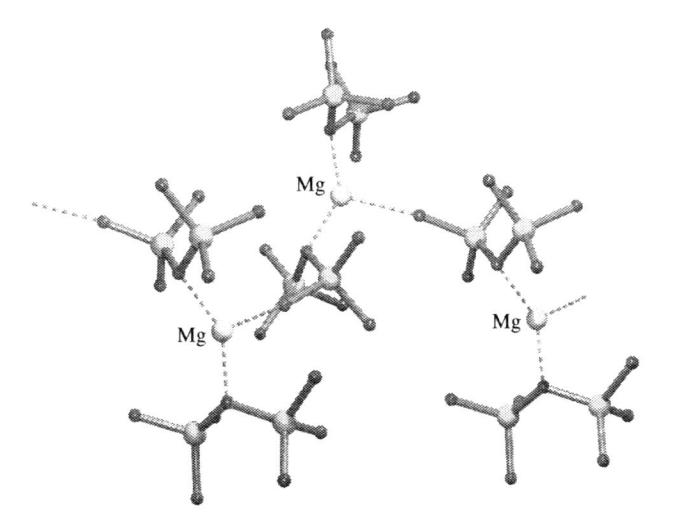

FIGURE 30. Part of the polymeric network of **50** in the solid state

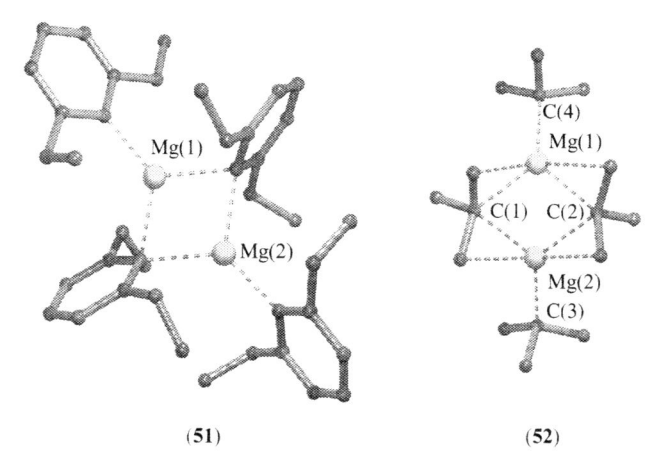

FIGURE 31. Molecular geometries of the dimeric diorganomagnesium compounds **51** and **52** in the solid state

C(2)–Mg(2) 2.2987(8) Å] as well as two t-Bu groups each terminally bonded to one magnesium atom [C(3)–Mg(2) 2.1483(8) and C(4)–Mg(1) 2.1424(9) Å] (Figure 31)[82]. The central four-membered C(1)–Mg(1)–C(2)–Mg(2) ring is folded, as is indicated by a C(1)–Mg(2)–Mg(1)–C(2) torsion angle of 140.45(4)°. Two methyl groups of each of the bridging t-Bu groups are rather close to a magnesium atom (Mg–C 2.489–2.542 Å). This distance is considerably less than the sum of the Van der Waals radii (3.4 Å) and points to agostic interactions with these methyl groups. It has been proposed that these agostic interactions promote β-hydrogen elimination and thus are responsible for the low thermal stability of **52**[82].

The simple diorganomagnesium compounds Me$_2$Mg (**53**) and Et$_2$Mg (**54**) are non-volatile solids, in strong contrast to their zinc analogs, which are low boiling liquids[29]. The structures of **53** and **54** in the solid state have been determined from X-ray powder diffraction data[21,22]. Both compounds form polymeric chains in the solid state (Figure 32) comprising a chain of magnesium atoms which are mutually connected by two bridging alkyl groups. As a consequence, each of the magnesium atoms has a tetrahedral coordination geometry. The observed Mg–C bond distances are 2.24(3) and 2.2(1) Å for **53** and **54**, respectively.

The structure of Ph$_2$Mg (**55**) in the solid state has been determined by a single-crystal X-ray diffraction study[83]. Like **53** and **54**, Ph$_2$Mg exists in the solid state as polymeric chains (Figure 33) in which two phenyl groups are symmetrically bridge-bonded [C–Mg 2.261(2) Å] between two magnesium atoms.

B. Diorganomagnesium Compounds Containing Multi-hapto-bonded Groups

The discovery and structural elucidation of ferrocene in 1951 and the subsequent development of metal-cyclopentadienyl chemistry started a new era in organometallic chemistry[84–86].

Soon after the first synthesis of bis(cyclopentadienyl)magnesium[87,88] (**56**), its structure in the solid state, based on X-ray powder diffraction data, was reported[89]. A more refined structure based on a single-crystal X-ray diffraction study was reported later[90]. The structure in the solid state is isostructural with that of ferrocene. The two parallel

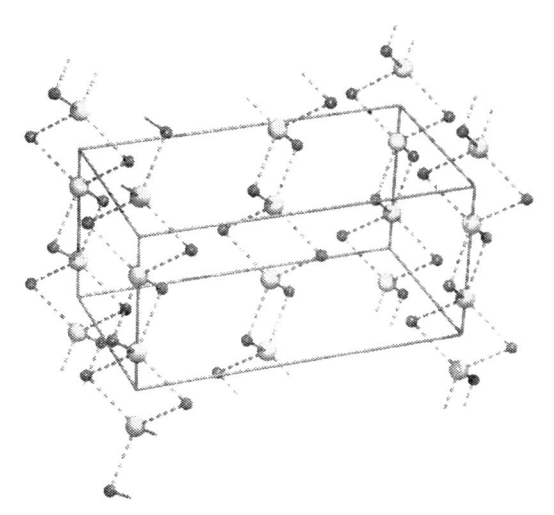

FIGURE 32. Unit cell content of **53** (space group I*bam*)

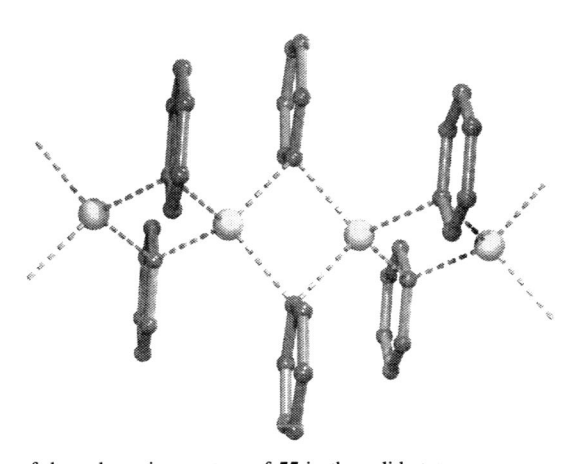

FIGURE 33. Part of the polymeric structure of **55** in the solid state

cyclopentadienyl rings each are η^5-bonded to the magnesium atom (Figure 34) with almost identical bond distances [average C–Mg 2.304(8) Å]. The two cyclopentadienyl rings adopt a staggered conformation, in contrast to the eclipsed conformation found for the structure of **56** in the gas phase, obtained from a gas-phase electron diffraction study[91].

Various substituted cyclopentadienylmagnesium compounds, $(t\text{-BuC}_5\text{H}_4)_2\text{Mg}^{92}$ (**56**), $[1,2,4\text{-}(Me_3Si)_3C_5H_2]_2Mg^{93}$ (**57**), $[1,2,4\text{-}(t\text{-Bu})_3C_5H_2]_2Mg^{94}$ (**58**), $(Me_4C_5H)_2Mg^{95}$ (**59**), $(t\text{-BuMe}_4C_5)_2Mg^{95}$ (**60**), $[(3\text{-butenyl})Me_4C_5]_2Mg^{96}$ (**61**) and $(Me_5C_5)_2Mg^{97}$ (**62**), have been prepared and were structurally characterized by X-ray crystallography. All compounds have a basic structural motif that is identical to **56**, but the conformation of the cyclopentadienyl rings is such that steric interference is minimal. A slight deviation from

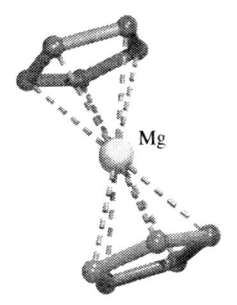

FIGURE 34. Molecular geometry of **56** in the solid state

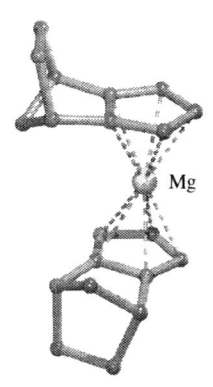

FIGURE 35. Molecular geometry of **61** in the solid state

the linear structure to a slightly bent structure is observed for **57**, **58** and **60**, due to the presence of bulky substituents.

The *exo,exo*-bis(iso-dicyclopentadienyl)magnesium metallocene (**61**) has been prepared by reacting Bu_2Mg with iso-dicyclopentadiene. Its structure in the solid state has been determined by X-ray crystallography (Figure 35)[98]. The two cyclopentadienyl rings are η^5-bonded to magnesium with bond distances that range from 2.314(1) to 2.347(1) Å and adopt a staggered conformation.

The substituted cyclopentadienylmagnesium compounds **62a** and **62b** have been prepared from the corresponding fulvenes (equation 8) and were structurally characterized in the solid state by X-ray crystallography[99]. The structures are, as expected, (η^5-bonded cyclopentadienyl groups) for bis(cyclopentadienyl)magnesium compounds.

$$(8)$$

(**62a**) $R^1 = Me$, $R^2 = Ph$
(**62b**) $R^1 = R^2 = c\text{-}Pr$

The magnesium atom of compound **62a** lies on a crystallographic inversion center, and consequently the substituents are in *anti*-configuration. The structure of **62b** shows an eclipsed conformation and leads to steric repulsion between the dicyclopropylmethyl groups (Figure 36). As a consequence a slight deviation from linear structure is observed.

Bis(indenyl)magnesium (**63**) has been prepared by the thermal decomposition of indenylmagnesium bromide, and its structure in the solid state has been established by X-ray crystallography[100]. Instead of the expected sandwich-type compound, a structure was found consisting of an infinite polymeric arrangement of which the repeating unit contains two magnesium atoms and four indenyl anions with two types of bonding modes (Figure 37). To each of the magnesium atoms an indenyl anion is η^5-bonded with its five-membered ring. The bond distances Mg(1)–C range from 2.31(1) to 2.54(1) Å and

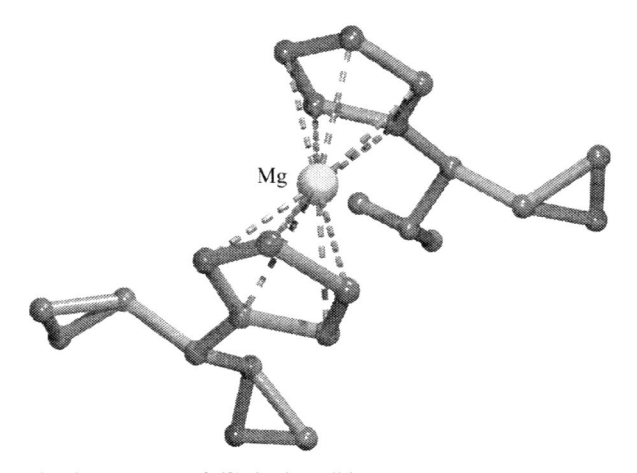

FIGURE 36. Molecular geometry of **62b** in the solid state

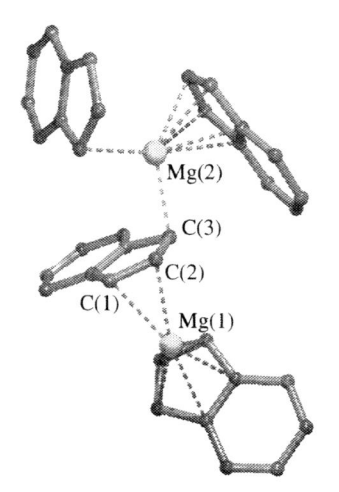

FIGURE 37. Molecular geometry of the repeating unit in polymeric **63**

Mg(2)–C range from 2.26(1) to 2.60(1) Å. One of the indenyl anions acts as bridge between the two magnesium atoms and is η^2-bonded to Mg(1) [Mg(1)–C(1) 2.40(1) and Mg(1)–C(2) 2.44(1) Å] and η^1-bonded to Mg(2) [Mg(2)–C(3) 2.26(1) Å]. In a similar bridging mode an indenyl anion is linking the repeating units.

Reaction of [Cp(Me)Mg(OEt$_2$)]$_2$ with phenylacetylene affords tetrameric [CpMgC≡CPh]$_4$ (**64**) (equation 9). Its structure in the solid state has been established by X-ray crystallography[101] and it is the only example of a heteroleptic diorganomagnesium compound for which the structure in the solid state is known. The structure of **64** has a heterocubane structure with alternating four magnesium atoms and four terminal carbon atoms of the acetylenic group at the corners of the cube. To each of the magnesium atoms a Cp group is η^5-bonded. The structure is shown schematically in equation 9. The bond distances between the terminal acetylenic carbon atoms and the magnesium atoms vary in a small range from 2.249(2) to 2.348(2) Å, resulting in an almost perfect cube.

$$2[\text{Cp(Me)Mg(OEt}_2)]_2 \;+\; 4\,\text{PhC}\equiv\text{CH} \xrightarrow{-2\text{MeH}} \quad (\textbf{64}) \qquad (9)$$

The structure of cyclopentadienyl(neopentyl)magnesium (**65**) in the gas phase has been determined by gas-phase electron diffraction[102]. The Mg–C bond distances of the η^5-bonded cyclopentadienyl group are 2.328(7) Å while the Mg–C bond distance of the neopentyl group was found to be 2.12(2) Å.

The structures in the solid state of bis(1-methylboratabenzene)magnesium (**66a**) and bis[3,5-dimethyl-1-(dimethylamino)boratabenzene]magnesium (**66b**) were determined by X-ray crystallography[103]. Both **66a** (Figure 38) and **66b** are typical sandwich structures and have common structural features. The magnesium atoms are located at crystallographic inversion centers, which implies coplanarity of the rings and an antiperiplanar arrangement with respect to the exocyclic substituents. The bond distances of the η^6-bonded boratabenzene ring to magnesium are Mg–C(1) 2.359(2), Mg–C(2) 2.422(2), Mg–C(3) 2.453(2), Mg–C(4) 2.420(2), Mg–C(5) 2.361(2) and Mg–B 2.436(2) Å.

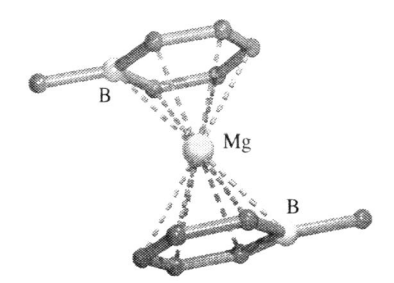

FIGURE 38. Molecular geometry of **66a** in the solid state

C. Diorganomagnesium Compounds Containing Intramolecularly Coordinating Substituents

In the early days of the development of organometallic chemistry it was thought that in many cases the metal–σ-carbon bond would be intrinsically unstable, especially in transition-metal organic compounds. Thermally induced homolytic cleavage of the metal–carbon bond and β-hydrogen elimination are the two most important pathways by which decomposition of organometallic compounds may occur. Several approaches have been put forward to suppress such decomposition pathways, e.g. the use of organic groups lacking β-hydrogen atoms, the introduction of bulky (often trimethylsilyl-containing) substituents and the use of organic groups containing a functionalized substituent capable of coordinating to the metal. The isolation and structural characterization of $(Me_3SiCH_2)_4Cu_4$[104] and $(2-Me_2NCH_2C_6H_4)_4Cu_4$[105] are clear examples of these approaches and represent the first examples of organocopper compounds sufficiently stable to allow their structural characterization by X-ray crystallography. In $(2-Me_2NCH_2C_6H_4)_4Cu_4$ the monoanionic, potentially bidentate $2-Me_2NCH_2C_6H_4$ ligand stabilizes the organocopper compound via intramolecular coordination of the nitrogen to copper. This particular ligand has been used in the early days to stabilize a variety of organometallic compounds. When other ligand skeletons and also other heteroatom-functionalized substituents capable of intramolecular coordination are included, several thousands of these organometallic derivatives, covering almost the whole periodic system of the elements, have been structurally characterized[23].

It is rather surprising that only a few diorganomagnesium compounds have been reported in which intramolecular coordination of a heteroatom-containing substituent is present. The synthesis of $(2-Me_2NCH_2C_6H_4)_2Mg$ (**67**) was reported. It has been used in a study on the influence of the presence of potentially intramolecular coordinating substituents on Schlenk equilibria[106]. However, it has never been structurally characterized.

The monoanionic potentially bidentate 2-[(dimethylamino)methyl]ferrocenyl ligand has coordinating properties similar to that of the 2-[(dimethylamino)methyl]phenyl ligand, and has also been used to stabilize a variety of organometallic derivatives. Bis{[(2-dimethylamino)methyl]ferrocenyl}magnesium (**68**) has been synthesized and was structurally characterized in the solid state by X-ray crystallography.

The molecular structure of **68**, crystallized from a solution containing THF and Et$_2$O, comprises two C,N-chelate bonded 2-[(dimethylamino)methyl]ferrocenyl groups [Mg–C 2.151(2) and 2.160(2) Å, and Mg–N 2.421(2) and 2.419(2) Å] (Figure 39). In addition, a THF molecule is coordinated [Mg–O 2.077(2) Å] to the magnesium atom resulting in five-coordinate magnesium. Based on the bond angles around magnesium the coordination geometry shows a 63% distortion from a trigonal bipyramid (with the carbon atoms and the oxygen atom in the equatorial plane) towards a square pyramid along the Berry pseudo-rotation coordinate[107]. When crystallization was performed in the absence of THF, the diethyl ether adduct was isolated, having structural features that are very similar to those of **68**.

An aggregate of bis{[(2-dimethylamino)methyl]ferrocenyl}magnesium with two molecules of LiBr (**69**) has been isolated and structurally characterized (Figure 40)[108]. This compound is a nice illustration of the capability of diorganomagnesium compounds to aggregate with other metal salts. In **69**, each of the 2-[(dimethylamino)methyl]ferrocenyl groups is slightly asymmetrically bridge-bonded with its carbon atom between magnesium and lithium [Mg–C 2.169(4) and 2.167(4) Å, and Li–C 2.390(8) and 2.311(8) Å]. Also, the bromine atoms are bridge-bonded between magnesium and lithium [Mg–Br 2.600(1) and 2.605(1), and Li–Br 2.508(7) and 2.493(7) Å] leading to a distorted tetrahedral coordination geometry at magnesium. The nitrogen atoms of the (dimethylamino)methyl substituents are each coordinating to a lithium atom [Li–N 2.074(8) and 2.065(7) Å]

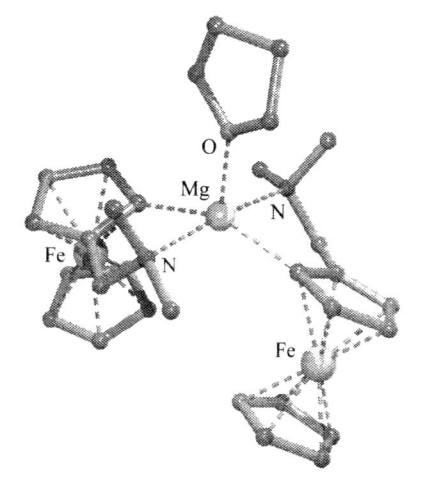

FIGURE 39. Molecular geometry of **68** in the solid state

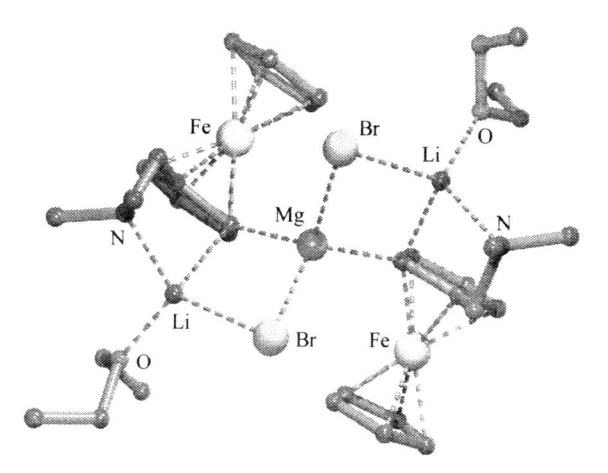

FIGURE 40. Molecular geometry of **69** in the solid state

while coordination saturation at each lithium atom is reached by the coordination of an additional diethyl ether molecule.

The potentially bidentate α-(2-pyridyl)-α,α-bis(trimethylsilyl)methyl monoanionic ligand also has been used in a variety of organometallic derivatives. Its magnesium derivative bis[α-(2-pyridyl)-α,α-bis(trimethylsilyl)methyl]magnesium (**70**) has been structurally characterized by X-ray crystallography (Figure 41)[109].

In **70**, the two α-(2-pyridyl)-α,α-bis(trimethylsilyl)methyl ligands are *C,N*-chelate bonded to magnesium [Mg–C 2.21(9), Mg–N 2.13(1) Å]. As a consequence of the four-membered chelate rings the coordination geometry at magnesium is distorted from tetrahedral, as is indicated by the large C–Mg–C and N–Mg–N bond angles of 157.0(7)° and 117(4)° respectively, and the acute C–Mg–N angle of 67.3(2)°.

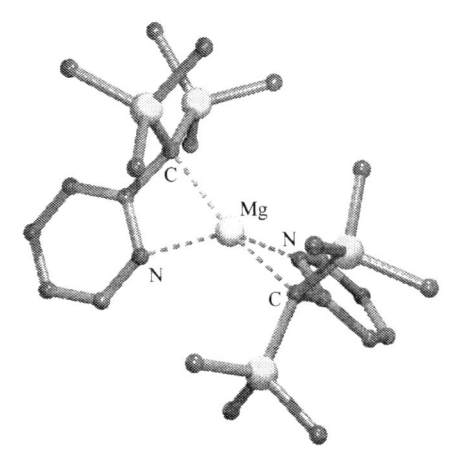

FIGURE 41. Molecular geometry of **70** in the solid state

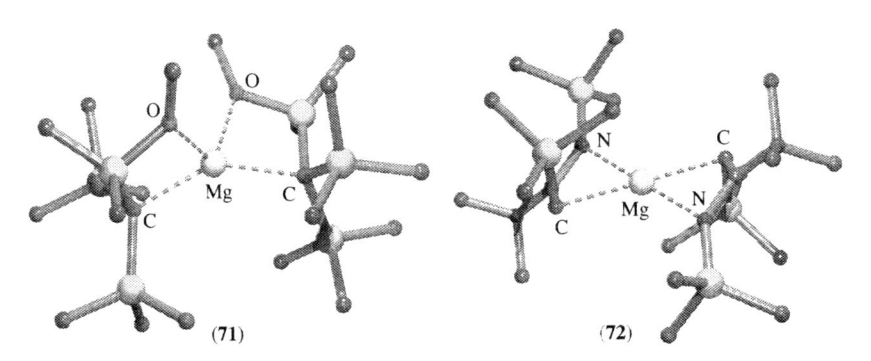

FIGURE 42. Molecular geometries of **71** and **72** in the solid state

A similar distortion of the tetrahedral coordination geometry at magnesium was observed in [MeOSi(Me)$_2$C(SiMe$_3$)$_2$C]$_2$Mg (**71**) in which two *C,O*-chelating ligands are forming four-membered chelate rings (Figure 42)[110].

An unusual planar coordination geometry at magnesium has been observed in [Me$_3$SiN = C(*t*-Bu)CHSiMe$_3$]$_2$Mg (**72**) in which two *C,N*-chelating ligands are present (Figure 42)[111]. However, the observed bond distances [Mg–C(1) 2.284(4), Mg–C(2) 2.408(4) and Mg–N 2.084(3) Å] suggest that the ligand binds rather in an aza-allyl type of manner than in a *C,N*-chelate bonding mode.

A heteroleptic diorganomagnesium compound [(Me$_2$N(Me)$_2$Si)(Me$_3$Si)$_2$C](*n*-Bu)Mg (THF) (**73**) has been synthesized and characterized by X-ray crystallography[112]. Its structure comprises one *C,N*-chelate bonded (Me$_2$N(Me)$_2$Si)(Me$_3$Si)$_2$C group [Mg–C 2.241(2) and Mg–N 2.203(2) Å], one σ-bonded *n*-butyl group [Mg–C 2.130(3) Å] and an additional coordinating THF molecule [Mg–O 2.069(2) Å] to complete a distorted tetrahedral coordination geometry at magnesium.

An elegant synthetic pathway to bis[(3-(dialkyl)aminobutyl]magnesium and bis[4-(dialkylamino)butyl]magnesium compounds was developed involving the addition reaction of dialkylallylamines and 1-dialkylamino-3-alkenes to highly reactive MgH$_2$[113] in

1. Structural organomagnesium chemistry

the presence of catalytic amounts (1mol%) of $ZrCl_4$[114]. According to this procedure, compounds **74a–74i** and **75a–75g** have been prepared (equation 10). ^{1}H and ^{13}C NMR spectroscopic studies in solutions indicated that in all these compounds the nitrogen atoms are involved in intramolecular coordination.

$$MgH_2 + 2 \diagup\diagdown(CH_2)_n\diagup NR^1R^2 \xrightarrow[ZrCl_4]{THF} [R^1R^2N(CH_2)_{n+2}]_2Mg \quad (10)$$

(**74a**) $n = 1, R^1 = R^2 = Me$ (**75a**) $n = 2, R^1 = R^2 = Me$
(**74b**) $n = 1, R^1 = R^2 = Et$ (**75b**) $n = 2, R^1 = R^2 = Et$
(**74c**) $n = 1, R^1 = R^2 = n\text{-}Pr$ (**75c**) $n = 2, R^1 = R^2 = n\text{-}Pr$
(**74d**) $n = 1, R^1 = R^2 = i\text{-}Pr$ (**75d**) $n = 2, R^1 = R^2 = n\text{-}Bu$
(**74e**) $n = 1, R^1 = R^2 = n\text{-}Bu$ (**75e**) $n = 2, R^1 = Me, R^2 = Et$
(**74f**) $n = 1, R^1 = Me, R^2 = Et$ (**75f**) $n = 2, R^1 = Me, R^2 = n\text{-}Bu$
(**74g**) $n = 1, R^1 = Me, R^2 = n\text{-}Bu$ (**75g**) $n = 2, R^1 = Me, R^2 = c\text{-}Hex$
(**74h**) $n = 1, R^1 = Me, R^2 = c\text{-}Hex$
(**74i**) $n = 1, R^1 = Me, R^2 = Ph$

Making use of the same procedure, the ether-functionalized diorganomagnesium compounds **76a–76f** were prepared from 3-butenyl ethers and MgH_2 (equation 11)[114]. It should be noted that the addition reaction of allyl ethers to MgH_2 failed because in that case ether cleavage by MgH_2 becomes a competing reaction. Also, for these compounds intramolecular O–Mg coordination in solution was established by NMR spectroscopic studies.

$$MgH_2 + 2 \diagup\diagdown\diagup OR \xrightarrow[ZrCl_4]{THF} [RO(CH_2)_4]_2Mg \quad (11)$$

(**76a**) R = Me
(**76b**) R = Et
(**76c**) R = n-Pr
(**76d**) R = n-Bu
(**76e**) R = n-Pent
(**76f**) R = n-Hex

An X-ray crystal-structure determination of **76a** confirmed that intramolecular O–Mg coordination is present in the solid state (Figure 43). Both 4-methoxybutyl ligands are C,O-chelate bonded to magnesium [Mg–C 2.144(4) and Mg–O 2.071(3) Å]. The bond angles around magnesium [C–Mg–C' 140.2(2)°, O–Mg–O' 96.4(1)°, C–Mg–O 95.7(1)° and C–Mg–O' 110.8(1)°] indicate that the coordination geometry at magnesium is distorted from the ideal tetrahedral geometry.

The above-mentioned functionalized diorganomagnesium compounds undergo clean redistribution reactions with Et_2Mg to form the heteroleptic diorganomagnesium compounds. For example, reaction of $[Me_2N(CH_2)_3]_2Mg$ (**74a**) with Et_2Mg gives $Me_2N(CH_2)_3$ MgEt (**77**) in quantitative yield[115]. An X-ray crystal-structure determination of **77a** showed this compound to exist as a centrosymmetric dimer in the solid state (Figure 43). The two dimethylaminopropyl groups are η^2-C bridge-bonded between two magnesium atoms [Mg–C 2.294(2) and 2.273(2) Å] while both nitrogen atoms are coordinated to the magnesium atoms [Mg–N 2.181(2) Å]. To each of the magnesium atoms an ethyl group is σ-bonded [Mg–C 2.142(3) Å]. Also, the structure of heteroleptic $MePhN(CH_2)_3MgEt$

(76a)   (77a)

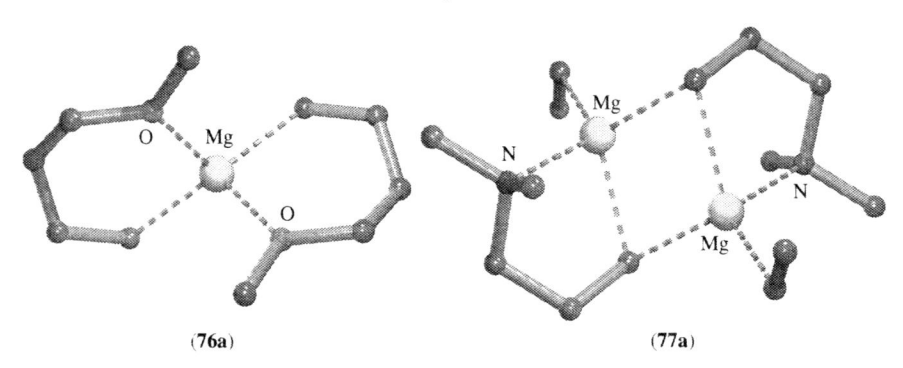

FIGURE 43. Molecular geometries of **76a** and **77a** in the solid state

(**77b**) in the solid state was established by X-ray crystallography. Its strucural features are similar to those of **77a**.

A remarkable structure was found for bis(*ortho*-anisyl)magnesium which crystallizes from THF as a dimeric bis-THF adduct (**78**) (Figure 44)[116]. This aggregate contains four *ortho*-anisyl groups with three different bonding modes. One anisyl group is μ^2-bridge-bonded between the two magnesium atoms [C(1)–Mg(1) 2.327(6) and C(1)–Mg(2) 2.305(6) Å] while the oxygen substituent is intramolecularly coordinated to one of these magnesiums [O–Mg(2) 2.166(4) Å]. A second anisyl group is σ-bonded to Mg(1) [C(2)–Mg(1) 2.199(7) Å] and the oxygen of the anisyl functionality coordinates to Mg(2) [O–Mg(2) 2.056(5) Å]. The two other *ortho*-anisyl groups are σ-bonded via C_{ipso} to different magnesium atoms [C–Mg(1) 2.147(7) and C–Mg(2) 2.132(6) Å] while the oxygen substituents are not involved in coordination to magnesium. Finally, to each of the magnesium atoms a THF molecule is coordinated, resulting in one four-coordinate magnesium atom [Mg(1)] and one five-coordinate magnesium atom [Mg(2)]. The rather strange structural motif present in **78** has been explained in terms of an intramolecular 'ate'-type of structure in which Mg(2) has a formally partial negative charge and Mg(1) has a formally partial positive charge.

A pseudo-trigonal-bipyramidal arrangement at magnesium was observed in the crystal structure of magnesacycle (**79**) (Figure 45)[106]. The two carbon atoms [C(1) and C(2)] of the σ-bonded aryl groups and the oxygen atom [O(2)] of one of the coordinating THF molecules lie in the equatorial plane. The intramolecular coordinating ether functionality and the oxygen atom of the other coordinating THF molecule are at the axial positions. As expected, the C-O distances of the axially bonded oxygen atoms [Mg–O(1) 2.242(4) and Mg–O(3) 2.221(4) Å] are significantly longer than of the equatorial one [Mg–O(2) 2.095(3) Å].

An X-ray crystal-structure determination of bis[1,3-bis-{(dimethoxy)methyl}phenyl]magnesium (**80**) revealed a distorted octahedral coordination geometry at magnesium because all four methoxy substituents are involved in intramolecular Mg–O coordination (Figure 45)[106]. The Mg—C bond distances are very short [Mg–C 2.093(4) and 2.105(4) Å], but are compensated for by relatively long Mg—O bond distances (average 2.315 Å). The C–Mg–C bond angle [173.4(2)°] deviates only slightly from linear.

An unprecedented metallation was observed when 1,3-xylyl crown ethers are reacted with diarylmagnesium compounds. Reaction of 1,3-xylene-15-crown-4 with diphenylmagnesium gives in quantitative yield 2-(phenylmagnesio)-1,3-xylene-15-crown-4 (**81**), the structure of which was established by an X-ray crystal-structure determination

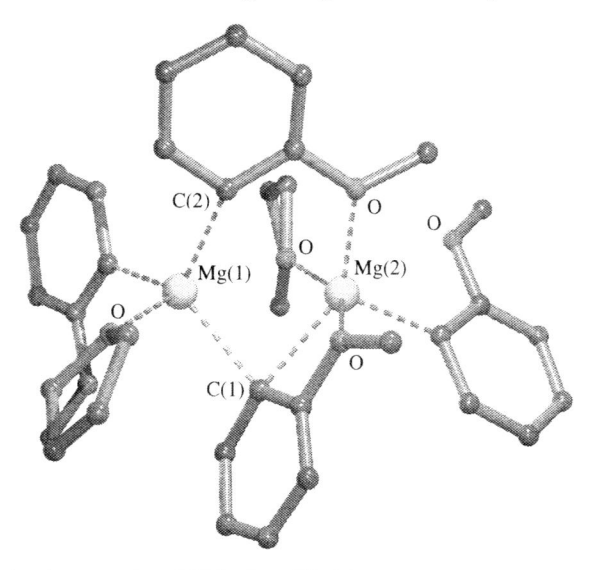

FIGURE 44. Molecular geometry of **78** in the solid state

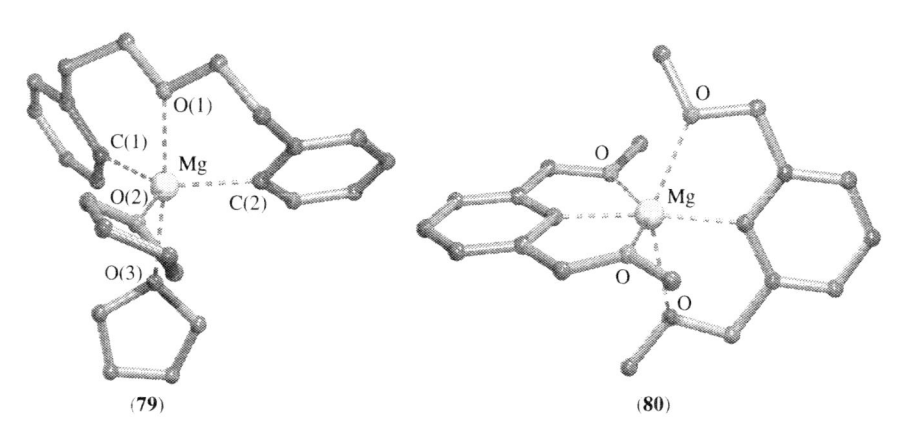

FIGURE 45. Molecular geometries of **79** and **80** in the solid state

(Figure 46)[117]. The Mg–C distances [Mg–C(1) 2.127(4) and Mg–C(2) 2.154(4) Å] are as expected for aryl groups σ-bonded to magnesium. All four oxygen atoms of the crown are involved in intramolecular coordination with Mg–O bond distances ranging from 2.183(3) to 2.619(3) Å, leading to a six-coordinate magnesium atom. Similarly, 2-[(4-*tert*-butylphenyl)magnesio]-1,3-xylene-18-crown-5 (**82**) has been prepared and structurally characterized in the solid state (Figure 46)[118]. Compound **82** has structural features similar to those of **81**, but only four of the five oxygen atoms of the crown are involved in intramolecular coordination to magnesium. The formation of compounds **81** and **82** has been explained by a mechanism involving arylmagnesium cations encapsulated in the crown and tris[aryl]magnesiate anions[118].

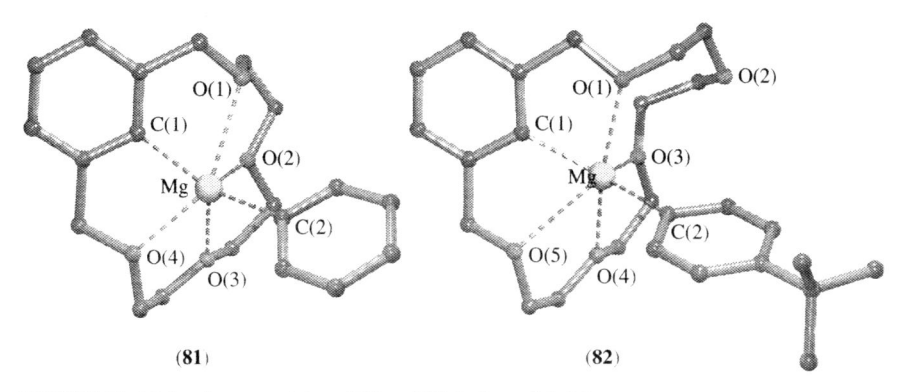

FIGURE 46. Molecular geometries of **81** and **82** in the solid state

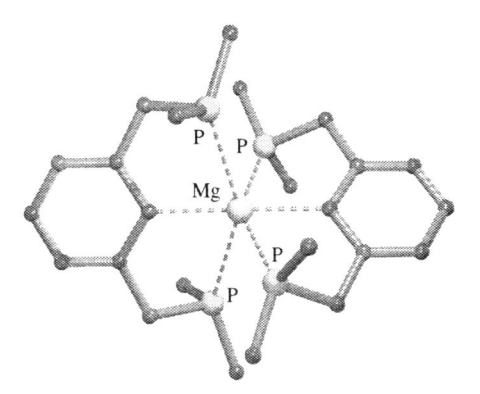

FIGURE 47. Molecular geometry of **83** in the solid state

Bis[1,3-bis{(dimethylphosphino)methyl}phenyl]magnesium (**83**) is the only example of a diorganomagnesium compound in which phosphorus to magnesium coordination in the solid state is established unambiguously by X-ray crystallography. The molecular geometry of **83** in the solid state comprises a centrosymmetric molecule in which the two aryl groups are σ-bonded to magnesium [Mg–C 2.216(1) Å] and all four phosphorus atoms are involved in intramolecular coordination [Mg–P 2.770(1) and 2.761(1) Å] (Figure 47)[119]. The coordination geometry at magnesium is a distorted octahedral one, with only a slight deviation of the C–Mg–C bond angle [178.10(8)°] from linear.

D. Donor–Acceptor Complexes of σ-Bonded Diorganomagnesium Compounds

As has been outlined before, the preferred coordination geometry at magnesium in organomagnesium compounds is tetrahedral, although also organomagnesium compounds are known with either lower or higher coordination numbers.

The only diorganomagnesium compound with three-coordination at magnesium, for which the structure was established by X-ray crystallography, is [(Me$_3$Si)$_2$CH]$_2$Mg(OEt$_2$) (**84**) (Figure 48)[120]. Probably, the combination of two sterically demanding (Me$_3$Si)$_2$CH groups and a rather bulky diethyl ether molecule in proximity to the magnesium atom

(84) (85)

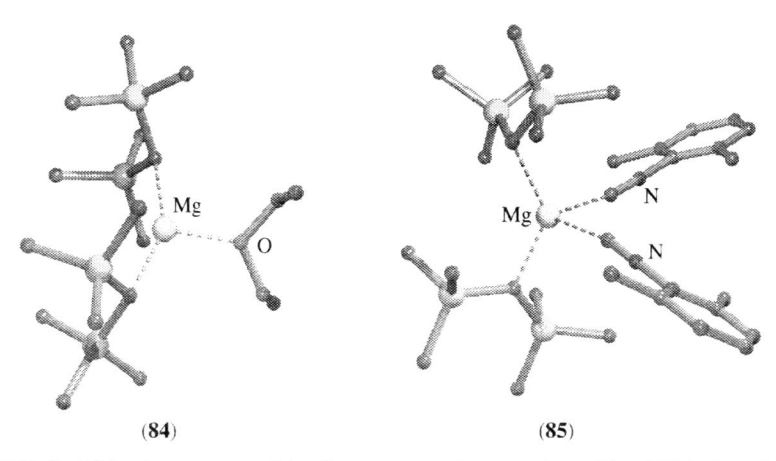

FIGURE 48. Molecular geometry of the diorganomagnesium complexes **84** and **85** in the solid state

prevents the coordination of a second diethyl ether molecule. Although in **84** the C–Mg–C bond angle is rather large (148.45°), the sum of the bond angles around magnesium is within experimental error 360°, indicating a planar trigonal coordination geometry.

The same diorganomagnesium compound is capable of coordinating two 2,6-xylylisocyanide molecules, thus forming [(Me$_3$Si)$_2$CH]$_2$Mg(CNC$_6$H$_3$Me$_2$-2,6)$_2$ (**85**) with a distorted tetrahedral coordination geometry at magnesium (Figure 48)[121]. The C–Mg–C bond angle of the bonding carbon atoms [Mg–C 2.148(7) and 2.138(9) Å] of the (Me$_3$Si)$_2$CH groups in **85** has narrowed to 128.40(7)° compared to that in **84**. However, the C–Mg–C bond angle of carbon atoms of the coordinating isocyanide groups [Mg–C 2.306(9) and 2.307(10) Å] is extremely acute with 88.79(2)°.

The usual, distorted, tetrahedral coordination geometry at magnesium has been established by X-ray crystallographic studies for a series of diorganomagnesium complexes **86–92** containing amine ligands. The relevant structural features of these complexes are summarized in Table 1. Notable are the acute N–Mg–N angles in the TMEDA complexes **86–91**; probably this is a consequence of the small bite angle of the chelating TMEDA ligand.

Like amines, ethers like THF and diethyl ether are also capable of coordinating to the magnesium atom of diorganomagnesium compounds to form 1:2 adducts causing a tetrahedral coordination geometry at magnesium. The structures in the solid state of a series of these adducts (**93–99**) have been unambiguously determined by X-ray crystallography. The relevant structural data of these compounds are compiled in Table 2. With the exception of the THF adduct of bis(*p*-tolyl)magnesium (**94**), *vide infra*, these compounds are discrete monomeric species. Notable are the relatively large C–Mg–C bond angles, which are compensated by acute O–Mg–O angles.

The X-ray crystal-structure determination of the THF adduct of bis(*p*-tolyl)magnesium showed that the unit cell contains two different molecules, a monomer (4-MeC$_6$H$_4$)$_2$Mg(THF)$_2$ (**94a**) and a dimeric molecule (4-MeC$_6$H$_4$)$_4$Mg$_2$(THF)$_2$ (**94b**) (Figure 49)[83].

The structure of monomer **94a** is straightforward and isostructural with those of **93** and **95–97**. In dimer **94b** two *p*-tolyl groups are bridge-bonded with the C$_{ipso}$ atoms between two magnesium atoms [Mg–C 2.245(7) and 2.313(7) Å]. To each of the magnesium atoms a *p*-tolyl group is σ-bonded [Mg–C 2.130(7) Å] and an additional THF molecule is coordinate bonded [Mg–O 2.020(5) Å], to give four-coordinate magnesium atoms. The

TABLE 1. Relevant structural features of diorganomagnesium complexes **86–92**

Compound	Mg–C (Å)	Mg–N (Å)	C–Mg–C (°)	N–Mg–N (°)	Reference
Me$_2$Mg(TMEDA) (**86**)	2.166(6) 2.166(6)	2.257(6) 2.227(6)	130.0(4)	81.5(3)	122
Ph$_2$Mg(TMEDA) (**87**)	2.167(5) 2.167(5)	2.205(5) 2.199(5)	119.2(1)	82.5(1)	123
Bn$_2$Mg(TMEDA) (**88**)	2.169(2) 2.169(2)	2.192(2) 2.207(2)	117.12(7)	83.36(5)	124
Et$_2$Mg(TMEDA) (**89**)	2.163(6) 2.137(6)	2.236(5) 2.237(7)	127.7(3)	82.7(2)	125
s-Bu$_2$Mg(TMEDA) (**90**)	2.181(3) 2.181(3)	2.252(3) 2.252(3)	133.6(3)	81.0(2)	126
(Ph$_2$PCH$_2$)$_2$Mg(TMEDA) (**91**)	2.171(4) 2.171(4)	2.226(4) 2.226(4)	130.0(2)	82.3(2)	127
Me$_2$Mg(quin)$_2$ [a] (**92**)	2.163(9) 2.224(8)	2.231(6) 2.247(6)	129.0(3)	108.2(2)	128

[a] Quinuclidine (1-azabicyclo[2.2.2]octane).

TABLE 2. Relevant structural features of diorganomagnesium ether complexes **93–99**

Compound	Mg–C (Å)	Mg–O (Å)	C–Mg–C (°)	O–Mg–O (°)	Reference
Ph$_2$Mg(THF)$_2$ (**93**)	2.132(8) 2.126(7)	2.050(5) 2.031(6)	124.4(3)	96.7(2)	83
(4-MeC$_6$H$_4$)$_2$Mg(THF)$_2$ (**94a**)	2.181(3) 2.181(3)	2.252(3) 2.252(3)	133.6(3)	81.0(2)	83
(2,4,6-Me$_3$C$_6$H$_2$)$_2$Mg(THF)$_2$ (**95**)	2.182(3) 2.165(3)	2.067(3) 2.079(3)	118.8(1)	88.4(1)	58
(2,4,6-i-Pr$_3$C$_6$H$_2$)$_2$Mg(THF)$_2$ (**96**)	2.179(3) 2.177(3)	2.107(2) 2.110(2)	123.1(1)	87.1(1)	58
(2-C$_2$H$_3$C$_6$H$_4$)$_2$Mg(THF)$_2$ (**97**)	2.14(1) 2.14(1)	2.044(8) 2.027(8)	127.8(5)	91.2(4)	129
Ph[(Me$_3$Si)$_3$Si]Mg(THF)$_2$ (**98**)	2.150(4) 2.650(1) [a]	2.051(3) 2.059(3)	128.2(1) [b]	95.4(1)	130
[Ph(Me)HC]$_2$Mg(OEt$_2$)$_2$ (**99**)	2.195(1) 2.195(1)	2.058(1) 2.058(1)	122.2(1)	93.8(1)	131

[a] Mg–Si bond length.
[b] C–Mg–Si bond angle.

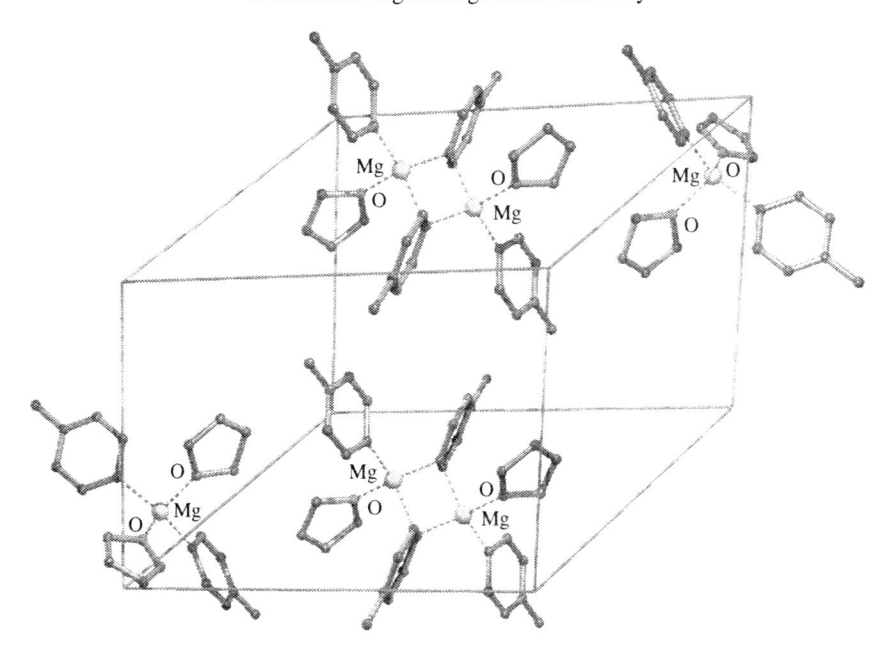

FIGURE 49. Unit-cell contents of **94** in space-group $P\bar{1}$

acute Mg–C–Mg bond angle of 77.5(2)° is in the range expected for bridging three-center two-electron bonded aryl groups.

It should be noted that in the solid-state structure of **99** the two chiral centers within an individual molecule have identical configurations, either both R or both S, although **99** was prepared from racemic starting material[131]. However, as a requirement of the space-group symmetry ($C2/c$) both enantiomers are present in 1:1 molar ratio in the crystal lattice.

Diorganomagnesium compounds also form complexes with bis-donor-atom ligands that are not capable of forming chelates. When dimethylmagnesium is crystallized from a THF solution that contains DABCO, a complex $(Me_2Mg)_2(DABCO)(THF)_2$ (**100**) is obtained. An X-ray crystal-structure determination showed that this complex consists of two $Me_2Mg(THF)$ moieties between which a DABCO molecule is N,N'-bridge bonded [Mg–N (2.208(3) Å] (Figure 50)[132]. As a result the magnesium atoms have a slightly distorted tetrahedral coordination geometry.

In many cases diorganomagnesium dioxane complexes, $R_2Mg(dioxane)$, have been obtained as unwanted side-products. Only for two of these, $Et_2Mg(dioxane)$[133] (**101**) and neo-$Pent_2Mg(dioxane)$[134] (**102**), were the structures in the solid state determined by X-ray crystallography. Both compounds exist in the solid state as polymeric chains in which dioxane molecules are O,O'-bridge bonded between the Et_2Mg units in **101** [Mg–O 2.077(2) and 2.084(2) Å], and the neo-$Pent_2Mg$ units in **102** [Mg–O 2.132(1) Å], resulting in tetrahedral coordinate magnesium atoms. The structure of **101** is shown in Figure 50.

During studies on bifunctional organomagnesium compounds[135,136], the structures in the solid state of several such compounds were determined. Association measurements of magnesiacyclohexane in THF solution indicated that this compound is in equilibrium with a dimer. An X-ray crystal-structure determination of the product that crystallizes from

(100) (101)

FIGURE 50. Molecular geometry of 100 in the solid state and part of the polymeric chain of 101 in the solid state

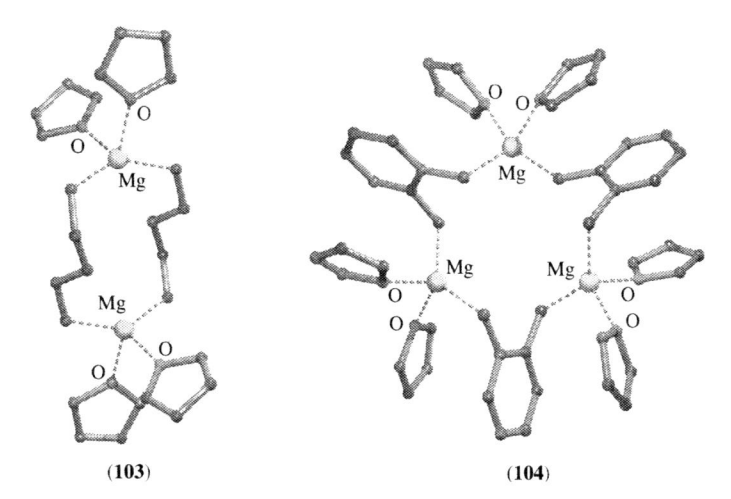

(103) (104)

FIGURE 51. Bifunctional cyclic organomagnesium compounds 103 and 104 in the solid state

such solutions indicated the presence of a cyclic dimer, the 1,7-dimagnesiocyclododecane tetra THF complex (103) in the solid state (Figure 51)[137,138]. To each of the magnesium atoms in the twelve-membered ring two additional THF molecules are coordinated to complete a tetrahedral coordination geometry at the magnesium atoms.

The o-xylidenemagnesium bis THF complex (104) exists in the solid state as a cyclic trimer (Figure 51)[139]. Each of the benzylic carbon atoms of the xylidene moieties is σ-bonded to a magnesium atom, thus forming a nine-membered ring. To each of the magnesium atoms two THF molecules are coordinated to give a tetrahedral coordination geometry.

The structures in the solid state of the 1,2-phenylenemagnesium THF complex (105), the 1,8-naphthalenediylmagnesium THF complex (106) and the cis-diphenylvinylenemagnesium THF complex (107) were determined by X-ray crystallography[140]. All three

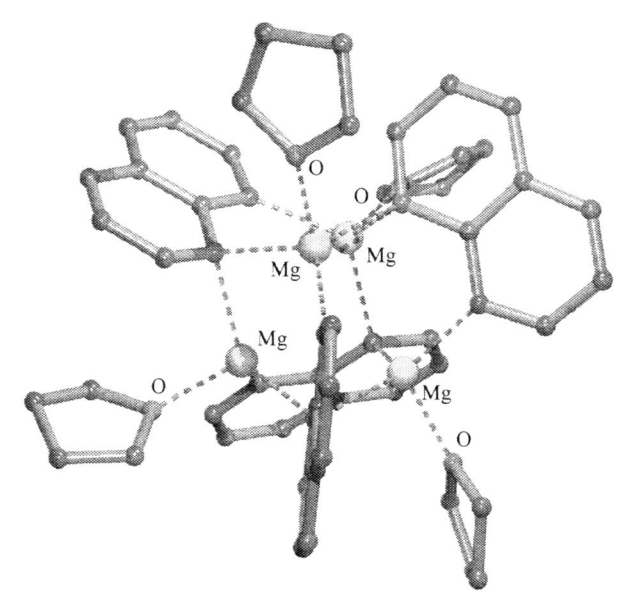

FIGURE 52. Molecular geometry of the tetrameric aggregate **106**

compounds are tetrameric aggregates with similar structural features. The core of these compounds consists of a tetrahedron of four magnesium atoms, arranged in a similar fashion as the Li_4 core of many tetrameric organolithium compounds[141]. Above each face of the tetrahedron an organic fragment is positioned that is bonded with two carbon atoms to three magnesium atoms. One of the carbon atoms is σ-bonded to one magnesium atom and the other carbon atom bridges two magnesium atoms via a three-center two-electron bond. To the top positions of each of the magnesium atoms a THF molecule is coordinated to complete a distorted tetrahedral coordination geometry at each of the magnesium atoms. As an illustrative example the structure of **106** is shown in Figure 52.

Tridentate nitrogen- or oxygen-containing ligands form complexes with diorganomagnesium compounds in which the magnesium atom is five-coordinate. The X-ray crystal-structure determination of $Me_2Mg(PMDTA)$ (**108**)[125,132] shows that in the solid state the magnesium atom exhibits a trigonal bipyramidal coordination geometry with the methyl groups [Mg−C 2.173(4) and 2.191(4) Å] and the central nitrogen atom [Mg−N 2.381(3) Å] at equatorial positions (Figure 53). The terminal nitrogen atoms occupy the axial positions. The sum of the bond angles in the equatorial plane is 360° within experimental error, but the N−Mg−N bond angle [138.3(1)°] between the axial nitrogen atoms deviates considerably from linear as a consequence of the presence of two five-membered chelate rings.

Bis(4-*tert*-butylphenyl)magnesium forms a complex with diglyme, complex **109**, in which all three oxygen atoms of the diglyme ligand are involved in coordination to magnesium. Like in **108**, in **109** the magnesium atom has a trigonal bipyramidal coordination geometry with the bonding carbon atoms and the central oxygen atom of the diglyme ligand in equatorial positions (Figure 53)[142]. Also, in this complex the bond angles in the equatorial plane add up to 360°. The terminal oxygen atoms of the diglyme ligand are at the axial sites, but as a consequence of the presence of the two five-membered chelate rings the O−Mg−O bond angle [141.97(9)°] deviates considerably from linear.

(108)   (109)

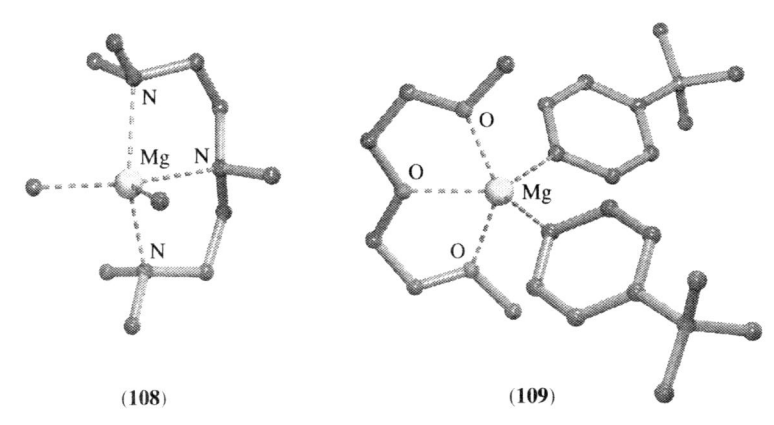

FIGURE 53. Molecular geometry of five-coordinate organomagnesium complexes **108** and **109** in the solid state

The structure of bis(4-*tert*-butylphenyl)magnesium tetraglyme (**110**) in the solid state has also been determined by X-ray crystallography[142]. Only the terminal and the next two oxygen atoms of the tetraglyme ligand are involved in coordination to magnesium. The asymmetric unit of **110** contains three crystallographically independent molecules in which the environments around the magnesium atoms are similar to that observed in **109** but differ in the orientation of the uncomplexed tail.

The crown ether 1,3,16,18-dixylylene-30-crown-8 is capable of forming a complex (**111**) with two molecules of bis(4-*tert*-butylphenyl)magnesium of which the structure in the solid state was determined by X-ray crystallography[142]. To each of the magnesium atoms three oxygen atoms of the crown ether are coordinated. Also, in this compound the environment around the magnesium atoms is similar to that observed in **109**. It has been suggested that the formation of complexes like **111** is the initial step in the formation of diorganomagnesium–rotaxane-type compounds, *vide infra*.

The structures in the solid state of the bis(thiomethyl)magnesium compounds $(MeSCH_2)_2$ $Mg(THF)_3$ (**112**) and $(PhSCH_2)_2Mg(THF)_3$ (**113**) have been determined by X-ray crystallography[143]. Because these compounds are isostructural, only details of **112** are given here. The overall structural geometry comprises a trigonal bipyramidal coordination geometry of the magnesium atom (Figure 54). The two bonding carbon atoms [Mg–C 2.178(3) and 2.191(3) Å] and an oxygen atom of one of the coordinating THF molecules [Mg–O 2.095(2) Å] are at the equatorial positions. The oxygen atoms of the two other coordinating THF molecules [Mg–O 2.178(2) and 2.185(2) Å] are at the axial sites. The sum of the bond angles in the equatorial plane is 360°, but the O–Mg–O bond angle [163.40(8)°] between the axial oxygen atoms slightly deviates from linear.

An octahedral coordination geometry at magnesium was observed in the solid-state structures of the magnesium acetylides $(PhC≡C)_2Mg(TMEDA)_2$ (**114a**)[144] (Figure 55) and $(t\text{-}BuC≡C)_2Mg(TMEDA)_2$ (**114b**)[49]. In **114a** the bonding carbon atoms [Mg–C 2.176(6) and 2.200(6) Å] are *trans* positioned in a perfect linear arrangement (C–Mg–C 180°). Also, the C(1)–Mg–N [89.4(2)°] and C(2)–Mg–N [90.6(2)°] bond angles are in agreement with an almost perfect octahedral coordination geometry. Only the N–Mg–N bond angle [80.4(2)°] of the nitrogen atoms in one TMEDA molecule is less than 90° as a consequence of the bite angle of the TMEDA ligand, but that is compensated by a larger N–Mg–N bond angle between the nitrogen atoms of the two TMEDA molecules. The structural features of **114b** are closely related to those of **114a**.

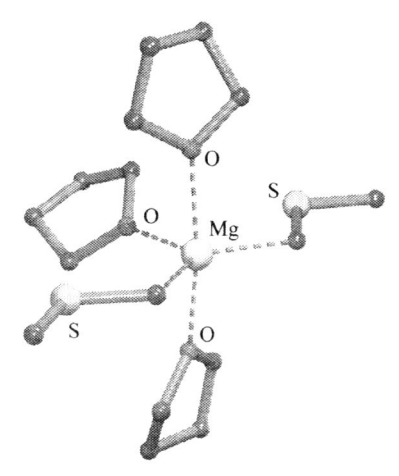

FIGURE 54. Molecular geometry of **112** in the solid state

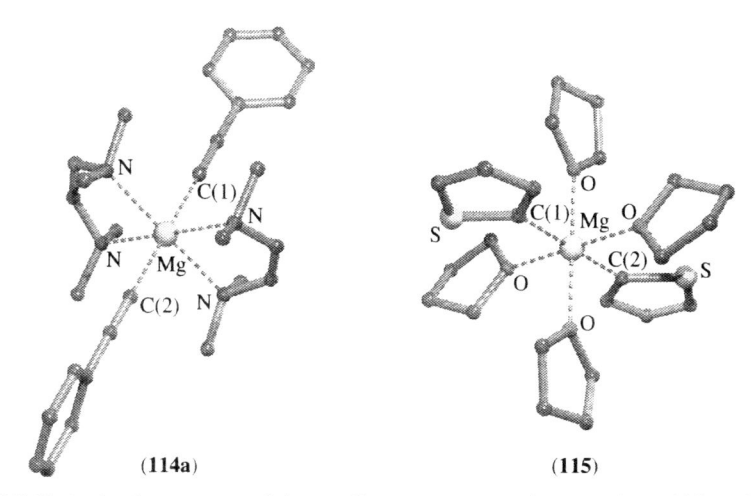

(**114a**)   (**115**)

FIGURE 55. Molecular geometry of six-coordinate organomagnesium complexes **114a** and **115** in the solid state

Bis(2-thienyl)magnesium crystallizes from THF as a complex (2-thienyl)$_2$Mg(THF)$_4$ (**115**) containing four-coordinated THF molecules. An X-ray crystal-structure determination showed that the bonded thienyl groups, like the acetylenic groups in **114a**, are *trans* positioned [C(1)–Mg–C(2) 180°] (Figure 55)[145]. Also, the other bond angles around magnesium deviate less than 0.2° from the ideal octahedral values.

When diphenylmagnesium is crystallized from a solution containing 1,3-xylyl-18-crown-5, an X-ray crystal-structure determination showed the formation of rotaxane **116** (Figure 56)[146]. Only four of the five oxygen atoms of the crown are involved in coordination to magnesium, two with a relatively short bond distance [2.204(3) and 2.222(4) Å] and two with a longer bond distance [2.516(4) and 2.520(4) Å]. The C(1)–Mg–C(2)

(116)   **(117)**

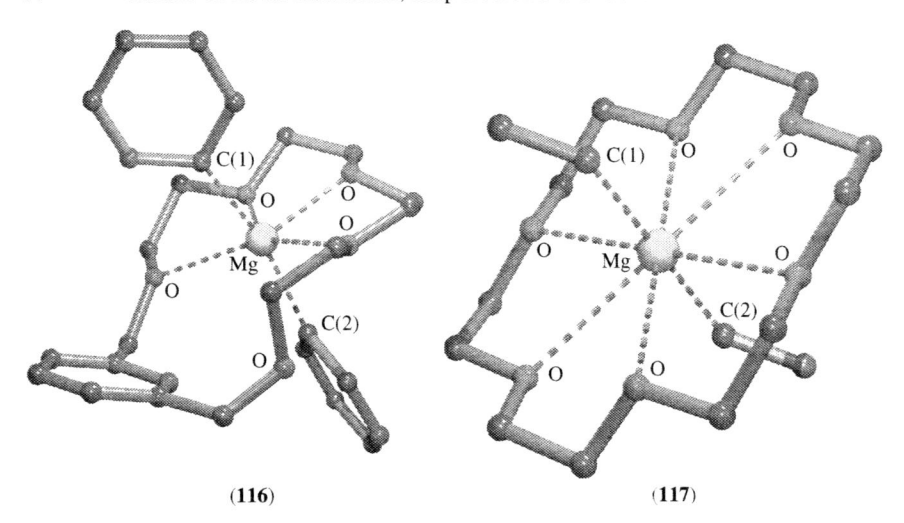

FIGURE 56. Molecular geometries in the solid state of the rotaxane complexes **116** and **117**

angle [163.8(2)°] deviates considerably from linear, pointing to a distorted octahedral coordination geometry at magnesium. An isostructural rotaxane was obtained from bis(*p-tert*-butylphenyl)magnesium and 1,3-xylyl-18-crown-5[142].

Diethylmagnesium and 18-crown-6 form a complex $Et_2Mg(18$-C-$6)$ (**117**) that, according to an X-ray crystal-structure determination, also has a rotaxane structure (Figure 56)[147]. The C(1)–Mg–C(2) arrangement is perfectly linear. At first sight it seems that all six oxygen atoms are involved in bonding to magnesium, although with extreme long bond lengths ranging from 2.767(1) to 2.792(2) Å. As an extreme, **117** might be regarded as a clathrate, having a linear Et_2Mg encapsulated within a crown ether, but bonded weakly to its oxygen atoms.

E. Magnesium Anthracene Compounds

Although the formation of magnesium anthracene was discovered in 1965 and mentioned in a patent[148], the chemistry of magnesium anthracene systems began to develop thirteen years later, triggered by the discovery of a catalyst system for the hydrogenation of magnesium under mild conditions[113].

Magnesium anthracene compounds attracted broad interest because of their versatile applications in synthesis and their ability to catalyze reactions of metallic magnesium. In the presence of a catalytic amount of anthracene, magnesium can be hydrogenated to a highly reactive form of magnesium hydride. This magnesium hydride is an excellent reducing agent for transition-metal salts, and can be used for the preparation of Grignard compounds that are inaccessible otherwise. Another application involves a MgH_2–Mg system, available via phase-transfer catalysis of magnesium, that can be used for chemical synthesis and is, moreover, an outstanding medium for hydrogen storage. These, and other topics of the magnesium anthracene system, have been reviewed[149–152].

Magnesium anthracene $C_{14}H_{10}Mg(THF)_3$ (**118**) can be prepared in high yield from the reaction of metallic magnesium and anthracene in THF (equation 12)[153]. Kinetic measurements showed that a reversible temperature-dependent equilibrium exists between anthracene, magnesium and **118**, the latter being favored at lower temperatures. This equilibrium opened a way to the preparation of elemental magnesium in a finely dispersed, very active form, by raising the temperature of a solution of **118**. Another method for the preparation of highly active magnesium involves heating of solid **118** to 200 °C in high vacuum to remove the THF and anthracene, leaving the highly active elemental magnesium as a black pyrophoric powder with a specific surface area of 60–110 m^2 g^{-1} [154].

$$Mg + C_{14}H_{10} \rightleftharpoons \underset{(118)}{C_{14}H_{10}Mg(THF)_3} \xrightleftharpoons[]{C_{14}H_{10}, -30\,°C} (C_{14}H_{10})_2Mg(THF)_6 \quad (12)$$

$$\downarrow \begin{array}{c} 3MgCl_2 \\ 20\,°C \end{array}$$

$$[Mg_2Cl_3(THF)_6]^+ \; 2\,[C_{14}H_{10}]^{\bullet-}$$

(119)

The molecular structures in the solid state of **118**[155] and its 1,4-dimethyl derivative[156] were determined by X-ray crystallography and appeared to be isostructural. In **118** the magnesium atom is bound to C(9) and C(10) with rather long bond distances [2.25(1) and 2.33(1) Å] (Figure 57). Due to the loss of aromaticity in the central ring of the anthracene skeleton, the molecule is folded (26.6°). As the result of three additional coordinating THF molecules [Mg–O 2.059(7), 2.028(8) and 2.091(8) Å], the magnesium atom is five-coordinate.

At −30 °C in THF, in the presence of anthracene a single-electron transfer from **118** to anthracene occurs, with the formation of insoluble (C$_{14}$H$_{10}$)$_2$Mg(THF)$_6$ (equation 12)[156]. A further reaction with MgCl$_2$ affords the radical anion complex [Mg$_2$Cl$_3$(THF)$_6$]$^+$ [C$_{14}$H$_{10}$]$^{\bullet-}$ (**119**). An X-ray crystal-structure determination of **119** clearly shows the presence of anthracene radical anions as distinct species in the crystal lattice (Figure 58)[156]. The bond lengths and the deformation of the electron density of the anthracene radical anion clearly show that in **119** the LUMO is occupied by one electron[156].

Also, the structure in the solid state of 9,10-bis(trimethylsilyl)anthracene magnesium was determined by X-ray crystallography[157]. Its structural features are similar to those of

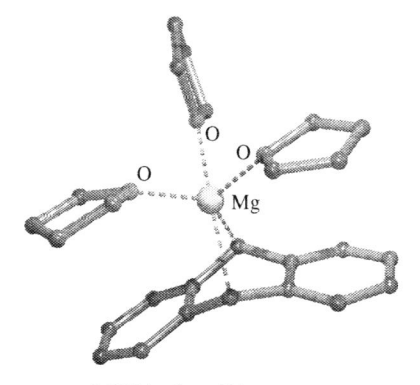

FIGURE 57. Molecular geometry of **118** in the solid state

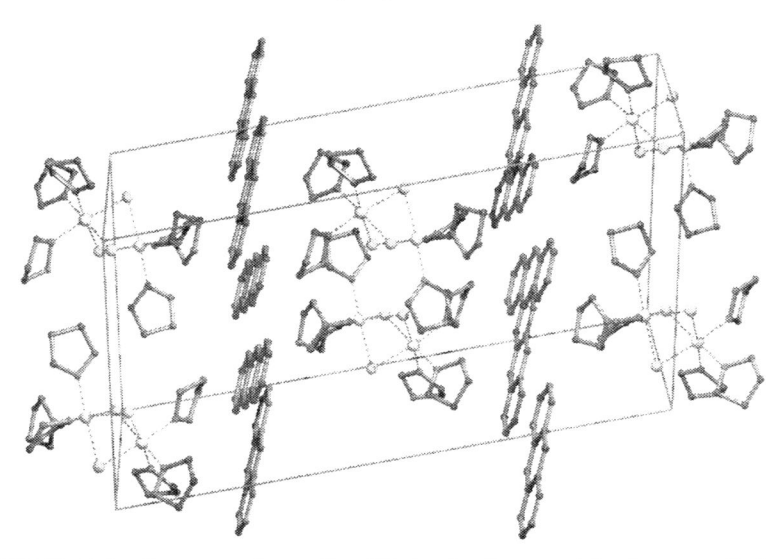

FIGURE 58. Unit-cell contents of crystalline **119** showing the separated anthracene radical anions and [$Mg_2Cl_3(THF)_6$]$^+$ cations

118, but, probably as a result of the sterically demanding Me_3Si groups, only two THF molecules are coordinated to magnesium, giving a tetrahedral coordinated magnesium atom. When this compound is crystallized from THF in the presence of TMEDA, an X-ray crystal-structure determination of the crystalline material (**120**) revealed an asymmetric unit that contains two molecules, one with two coordinating THF molecules and one with a chelate-bonded TMEDA molecule (Figure 59)[158].

Reaction of **118** with ethylene under a pressure of 60 bar gives a new magnesia-cyclic product (**121**) of which the structure in the solid state was established by X-ray

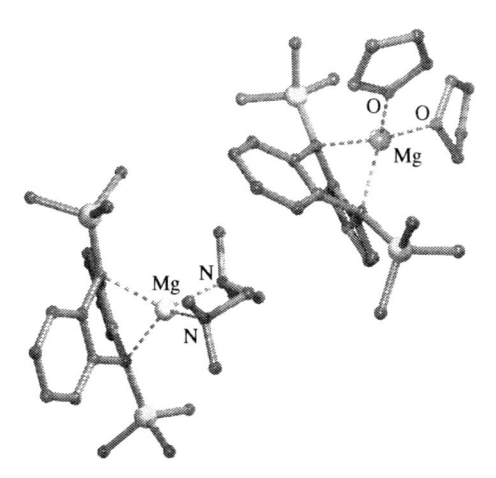

FIGURE 59. The asymmetric unit of **120**

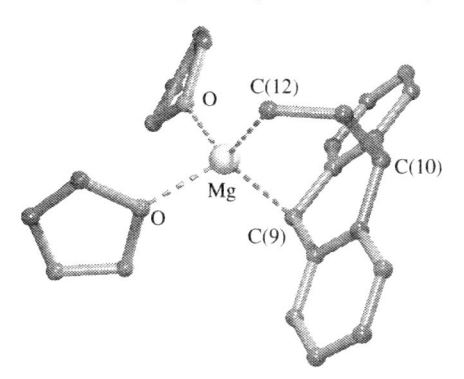

FIGURE 60. Molecular geometry of **121** in the solid state

crystallography (Figure 60)[159]. In **121**, the magnesium atom is bound to an anthracene carbon atom C(9) [Mg—C(9) 2.204(5) Å] and to the terminal carbon atom of the inserted ethylene molecule [Mg—C(12) 2.110(6) Å]. Two THF molecules are coordinated to magnesium, giving it a distorted tetrahedral coordination geometry.

F. Donor–Acceptor Complexes of Diorganomagnesium Compounds with Multi-hapto-bonded Groups

Bis(cyclopentadienyl)magnesium reacts with a variety of primary and secondary alkylamines to give the corresponding 1:1 adducts in high yields[160, 161]. The structures in the solid state of three of these, i.e. Cp$_2$Mg[H$_2$NCH(CHMe$_2$)$_2$] (**122**), Cp$_2$Mg[H$_2$N(c-C$_6$H$_{11}$)] (**123**) and Cp$_2$Mg[HN(i-Pr)(CH$_2$Ph)] (**124**), were determined by X-ray crystallography. In **122** (Figure 61), one of the cyclopentadienyl groups is η^5-bonded to magnesium with Mg—C bond distances ranging from 2.407(4) to 2.414(4) Å while the other cyclopentadienyl group is η^2-bonded [Mg—C 2.380(4) and 2.301(3) Å]. The nitrogen–magnesium bond length [2.112(3) Å] is in the range as expected for nitrogen-to-magnesium coordination bonds. Likewise, adduct **123** contains a η^5-bonded and a η^2-bonded cyclopentadienyl group, but in **124** both cyclopentadienyl groups are η^5-bonded to magnesium. In the latter compound, a somewhat longer N–Mg distance was found [Mg—N 2.210(4) Å]. Molecular

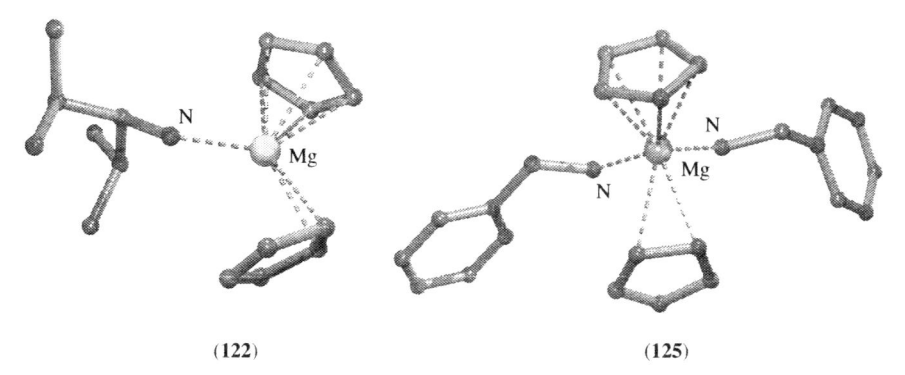

(**122**)   (**125**)

FIGURE 61. Molecular geometries of amine complexes **122** and **125** in the solid state

orbital calculations and infrared spectroscopic studies of these compounds suggest that a NH–hydrogen–$(C_5H_5)^-$ interaction is involved in the stabilization of these complexes.

In contrast to the other Cp_2Mg amine complexes, which were purified by sublimation (110 °C/0.05Torr), the benzylamine complex appeared to be thermally unstable under these conditions and disproportionates into Cp_2Mg and the bis-benzylamine complex $Cp_2Mg(H_2NCH_2Ph)_2$ (**125**). The structure of **125** in the solid state was established by an X-ray crystal-structure determination (Figure 61). Its molecular geometry comprises one cyclopentadienyl group η^5-bonded and one cyclopentadienyl group η^2-bonded to magnesium as well as two coordinating benzylamine molecules [Mg–N 2.146(3) and 2.156(2) Å].

When Cp_2Mg is reacted with t-BuNH$_2$ and the product subsequently recrystallized from THF, complex $Cp_2Mg(H_2NBu-t)(THF)$ (**126**) is obtained which contains N-coordinated t-BuNH$_2$ and O-coordinated THF molecules[162]. An X-ray crystal-structure determination clearly showed the presence of one η^5-bonded- and one η^2-bonded cyclopentadienyl group, a coordinating t-BuNH$_2$ molecule [Mg–N 2.140(2) Å] and a coordinating THF molecule [Mg–O 2.067(2) Å].

When Cp_2Mg is crystallized from neat THF, its bis-THF adduct $Cp_2Mg(THF)_2$ (**127**) is obtained[163]. Its X-ray crystal-structure determination shows, apart from the two coordinated THF molecules [Mg–O 2.088(2) and 2.098(2) Å], one cyclopentadienyl group which is η^5-bonded to magnesium. The distances between magnesium and the carbon atoms of the other cyclopentadienyl group, one being 2.282(2) Å and the next closest one 2.736(2) Å, suggest that this cyclopentadienyl group is η^1-bonded to magnesium.

Bis(indenyl)magnesium has a polymeric structure in the solid state, the details of which have been discussed in a previous section. When it is recrystallized from THF, a discrete monomeric bis-THF adduct, indenyl$_2$Mg(THF)$_2$ (**128**), is obtained[164]. An X-ray crystal-structure determination shows that the magnesium atom has one relatively short bond [2.256(3) Å] with C(1) of each of the indenyl groups, but also interactions with the two adjacent carbon atoms at a much longer distance [Mg–C 2.723(3) and 2.738(3)]. This bonding mode was described as intermediate between η^1 and η^3.

Several *ansa*-magnesocene complexes have been prepared and structurally characterized by X-ray diffraction[165].

The tetramethylethanediyl-bridged, t-butyl-substituted bis-cyclopentadienyl complex $Me_4C_2(3-t$-$BuC_5H_3)_2Mg(THF)$ (**129**) is present in the solid state as a *meso*-diastereoisomer. It has a structure in which both cyclopentadienyl groups are η^5-bonded to magnesium while only one THF molecule is coordinated to magnesium (Figure 62)[165].

For the Me$_2$Si-bridged analog Me$_2$Si(3-t-BuC$_5$H$_3$)$_2$Mg(THF)$_2$ (**130**) a different bonding situation is observed at magnesium (Figure 62). One of the cyclopentadienyl groups is η^5-bonded to magnesium with three relatively short Mg–C bond distances (2.34–2.41 Å) and two longer ones (2.64–2.68 Å). The other cyclopentadienyl group is η^1-bonded to magnesium (Mg–C 2.36 Å). The different bonding modes of the cyclopentadienyl rings result in a large dihedral angle of 77° between the planes of both cyclopentadienyl rings. As a result of the widened coordination gap two THF molecules are coordinated to the magnesium atom.

In the *ansa*-C$_2$H$_4$-bridged bis-indenyl complex C$_2$H$_4$(indenyl)$_2$Mg(THF)$_2$ (**131**)[165] (Figure 63), the bonding of the five-membered rings to magnesium is similar to that observed in (indenyl)$_2$Mg(THF)$_2$ (**128**), *vide supra*. Based on the observed bond distances to magnesium of the carbon atoms in the five-membered rings, the bonding mode is regarded to be intermediate between η^1 and η^3. Like in (indenyl)$_2$Mg(THF)$_2$, two THF molecules are coordinated to magnesium.

Because no crystals suitable for an X-ray crystal-structure determination could be obtained for the Me$_2$Si-bridged analog of **131**, the structure of its alkyl-substituted

(129) (130)

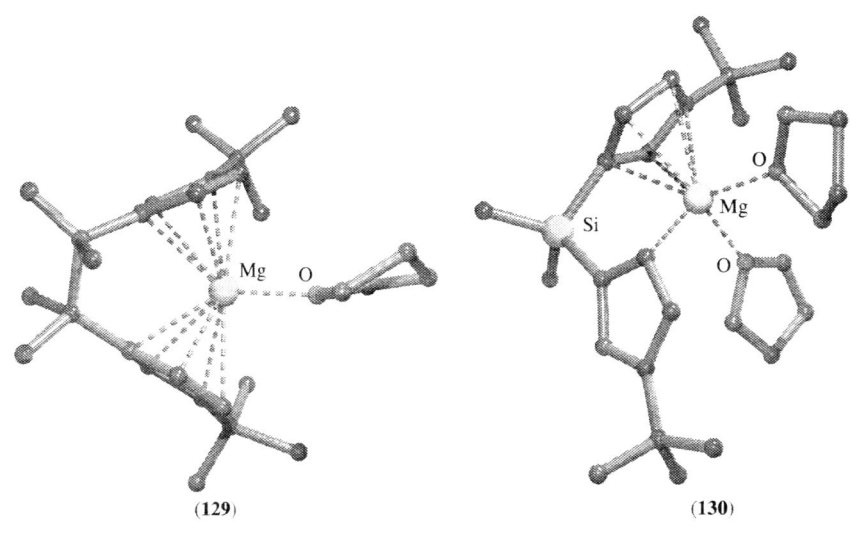

FIGURE 62. Structures of *ansa*-magnesocene complexes **129** and **130** in the solid state

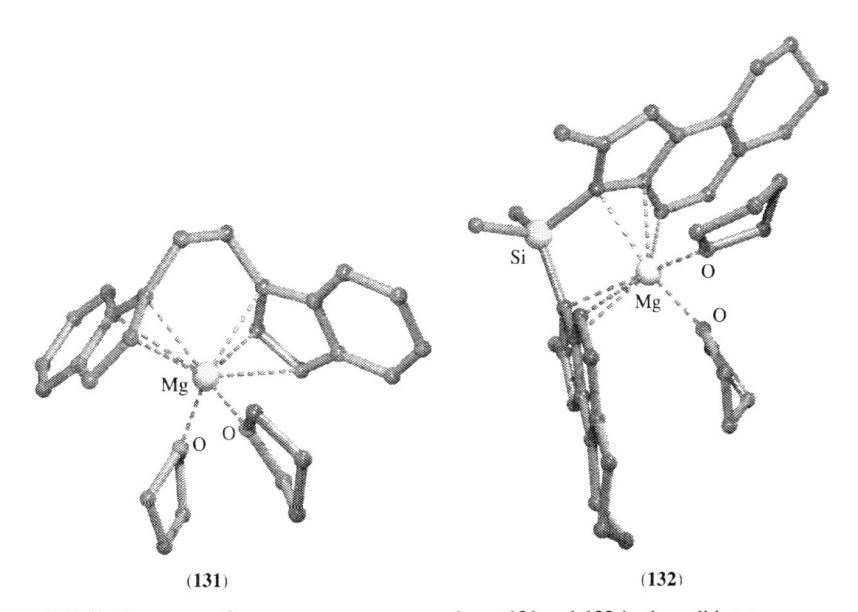

(131) (132)

FIGURE 63. Structures of *ansa*-magnesocene complexes **131** and **132** in the solid state

derivative Me$_2$Si(2-Me-6,7,8,9-tetrahydro-benz[e]indenyl)$_2$Mg(THF)$_2$ (**132**)[165] (Figure 63) was investigated. The indenyl–magnesium binding in the latter differs substantially from that observed in **131**. In **132** both indenyl groups are bound in an exocyclic η^3-geometry to magnesium. The magnesium atom is closest to a C$_3$-fragment comprising the bridgehead and the adjacent angular position as well as the neighboring carbon atom in

(133)   (135)

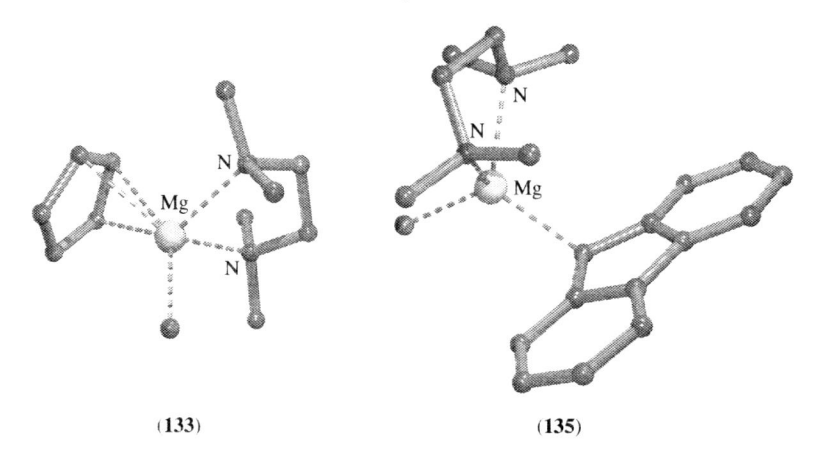

FIGURE 64. Molecular geometries of the heteroleptic organomagnesium TMEDA complexes **133** and **135** in the solid state

the aromatic six-membered ring. Coordination saturation at magnesium is reached by the additional coordination of two THF molecules.

A straightforward protonolysis reaction of Me$_2$Mg(TMEDA) with the carbon-acids cyclopentadiene, indene and fluorene affords the corresponding heteroleptic organomagnesium TMEDA complexes Me(Cp)Mg(TMEDA) (**133**), Me(indenyl)Mg(TMEDA) (**134**) and Me(fluorenyl)Mg(TMEDA) (**135**), respectively, in quantitative yield. The structures in the solid state of these complexes have been determined by X-ray crystallography[166]. In **133** the cyclopentadienyl group is η^3-bonded to magnesium [Mg–C 2.351(3), 2.488(3) and 2.488(3) Å] while the methyl group is σ-bonded to magnesium (Figure 64). The TMEDA molecule is N,N'-chelate bonded to magnesium [Mg–N 2.256(2) and 2.290(2) Å].

The structure in the solid state of **134** is closely related to that of **133**. Likewise, in **134** the indenyl group is η^3-bonded to magnesium.

In the fluorenyl derivative **135** the methyl group is σ-bonded and the TMEDA molecule N,N'-chelate bonded to magnesium (Figure 64). The fluorenyl group, however, is η^1-bonded to magnesium [Mg–C 2.273(2) Å], resulting in a distorted tetrahedral coordination geometry at the magnesium atom.

Heteroleptic (isodicyclopentadienyl)(butyl)magnesium TMEDA complex (**136**) was prepared via a quantitative redistribution reaction from its parent magnesocene, *exo,exo*-bis(isodicyclopentadienyl)magnesium (**61**) (*vide supra*) and Bu$_2$Mg in the presence of TMEDA[98]. An X-ray crystal-structure determination of **136** showed that the magnesium atom is positioned on the *exo* face and interacts in a η^5-manner with the cyclopentadienyl ring with Mg–C bond distances ranging from 2.439(4) to 2.545(4) Å (Figure 65). The butyl group is σ-bonded to magnesium [Mg–C 2.145(4) Å] and the two nitrogen atoms of the chelate-bonded TMEDA molecule complete the coordination sphere of magnesium.

Tetrameric [CpMgC≡CPh]$_4$ (**64**) (*vide supra*) has a heterocubane structure and deaggregates in THF solution to a dimeric THF complex [Cp(PhC≡C)Mg(THF)]$_2$ (**137**)[101]. The overall structural geometry of **137** comprises two magnesium atoms between which two terminal acetylenic carbon atoms are bridge-bonded in a slightly asymmetric manner [Mg–C 2.185(3) and 2.266(3) Å] (Figure 65). To each of the magnesium atoms a cyclopentadienyl ring is η^5-bonded and coordination saturation at magnesium is reached by the coordination of a THF molecule.

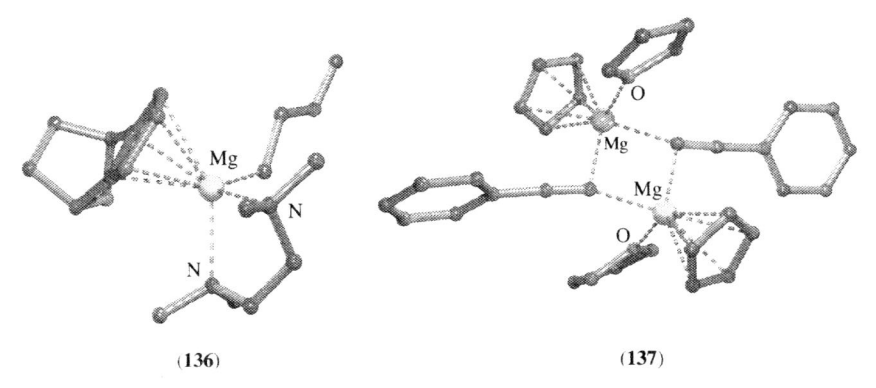

FIGURE 65. Molecular geometries of the heteroleptic organomagnesium complexes **136** and **137**

So far, only multi-hapto-bonded groups have been considered that contain a cyclic unsaturated functionality. It appeared that linear conjugated unsaturated functionalities are also capable of being involved in such multi-hapto interactions.

The 1,4-bis(phenyl)-2-butene-1,4-diylmagnesium tris-THF complex (**138**) has been prepared from activated magnesium and 1,4-diphenyl-1,3-butadiene, and its structure in the solid state was determined by X-ray crystallography[167]. Its molecular geometry comprises a Mg(THF)$_3$ unit (average Mg–O 2.12 Å) that has a η^4-interaction with the butene skeleton. The butene skeleton adopts a *s-cis* geometry in which the four central carbon atoms lie in one plane, with the magnesium atom positioned 1.71 Å above this plane (Figure 66). The bond distances between magnesium and the C(1) and C(4) atoms of the butene moiety, 2.32 and 2.26 Å, respectively, are shorter than the Mg–C(2) and Mg–C(3) interactions of 2.56 and 2.52 Å.

The structure in the solid state of the 1,4-bis(trimethylsilyl)-2-butene-1,4-diylmagnesium TMEDA complex (**139**) shows similarities with that of **138** (Figure 66)[168]. Also, in **139** the butene skeleton has a η^4-interaction with magnesium with shorter Mg–C(1) and Mg–C(4) bonds [2.200(9) and 2.191(9) Å, respectively] and longer Mg–C(2) and Mg–C(3) bonds [2.381(8) and 2.399(8) Å, respectively]. Instead of the three coordinating THF molecules in **138**, in **139** a *N,N'*-chelate bonded TMEDA molecule is present.

Another type of ligand, capable of forming multi-hapto interactions with metals, are boron and boron–carbon cage compounds of which in particular the carboranes have been used extensively in organometallic chemistry[169]. The structures in the solid state of a few magnesacarboranes have been determined by X-ray crystallographic studies.

An X-ray crystal-structure determination of the magnesocarborane, *closo*-1-Mg(THF)$_3$-2,4-(Me$_3$Si)$_2$-2,4-C$_2$B$_4$H$_4$ (**140**), showed that this compound exists in the solid state as a discrete monomer[170]. The molecular geometry of **140** comprises a Mg(THF)$_3$ moiety, of which the magnesium atom is located at an apical position above an open pentagonal face of the C$_2$B$_4$ cage (Figure 67). The observed Mg–C and Mg–B bond distances [Mg–C 2.390(3) and 2.429(3) Å, Mg–B 2.452(3), 2.404(3) and 2.381(4) Å] are indicative for η^5-bonding to magnesium.

The molecular geometry of the magnesacarborane *closo*-1-Mg(TMEDA)-2,3-(Me$_3$Si)$_2$-2,3-C$_2$B$_4$H$_4$ (**141**) in the solid state shows some similarities with that of **140**. In this case, a Mg(TMEDA) moiety is η^5-bonded to the open pentagonal face of the C$_2$B$_4$ cage (Figure 67)[171,172]. However, in contrast to **140**, here the molecule is dimeric via an interaction of the magnesium atoms with the unique boron atoms of the neighboring C$_2$B$_4$H$_4$ cage involving two Mg–H–B bridges.

(138)   (139)

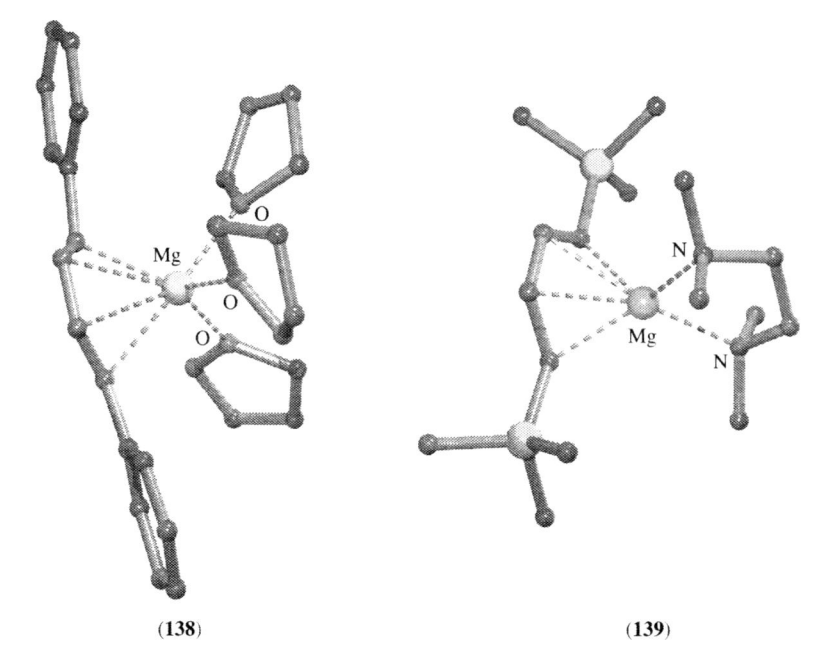

FIGURE 66. Molecular geometries of the 2-butene-1,4-diylmagnesium complexes **138** and **139**

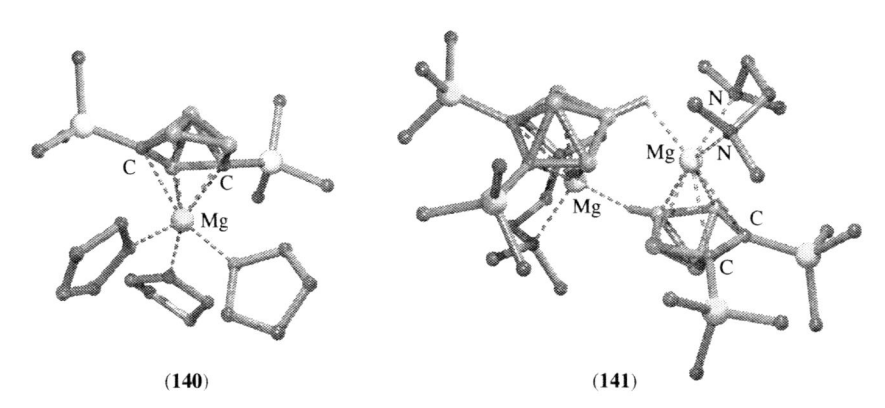

(140)   (141)

FIGURE 67. Molecular geometry of the magnesium $C_2B_4H_4$ carboranes **140** and **141**

Also, magnesacarboranes with a $C_4B_8H_8$ cage have been prepared and were structurally characterized in the solid state by X-ray diffraction studies. The magnesacarborane $(THF)_2Mg(Me_3Si)_4C_4B_8H_8$ (**142**) has been prepared from *nido*-2,4,6,12-$(Me_3Si)_4$-2,4,6,12-$C_4B_8H_8$ and metallic magnesium in THF. Its molecular geometry in the solid state comprises a $Mg(THF)_2$ unit that is η^4-bonded to four adjacent atoms, two carbon [Mg–C 2.315(10) and 2.326(9) Å] and two boron atoms [Mg–B 2.393(12) and 2.402(11)], of a seven-membered open face of the carborane cage (Figure 68)[173].

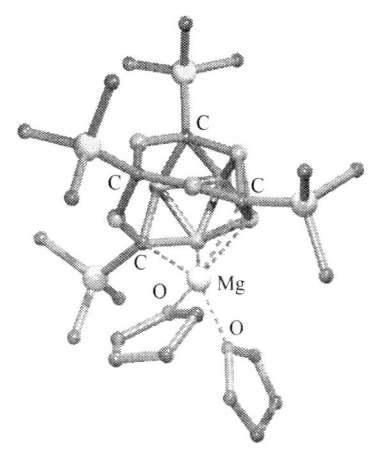

FIGURE 68. Molecular geometry of magnesacarborane **142**

The boron-substituted analogs $(THF)_2Mg(Me_3Si)_4(t\text{-}BuB)C_4B_7H_7$ (**143**) and $(THF)_2Mg(Me_3Si)_4(MeB)C_4B_7H_7$ (**144**) have also been prepared and structurally characterized in the solid state[174]. Their structures are essentially the same as that of **142**, except that one of the borons in **143** and **144** is alkylated.

So far, only diorganomagnesium complexes have been considered in which the heteroatom involved in coordination to magnesium is either nitrogen or oxygen. The only other functional group forming a coordination bond to magnesium in organomagnesium compounds that have been structurally characterized are carbenes.

Bis(pentamethylcyclopentadienyl)magnesium reacts smoothly with 1,3,4,5-tetramethylimidazol-2-ylidene to form carbene complex **145**[175]. An X-ray crystal-structure determination unambiguously showed that the ligand is bound to magnesium via its carbene carbon atom [Mg–C 2.194(2) Å] (Figure 69). Compared to $(Me_5C_5)_2Mg$ itself, in which both cyclopentadienyl groups are η^5-bonded to magnesium, the bonding mode of the cyclopentadienyl groups in **145** has changed. One of these is still η^5-bonded to magnesium with average Mg–C distances of 2.46(8) Å, but the other one exhibits a 'slipped'

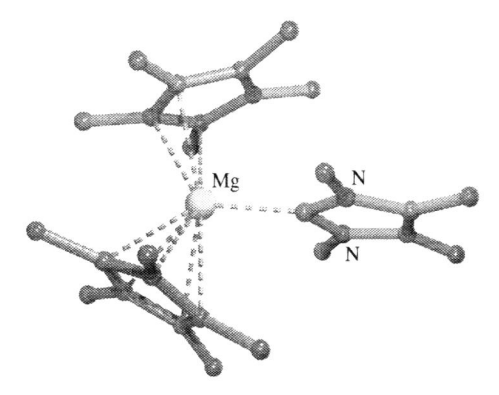

FIGURE 69. Molecular geometry of carbene complex **145**

geometry with three carbon atoms closest to magnesium at distances of 2.309(3), 2.465(2) and 2.605(2) Å. Together with the observed C—C bond distances in this ring, these values suggest a bonding mode of this cyclopentadienyl ring with magnesium that is intermediate between η^3 and η^1 (σ-bonded).

A similar carbene complex was obtained from the reaction of $(Me_4C_5H)_2Mg$ with 1,3-di-isopropyl-4,5-dimethylimidazol-2-ylidene, with a structure in the solid state that is closely related to that of **145**[95].

IV. HETEROLEPTIC RMgY COMPOUNDS

A. Introduction

In heteroleptic monoorganomagnesium compounds RMgY an organic group is σ-bonded or, in some particular cases, multi-hapto bonded to magnesium via several carbon atoms. The other group, Y, is bound to magnesium via a heteroatom. Examples of the latter groups are: halogen atoms, oxygen-containing groups like alkoxides, nitrogen-containing groups like primary and secondary amides, and other groups functionalized with heteroatoms like sulfur and phosphorus.

For the synthesis of heteroleptic monoorganomagnesium compounds, three major pathways are available (Scheme 1).

$$R_2Mg + HY \longrightarrow RMgY + RH$$

$$R_2Mg + MgY_2 \rightleftharpoons 2\,RMgY$$

$$RX + Mg \longrightarrow RMgX$$

X = Cl, Br, I
Y = heteroatom-functionalized organic group

SCHEME 1. The pathways for the synthesis of heteroleptic organomagnesium compounds

The first route involves the protonolysis of one of the alkyl or aryl groups of dialkyl- or diarylmagnesium compounds by an organic molecule containing an acidic proton bound to a heteroatom. Examples of such acidic compounds are alcohols and primary and secondary amines. A nice illustration of this first route is the formation of organomagnesium amides from the reaction of enantiomerically pure N-(2-methylamino-2-phenylethyl)piperidine with Bu_2Mg or i-Pr_2Mg (equation 13). The n-butylmagnesium derivative (**146**) has been successfully applied in the enantioselective addition of butyl groups to aldehydes[176], and the iso-propylmagnesium compound (**147**) has been used for the enantioselective reduction of ketones[177].

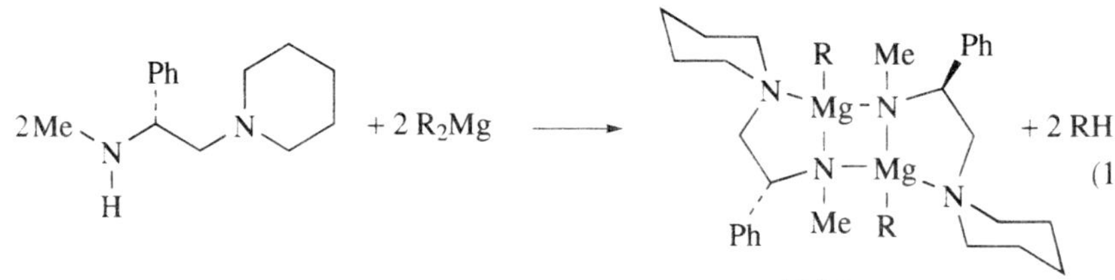

(**146**) R = n-Bu
(**147**) R = i-Pr

The second route (Scheme 1) is a redistribution reaction, in fact the Schlenk equilibrium. This route may be used in the reverse direction for the preparation of pure diorganomagnesium compounds from organomagnesium halides. Addition of a ligand, usually dioxane, that forms an insoluble complex with magnesium dihalide, shifts the Schlenk equilibrium completely to the left side and allows isolation of pure diorganomagnesium compounds from the remaining solution[178].

The third pathway (Scheme 1) for the preparation of heteroleptic monoorganomagnesium compounds, especially monoorganomagnesium halides, involves the reaction of an organic halide with metallic magnesium, the classical Grignard reaction (equation 14).

$$RX + Mg \xrightarrow{\text{solvent}} RMgX + \text{by-products} \qquad (14)$$

The reaction of the organic halide with magnesium is carried out in a non-protic polar solvent, usually diethyl ether or THF. Typical by-products are RR, RH and R(-H) (alkene), resulting from coupling and disproportionation reactions of the organic moiety. Also, by-products resulting from solvent attack are sometimes formed, but usually to a lesser extent.

Although the formation of Grignard reagents at first looks simple (equation 14), the mechanisms involved are still speculative despite about a hundred years of work and have been the subject of several reviews[179-182]. The mechanisms of the formation of Grignard reagents can be divided into two parts, an organic and an inorganic one. The organic mechanism traces the organic fragment R from RX to RMgX and by-products containing residues from R and occasionally the solvent. The inorganic part of the mechanism traces the Mg from metallic magnesium to RMgX and deals with surface films, inhibition, initiation and activation. It should be noted that the organic part of the mechanism has received far more attention than the inorganic part. Nowadays, there is overwhelming evidence that radicals play a major role in the formation of Grignard reagents. In the initial step, in which [RX$^{\bullet -}$] may be an intermediate or transition structure, both electron transfer and halogen-atom transfer may play a role (Scheme 2). For further details of these mechanistic studies, the reader is urged to consult the references cited.

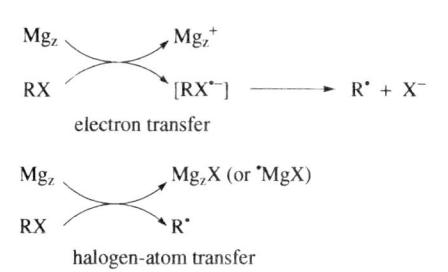

$$Mg_z \searrow \quad \nearrow Mg_z^+$$
$$RX \qquad [RX^{\bullet -}] \longrightarrow R^{\bullet} + X^-$$
$$\text{electron transfer}$$

$$Mg_z \searrow \quad \nearrow Mg_zX \text{ (or } ^{\bullet}MgX)$$
$$RX \qquad \qquad R^{\bullet}$$
$$\text{halogen-atom transfer}$$

SCHEME 2. The initial step in the formation of Grignard reagents

Due to the presence of an electronegative group directly bound to magnesium, the Lewis acidity of magnesium is enhanced and therefore heteroleptic monoorganomagnesium compounds readily form complexes with donor molecules. Moreover, these directly bound heteroatoms can act as multi-electron pair donors which facilitates aggregate formation in which these heteroatoms form four-electron three-center bridge-bonds between two or more magnesium atoms. The presence of the unavoidable Schlenk equilibrium in solutions of heteroleptic monoorganomagnesium compounds should also be taken into account. When crystalline material is harvested from such solutions for structural studies, the structures found in the solid state do not necessarily represent structures that actually are present (in a majority) in solution. Which particular compound, aggregated or not, crystallizes from solution is determined by several factors like thermodynamic stability, solubility and packing effects in the crystal lattice.

Based on the nature of the heteroatom directly bound to magnesium, the structures of heteroleptic monoorganomagnesium compounds can be divided into several sub-classes that will be discussed separately below. These sub-classes are: (i) monoorganomagnesium cations (i.e. compounds consisting of ion pairs), (ii) monoorganomagnesium halides, the Grignard reagents, (iii) monoorganomagnesium compounds with an oxygen atom σ-bonded to magnesium, (iv) monoorganomagnesium compounds with a nitrogen atom σ-bonded to magnesium and (v) monoorganomagnesium compounds containing anions σ-bonded to magnesium via other heteroatoms.

B. Monoorganomagnesium Cations

As has been outlined in the section on organomagnesiates (*vide supra*), in the presence of a 2,1,1-cryptand diorganomagnesium compounds undergo a disproportionation reaction, forming an organomagnesiate anion and a monoorganomagnesiate cation encapsulated in the 2,1,1-cryptand[55]. Likewise, a crystalline material containing [MeMg(15-C-5)]$^+$ cations and a linear polymeric chain in which the [Me$_5$Mg$_2$]$^-$ anion is the repeating unit has been isolated from a solution of Me$_2$Mg and 15-crown-5[57].

When a solution of heteroleptic MeMgCp is crystallized in the presence of the aza-crown 1,4,8,11-tetramethyl-1,4,8,11-tetraazacyclotetradecane (14-N-4) a crystalline material is obtained (**148a**) that, according to its X-ray crystal-structure determination, consists of isolated MeMg(14-N-4) cations and cyclopentadienyl anions in the crystal lattice (Figure 70)[183]. In the cationic MeMg(14-N-4)$^+$ fragment [Mg–C 2.136(7) Å] all four nitrogen atoms are involved in coordination to magnesium [Mg–N ranging from 2.208(6) to 2.252(5) Å] resulting in a penta-coordinate magnesium atom.

Similarly, crystallization of a solution containing Me$_2$Mg, Me$_2$Cd and 14-N-4 affords a crystalline product [MeMg(14-N-4)]$^+$ [Me$_3$Cd]$^-$ (**148b**)[184]. The cationic [MeMg(14-N-4)]$^+$ fragment is chemically identical with that of **148a** and shows only small differences in bond distances and angles.

When MeMgI is prepared in THF/DME as a solvent, a crystalline material [MeMg(DME)$_2$(THF)]$^+$ I$^-$ (**149**) was isolated in high yield. An X-ray crystal-structure deter-

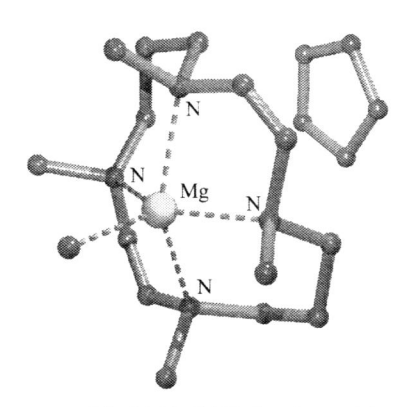

FIGURE 70. Molecular geometry of the MeMg(14-N-4) cation and the cyclopentadienyl anion **148a**

1. Structural organomagnesium chemistry 57

mination showed that in the solid state **149** consists of isolated [MeMg(DME)$_2$(THF)]$^+$ cations and iodide anions. In the cation two DME molecules are O,O-chelate bonded to magnesium and an additional coordinating THF molecule makes the magnesium atom octahedral coordinate (Figure 71)[185]. It should be noted that such octahedral arrangements are chiral. In fact, **149** crystallizes at $-20\,°$C as conglomerates in space group $P2_12_12_1$, i.e. the crystals are chiral. When crystallization is performed at much lower temperature ($-60\,°$C) a racemic phase crystallizes in space group Pbca in which as a requirement of space group symmetry both enantiomers, Δ-cis and Λ-cis, are present.

The only organomagnesium cation, containing a multi-hapto-bonded organic group, for which the structure in the solid state has been elucidated by X-ray crystallography, is [CpMg(PMDTA)]$^+$ [Cp$_2$Tl]$^-$ (**150**). This compound was obtained from the reaction of Cp$_2$Mg with CpTl in the presence of PMDTA. Its solid-state structure comprises isolated [CpMg(PMDTA)]$^+$ cations and [Cp$_2$Tl]$^-$ anions in the crystal lattice[186].

In the cation of **150** (Figure 72) the cyclopentadienyl group is η^5-bonded to magnesium with Mg–C distances of 2.40(3), 2.38(2), 2.44(3), 2.41(2) and 2.40(2) Å. The PMDTA molecule is chelate-bonded with its three nitrogen atoms to magnesium [Mg–N 2.21(2), 2.16(2) and 2.25(2) Å].

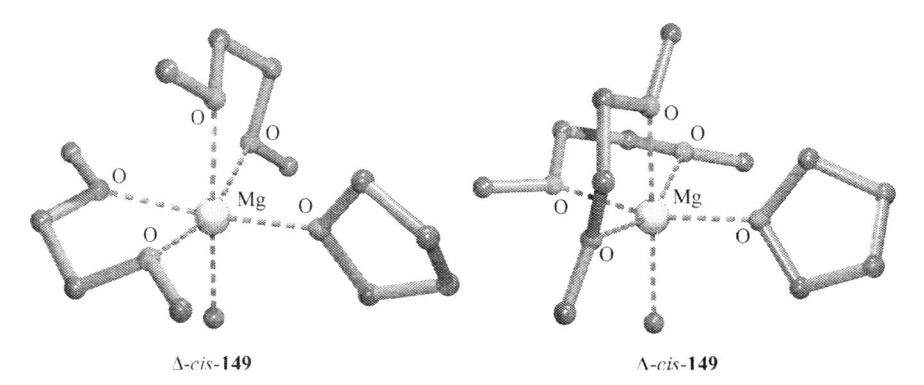

Δ-cis-**149** $\qquad\qquad\qquad$ Λ-cis-**149**

FIGURE 71. The two enantiomers of the chiral cation cis-[MeMg(DME)$_2$(THF)]$^+$

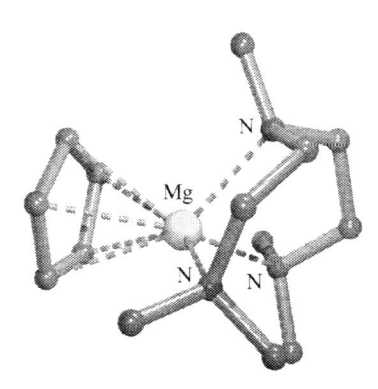

FIGURE 72. Molecular geometry of the [CpMg(PMDTA)]$^+$ cation of **150** in the solid state

C. Monoorganomagnesium Compounds RMgY with Y = halogen (Grignard Reagents)

One of the most fascinating and fundamental problems in organic chemistry concerns the constitution of Grignard reagents in ethereal solution. A closely related problem involves the mechanism of formation of Grignard reagents and the mechanism or mechanisms involved in the reaction of Grignard reagents with organic functional groups[3–5,11,13,14,187]. Structures in the solid state, obtained by X-ray crystallography from crystals grown from Grignard solutions, have helped to partly solve these problems. However, as mentioned before, the obtained structures are not necessarily representative for the bulk of the solution but only give an indication of what types of structures might be present in solution. Various structural motifs for Grignard reagents in the solid state have been observed: monomers, dimers and higher aggregates, with coordination numbers at magnesium varying from four to six. A particular type of Grignard reagents are those containing a heteroatom-functionalized group, which is capable of coordinating intramolecularly to magnesium. Those compounds can be regarded as containing a built-in coordinating solvent molecule.

The first Grignard compound that was structurally characterized in the solid state by X-ray crystallography was PhMgBr(OEt$_2$)$_2$ (**151**)[19,188]. It was unambiguously established that **151** exists in the solid state as a monomer with the phenyl group σ-bonded to magnesium. Furthermore, the bromine atom [Mg–Br 2.44(2) Å] and two oxygen atoms [Mg–O 2.01(4) and 2.06(4) Å] of two coordinating diethyl ether molecules are bonded to magnesium, giving it a distorted tetrahedral coordination geometry. Due to poor reflection data the location of the phenyl-carbon atoms could not be obtained exactly. Similarly, PhMgBr(THF)$_2$ (**152**) exists in the solid state as a discrete monomer, but also in this case reflection data were poor[189]. For EtMgBr(OEt$_2$)$_2$ (**153**) a more reliable data set was obtained, allowing a more detailed discussion of its structure in the solid state[20,190]. Compound **153** exists in the solid state as a monomer with four-coordinate magnesium as a result of bonding of the ethyl group [Mg–C 2.15(2) Å], the bromine atom [Mg–Br 2.48(1) Å] and two oxygen atoms [Mg–O 2.03(2) and 2.05(2) Å] of two coordinating diethyl ether molecules (Figure 73). With the exception of the C–Mg–Br bond angle [125.0(5)°] the bond angles around the magnesium atom are close to the ideal tetrahedral value, indicating an only slightly distorted tetrahedral coordination geometry at magnesium.

So far, the two other Grignard compounds known as having a monomeric structure in the solid state with four-coordinate magnesium are (Ph$_3$C)MgBr(OEt$_2$)$_2$ (**154**)[155]

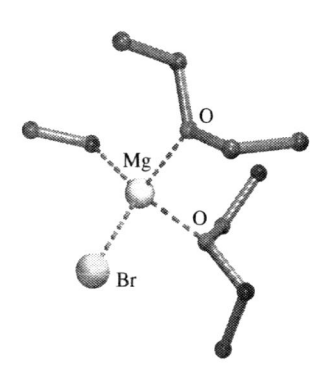

FIGURE 73. Molecular geometry of **153** in the solid state

and 2,6-Tip$_2$C$_6$H$_3$MgBr(THF)$_2$ (**155**)[191] (Tip = 2,4,6-i-Pr$_3$C$_6$H$_2$). The overall structural geometries of **154** and **155** with respect to the magnesium environment are closely related to that of **153**.

The structures of three monomeric organomagnesium bromides in which the magnesium center is penta-coordinate have been established in the solid state. These compounds are: MeMgBr(THF)(TMEDA) (**156**)[132], 9-bromo-9-[(bromomagnesium)methylene]fluorene tris-THF complex (**157**)[192] and MeMgBr(THF)$_3$ (**158**)[193].

The structure of **156** comprises a magnesium atom to which the methyl group and the bromine atom are σ-bonded [Mg–C 2.25(1) Å, Mg–Br 2.485(1) Å], the TMEDA ligand is N,N'-chelate bonded [Mg–N 2.246(2) and 2.334(3) Å] and an additional coordinating THF molecule [Mg–O 2.204(9) Å] completes five-coordination at magnesium (Figure 74). The magnesium atom has a trigonal bipyramidal coordination geometry with the carbon atom, the bromine atom and one of the nitrogen atoms of the TMEDA molecule at the equatorial sites. The other nitrogen atom and the oxygen atom of the coordinating THF molecule reside at the apical positions. The sum of the bond angles in the equatorial plane is 360°. The bond angle between magnesium and the apical bonded nitrogen and oxygen atom [N–Mg–O 166.5(5)°] deviates considerably from linear, but most probably is a consequence of the small bite angle of the TMEDA ligand.

In **157**, which represents a magnesium carbenoid compound, the magnesium atom has a trigonal bipyramidal coordination geometry with the carbon atom bound to magnesium [Mg–C 2.19(1) Å], the bromine atom [Mg–Br 2.517(3) Å] and an oxygen atom [Mg–O 2.045(7) Å] of one of the coordinating THF molecules at the equatorial positions (Figure 74).

Also, the magnesium atom in **158** has a trigonal bipyramidal geometry. However, due to disorder in the structure, details cannot be given.

A series of Grignard reagents has been recrystallized from neat dimethoxyethane (DME) resulting in compounds in which, according to X-ray crystal-structure determinations, the magnesium atoms have an octahedral coordination environment, due to the O,O'-chelate bonding of both DME molecules. The following compounds have been isolated and structurally characterized: n-PrMgBr(DME)$_2$ (**159**)[61], (allyl)MgBr(DME)$_2$ (**160**)[61], i-PrMgBr(DME)$_2$ (**161**)[61], (vinyl)MgBr(DME)$_2$ (**162**)[145], (2-thienyl)MgBr(DME)$_2$ (**163**)[145]

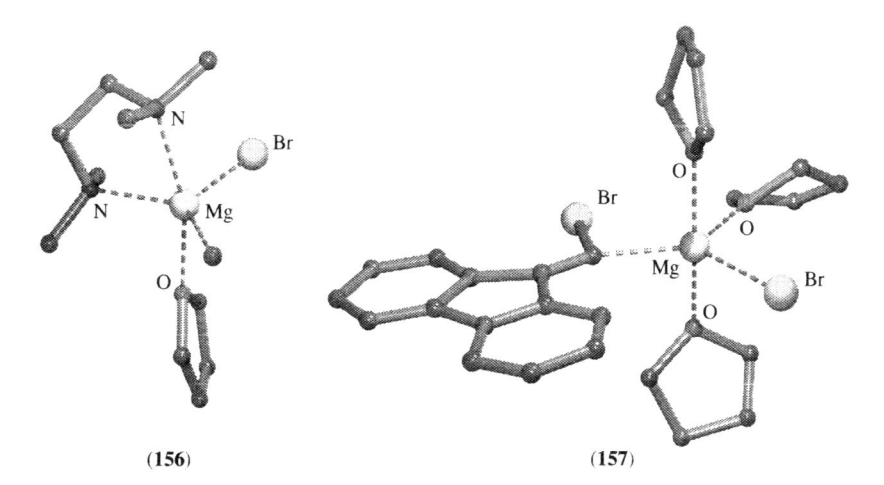

(**156**)   (**157**)

FIGURE 74. Molecular geometries of **156** and **157** in the solid state

Δ-enantiomer   Λ-enantiomer

FIGURE 75. The Δ- and Λ-enantiomers of *cis*-octahedral Grignard compounds

and *p*-TolMgBr(DME)$_2$ (**164**)[185]. In all compounds the organic group and the bromine atom are in *cis*-position. It should be noted that in such *cis*-octahedral arrangements the magnesium atom is chiral and thus a Δ- and Λ-enantiomer of the Grignard compound exists (Figure 75).

Upon crystallization of such chiral compounds two things might happen: (*i*) the material crystallizes as a racemate, i.e. the crystal contains both the Δ- and Λ-enantiomer, and (*ii*), the material crystallizes as a conglomerate, i.e. the crystals are chiral, one particular crystal contains only one of the enantiomers. X-ray crystal-structure determinations of **159–163** revealed centrosymmetric space groups, and thus these solid-state materials are by definition racemic. Surprisingly, crystallization of **164** at −20 °C yielded crystals which, according to an X-ray crystal-structure determination, have the a-centric spacegroup $P2_12_12_1$, and moreover, the asymmetric unit contains one molecule. This observation is a proof that **164** crystallizes as a conglomerate. Making use of special seeding techniques, both enantiomers of crystalline **164** could be isolated enantiomerically pure. Reaction of this enantiopure Grignard reagent, at −70 °C in DME as the solvent, with butyraldehyde afforded the corresponding alcohol with e.e. values of up to 22%.

Another approach to induce enantioselectivity during reactions with Grignard reagents is the use of chiral additives, usually chiral compounds that form a complex with the Grignard reagent. The structures in the solid state of the following complexes containing a chiral ligand have been determined: *t*-BuMgCl[(−)-sparteine] (**165**)[194], EtMgBr[(−)-sparteine] (**166**)[195], EtMgBr[(−)-α-isosparteine (**167**)[196] and EtMgBr[(+)-6-benzylsparteine] (**168**)[197]. These four compounds are discrete monomers with the organic group and the halogen atom σ-bonded to magnesium and the sparteine ligand N,N'-chelate bonded to magnesium resulting in a tetrahedral coordination geometry at magnesium. As an example

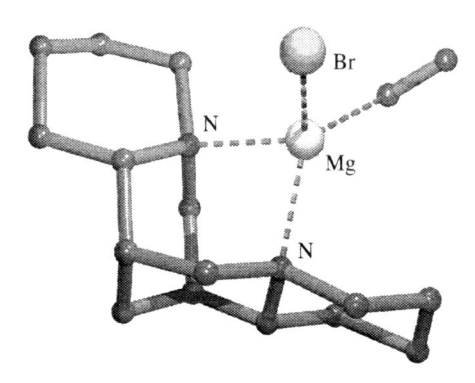

FIGURE 76. Molecular geometry of **166** in the solid state

the structure of **166** is shown (Figure 76). Complex **165** has been successfully applied as a catalyst for the selective asymmetric polymerization of racemic methacrylates[197].

1,1-Di-Grignard reagents are valuable synthons in both organic and organometallic chemistry[136]. The only 1,1-di-Grignard reagent for which the structure in the solid state was unambiguously established by X-ray crystallography is $(Me_3Si)_2C[MgBr(THF)_2]_2$ (**169**)[198]. Its structure comprises two almost identical $MgBr(THF)_2$ units σ-bonded [Mg–C 2.10(4) and 2.14(4) Å] to the central carbon atom (Figure 77). Each of the magnesium atoms has a slightly distorted tetrahedral coordination geometry. Its relative unreactivity was explained in terms of an effective shielding of the central carbon atom from attack by electrophiles by the two bulky Me_3Si groups and two bulky $MgBr(THF)_2$ units.

In the solid state, the Grignard complexes **170–177** (Table 3) all have a dimeric structural motif, via two bridging halogen atoms between the two magnesium atoms, as shown schematically in Figure 78.

The basic structure of these compounds consists of a central flat, four-membered Mg–X–Mg–X ring in which the halogen atoms are symmetrically bridge-bonded between the magnesium atoms. To each of the magnesium atoms is σ-bonded one organic group and a donor molecule which is coordinated via its heteroatom, resulting in a tetrahedral coordination geometry at the magnesium atoms. In principle two geometrically different isomers are possible, one with the organic groups approaching the four-membered ring from opposite sides, and one approaching the four-membered ring from the same side. So far only structures are known in which the organic groups, and consequently the coordinating donor molecules, are at opposite sides.

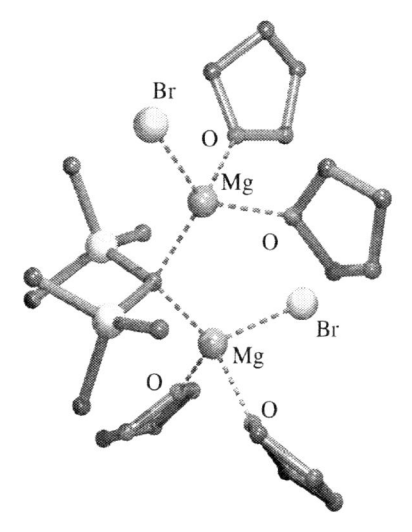

FIGURE 77. Molecular geometry of the 1,1-di-Grignard reagent **169** in the solid state

FIGURE 78. Schematic structural motif of the Grignard complexes **170–177** in the solid state

TABLE 3. Dimeric Grignard complexes **170–177**

Compound	R	X	L	Reference
170	Et	Br	$i\text{-Pr}_2\text{O}$	199
171	Et	Br	Et_3N	200
172	$(\text{Me}_3\text{Si})_2\text{CH}$	Br	Et_2O	120
173	$(\text{Me}_3\text{Si})_3\text{C}$	I	Et_2O	112
174	$(\text{PhMe}_2\text{Si})(\text{Me}_3\text{Si})_2\text{C}$	I	Et_2O	112
175	$(\text{PhMe}_2\text{Si})(\text{Me}_3\text{Si})\text{CH}$	Br	Et_2O	201
176	$2,6\text{-Mes}_2\text{C}_6\text{H}_3$	Br	THF	202
177	9-Anthryl	Br	$n\text{-Bu}_2\text{O}$	203

When allylmagnesium chloride is crystallized from THF in the presence of TMEDA, a dimeric complex [(allyl)MgCl(TMEDA)]$_2$ (**178**) is obtained. Its X-ray crystal-structure determination showed that in this complex two chlorine atoms are bridging between two magnesium atoms in a rather asymmetric way [Mg–Cl 2.400(1) and 2.694(1) Å]^{204}. To each of the magnesium atoms one allyl group is η^1-bonded via its terminal carbon atom [Mg–C 2.179(3) Å] and one TMEDA molecule N,N'-chelate bonded [Mg–N 2.211(2) and 2.285(2) Å], resulting in penta-coordinate magnesium atoms.

Crystallization of a series of monoorganomagnesium chlorides afforded crystalline materials with the formulation R$_2$Mg$_4$Cl$_6$(THF)$_6$; R = Et (**179**)^{205}, R = Me (**180**)^{60}, R = t-Bu (**181**)^{60} and R = benzyl (**182**)^{60}. X-ray structure determinations of these compounds show that they exist as complex aggregates shown schematically in Figure 79. The four magnesium atoms are linked via chloride bridges, four of which are bridge-bonded between two magnesium atoms and two are bridging between three magnesium atoms (Figure 79). The central two magnesium atoms have an octahedral coordination geometry, due to interaction with four chlorine atoms and two coordinating THF molecules in cis-position. The other two magnesium atoms have trigonal bipyramidal coordination geometry with the organic group and two chlorine atoms at equatorial positions and one chlorine atom and one coordinating THF molecule at the apical sites.

It should be noted that the ratio of organic group to magnesium to halide is not 1:1:1 as in the general formulation of Grignard reagents RMgX. In fact these aggregates contain an additional MgCl$_2$ molecule, which is always present in solutions of Grignard reagents due to the Schlenk equilibrium. This implies that if such a type of structures is present in solution, also other (aggregated) species having different stoichiometries must be present.

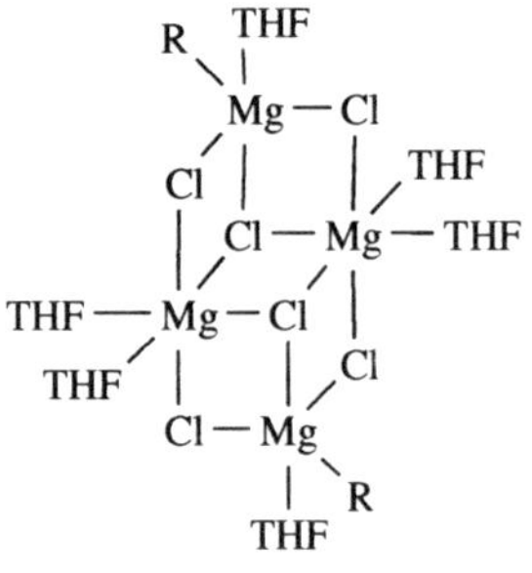

```
         R   THF
          \  |
           Mg — Cl
         /  |    |  THF
       Cl   |    | /
        |   Cl — Mg — THF        (179) R = Et
        |  /    /                 (180) R = Me
 THF — Mg — Cl  |                 (181) R = t-Bu
      / |    | /Cl                (182) R = benzyl
   THF  |    |/
         Cl — Mg
              | \R
             THF
```

FIGURE 79. Schematical representation of the structure in the solid state of the aggregated Grignard complexes **179–182**

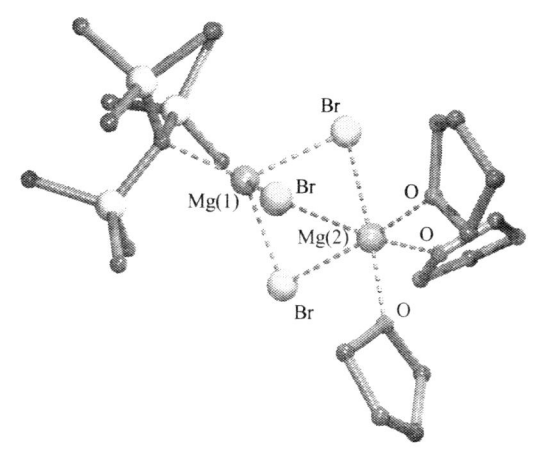

FIGURE 80. Molecular geometry of **183** in the solid state

According to an X-ray crystal-structure determination the crystalline material obtained from a solution of [(Me$_3$Si)$_3$C]MgBr in THF appeared to be [(Me$_3$Si)$_3$C]Mg$_2$Br$_3$(THF)$_3$ (**183**)[206]. Its molecular geometry comprises two magnesium atoms between which three bromine atoms are bridge-bonded. To one of the magnesium atoms the (Me$_3$Si)$_3$C group is σ-bonded [Mg–C 2.16(3) Å] while to the other magnesium atom three additional THF molecules are coordinated [Mg–O 2.09(2), 2.06(2) and 2.04(2) Å]. Consequently, one of the magnesium atoms is four-coordinate whereas the other one is six-coordinate (Figure 80). It is notable that the Br–Mg distances to the four-coordinate magnesium atom [average Mg(1)–Br 2.571(9) Å] are considerably shorter that those to the six-coordinate magnesium atom [Mg(2)–Br 2.741(9) Å]. It has been suggested, based on cold-spray ionization mass spectroscopy, that species having a structure similar to that of **183** are the predominant ones in THF solutions of Grignard reagents[60].

Grignard reagents are also capable of aggregating with other metal salts like LiBr. The structure of [(PhMe$_2$Si)(Me$_3$Si)$_2$C]MgBr$_2$Li(THF)(TMEDA) (**184**) was elucidated by X-ray crystallography[112]. In the structure of **184** two bromine atoms are symmetrically bridge-bonded between magnesium and lithium [Mg–Br 2.530(3) Å and Li–Br 2.507(13) Å] (Figure 81). The (PhMe$_2$Si)(Me$_3$Si)$_2$C group is σ-bonded [Mg–C 2.186(8) Å] to magnesium and a THF molecule is coordinating to magnesium [Mg–O 2.056(5) Å]. To attain a tetrahedral coordination geometry at lithium a TMEDA molecule is N,N'-chelated bonded to this lithium atom. A similar structure has been found in the solid state for [(Me$_3$Si)$_3$C]MgBr$_2$Li(THF)$_3$ (**185**). In **185**, instead of the chelate-bonded TMEDA molecule two THF molecules are coordinate to the lithium atom[207].

A few organomagnesium halides containing a monoanionic, C,N-chelating ligand have been structurally characterized by X-ray crystallographic studies. A discrete monomeric structure was found for (2-PySiMe$_2$)(Me$_3$Si)$_2$CMgBr(THF) (**186**)[208] (Figure 82). The (2-PySiMe$_2$)(Me$_3$Si)$_2$C monoanionic ligand forms a five-membered chelate ring with magnesium via a σ-carbon–magnesium bond [Mg–C 2.189(9) Å] and a coordinate bond of the pyridyl nitrogen atom with magnesium [Mg–N 2.097(9) Å]. A tetrahedral coordination geometry at magnesium is reached by a Mg–Br bond and an additional coordinating THF molecule.

The dimeric organomagnesium halide complexes **187–189** (Figure 82) were obtained from the reaction of (2-Py)(SiMe$_3$)$_2$C-Sb=C(SiMe$_3$)(2-Py) with Et$_2$Mg in THF in the

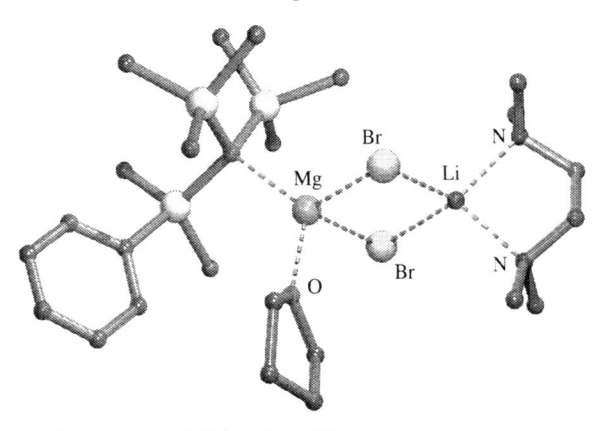

FIGURE 81. Molecular geometry of **184** in the solid state

(**186**)

(**187**) $X^1 = X^2 = Br$
(**188**) $X^1 = X^2 = Cl$
(**189**) $X^1 = Cl; X^2 = OEt$

(**190**)

FIGURE 82. Schematic structures of organomagnesium halides containing monoanionic C,N-chelating ligands

presence of Br^-, Cl^- and EtO^-, respectively[209]. These three complexes have closely related structures consisting of two (2-Py)(Me$_3$Si)$_2$CMg(THF) moieties linked via two bridging halogen atoms in **187** and **188**, or a chloride and an ethoxy bridge in **189**. The (2-Py)(Me$_3$Si)$_2$C monoanionic, C,N-chelating ligand forms a four-membered chelate ring with magnesium. Dimerization occurs via two symmetrically bridge-bonded halogen atoms in **187** and **188** and one bridge-bonded chlorine atom and one bridging ethoxy group in **189**. As a consequence the magnesium atoms are penta-coordinate, in **187** and **188** close to square pyramidal and in **188** distorted trigonal bipyramidal with the σ-bonded carbon atom, the oxygen atom of the ethoxy group and the oxygen atom of the coordinating THF molecule at the equatorial positions.

In the solid state the Grignard reagent Me$_2$N(CH$_2$)$_3$MgCl aggregates with MgCl$_2$ to a complex structure [Me$_2$N(CH$_2$)$_3$Mg$_2$Cl$_3$(THF)$_2$]$_2$ (**190**) (Figure 82)[210]. A similar overall structural motif has been found for monoorganomagnesium halides **179–182**, vide supra. In **190** the monoanionic, C,N-chelating Me$_2$N(CH$_2$)$_3$ group forms a five-membered chelate ring with magnesium via a σ-bonded carbon atom [Mg–C 2.146(9) Å] and a coordination bond with nitrogen [Mg–N 2.23(1) Å].

1. Structural organomagnesium chemistry

The Grignard reagent 2-pyridylmagnesium bromide crystallizes from THF as a dimeric complex (2-Py)$_2$Mg$_2$Br$_2$(THF)$_3$ (**191**). Its structure in the solid state comprises two magnesium atoms between which two 2-pyridyl groups are bridge-bonded via a σ-carbon–magnesium bond [Mg–C 2.149(3) Å] and a nitrogen–magnesium coordination bond [2.129(3) Å] (Figure 83)[211]. To each of the magnesium atoms one bromine atom is bonded [Mg–Br 2.4887(9) Å] and one THF molecule is coordinated. Finally, one additional THF molecule is bridge-bonded via its oxygen atom [Mg–O both 2.374(2) Å] between the two magnesium atoms. It here probably acts as a four-electron donor.

The structures of some Grignard reagents containing monoanionic, C,O-chelating ligands have been established in the solid state by X-ray crystallography. The 1-bromomagnesio-tris-THF derivative (**192**) of N-pivaloyl-tetrahydroisoquinoline crystallizes as a discrete monomer (Figure 84)[212]. Coordination of the carbonyl oxygen atom to magnesium [Mg–O 2.049(8) Å] results in the formation of a five-membered chelate ring. Three additional THF molecules are coordinate-bonded to the magnesium, resulting in an octahedral coordination geometry. Due to the geometry of the C,O-chelating ligand

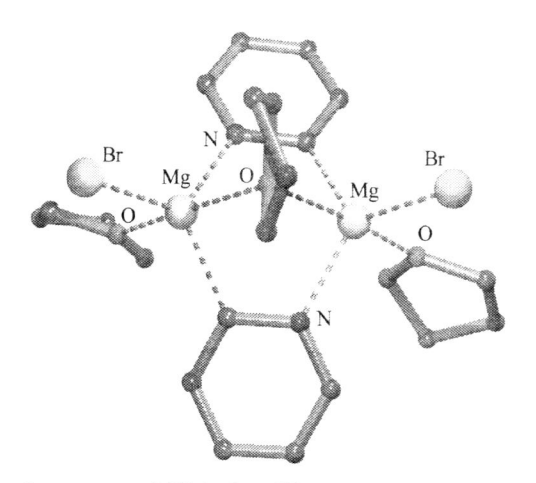

FIGURE 83. Molecular geometry of **191** in the solid state

(**192**)   (**193**)

FIGURE 84. Schematic representation of the molecular structures of **192** and **193** in the solid state

the magnesium-bonded carbon atom and the coordinating carbonyl–oxygen atom are in *cis*-position while the bromine atom is in *trans*-position with respect to this oxygen atom.

A systematic study of the structures in the solid state of 2-$CH_2(OCH_2CH_2)_n$OMe-substituted phenylmagnesium bromides ($n = 0-3$) has been carried out[213]. For the most simple compound ($n = 0$), i.e. 2-(methoxymethyl)phenylmagnesium bromide (**193**), a dimeric structure via two bridge-bonded bromine atoms in the solid state was found (Figure 84). The coordination geometry at the magnesium atoms is trigonal bipyramidal, with the magnesium-bonded carbon atom of the C,O-chelating ligand, a bromine atom and an additional coordinating THF molecule at the equatorial positions and the intramolecular coordinating oxygen atom and the other bromine atom at the axial sites. The bridge-bonding of the bromine atoms is such that an equatorial-bonded bromine atom in one half of the dimer occupies an axial site of the magnesium in the other half of the dimer, and *vice versa*. As might be expected the Mg–Br bond distance of equatorially-bonded bromine [Mg–Br 2.509(3) Å] is considerably shorter than that of an axially bonded one [Mg–Br 2.705(3) Å].

The 2-$CH_2(OCH_2CH_2)_n$OMe-substituted phenylmagnesium bromides with $n = 1$ (**194**), $n = 2$ (**195**) and $n = 3$ (**196**) are all discrete monomers in the solid state (Figure 85). In all three compounds the magnesium atom has an octahedral coordination geometry with the magnesium-bonded carbon atom and the coordinating benzylic oxygen atom in *cis*-position with respect to each other and the bromine atom in *trans*-position with respect to the coordinating benzylic oxygen atom. In **194–196** all the oxygen atoms of the substituents are involved in intramolecular coordination, but to complete six-coordination at magnesium in **194** two additional coordinating THF molecules are present, while in **195** only one additional THF molecule is required for that purpose.

In continuation of this study, the same authors investigated the structures in the solid state of crown-ether Grignard reagents. The structures of 2-(bromomagnesio)-1,3-xylyl-15-crown-4 (**197**)[214], 2-(bromomagnesio)-1,3-phenylene-16-crown-5 (**198**)[215] and 2-(bromomagnesio)-1,3-xylyl-18-crown-5 (**199**)[216] were determined by X-ray crystallography (Figure 86). In **197** all oxygen atoms of the crown are involved in coordination to magnesium, two with a relatively short Mg–O bond distance [Mg–O 2.12(1) and 2.13(1) Å] and two with a longer bond distance [Mg–O 2.33(1) and 2.49(1) Å]. The coordination sphere of magnesium may be considered as pentagonal-bipyramidal, with the bromine atom at the apex[214].

In **198** (Figure 86) the two phenolic oxygen atoms are not involved in coordination to magnesium, most probably because this would require the formation of two highly unfavorable four-membered chelate rings. Instead, an additional THF molecule is coordinating to the magnesium atom, giving it a distorted octahedral coordination geometry with the bromine atom *cis*-positioned with respect to the σ-bonded carbon atom and the coordinating THF molecule in a *trans*-position to the bromine atom.

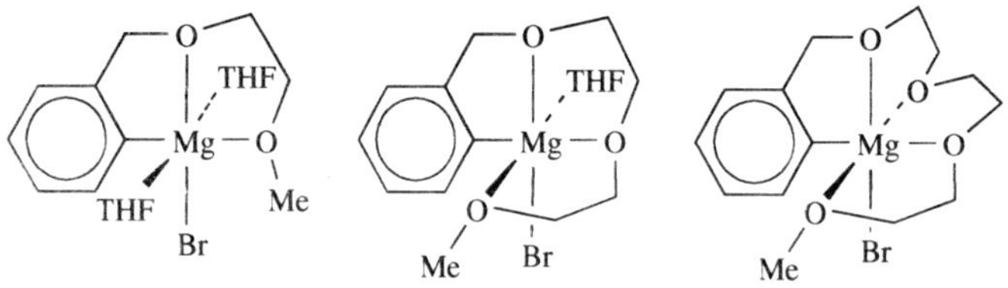

FIGURE 85. Schematic representation of the molecular structures of **194–196** in the solid state

(197) (198)

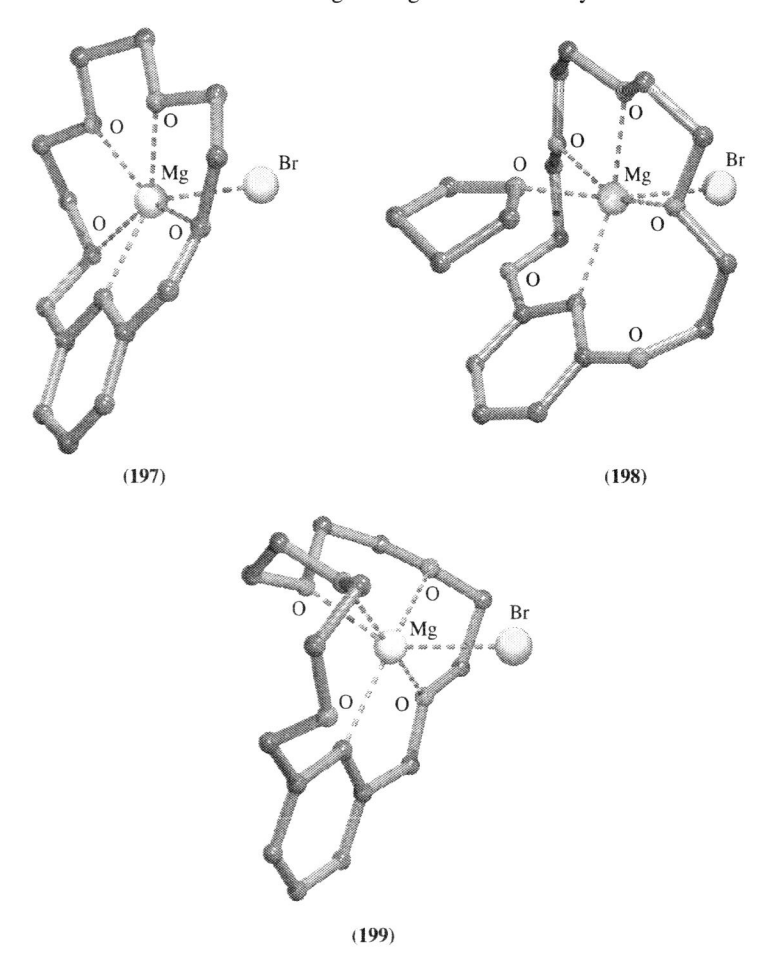

(199)

FIGURE 86. Molecular geometries of the crown-ether Grignard compounds **197–199** in the solid state

In the 18-crown-5 derivative **199** four of the five oxygen atoms of the crown-ether are involved in coordination to magnesium (Figure 86). Together with the σ-bonded carbon atom and the bromine atom this leads to a distorted octahedral coordination geometry at magnesium. Like in **198** the bromine atom is *cis*-positioned with respect to the σ-bonded carbon atom.

The structures in the solid state of a few cyclopentadienylmagnesium halide complexes have been determined by X-ray crystallography. The structures of CpMgBr(tetraethylethylenediamine) (**200**)[217] and 1,2,4-(Me$_3$Si)$_3$C$_5$H$_2$MgBr(TMEDA) (**201**)[93] show large similarities. The structure of **201** is shown (Figure 87). Both compounds are discrete monomers in which the cyclopentadienyl group is η^5-bonded to magnesium and the diamine ligand is N,N'-chelate bonded.

In the solid state cyclopentadienylmagnesium chloride exists as a dimer [CpMgCl (OEt$_2$)]$_2$ (**202**)[218] (Figure 87). The dimeric structure is caused by two symmetrically

(201)   (202)

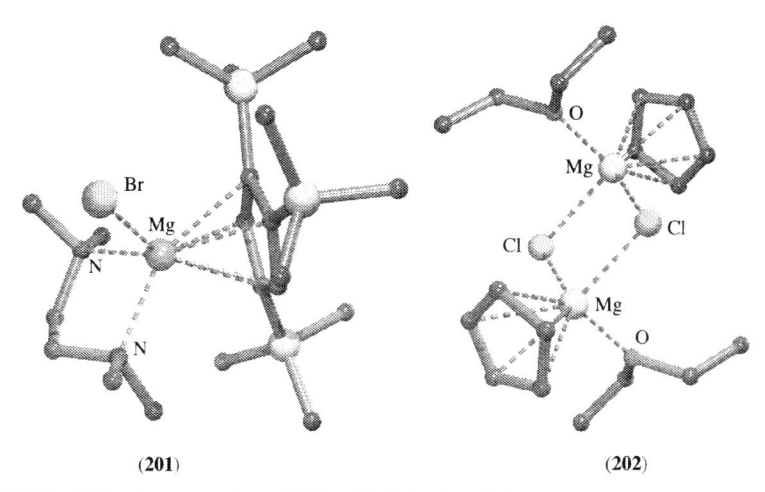

FIGURE 87. Molecular geometries of **201** and **202** in the solid state

bridge-bonded chlorine atoms between the two magnesium atoms [Mg–Cl 2.419(2) and 2.432(2) Å]. To each of the magnesium atoms one cyclopentadienyl group is η^5-bonded and one coordinating diethyl ether molecule completes the coordination sphere of magnesium. Also, the structures of [Cp*MgCl(OEt$_2$)]$_2$ (**203**)[218], [Cp*MgCl(THF)]$_2$ (**204**)[219] and [Cp*MgBr(THF)]$_2$ (**205**)[220] in the solid state were determined by X-ray crystallography. They have a similar dimeric structural motif as observed for **202**.

A cyclopentadienylmagnesium bromide containing a heteroatom-functionalized substituent at the cyclopentadienyl group has also been structurally characterized. When 1-[2-(dimethylamino)ethyl]-2,3,4,5-tetramethylcyclopentadienylmagnesium bromide is recrystallized from dichloromethane, dimeric [(Me$_2$N(CH$_2$)$_2$)Me$_4$C$_5$MgBr]$_2$ (**206**) (Figure 88) is obtained[221]. Its X-ray crystal-structure determination reveals a structure in the solid state

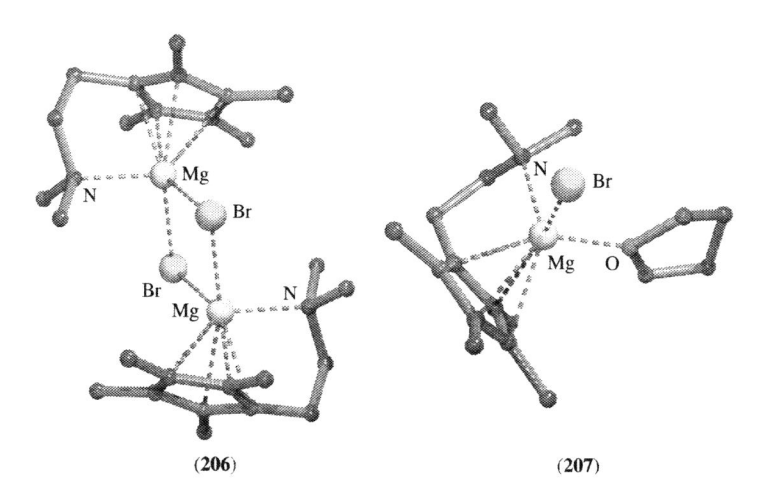

(206)   (207)

FIGURE 88. Molecular geometries of **206** and **207** in the solid state

1. Structural organomagnesium chemistry 69

that is closely related to the dimeric structure observed for **202**. However, instead of the coordinating diethyl ether molecule in **202**, in **206** the nitrogen atom of the Me$_2$NCH$_2$CH$_2$ substituent is coordinating intramolecularly to the magnesium atom.

When **206** is recrystallized from THF the dimeric structure is broken down to a discrete monomeric one (Me$_2$N(CH$_2$)$_2$)Me$_4$C$_5$MgBr(THF) (**207**), as was shown by an X-ray crystal-structure determination. In **207**, the substituted cyclopentadienyl group is η^5-bonded to magnesium while the nitrogen atom of the functional substituent is intramolecularly coordinated to magnesium. The bromine atom and an additional coordinating THF molecule complete the coordination at magnesium.

D. Monoorganomagnesium Compounds RMgY with Y = OR

The number of heteroleptic organomagnesium compounds RMgOR for which the structure in the solid state was established unambiguously by X-ray crystallography is rather limited, in contrast to the large number of structures known for the corresponding heteroleptic RZnOR congeners[29].

The structures of only two monomeric RMgOR complexes are known. The reaction of Et$_2$Mg with one equivalent of 2,6-di-*tert*-butylphenol in the presence of TMEDA affords a crystalline product with composition EtMgOC$_6$H$_3$Bu-*t*-2,6(TMEDA) (**208**)[189]. Its X-ray crystal-structure determination reveals a monomeric molecule with the ethyl group σ-bonded to magnesium [Mg–C 2.147(10) Å] and the phenoxy group also σ-bonded with a very short bond distance [Mg–O 1.888(5) Å] (Figure 89). A N,N'-chelate-bonded TMEDA molecule completes a tetrahedral coordination geometry at magnesium.

A similar reaction of *i*-Bu$_2$Mg with 2,6-di-*tert*-butylphenol in the presence of 18-crown-6 affords *i*-BuMgOC$_6$H$_3$Bu-*t*-2,6(18-crown-6) (**209**) as a crystalline solid[183]. An X-ray crystal-structure determination showed that this compound in the solid state also exists as a monomer with a σ-bonded *i*-butyl group and a σ-bonded phenoxy oxygen atom. Three adjacent oxygen atoms of the crown-ether are involved in coordination to magnesium, resulting in penta-coordination.

A dimeric structural motif, formed by bridge-bonding of two oxygen atoms between two magnesium atoms, has been observed in the solid-state structures of [*s*-BuMgOC$_6$H$_3$(Bu-*t*)$_2$-2,6]$_2$ (**210**)[222] and [*n*-HexMgOC$_6$H$_2$(Bu-*t*)$_2$-2,6-Me-4]$_2$ (**211**)[223] (Figure 90). In both compounds two phenoxy groups are symmetrically bridge-bonded between two magnesium atoms, while the organic group is σ-bonded to magnesium. As a result the magnesium atoms in **210** and **211** have a trigonal planar coordination geometry.

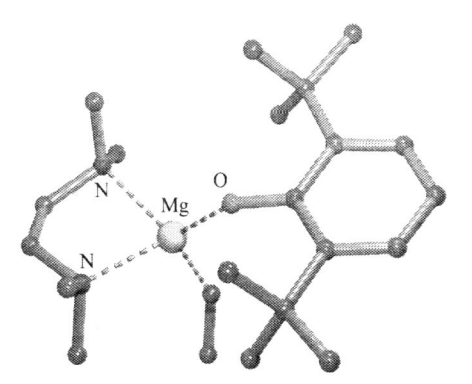

FIGURE 89. Molecular geometry of **208** in the solid state

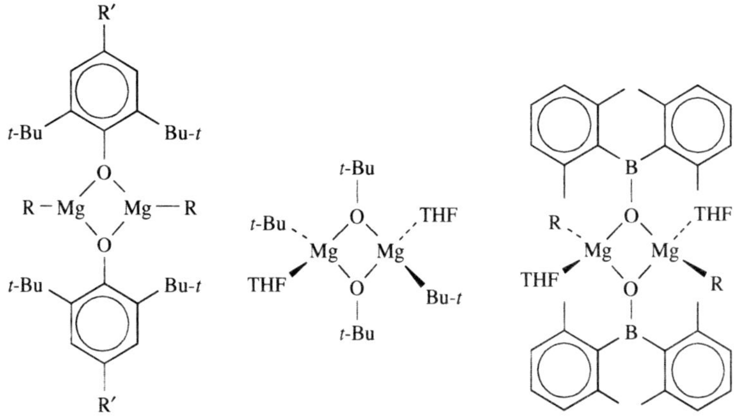

                                                                        R'
                                                                        |
                                                                    ┌───┴───┐
                                                              t-Bu──┤   ├──Bu-t
                                                                    └───┬───┘
                                                                        O
                                                                       / \
                                                                  R—Mg   Mg—R
                                                                       \ /
                                                                        O
                                                                    ┌───┴───┐
                                                              t-Bu──┤   ├──Bu-t
                                                                    └───┬───┘
                                                                        R'

(210) R = s-Bu; R' = H
(211) R = n-Hex; R' = Me

(212)

(213) R = n-Bu
(214) R = Me

FIGURE 90. Schematic representation of the structures of **210–214** in the solid state

Also, *t*-BuMgOBu-*t* exists in the solid state as a dimeric complex [*t*-BuMgOBu-*t*(THF)]₂ (**212**)[224]. Two *t*-BuO groups are symmetrically bridge-bonded between two magnesium atoms forming a flat four-membered Mg–O–Mg–O ring (Figure 90). To each of the magnesium atoms a *t*-Bu group is σ-bonded while a coordinating THF molecule completes a tetrahedral coordination geometry at magnesium.

Crystalline [*n*-BuMgOB(Mes)₂(THF)]₂ (**213**) was obtained from the reaction of *n*-Bu₂Mg with dimesityl boronic acid in THF. The corresponding methyl derivative [MeMgOB(Mes)₂(THF)]₂ (**214**) was prepared via a transmetallation reaction of the lithium salt of dimesityl boronic acid with MeMgCl in THF. For both compounds the structure in the solid state was determined by X-ray crystallography[225]. The basic structural motif of these compounds is identical to that of **212**; both are dimers, via bridge-bonding of the oxygen atoms of two dimesityl boronic acid anions between two magnesium atoms. An additional coordinating THF molecule completes a tetrahedral geometry at magnesium (Figure 90).

Methylmagnesium *tert*-butoxide exists in the solid state as a tetrameric aggregate [MeMgOBu-*t*]₄ (**215**). Its X-ray crystal-structure determination reveals a heterocubane structure with alternating magnesium and oxygen atoms at the corners of the cube (Figure 91)[226]. To each of the magnesium atoms one methyl group is σ-bonded.

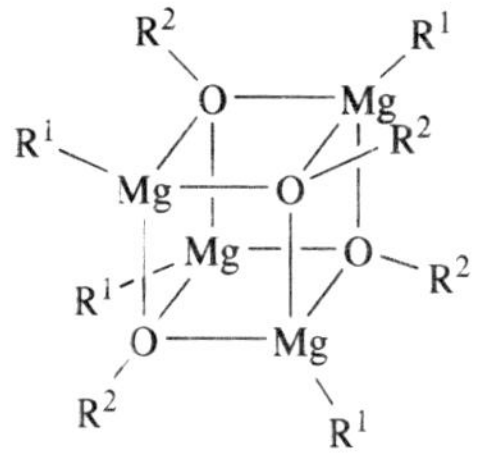

(215) R¹ = Me, R² = *t*-Bu
(216) R¹ = Cp, R² = Et

FIGURE 91. Schematic representation of the structures of **215** and **216** in the solid state

Like **215**, cyclopentadienylmagnesium ethoxide exists in the solid state as tetrameric [CpMgOEt]$_4$ (**216**) with a heterocubane structure (Figure 91)[227]. In **216** the cyclopentadienyl groups are η^5-bonded to the magnesium atoms.

E. Monoorganomagnesium Compounds RMgY with Y = NR$_2$

Despite the capability of anionic amide groups to form aggregates with metals via bridging nitrogen atoms, most of the monoorganomagnesium amides that have been structurally characterized in the solid state have a discrete monomeric structure.

The monoorganomagnesium amides **217**[228], **218**[228], **219**[229], **220**[224] and **221**[230] (Figure 92), derived from substituted anilines, have comparable structures in the solid state. They are all monomers with a tetrahedrally coordinated magnesium center formed by one σ-bonded organic group, one σ-bonded amido nitrogen atom and two coordinating heteroatoms. In **217–219** these are two THF molecules and in **220** a N,N'-chelate-bonded TMEDA molecule. In **217–220** the sum of the bond angles around the nitrogen atoms is 360° within experimental error, indicating that these nitrogen atoms are sp^2-hybridized. In **221** the carbazole skeleton is essentially flat, but the carbazole carbon C–N–Mg bond angles [both 108.2(2)°] indicate that the magnesium atom is bonded to a sp^3-hybridized nitrogen atom. In the latter compound the Mg–N bond distance [Mg–N 2.087(3) Å] is slightly longer than the Mg–N bond distances in **217–220** [2.040(3), 2.037(3), 2.027(4) and 2.004(2) Å, respectively].

Reaction of dialkylmagnesium compounds with 2,6-bis(imino)pyridines results in quantitative *N*-alkylation of the pyridine skeleton (equation 15).

The structures in the solid state of three of the initially formed organomagnesium amides, **222**, **223** and **224**, were determined by X-ray crystallography[231]. All three compounds are discrete monomers and have comparable structures of which that of **222** is shown (Figure 93). In **222** the ethyl group is σ-bonded to magnesium and interacts with the three nitrogen atoms of the *N*-alkylated 2,6-bis(imino)pyridine in a 'pincer-type'[232]

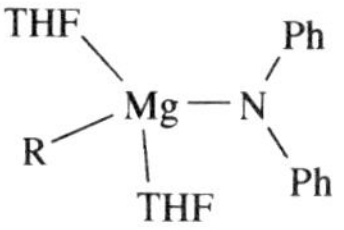

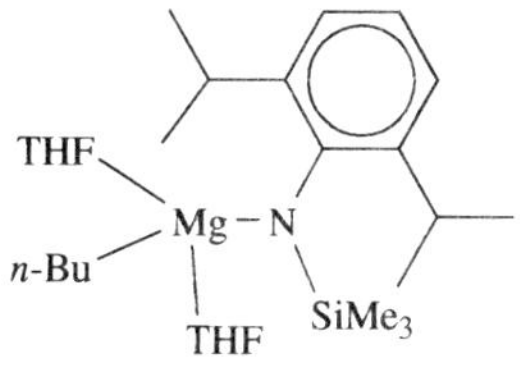

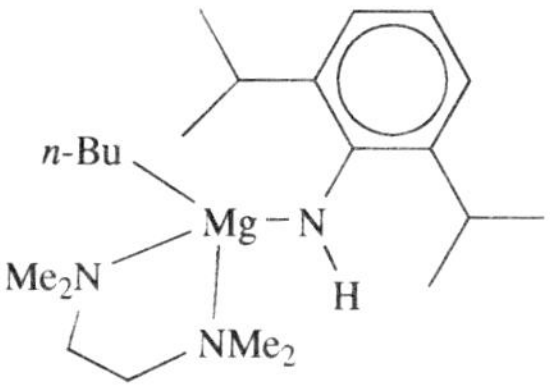

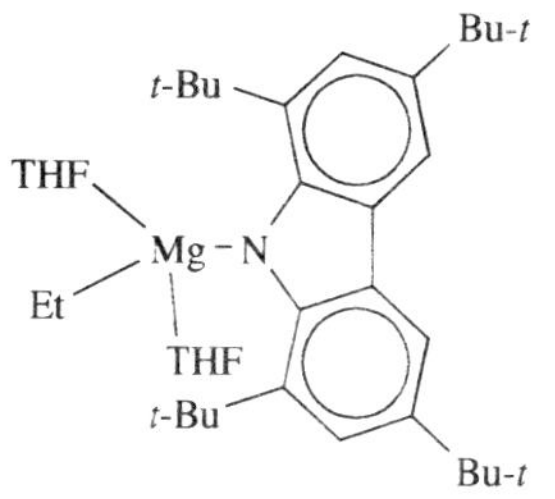

FIGURE 92. Schematic representation of the structures of **217–221** in the solid state

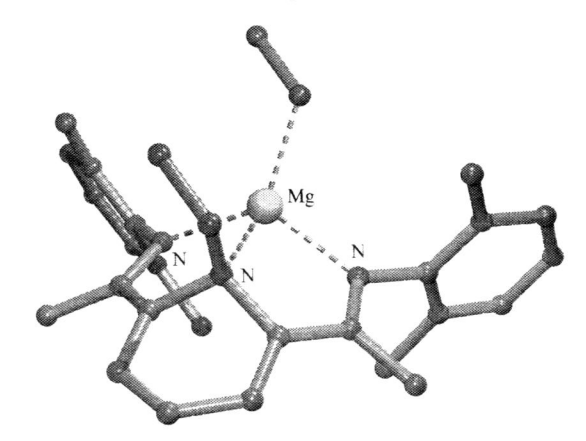

FIGURE 93. Molecular geometry of **222** in the solid state

fashion. Due to the rigidity of the monoanionic, tridentate ligand system the geometry around magnesium is severely distorted from tetrahedral.

$$\text{(pyridine-diimine)} + R_2Mg \longrightarrow \text{(chelate-Mg-R complex)} \quad (15)$$

(**222**) R = Et; R^1 = 2,6-$Me_2C_6H_3$
(**223**) R = i-Pr; R^1 = 2,6-$Me_2C_6H_3$
(**224**) R = i-Pr; R^1 = 2,6-$Et_2C_6H_3$

Reaction of Cp(Me)Mg(OEt$_2$) with 2,5-bis[(dimethylamino)methyl]pyrrole in diethyl ether results in selective protonolysis of the magnesium-bonded methyl group and results in the formation of the corresponding CpMg amide (**225**) (equation 16). An X-ray crystal-structure determination showed that **225**, of which the structure is shown schematically (equation 16), exists in the solid state as a monomer[233]. The cyclopentadienyl group is η^5-bonded to magnesium, while the pyrrole amido-nitrogen atom is σ-bonded to magnesium [Mg–N 2.043(1) Å]. Only one of the (dimethylamino)methyl substituents forms an intramolecular coordination bond to magnesium [Mg–N 2.225(2) Å]. An additional diethyl ether molecule coordinates to magnesium to complete the coordination saturation.

In a similar way Cp(Me)Mg(OEt$_2$) is capable of deprotonating N,N'-bis(2,4,6-trimethylphenyl)(*tert*-butyl)amidine to form the corresponding cyclopentadienylmagnesium amidinate complex (**226**) (equation 17). An X-ray crystal-structure determination of **226**, of which the structure is shown schematically (equation 17), showed that this compound also exists as a monomer in the solid state[234]. Like in **225** the cyclopentadienyl group is η^5-bonded to magnesium while the amidinate anion is N,N'-chelate bonded with almost equal Mg–N bond distances [Mg–N 2.090(2) and 2.097(2) Å]. Furthermore, an additional

THF molecule is coordinate-bonded to magnesium. Due to the small bite angle of the amidinate anion the N–Mg–N bond angle is very acute [N–Mg–N 63.3(1)°].

$$\text{(16)}$$

$$\text{(225)}$$

$$\text{(17)}$$

R = 2,4,6-Me$_3$C$_6$H$_2$

(226)

Various tris(pyrazolyl)borato alkylmagnesium derivatives have been prepared and were structurally characterized in the solid state. X-ray crystal-structure determinations of methylmagnesium tris(3-*tert*-butylpyrazolyl)hydroborate (**227**)[235, 236], isopropylmagnesium tris(3-*tert*-butylpyrazolyl)hydroborate (**228**)[236, 237] and trimethylsilylmethylmagnesium tris (3,5-dimethylpyrazolyl)hydroborate (**229**)[236, 238] show that they have comparable structures in the solid state (Figure 94). In these compounds the tris(pyrazolyl)hydroborate moiety acts as a monoanionic, tridentate ligand which is bonded with almost equal Mg–N bond distances to the alkylmagnesium moiety. Due to the small bite angle of the tripodal ligand, all N–Mg–N bond angles are close to 90°, and differ significantly from the ideal tetrahedral value. As a consequence the coordination geometry at magnesium is considerably distorted from tetrahedral.

The structures of methylmagnesium tris(3-*tert*-butylpyrazolyl)phenylborate (**230**) and ethylmagnesium tris(3-*tert*-butylpyrazolyl)phenylborate (**231**) have been determined by X-ray crystallography and are shown schematically (Figure 94)[239]. Their structures show large similarities with that of **227** and **228** and only differ in the presence of a boron-bonded phenyl group in **230** and **231**.

The X-ray crystal-structure determination of ethylmagnesium tris(3-phenylpyrazolyl) hydroborate (**232**)[240] (Figure 94) shows that the magnesium atom is penta-coordinate as the result of one σ-bonded ethyl group, three magnesium–nitrogen bonds with the tris(3-phenylpyrazolyl)hydroborate moiety and one additional, coordinating THF molecule.

Reaction of β-diketimines with dialkylmagnesium compounds in a 1:1 molar ratio affords the corresponding monoorganomagnesium β-diketiminates in high yield (Scheme 3). An alternative synthetic route involves deprotonation of the β-diketimine with *n*-BuLi and subsequent transmetallation of the initially formed lithium β-diketiminate with a suitable Grignard reagent. Extensive X-ray diffraction studies of the compounds obtained from these reactions have showed that, depending on the nature of the organic group bound to magnesium and the nature of the solvent used for the synthesis, three basic structural motifs, **A, B** and **C** (Scheme 3), are observed in the solid state for monoorganomagnesium β-diketiminates. These motifs are: (i), monomers in which the β-diketiminate anion is

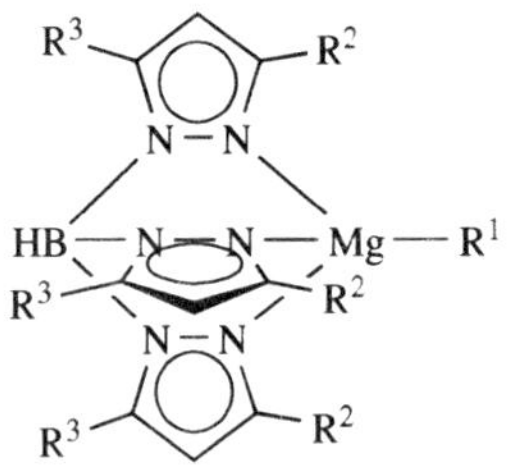

(227) R^1 = Me, R^2 = t-Bu, R^3 = H
(228) R^1 = i-Pr, R^2 = t-Bu, R^3 = H
(229) R^1 = CH_2SiMe_3, R^2 = R^3 = Me

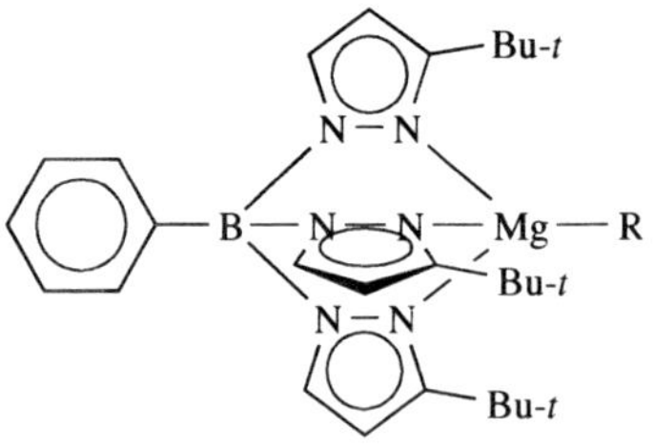

(230) R = Me
(231) R = Et

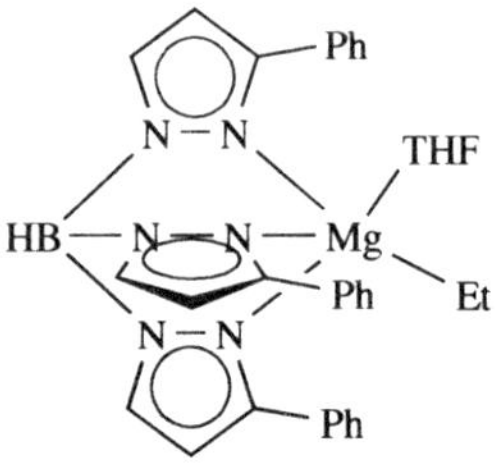

(232)

FIGURE 94. Schematic structures of the alkylmagnesium tris(pyrazolyl)borates 227–232

N,N'-chelate bonded to the alkylmagnesium moiety, with trigonal planar coordination geometry at magnesium, (ii) monomers in which the β-diketiminate anion is N,N'-chelate bonded to the alkylmagnesium moiety and an additional ligand is coordinating to the magnesium atom, giving it a tetrahedral coordination geometry, and (iii) dimers formed via bridging of two two-electron three-center bonded alkyl groups between two magnesium atoms, while a β-diketiminate anion is N,N'-chelate bonded to each magnesium atom.

The structures of the monoorganomagnesium β-diketiminates **233**[241], **234**[242], **235**[242] and **236**[243] are comparable. The N,N'-chelate bonding of the β-diketiminate anion with almost equal Mg—N bond distances to magnesium results in a six-membered MgN_2C_3 ring with all atoms located in one plane. The Mg—C bond of the σ-bonded alkyl group also lies in this plane. As a representative example the structure of **233** is shown (Figure 95).

The structures of the monoorganomagnesium β-diketiminates **237**[241], **238**[244], **239**[245], **240**[124], **241**[246], **242**[247], **243**[242] and **244**[242] show similar structural features. A distorted tetrahedral coordination geometry at magnesium is reached by a N,N'-chelate bonded β-ketiminate anion, a σ-bonded organic group and an additional coordinating solvent molecule, either THF or diethyl ether. In contrast to the flat six-membered MgN_2C_3 ring in **233–236**, this ring in **237–244** deviates considerably from planar and can best be described as having a distorted boat conformation with the magnesium atom at the front and the opposing carbon atom at the back. As an example the structure of **238** is shown (Figure 95).

The structures of the dimers **245**[244] (Figure 95) and **246**[248] show large similarities. The two halves of the dimers are linked via two symmetrically bridging two-electron three-center bonded alkyl groups. A β-diketiminate anion is N,N'-chelate bonded to each of the magnesium atoms giving them distorted tetrahedral coordination geometries. Also, in **245** and **246** the MgN_2C_3 ring deviates considerably from planar.

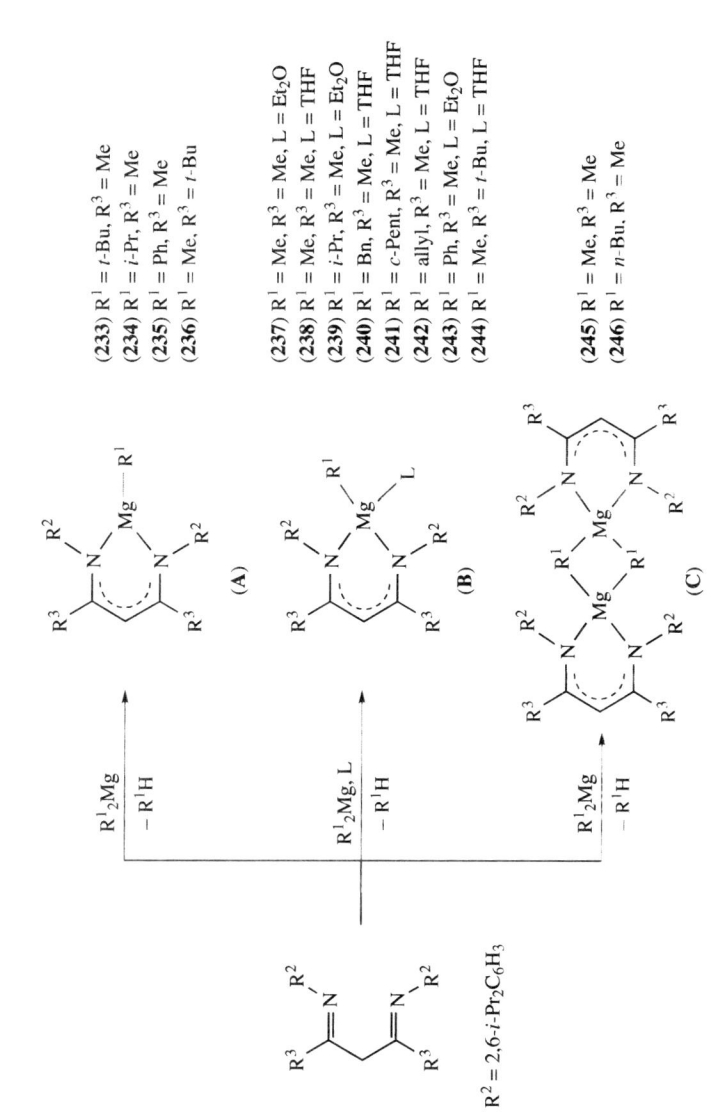

(233) $R^1 = t$-Bu, $R^3 = $ Me
(234) $R^1 = i$-Pr, $R^3 = $ Me
(235) $R^1 = $ Ph, $R^3 = $ Me
(236) $R^1 = $ Me, $R^3 = t$-Bu

(237) $R^1 = $ Me, $R^3 = $ Me, L = Et$_2$O
(238) $R^1 = $ Me, $R^3 = $ Me, L = THF
(239) $R^1 = i$-Pr, $R^3 = $ Me, L = Et$_2$O
(240) $R^1 = $ Bn, $R^3 = $ Me, L = THF
(241) $R^1 = c$-Pent, $R^3 = $ Me, L = THF
(242) $R^1 = $ allyl, $R^3 = $ Me, L = THF
(243) $R^1 = $ Ph, $R^3 = $ Me, L = Et$_2$O
(244) $R^1 = $ Me, $R^3 = t$-Bu, L = THF

(245) $R^1 = $ Me, $R^3 = $ Me
(246) $R^1 = n$-Bu, $R^3 = $ Me

$R^2 = 2,6$-i-Pr$_2$C$_6$H$_3$

SCHEME 3. Synthesis of organomagnesium β-diketiminates

(233) (238)

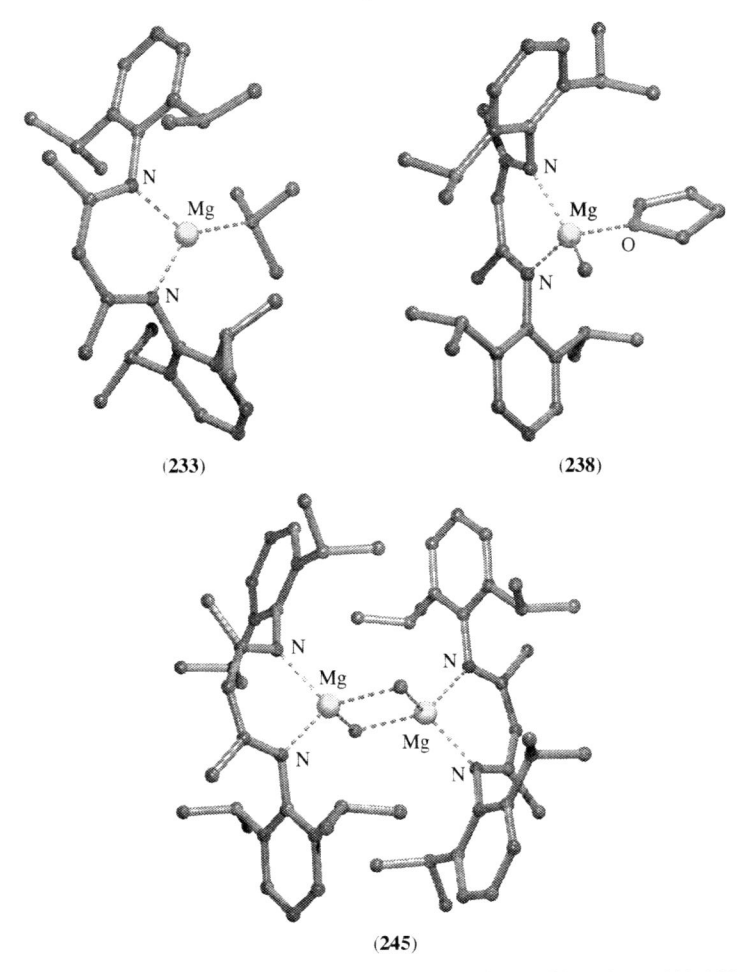

(245)

FIGURE 95. Molecular geometries of the monoorganomagnesium β-diketiminates **233, 238** and **245** in the solid state

In some of the monoorganomagnesium β-diketiminates having structural motif **B**, i.e. complexes **237–244**, the coordinating solvent molecule is relatively weakly bound and can be removed at reduced pressure. For example, when **243** is dried for a few hours in high vacuum and the residue is recrystallized from a non-coordinating solvent like toluene, crystalline **235** is obtained[242]. However, when the allylmagnesium β-diketiminate THF complex **242** is dried in vacuo it looses its coordinated THF molecule. An X-ray crystal-structure determination of the resulting product shows that instead of the anticipated monomeric structural motif **A**, this THF-free allylmagnesium β-diketiminate is a cyclic hexameric aggregate (**247**) in the solid state (Figure 96)[249]. In the twenty-four-membered ring structure the magnesium atoms are linked by bridging allyl groups in a very rare *trans-μ-η^1 : η^1* bonding mode. A β-diketiminate anion is N,N'-chelate bonded to each of the magnesium atoms to complete its coordination sphere.

1. Structural organomagnesium chemistry

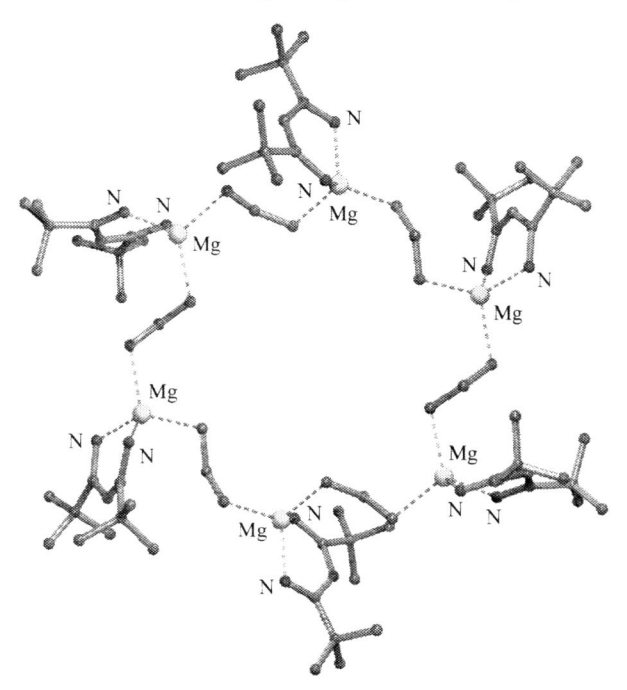

FIGURE 96. Molecular geometry of **247** in the solid state. The 2,6-i-Pr$_2$C$_6$H$_3$ substituents at the nitrogen atoms of the β-diketiminate anions are omitted for clarity

The reaction of isopropylmagnesium β-diketiminate **234** with 2′,4′,6′-trimethylacetophenone in an apolar solvent like toluene results in deprotonation of the ketone with the formation of an enolate (equation 18).

(**234**) $R^1 = 2,6\text{-}i\text{-Pr}_2\text{C}_6\text{H}_3$
            $R^2 = 2,4,6\text{-Me}_3\text{C}_6\text{H}_2$ (**248**)   (18)

An X-ray crystal-structure determination showed that this enolate exists in the solid state as a dimer (**248**) in which two enolate moieties are C,O-bridge-bonded [Mg−O 1.908(2) Å and Mg−C 2.318(3) Å] between two magnesium β-diketiminate units, resulting in distorted tetrahedral coordination geometries at the magnesium atoms[250]. The structure of **248** is shown schematically (equation 18). Such a C,O-bridge bonding mode for enolates is rather rare, but has also been observed in the Reformatski reagent [t-BuO$_2$CCH$_2$ZnBr(THF)]$_2$[251].

The cyclopentadienyl β-diketiminate **249** and its 4-*tert*-butylpyridine adduct **250** have been prepared and structurally characterized (equation 19)[252]. An X-ray crystal-structure determination of **249** showed that the cyclopentadienyl group is η^5-bonded to magnesium. On the basis of the observed bonding parameters of magnesium with the β-diketiminate skeleton [Mg–N 2.006(2) and 2.021(2) Å, Mg–C$_\alpha$ 2.729(3) and 2.826(3) Å and Mg–C$_\beta$ 2.689(3) Å] this bonding is described in terms of a π-interaction. However, in **250** the β-diketiminate is N,N'-chelate bonded to magnesium.

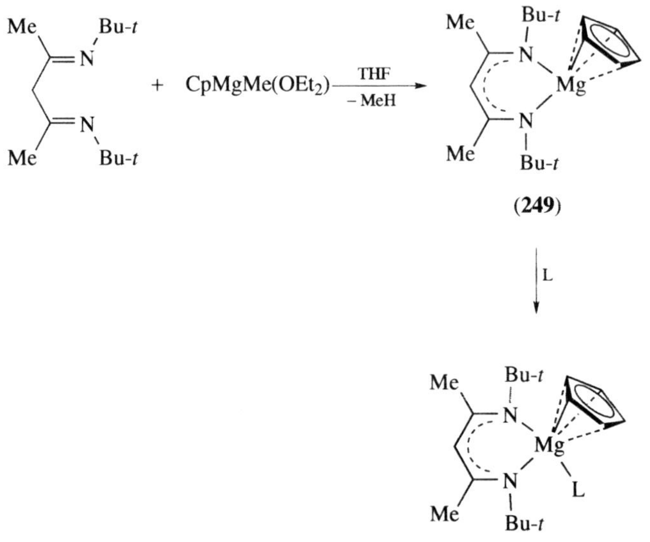

(19)

The hybrid boroamidinate/amidinate ligand as present in the methylmagnesium complex **251** (Figure 97) is isoelectronic with the β-diketiminate skeleton[253]. The X-ray crystal-structure determination of **251** shows that the boroamidinate/amidinate anion adopts a similar N,N'-chelate bonding as observed in organomagnesium β-diketiminates. The structure of **251** is shown schematically (Figure 97).

However, in the donor-ligand-free *tert*-butylmagnesium derivative (**252**), the same boroamidinate/amidinate ligand system adopts an entirely different bonding mode. An X-ray crystal-structure determination of **252** showed that all three nitrogen atoms of

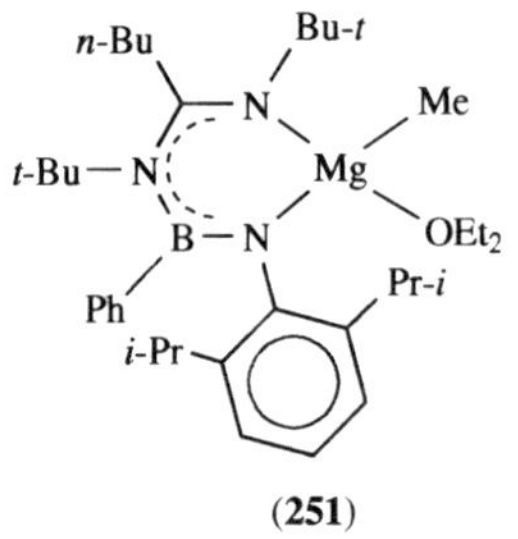

FIGURE 97. Schematic representation of the structure of **251** in the solid state

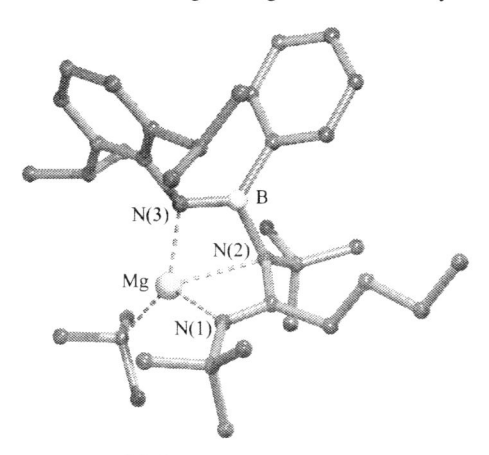

FIGURE 98. Molecular geometry of **252** in the solid state

the boroamidinate/amidinate anion are involved in bonding to magnesium [Mg–N(1) 2.004(2) Å, Mg–N(2) 2.390(2) Å and Mg–N(3) 2.080(2) Å]. Together with the σ-bonded *tert*-butyl group this leads to a distorted tetrahedral coordination geometry at magnesium (Figure 98).

From the reaction of the radical anion of 1,2-bis[(2,6-diisopropylphenyl)imino]acenaphthene (dpp-bian) with i-PrMgCl, the persistent radical complex isopropylmagnesium dpp-bian (**253**) was isolated in yields up to 60% (equation 20)[254]. An X-ray crystal-structure determination of **253** showed that the magnesium atom has distorted tetrahedral coordination geometry as the result of the σ-bonded isopropyl group, one coordinate-bonded diethyl ether molecule and N,N'-chelate bonding of the dpp-bian radical anion. The radical anionic character of the dpp-bian moiety is indicated by the relatively long Mg–N bond distances [Mg–N 2.120(2) and 2.103(2) Å].

$$R = 2,6\text{-}i\text{-}Pr_2C_6H_3 \quad (20)$$

(**253**)

The first step in the reaction of Me$_2$Mg with bulky α-diimine ligands is the formation of a complex in which the α-diimine is N,N'-chelate bonded to Me$_2$Mg (Scheme 4)[255, 256]. The second step is a single electron transfer (SET) resulting in radical-pair formation. From this point two pathways are possible. The first pathway involves escape of a methyl radical from the solvent cage resulting in a methylmagnesium diimine radical that subsequently dimerizes to **254a**. The second pathway involves transfer of a methyl radical to the diimine skeleton resulting in an imino-amide ligand bonded to magnesium which subsequently dimerizes to **254b**. At low temperature the methyl radical-transfer reaction predominates while at room temperature the dimerized radical is the major product. It should be noted that similar radical processes have been observed in the reaction of dialkylzinc compounds with α-diimines[257, 258].

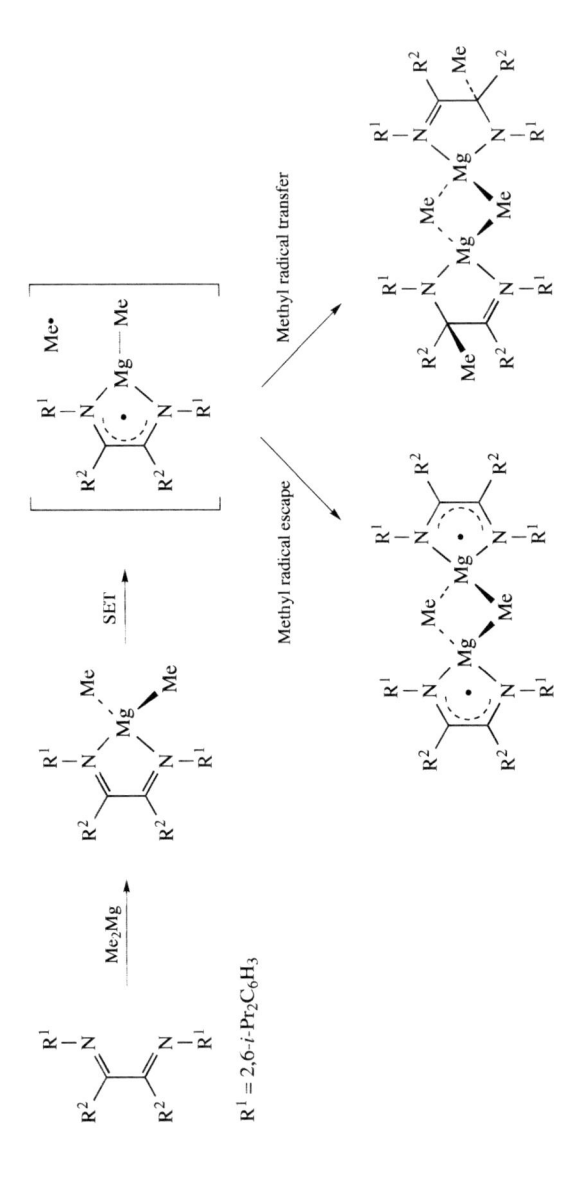

SCHEME 4. The radical mediated processes in the reaction of Me_2Mg with α-diimines

(254a) R^2R^2 = 1,8-naphthdiyl

(254b) R^2 = Me

$R^1 = 2,6$-i-$Pr_2C_6H_3$

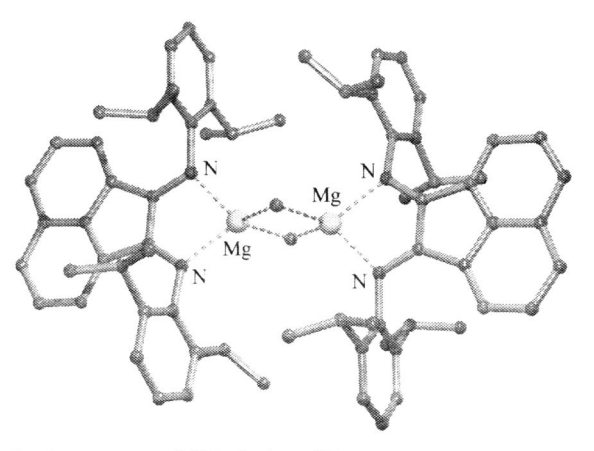

FIGURE 99. Molecular geometry of **254a** in the solid state

The structures of **254a** and **254b** in the solid state were established by X-ray crystal-structure determinations[255,256]. The structure of **254a** (Figure 99) comprises a symmetric dimer in which two methyl groups are two-electron three-center bonded between two magnesium atoms [Mg–C 2.263(5) Å] and a reduced dpp-bian ligand N,N'-chelate bonded to each magnesium atom [Mg–N 2.066(5) and 2.065(4) Å]. As a result both magnesium atoms have a tetrahedral coordination geometry. An X-ray crystal-structure determination of **254b** clearly shows its dimeric structure via two bridging two-electron three-center bonded methyl groups between the magnesium atoms. However, the methyl groups at the diimine skeleton are crystallographically disordered.

The structures of a series of organomagnesium amides, derived from secondary amines, have been determined. These compounds, [n-BuMg(TMP)]$_2$ (**255**)[259], [t-BuMg(TMP)]$_2$ (**256**)[224], [t-BuMgN(Bn)$_2$]$_2$ (**257**)[224], [t-BuMgN(Pr-i)$_2$]$_2$ (**258**)[224], [t-BuMgN(c-Hex)$_2$]$_2$ (**259**)[224], [t-BuMgN(SiMe$_3$)$_2$]$_2$ (**260**)[224] and [s-BuMgN(SiMe$_3$)$_2$]$_2$ (**261**)[260], have in common that they exist as dimers in the solid state. The amido nitrogen atoms are bridging in a symmetric way between the two magnesium atoms, forming a flat four-membered N–Mg–N–Mg ring. An organic group is σ-bonded to each of the magnesium atoms giving them a trigonal planar coordination geometry. The structures of these compounds are shown schematically (Figure 100).

The acetylenic organomagnesium amides [Me$_3$SiC≡CMgN(Pr-i)$_2$(THF)]$_2$ (**262**)[228] and [PhC≡CMgN(Pr-i)$_2$(THF)]$_2$ (**263**)[228] (Figure 100) exist in the solid state as dimers. Their structures comprise a central flat N–Mg–N–Mg four-membered ring as the result of two bridging amide nitrogen atoms between the two magnesium atoms. To each of the magnesium atoms an acetylenic group is σ-bonded and an additional THF molecule is coordinate-bonded, giving the magnesium atoms a tetrahedral coordination geometry. The magnesium-bonded acetylenic groups and the oxygen atoms of the coordinating THF molecules are pairwise located in *anti*-position with respect to the N–Mg–N–Mg plane.

The structures of the organomagnesium amides [t-BuMgNHBu-t(THF)]$_2$ (**264**)[162] and [MeMgNHSi(Pr-i)$_3$(THF)]$_2$ (**265**)[261], derived from primary amines, were determined by X-ray crystallography. The structure of **264** is closely related to those of **262** and **263**. The two amide nitrogen atoms are symmetrically bridge-bonded between the two magnesium atoms, resulting in a central flat, four-membered N–Mg–N–Mg ring (Figure 101). Like in **263** and **264**, the magnesium-bonded *tert*-butyl groups and the oxygen atoms of the coordinating THF molecules are pairwise located in *anti*-position with respect to the

(255) R = n-Bu
(256) R = t-Bu

(257) R¹ = t-Bu, R² = Bn
(258) R¹ = t-Bu, R² = i-Pr
(259) R¹ = t-Bu, R² = c-Hex
(260) R¹ = t-Bu, R² = Me₃Si
(261) R¹ = s-Bu, R² = Me₃Si

(262) R = Me₃SiC≡C
(263) R = PhC≡C

FIGURE 100. Schematic representation of the structures of organomagnesium amides **255–263** derived from secondary amines

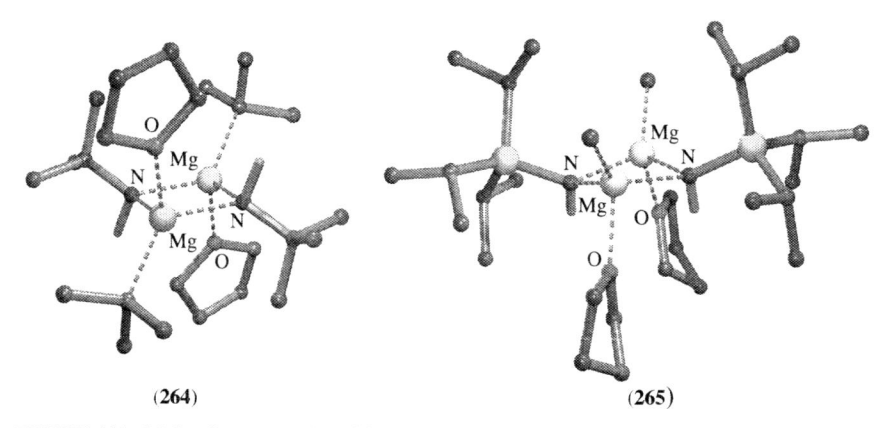

(264)   (265)

FIGURE 101. Molecular geometries of **264** and **265** in the solid state

N–Mg–N–Mg plane. A similar pairwise *anti*-position is observed for the nitrogen-bonded *tert*-butyl groups and the amide-hydrogen atoms.

The structure of **265** also comprises a central four-membered N–Mg–N–Mg ring as the result of bridging amide nitrogen atoms between the magnesium atoms. However, this ring is slightly folded (14.8°). The magnesium-bonded methyl groups and the oxygen atoms of the coordinating THF molecules show a pairwise *syn*-arrangement, as is also observed for the triethylsilyl substituent and the amide hydrogen atoms (Figure 101).

A remarkable structure in the solid state was found for the ethylmagnesium amide derived from the primary amine 2,6-diisopropylaniline. An X-ray crystal-structure determination showed that this compound exists as a cyclic dodecamer [EtMgN(H)C₆H₃(Pr-*i*)₂-2,6]₁₂ (**266**) in the solid state[162]. The cycle consists of twelve magnesium atoms between each of which one amide nitrogen atom is bridge-bonded and one ethyl group is two-electron three-center bridge-bonded resulting in a local N–Mg–N–Mg four-membered ring (Figure 102). The average Mg–C distance is 2.21(2) Å, and the average Mg–N distance is 2.086(10) Å. The ethyl groups are all disposed toward the interior of the cycle and the bulkier 2,6-*i*-Pr₂C₆H₃ substituents are all pointing outward.

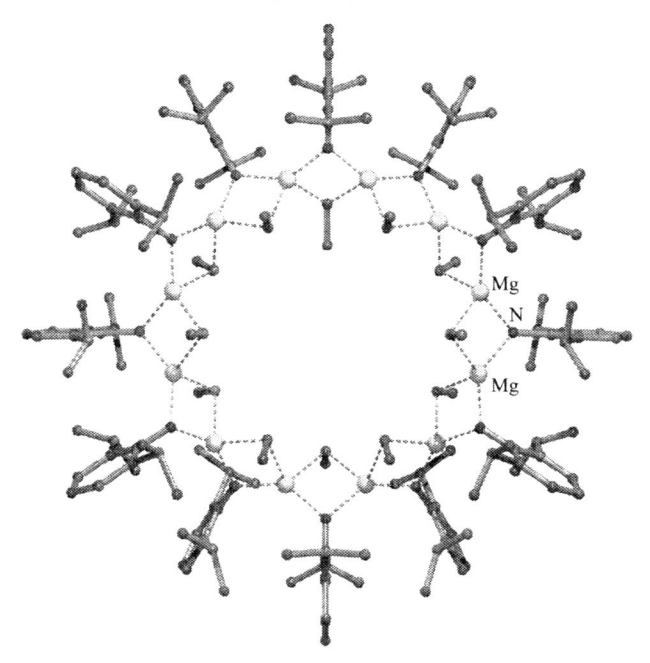

FIGURE 102. Molecular geometry of dodecameric **266** in the solid state

The solid-state structures of four cyclopentadienylmagnesium amides, [CpMgNPh$_2$]$_2$ (**267**), [CpMgN(H)CH(i-Pr)$_2$]$_2$ (**268**), [CpMgN(H)C$_6$H$_3$(Pr-i)$_2$-2,6]$_2$ (**269**) and [CpMgNBn(i-Pr)]$_2$ (**270**), have been determined[233]. The structures of these compounds are closely related and consist of a central flat four-membered N–Mg–N–Mg ring as the result of two bridging amide-nitrogen atoms between two magnesium atoms. In all compounds the cyclopentadienyl group is bonded in a η^5-fashion to magnesium. In **268** and **269** the nitrogen substituents and the amide-hydrogen atoms adopt an *anti*-configuration with respect to the N–Mg–N–Mg plane, like the benzyl-nitrogen and isopropyl-nitrogen substituents in **270**. As an example the structure of **268** is shown (Figure 103).

The monoorganomagnesium amides [MeMgN(Me)CH$_2$CH$_2$NMe$_2$]$_2$ (**271**)[262], [n-BuMgN(Bn)CH$_2$CH$_2$NMe$_2$]$_2$ (**272**)[263] and [n-BuMgN(Me)CH$_2$CH(Ph)N(CH$_2$)$_5$]$_2$ (**273**)[176] derived from N,N',N'-trisubstituted ethylenediamines have closely related structures in the solid state, of which the structures are shown schematically (Figure 104). These structures comprise a flat four-membered N–Mg–N–Mg ring formed via two bridging amide-nitrogen atoms between two magnesium atoms. The tertiary nitrogen atoms are intramolecularly coordinate-bonded to the magnesium atoms, one lying above the N–Mg–N–Mg plane and the other one below that plane. Consequently, the magnesium-bonded organic groups are in *anti*-position with respect to this plane. The chiral derivative **273** has been successfully applied in the enantioselective alkylation of aldehydes[176].

F. Monoorganomagnesium Compounds RMgY with Y = SR or PR$_2$

The structures of only a very few heteroleptic monoorganomagnesium compounds with a magnesium–heteroatom bond with heteroatoms other than halogen, oxygen or nitrogen have been determined.

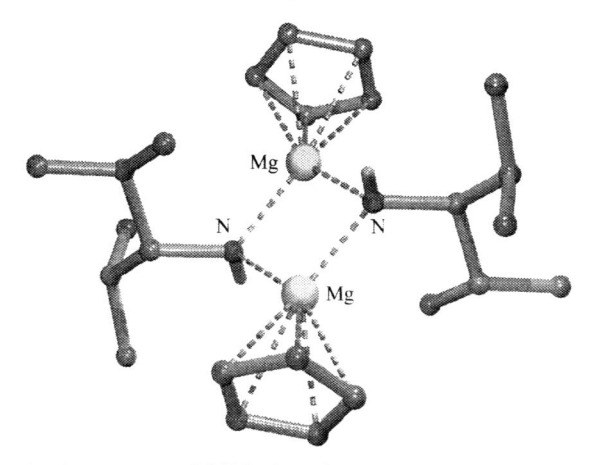

84   Johann T. B. H. Jastrzebski, Jaap Boersma and Gerard van Koten

FIGURE 103. Molecular geometry of **268** in the solid state

(**271**) $R^1 = R^2 = Me$
(**272**) $R^1 = n\text{-Bu}, R^2 = Bn$

(**273**)

FIGURE 104. Schematic representation of the structures of **271–273** in the solid state

Cyclopentadienylmagnesium *tert*-butylthiolate exists as a tetrameric aggregate [CpMgSBu-*t*]$_4$ (**274**) in the solid state. An X-ray crystal-structure determination revealed a heterocubane structure with magnesium and sulfur atoms at the corners (Figure 105)[264]. The Mg—S bond distances within the cube vary in a narrow range of 2.584(2) to 2.602(2) Å, indicating that the shape of the cube is close to perfect. To each of the magnesium atoms a cyclopentadienyl group is bonded in a η^5-fashion.

THF effectively breaks down tetrameric **274** to a dimeric THF complex [CpMgSBu-*t*(THF)]$_2$ (**275**). An X-ray crystal-structure determination of **275** showed a central four-membered S—Mg—S—Mg ring as the result of two symmetrically bridging sulfur atoms [Mg—S 2.503(1) and 2.504(1) Å] between two magnesium atoms (Figure 105)[264]. This four-membered ring is slightly folded, as indicated by the sum of the bond angles within this ring (243.93°). The cyclopentadienyl groups are η^5-bonded to magnesium and an additional THF molecule is coordinate-bonded to each magnesium atom.

The reaction of the secondary phosphine 2-MeOC$_6$H$_4$PHCH(SiMe$_3$)$_2$ with *s*-Bu$_2$Mg gives heteroleptic [*s*-BuMgP(C$_6$H$_4$OMe-2)(CH(SiMe$_3$)$_2$)]$_2$ (**276**). An X-ray crystal-structure determination revealed a dimeric complex with a flat four-membered P—Mg—P—Mg ring as the result of two slightly asymmetric bridging phosphido groups between the magnesium atoms [Mg—P 2.5760(8) and 2.5978(8) Å] (Figure 106)[265]. To each of

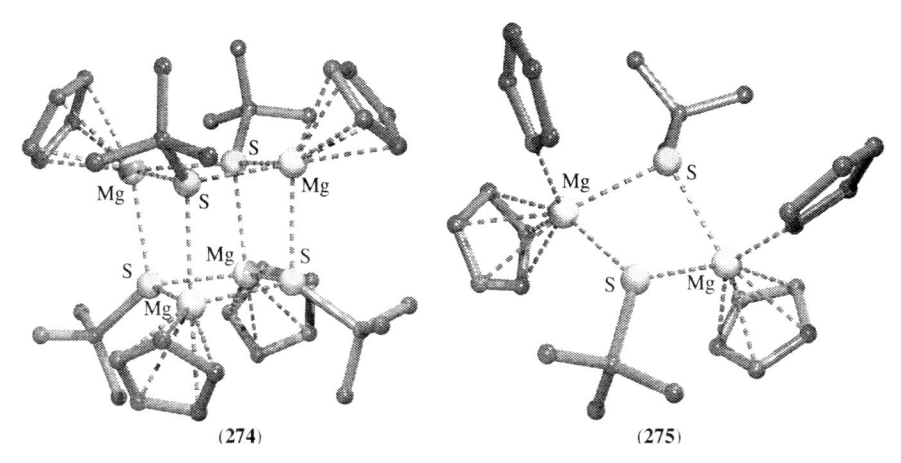

FIGURE 105. Molecular geometries of **274** and **275** in the solid state

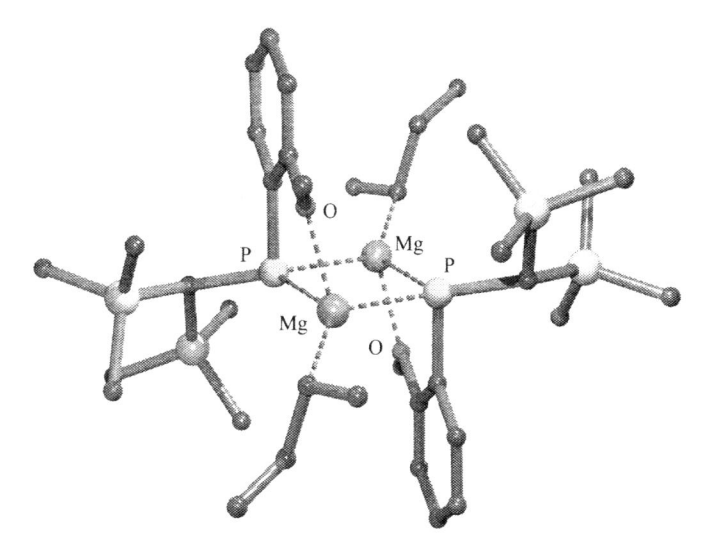

FIGURE 106. Molecular geometry of **276** in the solid state

the magnesium atoms a *s*-butyl group is σ-bonded. The magnesium atoms are four-coordinate as the result of intramolecular coordination of the oxygen atoms of methoxy substituents, one approaching a magnesium atom from above the P–Mg–P–Mg plane and the other approaching the other magnesium atom from below that plane. The resulting five-membered chelate rings are almost planar.

V. MIXED ORGANOMAGNESIUM TRANSITION-METAL COMPOUNDS

In this section, structures of compounds are described that contain both an organomagnesium moiety and a transition-metal-containing part. These moieties are linked via either

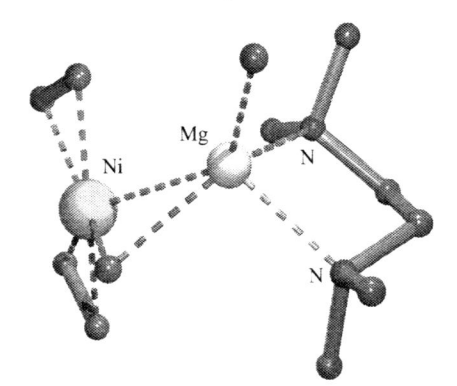

FIGURE 107. Molecular geometry of **277** in the solid state

a direct magnesium to transition-metal bond or via bridge-bonded atoms, like hydrogen, carbon and halogen, between magnesium and the transition metal.

The reaction of tris(ethylene)nickel(0) with R_2Mg in the presence of donor molecules like Et_2O, THF, dioxane and TMEDA, at -10 °C, gives crystalline materials with the formulation $R_2MgL_2Ni(C_2H_4)_2$[266,267]. The structure of one of these complexes, Me_2Mg(TMEDA)Ni$(C_2H_4)_2$ (**277**), was determined by X-ray crystallography (Figure 107)[266]. In **277** one methyl group is two-electron three-center bridge-bonded between magnesium and nickel [Mg-C 2.294(3) Å and Ni-C 2.031(3) Å]. The other methyl group is σ-bonded to magnesium [Mg-C 2.150(3) Å]. A N,N'-chelate bonded TMEDA molecule [Mg-N 2.252(2) and 2.264(2) Å] completes the coordination sphere of magnesium. Two ethylene molecules are π-bonded to nickel. The Mg-Ni bond distance of 2.615(1) Å indicates that a bonding interaction exists between these metals.

An X-ray crystal-structure determination of the copper–magnesium cluster compound $Ph_6Cu_4Mg(OEt_2)$ (**278**) shows that it comprises a central core of five metal atoms in a trigonal bipyramidal arrangement, with the magnesium atom at the axial position (Figure 108)[268]. The six phenyl groups bridge across the equatorial–axial edges of the trigonal bipyramid via two-electron three-center bonds. Coordination saturation at magnesium is reached by the additional coordination of a diethyl ether molecule.

Instead of the anticipated metalla-cyclobutane, reaction of the 1,1-di-Grignard reagent $Me_3SiCH(MgBr)_2$ with Cp_2ZrCl_2 gives an unexpected product. An X-ray structure determination showed the formation of a Tebbe-type spiro-organomagnesium compound $[Cp_2(Me_3SiCH)ZrBr]_2Mg$ (**279**)[269]. Its structure (Figure 109) comprises two Cp_2Zr moieties, each linked to a central magnesium atom via a bridging Me_3SiCH group [Mg-C 2.188(8) Å and Mg-Zr 2.147(7) Å] and a bridging bromine atom [Mg-Br 2.672(3) Å and Zr-Br 2.723(1) Å]. The coordination geometry at magnesium is distorted tetrahedral with the smallest angle being C-Mg-Br [92.2(2)°], which is a consequence of the four-membered C-Mg-Br-Zr ring.

Reduction of $(Me_3Si(Me)_4C_5)_2ZrCl_2$ with metallic magnesium in THF affords as the major product a mixed zirconium–magnesium hydride $[(Me_3Si(Me)_4C_5)(CH_2Me_2Si(Me)_3C_5(CH_2))ZrH_2Mg]_2$ (**280**) in which one of the substituted cyclopentadienyl groups at zirconium became doubly activated by abstraction of one hydrogen atom from the trimethylsilyl group and one hydrogen atom from the adjacent methyl group[270]. An X-ray crystal-structure determination showed that **280** is a centrosymmetric dimer as the result of two-electron three-center bridge-bonding of two methylene groups, generated from the

1. Structural organomagnesium chemistry

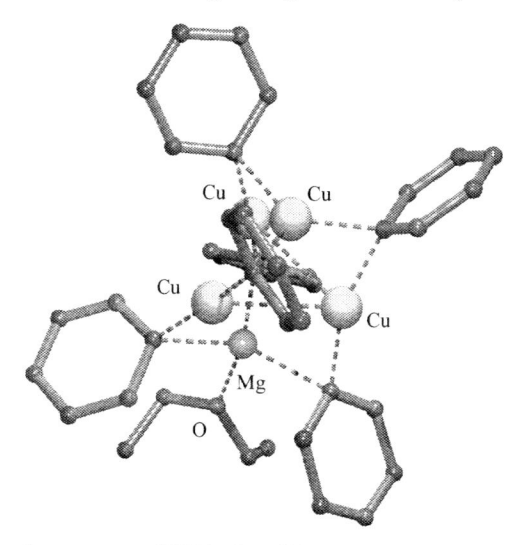

FIGURE 108. Molecular geometry of **278** in the solid state

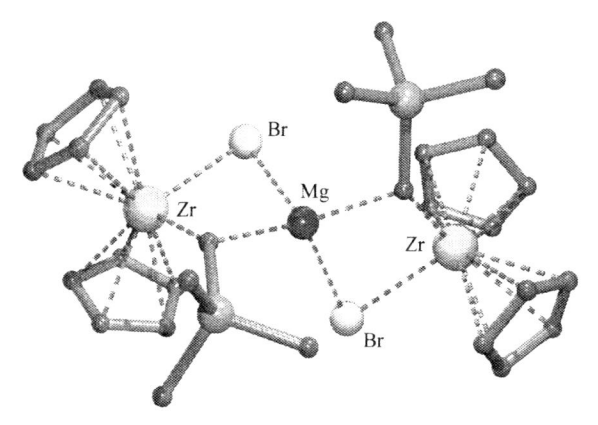

FIGURE 109. Molecular geometry of **279** in the solid state

trimethylsilyl substituents, between two magnesium atoms [Mg–C 2.218(9) and 2.255(9) Å] (Figure 110). Between each of the magnesium atoms and its adjacent zirconium atom two hydrogen atoms are bridge-bonded [Mg–H 1.81(5) and 1.86(4) Å and Zr–H 1.89(6) and 1.92(4) Å] resulting in a tetrahedral coordination geometry at magnesium. EPR studies showed that **280** is contaminated with a product having a structure that is closely related to that of **280**, but lacks the activation of the methyl group adjacent to the activated trimethylsilyl substituent. Consequently, in this complex the zirconium atoms have a trivalent oxidation state. In fact, the only isolable product from the reaction with the titanium analog is such a dimeric complex with trivalent titanium[271].

When (PhMe$_4$C$_5$)$_2$TiCl$_2$ is reduced with magnesium in THF three main products are formed. They are the diamagnetic doubly 'tucked-in' titanocene complex (PhMe$_4$C$_5$)

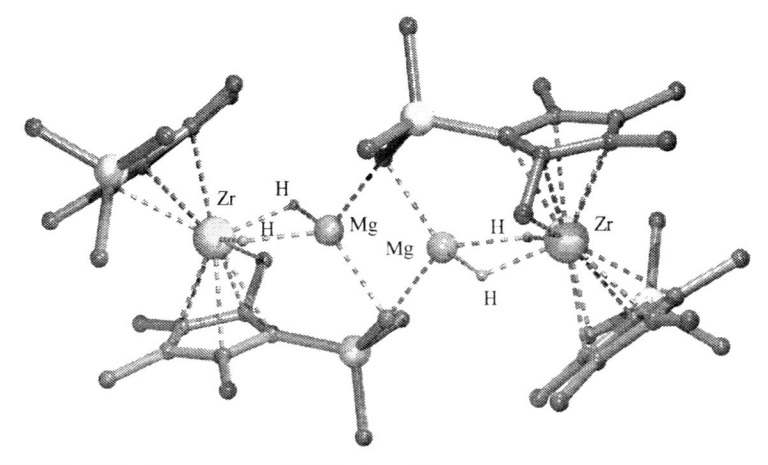

FIGURE 110. Molecular geometry of **280** in the solid state

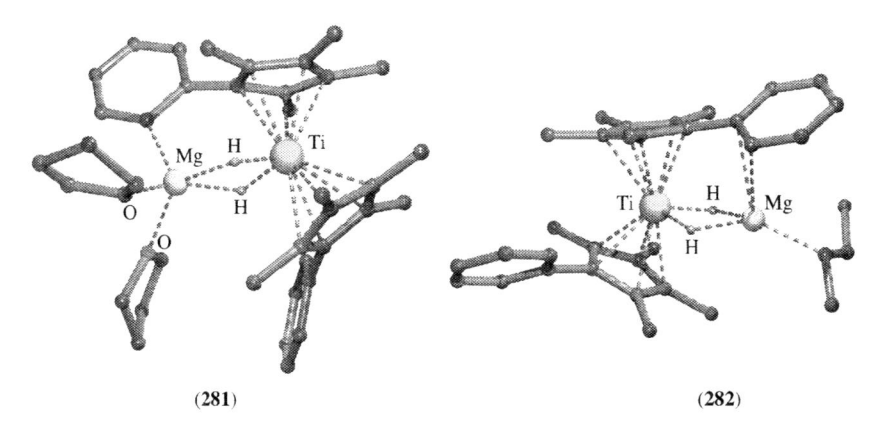

(**281**) (**282**)

FIGURE 111. Molecular geometries of **281** and **282** in the solid state

(PhMe$_2$(CH$_2$)$_2$C$_5$Ti, the paramagnetic trinuclear Ti–Mg–Ti hydride-bridged complex [(PhMe$_4$C$_5$)$_2$Ti(μ-H)$_2$]$_2$Mg and the paramagnetic binuclear titanocene hydride–magnesium hydride complex (PhMe$_4$C$_5$)[(2-C$_6$H$_4$)Me$_4$PhC$_5$]Ti(μ-H)$_2$Mg(THF)$_2$ (**281**). Of the latter complex the structure was determined by X-ray crystallography (Figure 111)[272,273]. Its structure comprises a (PhMe$_4$C$_5$)$_2$Ti moiety in which both cyclopentadienyl groups are η^5-bonded to titanium. The magnesium atom is linked to titanium via two bridge-bonded hydrogen atoms [Ti–H 1.72(3) and 1.78(3) Å, Mg–H 1.97(3) and 1.99(3) Å]. It appears that one of the phenyl groups was metallated in its 2-position by forming a σ-bond with magnesium [Mg–C 2.144(2) Å], resulting in an additional bridge between titanium and magnesium. The coordination sphere at magnesium is completed by two additional coordinating THF molecules, resulting in penta-coordinate magnesium.

When the same reaction was performed using an excess of n-Bu$_2$Mg as the reducing agent, essentially the same products are formed as from the reduction with metallic magnesium, *vide supra*. However, from this reaction mixture a by-product (PhMe$_4$C$_5$)$_2$Ti

(μ-H)$_2$Mg(2-buten-2-yl) (**282**) was isolated in low yield. Its X-ray crystal-structure determination showed a similar titanocene-type structure as observed for **281** with two bridge-bonded hydrogen atoms between titanium and magnesium [Ti–H 1.87(3) and 1.77(3) Å, Mg–H 1.85(3) and 1.85(3) Å] (Figure 111)[273]. One of the phenyl groups has a π-type interaction with magnesium, as is indicated by the distances between its C$_{ipso}$ and the adjacent carbon atom and the magnesium atom of 2.657(4) and 2.644(4) Å, respectively. The presence of a σ-bonded 2-buten-2-yl group at magnesium [Mg–C 2.123(4) Å] implies a hydrogen transfer from the butyl group into the titanium–magnesium bond.

When Cp$_2$TiCl$_2$ or its methyl-substituted derivative (MeH$_4$C$_5$)$_2$TiCl$_2$ are reduced with magnesium in THF, in the presence of bis(trimethylsilyl)acetylene, a mixture of products is obtained. Two of these appeared to be the mixed titanium–magnesium complexes CpTiMgCp(Me$_3$SiCCSiMe$_3$)$_2$ (**283**)[274] and its methyl-substituted derivative (**284**). X-ray crystal-structure determinations of **283** and **284** showed that they have similar structures, of which that of **284** is shown (Figure 112)[274, 275]. In **284** two Me$_3$SiCCSiMe$_3$ moieties are bridge-bonded in a μ-η^2-η^2 fashion between magnesium and titanium, while to both titanium and magnesium a methylcyclopentadienyl group is η^5-bonded. The C–C bond lengths of 1.31(1) Å in the Me$_3$SiCCSiMe$_3$ moiety and C–C–Si bond angles of average 140° indicate a change of the hybridization from sp to sp^2 of these carbon atoms. The observed Mg–Ti distance of 2.776(2) Å indicates the presence of a bonding interaction between these metals.

Another product isolated from the above-mentioned reaction mixture is (MeH$_4$C$_5$)TiMgCl$_2$Mg(C$_5$H$_4$Me)(Me$_3$SiCCSiMe$_3$)$_2$ (**285**)[275]. An X-ray crystal-structure determination showed an almost perfect linear Ti–Mg–Mg arrangement with two Me$_3$SiCCSiMe$_3$ moieties bridge-bonded in a μ-η^2-η^2 fashion between magnesium and titanium while a methylcyclopentadienyl group is η^5-bonded to titanium (Figure 112). Between the two magnesium atoms two chlorides are bridge-bonded in a slightly asymmetric way. The Mg–Cl bond distances to the terminal magnesium atom [Mg–Cl 2.440(5) and 2.439(5) Å]

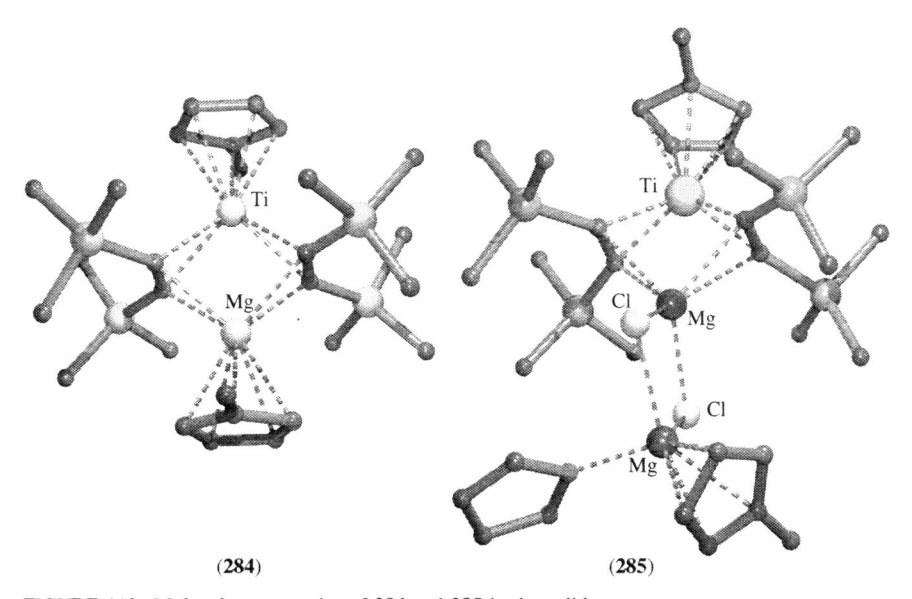

(**284**)   (**285**)

FIGURE 112. Molecular geometries of **284** and **285** in the solid state

90   Johann T. B. H. Jastrzebski, Jaap Boersma and Gerard van Koten

are slightly longer than those to the central magnesium atom [Mg–Cl 2.340(4) and 2.352(4) Å]. To the terminal magnesium atom a methylcyclopentadienyl group is η^5-bonded and coordination saturation is completed by an additional coordinating THF molecule. Also, in this compound the observed Mg–Ti distance of 2.763(4) Å indicates a bonding interaction between these two metals.

It has been shown that one of the μ^3-bridging hydrogen atoms in the Cp*$_3$Ru$_3$H$_5$ cluster can be easily replaced by a main group organometallic fragment like MeGa, EtAl i-PrMg or EtZn[276]. The X-ray crystal-structure determination of the product Cp*$_3$Ru$_3$Mg(i-Pr)H$_4$ (**286**) obtained from the reaction of Cp*$_3$Ru$_3$H$_5$ with i-Pr$_2$Mg shows that the main

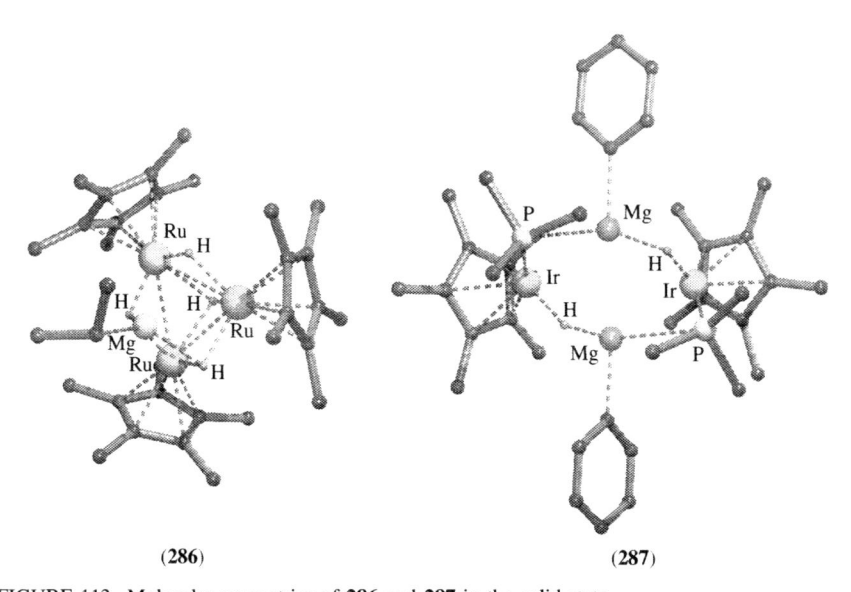

(**286**)   (**287**)

FIGURE 113. Molecular geometries of **286** and **287** in the solid state

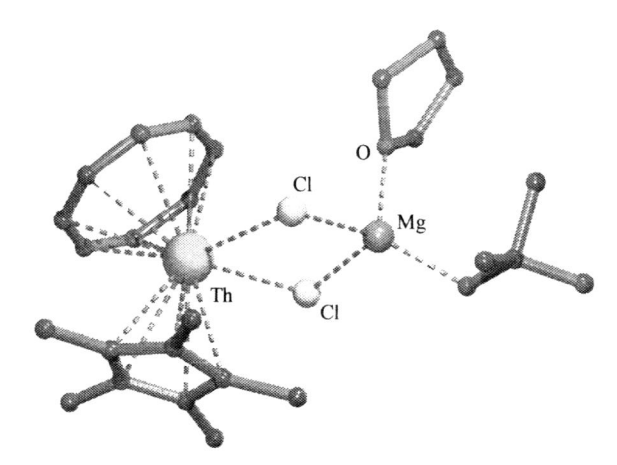

FIGURE 114. Molecular geometry of **288** in the solid state

structural features of the originating Cp*$_3$Ru$_3$H$_5$ cluster are retained, but one of the μ^3-bridging hydrogen atoms is replaced by a μ^3-bridging i-PrMg group (Figure 113), with almost equal Mg–Ru distances of 2.7487(13), 2.8007(12) and 2.7715(13) Å.

The reaction of Cp*IrH$_2$(PMe$_3$) with Ph$_2$Mg gives a product formulated as Cp*IrH(PMe$_3$)MgPh. An X-ray crystal-structure determination showed this compound to be a dimer [Cp*IrH(PMe$_3$)MgPh]$_2$ (**287**) in which two Cp*IrH(PMe$_3$) moieties are connected by the two PhMg groups, on one side to the phosphorus atom and on the other side to the hydride (Figure 113)[277]. The geometry at the magnesium atoms is trigonal planar, and the distances to iridium are slightly different [Mg–Ir 2.669(2) and 2.748(2) Å].

Instead of the anticipated transmetallation product, the reaction of Cp*(C$_8$H$_8$)ThCl with t-BuCH$_2$MgCl in THF gives a complex Cp*(C$_8$H$_8$)ThCl$_2$Mg(CH$_2$Bu-t)(THF) (**288**). An X-ray crystal-structure determination showed that **288** contains a Cp*(C$_8$H$_8$)Th moiety with a η^8-bonded cyclooctatetraenyl group and a η^5-bonded pentamethylcyclopentadienyl group (Figure 114)[278]. Two chlorides are bridge-bonded between thorium and magnesium while a neopentyl group is σ-bonded to magnesium and an additional THF molecule completes the coordination sphere at magnesium.

VI. CONCLUSIONS

In this chapter it has become clear that knowledge about the structures of organomagnesium compounds both in the solid state and in solution is often a pre-requisite for a better understanding of the reaction pathways involved in reactions of organomagnesium compounds. For the design of new synthetic pathways for the synthesis of new organic products this knowledge is of particular importance.

In contrast to their zinc analogs, simple dialkyl- and diaryl-magnesium compounds are, with a very few exceptions, not simple monomeric molecules. Due to the strong tendency of magnesium to extend its coordination number to usually four or even higher, these compounds form aggregates via multi-center, usually two-electron three-center, bonded organic groups. The only exceptions are a few diorganomagnesium compounds bearing very bulky substituents that prevent multi-center bonding for steric reasons. An example is [(Me$_3$Si)$_3$C]$_2$Mg that has a linear monomeric structure in the solid state.

In the presence of Lewis bases, diorganomagnesium compounds form complexes with one or two donor molecules. The usually observed coordination number for magnesium is four in complexes where magnesium has a tetrahedral coordination geometry. When a diorganomagnesium compound and/or ligand contains sterically demanding groups, complexes with one donor molecule are formed in which the magnesium is trigonal planar coordinate. However, higher coordination numbers are also observed, especially in complexes with multidentate donor ligands.

Various structural motifs are observed in the solid-state structures of heteroleptic organomagnesium compounds RMgY. In these compounds Y is either a halide or a heteroatom-containing group. In a few exceptional cases this group is σ-bonded to magnesium, resulting in monomeric heteroleptic organomagnesium compounds. Usually, such groups form multi-center bonds in which the group Y is either μ^2- or μ^3-bridge-bonded between two and three magnesium centers, respectively. Consequently, such bridge-bonding gives rise to the formation of aggregated structures.

In solutions containing RMgY species, the possible existence of a Schlenk equilibrium between RMgY and both R$_2$Mg and MgY$_2$ should always be considered. Moreover, equilibria between various aggregated species cannot be excluded. It should be noted that the formation of crystalline material, e.g. for structural studies in the solid state, may well be influenced by factors such as differences in solubilities of the various aggregates in solution and packing effects in the crystal lattice. As a consequence, care should be taken in drawing conclusions about structures in solution from data obtained from solid-state structural investigations (e.g. X-ray crystallography). It is perhaps prudent to regard these solid-state structures as resting states and to realize that they may represent only one of many structural forms present in solution.

VII. REFERENCES

1. P. Barbier, *Compt. Rend. Acad. Sci.*, **128**, 111 (1899).
2. V. Grignard, *Compt. Rend. Acad. Sci.*, **130**, 1322 (1900).
3. W. E. Lindsell, in *Comprehensive Organometallic Chemistry* (Eds. G. Wilkinson, F. G. A. Stone and E. W. Abel), Vol. 1, Pergamon, Oxford, 1982, pp. 155–252.
4. W. E. Lindsell, in *Comprehensive Organometallic Chemistry II* (Eds. E. W. Abel, F. G. A. Stone and G. Wilkinson), Vol. 1, Pergamon/Elsevier, Oxford, 1995, pp. 57–127.
5. T. P. Hanusa, in *Comprehensive Organometallic Chemistry III* (Eds. R. H. Crabtree and M. D. P. Mingos), Vol. 2, Elsevier, Oxford, 2006, pp. 67–152.
6. A. Baeyer and V. Villiger, *Chem. Ber.*, **35**, 1201 (1902).
7. V. Grignard, *Compt. Rend. Acad. Sci.*, **136**, 1260 (1903).
8. G. L. Stadnikov, *J. Russ. Phys.-Chem. Soc.*, **43**, 1235 (1912); *Chem. Abstr.*, **6**, 5482 (1912).
9. L. Thorp and O. Kamm, *J. Am. Chem. Soc.*, **36**, 1022 (1914).
10. R. Abegg, *Chem. Ber.*, **38**, 4112 (1905).
11. E. C. Ashby, *Q. Rev.*, **21**, 259 (1967).
12. W. Schlenk and W. Schlenk, Jr., *Chem. Ber.*, **62**, 920 (1929).
13. G. S. Silverman and P. E. Rakita (Eds.), *Handbook of Grignard Reagents*, Marcel Dekker, New York, 1996.
14. H. G. Richey, Jr. (Ed.), *Grignard Reagents, New Developments*, Wiley, Chichester, 2000.
15. P. R. Markies, O. S. Akkerman, F. Bickelhaupt, W. J. J. Smeets and A. L. Spek, *Adv. Organomet. Chem.*, **32**, 147 (1991).
16. T. S. Ertel and H. Bertagnolli, in *Grignard Reagents, New Developments* (Ed. H. G. Richey, Jr.), Wiley, Chichester, 2000, p. 329.
17. T. Holm and I. Crossland, in *Grignard Reagents, New Developments* (Ed. H. G. Richey, Jr.), Wiley, Chichester, 2000, p. 1.
18. C. Blomberg, in *Handbook of Grignard Reagents* (Eds. G. S. Silverman and P. E. Rakita), Marcel Dekker, New York, 1996, p. 249.
19. G. D. Stucky and R. E. Rundle, *J. Am. Chem. Soc.*, **85**, 1002 (1963).
20. L. J. Guggenberger and R. E. Rundle, *J. Am. Chem. Soc.*, **86**, 5344 (1964).
21. E. Weiss, *J. Organomet. Chem.*, **2**, 314 (1964).
22. E. Weiss, *J. Organomet. Chem.*, **4**, 101 (1965).
23. Cambridge Structural Database, release 5. 28, January 2007.
24. N. S. Poonia and A. V. Bajaj, *Chem. Rev.*, **79**, 389 (1979).
25. C. E. Holloway and M. Melnik, *J. Organomet. Chem.*, **465**, 1 (1994).
26. C. E. Holloway and M. Melnik, *Coord. Chem. Rev.*, **135/136**, 287 (1994).
27. H. L. Uhm, in *Handbook of Grignard Reagents* (Eds. G. S. Silverman and P. E. Rakita), Marcel Dekker, New York, 1996, p. 117.
28. F. Bickelhaupt, in *Grignard Reagents, New Developments* (Ed. H. G. Richey, Jr.), Wiley, Chichester, 2000, p. 299.
29. J. T. B. H. Jastrzebski, J. Boersma and G. van Koten, in *The Chemistry of Organozinc Compounds*, Part 1 (Eds. Z. Rappoport and I. Marek), Chap. 2, Wiley, Chichester, 2006, pp. 31–135.
30. J. A. Wanklyn, *Liebigs Ann. Chem.*, **108**, 67 (1858).
31. G. Wittig, F. J. Meyer and G. Lange, *Liebigs Ann. Chem.*, **571**, 167 (1951).
32. G. Wittig, *Angew. Chem.*, **70**, 65 (1958).
33. W. Tochtermann, *Angew. Chem., Int. Ed. Engl.*, **5**, 351 (1966).
34. G. E. Coates and J. A. Heslop, *J. Chem. Soc. A*, **514** (1968).
35. L. M. Seitz and T. L. Brown, *J. Am. Chem. Soc.*, **88**, 4140 (1966).
36. L. M. Seitz and T. L. Brown, *J. Am. Chem. Soc.*, **89**, 1602 (1967).
37. L. M. Seitz and B. F. Little, *J. Organomet. Chem.*, **18**, 227 (1969).
38. E. Weiss, *Angew. Chem., Int. Ed. Engl.*, **32**, 1501 (1993).

39. R. E. Mulvey, *Chem. Commun.*, 1049 (2001).
40. R. E. Mulvey, *Organometallics*, **25**, 1060 (2006).
41. K. Kitagawa, A. Inoue, H. Shinokubo and K. Oshima, *Angew. Chem., Int. Ed.*, **39**, 2481 (2000).
42. A. Inoue, K. Kitagawa, H. Shinokubo and K. Oshima, *J. Org. Chem.*, **66**, 4333 (2001).
43. S. Y. W. Lau, G. Hughes, P. D. O'Shea and I. W. Davies, *Org. Lett.*, **9**, 2239 (2007).
44. F. Mongin, A. Bucher, J. P. Bazureau, O. Bayh, H. Awad and F. Trécourt, *Tetrahedron Lett.*, **46**, 7989 (2005).
45. M. Hatano, T. Matsumura and K. Ishihara, *Org. Lett.*, **7**, 573 (2005).
46. D. B. Patterson and A. F. Halasa, *Macromolecules*, **24**, 1583 (1991).
47. J. L. Atwood and G. D. Stucky, *J. Am. Chem. Soc.*, **91**, 2538 (1969).
48. T. Greiser, J. Kopf, D. Thoennes and E. Weiss, *Chem. Ber.*, **114**, 209 (1981).
49. T. Greiser, J. Kopf and E. Weiss, *Chem. Ber.*, **122**, 1395 (1989).
50. B. Schubert and E. Weiss, *Chem. Ber.*, **117**, 366 (1984).
51. R. A. Layfield, T. H. Bullock, F. García, S. M. Humphrey and P. Schüler, *Chem. Commun.*, 2039 (2006).
52. P. C. Andrikopoulos, D. R. Armstrong, E. Hevia, A. L. Kennedy, R. E. Mulvey and C. T. O'Hara, *Chem. Commun.*, 1131 (2005).
53. D. Thoennes and E. Weiss, *Chem. Ber.*, **111**, 3726 (1978).
54. H. Viebrock, U. Behrens and E. Weiss, *Angew. Chem., Int. Ed. Engl.*, **33**, 1257 (1994).
55. E. P. Squiller, R. R. Whittle and H. G. Richey Jr., *J. Am. Chem. Soc.*, **107**, 432 (1985).
56. H. G. Richey Jr. and B. A. King, *J. Am. Chem. Soc.*, **104**, 4672 (1982).
57. A. D. Pajerski, M. Parvez and H. G. Richey Jr., *J. Am. Chem. Soc.*, **110**, 2660 (1988).
58. K. M. Waggoner and P. P. Power, *Organometallics*, **11**, 3209 (1992).
59. D. V. Graham, E. Hevia, A. R. Kennedy, R. E. Mulvey, C. T. O'Hara and C. Talmard, *Chem. Commun.*, 417 (2006).
60. S. Sakamoto, T. Imamoto and K. Yamaguchi, *Org. Lett.*, **3**, 1793 (2001).
61. M. Vestergren, J. Eriksson and M. Håkansson, *J. Organomet. Chem.*, **681**, 215 (2003).
62. N. D. R. Barnett, W. Clegg, A. R. Kennedy, R. E. Mulvey and S. Weatherstone, *Chem. Commun.*, 375 (2005).
63. M.-L. Hsueh, B.-T. Ko, T. Athar, C.-C. Lin, T.-M. Wu and S.-F. Hsu, *Organometallics*, **25**, 4144 (2006).
64. E. Hevia, K. W. Henderson, A. R. Kennedy and R. E. Mulvey, *Organometallics*, **25**, 1778 (2006).
65. G. C. Forbes, A. R. Kennedy, R. E. Mulvey, P. J. A. Rodger and R. B. Rowlings, *J. Chem. Soc., Dalton Trans.*, 1477 (2001).
66. P. C. Andrikopoulos, D. R. Armstrong, A. R. Kennedy, R. E. Mulvey, C. T. O'Hara, R. B. Rowlings and S. Weatherstone, *Inorg. Chim. Acta*, **360**, 1370 (2007).
67. D. R. Amstrong, A. R. Kennedy, R. E. Mulvey and R. B. Rowlings, *Angew. Chem., Int. Ed.*, **38**, 131 (1999).
68. P. C. Andrews, A. R. Kennedy, R. E. Mulvey, C. L. Raston, B. A. Roberts and R. B. Rowlings, *Angew. Chem., Int. Ed.*, **39**, 1960 (2000).
69. K. W. Henderson, A. R. Kennedy, R. E. Mulvey, C. T. O'Hara and R. B. Rowlings, *Chem. Commun.*, 1678 (2001).
70. P. C. Andrikopoulos, D. R. Armstrong, W. Clegg, C. J. Gilfillan, E. Hevia, A. R. Kennedy, R. E. Mulvey, C. T. O'Hara, J. A. Parkinson and D. M. Tooke, *J. Am. Chem. Soc.*, **126**, 11612 (2004).
71. E. Hevia, D. J. Gallagher, A. R. Kennedy, R. E. Mulvey, C. T. O'Hara and C. Talmard, *Chem. Commun.*, 2422 (2004).
72. P. C. Andrikopoulos, D. R. Armstrong, D. V. Graham, E. Hevia, A. R. Kennedy, R. E. Mulvey, C. T. O'Hara and C. Talmard, *Angew. Chem., Int. Ed.*, **44**, 3459 (2005).
73. E. Hevia, G. W. Honeyman, A. R. Kennedy, R. E. Mulvey and D. C. Sherrington, *Angew. Chem., Int. Ed.*, **44**, 68 (2005).
74. P. C. Andrikopoulos, D. R. Armstrong, E. Hevia, A. R. Kennedy and R. E. Mulvey, *Organometallics*, **25**, 2415 (2006).
75. R. A. Anderson and G. Wilkinson, *J. Chem. Soc., Dalton Trans.*, 809 (1977).
76. E. C. Ashby, L. Fernholt, A. Haaland, R. Seip and R. S. Smith, *Acta Chem. Scand.*, **A 34**, 213 (1980).

77. S. S. Al-Juaid, C. Eaborn, P. B. Hitchcock, C. A. McGeary and J. D. Smith, *J. Chem. Soc., Chem. Commun.*, 273 (1989).
78. S. S. Al-Juaid, C. Eaborn, P. B. Hitchcock, C. A. McGeary, K. Kundu and J. D. Smith, *J. Organomet. Chem.*, **480**, 199 (1994).
79. R. J. Wehmschulte and P. P. Power, *Organometallics*, **14**, 3264 (1995).
80. P. B. Hitchcock, J. A. K. Howard, M. F. Lappert, W.-P. Leung and S. A. Mason, *J. Chem. Soc., Chem. Commun.*, 847 (1990).
81. R. J. Wehmschulte, B. Twamley and M. A. Khan, *Inorg. Chem.*, **40**, 6004 (2001).
82. K. B. Starowieyski, J. Lewinski, R. Wozniak, J. Lipkowski and A. Chrost, *Organometallics*, **22**, 2458 (2003).
83. P. R. Markies, G. Schat, O. S. Akkerman, F. Bickelhaupt, W. J. J. Smeets, P. van der Sluis and A. L. Spek, *J. Organomet. Chem.*, **393**, 315 (1990).
84. T. J. Kealy and P. L. Pauson, *Nature*, **168**, 1039 (1951).
85. S. A. Miller, J. A. Tebboth and J. F. Tremaine, *J. Chem. Soc.*, 632 (1952).
86. P. Laszlo and R. Hoffman, *Angew. Chem., Int. Ed.*, **39**, 123 (2000).
87. F. Cotton and G. Wilkinson, *Chem. Ind.*, 307 (1954).
88. E. O. Fischer and W. Hafner, *Z. Naturforsch.*, **B9**, 503 (1954).
89. E. Weiss and E. O. Fischer, *Z. Anorg. Allg. Chem.*, **278**, 219 (1955).
90. W. Bünder and E. Weiss, *J. Organomet. Chem.*, **92**, 1 (1975).
91. A. Haaland, J. Lusztyk and D. P. Novak, *J. Chem. Soc., Chem. Commun.*, 54 (1974).
92. M. G. Gardiner, C. L. Raston and C. H. L. Kennard, *Organometallics*, **10**, 3680 (1991).
93. C. P. Morley, P. Jutzi, C. Krüger and J. M. Wallis, *Organometallics*, **6**, 1084 (1987).
94. F. Weber, H. Sitzmann, M. Schultz, C. D. Sofield and R. A. Andersen, *Organometallics*, **21**, 3139 (2002).
95. H. Schumann, J. Gottfriedsen, M. Glanz, S. Dechert and J. Demtschuk, *J. Organomet. Chem.*, **616–618**, 588 (2001).
96. H. Schumann, S. Schutte, H. J. Kroth and D. Lentz, *Angew. Chem., Int. Ed.*, **43**, 6208 (2004).
97. J. Vollet, E. Baum and H. Schnöckel, *Organometallics*, **22**, 2525 (2003).
98. O. Gobley, S. Gentil, J. D. Schloss, R. D. Rogers, J. C. Gallucci, P. Meunier, B. Gautheron and L. A. Paquette, *Organometallics*, **18**, 2531 (1999).
99. M. Westerhausen, N. Makropoulos, B. Wieneke, K. Karaghiosoff, H. Nöth, H. Schwenk-Kircher, J. Knizek and T. Seifert, *Eur. J. Inorg Chem.*, 965 (1998).
100. J. L. Atwood and K. D. Smith, *J. Am. Chem. Soc.*, **96**, 994 (1974).
101. A. Xia, J. Heeg and C. H. Winter, *Organometallics*, **22**, 1793 (2003).
102. R. A. Andersen, R. Blom, A. Haaland, B. E. R. Schilling and H. Volden, *Acta Chem. Scand.*, **A39**, 563 (1985).
103. X. Zheng, U. Englert, G. E. Herberich and J. Rosenplänter, *Inorg. Chem.*, **39**, 5579 (2000).
104. J. A. J. Jarvis, B. T. Kilbourn, R. Pearce and M. F. Lappert, *J. Chem. Soc., Chem. Commun.*, 475 (1973).
105. J. M. Guss, R. Mason, I. Søtofte, G. van Koten and J. G. Noltes, *J. Chem. Soc., Chem. Commun.*, 446 (1972).
106. P. R. Markies, R. M. Altink, A. Villena, O. S. Akkerman, F. Bickelhaupt, W. J. J. Smeets and A. L. Spek, *J. Organomet. Chem.*, **402**, 289 (1991).
107. R. R. Holmes, *Prog. Inorg. Chem.*, **32**, 119 (1984).
108. N. Seidel, K. Jacob, A. K. Fischer, C. Pietzsch, P. Zanello and M. Fontani, *Eur. J. Inorg. Chem.*, 145 (2001).
109. M. J. Henderson, R. I. Papasergio, C. L. Raston, A. H. White and M. F. Lappert, *J. Chem. Soc., Chem. Commun.*, 672 (1986).
110. C. Eaborn, P. B. Hitchcock, A. Kowalewska, Z.-R. Lu, J. D. Smith and W. A. Stańczyk, *J. Organomet. Chem.*, **521**, 113 (1996).
111. C. F. Caro, P. B. Hitchcock and M. F. Lappert, *Chem. Commun.*, 1433 (1999).
112. S. S. Al-Juaid, A. G. Avent, C. Eaborn, S. M. El-Hamruni, S. A. Hawkes, M. S. Hill, M. Hopman, P. B. Hitchcock and J. D. Smith, *J. Organomet. Chem.*, **631**, 76 (2001).
113. B. Bogdanović, S. Liao, M. Schwickardi, P. Sikorsky and B. Spliethoff, *Angew. Chem., Int. Ed. Engl.*, **19**, 818 (1980).
114. K. Angermund, B. Bogdanović, G. Koppetsch, C. Krüger, R. Mynott, M. Schwickardi and Y.-H. Tsay, *Z. Naturforsch.*, **B41**, 455 (1986).
115. B. Bogdanović, G. Koppetsch, C. Krüger and R Mynott, *Z. Naturforsch.*, **B41**, 617 (1986).

116. P. R. Markies, G. Schat, A. Villena, O. S. Akkerman, F. Bickelhaupt, W. J. J. Smeets and A. L. Spek, *J. Organomet. Chem.*, **411**, 291 (1991).
117. P. R. Markies, T. Nomoto, O. S. Akkerman, F. Bickelhaupt, W. J. J. Smeets and A. L. Spek, *Angew. Chem., Int. Ed. Engl.*, **27**, 1084 (1988).
118. P. R. Markies, T. Nomoto, G. Schat, O. S. Akkerman, F. Bickelhaupt, W. J. J. Smeets and A. L. Spek, *Organometallics*, **10**, 3826 (1991).
119. A. Pape, M. Lutz and G. Müller, *Angew. Chem., Int. Ed. Engl.*, **33**, 2281 (1994).
120. A. G. Avent, C. F. Caro, P. B. Hitchcock, M. F. Lappert, Z. Li and X.-H. Wei, *Dalton Trans.*, 1567 (2004).
121. C. F. Caro, P. B. Hitchcock, M. F. Lappert and M. Layh, *Chem. Commun.*, 1297 (1998).
122. T. Greiser, J. Kopf, D. Thoennes and E. Weiss, *J. Organomet. Chem.*, **191**, 1 (1980).
123. D. Thoennes and E. Weiss, *Chem. Ber.*, **111**, 3381 (1978).
124. P. J. Bailey, R. A. Coxall, C. M. Dick, S. Fabre, L. C. Henderson, C. Herber, S. T. Liddle, D. Loroño-González, A. Parkin and S. Parsons, *Chem. Eur. J.*, **9**, 4820 (2003).
125. H. Viebrock and E. Weiss, *J. Organomet. Chem.*, **464**, 121 (1994).
126. N. D. R. Barnett, W. Clegg, R. E. Mulvey, P. A. O'Neil and D. Reed, *J. Organomet. Chem.*, **510**, 297 (1996).
127. T. Rüffer, C. Bruhn and D. Steinborn, *Main Group Met. Chem.*, **24**, 369 (2001).
128. J. Toney and G. D. Stuckey, *J. Organomet. Chem.*, **22**, 241 (1970).
129. H. Erikson, M. Örtendahl and M. Håkansson, *Organometallics*, **15**, 4823 (1996).
130. W. Gaderbauer, M. Zirngast, J. Baumgartner, C. Marschner and T. D. Tilley, *Organometallics*, **25**, 2599 (2006).
131. U. Nagel and G. Nedden, *Chem. Ber./Recueil*, **130**, 535 (1997).
132. R. I. Yousef, B. Walfort, T. Rüffer, C. Wagner, H. Schmidt, R. Herzog and D. Steinborn, *J. Organomet. Chem.*, **690**, 1178 (2005).
133. R. Fischer, D. Walther, P. Gerbhardt and H. Görls, *Organometallics*, **19**, 2532 (2000).
134. M. Parvez, A. D. Pajerski and H. G. Richey Jr., *Acta Cryst.*, **C44**, 1212 (1988).
135. F. Bickelhaupt, *Angew. Chem., Int. Ed. Engl.*, **26**, 990 (1987).
136. F. Bickelhaupt, *Pure Appl. Chem.*, **62**, 699 (1990).
137. H. C. Holtkamp, C. Blomberg and F. Bickelhaupt, *J. Organomet. Chem.*, **19**, 279 (1969).
138. A. L. Spek, G. Schat, H. C. Holtkamp, C. Blomberg and F. Bickelhaupt, *J. Organomet. Chem.*, **131**, 331 (1977).
139. M. F. Lappert, T. R. Martin, C. L. Raston, B. W. Skelton and A. H. White, *J. Chem. Soc., Dalton Trans.*, 1959 (1982).
140. M. A. G. M. Tinga, G. Schat, O. S. Akkerman, F. Bickelhaupt, E. Horn, H. Kooijman, W. J. J. Smeets and A. L. Spek, *J. Am. Chem. Soc.*, **115**, 2808 (1993).
141. T. Stey and D. Stalke, in *The Chemistry of Organolithium Compounds*, Part 1 (Eds. Z. Rappoport and I. Marek), Chap. 2, Wiley, Chichester, 2004, pp. 47–120.
142. P. R. Markies, O. S. Akkerman, F. Bickelhaupt, W. J. J. Smeets and A. L. Spek, *Organometallics*, **13**, 2616 (1994).
143. D. Steinborn, T. Rüffer, C. Bruhn and W. Heinemann, *Polyhedron*, **17**, 3275 (1998).
144. B. Schubert, U. Behrens and E. Weiss, *Chem. Ber.*, **114**, 2640 (1981).
145. M. Vestergren, B. Gustafsson, Ö. Davidsson and M. Håkansson, *Angew. Chem., Int. Ed.*, **39**, 3435 (2000).
146. P. R. Markies, T. Nomoto, O. S. Akkerman, F. Bickelhaupt, W. J. J. Smeets and A. L. Spek, *J. Am. Chem. Soc.*, **110**, 4845 (1988).
147. A. D. Pajerski, G. L. BergStresser, M. Parves and H. G. Richey Jr., *J. Am. Chem. Soc.*, **110**, 4844 (1988).
148. H. E. Ramsden, U.S. Patent 3354190 (1967); *Chem. Abstr.*, **68**, 114744 (1968).
149. B. Bogdanović, *Angew. Chem., Int. Ed. Engl.*, **24**, 262 (1985).
150. B. Bogdanović, *Acc. Chem. Res.*, **21**, 261 (1988).
151. C. L. Raston, in *Grignard Reagents, New Developments* (Ed. H. G. Richey, Jr.), Wiley, Chichester, 2000, pp. 278–298.
152. B. Bogdanović, N. Janke, H.-G. Kinzelmann and U. Westeppe, *Chem. Ber.*, **121**, 33 (1988).
153. B. Bogdanović, S. Liao, R. Mynott, K. Schlichte and U. Westeppe, *Chem. Ber.*, **117**, 1378 (1984).
154. E. Bartmann, B. Bogdanović, N. Janke, S. Liao, K. Schlichte, B. Spliethoff, J. Treber, U. Westeppe and U. Wilczok,, *Chem. Ber.*, **123**, 1517 (1990).

155. L. M. Engelhardt, S. Harvey, C. L. Raston and A. H. White, *J. Organomet. Chem.*, **341**, 39 (1988).
156. B. Bogdanović, N. Janke, C. Krüger, R. Mynott, K. Schlichte and U. Westeppe, *Angew. Chem., Int. Ed. Engl.*, **24**, 960 (1985).
157. H. Lehmkuhl, A. Shakoor, K. Mehler, C. Krüger, K. Angermund and Y.-H. Tsay, *Chem. Ber.*, **118**, 4239 (1985).
158. T. Alonso, S. Harvey, P. C. Junk, C. L. Raston, B. W. Skelton and A. H. White, *Organometallics*, **6**, 2110 (1987).
159. B. Bogdanović, N. Janke, C. Krüger, K. Schlichte and J. Treber, *Angew. Chem., Int. Ed. Engl.*, **26**, 1025 (1987).
160. A. Xia, M. J. Heeg and C. H. Winter, *J. Am. Chem. Soc.*, **124**, 11264 (2002).
161. A. Xia, J. E. Knox, M. J. Heeg, H. B. Schlegel and C. H. Winter, *Organometallics*, **22**, 4060 (2003).
162. M. M. Olmstead, W. J. Grigsby, D. R. Chacon, T. Hascall and P. P. Power, *Inorg. Chim. Acta*, **251**, 273 (1996).
163. A. Jaenschke, J. Paap and U. Behrens, *Organometallics*, **22**, 1167 (2003).
164. H. Gritzo, F. Schaper and H.-H. Brintzinger, *Acta Cryst.*, **E60**, m1108 (2004).
165. H.-R. Damrau, A. Geyer, M.-H. Prosenc, A. Weeber, F. Schaper and H.-H. Brintzinger, *J. Organomet. Chem.*, **553**, 331 (1998).
166. H. Viebrock, D. Abeln and E. Weiss, *Z. Naturforsch.*, **49**, 89 (1994).
167. Y. Kai, N. Kanehisa, K. Miki, N. Kasai, K. Mashima, H. Yasuda and A. Nakamura, *Chem. Lett.*, 1277 (1982).
168. M. G. Gardiner, C. L. Raston, F. G. N. Cloke and P. B. Hitchcock, *Organometallics*, **14**, 1339 (1995).
169. R. N. Grimes, *Chem. Rev.*, **92**, 251 (1992).
170. C. Zheng, J.-Q. Wang, J. A. Maguire and N. S. Hosmane, *Main Group Met. Chem.*, **22**, 361 (1999).
171. N. S. Hosmane, D. Zhu, J. E. McDonald, H. Zhang, J. A. Maguire, T. G. Gray and S. C. Helfert, *J. Am. Chem. Soc.*, **117**, 12362 (1995).
172. N. S. Hosmane, D. Zhu, J. E. McDonald, H. Zhang, J. A. Maguire, T. G. Gray and S. C. Helfert, *Organometallics*, **17**, 1426 (1998).
173. N. S. Hosmane, H. Zhang, Y. Wang, K.-J. Lu, C. J. Thomas, M. B. Ezhova, S. C. Helfert, J. D. Collins, J. A. Maguire, T. C. Gray, F. Baumann and W. Kaim, *Organometallics*, **15**, 2425 (1996).
174. N. S. Hosmane, H. Zhang, J. A. Maguire, Y. Wang, T. Demissie, T. J. Colacot, M. B. Ezhova, K.-J. Lu, D. Zhu, T. C. Gray, S. C. Helfert, S. N. Hosmane, J. D. Collins, F. Baumann, W. Kaim and W. N. Lipscomb, *Organometallics*, **19**, 497 (2000).
175. A. J. Arduengo III, F. Davidson, R. Krafczyk, W. J. Marshall and M. Tamm, *Organometallics*, **17**, 3375 (1998).
176. K. H. Yong, N. J. Taylor and J. M. Chong, *Org. Lett.*, **4**, 3553 (2002).
177. K. H. Yong and J. M. Chong, *Org. Lett.*, **4**, 4139 (2002).
178. K. C. Cannon and G. R. Krow, in *Handbook of Grignard Reagents* (Eds. G. S. Silverman and P. E. Rakita), Marcel Dekker, New York, 1996, p. 271.
179. H. M. Walborsky, *Acc. Chem. Res.*, **23**, 286 (1990).
180. J. F. Garst, *Acc. Chem. Res.*, **24**, 95 (1991).
181. C. Walling, *Acc. Chem. Res.*, **24**, 255 (1991).
182. J. F. Garst and F. Ungváry, in *Grignard Reagents, New Developments* (Ed. H. G. Richey, Jr.), Wiley, Chichester, 2000, p. 185.
183. A. D. Pajerski, E. P. Squiller, M. Parvez, R. R. Whittle and H. G. Richey Jr., *Organometallics*, **24**, 809 (2005).
184. H. Tang, M. Parvez and H. G. Richey Jr., *Organometallics*, **15**, 5281 (1996).
185. M. Vestergren, J. Eriksson and M. Håkansson, *Chem. Eur. J.*, **9**, 4678 (2003).
186. D. R. Armstrong, R. Herbst-Irmer, A. Kuhn, D. Moncrieff, M. A. Paver, C. A. Russell, D. Stalke, A. Steiner and D. S. Wright, *Angew. Chem., Int. Ed. Engl.*, **32**, 1774 (1993).
187. F. Bickelhaupt, *J. Organomet. Chem.*, **475**, 1 (1994).
188. G. Stucky and R. E. Rundle, *J. Am. Chem. Soc.*, **86**, 4825 (1964).
189. F. A. Schröder, *Chem. Ber.*, **102**, 2035 (1969).
190. L. J. Guggenberger and R. E. Rundle, *J. Am. Chem. Soc.*, **90**, 5375 (1968).

1. Structural organomagnesium chemistry 97

191. C.-S. Hwang and P. P. Power, *Bull. Korean Chem. Soc.*, **24**, 605 (2003).
192. G. Boche, K. Harms, M. Marsch and A. Müller, *J. Chem. Soc., Chem. Commun.*, 1393 (1994).
193. M. Vallino, *J. Organomet. Chem.*, **20**, 1 (1969).
194. H. Kageyama, K. Miki, Y. Kai, N. Kasai, Y. Okamoto and H. Yuki, *Bull. Chem. Soc. Jpn.*, **56**, 2411 (1983).
195. H. Kageyama, K. Miki, N. Tanaka, N. Kasai, Y. Okamoto and H. Yuki, *Bull. Chem. Soc. Jpn.*, **56**, 1319 (1983).
196. H. Kageyama, K. Miki, Y. Kai, N. Kasai, Y. Okamoto and H. Yuki, *Acta Cryst.*, **B38**, 2264 (1982).
197. H. Kageyama, K. Miki, Y. Kai, N. Kasai, Y. Okamoto and H. Yuki, *Bull. Chem. Soc. Jpn.*, **57**, 1189 (1984).
198. M. Hogenbirk, G. Schat, O. S. Akkerman and F. Bickelhaupt, *J. Am. Chem. Soc.*, **114**, 7302 (1992).
199. A. L. Spek, P. Voorbergen, G. Schat, C. Blomberg and F. Bickelhaupt, *J. Organomet. Chem.*, **77**, 147 (1974).
200. J. Toney and D. Stucky, *J. Chem. Soc., Chem. Commun.*, 1168 (1967).
201. F. Antolini, P. B. Hitchcock, M. F. Lappert and X.-H. Wei, *Organometallics*, **22**, 2505 (2003).
202. J. J. Ellison and P. P. Power, *J. Organomet. Chem.*, **526**, 263 (1996).
203. H. Bock, K. Ziemer and C. Näther, *J. Organomet. Chem.*, **511**, 29 (1996).
204. M. Marsch, K. Harms, W. Massa and G. Boche, *Angew. Chem., Int., Ed. Engl.*, **26**, 696 (1987).
205. J. Toney and G. D. Stucky, *J. Organomet. Chem.*, **28**, 5 (1971).
206. S. S. Al-Juaid, C. Eaborn, P. B. Hitchcock, A. J. Jaggar and J. D. Smith, *J. Organomet. Chem.*, **469**, 129 (1994).
207. N. H. Buttrus, C. A. Eaborn, M. N. A. El-Kheli, P. B. Hitchcock, J. D. Smith, A. C. Sullivan and K. Tavakkoli, *J. Chem. Soc., Dalton Trans.*, 381 (1988).
208. S. S. Al-Juaid, C. Eaborn, P. B. Hitchcock, M. S. Hill and J. D. Smith, *Organometallics*, **19**, 3224 (2000).
209. P. C. Andrews, M. Brym, C. Jones, P. C. Junk and M. Kloth, *Inorg. Chim. Acta*, **359**, 355 (2006).
210. U. Castellato and F. Ossola, *Organometallics*, **13**, 4105 (1994).
211. A. V. Churakov, D. P. Krut'ko, M. V. Borzov, R. S. Krisanov, S. A. Belov and J. A. K. Howard, *Acta Cryst.*, **E62**, m1094 (2006).
212. D. Seebach, J. Hansen, P. Seiler and J. M. Gromek, *J. Organomet. Chem.*, **285**, 1 (1985).
213. P. R. Markies, G. Schat, S. Griffioen, A. Villena, O. S. Akkerman, F. Bickelhaupt, W. J. J. Smeets and A. L. Spek, *Organometallics*, **10**, 1531 (1991).
214. P. R. Markies, O. S. Akkerman, F. Bickelhaupt, W. J. J. Smeets and A. L. Spek, *J. Am. Chem. Soc.*, **110**, 4284 (1988).
215. I. D. Kostas, G.-J. M. Gruter, O. S. Akkerman, F. Bickelhaupt, H. Kooijman, W. J. J. Smeets and A. L. Spek, *Organometallics*, **15**, 4450 (1996).
216. P. R. Markies, A. Villena, O. S. Akkerman, F. Bickelhaupt, W. J. J. Smeets and A. L. Spek, *J. Organomet. Chem.*, **463**, 7 (1993).
217. C. Johnson, J. Toney and G. D. Stucky, *J. Organomet. Chem.*, **40**, C11 (1972).
218. C. Dohmeier, D. Loos, C. Robl and H. Schnöckel, *J. Organomet. Chem.*, **448**, 5 (1993).
219. R. E. Cramer, P. N. Richmann and J. W. Gilje, *J. Organomet. Chem.*, **408**, 131 (1991).
220. J. Vollet, J. R. Hartig and H. Schnöckel, *Angew. Chem., Int. Ed.*, **43**, 3186 (2004).
221. P. Jutzi, J. Kleimeier, T. Redeker, H.-G. Stammler and B. Neumann, *J. Organomet. Chem.*, **498**, 85 (1995).
222. K. W. Henderson, G. W. Honeyman, A. R. Kennedy, R. E. Mulvey, J. A. Parkinson and D. C. Sherrington, *Dalton Trans.*, 1365 (2003).
223. J. Gromada, A. Montreux, T. Chenal, J. W. Ziller, F. Leising and J.-F. Carpentier, *Chem. Eur. J.*, **8**, 3773 (2002).
224. B. Conway, E. Hevia, A. R. Kennedy, R. E. Mulvey and S. Weatherstone, *Dalton Trans.*, 1532 (2005).
225. S. C. Cole, M. P. Coles and P. B. Hitchcock, *Organometallics*, **23**, 5159 (2004).
226. M. M. Sung, C. G. Kim, J. Kim and Y. Kim, *Chem. Mater.*, **14**, 826 (2002).
227. H. Lehmkuhl, K. Mehler, R. Benn, A. Rufińska and C. Krüger, *Chem. Ber.*, **119**, 1054 (1986).

228. K.-C. Yang, C.-C. Chang, J.-Y. Huang, C.-C. Lin, G.-H. Lee, Y. Wang and M. Y. Chiang, *J. Organomet. Chem.*, **648**, 176 (2002).
229. W. Vargas, U. Englich and K. Ruhlandt-Senge, *Inorg. Chem.*, **41**, 5602 (2002).
230. N. Kuhn, M. Schulten, R. Boese and D. Bläser, *J. Organomet. Chem.*, **421**, 1 (1991).
231. I. J. Blackmore, V. C. Gibson, P. B. Hitchcock, C. W. Rees, D. J. Williams and A. J. P. White, *J. Am. Chem. Soc.*, **127**, 6012 (2005).
232. D. Morales-Morales and C. M. Jensen (Eds.), *The Chemistry of Pincer Compounds*, Elsevier, Amsterdam, 2007.
233. A. Xia, M. J. Heeg and C. H. Winter, *Organometallics*, **21**, 4718 (2002).
234. A. Xia, M. El-Kaderi, M. J. Heeg and C. H. Winter, *J. Organomet. Chem.*, **682**, 224 (2003).
235. R. Han, A. Looney and G. Parkin, *J. Am. Chem. Soc.*, **111**, 7276 (1989).
236. R. Han and G. Parkin, *Organometallics*, **10**, 1010 (1991).
237. R. Han and G. Parkin, *J. Am. Chem. Soc.*, **112**, 3662 (1990).
238. R. Han and G. Parkin, *Polyhedron*, **9**, 2655 (1990).
239. J. L. Kisko, T. Fillebeen, T. Hascall and G. Parkin, *J. Organomet. Chem.*, **596**, 22 (2000).
240. M. H. Chisholm, N. W. Eilerts, J. C. Huffman, S. S. Iyer, M. Pacold and K. Phomphrai, *J. Am. Chem. Soc.*, **122**, 11845 (2000).
241. V. C. Gibson, J. A. Segal, A. J. P. White and D. J. Williams, *J. Am. Chem. Soc.*, **122**, 7120 (2000).
242. A. P. Dove, V. C. Gibson, P. Hormnirun, E. L. Marshall, J. A. Segal, A. J. P. White and D. J. Williams, *Dalton Trans.*, 3088 (2003).
243. P. J. Bailey, R. A. Coxall, C. M. Dick, S. Fabre and S. Parsons, *Organometallics*, **20**, 798 (2001).
244. P. J. Bailey, C. M. E. Dick, S. Fabre and S. Parsons, *J. Chem. Soc., Dalton Trans.*, 1655 (2000).
245. J. Prust, K. Most, I. Müller, E. Alexopoulos, A. Stasch, I. Usón and H. Roesky, *Z. Anorg. Allg. Chem.*, **627**, 2032 (2001).
246. K. H. D. Ballem, K. M. Smith and B. O. Patrick, *Acta Cryst.*, **E60**, m408 (2004).
247. L. F. Sánchez-Barbara, D. L. Hughes, S. M. Humphrey and M. Bochman, *Organometallics*, **25**, 1012 (2006).
248. H. Hao, H. W. Roesky, Y. Ding, C. Cui, M. Schormann, H.-G. Schmidt, M. Noltemeyer and B. Žemva, *J. Fluorine Chem.*, **115**, 143 (2002).
249. P. J. Bailey, S. T. Liddle, C. A. Morrison and S. Parsons, *Angew. Chem., Int. Ed.*, **40**, 4463 (2001).
250. A. P. Dove, V. C. Gibson, E. L. Marshall, A. J. P. White and D. J. Williams, *Chem. Commun.*, 1208 (2002).
251. J. Dekker, P. H. M. Budzelaar, J. Boersma, G. J. M. van der Kerk and A. L. Spek, *Organometallics*, **3**, 1403 (1984).
252. H. M. El-Kaderi, A. Xia, M. J. Heeg and C. H. Winter, *Organometallics*, **23**, 3488 (2004).
253. T. Chivers, C. Fedorchuk and M. Parvez, *Organometallics*, **24**, 580 (2005).
254. I. L. Fedushkin, A. A. Skatova, M. Hummert and H. Schumann, *Eur. J. Inorg. Chem.*, 1601 (2005).
255. P. J. Baily, R. A. Coxall, C. M. Dick, S. Fabre, S. Parsons and L. J. Yellowlees, *Chem. Commun.*, 4563 (2005).
256. P. J. Baily, C. M. Dick, S. Fabre, S. Parsons and L. J. Yellowlees, *Dalton Trans.*, 1602 (2006).
257. G. van Koten, J. T. B. H. Jastrzebski and C. Vrieze, *J. Organomet. Chem.*, **250**, 49 (1983).
258. M. Kaupp, H. Stoll, H. Preuss, W. Kaim, T. Stahl, G. van Koten, E. Wissing, W. J. J. Smeets and A. L. Spek, *J. Am. Chem. Soc.*, **113**, 5606 (1991).
259. E. Hevia, A. R. Kennedy, R. E. Mulvey and S. Weatherstone, *Angew. Chem., Int. Ed.*, **43**, 1709 (2004).
260. L. M. Engelhardt, B. S. Jolly, P. C. Junk, C. L. Raston, B. W. Skelton and A. H. White, *Aust. J. Chem.*, **39**, 1337 (1986).
261. M. Westerhausen, T. Bollewein, N. Makropoulos and H. Piotroski, *Inorg. Chem.*, **44**, 6439 (2005).
262. V. R. Magnuson and G. D. Stucky, *Inorg. Chem.*, **8**, 1427 (1969).
263. K. W. Henderson, R. E. Mulvey, W. Clegg and P. O'Neil, *J. Organomet. Chem.*, **439**, 237 (1992).
264. A. Xia, M. J. Heeg and C. H. Winter, *J. Organomet. Chem.*, **669**, 37 (2003).

265. S. Blair, K. Izod, W. Clegg and R. W. Harrington, *Eur. J. Inorg. Chem.*, 3319 (2003).
266. W. Kaschube, K.-R. Pörschke, K. Angermund, C. Krüger and G. Wilke, *Chem. Ber.*, **121**, 1921 (1988).
267. G. Wilke, *Angew. Chem., Int. Ed. Engl.*, **27**, 185 (1988).
268. S. I. Khan, P. G. Edwards, H. S. H. Yuan and R. Bau, *J. Am. Chem. Soc.*, **107**, 1682 (1985).
269. M. Hogenbirk, G. Schat, F. J. J. de Kanter, O. S. Akkerman, F. Bickelhaupt, H. Kooijman and A. L. Spek, *Eur. J. Inorg. Chem.*, 2045 (2004).
270. M. Horáček, P. Štěpnička, J. Kubišta, K. Fejfarová, R. Gyepes and K. Mach, *Organometallics*, **22**, 861 (2003).
271. M. Horáček, J. Hiller, Y. Thewalt, M. Polášek and K. Mach, *Organometallics*, **16**, 4185 (1997).
272. V. Kupfer, U. Thewalt, M. Horáček, L. Petrusová and K. Mach, *Inorg. Chem. Commun.*, **2**, 540 (1999).
273. K. Mach, R. Gyepes, M. Horáček, L. Petrusová and J. Kubišta, *Collect. Czech. Chem. Commun.*, **68**, 1877 (2003).
274. V. Varga, K. Mach, G. Schmid and U. Thewalt, *J. Organomet. Chem.*, **454**, C1 (1993).
275. V. Varga, K. Mach, G. Schmid and U. Thewalt, *J. Organomet. Chem.*, **475**, 127 (1994).
276. M. Ohashi, K. Matsubara, T. Iizuka and H. Suzuki, *Angew. Chem., Int. Ed.*, **42**, 937 (2003).
277. J. T. Golden, T. H. Peterson, P. L. Holland, R. G. Bergman and R. A. Andersen, *J. Am. Chem. Soc.*, **120**, 223 (1998).
278. T. M. Gilbert, R. R. Ryan and A. P. Sattelberger, *Organometallics*, **8**, 857 (1989).

CHAPTER 2

The thermochemistry of organomagnesium compounds

JOEL F. LIEBMAN

Department of Chemistry and Biochemistry, University of Maryland, Baltimore County, 1000 Hilltop Circle, Baltimore, Maryland 21250, USA
Fax: +1 410 455 2608;   e-mail: jliebman@umbc.edu

TORKIL HOLM

Department of Chemistry, Technical University of Denmark, Building 201, DK-2800 Lyngby, Denmark
Fax: +45 45933968;   e-mail: th@kemi.dtu.dk

and

SUZANNE W. SLAYDEN

Department of Chemistry, George Mason University, 4400 University Drive, Fairfax, Virginia 22030, USA
Fax: +1 703 993 1055;   e-mail: sslayden@gmu.edu

I. INTRODUCTION: SCOPE AND DEFINITIONS	102
A. Thermochemistry	102
B. Sources of Data	102
C. Magnesium: A Metal Among Metals	102
D. Calorimetry of Organomagnesium Compounds	104
II. COMPOUNDS COMPOSED SOLELY OF MAGNESIUM AND CARBON	106
III. THE SCHLENK EQUILIBRIUM	107
IV. ORGANOMAGNESIUM HALIDES	109
A. Isomers and Homologous Series	110
B. Unsaturated Compounds	112
1. Formal protonation reactions	112
2. Enthalpies of hydrogenation	113
C. Organomagnesium Bromides Containing Heteroatoms	113

The chemistry of organomagnesium compounds
Edited by Z. Rappoport and I. Marek   © 2008 John Wiley & Sons, Ltd

V. DIORGANOMAGNESIUM COMPOUNDS	116
VI. ORGANOMAGNESIA AND RINGS	117
A. Cycloalkylmagnesium Halides	117
B. Magnesacycloalkanes and Their Dimers (Dimagnesacycloalkanes)	120
C. Other Magnesacycles	121
VII. MAGNESIUM SANDWICH SPECIES	122
A. Magnesocene (Bis(cyclopentadienyl) magnesium)	122
B. Neutral Magnesium Half-Sandwiches	123
C. Triple Decker (Club) Sandwiches	123
D. Cationic Sandwiches and Half-Sandwiches	124
VIII. MAGNESIUM COMPLEXES WITH CARBON MONOXIDE	125
IX. REFERENCES AND NOTES	126

I. INTRODUCTION: SCOPE AND DEFINITIONS

A. Thermochemistry

The current chapter is primarily devoted to the thermochemical properties of molar standard enthalpies of formation and of reaction, $\Delta_f H_m^\circ$ and $\Delta_r H_m^\circ$, often called the 'heat of formation', ΔH_f and 'heat of reaction', ΔH_r. We will only briefly discuss bond dissociation energies, Gibbs energy and complexation energies. This chapter foregoes discussion of other thermochemical properties such as entropy, heat capacity or excess enthalpy. Temperature and pressure are assumed to be 25 °C ('298 K') and 1 atmosphere or the nearly equal 1 bar (101,325 or 100,000 Pa) respectively. The energy units are kJ mol^{-1} where 4.184 kJ is defined to equal 1 kcal. Although our thermochemical preference is for the gas phase, we find that for many of the species discussed here, only solution phase data are available. We interpret the 'organomagnesium' in the title of this work to mean that the minimum requirement for a species to be included is that it have at least one magnesium atom and one carbon atom. And so there is a section on compounds consisting solely of magnesium and carbon. The remaining sections consider the traditional CHONS atom combinations in several manifestations as they are bonded to magnesium.

B. Sources of Data

Unreferenced enthalpies of formation for any organic species in the current chapter are taken from the now 'classic' thermochemical archives by Pedley and his coauthors[1]. Likewise, unreferenced enthalpies of formation for inorganic compounds come from the compilation of Wagman and his coworkers[2]. These thermochemical numbers are usually for comparatively simple and well-understood species where we benefit from the data evaluation performed by these authors rather than using the raw, but much more complete, set of data found in the recent, evolving, on-line NIST WebBook database[3]. All other thermochemical quantities come from sources explicitly cited in the chapter.

C. Magnesium: A Metal Among Metals

That magnesium is by far the most useful metal for preparing organometallic reagents to be used in syntheses is due to several factors. Although a highly electropositive metal, it is easily handled and stable in the atmosphere since it is protected by an invisible coating of oxide-hydroxide. It is non-toxic and presents no problems for the environment. Of the Group II metals, Ba, Sr, Ca, Mg, Be, Zn, Cd and Hg, the most electropositive Ba, Sr

2. The thermochemistry of organomagnesium compounds

and Ca have been little studied and the free metals and their alkyl compounds are rather inaccessible. Their reactions are similar to those of sodium but they are less reactive. Magnesium forms rather polar bonds to carbon which consequently possesses significant carbanionic character. Grignard reagents combine the virtues of being at the same time very reactive and very easy to prepare from metallic magnesium, which is unique among electropositive metals in being readily available and requiring little or no cleaning before use. Beryllium and its compounds are exceptionally toxic, and so discussion as useful reagents logically ends there.

Among hydrocarbylmetals formed from alkali metals, only hydrocarbyllithium compounds match the Grignard reagents in utility and reactivity. A choice will often exist between magnesium and lithium compounds for a given reaction, but magnesium is much easier and safer to handle and organomagnesium compounds furthermore have the advantage of being stable in ether solution while organic alkali compounds all attack ether and are handled in hydrocarbon solvents.

Metals more electronegative than magnesium, like beryllium, zinc, cadmium and mercury, form useful reagents for specific purposes, but the metals themselves are not sufficiently active to form organic derivatives under normal laboratory conditions and are unwanted in the environment since they are toxic. Aluminum compounds are useful for industrial purposes, but their use in the laboratory is insignificant in comparison with Grignard reagents.

Lest one forget and be complacent, organomagnesium species are high energy compounds as expressed in terms of the considerable exothermicity of many of their spontaneous reactions—those with water and/or air are perhaps best known. Almost all laboratory investigations of the chemistry of organomagnesium compounds have been with the homoleptic species R_2Mg, or with the classical Grignard reagents $RMgX$ with one hydrocarbyl (alkyl or aryl) R group and either chlorine, bromine or iodine attached as an X to the metal. Organomagnesium fluorides have been relatively ignored as they are more difficult to prepare than the related compounds with the other halogens[4]. These are plausible species in mixed metal fluorocarbon 'pyrolants', chemical sources of high temperatures (multi-thousand K) resulting from solid phase reactions of magnesium and fluorinated organic polymers.[5] That is, mechanically combined Mg and polymer are induced to chemically react presumably via the following schematic reactions (equations 1 and 2), shown here for polytetrafluoroethylene.

$$Mg(s) + -CF_2-CF_2-(s) \longrightarrow -CF_2-CF(MgF)-(s)$$
$$\longrightarrow MgF_2(s) + -CF=CF-(s) \quad (1)$$
$$Mg(s) + -CF=CF-(s) \longrightarrow -CF=C(MgF)-(s) \longrightarrow MgF_2(s) + 2C(s) \quad (2)$$

While the C−F bond is recognized as strong, the Mg−F bond is stronger. From enthalpy of formation data[2] per monomeric unit of C_2F_4, this reaction is exothermic by over 1400 kJ mol^{-1}.

Numerous other reactions are occasionally problematic because of unexpected heat evolution and temperature increase. Although not widely publicized, trifluoromethylphenyl chlorides and bromides are prone to explode during preparation of the Grignard reagent[6]. It was hypothesized that phenylethylene intermediates can polymerize in a runaway exothermic reaction[7], while loss of solvent contact and an excess of highly activated magnesium were shown to favor violent reactivity.[8] Fluorine-containing aryl Grignards are not the only culprits.[9] As such, there has been active industrial interest in safety hazards surrounding Grignard formation during scale-up, initiation and reagent addition[10].

D. Calorimetry of Organomagnesium Compounds

As with so many other classes of compounds, calorimetric measurements and derived thermochemical concepts were important in the early era of the study of organomagnesium compounds—and then largely ignored once the field gained maturity. For example, 100 years ago the interaction of amines with propylmagnesium iodide was discussed in terms of measured solvation energies, and the energies compared with those from the interaction of ethers with the same organometallic[11]. That ethers are less basic than amines and that oxonium ions and related salts are less stable than their ammonium counterparts, was used to suggest the solvation of Grignard reagents in terms of [Solvent−Mg−R]$^+$ I$^-$ ion pairs. These suggested structures presaged our modern understanding of solvent-stabilized molecular, rather than ionic complexes, in solution. Our current knowledge is that the C−Mg bond energy is very much the same for all primary alkyl groups attached to the magnesium—from observations on an extensive variety of other alkyl derivatives, we may now ask first how could it be otherwise, and then ask how could this entirely plausible result be experimentally demonstrated. Century-old experiments are relevant here as well. Direct calorimetric measurements[12] of the enthalpy of hydrolysis were made on three sets of isomeric pairs of R$_2$O•R′ MgI and RR′O•RMgI species in which the groups now recognized to be on oxygen and magnesium were interchanged. The reaction exothermicities were found to be nearly the same for the cases where R = Et, R′ = Pr; R = Et, R′ = Bu and R = Et, R′ = Pen.

Calorimetry is a discipline demanding exquisite experimental care, and is an art as well as a science: compared to many other areas of the chemical sciences, there are comparatively few new apprentices of this study. To aid future researchers interested in performing new experiments on the energetics of organomagnesium compounds, as well as historians of our science, we describe in considerable detail the earlier experiments performed by one of the authors (T.H.) but not included in the original publication.

Because of the high reactivity of Grignard reagents, calorimetric measurements require total exclusion of air and moisture and vacuum tight equipment must be used. The following three reactions (equations 3–5) have usually been studied: formation, protonation and reaction with bromine.

$$\text{RBr} + \text{Mg} \longrightarrow \text{RMgBr} \qquad (3)$$

$$\text{RMgBr} + \text{HBr} \longrightarrow \text{RH} + \text{MgBr}_2 \qquad (4)$$

$$\text{RMgBr} + \text{Br}_2 \longrightarrow \text{RBr} + \text{MgBr}_2 \qquad (5)$$

Protonation reagents such as water and alcohol have been used, but HBr is the preferred reagent because the reaction leads to well defined, ether-soluble products.

The use of a normal adiabatic calorimeter is not ideal when the reaction studied has an induction period as in reaction 3 or when a reaction has to be initiated by breaking an ampoule as in reaction 4 or 5. Much more convenient and reliable is the use of a steady-state heat flow calorimeter. The method used in References 13 and 14 is described here.

The calorimeter consisted of a 500-mL flask with an air-filled jacket, a magnetic stirrer and a manganin heating coil (Figure 1) . The calorimeter was closed with a B 29 standard taper rubber sealed adapter which fitted a Beckman thermometer (8), the leads (9) for the heating coil (4) and a glass capillary inlet (x) for the liquid or gaseous reactant. Internally, the inlet capillary had a 1.5-mm polyethylene tube leading to the bottom of the flask. Externally, this capillary was connected by a glass capillary either to a hydrogen bromide supply or to a Metrohm piston burette driven by a synchronous motor which delivered 20 mL/180 min.

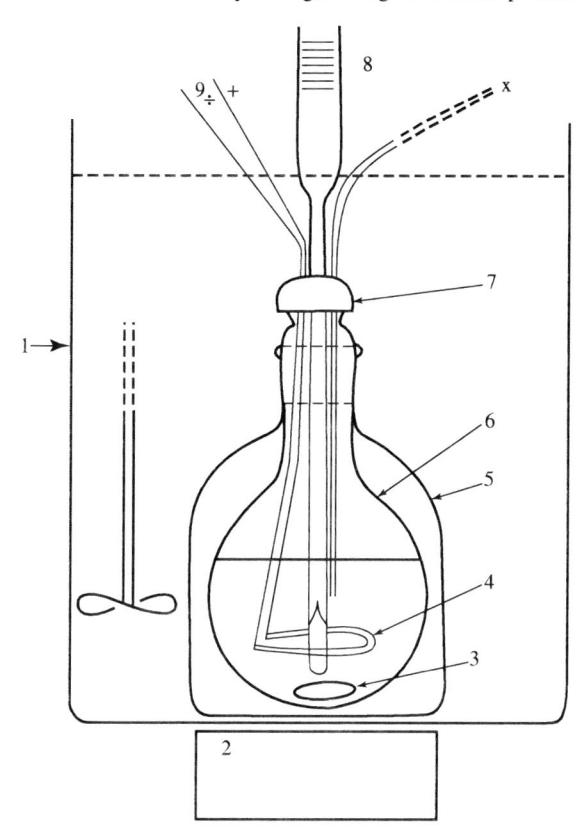

FIGURE 1. The steady-state heat flow calorimeter

For measurement of reaction 3 the calorimeter was filled with 15 g magnesium turnings and 400 mL of ether distilled from LiAlH$_4$. The calorimeter was mounted in a precision water thermostat (1) and the magnetic stirrer (2) was started. Pure alkyl bromide was pumped from the motorburette at the constant rate of 1.8517 μL s^{-1}. After the start of the reaction the addition was continued for 30–60 min. By adjusting the thermostat a steady state was obtained with a temperature in the calorimeter about 10 °C higher than in the thermostat, so that the reading of the Beckman thermometer was constant within ±0.002 °C. The addition of RBr was then stopped and the temperature was kept constant by leading an electric current through the heating coil using a precision constant current generator, 'Fluke 382 A' (not shown). The enthalpy of reaction is equal to the substituted electrical effect and, knowing the molarity of the pure alkyl bromide, the molar reaction enthalpy could be calculated.

Methyl bromide was kept in an ampoule at 0 °C and was displaced by the introduction of 1.8517 μ s^{-1} of mercury from the motorburette. The methyl bromide was passed through a 2-m stainless steel capillary heating coil which was placed in the thermostat water. In order to derive the enthalpy of reaction of liquid methyl bromide, the enthalpy of vaporization (23.0 kJ mol^{-1}) was subtracted from the value obtained for gaseous methyl bromide.

Measurement of the enthalpy of reaction 4 required a constant stream of HBr. This was obtained by placing an HBr cylinder in an ice bath and connecting the outlet to a glass capillary which allowed a stream of 18–20 μmol^{-1}. The exact value was found by leading the HBr stream into water for 100 s and titrating with sodium hydroxide. This determination was made before and after each experiment.

For addition of liquid bromine, a 5-mL piston burette was used delivering 0.4629 μL s^{-1}. In the calorimeter was placed 400 mL of a 0.4 M alkylmagnesium bromide in diethyl ether. The measuring procedure followed the same principles as used for HBr addition.

It was found that in the study of reaction 4, the most important source of error when using this calorimetry procedure was the change of vapor pressure in the calorimeter caused by the formation of gases. This resulted in a significant change in the heat transfer coefficient for the heat transfer for the calorimeter due to a change in the rate of reflux of the ether solvent from the uncovered walls. The error was almost eliminated by filling the calorimeter with ether leaving only 10% empty space. Errors were introduced also by assuming that gaseous alkanes dissolve in ether with evolution of the full enthalpy of vaporization. By measurements this was found to be true within experimental uncertainty for C_5 alkanes and higher, but incorrect for the lower alkanes. Corrections were made for C_1–C_4 alkanes. The results were usually reproducible to within ±1 kJ mol^{-1} when using liquid alkanes, and ±2.2 kJ mol^{-1} when using gaseous alkanes. The purity and the density of the alkyl bromides were the data given by the manufacturer and are estimated to be within ±0.5%.

II. COMPOUNDS COMPOSED SOLELY OF MAGNESIUM AND CARBON

In principle, there are many binary species that are composed solely of divalent Mg and C. Admittedly, such species characterized by carbon bonded to only magnesium or another carbon appear quite strange. Two such species would thus be the magnesium-containing 'too small' cyclopropyne, MgC_2 (**1**), and the cumulene, Mg_2C_3 (**2**), which is a bimetallic carbon suboxide mimic.

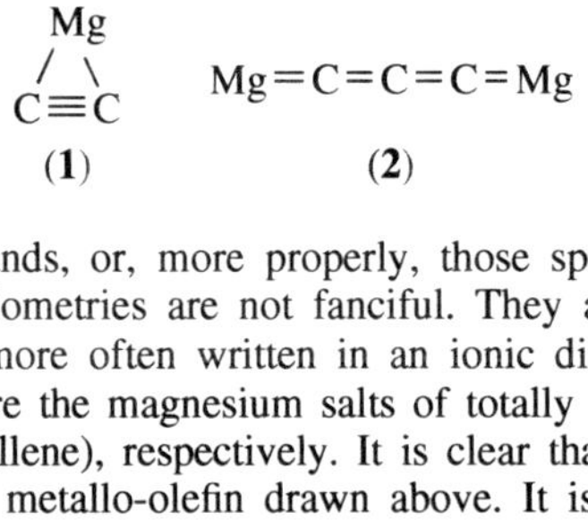

$$\begin{array}{c} Mg \\ / \ \backslash \\ C \equiv C \end{array} \qquad Mg=C=C=C=Mg$$

$$(1) \qquad \qquad (2)$$

However, these compounds, or, more properly, those species with the same Mg:C ratios and resulting stoichiometries are not fanciful. They are two of the best known magnesium carbides and more often written in an ionic dialect, as Mg^{2+} $(C_2)^{2-}$ and (Mg^{2+}) $(C_3)^{4-}$, i.e. they are the magnesium salts of totally deprotonated acetylene and propyne (or alternatively allene), respectively. It is clear that these species are not the covalent metallocycle and metallo-olefin drawn above. It is clear also that the totally ionic carbides also are inadequate representations since the isolated anions lie far above the corresponding neutrals and free electrons in energy[15].

These representations—essentially covalent molecule and totally ionized salt—presage the conflicting descriptions of the alternatives that can be drawn for the more conventional organomagnesium compounds that fill this chapter and the current volume. Indeed, the relative stability of the cycloalkyne ring description compared to the less-bonded (hypovalent) chain cumulene Mg=C=C recurs in the question of the general C_2X species with X chosen among other third row elements. For X = Na – Si, the ring is seemingly preferred over the chain and the opposite is found for X = P – Cl[16]. (For X = CH_2 the chain is seemingly the more stable, but not by so much that CH_2CC facilely automerizes into $CCCH_2$ by way of the parent cyclopropyne[17].) Additionally, crystalline MgC_2

2. The thermochemistry of organomagnesium compounds

has been described as having 'MgCCMgCCMg... chains ... [with] a weak interaction [2.510(1) Å] between Mg and the triple bond of the crossing chains above and below'[18]. Indeed, a crystallographic investigation of Mg_2C_3 described this species in terms of 'bridging of the C—C bonds by Mg ... reminiscent of polycenter, electron-deficient bonds'[19] and corresponding low ionicity.

However exotic are these species and however quixotic appear the attempts at a unique simple description, MgC_2 and Mg_2C_3 are well-known solids (see References 18 and 19 and citations therein) for which the enthalpies of formation of 84 and 71 kJ mol^{-1} are well-chronicled[2]. As no sublimation data are available from experiment or estimate, we are seemingly thwarted in any attempt to derive Mg—C bond energies from these data[20]. We remain optimistic in our understanding because, besides organic and organometallic chemists, materials scientists[18, 21] and astrochemists[22] have joined the hunt for new magnesium–carbon species and their understanding.

III. THE SCHLENK EQUILIBRIUM

In 1900, Victor Grignard[23] presented the reaction product from an alkyl halide, RX, and magnesium in ether as simply RMgX. He and contemporary workers were aware that ethyl ether was somehow built into the molecule and for a time an oxonium structure was suggested[24] that had no bond between carbon and magnesium. The modern concept of bonding between an anionic alkyl and a cationic magnesium was presented in 1905 by Abegg[25] and at the same time the possibility of alkyl–halogen exchange was suggested. In a footnote this author was the first to suggest an equilibrium as shown in equation 6.

$$2RMgX \rightleftharpoons R_2Mg + MgX_2 \qquad (6)$$

The equilibrium was 25 years later named after Schlenk and Schlenk[26], who found that magnesium halide precipitates from an ethereal Grignard reagent solution by addition of dioxane. They thought that by filtering and weighing the crystalline dioxanate precipitate it would be possible to determine the position of the equilibrium. This was not possible, however, because removal of magnesium halide led to an immediate readjustment of the equilibrium so that after addition of a sufficient amount of dioxane (>3 moles), all halide was removed leaving a solution of pure dialkylmagnesium.

For many years the equilibrium was formulated as in equation 7 and the monomer RMgBr was thought not to exist[27].

$$R_2Mg-MgX_2 \rightleftharpoons R_2Mg + MgX_2 \qquad (7)$$

Clarification of the problem was delayed several years after it was concluded by the use of isotopically labelled magnesium that magnesium–halogen exchange did not take place in the solution[28]. That this was incorrect was reported in 1963 when it was shown by the use of osmometric measurements that the monomeric EtMgBr is present in dilute solutions (<0.1 M) in THF[29] and diethyl ether[30], respectively. The osmometric measurements showed that at higher concentrations various loose aggregates form[31]. Aggregates are more apt to form in less polar solvents like diethyl ether than in more polar solvents like THF. Only alkylmagnesium fluorides are dimeric in THF[32]. The R group of the Grignard reagent likewise influences the degree of association. Organomagnesium molecules associate by halogen or/and alkyl bridges between magnesium atoms. Chlorine and fluorine are a better bridging ligands than either bromine and iodine, so alkyl magnesium chlorides and fluorides are dimeric over a wide concentration range.

Thermochemically, the association is not a major factor since the enthalpies of dilution of Grignard reagents are very small in diethyl ether as well as in THF[33, 34]. An explanation may be that bonding between molecules by means of halogen or alkyl bridging replace the coordinating ether molecules and that the enthalpies of coordination of the two types of bonding are nearly equivalent. Likewise, the association of Grignard reagents does not seem to have much influence on the position of the Schlenk equilibrium[33]. This is in accord with an equilibrium with the same number of entities on the two sides as in equation 6. Equation 7 represents an equilibrium that will be shifted to the right with higher dilution. The fact that the Schlenk equilibrium is almost independent of dilution must mean that equation 6 is a better description than equation 7 and that the tendency for association with solvent is, on an average, the same on both sides of equation 6.

Just as the association equilibria have little effect on the position of the Schlenk equilibrium, there has been no clear demonstration of a correlation between the association equilibria and the reactivity of the reagents. Plots of reaction rates versus concentration of Grignard reagents for various substrates often deviate from a straight line so that the reaction order is below 1 and even may approach zero[35]. This phenomenon has been shown not to correlate with an association of the reagent itself but rather with a complexation of the Grignard reagent with the substrate which occurs if the substrate has a Lewis basicity greater than that of the ether solvent[36]. With substrates of low basicity like methyl trifluoroacetate or benzonitrile the reaction order with respect to Grignard reagent is close to 1[37].

Although the position of equation 6 could not be determined by dioxane precipitation of magnesium halide, it was found that the position could be determined by thermometric titration[33,34,36,38]. Adding R_2Mg to a solution of $MgBr_2$ in ether led to an increase in temperature. The plot of added R_2Mg versus temperature gave both the enthalpy for complete reaction as well as the composition of the mixture and the equilibrium constant for equation 6. The Δt was positive in ether but was shown[34,36] to be negative in THF. Thermometric titration of dialkylmagnesium–magnesium bromide has been published for alkyl = methyl, ethyl, butyl and phenyl in both ether and THF as shown in Table 1.

The position of the Schlenk equilibrium has alternatively, and less accurately, been estimated by means of IR[39] and NMR spectra[40,41]. The latter method has confirmed the extreme rate of alkyl–halide exchange for alkyl = methyl and ethyl. Separate signals for dialkylmagnesium and alkylmagnesium halide were not discernable at room temperature, but for methyl separate signals appeared at $-80\,°C$ when the alkyl–halide exchange process was slowed down. For dimethylmagnesium, which is associated by bridging methyl

TABLE 1. Equilibrium constants ($K_{Schlenk}$) and enthalpies of reaction for the Schlenk reaction $R_2Mg + MgBr_2 \rightleftarrows 2RMgBr$

RMgBr	Solvent	$K_{Schlenk}$	ΔH_{rxn} (kJ mol^{-1})	Method[a]	Reference
MeMgBr	THF	3.5		IR	39
		4.0		NMR	40
	Et$_2$O	320		T	43
		455		C	32
EtMgBr	THF	5.09	25.5	C	34
	Et$_2$O	480, 484	-15.5	T	33
BuMgBr	THF	ca 9	14.2	T	36
	Et$_2$O	ca 1000	-13.4	T	36
		ca 1400		K	31
PhMgBr	THF	3.8	11.8	C	34
		4.0	13.4	NMR	41
	Et$_2$O	55, 62	-8.5	C	33

[a] T = thermometric titration; C = calorimetry; K = kinetics.

2. The thermochemistry of organomagnesium compounds

TABLE 2. Enthalpy of solvation of components of the Schlenk equilibrium in Et_2O and THF (kJ mol^{-1})[44]

Solvent	Et_2Mg	EtMgBr	$MgBr_2$
Et_2O	−24	−31	−32
THF	−36.6	−68.3	−109

groups, separate signals for terminal and for bridging methyl groups could be observed. Because crystallization took place, the signals were not useful for quantitative measurements. It was found that the alkyl exchange rate depended on both the solvent and on the alkyl group. The signals for di-t-butylmagnesium and t-butylmagnesium bromide in THF could be discerned at room temperature because of a slow alkyl exchange.

An estimate of the equilibrium may also be obtained by kinetic measurements since dialkylmagnesium is often 50–100 times more reactive than alkylmagnesium bromide[42]. Addition of $MgBr_2$ to a Grignard reagent converts R_2Mg to RMgBr. When comparing the reactivity of this manipulated Grignard reagent with the reactivity of both R_2Mg and 'normal' RMgBr, the content of R_2Mg and $K^{Schlenk}$ may be found. The non-basic methyl trifluoroacetate has a negligible reactivity toward BuMgBr and has been used for this type of estimation of the position of the Schlenk equilibrium[37]. The content of dibutylmagnesium in nominally 0.5 M butylmagnesium X was found to be 5%. This corresponds to an equilibrium constant of 1400 (not 400 as given in Reference 37). The reaction between methylmagnesium bromide and benzophenone has likewise been used to determine $K_{Schlenk}$ and the value found was in reasonable agreement with the value obtained by thermometric titration[43].

Of the three components of the Schlenk equilibrium, the electrophilicity decreases in the order $MgBr_2$ > RMgBr > R_2Mg. In diethyl ether the total bonding in 2 mol RMgBr is stronger than the total bonding in 1 mol each of R_2Mg and $MgBr_2$. For this reason, equilibrium 6 lies to the left in diethyl ether. The stronger solvation of especially $MgBr_2$ favors the right side of equilibrium 6 in a more solvating donor solvent like THF. The endothermic reaction between R_2Mg and $MgBr_2$ in THF is the result of an entropy-driven reaction leading to an almost statistical distribution of the three components. Approximate values of the enthalpy of solvation of the components of the Schlenk equilibrium are given in Table 2[44].

IV. ORGANOMAGNESIUM HALIDES

Holm determined the enthalpies of formation of a collection of hydrocarbylmagnesium bromides by reaction calorimetry with HBr in diethyl ether[13,14]. He also determined the enthalpies of formation in ethereal solution of the magnesium bromide salts of 20 Bronsted acids, HB, by measuring the enthalpies of reaction of the acid with pentylmagnesium bromide[45]. For those species that were reported in both studies (hydrocarbyl = phenylethynyl, phenyl, methyl, cyclopropyl, cyclopentyl, cyclohexyl), the enthalpies of formation were identical. The values are listed in Tables 3 and 4.

There is one other report in the literature of a measurement of the enthalpy of formation of an organomagnesium halide. The enthalpy of reaction of magnesium with methyl iodide in ether was calorimetrically determined as -273.6 ± 0.8 kJ mol^{-1} [46]. Using a recent enthalpy of formation for liquid methyl iodide of -13.6 ± 0.5 kJ mol^{-1} [47], the enthalpy of formation of methylmagnesium iodide is -287.2 kJ mol^{-1}. The exchange (equation 8) is thus 11.2 kJ mol^{-1} endothermic.

$$MeI + MeMgBr \longrightarrow MeBr + MeMgI \qquad (8)$$

TABLE 3. Enthalpies of reaction and enthalpies of formation of hydrocarbylmagnesium bromides (RMgBr) in ether solution (kJ mol^{-1})

R	ΔH_r RMgBr (soln)[a,b]	ΔH_f RMgBr(soln)[b]
Methyl	−274.5	−331.8
Vinyl	−294.1[c]	−264.4[c]
Ethyl	−299.2	−323.0
Allyl	−259.4	−265.7
n-Propyl		−360.7
i-Propyl	−305.9	−339.7
n-Butyl	−292.5	−378.2
i-Butyl	−289.1	−391.6
sec-Butyl	−305.9	−368.2
tert-Butyl	−306.7	−370.7
n-Pentyl		−406.7
1-Ethylpropyl	−306.3	−389.9
Neopentyl	−286.6	−430.1
n-Hexyl		−427.6
n-Heptyl		−452.3
n-Octyl		−478.6
Cyclopropyl	−282.8	−211.3
Cyclobutyl	−289.1	−229.7
Cyclopentyl	−291.6	−336.8
Cyclohexyl	−298.7	−380.3
Cycloheptyl	−299.6	−379.5
Cyclooctyl	−295.0	−395.4
Phenyl	−263.2	−208.4
Benzyl	−256.5	−252.3
4-Methylphenyl	−262.3	−244.8
4-Chlorophenyl	−260.2	−251.5
Phenylethynyl	−169.9	−69.5
Triphenylmethyl	−231.0	−120.5

[a] Enthalpies of reaction were determined for RMgBr(soln) + HBr(g) → RH(soln) + MgBr$_2$(soln).
[b] All values are from References 13 and 14. The experimental uncertainties are ±2.2 kJ mol^{-1}.
[c] In THF. It is expected that less heat would be evolved in ether solution.

There is one additional study on the enthalpy of hydrolysis of solid butylmagnesium chloride[48]. Additional calculations[3] result in a solid phase enthalpy of formation of −455.7 ± 2.0 kJ mol^{-1}.

A. Isomers and Homologous Series

We briefly discussed in an earlier volume the behavior of the isomeric and homologous organomagnesium bromides compared to the organolithium compounds as a means of furthering our understanding of the thermochemistry of the latter species[49]. Here, we will discuss only the magnesium compounds. Discussion of the cycloalkylmagnesium bromides is deferred to a later section in this chapter.

The linear correlation of enthalpies of formation with the number of carbon atoms is a useful and well-known feature of homologous series of functionalized organic compounds. The slope of the regression line for the gaseous n-alkanes (CH$_3$−(CH$_2$)$_x$−H), −20.6 kJ mol^{-1}, and the similar values of the slopes for other CH$_3$−(CH$_2$)$_x$−Z series is often called the 'universal methylene increment'[50]. In the liquid phase, the increment for the n-alkanes is −25.6 ± 0.1 kJ mol^{-1}. The most accurate determination of the increment

TABLE 4. Enthalpies of reaction and enthalpies of formation of organomagnesium bromides in ether solution (kJ mol^{-1})

HB	ΔH_r BMgBr [a,b]	ΔH_f BMgBr [c]
Methane	−15.1	−331.0
Cyclopropane	−6.7	−204.7
Cyclopentane	2.1	−336.2
Cyclohexane	9.2	−380.4
1,3-Cyclopentadiene	−148.5	−275.8
C_6H_6	−26.4	−210.6
$PhCH_3$	−33.1	−253.9
$PhC\equiv CH$	−125.9	−75.6
$CH_2(CN)_2$	−203.3	−59.7 [b]
CH_3NH_2	−130.5	−411.0
$c\text{-}C_6H_{11}NH_2$	−133.1	−514.0
$PhNH_2$	−153.1	−355.0
$(C_2H_5)_2NH$	−111.3	−448.2
$c\text{-}C_6H_{11}NHCH_3$	−122.6	−501.2
$c\text{-}(CH_2)_5NH$	−116.7	−436.4
Ph_2NH	−118.8	−192.0
$C_{11}H_{23}CONHCH_3$	−186.2	
CH_3OH	−219.7	−692.0
C_2H_5OH	−199.6	−721.0
$(CH_3)_2CHOH$	−193.3	−744.6
$(CH_3)_3COH$	−177.8	−770.2
CF_3CH_2OH	−199.6	−1365.2
PhOH	−202.5	−589.3
C_6F_5OH	−233.9	−1474.8
$C_2H_5CO_2H$	−251.0	−994.9
$C_{11}H_{23}CO_2H$	−243.1	−1214.2
CF_3CO_2H	−273.6	−1576.7
$C_{12}H_{25}SH$	−183.3	−744.6
PhSH	−178.2	−347.7

[a] Enthalpies of reaction are for $C_5H_{11}MgBr + B-H \rightarrow C_5H_{12} + B\text{-}MgBr$.
[b] Values are from Reference 45. The experimental uncertainties are ca 2–3%.
[c] Values calculated in this work unless otherwise noted. See text.

is from a dataset consisting of homologs of four or more carbons. The methyl derivative in most series deviates from the otherwise linear relationship.

There are seven n-alkylmagnesium bromides for which there are solution phase enthalpy of formation data, C_2–C_8. The methylene increment is -25.0 ± 0.8 kJ mol^{-1}, which is nearly identical to both the n-alkane series and the n-alkyl bromide series (-25.3 ± 0.4 kJ mol^{-1}). The methylmagnesium bromide enthalpy of formation is ca 9 kJ mol^{-1} more negative than that for ethylmagnesium bromide, even though MeMgBr has the smaller molecular weight. This is typical of a methyl group bonded to more electropositive atoms such as lithium, boron and aluminum. The enthalpies of formation of methyl derivatives bonded to atoms more electronegative than carbon also deviate from the correlation but in the opposite direction: they are typically less negative than for the ethyl derivatives. The magnitude of the gaseous methyl deviations can be correlated to the electronegativity of Z^{51}. For the three sec-alkylmagnesium bromides, the methylene increment is -25.1 ± 2.0 kJ mol^{-1}. There are only two sec-alkyl bromides to compare, isopropyl and sec-butyl, and the difference between their enthalpies of formation, and thus the methylene increment, is -24.3 kJ mol^{-1}.

In isomeric alkanes substituted with an electronegative atom, alkyl group branching at the carbon bonded to the substituent atom increases the thermodynamic stability in both the liquid and gaseous phases. For example, the stability order of the butyl bromides is t-Bu > sec-Bu > n-Bu. The increasing stability parallels the alkyl group carbocation stability. For alkyl groups bonded to metals, the alkyl group is more electronegative, and it might be expected that the stability order would be the opposite and thus parallel the alkyl group carbanion stability. Indeed, the enthalpy of formation of n-propylmagnesium bromide shows it to be more stable than the isomeric isopropylmagnesium bromide by ca 21 kJ mol^{-1}. Non-calorimetric corroboration of the relative stabilities is provided by the observation that in the presence of small amounts of $TiCl_4$, isopropylmagnesium bromide rearranges to n-propylmagnesium bromide[52]. Organomagnesia enthalpies of formation cannot track those of the parent hydrocarbon—after all, isopropyl hydride and n-propyl hydride must have the same enthalpy of formation since they are both propane, n- and sec-butyl hydrides must have the same enthalpy of formation since they are both n-butane, and isobutyl and $tert$-butyl hydride must have the same enthalpy of formation as they are both isobutane.

In the isomeric butyl series, the secondary butyl derivative is less stable than either of the primary butylmagnesium bromides. The carbon-branched isobutylmagnesium bromide is more stable than the n-butyl isomer in keeping with the usual observation that alkyl branching remote from the carbon bonded to the heteroatom increases the thermodynamic stability. Within the experimental uncertainties, sec-butyl- and t-butylmagnesium bromide have the same enthalpies of formation, which is the same as that observed for the corresponding alkyl lithiums. The explanation may be that there is a fortuitous cancellation of the stabilizing effects of carbon-branching in the tertiary group and of secondary-carbon bonded to metal.

A useful comparison is between the alkylmagnesium bromide and its corresponding hydrocarbon, as for the formal protonation reaction (equation 9). The average enthalpy of formation difference, $\delta \Delta H_f$, for primary R is 233.5 ± 5.1 kJ mol^{-1} and for secondary R it is 214.7 ± 3.3 kJ mol^{-1}. For the lone example of the tertiary butyl group, $\delta \Delta H$ is 217.2 kJ mol^{-1}, which is similar to that for secondary R, as expected. The larger endothermicity of the formal reaction is associated with the group of relatively more stable Grignard reagents.

$$RMgBr(soln) \longrightarrow RH \ (lq) \qquad (9)$$

B. Unsaturated Compounds

1. Formal protonation reactions

Just as there is a nearly constant $\delta \Delta H_f$ value for the enthalpy of the formal protonation reaction (equation 9) for Grignard reagents of similar structural type (primary vs. secondary, tertiary), we expect there to be a nearly constant (but different) $\delta \Delta H_f$ also for the various groups of unsaturated species. The enthalpies of formal reaction for the three aromatic Grignards are quite consistent, 259.0 ± 3.0 kJ mol^{-1}. The enthalpy of protonation of vinylmagnesium bromide is 304 kJ mol^{-1}. Considering that phenyl and vinyl species often exhibit similar thermochemistry, this latter value seems much too high. However, the reaction for vinylmagnesium bromide in THF is expected to be more exothermic than the same reaction in ether. The allyl- and benzylmagnesium bromides have nearly identical enthalpies of protonation: 267.4 and 264.7 kJ mol^{-1}, respectively. Using a liquid phase enthalpy of formation for triphenylmethane of 192.2 kJ mol^{-1} [53], the $\delta \Delta H_f$ is ca 314 kJ mol^{-1}. The enthalpy of reaction of the lone example of triple bond unsaturation, phenylethynylmagnesium bromide, is 353.0 kJ mol^{-1}. Again, the stable phenylethynyl and triphenylmethide carbanions have the most endothermic reaction enthalpies.

2. Enthalpies of hydrogenation

There are enthalpies of formation for several unsaturated organomagnesium bromides as well as for species that are their saturated counterparts. How do the enthalpies of the formal hydrogenation reaction (equation 10) of the organomagnesium bromides compare with those for the corresponding hydrocarbons?

$$>\!C\!=\!C\!< + H_2 \longrightarrow >\!CH\!-\!CH\!< \tag{10}$$

The unsaturated Grignard reagents are phenyl-, allyl- and vinylmagnesium bromide and their hydrogenated products are cyclohexyl-, n-propyl- and ethylmagnesium bromide. The calculated formal hydrogenation reaction enthalpies are -171.9, -95.0 and -58.6 kJ mol^{-1}, respectively. However, the last value for hydrogenation of vinylmagnesium bromide in THF rather than ether solution, must be corrected to ca -90 kJ mol^{-1} to account for the extremely high enthalpy of solution of MgBr$_2$ in THF. For the hydrocarbon counterparts, benzene/cyclohexane, propene/propane and ethene/ethane, the reaction enthalpies are -205.4, -123.4 and -136.2 kJ mol^{-1}, respectively. All of the Grignard reagents' reactions are less exothermic than those of the corresponding hydrocarbons. The lower exothermicity of hydrogenation of the phenyl and vinyl Grignard reagents indicates that there is a stabilizing interaction between the double bond electrons and the magnesium bonded to carbon. In the allyl case, stabilization takes place by resonance. In the phenyl and vinyl cases the carbon bonded to magnesium changes hybridization from sp^2 to sp^3. By this change, we go from a rather stable to a rather unstable Grignard reagent.

C. Organomagnesium Bromides Containing Heteroatoms

Enthalpies of formation of the magnesium bromide salts of the Bronsted acids are calculated from the measured enthalpies of the reaction of the Bronsted acid, HB, with pentylmagnesium bromide and the known enthalpies of formation of pentylmagnesium bromide and pentane according to equation 11. Because some of the HB enthalpies of formation have been revised and others newly measured since the original publication, the BMgBr enthalpies of formation have been recalculated and appear in Table 4[54].

$$HB + CH_3(CH_2)_4MgBr \longrightarrow BMgBr + CH_3(CH_2)_3CH_3 \tag{11}$$

There remain three Bronsted acids that have no liquid phase enthalpy of formation data that we know of: dodecanethiol, cyclohexyl methyl amine and N-methyl dodecanamide. Although the enthalpy of formation of 1-dodecanethiol has not been measured, there are experimental values available for other members of its homologous series, C_2–C_7, C_{10}. From a weighted least-squares analysis of the data from which a slope, -25.4, and an intercept, -23.3, are derived, the enthalpy of formation of dodecanethiol is -328.1 kJ mol^{-1} [55]. The enthalpy of formation of dodecanethiolate magnesium bromide is thus estimated as -744.6 kJ mol^{-1}. We can estimate the enthalpy of formation of cyclohexyl methyl amine by assuming equation 12 is thermoneutral.

$$c\text{-HexNH}_2 + Me_2NH \longrightarrow (c\text{-Hex})MeNH + MeNH_2 \tag{12}$$

From the archival enthalpies of formation of the other species, the enthalpy of formation of cyclohexyl methyl amine is -145.4 kJ mol^{-1}. From equation 11, the enthalpy of formation of the corresponding salt is -501.2 kJ mol^{-1}. Attempts to estimate an enthalpy of formation for N-methyldodecanamide reveals a paucity of data to work with, primarily for unsubstituted and N-methylamides[56]. There is much enthalpy of formation data for n-alkyl carboxylic acids, including dodecanoic acid. The methylene increment is -25.4 kJ mol^{-1}, typical of liquid phase enthalpies of formation. There are only two liquid enthalpies of formation for n-alkyl amides, butanamide and hexanamide, and the difference between them is -25.5 kJ mol^{-1} per $-CH_2-$group. A fairly accurate enthalpy of formation for dodecanamide of -550.9 kJ mol^{-1} could be derived from these data. However, for the only two enthalpies of formation for N-methyl-n-alkylamides in the liquid phase, N-methylacetamide and N-methylpropanamide, the difference is 6.9 kJ mol^{-1} which is extremely atypical for a methylene increment. However, the acetamide is the methyl derivative and so is expected to deviate from the other N-methylalkylamides. Furthermore, there is no liquid enthalpy of formation for either acetamide or propanamide upon which to base an estimate for N-methylation of any amide.

Earlier it was stated that within the set of hydrocarbylmagnesium bromides the enthalpy of formation difference, $\delta \Delta H_f$, for the magnesium compound and its corresponding hydrocarbon was slightly larger for the primary alkyl groups compared to the secondary and tertiary groups. This differentiation by $\delta \Delta H_f$ with respect to structure and stability would likewise be expected for sets of compounds with C–Mg, N–Mg, O–Mg and S–Mg bonds. Figure 2 shows a plot of the enthalpies of formation of the organomagnesium bromide species in Tables 3 and 4 vs. the enthalpies of formation of the corresponding protonated species. The data points for each bond type fall on separate straight lines, the slopes of which are close to 1 (C–Mg, 0.78; N–Mg, 0.98; O–Mg, 1.1; S–Mg, 1.0). Even though there are differences in structure within each set, the correlations (r^2) are quite good: C–MgBr, 0.98; N–MgBr, 0.98; O–MgBr, 0.99. There are only two data points for S–MgBr.

Within each bond-type group, further distinctions can be made, as mentioned earlier for the alkylmagnesium bromides. In Figure 2, the points corresponding to mono- and polyunsaturated hydrocarbyl groups all appear to the right of the points belonging to the saturated groups. Said differently, the $\delta \Delta H_f$ values are substantially larger for the species with unsaturated substituents, 257–359 kJ mol^{-1} vs. 211–248 kJ mol^{-1}. This difference indicates an extra stabilization for the Grignard reagent which is absent in the hydrocarbon. The negative charge can be better accommodated in such compounds by sp^2-inductive and/or resonance effects.

Within the O–MgBr group are at least three subcategories: the alkoxy and phenoxy, the carboxy and the fluorinated species. Within the alkoxy subgroup, the endothermic

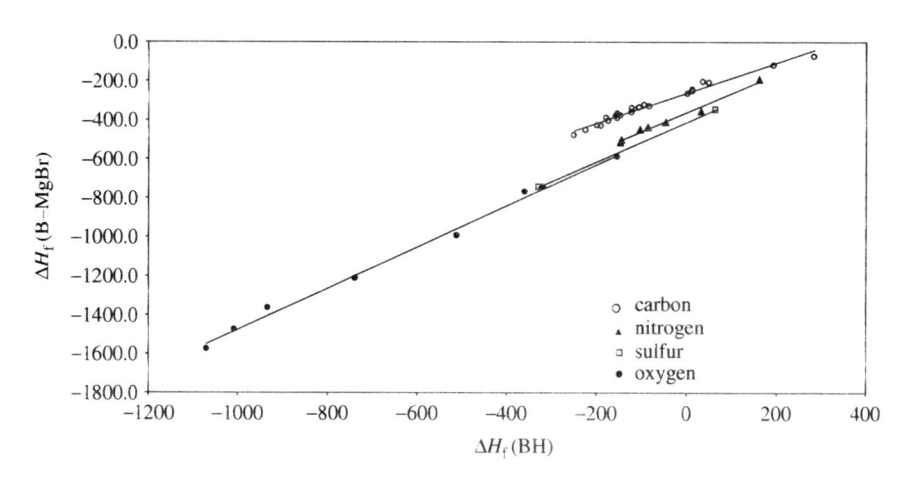

FIGURE 2. Enthalpies of formation of BMgBr vs. those of BH (kJ mol^{-1})

2. The thermochemistry of organomagnesium compounds

$\delta \Delta H_f$ from equation 9 is in the order t-BuO < i-PrO < EtO < MeO, which is the order of increasing stability of the alkoxides in solution. The two carboxy species are more endothermic than the alkoxy, and the trifluoroacetoxy is the most endothermic. While pentafluorophenoxy is more endothermic than phenoxy, trifluoroethoxy and ethoxy have identical endothermicities.

The N—MgBr group does not show much variation in $\delta \Delta H_f$ except for PhNHMgBr, which is comparatively endothermic. By comparison, Ph_2NMgBr is about the same as the saturated amine species.

The vertical distance between the lines may crudely be taken as the difference in bond energy between the B—H/B—MgBr bond types, at least for B atoms in the same row of the periodic table. From the plot, the bond strength increases in the order C—MgBr, N—MgBr, S—MgBr and O—MgBr, which is also expected from electronegativity differences. The bond strength to hydrogen increases in the order S—H, C—H, N—H and O—H. It has been suggested[45] that the very strong bond between oxygen and magnesium is due to back-donation of lone pairs on oxygen into empty orbitals on magnesium. This may be the case also with sulfur and less obviously with nitrogen. This is consistent from a consideration of the atomization energies of solid MgO, MgS and Mg_3N_2. From the enthalpies of formation of solid MgO, MgS and Mg_3N_2 and those of the gaseous atoms Mg, O, S and N, one can derive the enthalpies of atomization of the binary magnesium 'salts' to be 852, 773 and 1378 kJ mol^{-1}. Dividing these numbers by 2, 2 and 6, respectively (the number of 'bonds' per formula unit), results in 426, 382 and 230 kJ mol^{-1} for effective Mg—O, Mg—S and Mg—N bond strengths. Unfortunately, the enthalpy of formation of solid Mg_2C is not available from the literature—indeed, this seemingly simple binary species is still unknown—and so the remaining Mg—C bond strength cannot be derived for the final comparison.

One of the original goals in the determination of the enthalpy of reaction of Bronsted acids with pentylmagnesium bromide was to explain the relationship between the enthalpy and the pK_a of the acid[13, 14, 45]. For a set of hydrocarbons having disparate structures, the correlation coefficient, r^2, is 0.98. For the nitrogen-containing acids, again a group with disparate structures, there is an excellent correlation ($r^2 = >0.99$) if the aromatic aniline (phenylamine) and diphenylamine data are ignored. The oxygen-containing acid data show much scatter. For acids of the same acidity, the oxygen and nitrogen acid reactions are more exothermic than those for the carbon acids and the author assumes the cause is back-donation of lone pairs on the heteroatoms to empty orbitals on magnesium[45].

The only data omitted in this analysis are those for vinyl- (determined in THF) and cyclopentadienylmagnesium bromide. Including the data point for cyclopentadienyl in the analysis worsens the correlation. This may be caused by a difference in carbon bonding to the magnesium for cyclopentadienide compared to the other carbon—magnesium bonds. This bonding will be mentioned in a later section on magnesium sandwich compounds.

There is a report of calorimetrically-determined enthalpies of reaction of methyl- and ethylmagnesium bromides with some ketones in ether solution at 15 °C[57]. The reaction shown in equation 13, which results in an exotherm of -202.3 kJ mol^{-1}, produces an alkoxymagnesium bromide that also appears as a product of a different reaction in Table 4. Using the enthalpy of formation of MeMgBr[13, 14] and the liquid phase enthalpy of formation of acetone, the enthalpy of formation of the t-butoxymagnesium bromide is calculated as -783.5 kJ mol^{-1}. This is within 10 kJ mol^{-1} of the value reported in Table 4.

$$\text{MeMgBr} + \text{Me}_2\text{C=O} \longrightarrow t\text{-BuOMgBr} \qquad (13)$$

All the experimental enthalpies of the Grignard reaction appear in Table 5 along with the enthalpies of formation calculated using the same method as illustrated above. Unfortunately there are no liquid enthalpy of formation data for the halogenated ketones, nor are

TABLE 5. Enthalpies of reaction between ketones and Grignard reagents and calculated enthalpies of formation of alkoxymagnesium bromides (kJ mol^{-1})

Ketone	RMgBr	ΔH_{rxn} [a]	ΔH_f(R'OMgBr) [b]
Me$_2$CO	MeMgBr	−202.3	−782.2
MeCOEt	MeMgBr	−188.1	−793.2
Me$_3$CCOMe	MeMgBr	−162.1	−822.5
ClCH$_2$COMe	MeMgBr	−226.6	
BrCH$_2$COMe	MeMgBr	−233.7	
MeCOPh	MeMgBr	−184.9	−659.2
Me$_2$CO	EtMgBr	−222.6	−802.5
MeCOEt	EtMgBr	−209.2	−814.3
Me$_3$CCOMe	EtMgBr	−161.7	−822.1
ClCH$_2$COMe	EtMgBr	−240.2	
BrCH$_2$COMe	EtMgBr	−262.7	
MeCOPh	EtMgBr	−210.5	−684.8

[a] Enthalpies of reaction are from Reference 57.
[b] Enthalpies of formation of the alkoxymagnesium bromides are calculated from enthalpies of formation of the ketones from References 1 and 3 and of methyl and ethylmagnesium bromide from References 13 and 14. See text for discussion.

they easily estimated. The error in experimental measurements and assumptions can be understood by inspecting the two entries that both give the same alkoxymagnesium bromide product, although their calculated enthalpies of formation differ by *ca* 10 kJ mol^{-1}: Me$_2$CO + EtMgBr and MeEtCO + MeMgBr.

V. DIORGANOMAGNESIUM COMPOUNDS

The only dialkylmagnesium compound whose enthalpy of formation has been measured is dineopentyl magnesium[58]: (s) −236.8 ± 7.2 kJ mol^{-1} and (g) −74.3 ± 7.6 kJ mol^{-1}. Unfortunately there are no enthalpies of formation for any of its isomers or homologs. We cannot even calculate the enthalpy of the Schlenk equilibrium because, although the enthalpy of formation of neopentylmagnesium bromide is for the ether solution, that for dineopentylmagnesium is not, and there is no experimental value for the enthalpy of solution.

Organomagnesium compounds undergo fast intermolecular carbon–magnesium bond exchange in solution. One such process in THF solution, (equation 14) was studied by NMR line-shape analysis[59]:

$$(neo\text{-Pen})_2\text{Mg} + \text{Ph}_2\text{Mg} \longrightarrow 2(neo\text{-Pen})\text{MgPh} \quad (14)$$

The thermodynamic quantities for the reaction were found to be $\Delta H = 10.0$ kJ mol^{-1}, $\Delta S = 57.7$ eu and $\Delta G = -7.24$ kJ mol^{-1} at 298 K.

With knowledge of the enthalpies of formation of magnesium bromide and an alkylmagnesium bromide, and by using the data for the Schlenk reaction from Table 1, the enthalpy of formation of a dialkylmagnesium compound in ether solution may be calculated. In diethyl ether, the equilibrium equation 6 may be considered to be shifted to the side of the unsymmetrically substituted magnesium compound. Subtraction of the enthalpy of solution[44] gives the enthalpy of formation of the solvent-free components of the Schlenk equilibrium. The enthalpy of formation of MgBr$_2$ in diethyl ether is −559 ± 4 kJ mol^{-1} [13,14]. Only for ethyl- and butylmagnesium bromide in ether are all the enthalpy values available. The enthalpies of formation of diethylmagnesium

2. The thermochemistry of organomagnesium compounds   117

and dibutylmagnesium in diethyl ether are accordingly calculated to be -71.4 kJ mol^{-1} and -184.4 kJ mol^{-1}, respectively. Using the enthalpy of solvation for diethylmagnesium (the only solvation enthalpy available), its solid-phase enthalpy of formation is -47.4 kJ mol^{-1}. Is this a reasonable value? If we assume the methylene increment for the solid n-R$_2$Mg homologous series is at least -25 kJ mol^{-1}, then the enthalpy of formation of di-n-pentylmagnesium derived from the aforementioned value for diethylmagnesium is ca -197 kJ mol^{-1}. The stabilizing isomerization of n-pentylmagnesium bromide to neopentylmagnesium bromide is -23.4 kJ mol^{-1} and should be about the same as for isomerization of dipentylmagnesium. Twice that value yields an estimate of ca -244 kJ mol^{-1}, less than 10 kJ mol^{-1} different from the experimental measurement of the enthalpy of formation of dineopentylmagnesium.

VI. ORGANOMAGNESIA AND RINGS

Magnesium can be incorporated in cyclic compounds in two ways. The first is as an exocyclic divalent substituent, i.e. part of a species of the type RMgZ where R is a carbocyclic ring and Z is some univalent substituent. As discussed for acyclic organomagnesia, Z can be halide or hydrocarbyl (either cyclic or acyclic), and so again we consider Grignard reagents and diorganomagnesium compounds.

Alternatively, magnesium may be an endocyclic component of a ring, as found in a magnesacycle (or magnesiacycle). Recall that dicoordinate, divalent oxygen forms diverse, indeed nearly ubiquitous, heterocycles ranging from simple ethers such as the reactive oxiranes and the Grignard-'friendly' solvent THF, to biologically relevant sugars and nucleosides/tides. Dicoordinate oxygen with an unstrained bond angle of ca 105° is a natural ring component in that it mimics the tetracoordinate carbon that necessarily dominates the chemistry of organic rings. From simple models of molecular structure, dicoordinate, divalent magnesium is expected to be linear and so rings with $-$Mg$-$ might appear to mimic the generally highly strained cycloalkynes and cycloallenes with their linear multiple bond components rather than the saturated and considerably less strained cycloalkanes.

Perhaps, surprisingly, there are other structural types found for magnesium-containing rings. These, too, will be discussed. So, we now ask—what is found from the experimental literature, especially that which is of thermochemical consequence and direct interest in this chapter.

A. Cycloalkylmagnesium Halides

The enthalpies of formation of the cycloalkylmagnesium bromides that have been determined by reaction calorimetry are listed in Table 3[14]. As with other functionalized cycloalkanes and the cycloalkanes themselves, there is no regularity to these values with respect to carbon number as there are for their acyclic analogs because of the influence of ring strain on the enthalpies. Unfortunately, there are no enthalpies of formation for the bromocycloalkanes with which to compare these values; there are, however, enthalpies of formation for liquid phase cycloalkanes. Figure 3 is a plot of the enthalpies of formation for the cycloalkyl-MgBr vs. those for cycloalkyl-H. There is a linear relationship with $r^2 < 0.99$. Indeed, the enthalpies of formation of the cycloalkylmagnesium bromides were calculated from the enthalpies of formation of the cycloalkanes themselves by way of the protonation reaction (equation 15).

$$\text{RMgBr(soln)} + \text{HBr(g)} \longrightarrow \text{RH(soln)} + \text{MgBr}_2\text{(soln)} \quad (15)$$

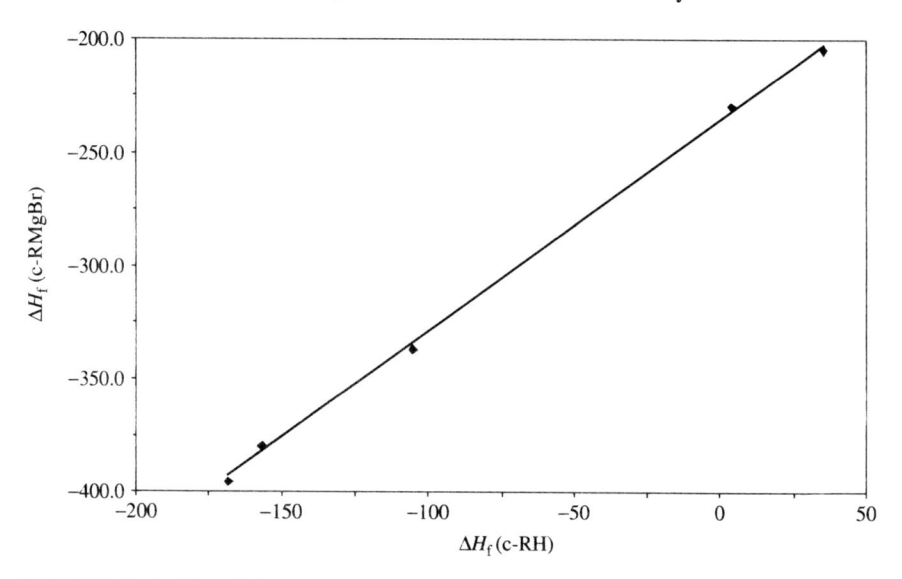

FIGURE 3. Enthalpies of formation of cycloalkylmagnesium bromides vs. those of cycloalkanes ($kJ\,mol^{-1}$)

The linear relationship thus demonstrates the near-constancy of the enthalpies of the protonation reaction of the secondary cycloalkylmagnesium bromides (-292.8 ± 6.3 $kJ\,mol^{-1}$).

Not surprisingly, the enthalpy of reaction for cyclopropylmagnesium bromide, -282.8 $kJ\,mol^{-1}$, is somewhat of an outlier, given the numerous anomalies associated with this small ring[60]. For example, cyclopropane is the most olefinic and most acidic of the cycloalkanes—which correctly suggests that cyclopropyl forms the most polar C—Mg bond and, accordingly, is the thermodynamically most stable cycloalkylmagnesium species.

Despite our earlier enunciated electronegativity and bond polarity logic, we must forego nearly all comparison with the free (uncomplexed) carbanions R^-. Unlike the rather stable cyclopropyl anion, the cyclobutyl and cyclopentyl ions[61] are unbound with regard to loss of their 'extra' electron. That is, the gas phase ionization process to form the radical from the carbanion, $R:^- \rightarrow R^{\bullet} + e^-$, is energetically favorable.

Nonetheless, it is telling that while allylMgBr is some 60 $kJ\,mol^{-1}$ more stable than its isomer cyclopropylMgBr (value from Table 3), at least in diethyl ether solution, the difference between liquid phase formation of the corresponding hydrocarbons, propene and cyclopropane, is only some 30 $kJ\,mol^{-1}$, favoring the former. This is consistent with allyl anion being more stable than cyclopropyl anion, a phenomenon generally ascribed to the significant resonance stabilization in the former. Presumably, at least some of that anionic resonance stabilization is still present—indeed, some 30 $kJ\,mol^{-1}$—in the derived organometallic, i.e. in the formally carbanionic part of the Grignard. However, despite their thermochemical proclivity, cyclopropyl Grignards seemingly do not rearrange, at least on the time scale of calorimetric investigations.

As shown by their reaction chemistry, cyclobutyl Grignards likewise do not rearrange to either their 3-butenyl or cyclopropylmethyl isomers; reaction of cyclobutylmagnesium chloride with benzoic acid results in almost quantitative yield of cyclobutane accompanied by only 1% 1-butene. In contrast, the cyclopropylmethylmagnesium chloride is *ca*

17 kJ mol^{-1} less stable than the 3-butenyl species at equilibrium (in refluxing ether) in terms of their free energies (and presumably, at least roughly in terms of enthalpies), while the corresponding bromides favor the acyclic species by *ca* 29 kJ mol^{-1}. There was no cyclobutane found with the isomeric butene and methylcyclopropane in the product mixture[62]. The average of these two differences is 23 kJ mol^{-1}. Applying an entropy correction of 16 eu at 298 K based on cyclopropane/propene, the enthalpy differences for the chloride and bromide are *ca* 1 and 13 kJ mol^{-1}, respectively. The enthalpy of formation difference between the corresponding liquid phase hydrocarbons, 1-butene and methylcyclopropane, is 22 kJ mol^{-1}.

An estimate of the enthalpies of formation of the magnesium species above is made as follows. The $\delta \Delta H_f$ introduced in an earlier section for the enthalpy of formation difference for equation 9 for the set of primary alkyl magnesium bromides in Table 3 (except allyl and benzyl) is 233.5 ± 5.1 kJ mol^{-1}. From the enthalpy of formation of liquid 1-butene (−20.5 kJ mol^{-1}), the enthalpy of formation of 3-butenylmagnesium bromide is thus −254 kJ mol^{-1}. The formal enthalpy of hydrogenation of 3-butenylmagnesium bromide to *n*-butylmagnesium bromide (equation 10) is then calculated as −124 kJ mol^{-1}, virtually identical to that for its butene/butane hydrocarbon counterpart, −126.1 kJ mol^{-1}. Since the double bond is remote from the C−Mg bond and there is no special stabilization of the Grignard, the hydrogenation enthalpies should be about the same. From the approximate difference between the equilibrium enthalpies for the cyclopropylmethyl-/3-butenylmagnesium bromide, 13 kJ mol^{-1}, the enthalpy of formation of cyclopropylmethylmagnesium bromide is −241 kJ mol^{-1}.

The enthalpies of formation of liquid methylcyclopropane and cyclobutane are quite close, −1.7 ± 0.6 and 3.7 ± 0.5 kJ mol^{-1}. How do the enthalpies of formation of their corresponding Grignards compare? The enthalpy of formation from Table 3 for cyclobutyl MgBr is −230 kJ mol^{-1}. The enthalpy difference between the cyclic C_4H_7MgBr isomers is thus *ca* 11 kJ mol^{-1}, which is not too different from that for the hydrocarbon counterparts, especially considering the uncertainties of the estimates used in this derivation.

Reaction chemistry and associated product analysis shows the free-energy difference between cyclobutylmethyl magnesium chloride and its more stable 4-pentenyl isomer to be *ca* 27 kJ mol^{-1} [63]. Using an unspecified entropic correction, the authors determined the difference in enthalpies of the isomers to be *ca* 9 kJ mol^{-1}. Again using the $\delta \Delta H_f$, above, the enthalpy of formation of 4-pentenylmagnesium bromide is −280.4 kJ mol^{-1}. The enthalpy of formation of cyclobutylmethylmagnesium bromide is accordingly -271.4 kJ mol^{-1}. The enthalpy difference between their hydrocarbon counterparts is 2.4 kJ mol^{-1}.

The experimental enthalpies of protonation[14] and the formal enthalpies of protonation, RMgBr → RH, are fairly constant for structurally similar species (R = cycloalkyl, primary alkyl) and would be expected to be constant also for the primary cycloalkylmethylmagnesium bromides. For the two examples just discussed, the formal enthalpies of protonation that are calculated using the derived enthalpies of formation for the cyclopropyl- and cyclobutylmethylmagnesium bromides are 262 and 235 kJ mol^{-1}, respectively. The mean value is thus 248 kJ mol^{-1}, which is very close to that expected for the formal protonation of other primary R groups.

Cyclopentylmethyl- and norbornylmethyl organometallic compounds are reportedly stable to ring cleavage[64]. Evidently, the ring strain associated with the small 3- and 4-membered rings is required for the reaction. However, *endo*- and *exo*-norbornenyl-5-methylmagnesium chlorides thermally interconvert with each other and with that of the ring-opened 4-allylcyclopentenylmagnesium chloride which is stabilized by allylic anion resonance[64]. These species apparently have comparable Gibbs energies. Now, how does this compare with the enthalpies of formation of corresponding hydrocarbons? The liquid phase enthalpy of formation of (*endo*)-5-methylnorborn-2-ene is 15.8 ± 1.1 kJ mol^{-1}.

The enthalpy of formation of allylcyclopentene is unknown, but accepting the value for allylcyclopentane (-66.1 ± 1.0 kJ mol^{-1}, from Reference 65) and assuming the same dehydrogenation enthalpy as for the parent carbocycles, cyclopentane/cyclopentene, of 109.5 ± 1.0 kJ mol^{-1}, we derive an enthalpy of formation of either allylcyclopentene isomer of ca 43 kJ mol^{-1}. This is a difference of 27 kJ mol^{-1} favoring the norbornene. However, correcting by ca 30 kJ mol^{-1} for the earlier enunciated resonance stabilization of an allyl anion results in the allylcyclopentene and norbornenyl Grignards having very nearly the same enthalpy of formation. This is consistent with the above putative thermoneutrality suggested for the two Grignards from their experimentally observed interconversion, although it must be acknowledged we have completely ignored entropic considerations in our analysis.

B. Magnesacycloalkanes and Their Dimers (Dimagnesacycloalkanes)

The magnesacycloalkanes, $(CH_2)_n Mg$ where $n = 4-6, 9$, have a strong tendency to dimerize in THF solution, as shown in equation 16[66, 67]. It would seem that relief of angle strain caused by the large C—Mg—C bond angle incorporated into the ring is the driving force for the dimerization of the smaller rings. However, magnesacyclodecane, which dimerizes to a 20-membered ring, should not be unduly strained[68].

$$2 \underset{Mg}{(CH_2)_n} \rightleftharpoons \underset{Mg}{\overset{Mg}{(CH_2)_n \quad (CH_2)_n}} \qquad (16)$$

Magnesacyclopentane, -heptane and -decane all completely dimerize, while magnesacyclohexane exists to a small extent as the monomer. The authors assert that the magnesacyclohexane monomer is only observable because of the highly dilute solution that shifts the equilibrium toward the monomer[67]. The enthalpy and entropy of the dimerization reaction for $n = 5$ were determined to be -48.0 ± 3 kJ mol^{-1} and 106.0 ± 10.0 J mol^{-1} deg^{-1}, respectively. The dimerization enthalpies for reactions when $n = 6, 9$ are more exothermic than 65 kJ mol^{-1}. This thermodynamic (and kinetic) proclivity to dimerization obviously is not shared by the corresponding carbocycles.

While no thermochemical data exist for any magnesacycloalkenes, there are some relevant data for their benzo analogs. It is interesting that 2,3-benzomagnesacyclohexene exists almost totally as a dimer, **3** (similar to its non-benzo analog, magnesacyclohexane), and 4,5-benzomagnesacycloheptene exists totally as monomer (shown as the dimer, **4**). While there is a mix of sp^2 and sp^3 carbon bonding to magnesium in the former, there can be only sp^3 bonding in the latter, and so this species would be expected to behave similarly to the other magnesacycles that also are sp^3-bonded and thus dimerize.

(3)   (4)

The reaction enthalpies[67] for the acetolysis reaction given in equation 17 are discussed as a measure of ring strain as compared to the strainless and monomeric diethylmagnesium.

$$[(CH_2)_{n-1}Mg]_2 + 4HOAc \longrightarrow 2[H(CH_2)_{n-1}H] + 2\,Mg(OAc)_2 \qquad (17)$$

The dimers of magnesacycloheptane and -decane are shown to be strain-free. Unlike the original authors (also, see Reference 69) we will not try to estimate strain energies for the various magnesacycles nor to interpret them. Besides the normal complications of medium-sized rings such as transannular repulsions of hydrogens, there are also electrostatic effects arising from the positively charged magnesium atoms and adjoining negative carbons. This may be compared to the enthalpy of formation of cyclic ethers, with negative oxygens and adjoining positive carbons, for which the dimerization and trimerization of dioxane to form 12-crown-4 (1,4,8,11-tetraoxacyclododecane) and 18-crown-6 (1,4,8,11,14-hexaoxacyclooctadecane) are essentially thermoneutral[70].

Having just mentioned negative oxygen and positive magnesium invites the question of rings containing both of these elements. We may expect strong dative, coordinate bonding between them. Indeed, this is found[71]: both 1-oxa-5-magnesacyclooctane and 1-oxa-6-magnesacyclodecane are found as monomers and have acetolysis enthalpies of -210.8 ± 3.8 and -212.6 ± 2.0 kJ mol^{-1}. The difference between these values and those for the rings with only magnesium is not that large—then it is to be remembered that these species also have interactions with the solvent THF which are weakened, if not replaced, upon intramolecular complexation.

C. Other Magnesacycles

Atomic magnesium has been shown to react with carbon dioxide and with ethylene to form small ring-containing products[72]. With CO_2 alone, a metastable 1:1 four-membered ring product $MgCO_2$ is found with magnesium bonded to both oxygens. With ethylene only, the monomeric (and unsolvated) 1:2 product, magnesacyclopentane is found. With the addition of both gaseous organics, a magnesalactone is formed suggesting that equation 18 is exothermic.

$$c\text{-}[(CH_2)_4Mg] + CO_2 \longrightarrow c\text{-}[CH_2CH_2COOMg] + C_2H_4 \qquad (18)$$

We note, however, that this is not the case for the corresponding carbocyclic reaction 19.

$$c\text{-}(CH_2)_5 + CO_2 \longrightarrow c\text{-}[(CH_2)_3COO] + C_2H_4 \qquad (19)$$

Indeed, it is endothermic by over 130 kJ mol^{-1}! This documents that the polar/ionic bond between magnesium and oxygen is exceptionally strong, a fact we already surmised by the vigor of the reaction of Grignard reagents with air and water.

Within the general description of ligand exchange[73], the relative stability of a variety of magnesium–olefin complexes/magnesacycles has been studied. For example, 1,4-diphenylbutadiene replaces the parent butadiene in equation 20 to form the pentacoordinated magnesium compound[74].

A variety of other olefins were studied: 1,6-diphenylhexatriene, anthracene and cyclooctatetraene also displace butadiene from its polymeric magnesium complex. Now should these olefin–magnesium species be viewed as magnesacycles? Or as contact ion pairs with olefin dianions? In any case, no enthalpies of hydrolysis are available, nor quantitation of stabilities by even equilibrium constants. We welcome this information.

VII. MAGNESIUM SANDWICH SPECIES

By the description "magnesium sandwich species" are meant compounds with the general structural formula $[(CH)_m]_n Mg_p$ (and their substituted and/or ionized derivatives) where $m \geqslant 3, n \geqslant 2$ and $p \geqslant 1$. In addition, it is tacitly assumed that the $(CH)_m$ rings are attached to the metal by at least three carbons, i.e., that they are η^k species with $k \geqslant 3$.

A. Magnesocene (Bis(cyclopentadienyl) magnesium)

The classic (if not classical) metal sandwich species bis(cyclopentadienyl)iron has the formula $[(CH)_5]_2 Fe$ ($Cp_2 Fe$) and the semisystematic name ferrocene, and so the related magnesium-containing species $[(CH)_5]_2 Mg$ ($Cp_2 Mg$) is often accompanied by the name magnesocene. We commence our discussion with this species.

There are two independent determinations of the solid-phase enthalpy of formation of magnesocene. The first measurement[75] of 66.9 ± 3.3 kJ mol^{-1} results from analysis of the hydrolytic reaction of magnesocene with aqueous H_2SO_4 (equation 21).

$$Cp_2Mg(s) + H_2SO_4(aq) \longrightarrow 2C_5H_6(lq) + Mg^{2+}(aq) + SO_4^{2-}(aq) \qquad (21)$$

From the solid-phase enthalpy of formation and the enthalpy of sublimation from the same source, the gas-phase enthalpy of formation is 135.1 ± 3.8 kJ mol^{-1}. By contrast, static bomb calorimetry[76] resulted in a value of 77 ± 3 kJ mol^{-1} for the enthalpy of formation value of the solid. The discrepancy of ca 10 kJ mol^{-1} may be ascribed to differences in the enthalpy of formation of the inorganic ancillary reference state species in the hydrolysis reaction (1N H_2SO_4 and 1:200 $MgSO_4$), the enthalpy of formation of the C_5H_6 product and 'foibles' of static as opposed to rotating bomb calorimetry. Let us accept a consensus value of ca 72 ± 5 kJ mol^{-1}. We note that the value of the sublimation enthalpy (uncontested from Reference 75) for magnesocene, 68.2 ± 1.3 kJ mol^{-1}, is very similar to that of other 'ocenes'. This is despite the considerable difference in their behavior otherwise, e.g. the ease of hydrolysis of magnesocene derivatives relative to the difficulty for those of ferrocene corroborates rather ionic ring–metal interactions in the former, and considerable covalency for the latter. Unfortunately, there are no data for the enthalpies of formation of correspondingly substituted magnesocenes and ferrocenes[77] with which one can further compare these at least formally related sandwich species.

Magnesocene does not appear to form a stable carbonyl complex. By contrast, there are seemingly stable 1:1 and 1:2 NH_3 complexes of magnesocene[78] with Mg–N bond energies of ca 25 kJ mol^{-1}, and indeed stable complexes of magnesocene with aliphatic (primary and secondary) amines have been crystallographically characterized[79]. There seems to be bonding between the N and the Mg and between the hydrogen of the ammonia or amines and the ring. There is also loss of hapticity of one of the cyclopentadienyl rings, i.e. one of the rings is coordinated by only two carbons as opposed to five for the other ring and in the uncomplexed magnesocene. Magnesocene is also complexed by a variety of other N and O (and P)—centered bases: NMR studies[80] suggest the order of increasing strength of complexation of di-isopropyl ether $\sim$ anisole $\sim$ triethylamine $<$ diethyl ether $<$ trimethylphosphine $<$ 1,4-dioxane $<$ 1,2-dimethoxyethane $<$ THF $< N, N', N'$-tetramethylethylenediamine. It is quite clear that the cyclopentadienyl rings are rather weakly attached to the magnesium core. For example, magnesocene reacts with DMSO and THF to form the totally dissociated salt $Mg(DMSO)_6^{2+}$ $(Cp^-)_2$ while THF forms the mixed $\eta^5 \eta^1$ complex dicoordinated by this ether[81].

There are few studies that address relative isomer stability of substituted magnesocenes. For example, the acid-catalyzed transalkylation (alkyl scrambling) studies of disubstituted benzenes[82] would destroy the organometallic of interest. One suggestive investigation—although it may reflect kinetic as opposed to thermodynamic effects—is the reaction of isodicyclopentadiene (4,5,6,7-tetrahydro-4,7-methano-2H-indene) with di-n-butylmagnesium to form the bis isodicyclopentadienyl complex in which both ligands are 'exo', i.e. it is the CH_2, and not the CH_2CH_2 bridge, that faces the Mg[83].

B. Neutral Magnesium Half-Sandwiches

By the description 'half-sandwich' we mean species of the type $[(CH)_m]_n Mg_p$ where n now equals 1. While there are no thermochemical data available for neutral cyclopentadienyl magnesium, CpMg, the lowest excited electronic state is known to be ca 242 kJ mol^{-1} above the ground state[84], which tells us that the Cp—Mg bond energy must be at least 242 kJ mol^{-1}. This value is within 1 kJ mol^{-1} (ca 100 cm^{-1}) for the excitation energy, and thus lower bound to the bond energies in the corresponding substituted methylcyclopentadienylmagnesium and nitrogen-containing pyrrolylmagnesium species. In that bond strength is often taken to relate to bond stretching frequencies and not bond energies, we note that the ring-Mg force constants are nearly the same as well: 112.6, 112.4 and 115.3 N m^{-1}. By contrast, that of magnesocene itself is 173 N m^{-1}, suggestive of stronger Cp—Mg bonding in the sandwich than half-sandwich compound. No spectroscopic or excitation data are available for the bisaza species, bis(pyrryl)magnesium for comparison, nor are there any enthalpy of formation or reaction measurements.

Affixing an R group to the Mg of our half-sandwiches results in the second class of species, e.g. CpMgR. It is quite clear that the reaction in equation 22

$$Cp_2Mg + R_2Mg \longrightarrow 2CpMgR \quad (22)$$

readily proceeds as written, e.g. for R = allyl[80] and neopentyl[85]. No relevant reaction or formation enthalpies are available, however, except for the solution phase difference of the η^1 and η^3-allyl (actually methallyl) derivative favoring the former by 54.4 kJ mol^{-1} [86]. Other CpMg derivatives are known but the associated thermochemistry is not available.

The species $(CH)_8Mg$, or we should say its THF 2.5-solvate[73,87,88], is readily formed from cyclooctatetraene and Mg[88]. The NMR spectrum shows eight equivalent ring atoms[87] and so suggests either the cyclooctatetraene dianion and Mg^{2+} salt[88] or a putative (and highly fluxional) solvated 'magnesacyclopentene' (or more properly magnesabicyclononatriene). However, there is no structural data for the η^8 open sandwich species and the enthalpy of formation of this simple and sensible half-sandwich, or tight ion pair, cannot even be estimated.

We note that both of the reactions (equations 23 and 24) (without additional solvating ligands explicitly being shown) proceed facilely[73,89].

$$-CH_2CH=CHCH_2-Mg- + (CH)_8 \longrightarrow CH_2=CHCH=CH_2 + (CH)_8Mg \quad (23)$$

$$C_{14}H_{10}(Mg) + (CH)_8 \longrightarrow C_{14}H_{10} + (CH)_8Mg \quad (24)$$

However, lacking thermochemical data on the other two butadiene and anthracene-related organomagnesia does not even allow us to deduce a bound for the enthalpy of formation of $(CH)_8Mg$[90].

C. Triple Decker (Club) Sandwiches

We know of no example wherein any $(Cp)_3Mg_2$ derivative or related species is known. Indeed, as documented crystallographically, magnesocene reacts with CpTl (thallocene) to form the $(Cp_2Tl)^-$ ion accompanied by a solvent-complexed CpMg$^+$ cation rather than a CpTlCpMgCp complex. (By contrast, CpLi produces a complexed 4-layer CpTlCpLi species[91].)

D. Cationic Sandwiches and Half-Sandwiches

We have already mentioned the formation of solvated $CpMg^+$ in the context of the solution phase reaction of magnesocene and thallocene. In this section are discussed aspects of the experimental gas-phase ion energetics of $CpMg^+$, Cp_2Mg^+ and related species.

The two ring-Mg bond energies in Cp_2Mg and $CpMg$ have been determined from electron impact measurements[92] for reactions 25–27.

$$Cp_2Mg \longrightarrow Cp_2Mg^+ + e \qquad (25)$$

$$Cp_2Mg \longrightarrow CpMg^+ + Cp + e \qquad (26)$$

$$Cp_2Mg \longrightarrow Mg^+ + 2Cp + e \qquad (27)$$

The energy thresholds (enthalpy of reactions) are roughly 8.0, 11.0 and 13.9 eV (772, 1061 and 1341 kJ mol^{-1}), respectively. From these values[93], the $CpMg^+$–Cp and Cp–Mg^+ bond dissociation energy values are very nearly the same, 289 and 280 kJ mol^{-1}. By contrast, there is the observation that the bond energy for $CpMg^+$–RH is meaningfully larger than that of Mg^+–RH and is consistent with the formal description of $CpMg^+$ as $Cp^- Mg^{2+}$ [94].

Although Cp_2Mg^+ does not undergo further ligation or reaction with RH, it does undergo a ligand exchange with HCN (equation 28)[95].

$$Cp_2Mg^+ + HCN \longrightarrow (CpMg-NCH)^+ + Cp \qquad (28)$$

The half-sandwich ion is also formed by direct clustering (with a third body M required) as in equation 29.

$$CpMg^+ + HCN + M \longrightarrow (CpMg-NCH)^+ + M \qquad (29)$$

With additional HCN molecules, additional clustering of $CpMg^+$ with this ligand is observed as opposed to proton transfer according to equations 30 and 31.

$$(CpMg-NCH)^+ + HCN \longrightarrow CpMg(NCH)_2^+ \qquad (30)$$

$$(CpMg-NCH)^+ + HCN \longrightarrow CpMgNC + HCNH^+ \qquad (31)$$

Analogous processes (some proceeding, some not) to the above ion–molecule reactions have been discussed for other ligands[96].

The Mg^+–C_6H_6 dissociation energy at 0 K was determined to be 134 ± 4 kJ mol^{-1} (1.39 ± 0.10 eV) using collision induced dissociation[97] and 112 kJ mol^{-1} by laser photodissociation[98]. Using the radiative association kinetics approach to ion cyclotron resonance spectrometry, the value was shown[99] to be the comparable 1.61 eV (155 kJ mol^{-1}). It was also shown that the binding of the second benzene to Mg^+, i.e. the Mg^+ (C_6H_6)–C_6H_6 bond energy, is less than 1.4 eV (135 kJ mol^{-1}).

The binding energy of $MgCl^+$ to a benzene was shown[99] to be >2.5 eV (*ca* 240 kJ mol^{-1}) and to a second benzene by less than 1.4 eV (*ca* 135 kJ mol^{-1}). Mesitylene might be expected to bind more strongly than benzene because of the electron donation from the three methyl groups; we are told[99] that the first and second bond energies to magnesium are greater than 2 eV (ca 190 kJ mol^{-1}). Based on these results and related ones for other alkaline earth species, the following rule was enunciated[99]: 'The qualitative picture is presented that MX$^+$ behaves as a metal ion center with the charge of a monovalent ion but the electronic character of a divalent alkaline earth cation.'

We close by noting that Mg$^+$ combines with multiple molecules of cyanoacetylene sequentially to form complexes of the generic formula Mg(HCCCN)$_n^+$ [95]. Of particular relevance to our discussion is the anomalous stability of the $n = 4$ species, suggested by the original authors to possibly be the Mg$^+$ complex of 1,3,5,7-tetracyanocyclooctatetraene. If so, this is the sole complex of the type $(CH)_n Mg^+$ with $n > 6$ or, more precisely, a tetracyano derivative thereof. We know of no experimental evidence for any $n = 7$ species with or without any additional ligands or any charge. Tropylium salts are reduced by Mg powder to form the radical dimer $(C_7H_7)_2$[100]. The sole species with the small $n = 3$ is $C_3H_3Mg^+$, seen experimentally as a product of the fragmentation of $(CH)_5Mg^+$ and quantum chemically suggested to be $CH_3C{\equiv}CMg^+$ and not a cyclopropenyl derivative[101].

VIII. MAGNESIUM COMPLEXES WITH CARBON MONOXIDE

Iron, as found in the porphyrin derivative hemoglobin, complexes CO to form a stable metal carbonyl. Iron also forms a variety of metal carbon monoxide derivatives such as the homoleptic Fe(CO)$_5$, Fe$_2$(CO)$_9$ and Fe$_3$(CO)$_{12}$, the anionic [Fe(CO)$_4$]$^{2-}$ and its covalent derivative Fe(CO)$_4$Br$_2$, [CpFe(CO)$_2$]$^-$ and its alkylated covalent derivatives CpFe(CO)$_2$-R with its readily distinguished π (and η^5) and σ (and η^1) iron carbon bonds. By contrast, Mg in its chlorin derivative chlorophyll, which very much resembles porphyrin, forms no such bonds with CO nor is there a rich magnesium carbonyl chemistry (if indeed, there is any at all).

This is not surprising—there are many textbook discussions of the difference between transition and main group elements. Consonant with this is the finding that (Cp*)$_2$Mg (the sandwich species alternatively called bis(pentamethylcyclopentadienyl)magnesium and decamethylmagnesocene) does not react with CO, unlike the corresponding species with Ca and some other metals and metalloids[102]. Indeed, in (Cp*)$_2$Mg, there is little room for another ligand around the central metal and Mg seems electronically satisfied.

Let us return to simple compounds and simple reactions involving Mg and CO. To begin with, consider the reaction 32.

$$Mg + CO \longrightarrow MgO + C \tag{32}$$

For solid Mg, MgO and C (and gaseous CO), this reaction is significantly exothermic: the reaction enthalpy is ca -491 kJ mol^{-1}. This is not surprising—Mg is more electropositive than C and so oxygen combines with the more metallic element. Indeed, combustion results from the aforementioned reaction (e.g. Reference 103). By contrast, this reaction is endothermic by an even more spectacular 697 kJ mol^{-1} when all of the species are in their gaseous phase[104]. The Mg—O bond in MgO, in any phase, is strong; the C—O bond in gaseous CO is stronger than any other bond in a gaseous molecule.

We know of no evidence for any discrete molecular species of the type Mg$_x$(CO)$_y$ that parallels mostly transition metal carbonyls. However, the related cations MgCO$^+$ and Mg(CO)$_2^+$ have been studied experimentally and quantum-chemically by gas phase ion chemists for which binding energies of ca 0.43 ± 0.06 [105, 106] and 0.40 ± 0.03 eV[105] (41.5 ± 5.8 and 38.6 ± 2.9 kJ mol^{-1}) were found. By contrast, Mg(CO)$^{2+}$ has a calculated binding energy[107] of ca 200 kJ mol^{-1} as befits the considerably stronger Lewis acidity of Mg^{2+} over that of Mg$^+$ [108]. Although there is an absence of neutral Mg$_x$(CO)$_y$ species, it is only for the 'ground state species' in that the excited state (involving the s$^1\pi^1$ Mg instead of s^2) of neutral Mg(CO)$_2$ has been calculated to be bound[109].

IX. REFERENCES AND NOTES

1. J. B. Pedley, R. D. Naylor and S. P. Kirby, *Thermochemical Data of Organic Compounds* (2nd edn.), Chapman & Hall, New York, 1986; J. B. Pedley, *Thermochemical Data and Structures of Organic Compounds*, Volume 1, Thermodynamics Research Center, College Station, 1994.
2. D. D. Wagman, W. H. Evans, V. B. Parker, R. H. Schumm, I. Halow, S. M. Bailey, K. L. Churney and R. L. Nuttall, 'The NBS tables of chemical thermodynamic properties: Selected values for inorganic and C_1 and C_2 organic substances in SI units', *J. Phys. Chem. Ref. Data*, **11** (Supplement No. 2) (1982).
3. P. J. Linstrom and W. G. Mallard (Eds.) *NIST Chemistry WebBook, NIST Standard Reference Database Number 69*, June 2005, National Institute of Standards and Technology, Gaithersburg MD, 20899 (http://webbook.nist.gov).
4. (a) E. C. Ashby and S. Yu, *J. Organomet. Chem.*, **29**, 339 (1971).
   (b) S. Yu and E. C. Ashby, *J. Org. Chem.*, **36**, 2123 (1971).
   (c) E. C. Ashby and J. Nackashi, *J. Organomet. Chem.*, **72**, 11 (1974).
5. (a) S. Cudzilo and W. A. Trzcinski, *Polish J. Appl. Chem.*, **45**, 25 (2001); *Chem. Abstr.*, **137**, 171930 (2001).
   (b) E-C. Koch, *Propellants, Explosives, Pyrotechnics*, **27**, 340 (2002).
6. I. C. Appleby, *Chem. Ind. (London)*, 120 (1971).
7. E. C. Ashby and D. M. Al-Fekri, *J. Organomet. Chem.*, **390**, 275 (1990).
8. J. L. Leazer Jr., R. Cvetovich, F-R. Tsay, U. Dolling, T. Vickery and D. Bachert, *J. Org. Chem.*, **68**, 3695 (2003).
9. T. Reeves, M. Sarvestani, J. J. Song, Z. Tan, L. J. Nummy, H. Lee, N. K. Yee and C. H. Senanayake, *Org. Proc. Res. Dev.* **10**, 1258 (2006).
10. (a) D. J. Ende, P. J. Clifford, D. M. DeAntonis, C. SantaMaria and S. J. Brenek, *Org. Proc. Res. Dev.*, **3**, 319 (1999).
    (b) H. D. Ferguson and Y. M. Puga, *J. Therm. Anal.*, **49**, 1625 (1997).
11. W. Tschelinzeff, *Ber. Dtsch. Chem. Ges.*, **40**, 1487 (1907); *Chem. Abstr.*, **1**, 8288 (1907).
12. (a) W. Tschelinzeff, *Compte Rendu*, **144**, 88 (1907).
    (b) V. Tshelintsev, *J. Russ. Phys.-Chem. Soc.*, **39**, 1019 (1908); *Chem. Abstr.*, **2**, 4832 (1908).
13. T. Holm, *J. Organomet. Chem.*, **56**, 87 (1973).
14. T. Holm, *J. Chem. Soc., Perkin Trans. 2*, 484 (1981).
15. H. D. B. Jenkins, *Z. Phys. Chem. (Munich)*, **194**, 165 (1996). This paper discusses the energetics of $(C_2)^{2-}$ and also that of C^{4-}. This study offers no reason to assume any particular stabilization for $(C_3)^{4-}$.
16. A. Largo, P. Redondo and C. Barrientos, *J. Am. Chem. Soc.*, **126**, 14611 (2004).
17. R. A. Seburg, E. V. Patterson, J. F. Stanton and R. J. McMahon, *J. Am. Chem. Soc.*, **119**, 5847 (1997).
18. P. Karen, Å. Kjekshus, Q. Huang and V. L. Karen, *J. Alloys Compd.*, **282**, 72 (1999); *Chem. Abstr.*, **130**, 176622 (1999).
19. H. Fjellvaag and P. Karen, *Inorg. Chem.*, **31**, 3260 (1992).
20. (a) The formal carbon-containing anions of MgC_2 and Mg_2C_3, the aforementioned C_2^{2-} and C_3^{4-}, are isoelectronic to CO and CO_2 and thereby to other 10 and 16 valence electron neutral and singly charged triatomics such as NO^+ and AlO^-, NO_2^+ and AlO_2^-. The atomization energies of these latter sets of diatomics and triatomics are simply related by E(triatomic)/E(diatomic) $\approx$ constant, cf. I. Tomaszkiewicz, G. A. Hope and P. A. G. O'Hare, *J. Chem. Thermodyn.*, **29**, 1031 (1997).
    (b) C. A. Deakyne, L. Li, W. Zheng, D. Xu and J. F. Liebman, *J. Chem. Thermodyn.*, **34**, 185 (2002).
    (c) C. A. Deakyne, L. Li, W. Zheng, D. Xu and J. F. Liebman, *Int. J. Quantum Chem.*, **95**, 713 (2003). However, we are convinced that none of this prepares us for the relevant multiply charged anionic species and/or extensive multicenter bonding.
21. K. Kobayashi and M. Arai, *Physica C*: Superconductivity and Its Applications (Amsterdam, Netherlands), **388–389**, 201 (2003).
22. See, for example:
    (a) D. E. Woon, *Astrophys. J.*, **456** (2 Pt 1), 602 (1996); *Chem. Abstr.*, **124**, 188082 (1996).

2. The thermochemistry of organomagnesium compounds

     (b) S. Itono, K. Takano, T. Hirano and U. Nagashima, *Astrophys J.*, **538** (2 Pt. 2), L163-5–(2000); *Chem. Abstr.*, **133**, 288140 (2000).
23. V. Grignard, *Compte Rendu*, **130**, 1322 (1900).
24. A. Baeyer and V. Villiger, *Ber.*, **35**, 1201 (1902).
25. R. Abegg, *Ber.*, **38**, 4112 (1905).
26. W. Schlenk and W. Schlenk Jr., *Ber.*, **62**, 920 (1929).
27. P. Jolibois, *Compte Rendu*, **155**, 353 (1912).
28. R. E. Dessy, G. S. Handler, H. Wotiz and C. Hollingsworth *J. Am. Chem. Soc.*, **79**, 3476 (1957).
29. E. C. Ashby and W. E. Becker, *J. Am. Chem. Soc.*, **85**, 118 (1963).
30. A. D. Vreugdenhill and C. Blomberg, *Recl. Trav. Chim. Pays-Bas*, **82**, 453 (1963).
31. E. C. Ashby and W. B. Smith, *J. Am. Chem. Soc.*, **86**, 4363 (1964).
32. E. C. Ashby, J. Laemmle and H. M. Neumann, *Acc. Chem. Res.*, **7**, 272 (1974).
33. M. B. Smith and W. F. Becker, *Tetrahedron*, **22**, 3027 (1966).
34. M. B. Smith and W. F. Becker, *Tetrahedron*, **23**, 4215 (1967).
35. S. G. Smith and G. Su, *J. Am. Chem. Soc.*, **86**, 2750 (1964).
36. T. Holm, *Acta Chem. Scand.*, **20**, 2821 (1966).
37. T. Holm, *Acta Chem. Scand.*, **21**, 2753 (1967).
38. M. B. Smith and W. F. Becker, *Tetrahedron Lett.*, 3843 (1965).
39. R. M. Salinger and H. S. Mosher, *J. Am. Chem. Soc.*, **86**, 1782 (1964).
40. G. E. Parris and E. C. Ashby, *J. Am. Chem. Soc.*, **93**, 1206 (1971).
41. (a) D. E. Evans and G. V. Fazakerley, *J. Chem. Soc. A*, 184 (1971).
     (b) D. E. Evans and G. V. Fazakerley, *Chem. Commun.*, 974 (1968).
42. T. Holm, *Tetrahedron Lett.*, 3329 (1966).
43. T. Holm, *Acta Chem. Scand.*, **23**, 579 (1969).
44. G. van der Wal, Thesis, Vrije Universiteit, Amsterdam, 1979.
45. T. Holm, *Acta Chem. Scand., Ser. B*, **37**, 797 (1983).
46. A. S. Carson and H. A. Skinner, *Nature*, **165**, 484 (1950).
47. A. S. Carson, P. G. Laye, J. B. Pedley and A. M. Welsby, *J. Chem. Thermodyn.*, **25**, 261 (1993).
48. V. G. Genchel, E. V. Evstigneeva and N. V. Petrova, *Zh. Fiz. Khim.*, **50**, 1909 (1976); *Chem. Abstr.*, **85**, 176675 (1976).
49. S. W. Slayden and J. F. Liebman, 'Thermochemistry of organolithium compounds', Chap. 3 in *The Chemistry of Organolithium Compounds* (Eds. Z. Rappoport and I. Marek), Wiley, Chichester, 2004.
50. J. D. Cox and G. Pilcher, *Thermochemistry of Organic and Organometallic Compounds*, Academic Press, London, 1970.
51. J. F. Liebman, J. A. Martinho Simões and S. W. Slayden, *Struct. Chem.*, **6**, 65 (1995).
52. G. D. Cooper and H. L. Finkbeiner *J. Org. Chem.*, **27**, 1493 (1962).
53. The enthalpy of formation for the solid triphenylmethane is from Reference 5. The enthalpy of fusion, 22.0 kJ mol^{-1} at 365 K, is taken from Reference 8.
54. The enthalpy of formation values for the liquid phase compounds are taken from Reference 1 when they are available. For the remaining compounds, data were obtained from Reference 3. The liquid phase enthalpy of formation of diphenyl amine of 132 kJ mol^{-1}, derived from combustion of the liquid, as reported in Reference 3 is seemingly incorrect with a too-small enthalpy of fusion, if the enthalpy of formation of the solid is accepted. The enthalpy of formation of the solid given by Reference 1, 130.2 ± 1.7 kJ mol^{-1}, is the average of two selected determinations. It is virtually the same as the average value of all known seven determinations in the 20th century, 129.8 ± 10 kJ mol^{-1}, as given in Reference 3. There are two enthalpies of sublimation: 89.1 as found in Reference 1 and 96.7 as found in Reference 3. Using the average value for both the above quantities, the enthalpy of formation of gaseous diphenyl amine is 222.6 kJ mol^{-1}. Finally, using the average of the two enthalpies of vaporization found in Reference 3, 62.6 kJ mol^{-1}, the enthalpy of formation for the liquid diphenyl amine is 160.0 kJ mol^{-1}. This value is used in the analysis.
55. J. F. Liebman, K. S. K. Crawford and S. W. Slayden, 'Thermochemistry of Organosulphur Compounds', Chap. 4 in *The Chemistry of Groups, Supplement S: The Chemistry of Sulphur-containing Functional Groups* (Eds. S. Patai and Z. Rappoport), Wiley, Chichester, 1993.

56. J. F. Liebman, H. Y. Afeefy and S. W. Slayden, 'The thermochemistry of amides', Chap. 4 in *The Amide Linkage: Structural Significance in Chemistry, Biochemistry, and Materials Science* (Eds. A. Greenberg, C. M. Breneman and J. F. Liebman), Wiley, New York, 2000.
57. A. V. Tuulmets, E. O. Parts and L. R. Ploom, *Zh. Obshch. Khim.*, **33**, 3124 (1963); *Chem. Abstr.*, **60**, 6714 (1964).
58. O. S. Akkerman, G. Schat, E. A. I. M. Evers and F. Bickelhaupt, *Recl. Trav. Chim. Pays-Bas*, **102**, 109 (1983).
59. G. Fraenkel and S. H. Yu, *J. Am. Chem. Soc.*, **96**, 6658 (1974).
60. See, for example, Z. Rappoport (Ed.), *The Chemistry of the Cyclopropyl Group*, Wiley, Chichester, 1987 and Vol. 2, Wiley, Chichester, 1995.
61. C. H. DePuy, S. Gronert, S. E. Barlow, V. M. Bierbaum and R. Damrauer, *J. Am. Chem. Soc.*, **111**, 1968 (1989).
62. D. J. Patel, C. L. Hamilton and J. D. Roberts, *J. Am. Chem. Soc.*, **87**, 5144 (1965).
63. E. A. Hill and H.-R. Ni, *J. Org. Chem.*, **36**, 4133 (1971).
64. E. A. Hill, K. Hsieh, K. Conroski, H. Sonnentag, D. Skalitzky and D. Gagas, *J. Org. Chem.*, **54**, 5286 (1989).
65. A. Labbauf and F. D. Rossini, *J. Phys. Chem.*, **65**, 476 (1961).
66. H. C. Holtkamp, G. Schat, C. Blomberg and F. Bickelhaupt, *J. Organomet. Chem.*, **240**, 1 (1982).
67. F. J. M. Freijee, G. Schat, O. S. Akkerman and F. Bickelhaupt, *J. Organomet. Chem.*, **240**, 217 (1982).
68. The bond angles in crystalline 1,6-dimagnesacyclodecane and 1,7-dimagnesacyclododecane, where each magnesium atom is coordinated with two THF molecules, are 128° and 142°, respectively. A. L. Spek, G. Schat, H. C. Holtkamp, C. Blomberg and F. Bickelhaupt, *J. Organomet. Chem.*, **131**, 331 (1977).
69. F. Bickelhaupt, *Pure Appl. Chem.*, **58**, 537 (1986).
70. The thermochemical data for 1,4-dioxane and 12-crown-4 come from our archival source; that for 18-crown 6 come from summing those from V. P. Vasil'ev, V. A. Borodin and S. B. Kopnyshev, *Russ. J. Phys. Chem. (Engl. Transl.)*, **66**, 585 (1992) (enthalpy of formation, solid) and G. Nichols, J. Orf, S. M. Reiter, J. Chickos and G. W. Gokel, *Thermochim. Acta*, **346**, 15 (2000) (enthalpy of sublimation).
71. F. J. M. Freijee, G. Van der Wal, G. Schat, O. S. Akkerman and F. Bickelhaupt, *J. Organomet. Chem.*, **240**, 229 (1982).
72. V. N. Solov'ev, E. V. Polikarpov, A. V. Nemukhin and G. B. Sergeev, *J. Phys. Chem. A*, **103**, 6721 (1999).
73. K. Mashima, Y. Matsuo, H. Fukumoto, K. Tani, H. Yasuda and A. Nakamura, *J. Organomet. Chem.*, **545–546**, 549 (1997).
74. Y. Kai, N. Kanehisa, K. Miki, N. Kasai, K. Mashima, H. Yasuda and A. Nakamura, *Chem. Lett.* 1277 (1982).
75. H. S. Hull, A. F. Reid and A. G. Turnbull, *Inorg. Chem.*, **6**, 805 (1967).
76. J. R. Chipperfield, J. C. R. Sneyd and D. E. Webster, *J. Organomet. Chem.*, **178**, 177 (1979).
77. In fact, we know of no enthalpy of formation data at all for substituted magnesocenes and only the following references for phase change enthalpies thereof: A. K. Baev, *Zh. Fiz. Khim.*, **78**, 1519 (2004); *Chem. Abstr.*, **142**, 392494 (2004) and A. I. Podkovyrov and A. K. Baev, *Izv. Vys.Ucheb. Zaved., Khim. Khim. Tekhn.*, **37**, 108 (1994); *Chem. Abstr.*, **122**, 239772 (1995). It has also been suggested that the monocyclopentadienyl amidinate species CpMg(HC(C(Me)(NBu-t)$_2$ has a comparable 'ease' of sublimation as magnesocene, H. M. El-Kaderi, A. Xia, M. Je. Heeg and C. H. Winter, *Organometallics*, **23**, 3488 (2004), suggestive of a comparable enthalpy of sublimation as well.
78. G. T. Wang and J. R. Creighton, *J. Phys. Chem. A*, **108**, 4873 (2004).
79. A. Xia, J. E. Knox, M. J. Heeg, H. Schlegel and C. H. Winter, *Organometallics*, **22**, 4060 (2003).
80. H. Lehmkuhl, K. Mehler, R. Benn, A. Rufinska and C. Krueger, *Chem. Ber.*, **119**, 1054 (1986).
81. A. Jaenschke, J. Paap and U. Behrens, *Organometallics*, **22**, 1167 (2003).
82. See, for example, the isomerization of 2- and 3-pentyl benzene and related dialkylated species, A. A. Pimerzin, T. N. Nesterova and A. M. Rozhnov, *J. Chem. Thermodyn.*, **17**, 641 (1985).

83. O. Gobley, S. Gentil, J. D. Schloss, R. D. Rogers, J. C. Gallucci, P. Meunier, B. Gautheron and L. A. Paquette, *Organometallics*, **18**, 2531 (1999).
84. E. S. J. Robles, A. M. Ellis and T. A. Miller, *J. Phys. Chem.*, **96**, 8791 (1992).
85. R. A. Andersen, R. Blom, A. Haaland, B. E. R. Schilling and H. Volden, *Acta Chem. Scand., Ser. A*, **39**, 563 (1985).
86. R. Benn, H. Lehmkuhl, K. Mehler and A. Rufinska, *J. Organomet. Chem.*, **293**, 1 (1985).
87. R. R. Muslukhov, L. M. Khalilov, A. A. Ibragimov, U. M. Dzhemilev and A. A. Panasenko, *Metalloorganicheskaya Khimiya*, **1**, 680 (1988); *Chem. Abstr.*, **110**, 95299 (1989).
88. H. Lehmkuhl, S. Kintopf and K. Mehler, *J. Organomet. Chem.*, **46**, C1 (1972).
89. T. Alonso, S. Harvey, P. C. Junk, C. L. Raston, B. Skelton and A. H. White, *Organometallics*, **6**, 2110 (1987).
90. Quantum chemical calculations on the valence isoelectronically related COT Ca, Sr and Ba complexes for which bonding energies of 89.8, 63.5 and 51.2 kcal mol^{-1} (376, 266 and 214 kJ mol^{-1}) were determined. Z. Gong, H. Shen, W. Zhu, X. Luo, K. Chen and H. Jiang, *Chem. Phys. Lett.*, **423**, 339 (2006).
91. D. R. Armstrong, R. Herbst-Irmer, A. Kuhn, D. Moncrieff, M. A. Paver, C. A. Russel, D. Stalke, A. Steiner and D. S. Wright, *Angew. Chem., Int. Ed. Engl.*, **32**, 1774 (1993).
92. G. M. Begun and R. N. Compton, *J. Chem. Phys.*, **58**, 2271 (1973).
93. We are encouraged in the use of this number because the photoelectron derived (vertical) ionization potential, nominally the same number as the first of the three electron impact derived values, is in fact very much the same value, namely 8.11 eV (783 kJ mol^{-1}), as the above value of 772 kJ mol^{-1}. See S. Evans, M. L. H. Green, B. Jewitt, A. F. Orchard and C. F. Pygall, *J. Chem. Soc., Faraday Trans. 2*, **68**, 1847 (1972).
94. R. K. Milburn, M. V. Frash, A. C. Hopkinson and D. K. Bohme, *J. Phys. Chem. A*, **104**, 3926 (2000).
95. R. K. Milburn, A. C. Hopkinson, and D. K. Bohme, *J. Am. Chem. Soc.*, **127**, 13070 (2005).
96. R. K. Milburn, V. Baranov, A. C. Hopkinson and D. K. Bohme, *J. Phys. Chem. A*, **103**, 6373 (1999).
97. A. Andersen, F. Muntean, D. Walter, C. Rue and P. B. Armentrout, *J. Phys. Chem. A*, **104**, 692 (2000).
98. K. F. Willey, C. S. Yeh, D. L. Robbins and M. A. Duncan, *J. Phys. Chem.*, **96**, 9106 (1992).
99. A. Gapeev and R. C. Dunbar, *J. Phys. Chem. A*, **104**, 4084 (2000).
100. K. Okamoto, K. Komatsu and H. Shingu, *Bull. Chem. Soc. Jpn.*, **42**, 3249 (1969).
101. J. Berthelot, A. Luna and J. Tortajada, *J. Phys. Chem. A*, **102**, 6025 (1998).
102. P. Selg, H. H. Brintzinger, M. Schultz and R. A. Andersen, *Organometallics*, **21**, 3100 (2002).
103. U. I. Gol'dshleger and E. Ya. Shafirovich, *Fizika Goreniya i Vzryva*, **36**, 67 (2000); *Chem. Abstr.*, **133**, 19400 (1985).
104. The enthalpy of formation of MgO(g) remains seriously contentious, cf. A. Lesar, S. Prebil and M. J. Hodoscek, *Chem. Inf. Comp. Sci.*, **42**, 853 (2002). However, our qualitative conclusions including our suggested adjectives remain unchanged.
105. A. Andersen, F. Muntean, D. Walter, C. Rue and P. B. Armentrout, *J. Phys. Chem. A*, **104**, 692 (2000).
106. R. C. Dunbar and S. Petrie, *J. Phys. Chem. A*, **109**, 1411 (2005).
107. S. Petrie, *J. Phys. Chem. A*, **106**, 7034 (2002).
108. When the Mg is suitably complexed, or, should we say, coordinated, such as found in 'polycrystalline MgO smoke', the binding of CO is much reduced, cf. the 11 kJ mol^{-1} suggested for this binding by G. Spoto, E. Gribov, A. Damin, G. Ricchiardi and A. Zecchina, *Surf. Sci.*, **540**, L605 (2003).
109. S. Sakai and S. Inagaki, *J. Am. Chem. Soc.*, **112**, 7961 (1990).

CHAPTER 3

NMR of organomagnesium compounds

PETER J. HEARD

School of Biological and Chemical Sciences, Birkbeck University of London, Malet Street, London, WC1E 7HX, UK
Fax: +0207 6316246; e-mail: p.heard@bbk.ac.uk

I. INTRODUCTION	131
II. CONCENTRATION OF REAGENT SOLUTIONS	132
III. ALKYL AND ARYL COMPOUNDS	133
A. Alkyl Compounds	133
B. Aryl Compounds	138
C. The Schlenk Equilibrium	140
IV. ALLYLIC AND VINYLIC COMPOUNDS	141
V. ALKOXIDE AND PEROXIDE COMPOUNDS	145
VI. CO-ORDINATION COMPLEXES	148
VII. ^{25}Mg NMR STUDIES	151
VIII. REFERENCES	153

I. INTRODUCTION

Since their discovery, the exact solution composition of Grignard compounds has been the subject of considerable debate, and given the power of NMR in elucidating chemical structure it is unsurprising that it was applied to the study of Grignard and other organomagnesium compounds from its earliest days. However, the complex nature of the solution behaviour of such compounds and the low magnetic field strengths then available often frustrated proper analysis of the data, and the first reported NMR studies were generally inconclusive[1]. Worse, the interpretation of early NMR spectra was often based on preconceived (and as it is now realized incorrect) notions as to the nature of the compounds in solution, so caution must be exercised when considering much of the pre-1970's data.

Although the advent of higher field NMR instruments and our increasing understanding of the solution behaviour of organomagnesium reagents have greatly improved the veracity of NMR studies, detailed NMR reports on such compounds remain relatively sparse. The

The chemistry of organomagnesium compounds
Edited by Z. Rappoport and I. Marek © 2008 John Wiley & Sons, Ltd

bulk of the literature that has been published was done so prior to the 1980's. The reasons for the paucity of reported NMR studies are probably three-fold: (i) organomagnesium reagents are generally highly sensitive, making the isolation of sufficiently pure samples problematic; (ii) different preparations can apparently give quite different NMR spectra; (iii) the exact solution behaviour depends on a number of factors, making it difficult to draw any general conclusions.

The weight of evidence, accumulated over many years of detailed studies using a combination of physicochemical techniques, including NMR, reveals that the solution composition of Grignard reagents is best represented by extended Schlenk equilibria (Scheme 1)[1,2]. However, because of the complexity of the solution behaviour, the vast majority of their NMR spectra are analysed on the basis of the basic equilibrium first proposed by Schlenk and Schlenk[3] (Scheme 2).

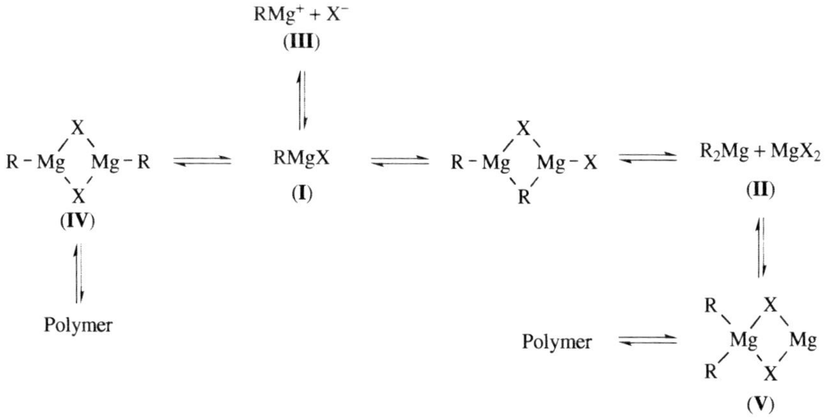

SCHEME 1. Extended Schlenk equilibria. Co-ordinated solvent molecules are omitted for clarity

SCHEME 2. Basic Schlenk equilibrium

In Et_2O, the fluorides and chlorides, RMgF and RMgCl, exist predominately as the halide-bridged symmetrical dimers $RMgX_2MgR$ (**IV**) (Scheme 1), whereas the Br and I analogues are best described as the monomeric RMgX species (**I**) at low concentrations (<0.1 mol dm^{-3}) and as linear, singly halide-bridged polymers at higher concentration.

In THF, monomeric RMgX (**I**) and R_2Mg (**II**) co-exist over a wide concentration range for the chlorides, bromides and iodides. The fluorides are present as the F-bridged dimers across the whole concentration range. Similarly, the alkoxide and aryloxide compounds, RMgOR', are present as the $R_2Mg(\mu\text{-}OR')_2Mg$ species in THF.

The predominate solution-state species in Et_2O and THF are summarized in Scheme 3.

II. CONCENTRATION OF REAGENT SOLUTIONS

The concentrations of organomagnesium reagent solutions have traditionally been determined by acid titration, but this method suffers from the disadvantage that it only provides

Solvent	X	Composition

Et$_2$O	F, Cl	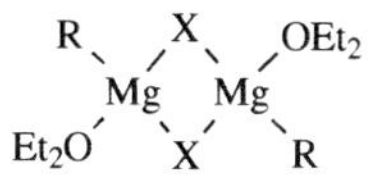 R X OEt$_2$
		  \ / \ /
		  Mg Mg
		  / \ / \
		Et$_2$O X R

	Br, I	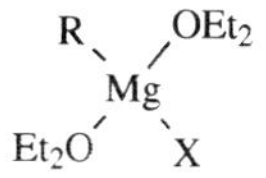 R OEt$_2$
		  \ /
		  Mg
		  / \
		Et$_2$O X
		< 0.1 mol dm^{-3}

		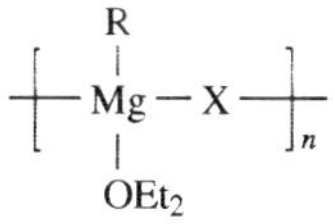 $\left[\begin{array}{c} R \\ \| \\ Mg-X \\ \| \\ OEt_2 \end{array} \right]_n$ > 0.1 mol dm^{-3}

THF	F, OR	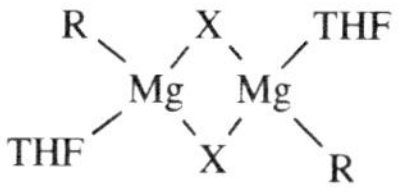 R X THF
		  \ / \ /
		  Mg Mg
		  / \ / \
		THF X R

	Cl, Br, I	2RMgX ⇌ MgX$_2$ + R$_2$Mg

SCHEME 3. Composition of Grignard compounds in solution

an estimation of the total basic content, which is also likely to include non-metal species. NMR, on the other hand, can provide a quick and convenient method for the direct determination of the concentration of organomagnesium species present, without the need to eliminate non-magnesium-containing bases. The NMR methodology relies on comparing the integrals of the reagent resonances with those of a suitable reference compound, of precisely known concentration. The accuracy of the method is limited only by the accuracy of the integration process, $ca \pm 5\%$. It is interesting to note that in the comparison of the two methods reported, the molarities determined by NMR were generally slightly higher than those estimated volumetrically[4]. Rather than NMR giving an overestimation, it seems more likely that the volumetric method slightly underestimates the concentration due to unavoidable decomposition of the reagents during the analysis.

III. ALKYL AND ARYL COMPOUNDS

A. Alkyl Compounds

At ambient temperature, most alkylmagnesium compounds display a single set of signals in their NMR spectra indicating that, if there is more than one solution-state species present, the organic groups are equivalent on the NMR time-scale. Although the positions of the NMR signals were shown to be both concentration and temperature dependent, early NMR studies failed to provide any direct evidence for the presence of more than a single species. However, with the development of higher field instruments, it has become possible to distinguish between R_2Mg and $RMgX$ species, and even between terminal and bridging groups in some associated species.

^1H and ^{13}C NMR chemical shift data for alkylmagnesium compounds are collected in Table 1[5–17], together with those for other selected organomagnesium compounds. The data presented in Table 1 should be interpreted with some caution. The chemical shifts are generally solvent, concentration and temperature dependent and different authors often quote different data for the same compounds because of the irreproducibility of the conditions. Unless otherwise noted in Table 1, the data are quoted at ambient temperature, under which conditions the compounds are generally undergoing rapid structural rearrangements: such data are therefore a weighted time-average of the various species present.

Despite the inherent limitations of the data, examination of Table 1 reveals several key features. The resonances of the α-hydrogen and α-carbon atoms are shifted significantly to lower frequency than those of the corresponding hydrocarbons[8]. The magnitude of the low frequency shift is generally greater in THF than in diethyl ether, in line with the relative strengths of the $Mg-O(THF)$ and $Mg-O(Et_2O)$ bonds (see below). Similar, though less marked trends are observed for the β-environments.

The first simultaneous observation by NMR of both $RMgX$ and R_2Mg species in solutions of Grignard reagents was made by Ashby and coworkers in 1969[18]. They showed that on cooling a diethyl ether solution of methylmagnesium bromide to ca $-100\,°C$, the ^1H NMR spectrum displayed signals due to both Me_2Mg and $MeMgBr$. The relative intensity of the signal due to Me_2Mg increased significantly on standing at low temperature, concomitant with the precipitation of $MgBr_2$, consistent with a gradual shift in the position of the Schlenk equilibrium. This first preliminary report was followed by a detailed NMR study of methylmagnesium compounds in both THF and diethyl ether solutions[5].

In diethyl ether at ca $-100\,°C$, Me_2Mg displays three signals in the ^1H NMR spectrum. The signal at -1.32 was assigned to bridging methyl groups of associated species (Scheme 1), while those at -1.74 and -1.70 were assigned to terminal methyl groups of the same associated species and to the methyl groups of solvated monomers, respectively. At the same temperature, the ^1H NMR spectrum of $MeMgBr$ displays signals at -1.55 ppm, which gradually disappears on standing, and at -1.70 ppm, assigned to $MeMgBr$ and Me_2Mg, respectively. Since the $RMgX$ species are known to be associated in diethyl ether solution, even at quite low concentrations, the observation of just a single '$RMgX$' species in the ^1H NMR spectrum indicates that the associated species are either indistinguishable from each other and/or that there is rapid exchange between them, even at low temperatures. Rapid halide exchange might certainly be expected and, although caution should be exercised when attempting to draw any inferences on the general nature of organomagnesium compounds, it is noteworthy in this context that NMR studies on aryl Grignard compounds (see below) indicate that halide exchange is significantly more rapid than aryl group exchange.

In THF solvent at $-76\,°C$, Me_2Mg displays two signals of widely different intensities at -1.83 (major) and -1.70 (minor) ppm, assignable to monomeric $Me_2Mg(thf)_n$ and to the terminal methyl groups of associated species, respectively. In the corresponding solution of the Grignard compound, $MeMgBr$, only signals due to Me_2Mg are observed, indicating that the Schlenk equilibrium is shifted much further towards Me_2Mg in THF than in diethyl ether.

On warming the solutions of both Me_2Mg and $MeMgBr$, the signals broaden and then coalesce giving a single, time-averaged signal at ambient temperature. In both cases the dynamic process involves alkyl group exchange. In Me_2Mg, exchange occurs between terminal and bridging methyl groups as a consequence of the reversible disassociation of associated species, while in the Grignard compound, the dynamic process also involves

TABLE 1. ^{1}H and ^{13}C NMR chemical shift data[a] for selected organomagnesium compounds

Compound	Solvent	δ^1H	δ^{13}C	Reference
Me$_2$Mg	Et$_2$O	−1.46 −1.74[b,c]		5
	THF	−1.76 −1.81[d]	−16.9	5 6
MeMgCl	THF	−1.72 −1.83[c]		5 5
MeMgBr	Et$_2$O THF	−1.55[c] −1.70 −1.85[c]	−16.3	5 5 6
MeMgI	Et$_2$O	−1.53	−14.5	7, 8
MeMgH	THF	−1.80 (−CH_3)		9
Et$_2$Mg	THF	ca −1.80 (−CH_2−) ca 1.15 (−CH_3)		9
EtMgBr	Et$_2$O		−2.9 (−CH$_2$−) 12.2 (−CH$_3$)	8
EtMgH	THF	−1.79 (−CH_2−) 1.15 (−CH_3)		9
n-PrMgBr	Et$_2$O		11.3 (MgCH$_2$−) 22.1 (−CH$_2$CH$_3$) 22.1 (−CH$_3$)	8
i-Pr$_2$Mg	THF	−0.75 (−CH−) 1.13 (−CH_3)	9.6 (−CH−) 26.3 (−CH$_3$)	10, 11
i-PrMgCl	THF	−0.44 (−CH−) 1.20 (−CH_3)	9.6 (−CH−) 26.3 (−CH$_3$)	11
i-PrMgBr			8.9 (−CH−) 22.9 (−CH$_3$)	8
n-BuMgBr	Et$_2$O		5.9 (MgCH$_2$−) 31.6 (MgCH$_2$$CH_2$−) 30.6 (−$CH_2CH_3$) 13.2 (−$CH_3$)	8
(t-BuCH$_2$)$_2$Mg	benzene	0.4 (−CH_3) 1.3 (−CH_2−)		12
n-BuCH(Cl)MgPr-i	THF[e]		68.8 (−CHCl−)	10
[n-BuCH(Cl)]$_2$Mg	THF[e]		69.4, 69.6 (−CHCl−)	10
EtCH(Me)CH$_2$MgBr	Et$_2$O	0.17 (Mg−CH_2−)		13
[EtCH(Me)CH$_2$]$_2$Mg	Et$_2$O THF	0.23 (Mg−CH_2−) 0.17 (Mg−CH_2−)		13
(PhCH$_2$)$_2$Mg	Et$_2$O		21.9 (−CH$_2$−) 115.9 (para−C) 123.2 (ortho−C)	8

(continued overleaf)

TABLE 1. (continued)

Compound	Solvent	δ^1H	δ^{13}C	Reference
			127.4 (*meta*−C)	
			155.2 (*ipso*−C)	
PhCH$_2$MgCl	Et$_2$O		22.2 (−CH$_2$−)	8
			116.1 (*para*−C)	
			123.4 (*ortho*−C)	
			127.3 (*meta*−C)	
			155.1 (*ipso*−C)	
PhCH$_2$CH(Br)MgPr-*i*	THF[e]		63.7 (−CHBr−)	10
			145.6 (*ipso*−C)	
[PhCH$_2$CH(Br)]$_2$Mg	THF[e]		66.7, 67.0 (−CHBr−)	10
			145.8, 146.2 (*ipso*−C)	
PhCH$_2$CH(I) MgPr-*i*	THF[e]		9.0 (−CHMe$_2$)	10
			26.5, 26.6 (−CH$_3$)	
			41.4 (−CHI−)	
			48.3 (PhCH$_2$−)	
			146.3 (*ipso*−C)	
[PhCH$_2$CH(I)]$_2$Mg	THF[e]		44.1, 46.7 (−CHI−)	10
			46.9, 47.3 (PhCH$_2$−)	
			145.8, 146.2 (*ipso*−C)	
Cp$_2$Mg	toluene	*ca* 5.75	103.8	14, 15
			107.7	
CpMgMe	THF	2.11 (CH_3)		7
		5.09 (Cp−H)		
CpMgCl	THF	*ca* 6.02		14
(1-MeC$_5$H$_4$)MgCl	THF		11.1 (−CH$_3$)	14
			101.6 (Cp−C)	
			104.1 (Cp−C)	
			116.1 (Cp−C)	
(1,3-Me$_2$C$_5$H$_3$)MgCl	THF		11.2 (−CH$_3$)	14
			101.4 (Cp−C)	
			105.2 (Cp−C)	
			114.4 (Cp−C)	
CpMgBr	THF		105.7	15
(MeSCH$_2$)$_2$Mg	THF	0.66 (−CH$_2$−)	12.9 (−CH$_2$−)	16
(PhSCH$_2$)$_2$Mg	THF	0.86 (−CH$_2$−)	4.9 (−CH$_2$−)	16
LMg^{13}CH$_3$[f]	benzene	−0.05		17
		($^1J_{CH}$ = 108Hz)		
LMgC≡CPh[f]	benzene	7.78 (*ortho*)	113.6 (Mg−CC)[g]	17
		7.15 (*meta*)	121.8 (Mg−CC)[g]	
		7.03 (*para*)	126.2 (*para*−C)	
			128.3 (*ortho*−C)	
			128.6 (*ipso*−C)	
			131.9 (*meta*−C)	

TABLE 1. (continued)

Compound	Solvent	δ^1H	δ^{13}C	Reference
LMgC≡CSi(Me)$_3$ [f]	benzene	0.40 (Si−CH_3)	1.35 (Si−CH$_3$) 120.0 (Mg−CC) [g] 146.7 (Mg−CC) [g]	17

[a] Data quoted in ppm relative to TMS. Spectra recorded at ambient temperature unless otherwise stated.
[b] Monomer (see text).
[c] Spectrum acquired at −100 °C.
[d] Spectrum acquired at −60 °C.
[e] Spectrum acquired at −78 °C.
[f] Ligand abbreviation: L = η^3-tris(3-*tert*-butylpyrazolyl)borate.
[g] Assignment may be reversed.

exchange between MeMgBr and Me$_2$Mg, i.e. the Schlenk equilibrium (Scheme 2). The rate of alkyl group exchange depends primarily on the nature of the organic group and the solvent.

Dynamic NMR studies have been used to probe organic group exchange in diorganomagnesium compounds[7]. Generally, rates of exchange are enhanced when the organic group is a good bridging ligand and reduced when bulky groups are present. The effect of the size of the alkyl group is particularly evident in PhCH$_2$CH(I)MgPr-*i*, which shows no tendency to disproportionate into (PhCH$_2$CHI)$_2$Mg and (*i*-Pr)$_2$Mg even in THF[10]. The exchange of the organic groups that takes place on mixing bis(3,3-dimethylbutyl)magnesium with bis(cyclopentadienyl)magnesium occurs with retention of configuration at the α-carbon atoms of bis(3,3-dimethylbutyl)magnesium: the rate of exchange is greater than the rate of inversion by a factor of 10^4-10^5 [7]. Such observations are consistent with a concerted exchange mechanism, with an alkyl-bridged intermediate (Figure 1). This mechanism is supported by the fact that rates of exchange are retarded in the presence of strongly co-ordinating solvents or chelating ligands, such as N,N,N',N'-tetramethylethylenediamine: binding of the donor groups inhibits association of the magnesium species.

In Grignard compounds, the halide exerts a secondary effect on the rate of alkyl group transfer. The rates are in the order RMgCl > RMgBr > RMgI, in accord with the relative ease of formation of halide bridges[19], suggesting that a mixed alkyl/halide bridged intermediate is involved in the exchange process.

Exchange between Mg-alkyl groups and the alkyl group of alkyl halides has also been long-suspected. The first direct evidence of such an exchange was demonstrated by ^{1}H NMR spectroscopy using ^{13}C-labelled methyl iodide. Han and Parkin[17] observed the appearance of a doublet at −0.05 ppm ($^1J_{CH}$ = 108 Hz) due to the Mg−^{13}CH$_3$ group of {η^3-HB(3-*t*-Bupz)$_3$}Mg^{13}CH$_3$ (3-*t*-Bupz = 3-*tert*-butylpyrazolyl) on mixing {η^3-HB(3-*t*-Bupz)$_3$}MgCH$_3$ and ^{13}CH$_3$I. Alkyl group exchange was also observed on mixing {η^3-HB(3-*t*-Bupz)$_3$}MgCH$_2$CH$_3$ with methyl iodide. Although no alkyl exchange was observed directly on mixing *n*-BuMgBr and *t*-BuBr, the NMR spectra of the reaction

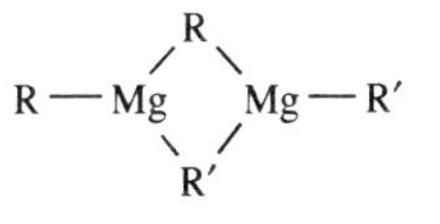

$$R-Mg\underset{R'}{\overset{R}{\diagup\hspace{-0.5em}\diagdown}}Mg-R'$$

FIGURE 1. Proposed intermediate in alkyl group exchange. Co-ordinated solvent molecules are omitted for clarity

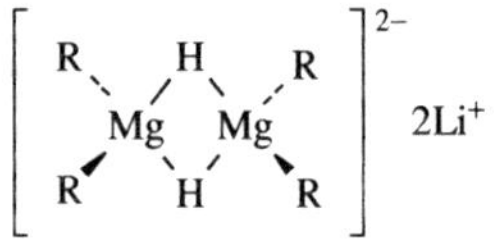

$$\left[\begin{array}{c} R \diagdown \diagup H \diagdown \diagup R \\ Mg \quad Mg \\ R \diagup \diagdown H \diagdown R \end{array}\right]^{2-} \quad 2Li^+$$

FIGURE 2. Proposed *ate* structure of the $LiMgR_2H$ compounds

mixture showed radical enhancement of signals assignable to isobutylene[20], formed from the disproportionation of *tert*-butyl radicals. The *tert*-butyl radicals are presumably formed as the result of a one-electron transfer from Mg, suggesting alkyl group exchange does indeed occur.

Reaction of R_2Mg (R = Me, Et, *i*-Pr, Bu) with activated MgH_2 yields the corresponding alkylmagnesium hydrides, RMgH, which display ^1H NMR spectra very similar to those of the dialkylmagnesium starting materials. Although no Mg—H signals are observed in the NMR, Mg—H stretching bands are observed in the IR spectra, clearly indicating the formation of the hydrides[9, 21]. Compounds of empirical formula $LiMgR_2H$ are obtained on addition of RLi to the alkylmagnesium hydrides, but their ambient-temperature ^1H NMR spectra in THF solution are indistinguishable from those of the alkylmagnesium hydrides from which they are derived and, importantly, their spectra are temperature independent. Taken together with molecular weight measurements, which suggest them to be strictly dimeric over a wide concentration range, the NMR data have been interpreted in terms of the hydrogen-bridged *ate* species shown in Figure 2[9].

The configurational stability of the metal-bonded carbon atom in organomagnesium compounds is of significant interest in terms of both our understanding of the structure and reactivity of such compounds, and more generally in gaining insights on the nature of the bonding of organic moieties to metals. If the β-carbon atom of the organic moiety is asymmetric or possesses a bulky substituent, the equivalence (chemical or magnetic) of the α-hydrogen atoms is broken and the NMR spectra become sensitive to the configuration at the α-carbon atom[13, 22-25].

Both the α-hydrogen atoms of 3,3-dimethylbutylmagnesium chloride and bis(3,3-dimethylbutyl)magnesium, for example[23], give rise to an AA' sub-spectrum of an AA'BB' spin system at *ca* $-55\,°C$ in diethyl ether solution. On warming, the signals collapse to an A_2 sub-spectrum of an A_2B_2 spin system. In these particular compounds the rate of inversion is much higher in the Grignard species, which reaches the fast exchange limit just above ambient temperature, than in the diorganomagnesium compound, which reaches the fast exchange regime only above $100\,°C$. This large difference in rate is not consistently observed, and in other organomagnesium compounds the rates of inversion are similar in the RMgX and R_2Mg species[13]. Rates of inversion also appear to vary for the same species, depending on the method of preparation, the solvent and concentration. This variation presumably reflects changes in the exact composition of the solution under investigation, and frustrates attempts to draw any general conclusions, for example on the mechanism of inversion, from the data.

B. Aryl Compounds

As with the alkyl Grignard compounds, most aryl Grignards display only a single set of NMR resonances due to the organic group at ambient temperature but, on cooling, signals assignable to RMgX and R_2Mg become apparent as aryl group exchange becomes slow on the NMR time-scale. Appreciable chemical shift differences are observed between the *ortho*-group ^1H NMR signals of the aryl Grignards and their corresponding diarylmagnesium compounds, enabling them to be distinguished unambiguously[19]. At $-65\,°C$ in THF

TABLE 2. ^{19}F chemical shift data[a] for the p-F in fluoroarylmagnesium compounds

Compound	δ^{19}F
$(C_6F_5)_2Mg$	97.8
	95.5[b]
C_6F_5MgBr	97.97[c], 98.12[c]
	95.64[c,d], 96.313[c,d]
C_6F_5MgI	95.30[c,e], 95.95[c,e]
C_6F_5MgCl	96.33[c,f], 96.97[c,f]
$(p\text{-}FC_6H_4)_2Mg$	56.42[d]
$p\text{-}FC_6H_4MgBr$	57.78
	55.77[c,d]
$p\text{-}FC_6H_4MgI$	55.62[c,f]

[a] Chemical shifts reported relative to benzotrifluoride in THF solvent at ambient temperature, unless otherwise stated. Data from Reference 26.
[b] In Et$_2$O.
[c] Shifts are concentration dependent. Data quoted at a concentration of ca 1.0 mol dm^{-3}.
[d] Shifts are concentration dependent. Data quoted at a concentration of ca 0.8 mol dm^{-3}.
[e] Shifts are concentration dependent. Data quoted at a concentration of ca 0.7 mol dm^{-3}.
[f] In toluene.

[3,5-^{2}H$_2$]-phenylmagnesium bromide, for example, displays two sets of *ortho*-hydrogen doublets of unequal intensity due to ArMgBr and Ar$_2$Mg. The chemical shift difference between the signals of the two species is ca 0.13 ppm, which compares with shift differences of <0.05 ppm between the alkyl signals in the corresponding alkylmagnesium compounds (Table 1).

The ^{19}F NMR spectra of fluoroarylmagnesium compounds have been studied in some detail[26]. The wider chemical shift range of ^{19}F, compared to that of ^{1}H, allows the various possible solution-state species to be distinguished readily. The *para*-fluorine resonances were found to be most sensitive to the chemical structure: p-F ^{19}F NMR data for selected fluoroarylmagnesium compounds are given in Table 2.

The presence of fluorine atoms on the phenyl ring reduces the rate of aryl group exchange and two sets of signals, due to ArMgX and Ar$_2$Mg species, are observed at ambient temperature in these compounds. However, these signals do undergo reversible broadening and coalesce at higher temperature giving a single, time-averaged signal. The same factors that govern the alkyl group exchange in alkylmagnesium compounds similarly govern rates of aryl group exchange, and an analogous (aryl-bridged) intermediate to that depicted in Figure 1 is presumed to be involved.

Despite the much greater range of chemical shifts and the slower rates of aryl group exchange (see below) only one set of signals assignable to 'ArMgX' species is observed in the ^{19}F NMR spectra of the Grignard compounds in diethyl ether, indicating a rapid equilibrium between associated species. The ^{19}F NMR spectrum of a mixture of C$_6$F$_5$MgBr and C$_6$F$_5$MgI at ambient temperature gives a single set of fluorine resonances at intermediate positions between those of the individual species. Since aryl group exchange is clearly slow at ambient temperature in these compounds, the observation of a single species is clearly indicative of rapid halide exchange.

In contrast to LiMgMe$_3$, which appears to dissociate to a mixture of MeLi and Me$_2$Mg in solution, NMR evidence suggests LiMgPh$_3$ remains intact[9]. The chemical shift difference between the centres of the *ortho* and *meta/para* multiplets (the latter being unresolved from each other) is 0.99 and 0.68 ppm for PhLi and Ph_2Mg, respectively, but is found to be 0.73 ppm in $LiMgPh_3$. That the chemical shift difference in $LiMgPh_3$ is not the weighted average of that found in PhLi and Ph_2Mg discounts the possibility that $LiMgPh_3$ xeists as a dynamic equilibrium mixture of PhLi and Ph_2Mg, and rather points towards a discrete species.

C. The Schlenk Equilibrium

Determining the position of the Schlenk equilibrium is clearly of key importance in understanding the reactivity of Grignard compounds and, provided the exchange rate can be slowed sufficiently, NMR can be used to determine populations of the various species present and the rates of exchange between them. Most data on the Schlenk equilibrium have been obtained in diethyl ether or THF, as Grignard reactions are generally performed in these solvents. Although the degree of aggregation of species is concentration dependent, particularly in diethyl ether, NMR spectra are usually analysed assuming only a basic Schlenk equilibrium (Scheme 2). The approximate equilibrium constants for selected Grignard compounds, determined by integration of their *static* NMR signals, are given in Table 3.

Since diorganomagnesium species are stronger Lewis acids than the corresponding Grignards, the Schlenk equilibrium generally lies further towards R_2Mg in stronger basic media. Thus diorganomagnesium species are generally more favoured in THF solution

TABLE 3. Schlenk equilibrium constants (K) for selected Grignard compounds

Compound[a]	Solvent	Temperature (°C)	K[b]	Reference
MeMgBr	THF	−85	1.1	19
MeMgBr(thf)	THF	−80	ca 0.1–0.2	6
MeMgBr(diglyme)	THF	−80	ca 0.1–0.2	6
MeMgBr(NEt$_3$)	THF	−80	ca 0.1–0.2	6
MeMgBr(tmeda)	THF	−80	ca 4	6
EtMgBr	THF	−60	ca 0.5	19
[3,5-^{2}H$_2$]C$_6$H$_3$MgBr	THF	−80	0.3	19
t-BuMgCl	THF (0.6 mol dm^{-3})	33	1.12	5
2-MeC$_6$H$_4$MgBr	THF	−50	2.3	19
(t-Bu-allyl)MgCl	THF		ca 50	27
(1,3-Me$_2$-allyl)MgCl	THF		ca 50	27
2-EtC$_6$H$_4$MgBr	THF	−40	4.0	19
2,6-Me$_2$C$_6$H$_3$MgBr	THF	−60	7.8	19
	Et$_2$O (0.3 mol dm^{-3})	−60	>400	19
2,6-Me$_2$C$_6$H$_3$MgCl	THF	−30	30.3	19
2,4,6-Me$_3$C$_6$H$_2$MgBr	THF	−40	12.3	19
2-CF$_3$C$_6$H$_4$MgBr	THF	−60	15.2	19
	Et$_2$O	−50	324	19
C$_6$F$_5$MgCl	Et$_2$O (0.7 mol dm^{-3})	22	16.0	19
C$_6$F$_5$MgBr	THF	22	2.0	19
	Et$_2$O (0.1–1.0 mol dm^{-3})	−55	4.0	19
C$_6$F$_5$MgI	Et$_2$O (0.85 mol dm^{-3})	22	7.8	19
4-FC$_6$H$_4$MgBr	Et$_2$O	−75	>1600	19
4-FC$_6$H$_4$MgI	Et$_2$O	−75	>1600	19
CpMgCl	THF (0.09 mol dm^{-3})	−75	54	14
CpMgBr	THF (0.20 mol dm^{-3})	−75	74	14

[a] Ligand abbreviations: diglyme = bis(2-methoxyethyl)ether; tmeda = tetramethylethylenediamine.
[b] The equilibrium constants given are for the formation of the Grignard species, i.e. $R_2Mg + MgX_2 \rightleftharpoons 2RMgX$.

than in Et$_2$O, irrespective of the nature of the R group or the halogen. Bulky organic groups, however, can restrict solvent co-ordination more in R$_2$Mg than in RMgX, thereby favouring RMgX.

Given the effect of the relative solvent basicity on the position of the Schlenk equilibrium, the affinity of particular solvents towards Mg is of importance, and has been investigated by NMR[28]. The chemical shifts of the organic moiety, particularly those on the α-carbon, have been shown to correlate with the co-ordinating ability of the solvent. Thus in more strongly basic solvents the NMR signals are generally shifted to lower frequency, consistent with a greater degree of charge separation between the Mg and α-carbon as a result of stronger Mg–solvent interactions. Based on ^{1}H NMR studies of EtMgBr and Et$_2$Mg, the preference for solvent co-ordination is in the order DME > THF > Et$_2$O > n-Bu$_2$O > Et$_3$N > i-Pr$_2$O. This trend is governed by both steric and electronic factors.

The number of co-ordinated solvent molecules is also of considerable interest. The magnesium atom has been shown typically to display co-ordination numbers of four or five in the solid state (see below) in organomagnesium compounds, depending on the nature of the magnesium moiety (i.e. organic group and/or halide atoms) and the donor groups. Although the situation is less clear in solution, the magnesium atom is probably co-ordinated by at least two or three solvent molecules. In many instances, the co-ordinated solvent molecules will be in rapid exchange with those in the bulk solution.

IV. ALLYLIC AND VINYLIC COMPOUNDS

The question of the solution structure of allylmagnesium compounds is an intriguing one and such compounds have been studied in more detail by NMR than any other organomagnesium species. ^{1}H and ^{13}C NMR data for selected allylmagnesium compounds are given in Tables 4 and 5, respectively.

Asymmetrically substituted allyl magnesium compounds often react to yield products derived from both the parent allyl halide and the corresponding allylic isomer, in varying relative yields[33,34]. Depending on the arrangement of the substituents, *cis* and *trans*

TABLE 4. ^{1}H NMR data[a] for allylmagnesium Grignards in Et$_2$O at ambient temperature

X	R^1	R^2	R^3	R^4	R^5	R^1	R^2	R^3	R^4	R^5	Reference
Cl	H	H	H	H	H	2.50	2.50	6.38	2.50	2.50	27
Cl	H	H	H	H	Me	ca 0.8	ca 0.8	ca 5.9	ca 4.5	ca 1.5	27
Cl	H	H	H	H	Et	0.79	0.79	5.94	4.56	2.09 ($-CH_2-$) 0.97 ($-CH_3$)	27
Cl	H	H	H	H	i-Pr	0.78	0.78	5.91	4.57	2.26 ($-CH-$) 0.96 ($-CH_3$)	27
Cl	H	H	H	H	t-Bu	0.71	0.71	5.95	4.73	0.98	27
Cl	H	H	H	Me	Me	0.57	0.57	5.58	1.55	1.55	27
Cl	Me	H	H	H	Me	1.76	2.82	6.20	2.82	1.76	27
Br	H	H	H	H	H	2.69	2.69	6.54	2.69	2.69	29
Br	H	H	Me	H	H	2.41	2.41	1.76	2.41	2.41	29

[a] Chemical shifts reported in ppm relative to TMS.

TABLE 5. ^{13}C NMR chemical shift data[a] for allylmagnesium compounds

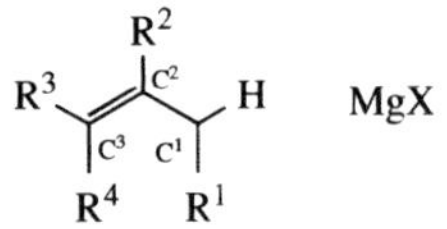

X	R^1	R^2	R^3	R^4	C^1	C^2	C^3	R^1	R^2	R^3	R^4	Reference
Cl	H	H	H	H	57.8	149.8	57.8					29
Cl	H	Me	H	H	26.7[b]	156.8[b]	57.8[b]					30
Cl	Me	H	Me	H	62.8	147.6	62.8	18.15		18.15		29
Br	H	H	H	H	57.5	149.5	57.5					31
Br	H	H	H	H	58.0	149.5	56.5					31
					57.3	137.3	57.3					32
Br	H	H	Me	H	[c]	141	102			15		31
					17.6[b]	141.5[b]	97.2[b]			12.5[b]		30
Br	H	Me	H	H	59.5	156.9	59.5		27.2			29
Br	H	H	Me	Me	ca 22	150	ca 92			[c]	[c]	31
					25.4[b,d]	131.8[b,d]	108.4[b,d]			16.7[b,d,e]	16.4[b,d,e]	30
allyl	H	H	H	H	57.2	149.4	57.2					29
					57.9[d]	148.7[d]	57.9[d]					29

[a] Chemical shifts reported in ppm relative to TMS. Spectra acquired at ambient temperature in THF except [b] and [d].
[b] Spectra recorded at $-78\,°C$.
[c] Not observed because of dynamic line broadening.
[d] In Et$_2$O.
[e] Assignment of R^3 and R^4 is arbitrary.

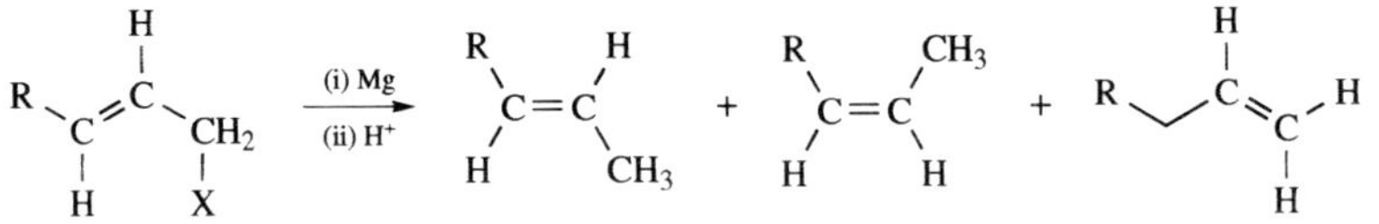

SCHEME 4. Possible stereochemistries of products from reactions proceeding via substituted allyl Grignard intermediates

isomers may also be observed (Scheme 4). Thus, besides the question of the position of the Schlenk equilibrium and the degree of aggregation, it is necessary to account for the observed patterns in reactivity.

The ^{1}H NMR spectra of allylmagnesium compounds display simple AX$_4$ patterns at temperatures as low as -80 to $-120\,°C$[35,36]. Such simple spectra can be interpreted in terms of either rapidly interconverting σ-bonded allylmagnesium species or an essentially ionic species, with rapid rotation about the C—C partial double bonds.

It is possible to distinguish indirectly between rapidly interconverting σ-bonded allylic and ionic species using the isotopic perturbation technique[31]. If allylmagnesium compounds exist as a pair of allylic isomers then, in the corresponding [1-^{2}H]allylmagnesium species (Figure 3), the concentration of **a** will be greater than that of **b**, irrespective of any exchange, because of the lower zero-point energy. Thus in the fast exchange regime, the average shift of C(1) will be moved towards that of the static shift of C(1) in isomer **a**, i.e. the exchange-averaged signal of C(1) will be shifted to higher frequency in the deuterium labelled analogue. Although the signal due to C(1) would also be expected to be shifted if the compounds existed as ionic species, any shift would be quite small. The

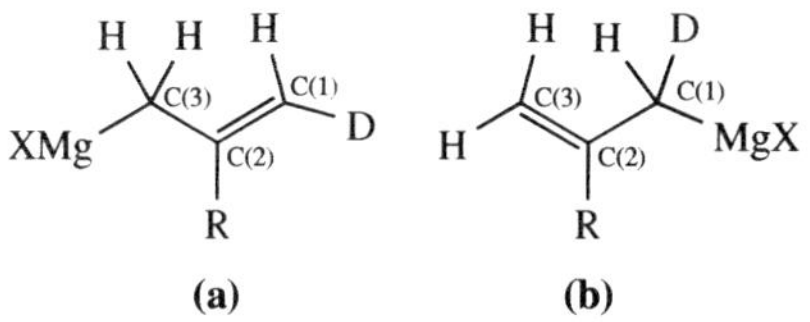

FIGURE 3. The two allylic isomers of [1-^{2}H]allylmagnesium compounds. **a** is favoured because of the lower zero-point energy

TABLE 6. Eyring activation parameters[a] for allylic exchange in allyl Grignards

Compound	$\Delta H^{\ddagger}$ (kcal mol^{-1})	$\Delta S^{\ddagger}$ (cal mol^{-1} K^{-1})	$\Delta G^{\ddagger}$ (298K) (kcal mol^{-1})
C$_3$H$_5$MgCl	5.5 (0.3)	−6.0 (1.3)	7.29
C$_3$H$_5$MgBr	5.88 (0.11)	−7.3 (0.5)	8.06
2-MeC$_3$H$_4$MgBr	9.7 (0.6)	7.0 (2.3)	7.61
1,3-Me$_2$C$_3$H$_3$MgCl[b]	6.6 (0.4)	−4.7 (1.3)	8.00

[a] Data from Reference 29. Obtained by ^{13}C NMR in Et$_2$O except for the last entry. Standard deviations are given in parentheses.
[b] In THF solvent.

actual change in the shift of C(1) in allylmagnesium bromide observed on deuteriation is ca 1.4–1.9 ppm in both diethyl ether and tetrahydrofuran solution, consistent with a dynamic equilibrium between the allylic isomers.

The advent of higher field NMR instruments, together with the larger chemical shift range of ^{13}C, has subsequently enabled the direct observation of σ-bonded allylic isomers[29, 32], allowing an estimation of the activation barrier to allylic exchange to be obtained[29] in allyl Grignards: activation parameters are given in Table 6. In contrast to the Grignard compounds, the ^{13}C spectra of bis(allyl)magnesium remain essentially temperature independent down to at least −95 °C, indicating that either (i) the barrier is significantly lower or (ii) a more ionic-type structure is preferred. Conversely, the barrier to the allyl rearrangement is significantly higher in (cyclopentadienyl)(2-methylallyl)magnesium and, at moderately low temperatures, their NMR spectra clearly show the allyl ligand to be σ-bonded[37]. ^{25}Mg NMR data (see below) are also in accord with σ-bonded species.

The dynamic equilibrium between allylic isomers accounts for the observation of *cis* and *trans* product isomers in reactions of substituted allylmagnesium compounds. Rapid rotation about the C−C single bonds in each allylic isomer gives rise to both *cis* and *trans* magnesium species (Scheme 5) that can go on to yield *cis* and/or *trans* products. This rapid rotation is clearly evidenced in the ^{1}H NMR spectra by the equivalence of the methylene hydrogens, i.e. both isomer interconversion and rapid C−C is necessary to account for the observation of an AX$_4$ spin pattern in the fast exchange regime. The exact constitution of the product mixture resulting from reaction of allylmagnesium compounds thus depends on the equilibrium populations of the various species and the relative kinetics for the reaction of each isomer. It is therefore difficult to make any generalizations regarding the likely composition of products. However, it is noteworthy that, in the absence of steric hindrance, allylmagnesium reagents have been shown to favour the *cis* configuration, while in the presence of bulky substituents this gives way to a *trans* preference[27].

The $^3J_{HH}$ spin coupling constants (Table 7) also provide valuable insight into the structure of allylmagnesium compounds[27]. Assuming the Karplus relationship holds, the magnitude of the coupling between the unique hydrogen, H^2, and the two equivalent

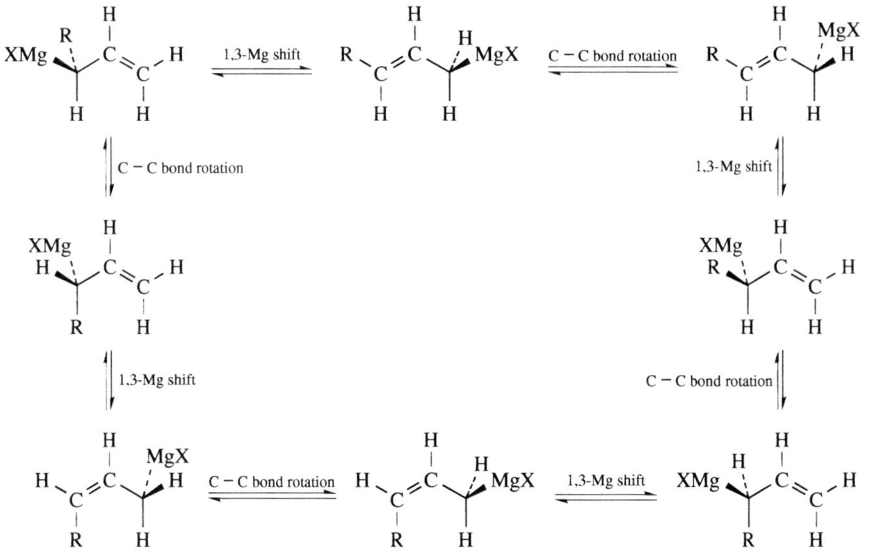

SCHEME 5. Dynamic equilibria in allylmagnesium compounds

TABLE 7. $^nJ_{HH}$ coupling constants[a] for allylmagnesium chlorides

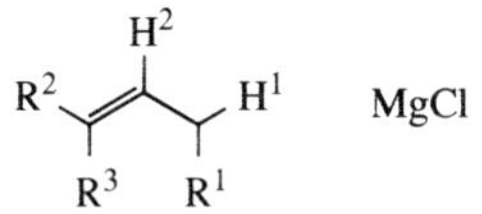

$$R^2\underset{R^3}{\overset{H^2}{\diagdown}}\!\!=\!\!\underset{R^1}{\diagup}\!\!H^1 \quad MgCl$$

R^1	R^2	R^3	$J_{H(1)H(2)}$[b]	$J_{H(1)R(2)}$[c]	$J_{H(2)R(2)}$	$J_{H(2)R(3)}$	$J_{R(2)R(3)}$
H	H	Me	9.6	1.2	11.7	1.5	6.4
H	H	Et	9.5	1.5	12.4	1.25	6.5
H	H	i-Pr	9.4	1.3	13.6	1.1	7.3(CH)
H	H	t-Bu	9.1	1.3	15.1		
Me	H	Me	11.0		11.0	ca 0.8	7.0

[a] Data from Reference 29. Recorded at 32 °C in Et$_2$O solvent.
[b] Where R^1 = H, $J_{R(1)H(2)} = J_{H(1)H(2)}$.
[c] Where R^1 = H, $J_{R(1)R(2)} = J_{H(1)R(2)}$.

allylic ($-CH_2-$) hydrogens suggests the presence of two rapidly exchanging, energetically equivalent conformers with dihedral angles of approximately 30° (Figure 4). Thus the $^3J_{HH}$ couplings are consistent with the magnesium being σ-bonded to an sp^3-hybridized carbon atom.

Although the interconversion of the allylic isomers remains rapid at temperatures as low as -80 °C, the Schlenk equilibrium is slowed sufficiently to enable signals due to both the allyl Grignard and bis(allyl)magnesium compounds to be observed at such temperatures in their ^{1}H NMR spectra. The NMR parameters for the bis(allyl)magnesium compounds are not very different from those of the Grignards, suggesting that they possess essentially

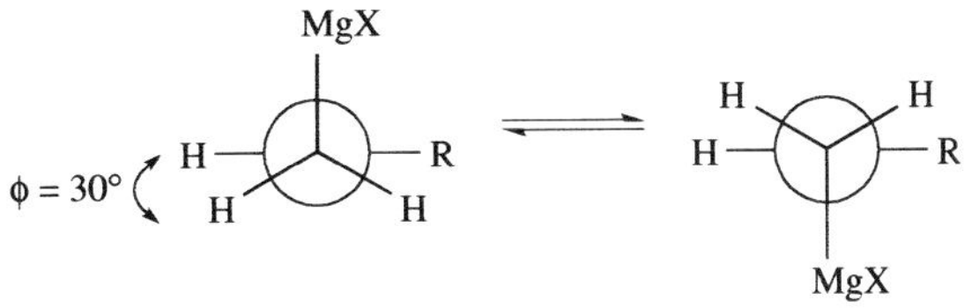

FIGURE 4. The magnitudes of $^nJ_{HH}$ coupling constants found in allylmagnesium compounds are consistent with the presence of two rapidly interconverting, equivalent conformers with dihedral angles of *ca* 30°

the same structural features. Substantial overlap of the signals due to the RMgX and R_2Mg species frustrates the evaluation of a good quantitative estimate of the Schlenk equilibrium constant, but a value of *ca* 50 (Table 3) has been estimated for both *tert*-butylallylmagnesium chloride and 1,3-dimethylallylmagnesium chloride[27].

Few NMR studies have been carried out on vinylmagnesium compounds. However, NMR has been used to probe the stereospecificity of vinyl Grignard formation, as this has important consequences on product stereochemistry[38,39]. Reaction of *cis-β*-bromostyrene with Mg in THF, followed by the addition of D_2O gave a 10:1 *cis:trans* ratio of *β*-[1-^{2}H]-styrene, indicating the reaction proceeds, essentially, with overall retention of configuration[38]. A similar result is obtained when the Grignard reagent is formed by the magnesium–halogen exchange reaction of *cis-β*-bromostyrene with butylmagnesium bromide[39]. Retention of stereochemistry is also generally observed when starting from *trans-β*-bromostyrene, although to a significantly lesser extent. The degree of retention is solvent dependent: retention is greater in THF than in Et_2O. The solvent effect has been shown clearly to exert itself in the formation of the Grignard, rather than in the subsequent reaction with D_2O.

V. ALKOXIDE AND PEROXIDE COMPOUNDS

The solution compositions of a number of methylmagnesium alkoxides have been studied in some detail by Ashby and coworkers using a variety of physicochemical methods, including ^1H NMR spectroscopy[40]. The NMR spectra displayed broad signals due to the Mg–CH_3 groups in the region −1 to −2 ppm (Table 8), which are strongly solvent, concentration and time dependent.

The solvent and concentration dependence of the spectra arises from changes in the degree of molecular association. Time-dependent NMR studies have shown that, in more strongly co-ordinating solvents such as THF, μ^2-alkoxide bridged dimers are favoured, but in weakly co-ordinating solvents such as diethyl ether, linear oligomers or μ^3-alkoxide bridged cubane-like tetramers (Figure 5) gradually form on standing. The nature of the alkoxide also affects the degree of association: bulky groups hinder association.

Variable temperature NMR studies on 1:1 mixtures of the alkoxides, RMgOR′, and Me_2Mg reveals methyl group exchange between the magnesium atoms. Exchange is rapid in the dimeric species, but slow in the tetrameric species, suggesting that there is no convenient mechanism in the latter case. Mixed alkyl/alkoxide bridged dimeric species are thus assumed to be intermediate in the exchange process: in the tetrameric species formation of such dimers first requires dissociation of the *cube*. Although there is no evidence of alkoxide group exchange in these compounds, the ^{13}C NMR spectrum of [*n*-BuMg(μ-OAr)]$_2${Ar = 2,6-(*t*-Bu)$_2$C$_6$H$_3$} in THF solution displays two distinct

TABLE 8. Selected ^{1}H and ^{13}C NMR data[a] for magnesium alkoxide and aryloxide compounds

Compound[b]	Structure[c]	Solvent	δ (^{1}H) (ppm)	δ (^{13}C) (ppm)	Reference
MeMgOBu-t	cubane	C$_6$D$_6$	1.47 [C(CH_3)$_3$] −0.66[MgCH_3]		40
		Et$_2$O	1.55 [C(CH_3)$_3$] −1.11[MgCH_3]		
		THF	1.55 [C(CH_3)$_3$] −1.12[MgCH_3]		
	oligomer	Et$_2$O	1.2 [C(CH_3)$_3$] −1.20[MgCH_3] −1.45[MgCH_3]		
	dimer	THF	1.20 [C(CH_3)$_3$] −1.60[MgCH_3]		
MeMgOPr-i	cubane	C$_6$D$_6$	1.30 [CH(CH_3)$_2$] −0.76[MgCH_3]		40
	cubane/oligomer	Et$_2$O	1.44 [CH(CH_3)$_2$] −1.30[MgCH_3]		
	dimer	THF	1.12 [CH(CH_3)$_2$] −1.66[MgCH_3]		
MeMgOPr-n	oligomer[d]	C$_6$D$_6$	−0.82[MgCH_3]		40
	cubane/oligomer	Et$_2$O	−1.33[MgCH_3]		
	dimer	THF	−1.70[MgCH_3]		
BuMgOAr^1	dimer	C$_6$D$_5$CD$_3$	−0.10[MgCH_2−] 1.37 [OArC(CH_3)$_3$]	7.06 [MgCH_2−] 32.80 [OArC(CH$_3$)$_3$] 35.14 [OАrC(CH$_3$)$_3$]	41
		THF	−1.70[MgCH_2−] 1.53 [OArC(CH_3)$_3$]	8.67 [MgCH_2−] 31.12 [OArC(CH$_3$)$_3$] 35.64 [OАrC(CH$_3$)$_3$]	
Mg(OAr^1)$_2$(thf)$_2$	monomer	C$_6$D$_5$CD$_3$	1.25 [−CH_2(thf)] 1.55 [OArC(CH_3)$_3$] 3.64 [−OCH_2(thf)]	24.86 [−CH$_2$(thf)] 31.88 [OArC(CH$_3$)$_3$] 35.53 [OArC(CH$_3$)$_3$] 70.75 [−OCH$_2$(thf)]	41
		THF	1.37 [OArC(CH_3)$_3$] 1.77 [−CH_2(thf)] 3.62 [−OCH_2(thf)]	26.59 [−CH$_2$(thf)] 32.24 [OArC(CH$_3$)$_3$] 36.01 [OАrC(CH$_3$)$_3$] 68.44 [−OCH$_2$(thf)]	
Mg(OAr^1)$_2$	dimer	C$_6$D$_5$CD$_3$	1.20 [OArC(CH_3)$_3$] 1.58 [OArC(CH_3)$_3$]	31.92 [OArC(CH$_3$)$_3$] 34.13 [OArC(CH$_3$)$_3$] 35.07 [OАrC(CH$_3$)$_3$] 36.06 [OАrC(CH$_3$)$_3$]	41
		THF	1.37 [OArC(CH_3)$_3$] 1.40 [OArC(CH_3)$_3$] 1.41 [OArC(CH_3)$_3$]	32.17 [OArC(CH$_3$)$_3$] 32.22 [OArC(CH$_3$)$_3$] 32.37 [OArC(CH$_3$)$_3$] 35.94 [OАrC(CH$_3$)$_3$] 36.00 [OАrC(CH$_3$)$_3$]	
HMgOAr^2	dimer	THF	1.19 [i-Pr−CH_3] 6.79 [*meta-H*] 6.92 [*para-H*]		42
HMgOAr^3	dimer	THF	1.39 [t-Bu−CH_3] 2.12 [Ph−CH_3]		42
HMgOCH$_2$CH$_2$Ph	dimer	THF	1.26 [−CH_2Ph] 6.92−7.36 [Ph−H]		42

TABLE 8. (continued)

Compound[b]	Structure[c]	Solvent	δ (^{1}H) (ppm)	δ (^{13}C) (ppm)	Reference
HMgOCPh$_3$	dimer	THF	7.19–7.36 [Ph–H]		42
L^{1}MgOEt	monomer	C$_6$D$_6$	1.72 [–OCH$_2$CH_3] 4.93 [–OCH_2CH$_3$]	35.7 [–OCH$_2$CH$_3$] 64.2 [–OCH$_2$CH$_3$]	17
L^{1}MgOPr-i	monomer	C$_6$D$_6$	1.65 [–OCH(CH_3)$_2$] 4.84 [–OCH(CH$_3$)$_2$]	30.2 [–OCH(CH$_3$)$_2$] 64.2 [–OCH(CH$_3$)$_2$]	17
L^{1}MgOBu-t	monomer	C$_6$D$_6$	1.75 [–OC(CH_3)$_3$]	35.7 [–OC(CH$_3$)$_3$] 68.1 [–OC(CH$_3$)$_3$]	17
L^{1}MgOPh	monomer	C$_6$D$_6$	6.96 [para-H] 7.28 [ortho-H] 7.52 [meta-H]	114.5 [para-C] 120.2 [meta-C] 129.0 [ortho-C] 163.3 [ipso-C]	17

[a] Data acquired at ambient temperature.
[b] Ligand abbreviations: OAr^1 = 2,6-di-*tert*-butylphenoxy; OAr^2 = 2,6-diisopropylbenzyl; OAr^3 = 2,6-di-*tert*-butyl-4-methylbenzyl; L^1 = η^3-tris(3-*tert*-butylpyrazolyl)borate.
[c] See text.
[d] Unsolvated.

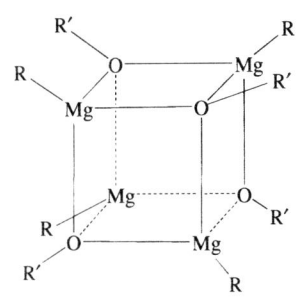

FIGURE 5. Cubane structure of the tetrameric compounds RMgOR'. The tetramers are thermodynamically favoured when the alkyl groups are small and the solvent is only weakly co-ordinating

(CH$_3$)$_3$C–resonances of widely different intensities at 31.1 (major) and 32.24 (minor) ppm, respectively[41]. The latter signal coincides with the (CH$_3$)$_3$C–resonance of Mg(μ-OAr)$_2$, indicating at least some disproportionation of [n-BuMg(μ-OAr)]$_2$ and implying slow alkoxide group exchange on the NMR chemical shift time-scale.

Dimeric alkoxy- and aryloxy-magnesium hydrides, HMgOR, are prepared by the reaction of activated MgH$_2$ with the appropriate Mg(OR)$_2$ compounds in THF[42]. Their NMR spectra display a single set of signals due to the alkoxide/aryloxide group (Table 8) but, as with the alkylmagnesium hydrides, no Mg–H resonance is observed. An Mg–H stretching band is, however, observed in the IR spectra. The complexes are dimeric and presumed to possess bridging hydrides, rather than bridging alkoxide groups, on steric grounds.

The oxidation of Grignard reagents with dioxygen, yielding alcohols, has long been known. The reaction is presumed to proceed via an alkylperoxide intermediate, ROOMgX. The first magnesium alkylperoxides reported, {η^3-HB(3-t-Bupz)$_3$}MgOOR (R=Me, Et, i-Pr, t-Bu), were prepared by the insertion of dioxygen into the Mg–C bond in {η^3-HB(3-t-Bupz)$_3$}MgR and characterized by ^{1}H and ^{17}O NMR spectra[17]. The ^{17}O NMR

TABLE 9. ^{17}O NMR data[a] for magnesium peroxide complexes

Compound[b]	δ(MgOOR)	δ(MgOOR)
LMgOOMe	427	102
LMgOOEt	407	130
LMgOOPr-i	373	159
LMgOOBu-t	323	183

[a] Data from Reference 17. Chemical shifts reported relative to H$_2$O.
[b] L = η^3-tris(3-*tert*-butylpyrazolyl)borate.

spectra display two well-separated signals in the regions 102—183 and 323—427 ppm (Table 9), assigned to the β- and α-oxygen atoms, respectively. Interestingly, the oxygen chemical shifts vary almost linearly with increasing steric bulk of the alkyl group. More recently, Bailey and coworkers reported the structure of the benzylperoxide complex HC{C(CH$_3$)NAr}$_2$MgOOCH$_2$Ph {Ar=2,6-(i-Pr)$_2$C$_6$H$_3$}, in which the peroxybenzyl moiety binds in an unusual μ-η^2 : η^1–O,O fashion in the solid state[43]. No ^{17}O NMR data were reported for the complex, so it is not possible to compare data with those for {η^3-HB(3-t-Bupz)$_3$}MgOOR complexes, in which the bonding mode of the alkylperoxide was not established.

VI. CO-ORDINATION COMPLEXES

The co-ordination complexes of organomagnesium reagents have been studied quite extensively, primarily with the aim of obtaining sufficiently stable adducts to permit their structural characterization. X-ray studies have revealed a range of co-ordination numbers from two to eleven for the magnesium atom in the solid state: unsurprisingly, the most commonly occurring co-ordination number is four[44]. The higher co-ordination numbers are found in η^5-cyclopentadienyl complexes, in which each carbon is considered to occupy a separate co-ordination site.

It is difficult to ascertain if the same co-ordination numbers are retained in solution. Solution NMR studies indicate small co-ordination shifts for the ligand resonances, suggesting relatively weak, and hence labile, metal–ligand bonds[6,45]. The lability of the ligands, particularly monodentate ones, is further illustrated by the fact that adducts of different ligands often give identical NMR spectra (Table 10): the ligands are presumably substituted by solvent molecules, yielding identical solution species. The degree of solvation is clearly a matter of conjecture in most instances, but it is not unreasonable to expect co-ordination numbers of four or five to predominate in solution as they do in the solid state.

Despite the greater Lewis acidity of R$_2$Mg species, the co-ordination induced ligand shifts are smaller in the diorganomagnesium compounds than in the analogous Grignards, indicating that they have a lower affinity for complex formation. A particularly interesting exception to the low propensity of R$_2$Mg compounds for complex formation is that when sparteine, which is used to treat arrythmic heart disorders, is the ligand in question. Sparteine forms a stable 1:1 adduct with bis(2-methylbutyl)magnesium in which the ligand has been shown by ^{1}H NMR to adopt a *cisoid* configuration (Figure 6), even at elevated temperatures[48].

3. NMR of organomagnesium compounds

TABLE 10. NMR data[a] for selected organomagnesium adducts

Complex	Solvent	δ^1H Mg–R	δ^1H ligand	δ^{13}C Mg–R	δ^{13}C ligand	Reference
MeMgBr(OEt$_2$)	THF	−1.71	1.10 3.38	−16.4	15.7 66.3	6
MeMgBr(thf)	THF	−1.70		−16.3		6
MeMgBr(diglyme)	THF	−1.73	3.33 3.52 3.61	−16.0	59.1 71.0 72.5	6
MeMgBr(NEt$_3$)	THF	−1.72	0.96 2.43	−16.5	12.6 47.3	6
MeMgBr(tmeda)	THF	−1.67	2.32 2.48	−15.5	47.0 57.7	6
MeMgBr(pmdta)	THF	−1.68	2.31 2.43 2.54 2.70	−13.2	44.2 45.8 57.4 57.6	6
Me$_2$Mg(pmdta)	THF	−1.80	2.25 2.46 2.57	−14.1	42.6 46.7 57.6	6
EtMgBr(teed)	benzene		0.76 ca 2.05 2.12			45
EtMgNPh$_2$(thf)$_2$	benzene	0.51 1.82	1.11 3.34 6.76– 7.19	1.39 14.32	25.59 69.39 117.7 121.68 130.6 157.02	46
i-PrMgNPh$_2$(thf)$_2$	benzene	0.26 1.81	1.17 3.38 6.77– 7.27	9.63 26.39	26.39 69.53 116.99 117.66 121.27 121.61 129.99 130.10 156.99 157.45	46
s-Bu$_2$Mg(tmeda)	benzene	0.05 1.44 1.75	1.53 2.36			47
p-FC$_6$H$_4$MgBr(teed)	benzene		0.82 ca 2.26 2.33			45
(PhCH$_2$)$_2$Mg(thf)$_2$	benzene	1.9 6.83 7.18 7.25	1.28 3.34	22.8 115.4 123.2 127.7 157.2	25.8 67.7	43

(*continued overleaf*)

TABLE 10. (continued)

Complex	Solvent	δ^1H Mg–R	δ^1H ligand	δ^{13}C Mg–R	δ^{13}C ligand	Reference
(PhCH$_2$)$_2$Mg(tmeda)	benzene	1.33 6.30 6.67 6.75	2.15 2.33	21.0 113.6 121.6 125.9 155.2	43.9 56.0	43
Mg(t-BuCH$_2$)$_2$(OEt$_2$)$_2$	benzene	0.30 1.45	0.97 3.46			12
Mg(t-BuCH$_2$)$_2$(tmeda)	benzene	0.07 1.59	1.74 1.97			12
Mg(PhCMe$_2$CH$_2$)$_2$(tmeda)	benzene	−0.08 1.08 7.3 6.8	1.06 1.32			12
CpMgBr(teed)	benzene		0.81 ca 2.20 2.26			45
Mg(OAr)$_2$(thf)$_2$	toluene	1.55 6.73 7.33	1.25 3.64	31.88 35.53 114.22 125.25 137.57 163.17	24.64 70.75	41
Mg(OAr)$_2$(tmeda)	toluene	1.55 6.71 7.33	1.55 2.05	32.53 35.83 114.28 125.57 137.48 163.05	32.53 57.18	41

[a] Data acquired at ambient temperature. Ligand abbreviations: diglyme = bis(2-methoxyethyl) ether; tmeda = tetramethylethylenediamine; pmdta = pentamethyldiethylenetriamine; teed = tetraethylethylenediamine; OAr = 2,6-di-tert-butylphenoxy.

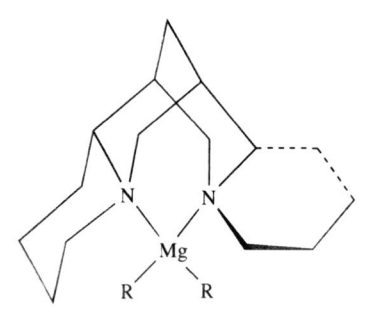

FIGURE 6. Dialkylmagnesium compounds form unusually strong adducts with sparteine. NMR data indicate that sparteine adopts *cisoid* configuration on co-ordination

VII. ^{25}Mg NMR STUDIES

Magnesium possesses a single NMR active nuclide, ^{25}Mg, which is only of limited utility owing to its low natural abundance and high quadrupole moment (Table 11)[49]. The large quadruople moment (and small magnetic moment) also gives rise to efficient quadrupolar relaxation effects in solution resulting in broad spectral lines which, coupled with the relatively narrow chemical shift range, further limit the utility of ^{25}Mg NMR studies on organomagnesium complexes.

Despite the obvious difficulties associated with the acquisition of good quality spectra, ^{25}Mg NMR has been usefully applied to the study of organomagnesium compounds[15,50,51]. The ^{25}Mg NMR parameters reported for organomagnesium complexes are listed in Table 12. Examination of Table 12 shows that the total solution chemical shift range is relatively narrow: −85 to +110 ppm. The η^5-cyclopentadienyl complexes resonate at significantly lower frequency than the σ-bonded alkyl and aryl compounds, the latter occurring between 56–110 ppm. Comparison of the chemical shift data for these compounds can provide useful additional information on the bonding between the organic moiety and the metal centre. The relatively high ^{25}Mg chemical shift (ca 70 ppm) observed for bis(allyl)magnesium, for example, is similar to that found in alkylmagnesium compounds, suggesting that the allyl moiety is indeed σ-bonding to magnesium, in agreement with more recent ^{1}H and ^{13}C variable-temperature NMR measurements.

Unsurprisingly, the ^{25}Mg chemical shifts of Grignard compounds are solvent, temperature and concentration dependent, in keeping with the effect of these variables on the position of the Schlenk equilibrium. Although the chemical shifts of $MgCl_2$ and $MgBr_2$ (Table 12) lie within the range found for organomagnesium compounds, they are sufficiently separated from those of the RMgX and R_2Mg (R=alkyl) compounds to allow the simultaneous observation of all three species. The ^{25}Mg NMR spectrum of EtMgBr (0.36 mol dm^{-3}; THF solution), for example, reveals the presence of three non-exchanging species, namely Et_2MgBr, Et_2Mg and $MgBr_2$. On warming, the spectra broaden and coalesce, and at 340 K a single exchange averaged signal is observed at ca 54 ppm[15].

Although often a hindrance to the acquisition of good quality spectra, the half-height line widths of the ^{25}Mg resonances are of diagnostic use. The degree of covalency in (cyclopentadienyl)magnesium compounds has been the subject of considerable conjecture, but the very narrow half-height line width (105 Hz) of the ^{25}Mg NMR signal of Cp_2Mg in non-polar solvents clearly suggests significant covalent character[50], despite the

TABLE 11. Magnesium-25 NMR parameters

Spin	5/2
Natural abundance (%)	10.13
Magnetogyric ratio (10^7 rad T^{-1} s^{-1})	−1.6370
Frequency^a (MHz)	6.120
Quadrupole moment (10^{-28} m^{-2})	0.22
Relative sensitivity^b	2.71×10^{-4}
Standard reference	Mg^{2+}_{aq}
Chemical shift range^c	ca 180 ppm

^{a} Relative to ^{1}H = 100 MHz.
^{b} Relative to ^{1}H.
^{c} Total range reported for organomagnesium complexes[50].

TABLE 12. Magnesium-25 NMR data[a] for organomagnesium compounds

Compound	δ^{25}Mg	$W_{1/2}$ (Hz)
MeMgBr	67.8	1900
EtMgBr	56.2	1100
allylMgBr	29.7	1000
CpMgBr	−26.8	60
^{R}CpMgBr(tmeda)[b]	−15.0	300
Et$_2$Mg	99.2	3200
n-Pr$_2$Mg(tmeda)[c]	110.0	1700
{CH$_2$CH(Me)CH$_2$}$_2$Mg	70.4	2000
Ph$_2$Mg·dioxane[c]	108	2800
Cp$_2$Mg[b]	−85.4	105
Cp$_2$Mg(thf)$_n$	−33.8	90
Cp*$_2$Mg	−78.3	350
^{R}Cp$_2$Mg	−82.0	550
^{R}Cp$_2$Mg(thf)$_n$	−36.7	250
CpMgEt	−4.0	1500
CpMg{CH$_2$CH(Me)CH$_2$}	−14.7	710
Cp*MgEt	−10.0	1300
(indenyl)MgEt[d]	26.6	900
^{R}CpMgMe(tmeda)	15.0	1100
CpMg·OEt$_2$	−34.1	160
MgCl$_2$	16.4	350
MgBr$_2$	14.1	36

[a] Data from Reference 50. Data acquired in THF solvent at ambient temperature, unless otherwise stated. Ligand abbreviations: ^{R}Cp = 1, 2, 4-tris(trimethylsilyl)cyclopentadienyl; tmeda = tetramethylethylenediamine. Data for MgCl$_2$ and MgBr$_2$ are given for comparison.
[b] In toluene solvent.
[c] At 353 K.
[d] At 340 K.

relatively long C−Mg distances observed in the solid state[52]. The narrow line widths displayed by (cyclopentadienyl)magnesium compounds has also permitted their co-ordination chemistry with a variety of N, O and P Lewis bases to be explored by ^{25}Mg NMR. Cp$_2$Mg forms tetrahedral co-ordination complexes of the type Cp$_2$MgB$_2$ which, in some cases, have been isolated from toluene solution. The ^{25}Mg chemical shifts of these adducts (Table 12) correlate closely with the ^{13}C chemical shifts of the cyclopentadienyl ring carbons (Figure 7). Although, as expected, the ^{25}Mg signals are the more sensitive, both the ^{25}Mg and ^{13}C chemical shifts move to higher frequency as the stability of the adduct increases[51]. The comparatively high ^{25}Mg shift of the THF adduct, [Cp$_2$Mg ·(thf)$_n$], is presumed to arise from the formation of a penta-coordinate species, rather than an unusually stable adduct.

The acquisition of solid-state NMR spectra of half-integer quadrupolar nuclei, such as ^{25}Mg, is particularly challenging and the first ^{25}Mg SSNMR study of an organomagnesium compound, Cp$_2$Mg, has only been reported within the last few years[52]. The ^{25}Mg MAS QCPMG NMR spectrum of Cp$_2$Mg displays a single second-order quadrupolar pattern with an isotropic shift of −91 ppm. The chemical shielding anisotropy is estimated to be less than 60 ppm. The spectrum is consistent with Cp$_2$Mg possessing local C_i symmetry, with the two cyclopentadienyl rings undergoing rapid rotation about the Cp(centroid)−Mg axis at ambient temperature.

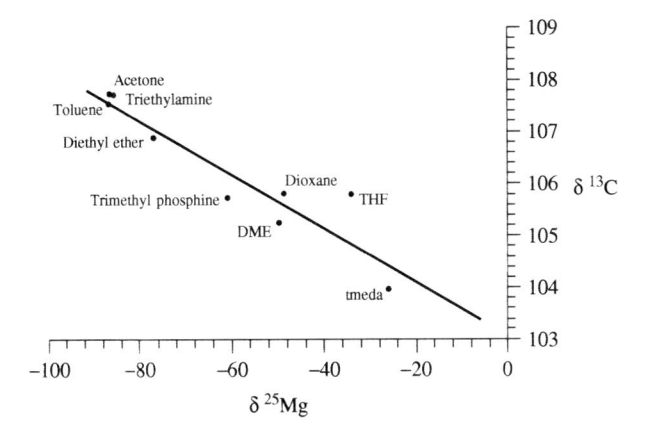

FIGURE 7. Correlation of the ^{25}Mg and ^{13}C NMR chemical shifts in bis(cyclopentadienyl) magnesium adducts. The ^{13}C shifts are those of the cyclopentadienyl ring carbons. The points on the line are the co-ordinating ligands

VIII. REFERENCES

1. E. C. Ashby, *Q. Rev.*, **21**, 259 (1967).
2. E. C. Ashby, *Pure Appl. Chem.*, **52**, 545 (1980).
3. W. Schlenk and W. Schlenk Jr., *Ber.*, **61B**, 720 (1928).
4. R. Jones, *J. Organomet. Chem.*, **18**, 15 (1969).
5. G. E. Paris and E. C. Ashby, *J. Am. Chem. Soc.*, **93**, 1206 (1971).
6. R. I. Yousef, B. Walfort, T. Ruffer, C. Wagner, H. Schmidt, R. Herzog and D. Steinborn, *J. Organomet. Chem.*, **690**, 1178 (2005).
7. H. O. House, R. L. Latham and G. M. Whitesides, *J. Org. Chem.*, **32**, 2481 (1967).
8. D. Leibfritz, B. O. Wagner and J. D. Roberts, *Liebigs Ann. Chem.*, **763**, 173 (1972).
9. E. C. Ashby and A. B. Goel, *Inorg. Chem.*, **17**, 322 (1978).
10. V. Schulze, R. Lowe, S. Fau and R. W. Hoffmann, *J. Chem. Soc., Perkin Trans. 2*, 463 (1998).
11. V. P. W. Bohm, V. Schulze, M. Bronstrup, M. Muller and R. W. Hoffmann, *Organometallics*, **22**, 2925 (2003).
12. R. A. Anderson and G. Wilkinson, *J. Chem. Soc., Dalton Trans.*, 809 (1977).
13. G. Fraenkel and D. T. Dix, *J. Am. Chem. Soc.*, **88**, 979 (1966).
14. W. T. Ford and J. B. Grutzner, *J. Org. Chem.*, **37**, 2561 (1972).
15. R. Benn, H. Lehmkul, K. Mehler and A. Rufinska, *Angew. Chem., Int. Ed. Engl.*, **23**, 534 (1984).
16. D. Steinborn, T. Ruffer, C. Bruhn and F. W. Heinemann, *Polyhedron*, **17**, 3275 (1998).
17. R. Han and G. Parkin, *J. Am. Chem. Soc.*, **114**, 748 (1992).
18. E. C. Ashby, G. Paris and F. Walker, *Chem. Commun.*, 1464 (1969).
19. D. F. Evans and G. V. Fazakerley, *J. Chem. Soc., A*, 184 (1971).
20. H. R. Ward, R. G. Lawler and T. A. Marzilli, *Tetrahdron Lett.*, 521 (1970).
21. E. C. Ashby and A. B. Goel, *Chem. Commun.*, 169 (1977).
22. G. M. Whitesides, F. Kaplan and J. D. Roberts, *J. Am. Chem. Soc.*, **85**, 2167 (1963).
23. G. M. Whitesides, M. Witanowski and J. D. Roberts, *J. Am. Chem. Soc.*, **87**, 2854 (1965).
24. G. M. Whitesides and J. D. Roberts, *J. Am. Chem. Soc.*, **87**, 4878 (1965).
25. M. Witanowski and J. D. Roberts, *J. Am. Chem. Soc.*, **88**, 737 (1966).
26. D. F. Evans and M. S. Khan, *J. Chem. Soc.*, A, 1643 (1967).
27. G. Westera, C. Blomberg and F. Bickelhaupt, *J. Organomet. Chem.*, **155**, C55 (1978).
28. D. A. Hutchison, K. R. Beck, R. A. Benkeser and J. B. Grutzner, *J. Am. Chem. Soc.*, **95**, 7075 (1973).
29. E. A. Hill, W. A. Boyd, H. Desai, A. Darki and L. Bivens, *J. Organomet. Chem.*, **514**, 1 (1996).

30. B. H. Lipshutz and C. Hackmann, *J. Org. Chem.*, **59**, 7437 (1994).
31. M. Schlosser and M. Stahle, *Angew. Chem., Int. Ed. Engl.*, **19**, 487 (1980).
32. R. Benn and A. Rufinska, *Organometallics*, **4**, 209 (1985).
33. R. H. DeWolfe and W. G. Young, *Chem. Rev.*, **56**, 753 (1956).
34. G. M. Whitesides, J. E. Nordlander and J. D. Roberts, *Discuss. Faraday Soc.*, **34**, 185 (1962).
35. J. E. Nordlander and J. D. Roberts, *J. Am. Chem. Soc.*, **81**, 1769 (1959).
36. H. E. Zieger and J. D. Roberts, *J. Org. Chem.*, **34**, 1976 (1969).
37. R. Benn, H. Lehmkuhl, K. Mehler and A. Rufinska, *J. Organomet. Chem.*, **293**, 1 (1985).
38. T. Yoshino and Y. Manabe, *J. Am. Chem. Soc.*, **85**, 2860 (1963).
39. T. Sugita, Y. Sakabe, T. Sasahara, M. Tsukada and K. Ichikawa, *Bull. Chem. Soc. Jpn.*, **57**, 2319 (1984).
40. E. C. Ashby, J. Nackashi and G. E. Parris, *J. Am. Chem. Soc.*, **97**, 3162 (1975).
41. K. W. Henderson, G. W. Honeyman, A. R. Kennedy, R. E. Mulvey, J. A. Parkinson and D. C. Sherrington, *J. Chem. Soc., Dalton Trans.*, 1365 (2003).
42. E. C. Ashby and B. Goel, *Inorg. Chem.*, **18**, 1306 (1979).
43. P. J. Bailey, R. A. Coxhall, C. M. Dick, S. Fabre, L. C. Henderson, C. Herber, S. T. Liddle. D. Lorono-Gonzalez, A. Parkin and S. Parsons, *Chem. Eur. J.*, **9**, 4820 (2003).
44. Cambridge Crystallographic Database.
45. D. F. Evans and M. S. Khan, *J. Chem. Soc., A*, 1648 (1967).
46. K.-C. Yang, C.-C. Chang, J.-Y. Huang, C.-C. Lin, G.-H. Lee, Y. Wang and M. Y. Chiang, *J. Organomet. Chem.*, **648**, 176 (2002).
47. N. D. R. Barnett, W. Clegg, R. E. Mulvey, P. A. O'Neil and D. Reed, *J. Organomet. Chem.*, **510**, 297 (1996).
48. G. Fraenkel, C. Cotterell, J. Ray and J. Russell, *Chem. Commun.*, 273 (1971).
49. R. K. Harris and B. E. Mann (Eds.), *NMR and the Periodic Table*, Academic Press, New York, 1978.
50. R. Benn and A. Rufinska, *Angew. Chem., Int. Ed. Engl.*, **25**, 861 (1986) and references cited therein.
51. H. Lehmkul, K. Mehler, R. Benn, A. Rufinska and C. Kruger, *Chem. Ber.*, **119**, 1054 (1986).
52. I. Hung and R. W. Schurko, *Solid State Nucl. Magn. Reson.*, **24**, 78 (2003).

CHAPTER 4

Formation, chemistry and structure of organomagnesium species in solvent-free environments

RICHARD A. J. O'HAIR

School of Chemistry, The University of Melbourne, Victoria 3010, Australia; Bio21 Molecular Science and Biotechnology Institute, The University of Melbourne, Victoria 3010, Australia; ARC Centre of Excellence in Free Radical Chemistry and Biotechnology
Fax: +61 3 9347-5180;   e-mail: rohair@unimelb.edu.au

I. INTRODUCTION	156
II. FORMATION OF ORGANOMAGNESIUM SPECIES IN SOLVENT-FREE ENVIRONMENTS	156
A. Reactions of Magnesium Atoms with Organic Substrates	157
1. Reactions of Mg with alkanes	157
2. Reactions of Mg with alkyl halides	157
3. Reactions of Mg with unsaturated organic substrates	158
4. Reactions of Mg with other organic substrates	159
B. Reactions of Magnesium Cations with Organic Substrates	160
1. Adduct-forming reactions of $Mg^{+\bullet}$ with alkanes, alkenes and other unsaturated species	160
2. Reactions of $Mg^{+\bullet}$ with alkyl halides	160
3. Reactions of $Mg^{+\bullet}$ with alcohols	161
4. Reactions of MgX^+ (X = Cl, O, OH)	162
5. Photoactivation reactions of complexes $Mg(L)^{+\bullet}$, where L = an alkane	162
6. Photoactivation reactions of complexes $Mg(L)^{+\bullet}$, where L = an organohalogen	163
7. Photoactivation reactions of complexes $Mg(L)^{+\bullet}$, where L = an alcohol or ether	165
8. Photoactivation reactions of complexes $Mg(L)^{+\bullet}$, where L = an amine	167

The chemistry of organomagnesium compounds
Edited by Z. Rappoport and I. Marek   © 2008 John Wiley & Sons, Ltd

      9. Photoactivation reactions of complexes Mg(L)$^{+\bullet}$, where L contains a
         C=X bond (X = O or S) 169
   C. Reactions of Magnesium Clusters with Organic Substrates 171
   D. Reactions of Magnesium Surfaces and Films with Organic Substrates . 172
   E. Gas-phase Fragmentation of Ligated Magnesium Ions to Yield
      Organomagnesium Ions 174
      1. Fragmentation reactions of Mg(L)$_n^{2+}$ complexes 175
      2. Fragmentation reactions of XMg(L)$_n^+$ complexes (where X = an
         anionic ligand) 176
      3. Fragmentation reactions of Mg(X)$_3^-$ complexes 178
III. BIMOLECULAR REACTIONS OF ORGANOMAGNESIUM IONS
    IN THE GAS PHASE 179
IV. UNIMOLECULAR REACTIONS OF ORGANOMAGNESIUM
    IONS IN THE GAS PHASE 182
V. STRUCTURES OF ORGANOMAGNESIUMS AND MAGNESIUM
    HALIDES IN THE GAS PHASE 183
VI. CONCLUDING REMARKS 184
VII. ACKNOWLEDGMENTS 184
VIII. REFERENCES 184

I. INTRODUCTION

Unlike other stable organoelement compounds (e.g. organosilicon[1] and organophosphorus[2]), which have been widely studied by mass-spectrometry-based techniques, only a handful of studies have examined organomagnesium species using mass spectrometry[3–6]. This may be due to the challenges of introducing these water- and oxygen-sensitive compounds into traditional EI sources of mass spectrometers. Recent studies using coldspray ionization[6,7] and matrix assisted laser desorption ionization (MALDI)[5] show promise for the analysis of organomagnesium compounds. Thus, unlike previous chapters in 'The Chemistry of Functional Groups' series that were solely devoted to mass spectrometry of organoelement species[1,2], a wider net is cast in this review to include studies relevant to the formation, chemistry, structure and mass spectrometry of organomagnesium species in the gas phase and related solvent-free environments (e.g. matrix conditions). These studies highlight the broad scientific interest in the interaction of magnesium species with organic molecules, which span the range from traditional organic and organometallic chemistry through to the role of magnesium in planetary atmospheres[8–11] and interstellar science[12–14]. Although the heavier congeners of magnesium are not reviewed here, some comparison of their reactivity is made where appropriate. Theoretical studies are not reviewed here, unless they are directly related to experimental work. Finally, experimental techniques are not reviewed in this chapter.

II. FORMATION OF ORGANOMAGNESIUM SPECIES IN SOLVENT-FREE ENVIRONMENTS

The formation of organomagnesium species such as Grignard reagents typically involves activating C−X bonds by magnesium metal. Since it is difficult to theoretically model in detail the interactions of an organic substrate with bulk magnesium metal, there is considerable interest in C−X bond activation in solvent-free environments using simpler magnesium species (e.g. magnesium atoms and ions; magnesium clusters or well-defined magnesium surfaces). In fact it has been argued that 'the active sites of a Mg surface are constituted by sets of clusters of highly variable reactivity rather than by a unique entity called metallic Mg'[15]. Key experimental and theoretical work on these 'idealized' systems are described in the next sections. Note that reactivity of a wide range of organic substrates is considered, including those that do not ultimately yield organomagnesium species.

A. Reactions of Magnesium Atoms with Organic Substrates

Magnesium atoms are readily formed via vaporization of magnesium metal using either thermal techniques or laser ablation. The reactions of magnesium atoms with a range of compounds have been the subject of several studies in both the gas phase[16] and using matrix isolation techniques[17]. Since both areas have been nicely reviewed[16, 17], here the focus is on key aspects of reactivity of Mg with organic substrates. Magnesium atoms in the ^1S ground state are generally unreactive towards organic substrates such as CH_4 due to repulsive interactions with bonding orbitals. In order to activate a bond in the organic substrate, photoactivation of one of the valence 3s electrons of Mg is required to generate the $^3P_1$ or $^1P_1$ excited-state. The outcome of reactions of excited-state Mg atoms with organic substrates is dependent on the medium. For example, gas-phase reactions with alkanes, RH, tend to produce $R^•$ + $MgH^{•}$ [16], while the insertion product, RMgH, can be observed in matrix environments[17]. In the next sections, the reactions of Mg atoms are described with alkanes, alkyl halides and other substrates.

1. Reactions of Mg with alkanes

Excited-state Mg atoms react with methane and other alkanes via H atom abstraction in the gas phase (equation 1). By studying the vibrational states of the $MgH^•$ product, information on the mechanism has been inferred[18]. It has been found that regardless of the alkane, RH (and thus the C−H bond strength), the vibrational state distributions are essentially identical. This suggests that long-lived vibrationally excited [RMgH]* complexes are not intermediates for equation 1 in the gas phase. The situation is quite different for excited-state Mg atoms reacting with methane under matrix conditions, where the insertion product (equation 2) is sufficiently stable for analysis via infrared spectroscopy[19, 20]. Calcium atoms have been shown to insert into the C−H bonds of cycloalkanes[21].

$$Mg^* + RH \longrightarrow MgH^• + R^• \qquad (1)$$

$$Mg^* + CH_4 \longrightarrow CH_3MgH \qquad (2)$$

2. Reactions of Mg with alkyl halides

Skell and Girard appear to have been the first to report on the formation of solvent-free Grignard reagents via the codeposition of alkyl halides and magnesium atoms under matrix conditions over 30 years ago[22]. They noted that these solvent-free Grignards reacted differently compared to solution-phase Grignard reagents. For example, the solvent-free Grignard formed from n-propyl iodide reacted with acetone via enolization rather than addition. For some time the precise nature and mechanism of formation of these solvent-free Grignards formed under matrix conditions was obscure. Although Skell and Girard claimed they were observing ground-state reactivity of Mg atoms and Ault later confirmed that Mg atoms could react with methyl halides under matrix conditions to form monomeric reagents, CH_3MgX[23], subsequent work by Klabunde and coworkers suggested that the reactivity was due to the presence of Mg clusters[24, 25]. Thus Imizu and Klabunde found that Mg atoms were 'totally inert' towards CH_3Br[24]. This is consistent with early theoretical calculations, which predicted a substantial activation barrier to form CH_3MgX

from reaction of Mg with CH_3X (31.3 kcal mol^{-1} for X = F; and 39.4 kcal mol^{-1} for X = Cl)[26]. In 1997, Solov'ev and coworkers revisited the formation of CH_3MgX (X = Cl and Br) using matrix isolation of reactions between evaporated Mg atoms and the methyl halide[27]. They concluded that the products were monomeric reagents. A year later Bare and Andrews examined the reactivity of laser-ablated Mg atoms with CH_3X (X = F, Cl, Br and I)[28]. Using a combination of IR spectroscopy, C, H and Mg isotopic labeling and DFT calculations, they identified the isolated monomeric CH_3MgX species as the primary product together with the following other products: MgX•, MgX_2, MgH•, MgH_2, CH_4, C_2H_6, CH_2X•, CH_3MgCH_3, $XMgCH_2$•, $MgCH_2$, CH_3MgH and $HMgCH_2X$. A key difference in their experiments is that laser ablation produces a portion of excited-triplet-state Mg atoms. They suggested that these excited-state atoms react with the methyl halide to form two different excited-state monomeric species, which arise from the expected C–X bond activation pathway (equation 3) as well as an unusual C–H bond activation pathway (equation 4). These may either relax to form the monomeric Grignard CH_3MgX (equation 5a) and C–H insertion product $HMgCH_2X$ (equation 6a), or can decompose via a range of pathways (equations 5b–e and 6b–c). The most recent theoretical calculations confirm the role of triplet states in the insertion reaction of Mg with CH_3Cl[29,30]. Finally, the magnesium carbene product, $MgCH_2$, has been examined in more detail[31].

$$Mg^* + CH_3X \longrightarrow CH_3MgX^* \qquad (3)$$

$$\longrightarrow HMgCH_2X^* \qquad (4)$$

$$CH_3MgX^* \longrightarrow CH_3MgX \qquad (5a)$$

$$\longrightarrow CH_3^\bullet + MgX^\bullet \qquad (5b)$$

$$\longrightarrow MgCH_2 + HX \qquad (5c)$$

$$\longrightarrow CH_3Mg^\bullet + X^\bullet \qquad (5d)$$

$$\longrightarrow CH_2MgX^\bullet + H^\bullet \qquad (5e)$$

$$HMgCH_2X^* \longrightarrow HMgCH_2X \qquad (6a)$$

$$\longrightarrow MgH^\bullet + CH_2X^\bullet \qquad (6b)$$

$$\longrightarrow MgCH_2 + HX \qquad (6c)$$

3. Reactions of Mg with unsaturated organic substrates

The reactions of Mg atoms and clusters with CO_2, ethylene and their mixtures have been examined using a combination of matrix isolation and theoretical calculations[32]. Products were characterized by IR and UV-visible techniques. One of the most interesting findings is that Mg appears to promote the formation of bonds between two ligands, thereby forming the five-membered rings **1** (between two ethylene ligands) and **2** (between one ethylene ligand and one CO_2 ligand).

4. Solvent-free organomagnesiums

The only identified reaction product of laser-ablated Mg atoms and acetylene under matrix isolation conditions is MgC≡CH$^{\bullet}$ [33]. It was suggested that this reaction involves the insertion of excited-state Mg into the H—C bond to form an excited complex (equation 7), which then decomposes via H atom loss (equation 8).

$$Mg^* + C_2H_2 \longrightarrow HMgC\equiv CH^* \tag{7}$$

$$HMgC\equiv CH^* \longrightarrow MgC\equiv CH^{\bullet} + H^{\bullet} \tag{8}$$

Mg atoms formed via laser ablation react with H≡CN to form MgN≡C$^{\bullet}$ rather than MgC≡N$^{\bullet}$ [34]. This suggests coordination at N to form an excited-state intermediate (equation 9) which decomposes via H atom loss (equation 10) rather than via a C—H insertion reaction (cf equations 7 and 8). Finally, the monomethyl magnesium radical, MgCH$_3^{\bullet}$, has been formed via reaction of laser-ablated magnesium metal reacting with either CH$_3$I or acetone[35]. Although the emphasis of this study was on an examination of the radical via ESR spectroscopy, a possible mechanism may involve C—X bond insertion via excited-state magnesium atoms to form an excited organomagnesium intermediate (cf equation 3) which then decomposes (cf equation 5d).

$$Mg^* + HC\equiv N \longrightarrow MgN\equiv CH^* \tag{9}$$

$$MgN\equiv CH^* \longrightarrow MgN\equiv C^{\bullet} + H^{\bullet} \tag{10}$$

4. Reactions of Mg with other organic substrates

Much less is known about the reactions of Mg atoms with other organic substrates. In fact it appears that there is only one gas-phase study in which the reactions of neutral organic substrates other than alkanes were studied. Thus as part of a systematic study of C—H bond activation by excited-state Mg atoms, Breckenridge and Umemoto studied a range of organic substrates including CH$_3$OH, (CH$_3$)$_2$O, (CH$_3$CH$_2$)$_2$O, CH$_3$NH$_2$ and (CH$_3$)$_4$Si[18]. All reacted via H atom abstraction (equation 1). In contrast, reaction of Mg atoms with CH$_3$OH under matrix conditions yields a Mg(CH$_3$OH) complex, which undergoes C—O bond activation to yield CH$_3$MgOH under conditions of UV-Vis irradiation (cf equation 5a)[36]. Interestingly, CH$_3$MgOH undergoes further reaction with Mg to yield CH$_3$MgOMgH arising from O—H bond activation.

Finally, two studies have reported on the reactions of carbocations with Mg atoms using mass spectrometry[37,38]. The types of products formed depend on the nature of the carbocation. The labeled methanium ion, CH$_4$D$^+$, reacts via proton transfer (equation 11), deuteron transfer (equation 12) and charge transfer (equation 13)[37]. The ethyl cation reacts via charge transfer (equation 14)[38] while the *tert*-butyl cation reacts via proton transfer (equation 15)[37]. In all cases there was no evidence for formation of an organomagnesium species.

$$CH_4D^+ + Mg \longrightarrow MgH^+ + CH_3D \tag{11}$$

$$\longrightarrow MgD^+ + CH_4 \tag{12}$$

$$\longrightarrow Mg^{+\bullet} + \text{products} \tag{13}$$

$$C_2H_5^+ + Mg \longrightarrow Mg^{+\bullet} + \text{products} \tag{14}$$

$$(CH_3)_3C^+ + Mg \longrightarrow MgH^+ + (CH_3)_2C=CH_2 \tag{15}$$

B. Reactions of Magnesium Cations with Organic Substrates

The bond activation reactions of monoatomic main group and transition metal cations have been widely studied for decades and have been the subject of several reviews[39–44]. Gas-phase monoatomic magnesium cations can readily be formed via a range of processes including electron impact on magnesium vapors[45] and magnesium organometallics[46] and laser ablation on magnesium metal[47]. The reactivity of $Mg^{+\bullet}$ is first described, followed by a discussion on the reactions of ligated magnesium ions and finally on the photoactivation of magnesium adduct ions.

1. Adduct-forming reactions of $Mg^{+\bullet}$ with alkanes, alkenes and other unsaturated species

Using the selected ion flow tube technique (SIFT), Bohme and coworkers have shown that thermalized $Mg^{+\bullet}$ reacts with alkanes (L) via simple adduct formation without bond activation (equation 16)[46]. Only single ligation was observed, and the efficiency of this reaction depends on the size of the alkane (L). Methane and ethane are unreactive. Larger alkanes become more reactive, with n-heptane reacting at the collision rate. DFT calculations reveal that while the binding energies can be strong (around 12–16 kcal mol^{-1} for n-pentane), interconversion of different $MgL^{+\bullet}$ isomers should be facile.

$$Mg^{+\bullet} + L \longrightarrow MgL^{+\bullet} \quad (16)$$

Several studies have examined the reactions of $Mg^{+\bullet}$ with unsaturated molecules. Under the lower pressure conditions of FT-ICR mass spectrometry, $Mg^{+\bullet}$ reacts with the polycyclic aromatic hydrocarbon, coronene, via a combination of radiative associative adduct formation (equation 16) and electron transfer (equation 17)[48]. The latter reaction is 8 times faster, consistent with it being exothermic. Adduct formation (equation 16) also readily occurs in reactions with C_{60}[49–51]. Theoretical calculations suggest that related radiative associative adduct formation of $Mg^{+\bullet}$ with cyanopolyenes and polyenes should be highly efficient[52,53].

$$Mg^{+\bullet} + L \longrightarrow Mg + L^{+\bullet} \quad (17)$$

The reactions of $Mg^{+\bullet}$ with cyanoacetylene are remarkable in that while $Mg^{+\bullet}$ is unreactive towards HCN, multiple ligation occurs for cyanoacetylene[54]. Furthermore, there is evidence from collision induced dissociation (CID) studies that ligand–ligand interactions occur. In fact, the $Mg(NC_3H)_4^{+\bullet}$ is especially stable, being resistant to CID. DFT calculations suggest that the semibulvalene-type structure, **4**, is around 12 kcal mol^{-1} more stable than the tetrahedral structure, **3**. These are reminiscent of the reactions of Mg atoms with ethylene to form **1**[32].

2. Reactions of $Mg^{+\bullet}$ with alkyl halides

$Mg^{+\bullet}$ reacts with alkyl halides in the gas phase via a range of substrate-dependent pathways[45,47]. Not all halides are reactive—examples of unreactive substrates include methyl chloride, vinyl chloride, trichloro and tetrachloro ethylene. Reaction with ethyl chloride proceeds via an elimination reaction (equation 18) followed by a displacement reaction (equation 19). For larger alkyl halides, such as isopropyl chloride, chloride abstraction also occurs (equation 20). For multiply halogenated substrates such as carbon tetrachloride, oxidative reactions occur (equations 21 and 22), although organometallic

(3)   (4)

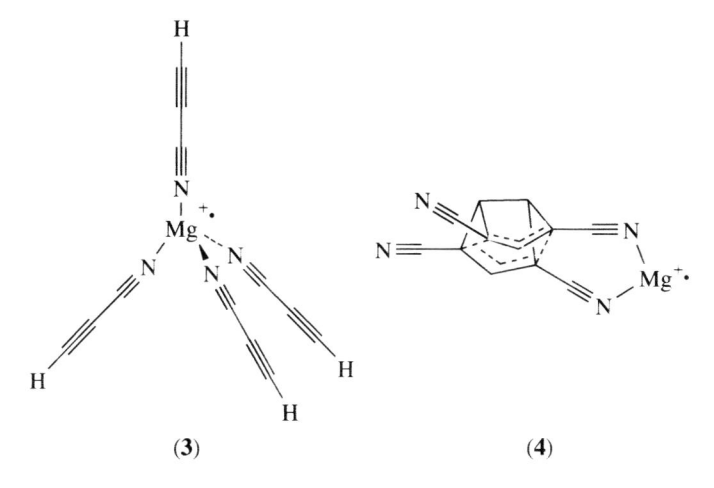

species are not found. Finally, the related calcium cation reacts with methyl fluoride via oxidation (equation 23)[55].

$$Mg^{+\bullet} + CH_3CH_2Cl \longrightarrow Mg(CH_2=CH_2)^{+\bullet} + HCl \qquad (18)$$

$$Mg(CH_2=CH_2)^{+\bullet} + CH_3CH_2Cl \longrightarrow Mg(CH_3CH_2Cl)^{+\bullet} + CH_2=CH_2 \qquad (19)$$

$$Mg^{+\bullet} + (CH_3)_2CHCl \longrightarrow (CH_3)_2CH^+ + MgCl^\bullet \qquad (20)$$

$$Mg^{+\bullet} + CCl_4 \longrightarrow CCl_2^{+\bullet} + MgCl_2 \qquad (21)$$

$$\longrightarrow MgCl^+ + CCl_3^\bullet \qquad (22)$$

$$Ca^{+\bullet} + CH_3F \longrightarrow CaF^+ + CH_3^\bullet \qquad (23)$$

3. Reactions of $Mg^{+\bullet}$ with alcohols

$Mg^{+\bullet}$ reacts with alcohols via either condensation or via elimination. The outcome is substrate dependent. Thus while n-BuOH reacts via condensation (equation 24), t-BuOH reacts via elimination (equation 25). No oxidative reactions (cf equations 21–23) are observed. Armentrout and coworkers[56] have examined the CID reactions of methanol adducts of $Mg^{+\bullet}$ using Xe as the collision gas and found competition between ligand loss (equation 26), ligand switching (equation 27) and C—O bond activation (equations 28 and 29).

$$Mg^{+\bullet} + CH_3CH_2CH_2CH_2OH \longrightarrow Mg(CH_3CH_2CH_2CH_2OH)^{+\bullet} \qquad (24)$$

$$Mg^{+\bullet} + (CH_3)_3COH \longrightarrow Mg(H_2O)^{+\bullet} + (CH_3)_2C=CH_2 \qquad (25)$$

$$Mg(CH_3OH)^{+\bullet} + Xe \longrightarrow Mg^{+\bullet} + CH_3OH + Xe \qquad (26)$$

$$\longrightarrow Mg(Xe)^{+\bullet} + CH_3OH \qquad (27)$$

$$\longrightarrow MgOH^+ + CH_3^\bullet + Xe \qquad (28)$$

$$\longrightarrow CH_3^+ + MgOH^\bullet + Xe \qquad (29)$$

4. Reactions of MgX$^+$ (X = Cl, O, OH)

MgCl$^+$ ions undergo anion abstraction reactions with organic halides (equation 30)[42,44] and nitric acid (equation 31).

$$MgCl^+ + CXCl_3 \longrightarrow XCCl_2^+ + MgCl_2 \quad X = H, Cl \quad (30)$$

$$MgCl^+ + HNO_3 \longrightarrow NO_2^+ + HOMgCl \quad (31)$$

The MgO$^{+\bullet}$ ion has significant radical character and reacts via electron transfer (equation 32)[57]. It is also a potent H atom acceptor, readily reacting with water via H atom abstraction (equation 33, X = HO)[53]. A recent combined experimental and theoretical study reveals that the MgO$^{+\bullet}$ ion readily activates the C–H bond of methane to yield MgOH$^+$ as the major product ion (equation 33, X = CH$_3$) as well as Mg$^{+\bullet}$ as a minor product ion via O atom insertion into the C–H bond (equation 34)[58].

$$MgO^{+\bullet} + Me_3N \longrightarrow Me_3N^{+\bullet} + MgO \quad (32)$$

$$MgO^{+\bullet} + HX \longrightarrow MgOH^+ + HX^{\bullet} \quad (33)$$

$$MgO^{+\bullet} + CH_4 \longrightarrow Mg^{+\bullet} + CH_3OH \quad (34)$$

The MgOH$^+$ ion is a weak acid, failing to react via proton transfer (equation 35) with even a strong base such as N,N,N',N'-tetramethyl-1,8-naphthalenediamine[53]. Although its reactions with organic reagents are largely unexplored, MgOH$^+$ reacts with nitric acid via HO$^-$ abstraction (cf equation 31)[42].

$$MgOH^+ + B \not\longrightarrow BH^+ + MgO \quad (35)$$

5. Photoactivation reactions of complexes Mg(L)$^{+\bullet}$, where L = an alkane

Intracomplex reactions in Mg(L)$^{+\bullet}$ complexes (where L = an organic molecule) have been studied for a wide range of organic molecules using gas-phase photodissociation spectroscopy experiments. These experiments offer a number of advantages since: (a) they start from a well-defined complex; (b) chemical reactivity is triggered by exciting Mg$^{+\bullet}$ electronically (the Mg$^{+\bullet}$ $3P \leftarrow 3S$ transition) via absorption of a photon in the UV-Vis region; (c) the presence of a positive charge means that ionic products can readily be detected via mass spectrometry; (d) the systems are often sufficiently small so that they are amenable to high-level theoretical calculations. While this area was reviewed in 1998[59], progress has been rapid and so in the next sections the C–X bond activation observed in these studies is described by class of organic molecule.

Cheng and coworkers have examined the photodissociation spectroscopy of MgCH$_4^{+\bullet}$ in detail[60]. The photofragmentation action spectrum has a broad featureless continuum ranging from 310 to 342 nm, with a maximum at 325 nm. In this region the channels observed are nonreactive (equation 36, ca 60%), H abstraction (equation 37, ca 7%) and CH$_3$ abstraction (equation 38, ca 33%). Recent theoretical calculations on the C–H bond activation in MgCH$_4^{+\bullet}$ reveal that the formation of the insertion intermediate, CH$_3$MgH$^{+\bullet}$, proceeds via a three-centered transition state[61].

$$MgCH_4^{+\bullet} + h\nu \longrightarrow Mg^{+\bullet} + CH_4 \quad (36)$$

$$\longrightarrow MgH^+ + CH_3^{\bullet} \quad (37)$$

$$\longrightarrow MgCH_3^+ + H^{\bullet} \quad (38)$$

The photodissociation spectroscopy of $CaCH_4^{+\bullet}$ contrasts that of $MgCH_4^{+\bullet}$, with complex rovibrational structure in the spectrum, but with no evidence for C–H bond activation[62]. Although not relevant to C–H bond activation, the photodissociation spectroscopy of the prototypical alkene complex, $MgC_2H_4^{+\bullet}$, exhibits a rich photochemistry arising from metal-centered transitions, ligand-centered transitions, and from charge transfer processes (which give rise to electron transfer to yield Mg and $C_2H_4^{+\bullet}$)[63].

6. Photoactivation reactions of complexes $Mg(L)^{+\bullet}$, where L = an organohalogen

The photoactivation of $Mg(L)^{+\bullet}$ complexes of organohalogens has been widely studied[64–74]. The photodissociation spectra of the methyl halide complexes, $Mg(XCH_3)^{+\bullet}$ (where X = F, Cl, Br and I), have been studied in great detail by Furuya and coworkers in two publications[66, 67]. Each of the four halides exhibits spectra with three absorption bands at the red and blue sides of the free $Mg^{+\bullet}$ $^2P \leftarrow {}^2 2S$ transition. These three absorption bands were assigned to the splitting of the $Mg^{+\bullet}$ $3p$ orbitals as a result of interaction with the methyl halide molecules. Six different fragment ions were produced including intermolecular bond breaking (evaporation) without (equation 39) and with charge transfer (equation 40), anion abstraction (equations 41 and 42) and oxidation (equations 43 and 44). Equations 39, 41 and 43 were observed for all four halides, equation 40 was observed for X = Cl, Br and I, equation 42 was only observed for the iodide, while equation 44 was observed when X = Cl and Br. The yields of each product channel depend on which of the three absorption bands was excited. Detailed theoretical calculations were carried out to explain the experimental data. Of interest is that the complex with connectivity $MgXCH_3^{+\bullet}$ is predicted to be more stable than the organomagnesium ion, $CH_3MgX^{+\bullet}$, in all cases (for X = F, by 12.4 kcal mol^{-1}; X = Cl, by 4.8 kcal mol^{-1}; X = Br, by 3.7 kcal mol^{-1}; X = I, by 3.9 kcal mol^{-1}). The photodissociation spectra of $Mg(FCH_3)^{+\bullet}$ complexes 'solvated' by up to three other methyl fluoride molecules are dominated by the formation of bare and solvated MgF^+ (cf equation 43)[68].

$$Mg(XCH_3)^{+\bullet} + h\nu \longrightarrow Mg^{+\bullet} + XCH_3 \qquad (39)$$

$$\longrightarrow XCH_3^{+\bullet} + Mg \qquad (40)$$

$$\longrightarrow CH_3^+ + MgX^{\bullet} \qquad (41)$$

$$\longrightarrow X^+ + MgCH_3^{\bullet} \qquad (42)$$

$$\longrightarrow MgX^+ + CH_3^{\bullet} \qquad (43)$$

$$\longrightarrow MgCH_3^+ + X^{\bullet} \qquad (44)$$

The photodissociation spectra of benzene and halobenzene complexes, $Mg(C_6H_5X)^{+\bullet}$ (where X = H, F, Cl and Br), are dominated by the formation of $Mg^{+\bullet}$ (cf equation 39), although charge transfer (cf equation 40) is observed for all cases[69]. Fluorobenzene gives MgF^+ as a unique product (cf equation 43). New fragmentation channels open up in the photodissociation spectra of polyfluorinated benzenes, $C_6H_{4-n}F_{2+n}$, 5–9[70, 71, 73]. Apart from formation of $Mg^{+\bullet}$ (cf equation 39) and MgF^+ (cf equation 43), benzyne radical cations, $C_6H_{4-n}F_n^{+\bullet}$, are formed (equation 45). These benzyne radical cations undergo further fragmentation reactions, the nature of which depends on the number of fluorines present. The $C_6H_{4-n}F_n^{+\bullet}$ ions of 5 and 7 fragment via loss of C_2H_2 and C_2HF respectively, while 6 fragments via competitive loss of C_2H_2 and C_2HF. Instead of fragmenting via C_2HX loss (where X = H or F), 8 yields CF^+, C_5H^+ and $C_5HF^{+\bullet}$ fragment ions, while

9 yields CF^+, C_5F^+, $C_5F_2^{+\bullet}$ and $C_5F_3^+$.

$$Mg(C_6H_{4-n}F_{2+n})^{+\bullet} + h\nu \longrightarrow C_6H_{4-n}F_n^{+\bullet} + MgF_2 \qquad (45)$$

(5a) *ortho*   (6)   (7)   (8)   (9)
(5b) *meta*
(5c) *para*

(10)   (11)   (12)

The same group has studied the photodissociation spectra of $Mg^{+\bullet}$ complexes of 2-fluoropyridine **10**[63] and the polyfluorinated pyridines **11** and **12**[64]. The photodissociation chemistry of the $Mg^{+\bullet}$ complexes of **10** is rich. Aside from $Mg^{+\bullet}$ formation (equation 39), anion abstraction (equation 41) and oxidation (equation 43), HF extrusion (equation 46) and reactions which result in the destruction of the aromatic ring are also observed. The latter include $FMgNC^{+\bullet}$ (equation 47) and FMgNC (equation 48) formation and extrusion of HCN (equation 49). Further substitution of fluorine onto the pyridine ring results in changes in the photodissociation chemistry. Thus the polyfluorinated pyridines **11** and **12** react via $Mg^{+\bullet}$ formation (equation 39), oxidation (equation 43) and dehydropyridine radical cation formation (cf equation 45)[64]. In the case of **11**, the resultant dehydropyridine radical cation undergoes further fragmentation via loss of HCN to give $C_4H_2^{+\bullet}$. Although the structures of some of these product ions and neutrals are not known from experiment, DFT calculations were preformed to suggest possible mechanisms. For example, all the fragments for the complex of **11** were rationalized as potentially arising from the initial N bound adduct **13** reacting via the processes shown in Scheme 1. Note that the radical cation structures, $C_5H_3N^{+\bullet}$, can include the dehydropyridines **14c**, **15c** and **16c** as well as open-chain isomers. Structures **14c**, **15c** and **16c** can arise from H atom migrations either within the initial Mg complex (e.g. processes **14a** → **15a** → **16a** in Scheme 1), or from subsequent H atom migrations within the $C_5H_3N^{+\bullet}$ product ion (e.g. processes **14c** → **15c** → **16c** in Scheme 1). The DFT calculations suggest that the experiments are likely to produce a mixture of $C_5H_3N^{+\bullet}$ isomers, but that extrusion of HCN is energetically preferred from **16c**.

$$Mg(C_5H_4FN)^{+\bullet} + h\nu \longrightarrow Mg(C_5H_3N)^{+\bullet} + HF \qquad (46)$$
$$\longrightarrow FMgNC^{+\bullet} + C_4H_4 \qquad (47)$$
$$\longrightarrow C_4H_4^{+\bullet} + FMgNC \qquad (48)$$
$$\longrightarrow Mg(C_4H_3F)^{+\bullet} + HCN \qquad (49)$$

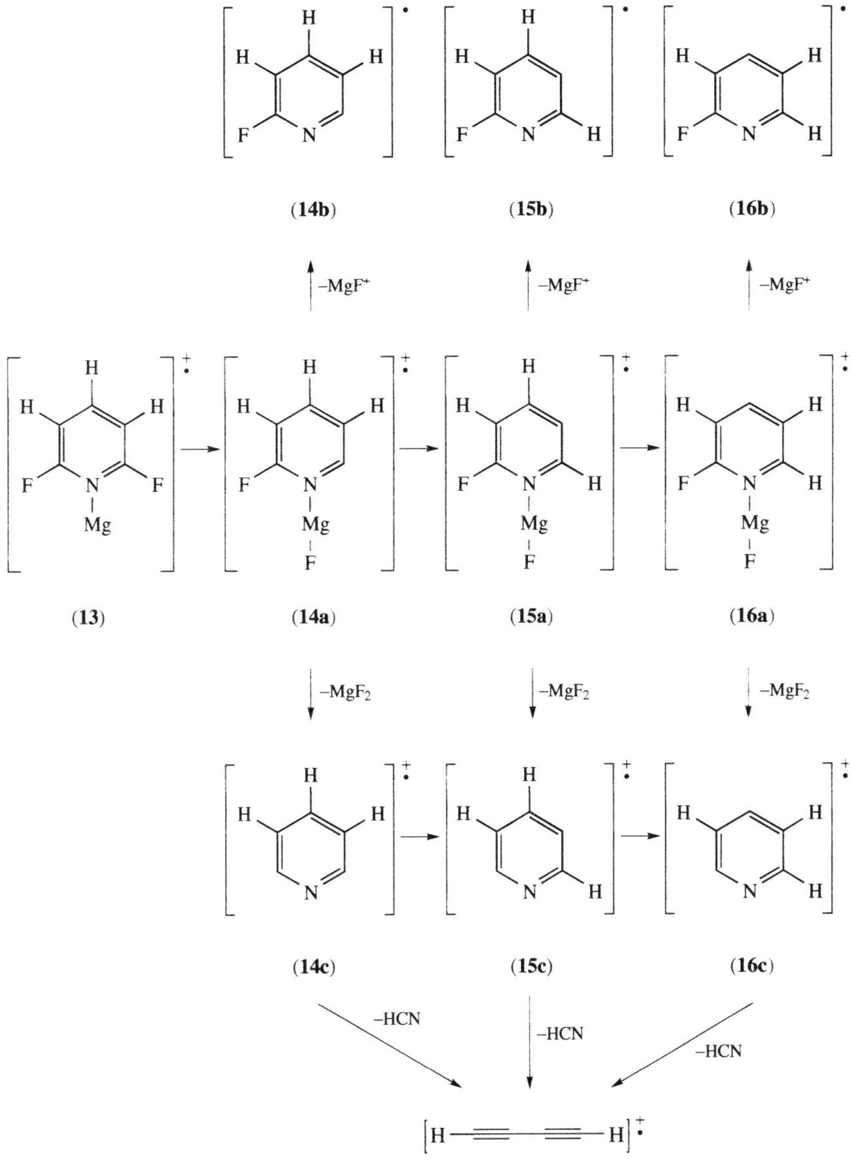

SCHEME 1

7. Photoactivation reactions of complexes Mg(L)$^{+\bullet}$, where L = an alcohol or ether

The photodissociation spectra of Mg(CH$_3$OH)$_n$$^{+\bullet}$ complexes has been studied as a function of cluster size, n[75]. For $n = 1$, the main reaction channels involve formation of Mg$^{+\bullet}$ (cf equation 39) and MgOH$^+$ (cf equation 43). Small amounts of CH$_3$$^+$ (cf equation 41) and $MgO^{+\bullet}$ are also observed. Although the neutral product(s) and mechanism associated with the formation of the latter ion are unknown, they may represent the reverse of the reaction involving C—H bond activation of methane by $MgO^{+\bullet}$ (equation 34). Larger clusters ($2 \leqslant n \leqslant 6$) undergo efficient dissociation at 350 nm to yield products arising from two competing pathways: (i) elimination of solvent, and (ii) an excited-state photoinduced reaction to yield $MgOH(CH_3OH)_m^+$ (the solvated equivalent of equation 43, where $m < n$). When $n \geqslant 6$, photodissociation is no longer efficient, suggesting the loss of the $Mg^{+\bullet}$ chromophore. Similar results were observed in the photodissociation spectra of $Mg(CH_3OD)_n^{+\bullet}$ complexes, although a unique loss of CH_3D was observed when $n = 2$[76].

The photoproducts of the $Mg(CF_3CH_2OH)^{+\bullet}$ complex have been examined using a combination of experiment and theory[77]. Apart from non-reactive formation of $Mg^{+\bullet}$ (cf equation 39), ionic products arise from the scission of the C—F bond (to yield MgF^+), as well as from the simultaneous rupture of two bonds. The latter include $MgOH_2^{+\bullet}$, CHF_2CO^+ and $CF_2CH_2^{+\bullet}$. The observed products are consistent with those arising from structure **17**, which was predicted to be the minimum energy structure based on *ab initio* calculations.

(17)

A similar formation of five-membered rings involving bidentate coordination to $Mg^{+\bullet}$ appears to be at the heart of the photofragmentation of the $Mg^{+\bullet}$ complexes of 2-methoxyethanol and 1,2-dimethoxyethane[78]. Aside from $Mg^{+\bullet}$ formation (cf equation 39), a range of photoproducts were identified. Interestingly, a significant number of the photoproducts are complexes of $Mg^{+\bullet}$ with neutral molecules such as H_2O, CH_2O and CH_3OH. Based on *ab initio* calculations, a key hydrogen shift mechanism is proposed to form intermediate **18** (Scheme 2, where X = H or CH_3). This carbon-centered radical intermediate then undergoes a range of competing fragmentation reactions. Two decomposition pathways of **18** which are common to the complexes of 2-methoxyethanol and 1,2-dimethoxyethane are shown in Scheme 2. The first involves H attack onto a carbon atom to yield **19**, which can then fragment via either loss of CH_2O or CH_3CH_2OX. The second involves H attack onto an oxygen atom to yield **20**, which can then fragment via loss of the cyclic ether.

The photoinduced reactions of the $Mg(CF_3OC_6H_5)^{+\bullet}$ complex have been examined using a combination of experiments and theory[79]. Four ionic products are observed: $Mg^{+\bullet}$ (cf equation 39), MgF^+ (cf equation 43), $C_6H_5^+$ (cf equation 41) and $CF_3OC_6H_5^{+\bullet}$ (cf equation 40). Other $Mg^{+\bullet}$ complexes of ethers that have been examined using photodissociation spectroscopy are those of 2-methoxyethanol and 1,2-dimethoxyethane described above (Scheme 2) and those of 1,3- (**21**) and 1,4- (**22**) dioxane[80]. The main ionic photoproduct of the complexes of **21** and **22** is $Mg^{+\bullet}$ (cf equation 39). While the complex of **21** gives only one other product, $Mg(O=CH_2)^{+\bullet}$, the complex of **22** gives a much richer range of ionic fragments including $MgOH^+$, $Mg(O=CH_2)^{+\bullet}$, $Mg(OCH=CH_2)^{+\bullet}$, $Mg(OCH_2CH_3)^{+\bullet}$, $C_2H_4^{+\bullet}$ and $C_3H_6O^{+\bullet}$. Based on theoretical calculations, the insertion complex **23** is the first key intermediate in the formation of $Mg(O=CH_2)^{+\bullet}$ from the complex of **21**. In a similar fashion, the insertion complex **24** is a key intermediate for

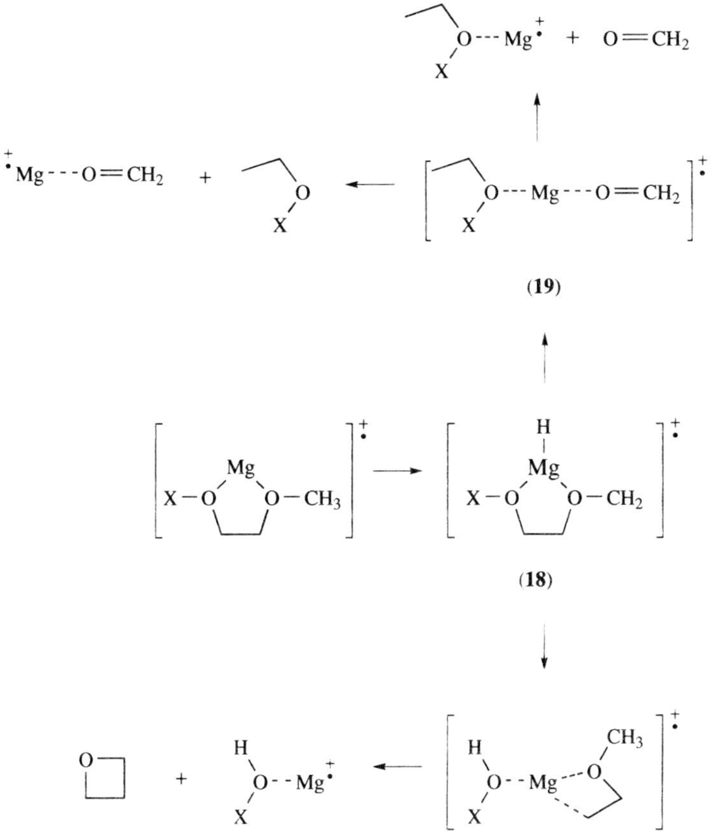

4. Solvent-free organomagnesiums

SCHEME 2

(19), (18), (20), (21), (22), (23), (24)

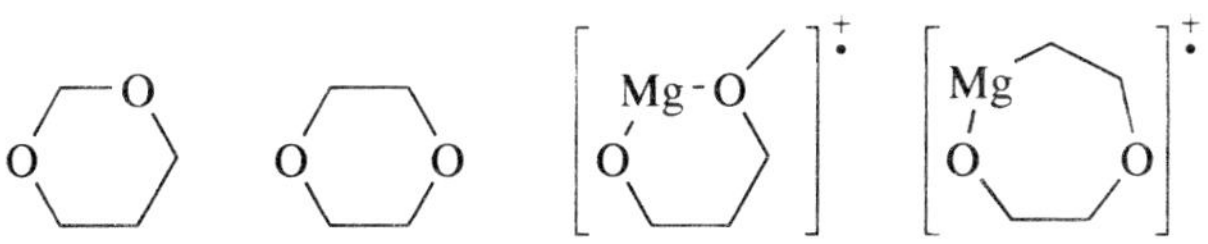

the formation of many of the products of the complex of **22**. Photoionization of neutral $Mg(O(CH_3)_2)_n$ clusters results in the formation of $Mg(O(CH_3)_2)_n^{+\bullet}$ clusters as well as $Mg(OCH_3)(O(CH_3)_2)_n^+$ ions arising from C—O bond activation[81].

8. Photoactivation reactions of complexes Mg(L)$^{+\bullet}$, where L = an amine

Photoactivation of Mg(amine)$^{+\bullet}$ complexes has been the subject of several studies[82-85]. The photofragmentation pathways are dependent on the structure of the amine. For methylamine, four processes are observed: $Mg^{+\bullet}$ formation (cf equation 39), electron transfer to form $CH_3NH_2^{+\bullet}$ (cf equation 40), C−N bond activation to yield $MgNH_2^+$ (cf equation 43) and C−H bond activation to form the immonium ion $CH_2=NH_2^+$ (equation 50 where R = H)[72]. Theoretical calculations suggest that this pathway involves hydrogen transfer from C to Mg via a four-centered transition state[74]. For dimethylamine, only $Mg^{+\bullet}$ (cf equation 39) and the immonium ion $CH_2=NHCH_3^+$ (equation 50 where R = CH_3) are observed[72]. For amines with larger alkyl groups, new reaction channels open up. Apart from forming $Mg^{+\bullet}$ (cf equation 39) and the immonium ion $CH_3CH=NHCH_2CH_3^+$ (cf equation 50), the $Mg(HN(CH_2CH_3)_2)^{+\bullet}$ complex fragments to eliminate a C_3H_7 radical (equation 51 where R = H)[73]. The $Mg^{+\bullet}$ complex of triethylamine fragments via $Mg^{+\bullet}$ formation (cf equation 39), electron transfer (cf equation 40) and elimination of a C_3H_7 radical (equation 51 where R = CH_3CH_2)[73]. The exact structure of the ionic product is uncertain. When R = H, DFT calculations suggest that the $Mg(HNCH_3)^+$ isomer is only about 1 kcal mol^{-1} more stable than the $Mg(H_2NCH_2)^+$ isomer. In a follow-up paper, DFT calculations were carried out to suggest a mechanism for reaction 51 (R = H). Scheme 3 highlights that this intriguing reaction involves multiple bond breaking and bond making. Thus, insertion of the Mg into the C−N bond yields the organometallic ion **25**, which then forms **26** via H transfer. CH_3 transfer occurs via the six-centered transition state **27**. Ultimately, the organometallic ion **28** is formed. The $Mg^{+\bullet}$ complexes of propylamine, isopropylamine, dipropylamine and diisopropylamine exhibit a rich photochemistry[75]. One of the most interesting sets of products occurs for the secondary amines and involves C−C bond coupling, which may occur via processes related to those shown in Scheme 3.

$$Mg(CH_3NHR)^{+\bullet} + h\nu \longrightarrow CH_2=NHR^+ + MgH^{\bullet} \qquad (50)$$

$$Mg(RN(CH_2CH_3)_2)^{+\bullet} + h\nu \longrightarrow Mg(R,N,C,H_3)^+ + C_3H_7^{\bullet} \qquad (51)$$

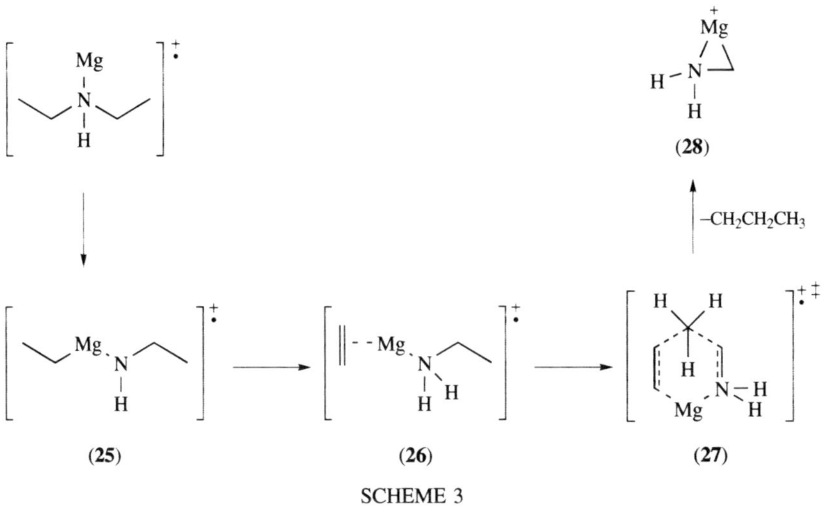

SCHEME 3

Photodissociation of the $Mg^{+\bullet}$ complex of pyridine yields two products: $Mg^{+\bullet}$ as the major product (cf equation 39) and $C_5H_5N^{+\bullet}$ via electron transfer (cf equation 40) as the minor channel[86]. The photodissociation spectra of $Mg(NCCH_3)_n^{+\bullet}$ complexes has been studied as a function of cluster size, n^{87}. For $n = 1$, there are two products: $Mg^{+\bullet}$ as the major product (cf equation 39) and $MgNC^+$ as the minor product (cf equation 43). Solvent evaporation is the sole reaction channel observed for all the other clusters, $n = 2-4$ (equation 52).

$$Mg(NCCH_3)_n{}^{+\bullet} + h\nu \longrightarrow Mg(NCCH_3)_{n-1}{}^{+\bullet} + CH_3CN \qquad (52)$$

9. Photoactivation reactions of complexes Mg(L)$^{+\bullet}$, where L contains a C=X bond (X = O or S)

The photodissociation spectra of the $Mg(L)^{+\bullet}$ complexes containing substrates with a C=O moiety have been studied for formaldehyde[88, 89], acetaldehyde[90, 91], acetic acid[92] and N,N-dimethylformamide[93]. Apart from $Mg^{+\bullet}$ formation (cf equation 39), these complexes share some common reactive channels (equations 53–56). Formaldehyde undergoes a significant amount of H abstraction (equation 53, X = Y = H) for the magnesium complex[78], which contrasts with the calcium complex[79]. For the acetaldehyde complex, the Mg attacks both the C–H bond (equations 53, where X = H and Y = CH$_3$) as well as the C–C bond (equations 54 and 56, where X = H and Y = CH$_3$)[80, 81]. Deuterium labeling confirms that the C–H bond attacked is the aldehydic C–H bond rather than the methyl C–H bond. The adduct of acetic acid fragments via formation of the following ions: $Mg^{+\bullet}$ (cf equation 39), $MgCH_3{}^+$ (equation 53 where X = CH$_3$ and Y = OH), $MgOH^+$ (equation 53) and CH_3CO^+ (equation 55). In addition, dehydration of acetic acid to form ketene and $MgOH_2^{+\bullet}$ is observed. When CH_3CO_2D is used, a minor yield of $MgOH^+$ is observed, suggesting activation of the C–H bond. When L = N,N-dimethylformamide the following photoproducts are observed: $Mg^{+\bullet}$ (cf equation 39), MgH^+ [equation 53 where X = H and Y = (CH$_3$)$_2$N] and (CH$_3$)$_2$NCO$^+$ (equation 55). Once again, deuterium labeling was used to confirm that the C–H bond attacked is the formyl C–H bond rather than the methyl C–H bond. The dimer complex, $Mg(HCON(CH_3)_2)_2{}^{+\bullet}$, simply undergoes solvent evaporation (cf equation 52)[93].

$$Mg(O=CXY)^{+\bullet} + h\nu \longrightarrow MgX^+ + YCO^{\bullet} \qquad (53)$$

$$\longrightarrow MgY^+ + XCO^{\bullet} \qquad (54)$$

$$\longrightarrow YCO^+ + MgX^{\bullet} \qquad (55)$$

$$\longrightarrow MgCXO^+ + Y^{\bullet} \qquad (56)$$

The photodissociation spectra of the $Mg(L)^{+\bullet}$ complexes of ethyl isocyanate[94, 95] and ethyl isothiocyanate[96] show some common photofragments. Aside from the ubiquitous formation of $Mg^{+\bullet}$ (cf equation 39), both ethyl isocyanate and ethyl isothiocyanate yield products from attack of the N–C single bond (equations 57 and 58). The ethyl isothiocyanate complex also yields MgS via equation 59. The photodissociation spectrum of the ethyl thiocyanate isomer was also examined and gave the products shown in equations 60–62. Thus each isomer gives a unique ionic product [MgS$^{+\bullet}$ for ethyl isothiocyanate (equation 59) vs $MgNC^+$ for ethyl thiocyanate (equation 62)] which allows their distinction. Finally, the Mg(ethyl isocyanate)$_n{}^{+\bullet}$ complexes simply undergo solvent evaporation for $n = 2$ and 3 (cf equation 52).

$$Mg(X=C=NCH_2CH_3)^{+\bullet} + h\nu \longrightarrow Mg(NCX)^+ + CH_3CH_2{}^{\bullet} \qquad (57)$$

$$\longrightarrow CH_3CH_2{}^+ + Mg(NCX)^{\bullet} \qquad (58)$$

$$\longrightarrow MgX^{+\bullet} + CH_3CH_2NC \qquad (59)$$

$$Mg(NCSCH_2CH_3)^{+\bullet} + h\nu \longrightarrow Mg(NCS)^+ + CH_3CH_2^\bullet \quad (60)$$
$$\longrightarrow CH_3CH_2^+ + Mg(NCS)^\bullet \quad (61)$$
$$\longrightarrow MgNC^+ + CH_3CH_2S^\bullet \quad (62)$$

The $Mg^{+\bullet}$ complexes of cytosine, thymine and uracil are the most complex system studied via photodissociation spectroscopy to date[97,98]. A complication for these systems is that these nucleobases can exist in various tautomeric forms and that complexation of a metal can change the stability order of the tautomers. DFT calculations located four tautomeric $Mg(cytosine)^{+\bullet}$ complexes, and three of these (**29**, **30**, and **31**) were suggested to be responsible for the four reactive photofragment ions **32**–**35** observed at a wavelength of 360 nm (Scheme 4)[97]. Related photofragmentation reactions were observed for the $Mg(thymine)^{+\bullet}$ and $Mg(uracil)^{+\bullet}$ complexes[98].

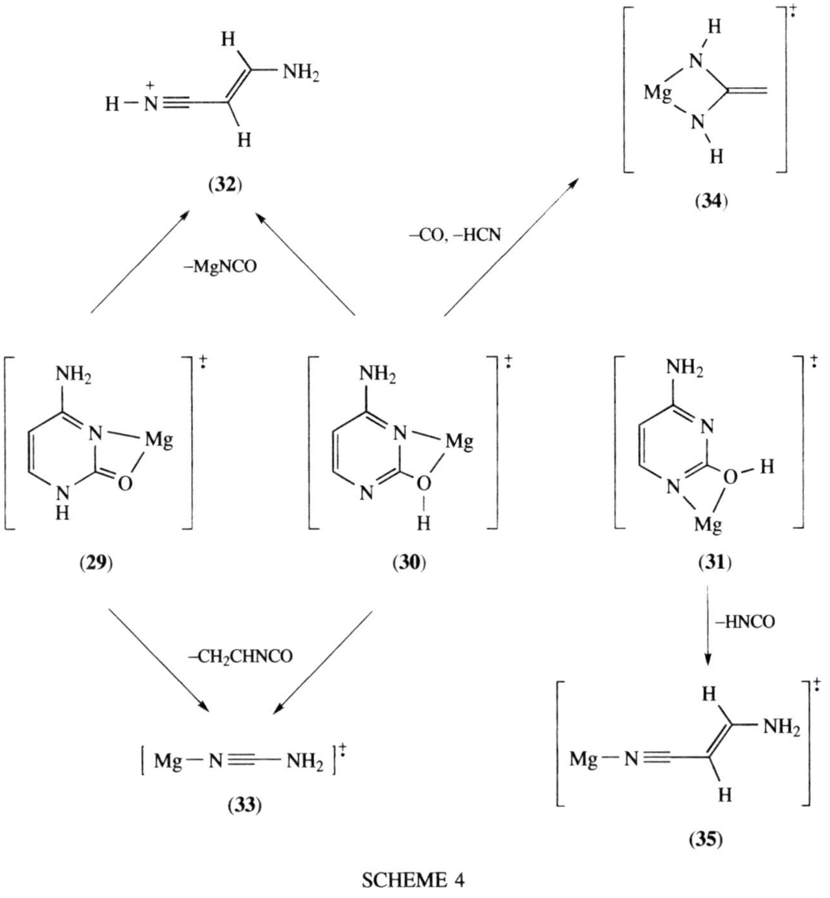

SCHEME 4

C. Reactions of Magnesium Clusters with Organic Substrates

Klabunde and coworkers compared the reactivity of metal atoms with metal clusters (metal = magnesium and calcium) with CH_3X (X = F, Cl, Br and I) under conditions of matrix isolation (Ar matrix)[24,25]. UV-Vis spectroscopy was used to monitor the reactivity of the metal clusters to form organometallic cluster reagents (equation 63). The general metal reactivity trends were: Ca_x ($x > 2$) ≈ Ca_2 > Mg_x ($x > 4$) ≈ Mg_4 > Mg_3 ≈ Mg_2 > Ca > Mg. The substrates reacted in the order: CH_3I > CH_3F > CH_3Br > CH_3Cl. The enhanced reactivity of the clusters is consistent with the early theoretical calculations, which predicted a greater stability of CH_3Mg_2X relative to the formation of CH_3MgX and an isolated Mg atom[99].

$$\text{Metal}_n + CH_3X \longrightarrow CH_3\text{Metal}_nX \tag{63}$$

Although cluster Grignard reagents are the proposed products of equation 63 (where Metal = Mg), the first spectroscopic evidence for the formation of the $PhMg_4X$ cluster Grignard reagents (X = F, Cl and Br) involved the assignment of their molecular weights via the formation of their protonated ions under conditions of MALDI MS[5]. These assignments were consistent with the stoichiometries of the hydrolysis reaction (equation 64). The cluster Grignard reagents undergo a number of interesting reactions including transmetallation (equation 65)[100,101] and catalysis of the isomerization of allylbenzene to β-methylstyrene[102].

$$C_6H_5Mg_4F + 7H_2O \longrightarrow C_6H_6 + MgF(OH) + 3Mg(OH)_2 + 3H_2 \tag{64}$$

$$C_6H_5Mg_4X + RY \longrightarrow RMg_4Y + C_6H_5X \tag{65}$$

Two studies have examined the formation of CH_3Mg_nCl cluster Grignard reagents via the use of theoretical methods[103,104]. Two different competing pathways were located for the reaction of the tetrahedral Mg_4 cluster with CH_3Cl (Scheme 5). The first involves formation of the transition state **36**, which yields the cluster **37**, in which the tetrahedral Mg_4 framework is maintained. The second involves the formation of the transition state **38**, which yields the cluster **39**, in which the Mg_4 framework is rhombic instead. Interestingly, while the activation energy for the first pathway is lower (18.4 kcal mol^{-1} for **36** versus 24.8 kcal mol^{-1} for **38**) the most stable product is that for the second pathway (-48.2 kcal mol^{-1} for **37** versus -51.5 kcal mol^{-1} for **39**)[92]. Jasien and Abbondondola found that the activation energy for transition states related to **38** decrease as the size of the magnesium cluster increases, with the lowest activation energy being 9.8 kcal mol^{-1} for the Mg_{21} cluster[103].

The sole gas-phase study on a cationic magnesium cluster examined the photodissociation spectrum of the $Mg_2(CH_4)^{+\bullet}$ complex[105]. $Mg_2^{+\bullet}$ is only a minor product (equation 66) while $Mg^{+\bullet}$ is the main ionic fragment and may arise via either of the processes shown in equations 67 and 68. The latter reaction is predicted to only be slightly more endothermic.

$$Mg_2CH_4^{+\bullet} + h\nu \longrightarrow Mg_2^{+\bullet} + CH_4 \tag{66}$$

$$\longrightarrow Mg^{+\bullet} + MgCH_4 \tag{67}$$

$$\longrightarrow Mg^{+\bullet} + Mg + CH_4 \tag{68}$$

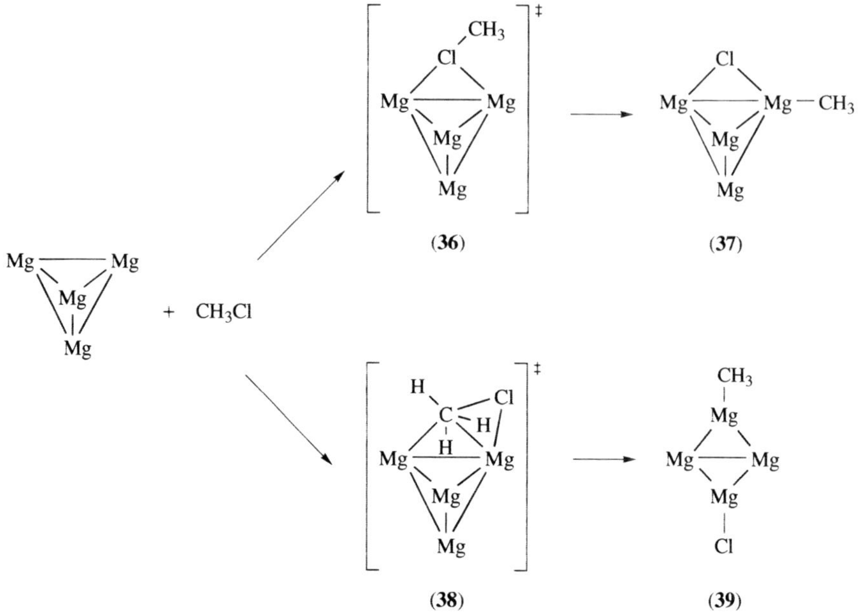

172   Richard A. J. O'Hair

SCHEME 5

D. Reactions of Magnesium Surfaces and Films with Organic Substrates

The formation of Grignard reagents is a complex heterogeneous process that involves surface chemistry, interfacial chemistry and solution-phase chemistry. Since this topic has been comprehensively reviewed[106, 107], here the chemistry of well-defined, clean Mg surfaces and Mg thin films is briefly discussed. Using X-ray photoelectron spectroscopy, Abreu and coworkers have examined the nature of the surface film formed when a clean Mg metal surface is subjected to pretreatments that simulate exposure to ambient environments[108]. They noted that as-received Mg metal contains a surface covered by a film mainly composed of magnesium hydroxide with smaller quantities of magnesium bicarbonate. These surface films slow down Grignard formation since the alkyl halide must bypass the surface hydroxide and bicarbonate sites to interact with Mg metal site(s). Nuzzo and Dubois used a Mg(0001) single-crystal surface to examine the chemisorption and subsequent decomposition of MeBr[109]. They found evidence for the formation of a surface bromide and gas-phase ethane. Although stable surface methyls were not observed even at $-150\,°C$, they suggested the mechanistic picture in Scheme 6, in which cleavage of the C—Br bond yields **40**. Rather than form the Grignard, **41**, the surface bromide **42** is formed. The role of surface modification was also examined. For example, while co-adsorbed dimethyl ether does not perturb the reactivity pattern, formation of either a thin surface bromide or a surface oxide passivates the surface to further reaction under the ultra-high-vacuum conditions of the experiments.

Gault and coworkers have used a specially constructed chamber interfaced to a mass spectrometer to examine the reactions of organic substrates with magnesium films[110–114]. In contrast to the chemistry of the pristine Mg(0001) surface, Gault found that alkyl halides were adsorbed irreversibly on Mg films to ultimately yield solid dull films of the organomagnesium 'RMgX' (R = Et, Me_2CH, n-Pr, n-Bu; X = Br, Cl)[110]. These

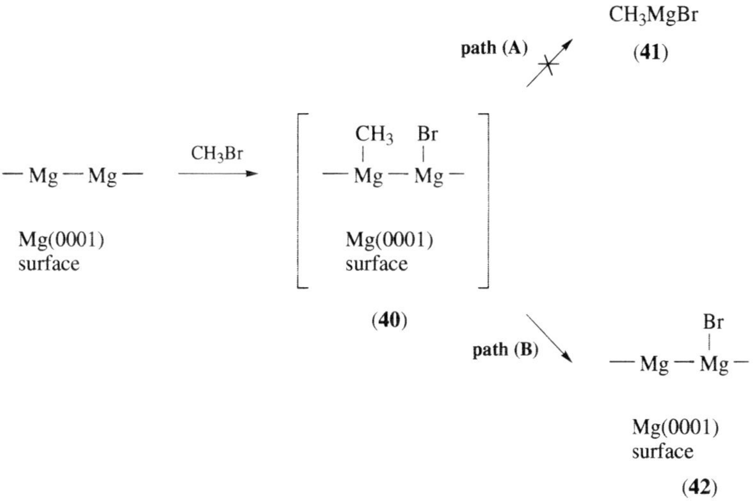

4. Solvent-free organomagnesiums

$$
\begin{array}{c}
\text{—Mg—Mg—} \\
\text{Mg(0001)} \\
\text{surface}
\end{array}
\xrightarrow{\text{CH}_3\text{Br}}
\left[
\begin{array}{c}
\text{CH}_3\quad \text{Br} \\
|\qquad\ | \\
\text{—Mg—Mg—} \\
\text{Mg(0001)} \\
\text{surface}
\end{array}
\right]
\begin{array}{c}
\xrightarrow{\text{path (A)}}\!\!\!\!\!/\;\; \text{CH}_3\text{MgBr}\quad(\mathbf{41}) \\
\\
\\
\xrightarrow{\text{path (B)}}
\begin{array}{c}
\text{Br} \\
| \\
\text{—Mg—Mg—} \\
\text{Mg(0001)} \\
\text{surface} \\
(\mathbf{42})
\end{array}
\end{array}
$$

(**40**)

SCHEME 6

films were soluble in diethyl ether and underwent all the reactions of solvated Grignard reagents. For example, the 'EtMgBr' film liberated C_2H_6 on treatment with water or alcohols and underwent the Grignard reaction with adsorbed carbonyl compounds. In separate experiments, Gault[110] and Lefrancois and Gault[111] examined the decomposition of the organomagnesium films at high temperature. For example, 'EtMgBr' decomposed at 180 °C with liberation of a mixture of saturated and unsaturated hydrocarbons. Ethylene was the main constituent of the unsaturated fraction along with butenes and hexenes. All hydrocarbons were formed initially, thus ruling out a chain reaction mechanism involving $CH_2=CH_2$. Decomposition of the 'EtMgBr' film in the presence of propene gave appreciable amounts of pentenes and some heptenes as immediate products, suggesting reactions between olefins and 'RMgX' or radicals produced during decomposition.

In a series of three papers, Gault[112,113] and Choplin and Gault[114] used the same apparatus to examine the self-hydrogenation of alkynes and dienes of Mg films. In the first report, Gault noted that when either 1-alkynes, 2-alkynes or 1,2-dienes are allowed to interact with a magnesium film in the absence of co-adsorbed hydrogen, the gaseous reaction products consisted of alkenes as well as isomers of the starting material[112]. In follow-up studies, the species which remained adsorbed on the Mg during the reaction were desorbed by quenching reactions with D_2O and characterized as the deuterated hydrocarbons. The structures of these hydrocarbons and the variation in their D distribution with temperature and reaction time are consistent with a mechanism consisting of two parallel processes:

(i) The dehydrogenation to the metallated species **43** which is stable at 373 K but is further dehydrogenated to the carbide, **44**, at 423 K. Indirect evidence for these intermediates was gained via reaction with D_2O, which yielded d_1 and d_4 propyne (Scheme 7).

(ii) The two-step hydrogenation of the reactant to propene. A possible mechanism to rationalize the experimental data is shown in Scheme 8. Reaction with a portion of the film containing magnesium hydride moieties, **45** (formed via processes related to Scheme 7), yields intermediate **46**. The desorption of propene can occur via either heating of **46** (which presumably involves a reductive elimination) or via reaction of **46** with D_2O. By

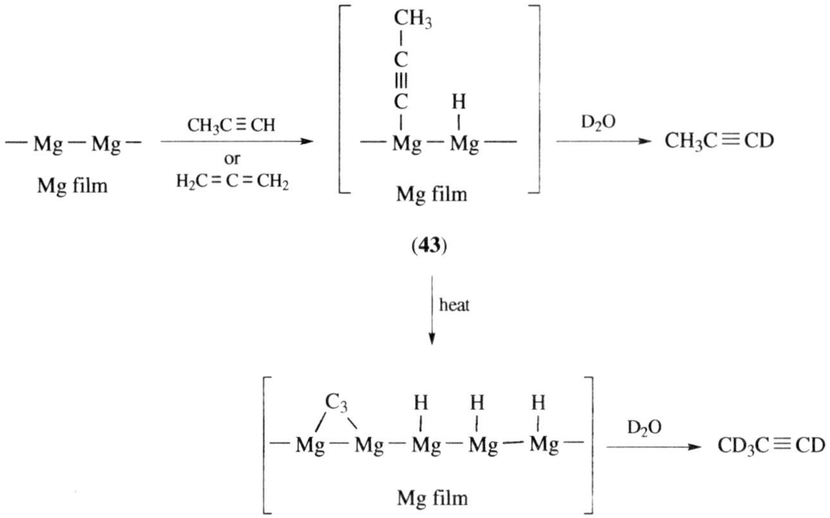

SCHEME 7

using microwave spectroscopy, Choplin and Gault were able to show that, regardless of the precursor (allene on propyne), the main product of the latter reaction is the (*E*)-1-deuteriopropene, **47**[114]. This suggests a four-centered transition state for the initial reaction to form **46**.

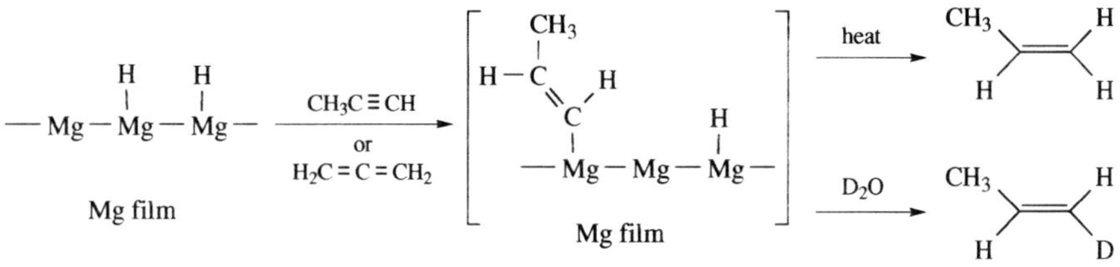

SCHEME 8

E. Gas-phase Fragmentation of Ligated Magnesium Ions to Yield Organomagnesium Ions

With the advent of electrospray ionization (ESI), it is now possible to study the gas-phase chemistry of Mg(II) species. These can be in three different charge states, depending on the type(s) of ligand(s), L, coordinated to the Mg center. For complexes containing only neutral ligand(s), the net charge on the complex is $+2$. If one ligand is monoanionic, the charge becomes $+1$. Mg complexes containing ligands with a total of three negative charges can be observed in the negative ion mode. The chemistry of each of these types of complexes is now described.

1. Fragmentation reactions of $Mg(L)_n^{2+}$ complexes

Kebarle and coworkers carried out some of the pioneering studies in this area by subjecting metal salts to ESI conditions using various solvents/ligands, L^{115}. In addition, Stace and coworkers have developed an alternative technique whereby solvated/ligated Mg atoms are subject to electron impact[116-122]. Both these techniques provide fundamental information on the inherent kinetic stability of $Mg(L)_n^{2+}$ complexes with respect to fragmentation. In addition, Kebarle and coworkers have determined the sequential binding energies of water, acetone and N-methylacetamide to Mg^{2+} [123]. Table 1 lists the stability of a range of $Mg(L)_n^{2+}$ complexes as a function of the ionization energy of the ligand. The criterion for stability is a kinetic one and relates to the smallest cluster number (n) for which a stable doubly charged cluster, $Mg(L)_n^{2+}$, is observed (defined as n_{min}). Since the second ionization energy of Mg is 15.03 eV, it is not surprising that $n_{min} > 1$ for most ligands.

Several studies have not only examined n_{min}, but have also identified the key fragmentation channels of $Mg(L)_n^{2+}$ complexes as a function of both the ligand structure as well as the number of ligands, n. The fragmentation channels can be divided into 5 main classes of reactions: loss of a neutral ligand (equation 69); metal charge reduction via electron transfer which yields the two complementary ions $Mg(L)_{n-1}^{+\bullet}$ and $L^{+\bullet}$ (equation 70); charge reduction via interligand proton transfer which yields the two complementary ions $Mg(L)_{n-2}(L-H)^+$ and $[L+H]^+$ (equation 71); ligand fragmentation via neutral loss (equation 72); and ligand fragmentation via loss of a cation (equation 73). Which fragmentation channel dominates depends upon both the type of ligand and its properties (such as ionization energy, which influences reaction 70) as well as the number of ligands[101]. A key difference in the stability and fragmentation reactions of the related $Ca(L)_n^{2+}$ complexes is that the lower second ionization energy of Ca allows ions with

TABLE 1. Stability of $Mg(L)_n^{2+}$ complexes in the gas phase as a function of ligand ionization energy (IE)

Ligand, L =	IE (eV) [a,b]	n_{min}	Reference
CO_2	13.78	2	122
H_2O	12.62	2	122
CH_3CN	12.20	1	122
Methanol	10.84	2	122
Ethanol	10.48	3	122
n-Propanol	10.22	3	122
Ammonia	10.07	2	122
Acetone	9.70	3	122
Acetamide	9.69	2	129
2-Butanone	9.52	2	122
Diethyl ether	9.51	2	122
Tetrahydrofuran	9.40	2	122
2-Pentanone	9.38	2	122
Pyridine	9.26	2	122
Dimethyl sulfoxide	9.10	1	126
n-Butylamine	8.73	2	122
Pyrrole	8.21	2	122
Pentan-2,4-dione	8.85	1	122
4-Hydroxy-4-methylpentan-2-one	unknown	1	128
Ethylene diamine	8.6	—	122

[a] IE is defined as: $M \to M^{+\bullet} + e^-$; $\Delta H = IE$.
[b] All IEs are taken from http://webbook.nist.gov.

low n to become more kinetically stable with respect to electron transfer (equation 70).

$$Mg(L)_n^{2+} \longrightarrow Mg(L)_{n-1}^{2+} + L \qquad (69)$$
$$\longrightarrow Mg(L)_{n-1}^{+\bullet} + L^{+\bullet} \qquad (70)$$
$$\longrightarrow Mg(L)_{n-2}(L-H)^+ + [L+H]^+ \qquad (71)$$
$$\longrightarrow Mg(L)_{n-1}(L-X)^{2+} + X \qquad (72)$$
$$\longrightarrow Mg(L)_{n-1}(L-Y)^+ + Y^+ \qquad (73)$$

Under conditions of CID, $Mg(L)_n^{2+}$, where L = methanol, fragment via several different pathways[120]. For larger clusters (e.g. $n = 10$), methanol loss dominates (equation 69). At $n = 4$, charge transfer products appear (equation 70) together with products from other fragment channels. The $n = 3$ cluster fragments via reactions that reduce the overall charge state to +1: charge transfer (equation 70), proton transfer (equation 71) and two ligand fragmentation channels [equation 71, where (L−Y) = OH and H]. $Mg(L)_n^{2+}$ clusters from larger alcohols such as n-propanol are less stable and undergo charge transfer more readily[121]. Recent theoretical calculations on $Mg(CH_3OH)^{2+}$ suggest that while there are three exothermic reaction channels (charge transfer, equation 70, and ligand fragmentation, equation 73, to form $MgOH^+$ and MgH^+), this cluster ion should be kinetically stable due to significant barriers to all three reaction channels[124].

When L is the commonly used solvent THF, a particularly stable $Mg(THF)_4^{2+}$ ion is formed in the gas phase. $Mg(THF)_3^{2+}$ undergoes ligand fragmentation via neutral loss (equation 72, $X = C_3H_6$) to yield $Mg(THF)_2(CH_2O)^{2+}$ [119]. There are interesting differences in the fragmentation of the CH_3-X bonds in the $Mg(L)_n^{2+}$ complexes of acetonitrile and dimethyl sulfoxide. Thus while the acetonitrile clusters tend to fragment via heterolytic cleavage to yield $MgCN(L)_{n-1}^+$ and CH_3^+ (equation 73)[125], the dimethyl sulfoxide fragment via CH_3 and CH_4 loss (equation 72)[126]. Destruction of the aromatic ring is observed for the $Mg(L)_n^{2+}$ complexes of pyridine[127]. For example, $MgCN(L)_{n-1}^+$ and $C_4H_5^+$ formation is observed (equation 73). The $Mg(L)_n^{2+}$ complexes of 4-hydroxy-4-methylpentan-2-one undergo a range of ligand fragmentation including C−C bond cleavage via a retro-aldol reaction[128].

One of the few studies to have combined experiment and theory has provided detailed mechanistic insights into the fragmentation reactions of the $Mg(L)_n^{2+}$ complexes of acetamide[129]. The $Mg(L)_3^{2+}$ complex fragments via neutral ligand loss (equation 69) in competition with interligand deprotonation (equation 71). The chemistry of the latter product, $Mg(L-H)(L)^+$, is described further below. The $Mg(L)_2^{2+}$ complex fragments solely via heterolytic amide bond cleavage (equation 71) to yield $Mg(NH_2)(L)^+$ and CH_3CO^+. DFT calculations on the $Mg(L)_2^{2+}$ complex reveal that neutral ligand loss (equation 69) is much more endothermic than interligand deprotonation (equation 71).

2. Fragmentation reactions of $XMg(L)_n^+$ complexes (where X = an anionic ligand)

Surprisingly few studies have thoroughly investigated the gas-phase chemistry of $XMg(L)_n^+$ complexes. An exception is the combined experimental and theoretical study on the fragmentation reactions of the $XMg(L)^+$ (where X = L−H and NH_2) and $Mg(L−H)^+$ complexes of acetamide (L = CH_3CONH_2)[129]. The $H_2NMg(L)^+$ complex fragments via ligand loss (equation 74) and NH_3 loss (equation 75). The $Mg(L−H)(L)^+$ complex fragments via ligand loss (equation 76), water (equation 77) and acetonitrile loss (equation 78). Finally, the Mg(L−H)$^+$ complex fragments via losses of HNCO (equation 79), MgO (equation 80) and acetonitrile loss (equation 81).

$$H_2NMg(L)^+ \longrightarrow H_2NMg^+ + L \tag{74}$$

$$\longrightarrow Mg(L-H)^+ + NH_3 \tag{75}$$

$$Mg(L-H)(L)^+ \longrightarrow Mg(L-H)^+ + L \tag{76}$$

$$\longrightarrow Mg(L-H)(CH_3CN)^+ + H_2O \tag{77}$$

$$\longrightarrow Mg(L-H)(H_2O)^+ + CH_3CN \tag{78}$$

$$Mg(CH_3CONH)^+ \longrightarrow MgCH_3^+ + HNCO \tag{79}$$

$$\longrightarrow CH_3CNH^+ + MgO \tag{80}$$

$$\longrightarrow MgOH^+ + CH_3CN \tag{81}$$

DFT calculations on the Mg(L−H)(L)$^+$ complex reveal how water and acetonitrile can be lost (Scheme 9). Thus intramolecular proton transfer tautomerizes the neutral acetamide ligand in **48** into the hydroxyimine form in **49**, which can then dissociate via another intramolecular proton transfer to yield the four-coordinate adduct **50**, which now contains both water and acetonitrile ligands. It is this complex that is the direct precursor to water and acetonitrile loss. Note that the reaction shown in Scheme 9 is a retro-Ritter reaction and involves fragmentation of the neutral rather than the anionic acetamide ligand, which is a bidentate spectator ligand.

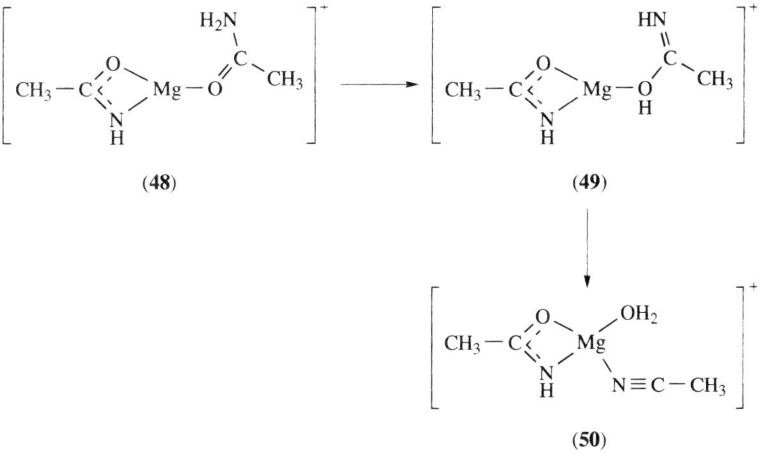

SCHEME 9

DFT calculations on the Mg(L−H)$^+$ complex also reveal how HNCO might be lost (Scheme 10). Thus the bidentate interaction of the acetamide ligand with Mg in **51** must be disrupted to yield either of the monodentate structures **52** or **54**. These intermediates can insert into the CH$_3$−C bond via four-centered transition states to yield the organometallic ions **53** or **55**, which can then lose HNCO to form MgCH$_3^+$. The DFT calculations reveal that path (A) of Scheme 10 is kinetically favored.

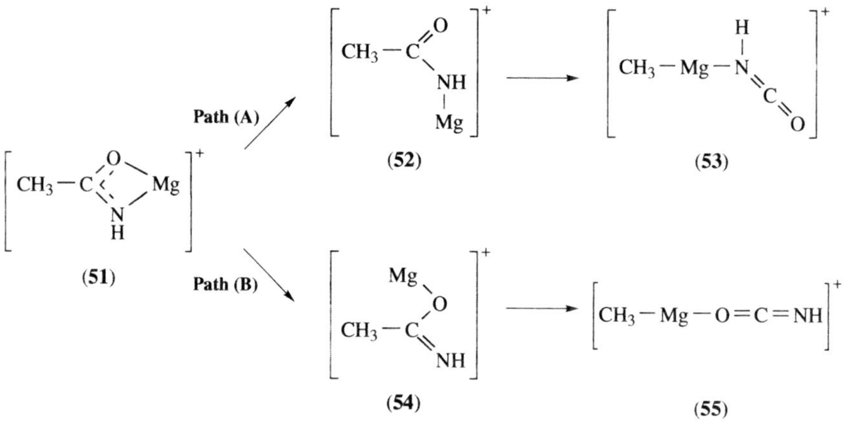

SCHEME 10

Wu and Brodbelt have studied the gas-phase fragmentation reactions of $HOMg(L)^+$ complexes of crown ethers and glymes[130]. A common loss involves units of C_2H_4O, which can either directly occur from the precursor ion, or can be triggered by an initial interligand reaction between HO^- and L. This latter reaction is illustrated in Scheme 11 for the complex of 12-Crown-4. Thus loss of H_2O from the initial adduct **56** yields the ring-opened complex **57**, which contains a coordinated alkoxide moiety, which can then lose an epoxide to form the related complex **58**.

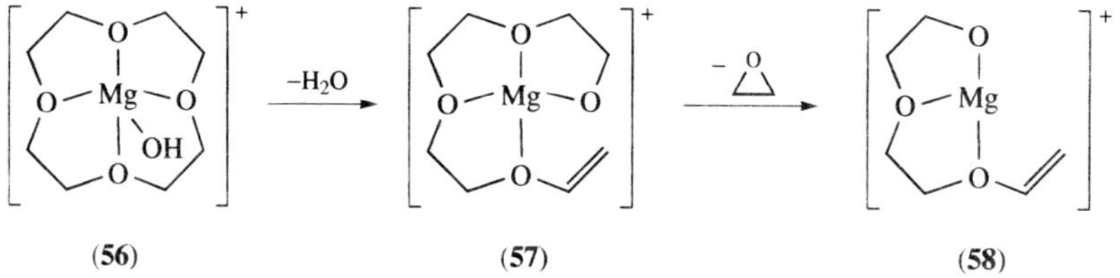

SCHEME 11

3. Fragmentation reactions of $Mg(X)_3^-$ complexes

The final class of Mg(II) ions which can readily be formed via ESI MS involves magnesate anions, which are formed by coordinating an anion to a Mg(II) salt. The gas-phase fragmentation reactions of $CH_3CO_2MgX_2^-$ (where X = Cl and CH_3CO_2) have been studied using a combination of CID experiments in a quadrupole ion trap in conjunction with DFT calculations[131]. Decarboxylation (equation 82) is the main reaction channel, with some acetate loss (equation 83) also being observed. DFT calculations reveal that the former reaction is less endothermic. The decarboxylation reactions yield the organomagnesates, $CH_3MgX_2^-$, and are reminiscent of the HNCO loss described above (equation 79 and Scheme 10). The DFT calculations also provide insights into the coordination modes of reactants and products for the decarboxylation reactions (Scheme 12). Generally, the carboxylate ligands bind in a bidentate fashion, while the chloride ions are monodentate.

4. Solvent-free organomagnesiums

When X = CH$_3$CO$_2$, the ground-state reactant structure is six-coordinate, **59**. The 'reactive' geometry for decarboxylation, **60**, requires cleavage of one of the Mg—O bonds. The product, **61**, is five-coordinate. In contrast, when X = Cl, the three-coordinate product, **63**, is directly formed via decarboxylation of the four-coordinate reactant **62**.

$$CH_3CO_2MgX_2^- \longrightarrow CH_3MgX_2^- + CO_2 \tag{82}$$

$$\longrightarrow CH_3CO_2^- + MgX_2 \tag{83}$$

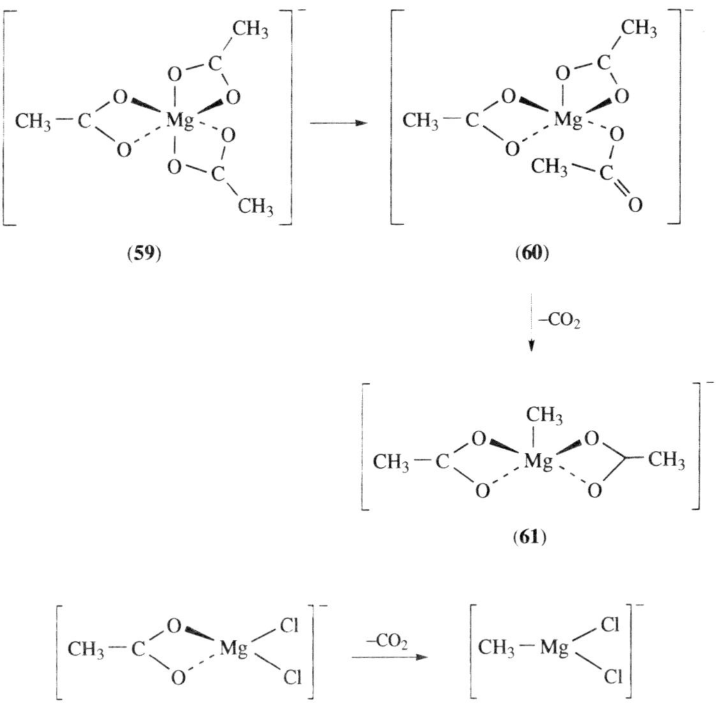

SCHEME 12

Related decarboxylation reactions have been used to synthesize magnesium hydride anions from formate anions[132] and organocalcium, organobarium and organostrontium metallates[133].

III. BIMOLECULAR REACTIONS OF ORGANOMAGNESIUM IONS IN THE GAS PHASE

Relatively few studies have examined the bimolecular reactivity of organomagnesium species in the gas phase. In terms of organomagnesium ions, this has largely been due to the fact that it has been difficult to generate organomagnesium cations via traditional electron ionization (EI). Not surprisingly, the bimolecular chemistry of organomagnesium cations has focused on the chemistry of $(c\text{-}C_5H_5)Mg^+$ and $(c\text{-}C_5H_5)_2Mg^{+\bullet}$, which are readily formed via EI on magnesocene. Bohme and coworkers have compared the reactivity of $Mg^{+\bullet}$, $(c\text{-}C_5H_5)Mg^+$ and $(c\text{-}C_5H_5)_2Mg^{+\bullet}$ towards alkanes[46] and a range of small inorganic ligands (H_2, NH_3, H_2O, N_2, CO, NO, O_2, CO_2, N_2O and NO_2)[134,135]. As noted in Section II.B.1, $Mg^{+\bullet}$ reacts with alkanes via ligand addition (equation 16). Single ligation of $Mg^{+\bullet}$ with $(c\text{-}C_5H_5)^\bullet$ substantially enhances the efficiency of subsequent ligation. Thus ligation is rapid with all the hydrocarbons investigated. In contrast, no reaction was observed between the alkanes and the full-sandwich magnesocene cation, $(c\text{-}C_5H_5)_2Mg^{+\bullet}$. $Mg^{+\bullet}$ was unreactive towards all the inorganic ligands except with ammonia, which was found to sequentially add up to 5 ligands (equation 84). The structures of the $Mg(NH_3)_n^{+\bullet}$ ions ($n = 1-4$) were probed via DFT calculations. In all cases, structures in which the NH_3 ligands are directly coordinated to the $Mg^{+\bullet}$ were more stable than other structures (such as those in which one NH_3 ligand is hydrogen bonded to a coordinated NH_3 ligand). Once again, the singly ligated $(c\text{-}C_5H_5)Mg^+$ complex substantially enhances the efficiency of ligation by inorganic ligands (equation 85). Thus initial ligation is rapid with all ligands except H_2, N_2 and O_2. The 'full-sandwich' magnesocene radical cation, $(c\text{-}C_5H_5)_2Mg^{+\bullet}$. does not undergo ligation. Instead, fast bimolecular ligand-switching reactions occur with NH_3 and H_2O, suggesting that these two ligands bind more strongly to $(c\text{-}C_5H_5)Mg^+$ than does $c\text{-}C_5H_5^\bullet$ itself (equation 86).

$$Mg(L)_n^{+\bullet} + L \longrightarrow Mg(L)_{n+1}^{+\bullet} \tag{84}$$

$$(c\text{-}C_5H_5)Mg(L)_n^+ + L \longrightarrow (c\text{-}C_5H_5)Mg(L)_{n+1}^+ \tag{85}$$

$$(c\text{-}C_5H_5)_2Mg^{+\bullet} + L \longrightarrow (c\text{-}C_5H_5)Mg(L)^+ + c\text{-}C_5H_5^\bullet \tag{86}$$

As noted in a previous review[136], one of the benefits of the quadrupole ion trap mass spectrometer is that ions are stored in the quadrupole ion trap and can be manipulated to undergo multiple stages of mass spectrometry associated with different types of reactions. Thus CID can be used to 'synthesize' organometallic ions via CID and their gas-phase reactivity can then be examined via subsequent ion–molecule reactions. We have used the decarboxylation reaction (equation 82) to synthesize organoalkaline earths, $[CH_3MetalX_2]^-$, and have studied their acid–base reactions with neutral acids, AH (equation 87), to establish how reactivity is controlled by the auxiliary ligand[131], the nature of the metal[133] and the substrate, AH[131]. In our first study on the organomagnesates $CH_3MgX_2^-$ (X = Cl and O_2CCH_3) we examined the influence of the auxiliary ligand and the substrate on reactivity[131]. We found that these $CH_3MgX_2^-$ ions exhibit some of the reactivity of Grignard reagents, reacting with acids, AH, via addition with concomitant elimination of methane to form $AMgX_2^-$ ions (equation 87, Metal = Mg), in direct analogy to the acid–base reactions of Grignard reagents. Kinetic measurements, combined with DFT calculations, provided clear evidence for an influence of the auxiliary ligand on reactivity of the organomagnesates $[CH_3MgX_2]^-$. Thus when X = O_2CCH_3, reduced reactivity towards water was observed. The DFT calculations suggest that this may arise from the bidentate binding mode of acetate, which induces overcrowding of the Mg coordination sphere in the transition state relative to the chloride organomagnesate (compare **64** and **65** of Scheme 13). Interestingly, there is a report in the literature on the enhanced selectivity (i.e. reduced reactivity) of solution-phase Grignard reagents processing carboxylate ligands instead of the traditional halides[137].

$$CH_3MetalX_2^- + AH \longrightarrow AMetalX_2^- + CH_4 \tag{87}$$

The substrate also plays a key role in the reactivity of the $[CH_3MgX_2]^-$ ions. This is illustrated dramatically for the reaction of aldehydes containing enolisable protons, which reacted via enolisation (equation 87), rather than via the Grignard reaction (equation 88).

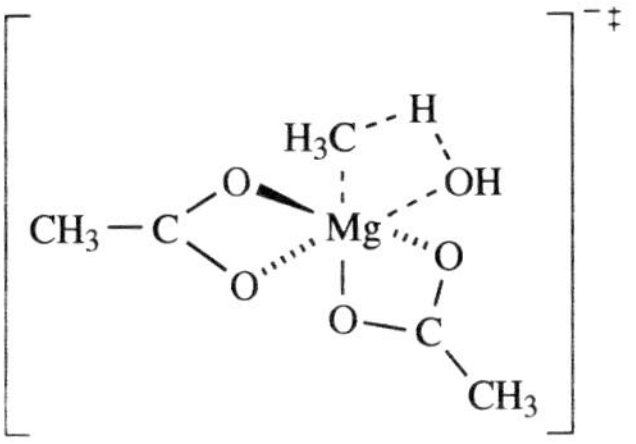

4. Solvent-free organomagnesiums

(64) 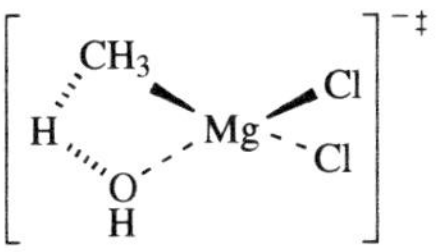 **(65)**

SCHEME 13

This is consistent with DFT calculations on $[CH_3MgCl_2]^-$, which reveal that the six-centered transition state for the enolisation reaction **66** is entropically favored over the four-centered transition state for the Grignard reaction **67** (Scheme 14).

$$CH_3MgX_2^- + RCHO \longrightarrow RCH(CH_3)OMgX_2^- \qquad (88)$$

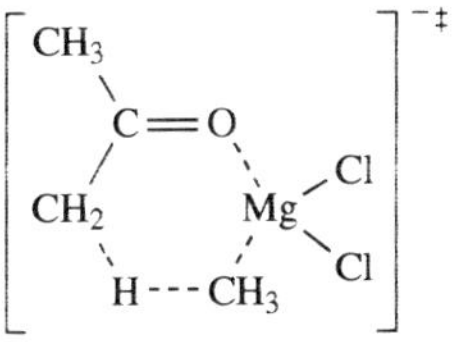

(66) 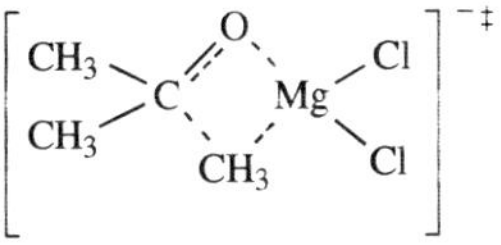 **(67)**

SCHEME 14

Interestingly, when acetic acid is the substrate, the $CH_3MgX_2^-$ ions (where $X = Cl$ or CH_3CO_2) complete a catalytic cycle for the decarboxylation of acetic acid (equation 89, Scheme 15)[131]. The first step is a metathesis reaction, in which a CH_3 ligand is switched for a carboxylato ligand (equation 87, Scheme 15). The second step is the rate-determining step (equation 82, Scheme 15) as it requires activation (under CID conditions) to induce decarboxylation of the magnesium acetate anion $CH_3CO_2MgX_2^-$, to reform the organometallic catalyst $CH_3MgX_2^-$. A similar catalytic cycle has been observed for decarboxylation of formic acid[132].

$$CH_3CO_2H \longrightarrow CH_4 + CO_2 \qquad (89)$$

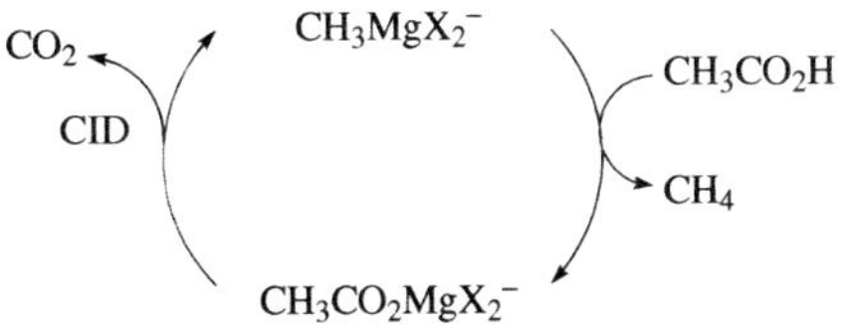

SCHEME 15

Each of the organoalkaline earths $[CH_3Metal(O_2CCH_3)_2]^-$ reacts with water via addition with concomitant elimination of methane to form the metal hydroxide $[HOMetal(O_2$

CCH$_3$)$_2$]$^-$ ions (equation 87), with a relative reactivity order of: [CH$_3$Ba(O$_2$CCH$_3$)$_2$]$^-$ ≈ [CH$_3$Sr(O$_2$CCH$_3$)$_2$]$^-$ > [CH$_3$Ca(O$_2$CCH$_3$)$_2$]$^-$ > [CH$_3$Mg(O$_2$CCH$_3$)$_2$]$^{-\,133}$. DFT calculations on the reaction exothermicities for these reactions generally supported the reaction trends observed experimentally, with [CH$_3$Mg(O$_2$CCH$_3$)$_2$]$^-$ being the least reactive.

IV. UNIMOLECULAR REACTIONS OF ORGANOMAGNESIUM IONS IN THE GAS PHASE

As noted in the introduction, few studies have examined the mass spectra of organomagnesium compounds. Of these, there are only three that have examined the unimolecular fragmentation reactions of organomagnesium ions. Under conditions of electron ionization, magnesocene yields the following ions in the positive ionization mode: the parent radical cation (equation 90), the monoligated ('half-sandwich') cation (equation 91) and the bare magnesium ion (equation 92)[4]. Note that Bohme and coworkers have studied the bimolecular reactivity of all these ions as described in Sections II and III above. The gas-phase fragmentation reactions of (c-C$_5$H$_5$)Mg$^+$ have been studied using a combination of metastable and CID experiments as well as DFT calculations[138]. Under metastable conditions, (c-C$_5$H$_5$)Mg$^+$ fragments via loss of a H atom (equation 93) and c-C$_5$H$_5$$^\bullet$ (equation 94). Collisional activation induces further fragmentation, resulting in the formation of (C$_3$H$_3$)Mg$^+$ and (C$_3$H$_2$)Mg$^{+\bullet}$. Electron ionization of magnesocene in the negative ionization mode yields the monoligated ('half-sandwich') anion (equation 95) and the cyclopentadienyl anion (equation 96)[4].

$$(c\text{-C}_5\text{H}_5)_2\text{Mg} + e^- \longrightarrow (c\text{-C}_5\text{H}_5)_2\text{Mg}^{+\bullet} + 2e^- \quad (90)$$

$$\longrightarrow (c\text{-C}_5\text{H}_5)\text{Mg}^+ + c\text{-C}_5\text{H}_5^\bullet + 2e^- \quad (91)$$

$$\longrightarrow \text{Mg}^{+\bullet} + 2c\text{-C}_5\text{H}_5^\bullet + 2e^- \quad (92)$$

$$(c\text{-C}_5\text{H}_5)\text{Mg}^+ \longrightarrow (\text{C}_5\text{H}_4)\text{Mg}^{+\bullet} + \text{H}^\bullet \quad (93)$$

$$\longrightarrow \text{Mg}^{+\bullet} + c\text{-C}_5\text{H}_5^\bullet \quad (94)$$

$$(c\text{-C}_5\text{H}_5)_2\text{Mg} + e^- \longrightarrow (c\text{-C}_5\text{H}_5)\text{Mg}^- + c\text{-C}_5\text{H}_5^\bullet \quad (95)$$

$$\longrightarrow c\text{-C}_5\text{H}_5^- + (c\text{-C}_5\text{H}_5)\text{Mg}^\bullet \quad (96)$$

The only study to have examined the composition and fragmentation reactions of a Grignard reagent via MS is that of Sakamoto and coworkers, who used a combination of coldspray ionization in conjunction with tandem mass spectrometry to evaluate the types of ions formed from a THF solution of 'CH$_3$MgCl'[6]. They noted the formation of [CH$_3$Mg$_2$Cl$_3$(THF)$_n$-H]$^+$ (where $n = 4$–6) under coldspray ionization. When [CH$_3$Mg$_2$Cl$_3$(THF)$_6$-H]$^+$ was subjected to CID, an envelope of [CH$_3$Mg$_2$Cl$_3$(THF)$_n$-H]$^+$ product ions was observed arising from sequential losses of up to 4 THF solvent molecules. Based on these results in conjunction with considerations of X-ray crystal structures of a range of organomagnesiums, structure **68** was suggested to be the core of the 'CH$_3$MgCl' Grignard in THF solution.

$$\text{CH}_3-\text{Mg}\underset{\underset{\text{Cl}}{\diagdown\,\diagup}}{\overset{\overset{\text{Cl}}{\diagup\,\diagdown}}{}}\text{Mg}$$

(**68**)

V. STRUCTURES OF ORGANOMAGNESIUMS AND MAGNESIUM HALIDES IN THE GAS PHASE

Gas-phase electron diffraction (GED) has been used to gain insights into the gas-phase structures of a range of inorganic and organometallic compounds. Since the area has been reviewed on several occasions[139–145], here the structures of key organomagnesiums[146–149] and magnesium halides[150–155] are briefly described. Scheme 16 shows the structures of all organomagnesium and magnesium halides studied via gas-phase electron diffraction to date, and includes key bond lengths. A key feature is that all monomers (**71**, **73–76**) are linear. The GED data require modeling of the structure to determine the best fit. For magnesocene, the best fit is for the eclipsed structure **69a** rather than the staggered structure, **69b**[146]. Permethylation of magnesocene results in a slight elongation of the Mg−C and ring C−C bonds (compare **69** and **70**) while replacement of a Cp ring with a neopentyl group decreases the Mg−Cp bond (compare **69** and **72**). The GED of the halides **73–76** were first studied 50 years ago[152, 154], and their structures have been refined over the years by further experimental and theoretical work. The most recent studies of Hargittai and coworkers on $MgCl_2$ (**74**)[150] and $MgBr_2$ (**75**)[151] were carried out on a GED instrument interfaced with a mass spectrometer, which allowed the vapor constitution to

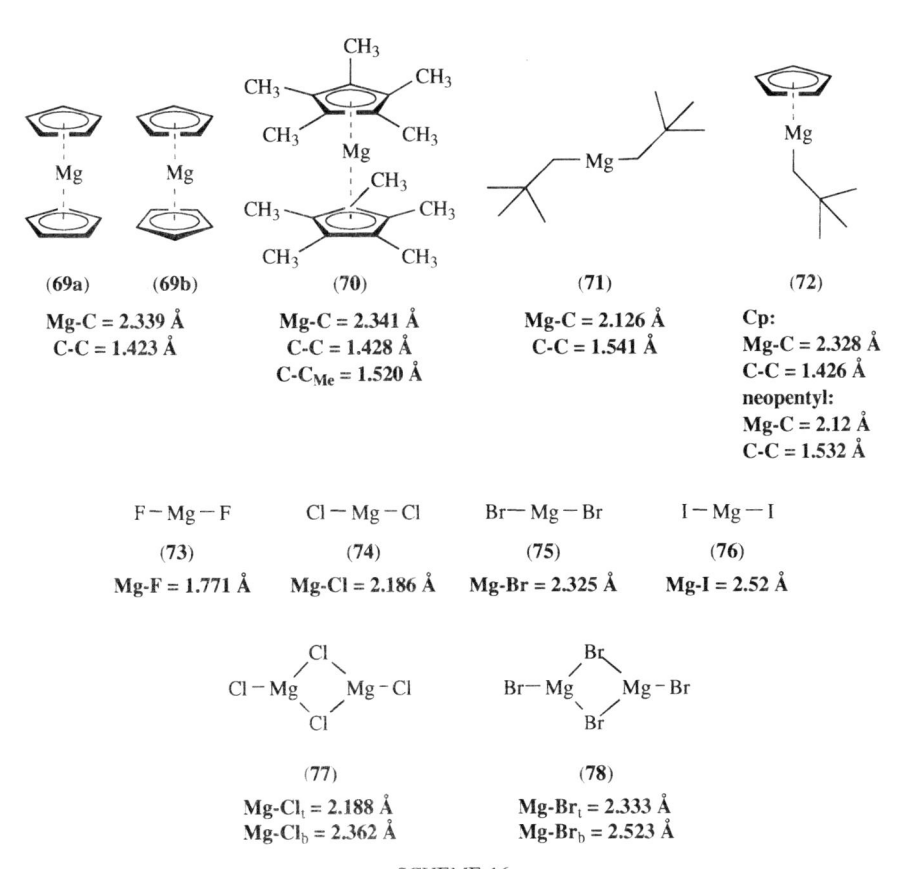

(69a)
Mg-C = 2.339 Å
C-C = 1.423 Å

(69b)

(70)
Mg-C = 2.341 Å
C-C = 1.428 Å
C-C$_{Me}$ = 1.520 Å

(71)
Mg-C = 2.126 Å
C-C = 1.541 Å

(72)
Cp:
Mg-C = 2.328 Å
C-C = 1.426 Å
neopentyl:
Mg-C = 2.12 Å
C-C = 1.532 Å

F−Mg−F
(73)
Mg-F = 1.771 Å

Cl−Mg−Cl
(74)
Mg-Cl = 2.186 Å

Br−Mg−Br
(75)
Mg-Br = 2.325 Å

I−Mg−I
(76)
Mg-I = 2.52 Å

(77)
Mg-Cl$_t$ = 2.188 Å
Mg-Cl$_b$ = 2.362 Å

(78)
Mg-Br$_t$ = 2.333 Å
Mg-Br$_b$ = 2.523 Å

SCHEME 16

be analyzed via MS. It was found that the vapor consisted of over 10% of the dimers, **77** and **78**. By carefully modeling the GED data, the structures of both the monomers and dimers were determined. The latter are interesting structures, directly relevant to the Schlenk equilibrium. Note that in both cases, the terminal Mg−X bond lengths (Mg−X_t) are shorter than the bridging Mg−X bond lengths (Mg−X_b). If the GED data do not take into account the presence of dimers, the Mg−X bond is overestimated. Thus the bond lengths for MgF_2 (**73**)[155] and MgI_2 (**76**)[152] shown in Scheme 16, which are derived from early data that were not modeled using dimer contributions, may be overestimates.

VI. CONCLUDING REMARKS

An attempt has been made to bring together a seemingly disparate set of studies on the formation, reactions and structures of organomagnesium species in solvent-free environments. Although the spectroscopy of such species has not been discussed, there have been several studies on the ESR, IR, UV-Vis and laser-induced fluorescence of organomagnesium species in the gas-phase, in matrices and in helium nanodroplets. Interested readers are referred to a number of recent reviews and articles[156−162]. Finally, given the advances in mass spectrometry, further studies on the gas-phase reactivity of organomagnesium ions are eagerly anticipated.

VII. ACKNOWLEDGMENTS

R.A.J.O thanks the Australian Research Council for financial support for studies on metal-mediated chemistry (Grant# DP0558430) and via the ARC Centres of Excellence program (ARC Centre of Excellence in Free Radical Chemistry and Biotechnology).

VIII. REFERENCES

1. N. Goldberg and H. Schwarz, in *The Chemistry of Organic Silicon Compounds*, Vol. 2, Part 2, (Eds. Z. Rappoport and Y. Apeloig), Wiley, Chichester, 1998, pp. 1105–1142.
2. R. A. J. O'Hair, in *The Chemistry of Organophosphorus Compounds*, Vol. 4, (Ed. F. R. Hartley), Wiley, Chichester, 1996, pp. 731–765.
3. H. O. House, R. A. Latham and G. M. Whitesides, *J. Org. Chem.*, **32**, 2481 (1967).
4. G. M. Bejun and R. N. Compton, *J. Chem. Phys.*, **58**, 2271 (1973).
5. L. A. Tjurina, V. V. Smirnov, G. B. Barkovskii, E. N. Nikolaev, S. E. Esipov and I. P. Beletskaya, *Organometallics*, **20**, 2449 (2001).
6. S. Sakamoto, T. Imamoto and K. Yamaguchi, *Org. Lett.*, **3**, 1793 (2001).
7. K. Yamaguchi, *J. Mass Spectrom.*, **38**, 473 (2003).
8. E. E. Ferguson, B. R. Rowe, D. W. Fahey and F. C. Fehsenfeld, *Planet. Space. Sci.*, **29**, 479 (1981).
9. S. Petrie, *Icarus*, **171**, 199 (2004).
10. S. Petrie, *Environ. Chem.*, **2**, 25 (2005).
11. S. Petrie and R. C. Dunbar, *AIP Conf. Proc.*, **855**, 272 (2006).
12. S. Petrie, *Aust. J. Chem.*, **56**, 259 (2003).
13. S. Petrie and D. K. Bohme, *Top. Curr. Chem.*, **225**, 37 (2003).
14. R. C. Dunbar and S. Petrie, *AIP Conf. Proc.*, **855**, 281 (2006).
15. E. Peralez, J.-C. Negrel, A. Goursot and M. Chanon, *Main Group Met. Chem.*, **21**, 69 (1998).
16. W. H. Breckenridge, *J. Phys. Chem.*, **100**, 14840 (1996).
17. H.-J. Himmel, A. J. Downs and T. M. Greene, *Chem. Rev.*, **102**, 4191 (2002).
18. W. H. Breckenridge and H. Umemoto, *J. Chem. Phys.*, **81**, 3852 (1984).
19. J. G. McCaffrey, J. M. Parnis, G. A. Ozin and W. H. Breckenridge, *J. Phys. Chem.*, **89**, 4945 (1985).
20. T. M. Greene, D. V. Lanzisera, L. Andrews and A. J. Downs, *J. Am. Chem. Soc.*, **120**, 6097 (1998).

21. K. Mochida, K. Kojima and Y. Yoshida, *Bull. Chem. Soc. Jpn*, **60**, 2255 (1987).
22. P. S. Skell and J. E. Girard, *J. Am. Chem. Soc.*, **94**, 5518 (1972).
23. B. S. Ault, *J. Am. Chem. Soc.*, **102**, 3480 (1980).
24. Y. Imizu and K. J. Klabunde, *Inorg. Chem.*, **23**, 3602 (1984).
25. K. J. Klabunde and A. Whetten, *J. Am. Chem. Soc.*, **108**, 6529 (1986).
26. S. R. Davis, *J. Am. Chem. Soc.*, **113**, 4145 (1991).
27. V. N. Solov'ev, G. B. Sergeev, A. V. Nemukhin, S. K. Burt and I. A. Topol, *J. Phys. Chem. A*, **101**, 8625 (1997).
28. W. D. Bare and L. Andrews, *J. Am. Chem. Soc.*, **120**, 7293 (1998).
29. A. A. Tulub, *Russ. J. Gen. Chem.*, **72**, 886 (2002).
30. A. V. Tulub, V. V. Porsev and A. Tulub, *Dokl. Phys. Chem.*, **398**, 241 (2004).
31. W. D. Bare, A. Citra, C. Trindle and L. Andrews, *Inorg. Chem.*, **39**, 1204 (2000).
32. V. N. Solov'ev, E. V. Polikarpov, A. V. Nemukhin and G. B. Sergeev, *J. Phys. Chem. A*, **103**, 6721 (1999).
33. C. A. Thompson and L. Andrews, *J. Am. Chem. Soc.*, **118**, 10242 (1996).
34. D. V. Lanzisera and L. Andrews, *J. Phys. Chem. A*, **101**, 9666 (1997).
35. A. J. McKinley and E. Karakyriakos, *J. Phys. Chem. A*, **104**, 8872 (2000).
36. Z. Huang, M. Chen, Q. Liu and M. Zhou, *J. Phys. Chem. A*, **107**, 11380 (2003).
37. P. L. Po and R. F. Porter, *J. Am. Chem. Soc.*, **99**, 4922 (1977).
38. G. I. Gellene, N. S. Kleinrock and R. F. Porter, *J. Chem. Phys.*, **78**, 1795 (1983).
39. P. B. Armentrout, *Top. Organomet. Chem.*, **4**, 1 (1999).
40. K. Eller and H. Schwarz, *Chem. Rev.*, **91**, 1121 (1991).
41. K. Eller, *Coord. Chem. Rev.*, **126**, 93 (1993).
42. B. S. Freiser, *Acc. Chem. Res.*, **27**, 353 (1994).
43. B. S. Freiser, *J. Mass Spectrom.*, **31**, 703 (1996).
44. L. Operti and R. Rabezzana, *Mass Spectrom. Rev.*, **25**, 483 (2006).
45. B. R. Rowe, D. W. Fahey, E. E. Ferguson and F. C. Fehsenfeld, *J. Chem. Phys.*, **75**, 3325 (1981).
46. R. K. Milburn, M. V. Frash, A. C. Hopkinson and D. K. Bohme, *J. Phys. Chem. A*, **104**, 3926 (2000).
47. J. S. Uppal and R. H. Staley, *J. Am. Chem. Soc.*, **104**, 1229 (1982).
48. B. P. Pozniak and R. C. Dunbar, *J. Am. Chem. Soc.*, **119**, 10439 (1997).
49. M. Welling, R. I. Thompson and H. Walther, *Chem. Phys. Lett.*, **253**, 37 (1996).
50. M. Welling, H. A. Schuessler, R. I. Thompson and H. Walther, *Int. J. Mass Spectrom. Ion Proc.*, **172**, 95 (1998).
51. R. I. Thompson, M. Welling, H. A. Schuessler and H. Walther, *J. Chem. Phys.*, **116**, 10201 (2002).
52. S. Petrie and R. C. Dunbar, *J. Phys. Chem. A*, **104**, 4480 (2000).
53. R. C. Dunbar and S. Petrie, *Astrophys. J.*, **564**, 792 (2002).
54. R. K. Milburn, A. C. Hopkinson and D. K. Bohme, *J. Am. Chem. Soc.*, **127**, 13070 (2005).
55. X. Zhao, G. K. Koyanagi and D. K. Bohme, *J. Phys. Chem. A*, **110**, 10607 (2006).
56. A. Andersen, F. Muntean, D. Walter, C. Rue and P. B. Armentrout, *J. Phys. Chem. A*, **104**, 692 (2000).
57. L. Operti, E. C. Tews, T. J. MacMahon and B. S. Freiser, *J. Am. Chem. Soc.*, **111**, 9152 (1989).
58. D. Schroeder and J. Roithova, *Angew. Chem., Int. Ed.*, **45**, 5705 (2006).
59. P. D. Kleiber and J. Chen, *Int. Rev. Phys. Chem.*, **17**, 1 (1998).
60. Y. C. Cheng, J. Chen, L. N. Ding, T. H. Wong, P. D. Kleiber and D.-K. Liu, *J. Chem. Phys.*, **104**, 6452 (1996).
61. W. Guo, T. Yuan, X. Chen, L. Zhao, X. Lu and S. Wu, *J. Mol. Struct. (THEOCHEM)*, **764**, 177 (2006).
62. J. Chen, Y. C. Cheng and P. D. Kleiber, *J. Chem. Phys.*, **106**, 3884 (1997).
63. J. Chen, T. H. Wong and P. D. Kleiber, *Chem. Phys. Lett.*, **279**, 185 (1997).
64. X. Yang, Y. Hu and S. Yang, *J. Phys. Chem. A*, **104**, 8496 (2000).
65. X. Yang, Y. Hu and S. Yang, *Chem. Phys. Lett.*, **322**, 491 (2000).
66. A. Furuya, F. Misaizu and K. Ohno, *J. Chem. Phys.*, **125**, 094309/1 (2006).
67. A. Furuya, F. Misaizu and K. Ohno, *J. Chem. Phys.*, **125**, 094310/1 (2006).
68. X. Yang, H. Liu and S. Yang, *J. Chem. Phys.*, **113**, 3111 (2000).

69. X. Yang, K. Gao, H. Liu and S. Yang, *J. Chem. Phys.*, **112**, 10236 (2000).
70. H.-C. Liu, X.-H. Zhang, Y.-D. Wu and S. Yang, *Phys. Chem. Chem. Phys.*, **7**, 826 (2005).
71. H.-C. Liu, C.-S. Wang, W. Guo, Y.-D. Wu and S. Yang, *J. Am. Chem. Soc.*, **124**, 3794 (2002).
72. H.-C. Liu, S. Yang, X.-H. Zhang and Y.-D. Wu, *J. Am. Chem. Soc.*, **125**, 12351 (2003).
73. H. Liu, X. Zhang, C. Wang, W. Guo, Y. Wu and S. Yang, *J. Phys. Chem. A*, **108**, 3356 (2004).
74. H.-C. Liu, X.-H. Zhang, C. Wang, Y.-D. Wu and S. Yang, *Phys. Chem. Chem. Phys.*, **9**, 607 (2007).
75. M. R. France, S. H. Pullins and M. A. Duncan, *Chem. Phys.*, **239**, 447 (1998).
76. J. I. Lee, D. C. Sperry and J. M. Farrar, *J. Chem. Phys.*, **114**, 6180 (2001).
77. W. Guo, H. Liu and S. Yang, *J. Chem. Phys.*, **116**, 9690 (2002).
78. H. Liu, J. Sun and S. Yang, *J. Phys. Chem. A*, **107**, 5681 (2003).
79. Y. Hu, H. Liu and S. Yang, *Chem. Phys.*, **332**, 66 (2007).
80. H. Liu, Y. Hu, S. Yang, W. Guo, Q. Fu and L. Wang, *J. Phys. Chem. A*, **110**, 4389 (2006).
81. B. Soep, M. Elhanine and C. P. Schulz, *Chem. Phys. Lett.*, **327**, 365 (2000).
82. W. Guo, H. Liu and S. Yang, *J. Chem. Phys.*, **117**, 6061 (2002).
83. H. Liu, Y. Hu, S. Yang, W. Guo, X. Lu and L. Zhao, *Chem. Eur. J.*, **11**, 6392 (2005).
84. W. Guo, X. Lu, S. Hu and S. Yang, *Chem. Phys. Lett.*, **381**, 109 (2003).
85. W. Guo, H. Liu and S. Yang, *J. Chem. Phys.*, **116**, 2896 (2002).
86. W. Guo, H. Liu and S. Yang, *Int. J. Mass Spectrom.*, **226**, 291 (2003).
87. H. Liu, W. Guo and S. Yang, *J. Chem. Phys.*, **115**, 4612 (2001).
88. W. Y. Lu, T. H. Wong, Y. Sheng and P. D. Kleiber, *J. Chem. Phys.*, **117**, 6970 (2002).
89. W. Y. Lu, T. H. Wong, Y. Sheng and P. D. Kleiber, *J. Chem. Phys.*, **118**, 6905 (2003).
90. W. Y. Lu and P. D. Kleiber, *Chem. Phys. Lett.*, **338**, 291 (2001).
91. W. Y. Lu and P. D. Kleiber, *J. Chem. Phys.*, **114**, 10288 (2001).
92. Y. Abate and P. D. Kleiber, *J. Chem. Phys.*, **125**, 184310 (2006).
93. H. Liu, Y. Hu and S. Yang, *J. Phys. Chem. A*, **107**, 10026 (2003).
94. J.-L. Sun, H. Liu, K.-L. Han and S. Yang, *J. Chem. Phys.*, **118**, 10455 (2003).
95. J.-L. Sun, H. Liu, H.-M. Yin, K.-L. Han and S. Yang, *J. Phys. Chem. A*, **108**, 3947 (2004).
96. Y. Hu, H. Liu and S. Yang, *J. Chem. Phys.*, **120**, 2759 (2004).
97. J.-L. Sun, H. Liu, H.-M. Wang, K.-L. Han and S. Yang, *Chem. Phys. Lett.*, **392**, 285 (2004).
98. H. Liu, J.-L. Sun, Y. Hu, K.-L. Han and S. Yang, *Chem. Phys. Lett.*, **389**, 342 (2004).
99. P. G. Jasien and C. E. Dykstra, *J. Am. Chem. Soc.*, **105**, 2089 (1983).
100. L. A. Tjurina, V. V. Smirnov and I. P. Beletskaya, *J. Mol. Catal. A*, **182–183**, 395 (2002).
101. L. A. Tjurina, V. V. Smirnov, D. A. Potapov, S. A. Nikolaev, S. E. Esipov and I. P. Beletskaya, *Organometallics*, **23**, 1349 (2004).
102. D. A. Potapov, L. A. Tjurina and V. V. Smirnov, *Russ. Chem. Bull.*, **54**, 1185 (2005).
103. P. G. Jasien and J. A. Abbondondola, *J. Mol. Struct. (THEOCHEM)*, **671**, 111 (2004).
104. V. V. Porsev and A. V. Tulub, *Dokl. Phys. Chem.*, **409**, 237 (2006).
105. Y. C. Cheng, J. Chen, P. D. Kleiber and M. A. Young, *J. Chem. Phys.*, **107**, 3758 (1997).
106. J. F. Garst and M. P. Soriaga, *Coord. Chem. Rev*, **248**, 623 (2004).
107. J. F. Garst and F. Ungvary, 'Mechanisms of Grignard reagent formation', in *Grignard Reagents: New Developments*; (Ed. G. H. Richey, Jr.), Wiley, Chichester, 2000, p. 185.
108. J. B. Abreu, J. E. Soto, A. Ashley-Facey, M. P. Soriaga, J. F. Garst and J. L. Stickney, *J. Colloid Interface Chem.*, **206**, 247 (1998).
109. R. G. Nuzzo and L. H. Dubois, *J. Am. Chem. Soc.*, **108**, 2881 (1986).
110. Y. Gault, *Tetrahedron Lett.*, 67 (1966); *Chem. Abstr.*, **64**, 75247 (1966).
111. M. Lefrancois and Y. Gault, *J. Organomet. Chem.*, **16**, 7 (1969); *Chem. Abstr.*, **70**, 68449 (1969).
112. Y. Gault, *J. Chem. Soc., Chem. Commun.*, 478 (1973).
113. Y. Gault, *J. Chem. Soc., Faraday Trans.*, **74**, 2678 (1978).
114. A. Choplin and Y. Gault, *J. Organomet. Chem.*, **179**, C1 (1979).
115. A. T. Blades, P. Jayaweera, M. G. Ikonomou and P. Kebarle, *J. Chem. Phys.*, **92**, 5900 (1990).
116. A. J. Stace, *J. Phys. Chem. A*, **106**, 7993 (2002).
117. A. Stace, *Science*, **294**, 1292 (2001).
118. A. J. Stace, *Phys. Chem. Chem. Phys.*, **3**, 1935 (2001).
119. M. P. Dobson and A. J. Stace, *Chem. Commun.*, 1533 (1996).
120. C. A. Woodward, M. P. Dobson and A. J. Stace, *J. Phys. Chem. A*, **101**, 2279 (1997).
121. C. A. Woodward, M. P. Dobson and A. J. Stace, *J. Phys. Chem.*, **100**, 5605 (1996).

4. Solvent-free organomagnesiums 187

122. N. Walker, M. P. Dobson, R. R. Wright, P. E. Barran, J. N. Murrell and A. J. Stace, *J. Am. Chem. Soc.*, **122**, 11138 (2000).
123. M. Peschke, A. T. Blades and P. Kebarle, *J. Am. Chem. Soc.*, **122**, 10440 (2000).
124. A. M. El-Nahas, S. H. El-Demerdash and E.-S. E. El-Shereefy, *Int. J. Mass Spectrom.*, **263**, 267 (2007).
125. A. A. Shvartsburg, J. G. Wilkes, J. O. Lay and K. W. M. Siu, *Chem. Phys. Lett.*, **350**, 216 (2001).
126. A. A. Shvartsburg and J. G. Wilkes, *J. Phys. Chem. A*, **106**, 4543 (2002).
127. A. A. Shvartsburg, *Chem. Phys. Lett.*, **376**, 6 (2003).
128. A. A. Shvartsburg and J. G. Wilkes, *Int. J. Mass Spectrom.*, **225**, 155 (2003).
129. T. Shi, K. W. M. Siu and A. C. Hopkinson, *Int. J. Mass Spectrom.*, **255–256**, 251 (2006).
130. H.-F. Wu and J. S. Brodbelt, *J. Am. Chem. Soc.*, **116**, 6418 (1994).
131. R. A. J. O'Hair, A. K. Vrkic and P. F. James, *J. Am. Chem. Soc.*, **126**, 12173 (2004).
132. G. N. Khairallah and R. A. J. O'Hair, *Int. J. Mass Spectrom.*, **254**, 145 (2006).
133. A. P. Jacob, P. F. James and R. A. J. O'Hair, *Int. J. Mass Spectrom.*, **255–256**, 45 (2006).
134. R. K. Milburn, V. Baranov, A. C. Hopkinson and D. K. Bohme, *J. Phys. Chem. A*, **103**, 6373 (1999).
135. R. K. Milburn, V. I. Baranov, A. C. Hopkinson and D. K. Bohme, *J. Phys. Chem. A*, **102**, 9803 (1998).
136. R. A. J. O'Hair, *Chem. Commun.*, 1469 (2006).
137. M. T. Reetz, N. Harmat and R. Mahrwald, *Angew. Chem.*, **104**, 333 (1992).
138. J. Berthelot, A. Luna and J. Tortajada, *J. Phys. Chem. A*, **102**, 6025 (1998).
139. A. Haaland, *Top. Curr. Chem.*, 1 (1975).
140. P. R. Markies, O. S. Akkerman, F. Bickelhaupt, W. J. J. Smeets and A. L. Spek, *Adv. Organomet. Chem.*, **32**, 147 (1991).
141. D. W. H. Rankin and H. E. Robertson, *Spect. Prop. Inorg. Organomet. Compd.*, **38**, 348 (2006).
142. I. Hargittai, *Struct. Chem.*, **16**, 1 (2005).
143. M. Hargittai, *Struct. Chem.*, **16**, 33 (2005).
144. M. Hargittai, *Chem. Rev.*, **100**, 2233 (2000).
145. M. Hargittai, *Coord. Chem. Rev*, **91**, 35 (1988).
146. A. Haaland, J. Lusztyk, D. P. Novak, J. Brunvoll and K. B. Starowieyski, *J. Chem. Soc., Chem. Commun.*, 54 (1974).
147. R. A. Andersen, R. Blom, J. M. Boncella, C. J. Burns and H. V. Volden, *Acta Chem. Scand., Ser. A*, **A41**, 24 (1987).
148. E. C. Ashby, L. Fernholt, A. Haaland, R. Seip and R. S. Smith, *Acta Chem. Scand., Ser. A*, **A34**, 213 (1980).
149. R. A. Andersen, R. Blom, A. Haaland, B. E. R. Schilling and H. V. Volden, *Acta Chem. Scand., Ser. A*, **A39**, 563 (1985).
150. J. Molnar, C. J. Marsden and M. Hargittai, *J. Phys. Chem.*, **99**, 9062 (1995).
151. B. Reffy, M. Kolonits and M. Hargittai, *J. Phys. Chem. A*, **109**, 8379 (2005).
152. P. A. Akishin and V. P. Spiridonov, *Zh. Fiz. Khim.*, **32**, 1682 (1958); *Chem. Abstr.*, **53**, 4621 (1959).
153. V. V. Kasparov, Y. S. Ezhov and N. G. Rambidi, *Zh. Strukt. Khim.*, **20**, 260 (1979); *Chem. Abstr.*, **91**, 30922 (1979).
154. P. A. Akishin, V. P. Spiridonov, G. A. Sobolev and V. A. Naumov, *Zh. Fiz. Khim.*, **31**, 461 (1957); *Chem. Abstr.*, **51**, 84261 (1957).
155. V. V. Kasparov, Y. S. Ezhov and N. G. Rambidi, *Zh. Strukt. Khim.*, **21**, 41 (1980). *Chem. Abstr.*, **93**, 86015 (1980).
156. M. S. Beardah and A. M. Ellis, *J. Chem. Tech. Biotech.*, **74**, 863 (1999).
157. A. M. Ellis, *Int. Rev. Phys. Chem.*, **20**, 551 (2001).
158. M. A. Duncan, *Ann. Rev. Phys. Chem.*, **48**, 69 (1997).
159. N. R. Walker, R. S. Walters and M. A. Duncan, *New J. Chem.*, **29**, 1495 (2005).
160. F. Dong and R. E. Miller, *J. Phys. Chem. A*, **108**, 2181 (2004).
161. D. T. Moore and R. E. Miller, *J. Phys. Chem. A*, **108**, 9908 (2004).
162. P. L. Stiles, D. T. Moore and R. E. Miller, *J. Chem. Phys.*, **121**, 3130 (2004).

CHAPTER **5**

Photochemical transformations involving magnesium porphyrins and phthalocyanines

NATALIA N. SERGEEVA and MATHIAS O. SENGE

School of Chemistry, SFI Tetrapyrrole Laboratory, Trinity College Dublin, Dublin 2, Ireland
Fax: +353-1-608-8536; e-mail: sengem@tcd.ie

I. INTRODUCTION	190
A. Abbreviations	190
B. General Introduction	190
II. BASIC PHOTOCHEMISTRY OF PORPHYRINS	192
A. General Concepts and Theoretical Background	192
B. Stability	193
III. PHOTOSYNTHESIS	194
IV. ELECTRON TRANSFER SYSTEMS	196
A. Introduction	196
B. Donor–Acceptor Electron Transfer Compounds	196
C. Heteroligand Systems	201
V. PHOTOCHEMICAL REACTIONS	206
A. Porphyrins	206
B. Photoinduced Ring-opening Reactions	207
C. Reactions of Chlorophyll	209
VI. APPLIED PHOTOCHEMISTRY	212
VII. ACKNOWLEDGMENT	214
VIII. REFERENCES	214

The chemistry of organomagnesium compounds
Edited by Z. Rappoport and I. Marek   © 2008 John Wiley & Sons, Ltd

I. INTRODUCTION

A. Abbreviations

bchl	bacteriochlorophyll	OEP	2,3,7,8,12,13,17,18-octaethylporphyrinato
chl	chlorophyll	P	porphyrin
D–B–A	donor–bridge–acceptor systems	PET	photoinduced electron transfer
ET	electron transfer	PS	photosensitizer
MV	methyl viologen	TPP	5,10,15,20-tetraphenylporphyrinato

B. General Introduction

The photochemistry of true organomagnesium compounds remains almost completely unexplored. A literature search in preparation of this work found only a few scattered examples of photochemical studies, mostly in relation to Grignard reactions[1] and 1,3-diketonate chelates[2,3]. Similar to the situation with organozinc compounds[4] magnesium tetrapyrrole chelates, i.e. magnesium porphyrins **1**, 5,10,15,20-tetraazaporphyrins (porphyrazines) **2** and phthalocyanines **3** have found more interest. This is primarily related to the biological relevance of magnesium porphyrins in nature, notably in photosynthesis and electron transfer, and we will focus on this aspect in this review. Outside this area not many 'true' photochemical studies have been performed with magnesium tetrapyrroles. Nevertheless, even in this area the body of available literature is limited and we only use selected examples to highlight the state of the art of this field. A description of syntheses, methodology or electron transfer reactions is outside the purview of this work and the present work can only give a broad overview and selected examples of studies in this area.

Chlorophylls (chl) and the related bacteriochlorophylls (bchl) are the ubiquitous pigments of photosynthetic organisms and the predominant class of magnesium tetrapyrroles in nature. As such they share common structural principles and functions. They are either involved in light harvesting (exciton transfer) as antenna pigments or charge separation (electron transfer) as reaction center pigments. The best-known pigment is chl a, **4**, which occurs in all organisms with oxygenic photosynthesis. In higher plants it is accompanied in a 3:1 ratio by chl b, **6**, where the 7-methyl group has been oxidized to a formyl group. Both compounds typically consist of the tetrapyrrole moiety and a C-20 terpenoid alcohol, phytol. Most compounds are magnesium chelates, but the free base of chl a, pheo a **5**, is also active in electron transfer. Chl a and b can be obtained easily from plants or algae and their synthetic chemistry has mainly targeted total syntheses and medicinal application in photodynamic therapy (PDT)[5,6].

However, many other similar photosynthetic pigments occur in nature[4-7]. All share either a phytochlorin **7** or a 7,8-dihydrophytochlorin framework and by now about one hundred related pigments have been isolated[8]. For example, such compounds include chl d **8** from Rhodophytes, the bchls c (**9**), d and e (which are chlorins **12** and show significant variability in their peripheral groups) from Chlorobiaceae and Chloroflexaceae, and bchl a (**10**) and b (true bacteriochlorins **13**) found in Rhodospirillales. Other natural pigments are chl c, bchl g and many of these are esterified with different isoprenoid alcohols. Chemically related chlorins have also been found in many oxidoreductases, marine sponges, tunicates and in *Bonella viridis*. The deep-see dragon fish *Malacosteus niger* even utilizes a chl derivative as a visual pigment[9]. Most of these are believed to be derived from chl and then processed by the plant or animal.

(**4**) M = Mg, chl a
(**5**) M = 2H, pheo a

(**6**) chl b

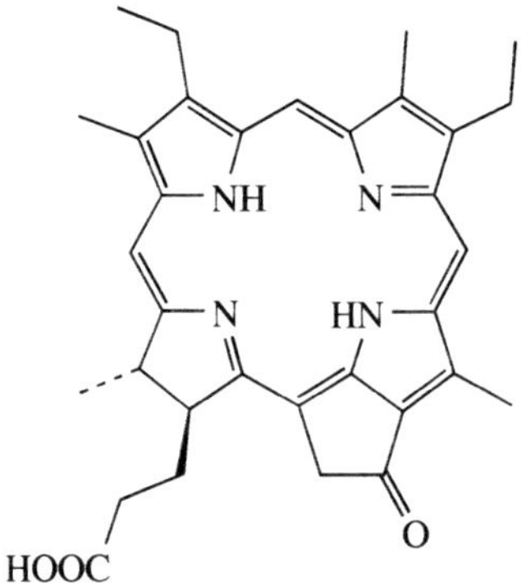

(7) phytochlorin

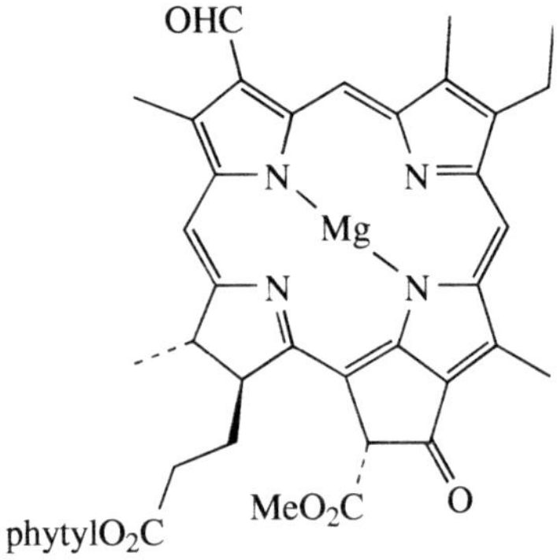

(8) chl d

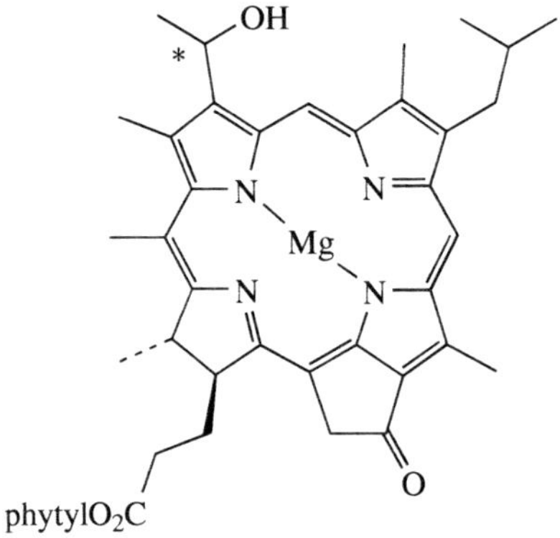

(9) bchl c

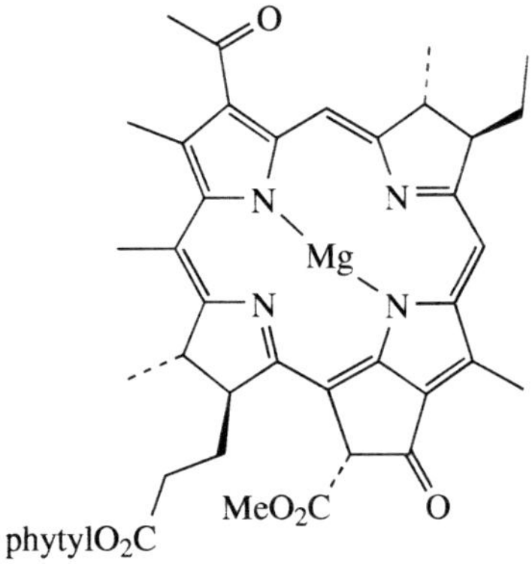

(10) bchl a

II. BASIC PHOTOCHEMISTRY OF PORPHYRINS
A. General Concepts and Theoretical Background

Chls and all tetrapyrroles are heteroaromatic compounds and the aromatic character of the underlying tetrapyrrole moiety and the reactivity of the functional groups in the side chains govern their chemistry. Three different classes of tetrapyrroles, differentiated by their oxidation level, occur in nature: porphyrins (**11**, e.g. hemes), chlorins (**12**, e.g. chls) and bacteriochlorins (**13**, e.g. bchls). As a cyclic tetrapyrrole with a fused five-membered ring, the overall reactivity of chl is that of a standard phytochlorin **7**. Such compounds are capable of coordinating almost any known metal with the core nitrogen atoms. Together with the conformational flexibility of the macrocycle and the variability of its side chains, this accounts for their unique role in photosynthesis and applications[10,11].

Tetrapyrroles contain an extended π-conjugated system which is responsible for their use in a wide range of applications ranging from technical (pigments, catalysts, photoconductors) to medicinal (photodynamic therapy) uses. The electronic absorption spectra are governed by the aromatic 18 π-electron system and typically consist of two main bands. In phthalocyanines the Q band around 660–680 nm is the most intense one accompanied by a weaker Soret band near 340 nm[12]. In porphyrins the situation is reversed with an

(**11**) porphyrin   (**12**) chlorin   (**13**) bacteriochlorin

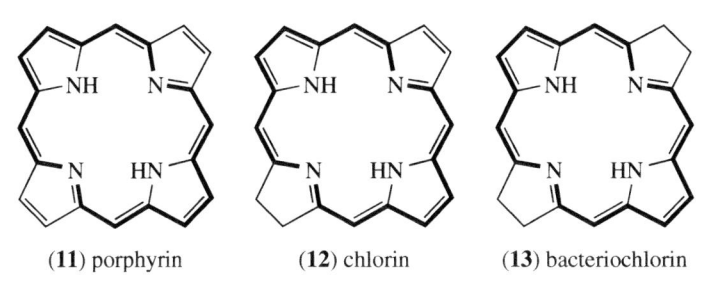

intense Soret band around 380–410 nm and weaker Q bands in the 550–650 nm region. The position and intensity of the absorption bands are affected by the central metal, axial ligands, solvation, substituents and their regiochemical arrangement, and aggregation. The theoretical background has been widely reviewed and established[13] in pioneering works by Gouterman[14] and Mack and Stillman[15]. The spectral characteristics depend strongly on the substituent pattern. By now almost all possible combinations of electron-donating, electron-withdrawing or sterically demanding groups have been prepared[11].

Magnesium(II) tetrapyrroles behave like most other organic chromophores. Absorption of light will lead to the rapid formation of the lowest excited singlet state by promotion of an electron from the HOMO to the LUMO. The excited state can then either relax to the ground state via radiative (fluorescence) or nonradiative processes (internal conversion of vibrational relaxation). Another possibility is intersystem crossing to form a triplet state which again can relax either via radiative (phosphorescence) or nonradiative processes. In our context, both excited-state types can take part in photochemical reactions and, in the presence of donor or acceptor units, energy transfer or electron transfer between the chromophores, can compete with these processes[16]. In addition, metallo(II) porphyrins and phthalocyanines may form ions upon illumination. These are either anion or π-cation radicals that undergo further photochemical reactions[17,18].

B. Stability

Although porphyrins and especially phthalocyanines are stable compounds, both will undergo photooxidative degradation or photoexcited ET reactions[19–21]. An additional problem with magnesium complexes is their low stability in aqueous solution, as they demetallate quite easily. This is one of the main reasons that many photochemical studies targeted at modeling the natural situation use the more stable zinc(II) complexes. In addition, past years have seen increasing evidence that both Mg(II) and Zn(II) chlorophylls do exist in nature.

Photosynthetic organisms that utilize chls or bchls containing metals other than Mg were unknown for a long time[22]. By now it has been satisfactorily demonstrated that a novel purple pigment occurring in a group of obligatory aerobic bacteria is in fact a zinc-chelated bchl (Zn-bchl)[23–25]. The natural occurrence of Zn-bchl *a* has been proven for a limited group of aerobic acidophilic proteobacteria, including species of the genus *Acidiphilium*. The major photopigment in *Acidiphilium* was first identified tentatively as Mg-bchl *a* on the basis of preliminary spectral analyses[26,27]. However, more detailed studies revealed that all previously known species of *Acidiphilium* contained Zn-bchl *a* as the major photopigment and showed *Acidiphilium* to be a photosynthetic organism[23,25].

The naturally occurring Zn-bchl *a* and Mg-bchl *a* show large structural similarities and have very similar physicochemical characteristics[28]. Likewise, Zn-chl *a* exhibits features similar to Mg-chl *a* with regard to redox potential and absorption maxima in organic solvents. The light-harvesting efficiency of Zn-chl *a* and Mg-chl *a* are very similar although the fluorescence quantum yield of the former is lower than that of the latter. Compared to other chlorophyll-type pigments Zn-bchl *a* is much more stable towards acid. For example, the rate of pheophytinization for Zn-bchl *a* is 10^6-fold slower than for Mg-bchl a^{29}. In fact, it is difficult to fully demetallate Zn-bchl *a* to bacteriopheophytin (bPhe) by treatment with 1N HCl, which is commonly used for pheophytinization of Mg-bchl and Mg-chl. Due to the chemical stability of Zn-(b)chl *a* and their photo- and electrochemical similarities with Mg-(b)chl *a*, Zn-(b)chls are an alternative pigment for photosynthesis. Thus, it is not surprising that they have been used along with magnesium porphyrins in studies of artificial photosynthetic systems[30,31].

III. PHOTOSYNTHESIS

The natural photosynthetic process is a rather complex biochemical system that primarily relies on the light absorption by organic chromophores, followed by generation of reduction equivalents and ATP. The main photosynthetic pigments are chlorophylls that have very strong absorption bands in the visible region of the spectrum. Together with accessory pigments (carotenoids and open-chain tetrapyrroles) the various photosynthetic pigments complement each other in absorbing sunlight. Photosynthetic bacteria mostly contain bacteriochlorophylls with absorption maxima shifted towards the bathochromic region compared to chlorin-based pigments.

In its simplest form photosynthesis can be envisaged as the absorption of light through pigments arranged in a light-harvesting complex. These antenna systems permit an organism to increase greatly the absorption cross section for light and the use of light harvesting complexes with different pigments allows for a more efficient process through absorption of more photons and a more efficient use of the visible spectrum. The antenna pigments funnel the excitation energy through exciton transfer to a closely coupled pair of (b)chl molecules in the photochemical reaction center (Figure 1). The reaction center is an integral membrane pigment-protein that carries out light-driven electron transfer reactions. The excited (bacterio)chlorophyll molecule transfers an electron to a nearby acceptor molecule, thereby creating a charge separated state consisting of the oxidized chlorophyll and reduced acceptor.

After the initial electron transfer event, a series of electron transfer reactions takes place that eventually stabilizes the stored energy in reduction equivalents and ATP. Higher plants have two different reaction center complexes that work together in sequence, with the reduced acceptors of one photoreaction (photosystem II) serving as the electron donor for photosystem I. Here, the ultimate electron donor is water, liberating molecular oxygen, and the ultimate electron acceptor is carbon dioxide, which is reduced to carbohydrates. More simple and evolutionary older types of photosynthetic organisms contain only a single photosystem, either similar to photosystem II or photosystem I[32-35]. A simplified scheme of the complex photosynthetic apparatus is shown in an adaptation of the Z-scheme in Figure 2. The Z-scheme illustrates the two light-dependent reactions in photosynthetic systems of higher plants and exemplifies that two photosystems function in sequence to convert solar energy into chemical energy.

In chemical terms the photoinduced electron transfer results in transfer of an electron across the photosynthetic membrane in a complex sequence that involves several donor–acceptor molecules. Finally, a quinone acceptor is reduced to a semiquinone and subsequently to a hydroquinone. This process is accompanied by the uptake of two protons from the cytoplasma. The hydroquinone then migrates to a cytochrome bc complex, a proton pump, where the hydroquinone is reoxidized and a proton gradient is established via transmembrane proton translocation. Finally, an ATP synthase utilizes the proton gradient to generate chemical energy. Due to the function of tetrapyrrole-based pigments as electron donors and quinones as electron acceptors, most biomimetic systems utilize some

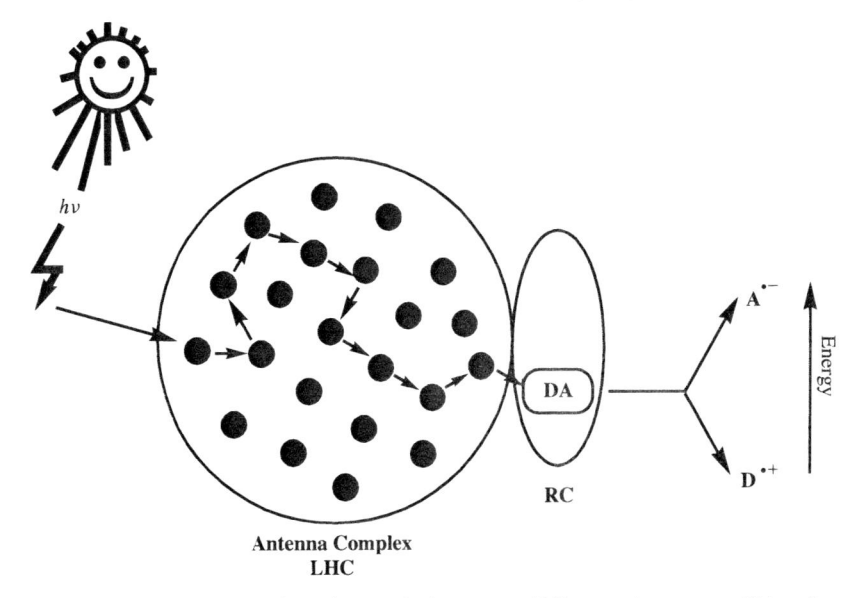

FIGURE 1. General scheme of a photosynthetic system (RC = reaction center, DA = donor–acceptor complex, LHC = light-harvesting complex)

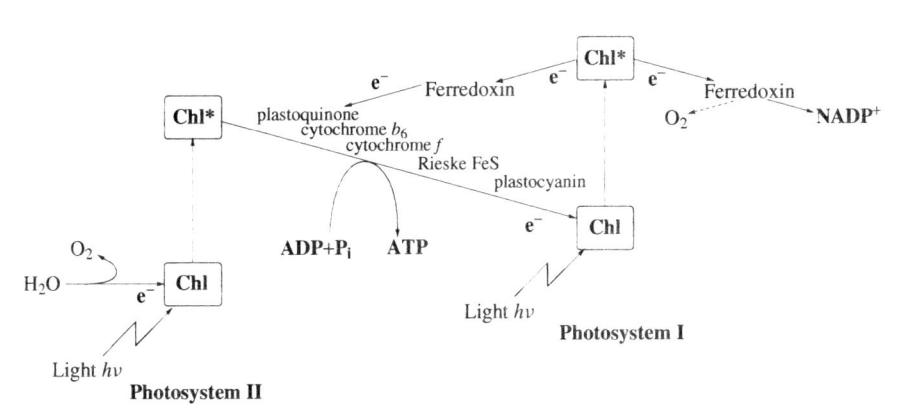

FIGURE 2. Simplified Z-scheme of the photosynthetic apparatus in higher plants

kind of donor–acceptor construct to model the natural photosynthetic process (Figure 3). Variation of the components (donor, bridge, linking group, acceptor), their spatial relationship, solvents and environmental factors then serves to modulate and optimize the physicochemical properties. Several thousand systems of this general type have been prepared and used for investigation of the photoinduced electron transfer (PET) and numerous reviews have been published in this area[16,36]. Most of the available literature on ET studies in donor–acceptor compounds focuses on porphyrins. Phthalocyanine building blocks have been used less often, a result of their low solubility and the lack of appropriate synthetic methodologies to selectively introduce functional groups or for the synthesis

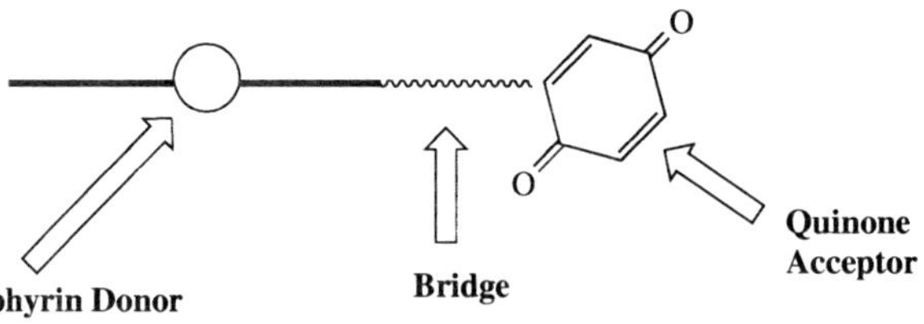

FIGURE 3. Schematic view of a biomimetic electron transfer compound

of unsymmetrically substituted derivatives. An overview of the various synthetic and structural principles to model the components of the photosynthetic apparatus has been given in the relevant chapter on zinc(II) porphyrins[4].

IV. ELECTRON TRANSFER SYSTEMS
A. Introduction

Studies on photoinduced energy and electron transfer in supramolecular assemblies have witnessed a rapid growth in the past decade. These studies were focused on the mechanistic details of light-induced chemical processes. One of these aims of photoinduced electron transfer studies in molecular systems is to produce a long-lived charge-separated state to mimic photosynthesis. Recently, the development of novel photochemically active systems has focused on polychromophoric, dendritic, supramolecular systems and novel materials. Researchers attempt to generate systems with ultrafast charge transfer and charge recombination applicable as light-induced switches or with a long-lived charge-separated state for solar energy generation[37]. These studies have yielded an expanding body of information on porphyrin/phthalocyanine dyads, their design and energy, exciton and charge transfer properties. Incorporation of these systems into larger architectures now offers the possibility for applications in molecular photonics, electronics, solar energy conversion and quantum optics.

The simplest covalently linked systems consist of porphyrin linked to electron acceptor or donor moiety with appropriate redox properties as outlined in Figure 1. Most of these studies have employed free base, zinc and magnesium tetrapyrroles because the first excited singlet state is relatively long-lived (typically 1–10 ns), so that electron transfer can compete with other decay pathways. Additionally, these pigments have relatively high fluorescence quantum yields. These tetrapyrroles are typically linked to electron acceptors such as quinones, perylenes[38–40], fullerenes[41,42], acetylenic fragments (**14**, **15**) and aromatic spacers[43–46] and other tetrapyrroles (e.g. boxes and arrays).

The basic photochemistry of magnesium tetrapyrroles is similar to other tetrapyrroles. Magnesium porphyrins[47–49] and phthalocyanines[50–53] may form cation radicals and ions via the triplet state upon illumination. For (phthalocyaninato)magnesium both photochemical oxidations and reductions have been shown. In the presence of carbon tetrabromide as an irreversible electron acceptor the mechanism proceeds via the radical cation[54]. The suggested mechanism for the photochemical oxidation is through the lowest lying triplet state of the phthalocyanine and is thought to be similar to that of porphyrins such as (2,3,7,8,12,13,17,18-octaethylporphyrinato)magnesium and (5,10,15,20-tetraphenylporphyrinato)magnesium.

B. Donor–Acceptor Electron Transfer Compounds

Biomimetic systems comprised of porphyrins and quinones have been studied extensively with regard to their electron transfer and charge transfer properties. Porphyrin–

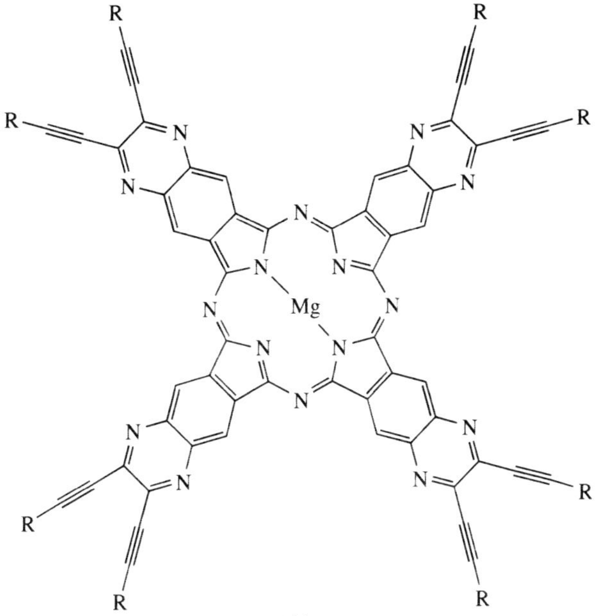

(14)

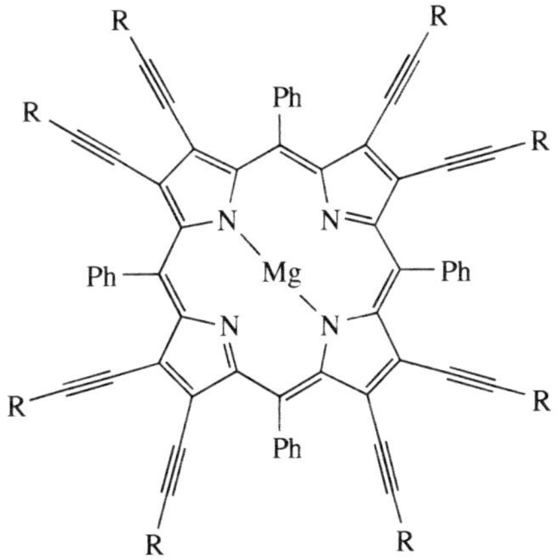

(15)

quinone (PQ) model systems, in which the quinone is fused directly to the porphyrin periphery, therefore have a special relevance for the fundamental understanding of rapid biological electron transfer reactions. Although the importance of these compounds as structurally simple models with large electronic donor–acceptor coupling has long been recognized, only few examples of magnesium-containing systems have been reported so far.

Many spectroscopic methods have been employed for the investigation of such systems[55–59]. For example, wide-band, time-resolved, pulsed photoacoustic spectroscopy was employed to study the electron transfer reaction between a triplet magnesium porphyrin and various quinones in polar and nonpolar solvents[55]. Likewise, ultrafast time-resolved anisotropy experiments with [5-(1,4-benzoquinonyl)-10,15,20-triphenylporphyrinato]magnesium **16** showed that the photoinduced electron transfer process involving the locally-excited MgP*Q state is solvent-independent, while the thermal charge recombination reaction is solvent-dependent[56, 57]. Recently, several examples of quinone–phthalocyanine systems have also been reported[58, 59].

(**16**)

Viologen (4,4′-bipyridyl) derivatives are attractive electron-accepting units for tetrapyrrole-containing dyads and more complex donor–acceptor systems as they can be easily reduced, conveniently linked to other molecules via N-alkylation of precursors, and can be used to vary the solubility in polar solvents by virtue of their charged nature. Based on the fact that the viologen radical monocation absorbs in the visible region, they can be used as convenient charge-separation indicators. As a result, a number of magnesium porphyrin[60, 61]/phthalocyanine[62–64]–viologen systems have been studied. Typically, excitation of a porphyrin–viologen dyad **17** leads to the porphyrin first excited singlet state, which can than induce photoelectron transfer to the viologen or undergo intersystem crossing to yield the porphyrin triplet state. As viologen is easily reduced, the porphyrin triplet state may also act as an electron donor in these systems.

A different strategy involves using a transition metal center linked to an organic chromophore. This greatly expands the number of electron/energy transfer reactions that can take place within the assembly compared to pure organic or inorganic-organometallic systems. Covalently linking metal complexes to porphyrins yields a cornucopia of candidates for photosynthesis-related studies. Again, only a few examples of photoinduced processes based on magnesium phthalocyanine[65] and porphyrins[66, 67] have been reported so far (e.g. **18, 19**).

For example, in 1963 the photochemistry of magnesium phthalocyanine with coordinated uranium cations was studied in pyridine and ethanol and indicated the occurrence of PET to the uranium complex[65]. A rapid photoinduced electron transfer (2–20 ps) followed by an ultrafast charge recombination was shown for various zinc and magnesium porphyrins linked to a platinum terpyridine acetylide complex[66]. The results indicated the electronic interactions between the porphyrin subunit and the platinum complex, and underscored the potential of the linking *para*-phenylene bisacetylene bridge to mediate a rapid electron transfer over a long donor–acceptor distance.

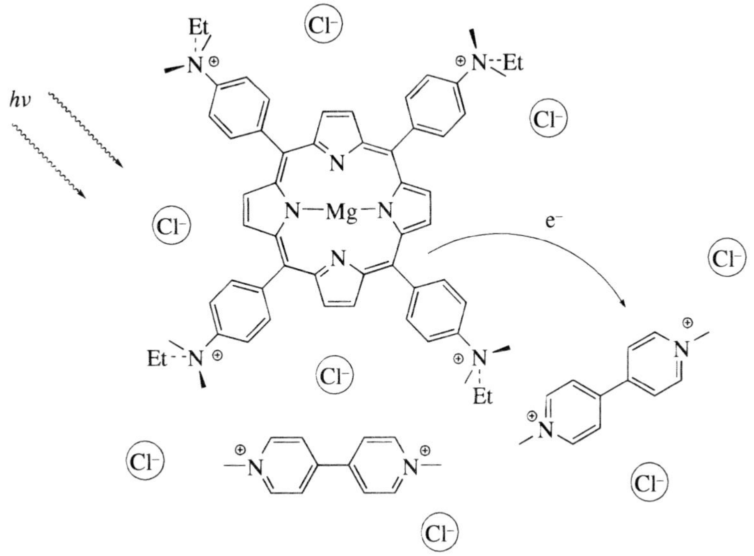

(17)

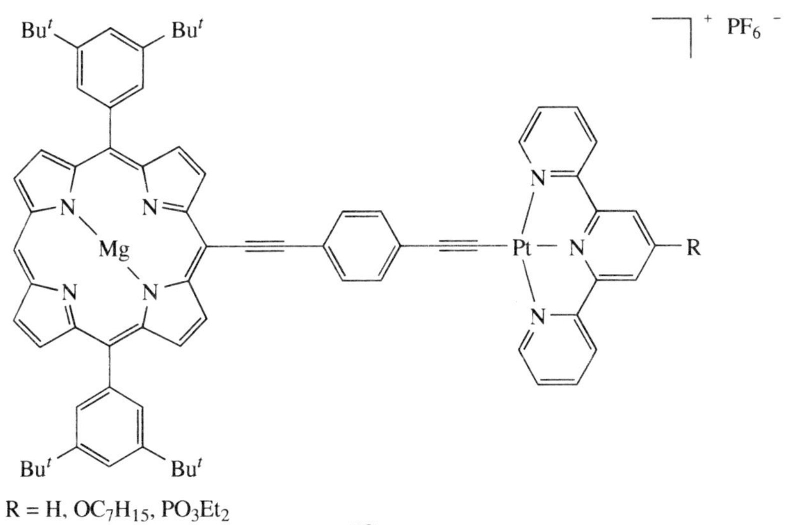

(18)

R = H, OC$_7$H$_{15}$, PO$_3$Et$_2$

Complexes of rhenium(bipyridine)(tricarbonyl)(picoline) units linked covalently to magnesium tetraphenylporphyrins via an amide bond between the bipyridine and one phenyl substituent of the porphyrin **19** exhibited no signs of electronic interaction between the Re(CO)$_3$(bpy) units and the metalloporphyrin units in their ground states. However, emission spectroscopy revealed a solvent-dependent quenching of porphyrin emission upon irradiation into the long-wavelength absorption bands localized on the porphyrin.

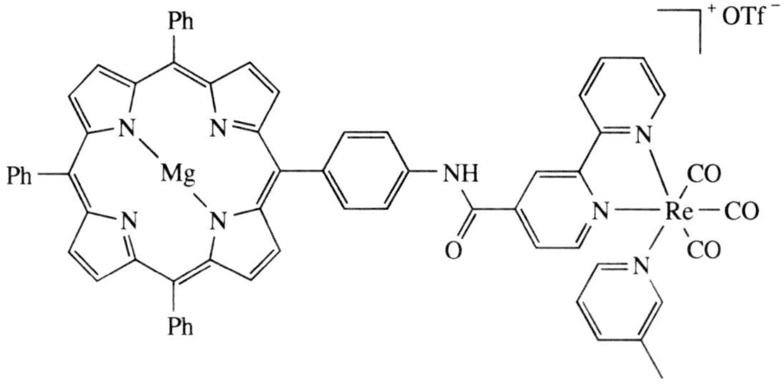

(19)

The presence of the charge-separated state involving electron transfer from Mg(II)TPP to Re(bpy) was shown by time-resolved IR spectroscopy[67].

The system is reversible in the absence of an added electron donor but undergoes irreversible reaction at the reduced rhenium bipyridine center in the presence of added triethylamine. The observation of reaction at the rhenium site upon excitation in the absorption band of the metalloporphyrin site is compatible with an ultrafast back electron transfer, provided that the triethylamine coordinated to the magnesium prior to absorption and that the electron transfer from the metalloporphyrin to the bipyridine was followed rapidly by irreversible electron transfer from the triethylamine to the metalloporphyrin. The experiments graphically demonstrated the benefits of the incorporation of carbonyl ligands at the electron acceptor as they allowed a tracking of the sequence of charge separation and back electron transfer via time-resolved IR data[67].

Fullerenes are currently enjoying considerable attention as acceptor groups in ET compounds[68, 69]. Fullerenes can accept up to six electrons, exhibit small reorganization

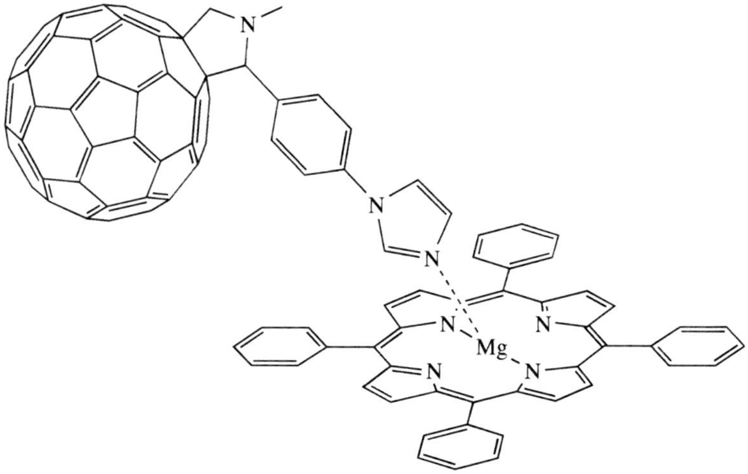

(20)

energies while photoinduced charge separation is accelerated and charge recombination is slowed. Thus, relatively long-lived charge-separated states are obtained without a special environment such as an apoprotein[70]. A recently described system consisting of a ferrocene, two porphyrins and one C_{60} unit exhibited a lifetime of 1.6 s (!), comparable to bacterial photosynthetic reaction centers. The quantum yields for charge separation in complex biomimetic systems can reach unity. Recent advances in their synthetic methodologies allow one to functionalize fullerenes and link them to other pigments.

Several self-assembled donor–acceptor systems containing fullerenes as three-dimensional electron acceptors and porphyrins as electron donors have been described. Noncovalently and covalently linked Mg porphyrin–fullerene dyads have been synthesized and investigated spectroscopically[41,42]. For example, a covalently linked magnesium porphyrin–fullerene (MgP–C_{60}) dyad with a flexible ethylene dioxide bridge[41] was compared to a self-assembled noncovalently linked dyad (MgP···C_{60}Im, **20**). In the latter, axial coordination of an imidazole (Im) functionalized fullerene[42] to the magnesium porphyrin was used for bonding. Significant increases in the lifetime of the charge-separated states were observed upon coordinating nitrogenous axial ligands to the latter.

Perylene-linked systems represent another class of useful compounds for PET studies. Classic cases are **21** and **22**. They represent a family of closely related bichromophoric systems with properties designed to utilize PET strategies[38–40].

C. Heteroligand Systems

Photoinduced ET between metalloporphyrins and free bases in dimeric, trimeric and oligomeric porphyrin systems has been studied extensively. Depending on the choice of the donor and acceptor unit, electron transfer from either the singlet or triplet states can be observed. Electron transfer studies in systems based on heterodimers with covalent or electrostatic bonds is of particular interest as it relates directly to the special pair of the reaction center chlorophylls. For systems such as the magnesium–free-base porphyrin, heterodimer **23** EPR spectroscopy has been shown to be an essential analytical tool that provides information not available from optical studies. It provides details on the magnetic interactions and spin dynamics of states with different multiplicities, such as doublets, triplets and charge-transfer states. The communication between these states strongly depends on the temperature and the solvent, and the EPR results established the existence of the radical species deduced in ps optical experiments and the corresponding theoretical calculations[71–73].

Using series of conformationally restricted magnesium–free-base hybrid arrays bridged linearly via aryl-spacers to form di- (**24**) or trimeric porphyrins, the intramolecular electron-transfer reactions from the singlet excited state of the distal doubly strapped free-base porphyrin to the pyromellitimide acceptor (Plm) was studied by time-resolved ps fluorescence and transient absorption spectroscopy[74–76]. The electron transfer was more effective in magnesium–porphyrin bridged models than in the related zinc–porphyrin bridged ones, indicating that the past reliance on the use of zinc-based biomimetic models is not always sufficient. Remarkably, the electron transfer over two porphyrins proceeded with rates almost similar to those for the ET over one porphyrin regardless of the bridging metalloporphyrin.

A simple method has been developed to construct a variety of molecular architectures containing free base–magnesium or magnesium–metalloporphyrin systems consisting of two to nine porphyrin units. Compound **25** is a typical example for such a compound that are important for studying the electronic communication in multichromophoric systems.[45,46]

(21)

R = *t*-Bu

(22)

R=*t*-Bu

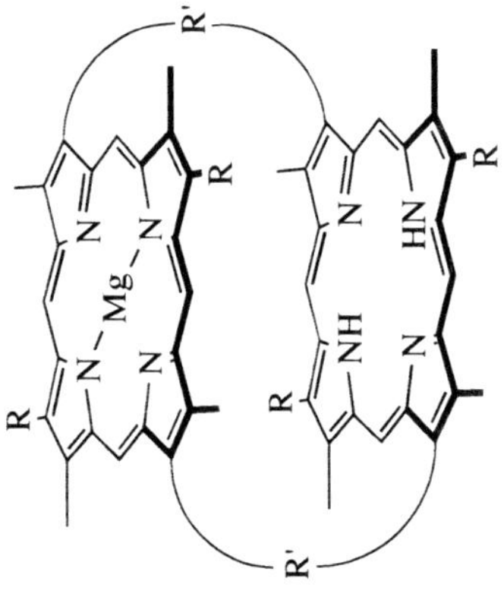

R = *n*-C$_8$H$_{17}$, R' = CH$_2$CON(*n*-Bu)CH$_2$CH$_2$

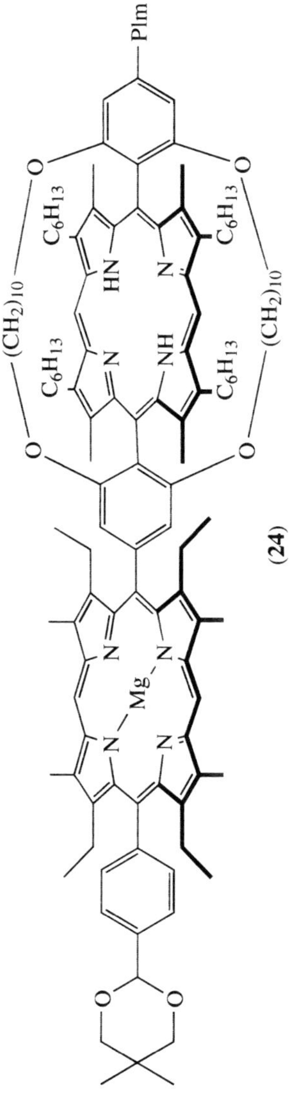

(23)

(24)

Plm:

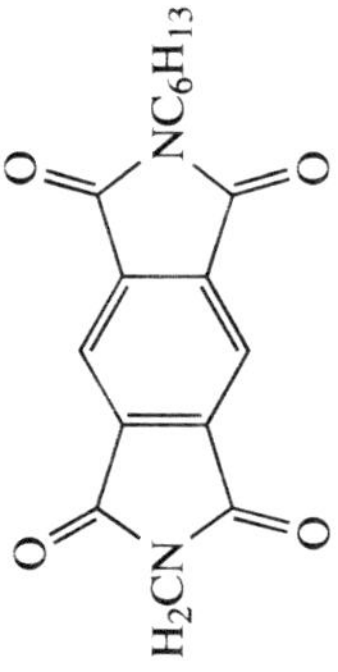

(25)

An example for a study involving dimeric systems linked through noncovalent bonds used (5,10,15,20-tetrakis(4-sulfonatophenyl)porphyrinato)zinc(II) and (5,10,15,20-tetrakis(4-N,N,N-trimethylanilinium)porphyrinato)magnesium(II) with complementary charge and results in dimerization in solution. Continuous-wave time-resolved EPR spectroscopy demonstrated that intramolecular electron and/or energy transfer in electrostatically bound metalloporphyrin dimers can be controlled via simple metal and substituent effects. Although the metal constituents are identical in these two dimers, it was the peripheral charged substituents that governed the fate of the electron transfer, whereas the energy transfer is controlled via the metal substituents[77,78].

V. PHOTOCHEMICAL REACTIONS

A. Porphyrins

As lipophilic pigments where the (b)chls are embedded in natural systems in apoproteins, photosynthesis in general is a transmembrane process. Thus, PET reactions in lipid membranes have been investigated extensively. Many reports have been published on photoinitiated (where the photoinitiated species acts as a catalysts to mediate thermodynamically favored reactions) and photodriven (where some of the light energy is converted into the products) processes[79]. A typical example are Mg(II)OEP-sensitized electron transfer reactions across lipid bilayer membranes[80]. The reaction mechanism involved a reduction of photoexcited Mg(II)OEP at the reducing (ascorbate) side of the bilayer with the charge carrier most likely being a neutral protonated Mg(II)OEP anion. Thus, the magnesium porphyrin participated as a sensitizer and a transmembrane redox mediator.

More detailed data are available on Mg-substituted horseradish peroxidases. This system can form stable porphyrin π-cation radicals in the presence of oxidants[81,82] and photooxidation and reduction occur through direct reaction of the excited-state porphyrins with oxidants and reductants, respectively. In general, porphyrins appear to be photooxidized both via electron transfer and $^1O_2$ mechanisms. Thus, photoirradiation of the Mg-substituted horseradish peroxidase under aerobic conditions results in two simultaneously occurring reactions. A porphyrin π-cation radical is generated through electron transfer from excited porphyrin to O_2 and a so-called 448-nm compound via a singlet oxygen mechanism. A species with an absorption band at 448 nm was first formed upon irradiation and was then converted in the dark to a final product with a band at 489 nm (Figure 4). The conversion of the 448 nm compound to the 489 nm compound seems

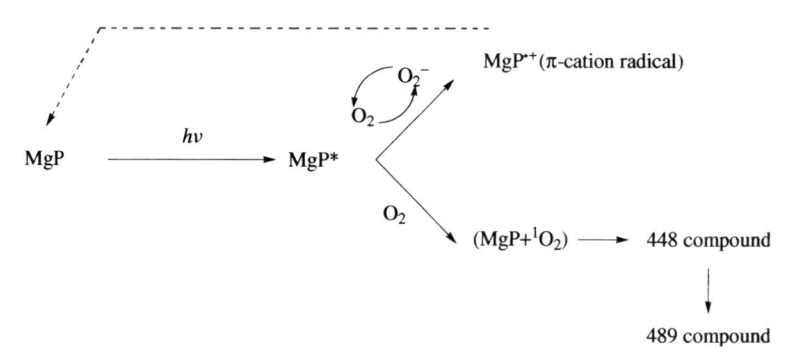

$$MgP \xrightarrow{h\nu} MgP^* \begin{array}{c} \xrightarrow{O_2, \, O_2^-} MgP^{\cdot+} (\pi\text{-cation radical}) \\ \xrightarrow{O_2} (MgP + {}^1O_2) \longrightarrow 448 \text{ compound} \\ \downarrow \\ 489 \text{ compound} \end{array}$$

FIGURE 4. Scheme for the photooxidation of porphyrin (P) by Mg horseradish peroxidase

to be an isomerization reaction which requires a high activation energy, probably due to structural restrictions in the heme crevice. Noteworthy is that the 448 and 489 nm compounds both form the same chlorine-type hydroporphyrin with an intense band at 712 nm upon the addition of ascorbate.

B. Photoinduced Ring-opening Reactions

The formation of long-lived excited states of chlorophyll and its function as energy-storage and catalytic material in photosensitization reactions has been postulated for some time on the basis of indirect evidence. For example, in the 1930s Rabinowitch and Weiss performed spectrophotoelectrochemical studies on the reversible oxidation and reduction of chl[83, 84]. An ethylchlorophyllide solution was reversibly oxidized by $FeCl_3$ to a yellow, unstable intermediate from which the green solution was regenerated by reduction with $FeCl_2$. The oxidation was greatly favored by illumination and the equilibrium was shifted by light towards the yellow form. The nature of the reversible reaction with Fe^{3+} was considered to be an oxidation in which Fe^{3+} was reduced to Fe^{2+} and chlorophyll was oxidized to a chl cation or a dehydrochlorophyll species.

About a decade later Calvin and coworkers[85-87] reported that the photochemical reactions of simple chlorins in the presence of either oxygen or various *ortho/para* quinones led to the corresponding porphyrins and unidentified products. Based on kinetic experiments, they proposed a mechanism for the photochemical oxidation of 5,10,15,20-tetraphenylchlorin (H_2TPC) and β-naphthoquinone involving the triplet state of the chlorin molecule as an intermediate[85]. The production of ions P+ and P− from the first excited triplet state (**T**) of Mg(II)OEP (**P**) predominantly involves triplet–triplet annihilation. Evidence was obtained indicating that the reaction of T with ground-state P is not a significant source of ions. On the other hand, the two triplets initially can combine to form an excited charge-transfer complex. The relationship between the multiplicity of this charge-transfer complex and triplet quenching, delayed fluorescence and ion formation is illustrated in Figure 5. Less extensive experiments were carried out with Mg(II)TPC due to its instability. However, the data obtained confirmed the existence of a phosphorescence state[86]. The photooxidation rates for the magnesium chlorins were significantly lower (almost 8 times) than for the corresponding zinc complexes.

The magnesium and zinc complexes of TPC can be photooxidized using quinones as hydrogen acceptors. More detailed studies showed that the reaction between quinones and Zn(II)TPC resulted in the formation of Zn(II)TPP[86]. Subsequent work showed that Mg(II)TPC and Zn(II)TPC can be photooxidized by molecular oxygen and *o/p*-quinones. Oxygen is reduced to hydrogen peroxide with a concomitant reduction of quinones to hydroquinones. However, oxygen differs from quinones, as the primary formation of oxidation to porphyrins here is followed by secondary reactions[87]. This second reaction involves H_2O_2 that can react either directly or as an initiator of Haber–Weiss processes and resulted in the formation of unidentified products[87] similar to those obtained by 'bleaching' of chlorophyll in the presence of oxygen[83, 84].

Subsequent work in this area clarified some aspects of the photooxidation of magnesium porphyrins. Barrett found no alteration in the spectral and chromatographic properties of nonfluorescent protoporphyrin complexes of Fe, Ni, Co, Cu and Ag upon irradiation[88]. However, irradiation of (protoporphyrinato dimethyl ester)magnesium(II) **26** in various organic solvents resulted in rapid photooxygenation to green-brown products that did not contain magnesium. Moreover, spectroscopic data indicated an interruption of the aromatic ring system, showed no fluorescence and the appearance of a strong band at 1680 cm^{-1}

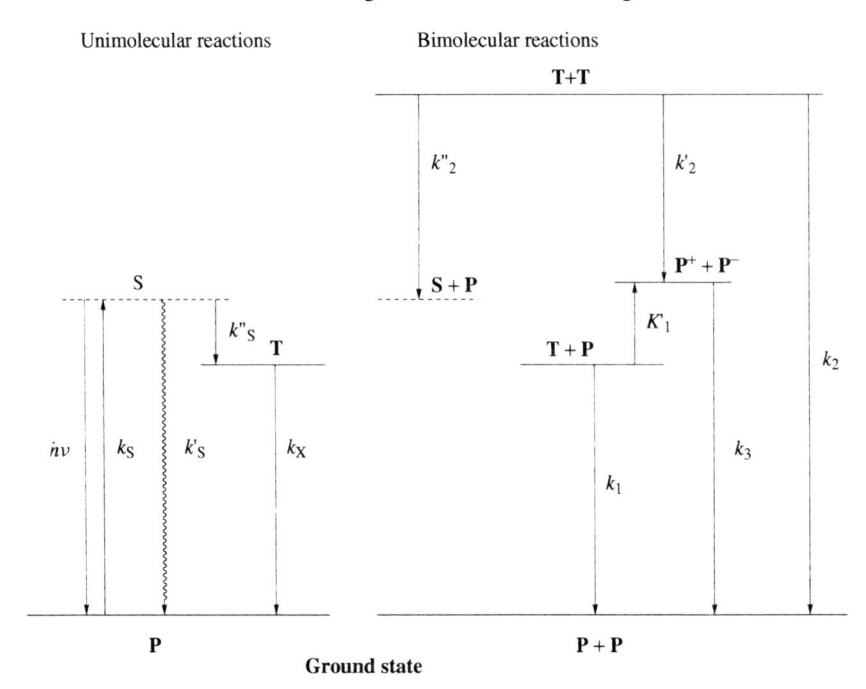

FIGURE 5. Diagram of the triplet decay mechanism

(CCl$_4$) in the IR spectrum of the newly formed compound. This green pigment was very photolabile and quickly decomposed to yield 15,16-hydrobiliverdin **27**. Magnesium porphyrins without vinyl side-chains were photooxidized to similar green compounds with the band at 1680 cm^{-1}, confirming no oxidation of the vinyl group. The UV/vis spectra of these green products were similar to those of the phlorins obtained by photoreduction of uro-, copro- and hematoporphyrins[89]. Thus, photooxidation of magnesium protoporphyrins resulted in the formation of 15,16-hydrobiliverdins upon ring cleavage. This is in contrast to the enzymatic breakdown of heme which proceeds through biochemical transformations via biliverdin **28** towards the phycobilins[90–92].

The reaction of porphyrin ligands with molecular oxygen is related to catabolic processes of naturally occurring porphyrins and drugs and is of great importance. Various metalloporphyrins, particularly the chlorophylls present in photosynthetic organisms, can be rapidly destroyed by light and oxygen. In fact, without the presence of photoprotective pigments such as carotenes, no natural chlorophyll-based photosynthetic system would be stable. First studies on the photooxygenation of Mg(II)OEP (**29**),[93–95] Mg(II)TPP[96,97] and Mg(II)protoporphyrin[88] and Mg(II)(tetrabenzoporphyrin)[98,99] were reported in the 1970s and 1980s. For example, when Mg(II)OEP was exposed to visible light in the presence of air in benzene solution, spectroscopic examinations showed that the porphyrin was quantitatively converted into a chromophore with an intense absorption band above 800 nm[88,93,94]. This reaction proceeded uniformly and no intermediates with lifetimes of more than 10 s occurred. The primary product was an open-chain magnesium

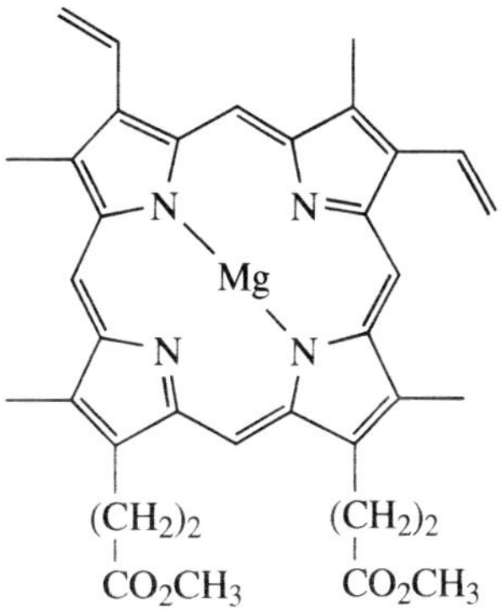

(26)

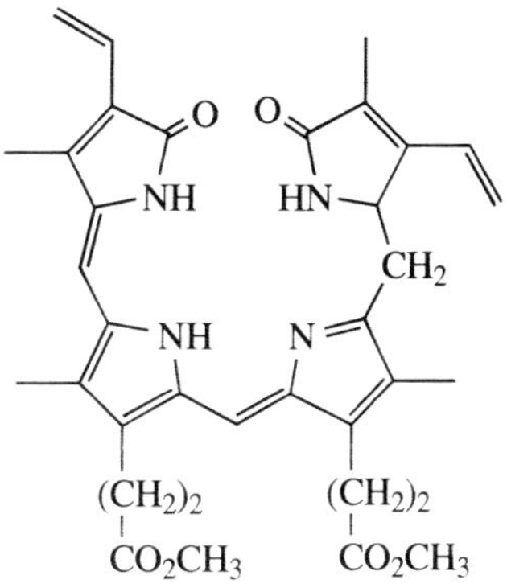

(27)

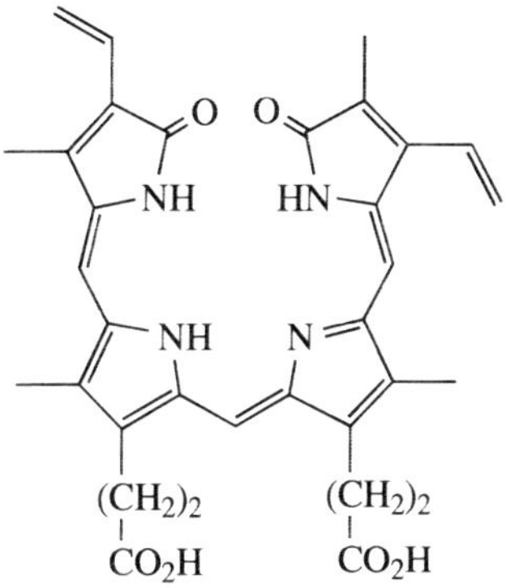

(28)

formylbiliverdin complex **30** that can be easily demetallated to the formylbiliverdin (Scheme 1).

A similar photooxidation pathway was found for Mg(II)TPP. It reacted readily with molecular oxygen to give the corresponding 15,16-dihydrobiliverdin, similar to the one shown for Mg(II)OEP in Scheme 1. Further studies have proposed that the photooxygenation of metallo-*meso*-tetrasubstituted porphyrins proceeds via a one-molecule mechanism involving only one oxygen molecule. Most likely, the first intermediates formed upon photooxygenation are short-lived peroxides. Such compounds are very unstable and a possible dioxetane structure is shown in formula **31**.

C. Reactions of Chlorophyll

The most obvious chemical reaction involving chlorophyll is the chlorophyll breakdown in fall and during senescence. This process involves annually more than 10^9 tons of chlorophyll and, despite its obvious prominence in the natural beauty of the fall season,

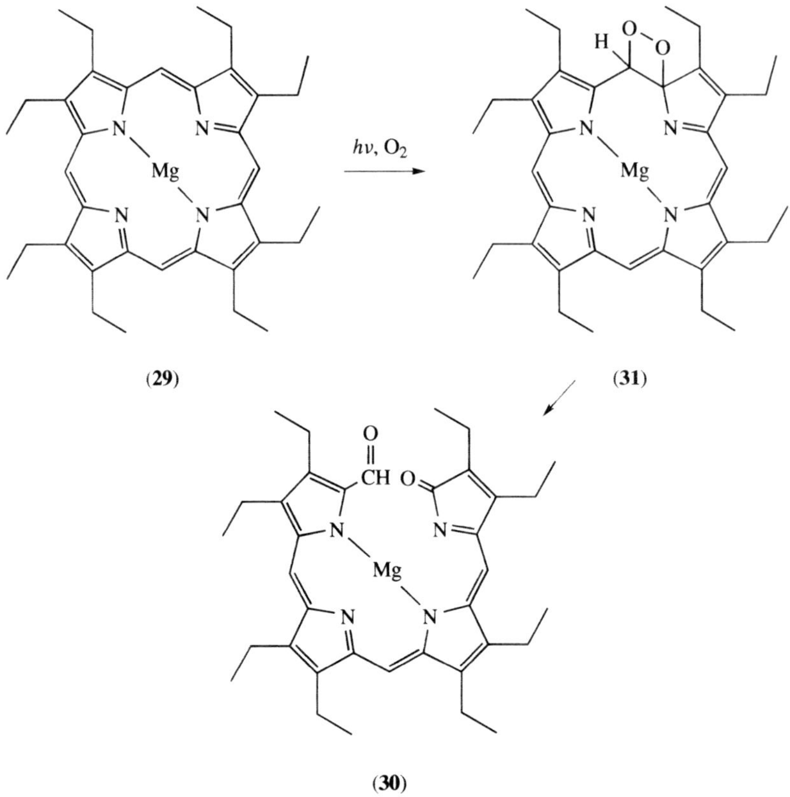

(29) → hv, O₂ → (31) → (30)

SCHEME 1. Photooxygenation of Mg(II)OEP

and its mechanism remained unknown until about 20 years ago[100]. Work by the groups of Kräutler, Matile and Gossauer showed that the central step is a ring-opening reaction at the 5-position[101–103]. This is in contrast to the situation encountered for heme, which is oxidatively cleaved at the 20-position. As shown in Scheme 2, the crucial steps during chl a degradation are the conversion of chl a into pheophorbide a (**5**), followed by enzymatic transformation into the bilinone **32**. During this step the macrocycle undergoes oxidative C5 ring-opening, incorporates two oxygen atoms (the CHO one from O_2) and is saturated at the 10-position. This reaction is catalyzed by a monooxygenase and the red compound **32** is further converted to the still fluorescing compound **33** and finally into the nonfluorescing derivative **34**, along with some changes in the side chains directed to increase the hydrophilicity of the breakdown products. Chl b **6** is first converted into chl a **4** and then subjected to the same reactions. Note that this is an enzymatic process, not a simple photochemical reaction, and should not be confused with the photooxidative ring-opening reactions.

A second reaction involves the chlorin-to-porphyrin conversion. Any chlorin which has hydrogen atoms at the sp^3-hybridized centers of the reduced ring can be oxidized to the respective porphyrin. Oxidation may be achieved by various oxidants including oxygen[104]. Likewise, reductions to hydroporphyrins and other reactions of the macrocycle are possible. However, most of these are of interest only for the specialist. Under

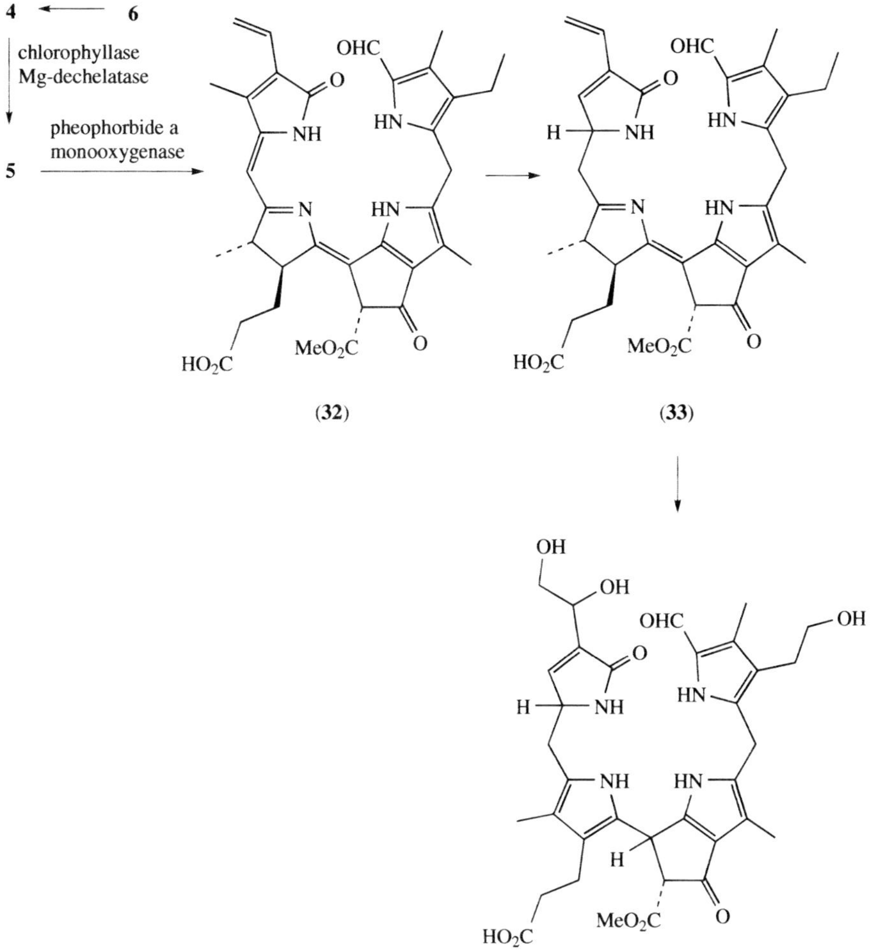

SCHEME 2. Chlorophyll breakdown during senescence

appropriate conditions photochemical reductions, notably the Krasnovskii reduction to **35**, can occur[105].

Like porphyrins, chls undergo photooxygenation[106,107]. Chlorophylls are potent photosensitizers and will produce singlet oxygen in the presence of air or triplet oxygen[104]. Thus, chls can undergo self-destruction (Figure 6). The chemistry of this photooxygenation is very involved and differs somewhat for individual types of (b)chls[103,105,108,109]. While being partially responsible for the low stability of chls in solution[110], and for unwanted side reactions in food stuff[111], the same reaction also offers potential for future applications. Chl and derivatives thereof may be used as photosensitizers to affect desired chemical transformations and they have been utilized for applications in photodynamic therapy (PDT)[112].

(4) (35)

SCHEME 3. Formation of the Krasnovski photoproduct of chl

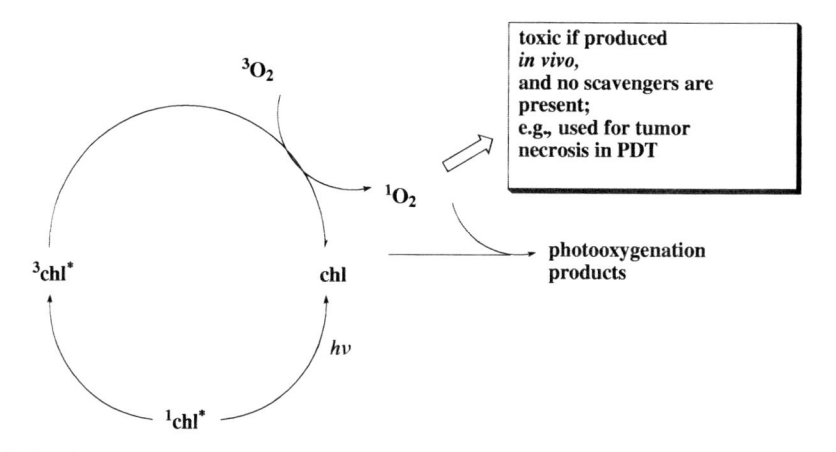

FIGURE 6. Photooxidation reactions of chl

VI. APPLIED PHOTOCHEMISTRY

The low stability of the magnesium porphyrins has precluded most potential applications. Other metallotetrapyrroles have found industrial uses for oil desulfurization, as photoconducting agents in photocopiers, deodorants, germicides, optical computer disks, semiconductor devices, photovoltaic cells, optical and electrochemical sensing, and molecular electronic materials. A few scattered examples of the use of Mg porphyrins in nonlinear optical studies have appeared[113,114], and magnesium phthalocyanines have been used in a few studies as semiconductor or photovoltaic materials[115–117]. One of the few

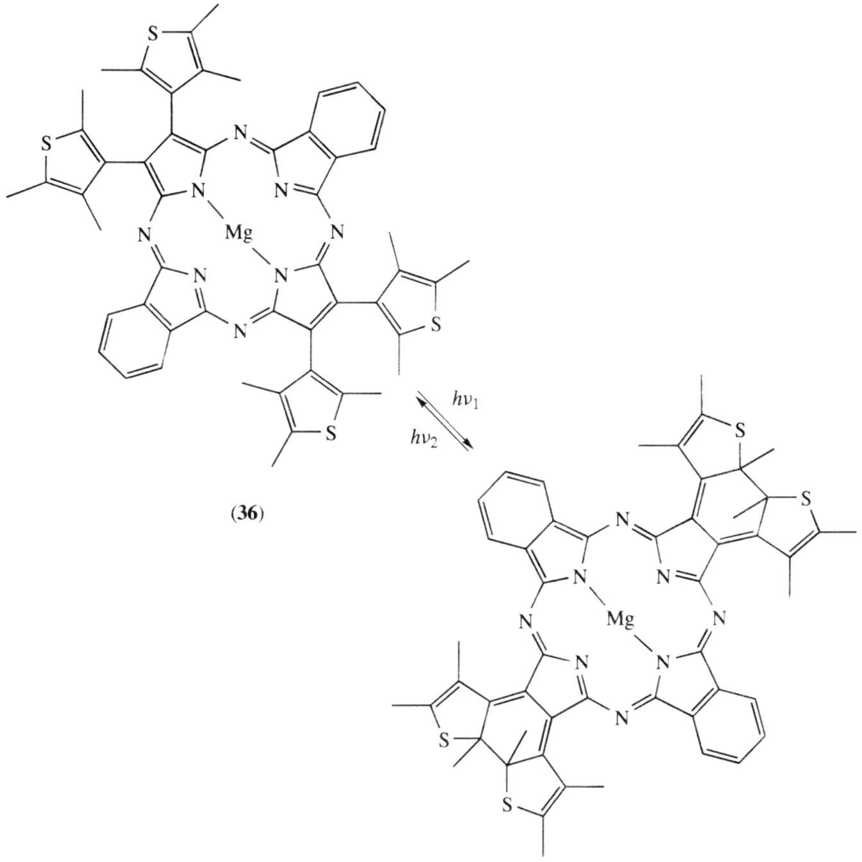

SCHEME 4. Photochemical interconversion of a magnesium(II) phthalocyanine

true photochemical reactions described for Mg(II) phthalocyanines involved a wavelength-dependent photocyclization of **36** to **37**. Together with the back reaction, the system shown in Scheme 4 was developed to act as a photochromic readout system.[118]

Like the porphyrins, the phthalocyanines can undergo photooxidation and act as photosensitizers for the production of singlet oxygen[119-122]. One of the few chemical synthetic applications was the acceleration of the autoxidation of cumene and photooxidation of pinenes[123].

Despite their low (photo)stability, chlorophylls, or rather their derivatives, have found some applications, especially in the nutrition industry. In Europe the food additive E140 is chl, and E141 is chlorophyllin (a semisynthetic sodium/copper derivative of chlorophyll), and they are used in cakes, beverages, sweets, icecream etc. As color No 125 they find applications in toothpaste, as a soap pigment and in shampoos. The older literature also describes its use in candles[124] and as a lipophilic oil bleaching additive (to neutralize the yellow color of oils in food stuff or giving them a greener touch)[125].

Nevertheless, this is a somewhat misleading statement as in most cases chl is used in the form of chlorophyllin and metal complexes thereof. Chlorophyllin is an inhomogeneous water-soluble material. It is prepared by saponification of the phytyl side chain with NaOH and exchange of the central magnesium atom against copper (or other metals). The harsh reaction conditions (and the use of the natural chl a/b mix) results in the formation of a mixture of chemical compounds. Most prominent constituents are derivatives $3^1,3^2$-didehydrorhodochlorin, pheophorbide salts and the typical allomerization products[126,127]. Chlorophyllin is a stable pigment with intense light green to dark blue-green color. Related formulations are sodium zinc chlorophyllin, chlorophyll paste, oil-soluble chlorophyll and sodium magnesium chlorophyllin.

Besides the traditional use of chlorophyll and its derivatives as pigments, early investigations on the medicinal use of chlorophyllin in the 1940s led to a first boom in chlorophyll use and initiated more serious investigation of medicinal applications. During those times it was used in bathroom tissue, diapers, chewing gum, bed sheets, shoe liners, toothpaste[128] and other daily products, mostly as an antiodorant. Chlorophyll preparations are still available as over-the-counter (OTC) medicine to reduce fecal odor due to incontinence or to reduce odor from a colostomy or ileostomy. Other applications involved use in wound healing, germ killing and the treatment of infections and inflammations (use of bandages, antiseptic ointments, surgical dressings). Despite these sometime dubious applications all outside the area of photochemistry, there is growing evidence for a medicinal use of chlorophylls. Antimutagenic effects, both *in vitro* and in animal models, have been proven, notably against aflatoxins. Likewise, there are indications for an anticarcinogenic role[128,129]. For example, an animal study showed inhibition of dioxin absorption and increased fecal excretion of dioxin[130]. At the very least, these results indicate the need for further research and offer the promise of some future chl applications[131,132].

Photodynamic therapy presents the one clearly established medicinal application of chlorophyll derivatives to date[133,134]. This method relies on the selective accumulation of a tetrapyrrole photosensitizer in target tissue where it can be activated with light to produce toxic singlet oxygen resulting in, e.g., tumor necrosis as outlined in Figure 6. Several porphyrin-based compounds have been approved for medicinal applications and others are in Phase-2 trials. Among these tetrapyrroles, chlorophyll derivatives are currently under active investigation and show great promise[112]. Due to the low stability, most cases of chlorin-type hydroporphyrins used in clinical studies and applications are free-base tetrapyrroles. The use of chlorophyll derivatives in technical applications is still in the early developmental stage. Topics of current interest are both solar energy conversion and hydrogen production[135]. As in the case of the PET model compounds, most studies in these cases do use the zinc(II) and not the magnesium(II) derivatives[136,137]. More specialized works have treated emerging trends and contemporary approaches and a number of reviews dealt with chlorophyll chemistry and related chromophore systems[138,139].

VII. ACKNOWLEDGMENT

Writing of this chapter was made possible by generous funding from Science Foundation Ireland (Research Professorship—SFI 04/RP1/B482).

VIII. REFERENCES

1. P. C. Harvey, C. L. Junk, C. L. Colin and G. Salem, *J. Org. Chem.*, **53**, 3134 (1988).
2. B. Marciniak and G. L. Hug, *J. Photochem. Photobiol. A: Chem.*, **78**, 7 (1994).
3. B. Marciniak and G. L. Hug, *Coord. Chem. Rev.*, **159**, 55 (1997).

4. M. O. Senge and N. N. Sergeeva, in *The Chemistry of Organozinc Compounds* (Eds. Z. Rappoport and I. Marek), Wiley, Chichester, 2006, pp. 395–419.
5. H. Scheer, in *Chlorophylls* (Ed. H. Scheer), CRC Press, Boca Raton, 1991, pp. 3–30.
6. M. O. Senge, *Chem. Unserer Zeit*, **26**, 86 (1992).
7. M. O. Senge and J. Richter, in *Biorefineries-Industrial Processes and Products* (Eds. B. Kamm, P. R. Gruber and M. Kamm), Vol. 2, Wiley-VCH, Weinheim, 2006, pp. 325–343.
8. M. O. Senge, A. Wiehe and C. Ryppa, in *Chlorophylls and Bacteriochlorophylls* (Eds. B. Grimm, R. J. Porra, W. Rüdiger and H. Scheer), Springer, Berlin, 2006, pp. 27–37.
9. R. H. Douglas, J. C. Partridge, K. Dulai, D. Hunt, C. W. Mullineaux, A. Y. Tauber and P. H. Hynninen, *Nature*, **393**, 423 (1998).
10. M. O. Senge, *J. Photochem. Photobiol. B: Biol.*, **16**, 3 (1992).
11. K. M. Kadish, K. M. Smith and R. Guilard, *The Porphyrin Handbook*, Vol. 1–20, Academic Press, San Diego, CA, 2000.
12. T. Nyokong and H. Isago, *J. Porphyrins Phthalocyanines*, **8**, 1083 (2004).
13. K. M. Kadish, K. M. Smith and R. Guilard, *The Porphyrin Handbook*, Vol. 17, Vol. 19, Academic Press, San Diego, CA, 2000.
14. M. Gouterman, in *The Porphyrins* (Ed. D. Dolphin), Vol. 3, Academic Press, New York, 1978, p. 1.
15. J. Mack and M. J. Stillman, in *The Porphyrin Handbook* (Eds. K. M. Kadish, K. M. Smith and R. Guilard), Vol. 16, Academic Press, San Diego, CA, 2003, pp. 43–116.
16. D. Gust and T. A. Moore, in *The Porphyrin Handbook* (Eds. K. M. Kadish, K. M. Smith and R. Guilard), Vol. 8, Academic Press, San Diego, CA, 2000, pp. 153–190.
17. S. G. Ballard and D. Mauzerall, *Biophys. J.*, **24**, 335 (1978).
18. J. Mack and M. J. Stillman, *J. Am. Chem. Soc.*, **116**, 1292 (1994).
19. S. P. Keizer, W. Han and M. J. Stillman, *Inorg. Chem.*, **41**, 353 (2002).
20. K. Lang, D. M. Wagnerova and J. Brodilova, *J. Photochem. Photobiol. A: Chem.*, **72**, 9 (1993).
21. J. D. Spikes, J. E. van Lier and J. C. Bommer, *J. Photochem. Photobiol. A: Chem.*, **91**, 193 (1995).
22. A. Hiraishi and K. Shimada, *J. Gen. Appl. Microbiol.*, **47**, 161 (2001).
23. A. Hiraishi, K. V. P. Nagashima, K. Matsuura, K. Shimada, S. Takaichi, N. Wakao and Y. Katayama, *Int. J. Syst. Bacteriol.*, **48**, 1389 (1998).
24. N. Wakao, A. Hiraishi, K. Shimada, M. Kobayashi, S. Takaichi, M. Iwaki and S. Itoh, in *The Phototrophic Prokaryotes* (Ed. G. A. Peschek), Kluwer Academic/Plenum Publishers, New York, 1999, p. 745–753.
25. N. Wakao, N. Yokoi, N. Isoyama, A. Hiraishi, K. Shimada, M. Kobayashi, H. Kise, S. Iwaki, S. Itoh, S. Takaichi and Y. Sakurai, *Plant Cell Physiol.*, **37**, 889 (1996).
26. N. Wakao, T. Shiba, A. Hiraishi, M. Ito and Y. Sakurai, *Curr. Microbiol.*, **27**, 277 (1993).
27. N. Kishimoto, F. Fukaya, K. Inagaki, T. Sugio, H. Tanaka and T. Tano, *FEMS Microbiol. Ecol.*, **16**, 291 (1995).
28. H. Scheer and G. Hartwich, in *Anoxygenic Photosynthetic Bacteria* (Eds. R. E. Blankenship, M. T. Madigan and C. E. Bauer), Kluwer Academic Publishers, Dordrecht, 1995, pp. 649–663.
29. M. Kobayashi, M. Yamamura, M. Akiyama, H. Kise, K. Inoue, M. Hara, S. Takaichi, N. Wakao, K. Yahara and T. Watanabe, *Anal. Sci.*, **14**, 1149 (1998).
30. A. Osuka, S. Nakajima, K. Maruyama, N. Mataga, T. Asahi, I. Yamazaki, Y. Nishimura, T. Ohno and K. Nozaki, *J. Am. Chem. Soc.*, **115**, 4577 (1993).
31. M. R. Wasielewski and M. P. Niemczyk, *J. Am. Chem. Soc.*, **106**, 5043 (1984).
32. K. M. Kadish, K. M. Smith and R. Guilard, *The Porphyrin Handbook*, Vol. 13, Academic Press, San Diego, CA, 2000.
33. A. S. Raghavendra, *Photosynthesis*, Cambridge University Press, Cambridge, 1998.
34. R. van Grondelle and V. I. Novoderezhkin, *Phys. Chem. Chem. Phys.*, **8**, 793 (2006).
35. D. Gust, T. A. Moore and A. L. Moore, *Acc. Chem. Res.*, **34**, 40 (2001).
36. M. R. Wasielewski, *Chem. Rev.*, **92**, 435 (1992).
37. Y. Iseki and S. Inoue, *J. Chem. Soc., Chem. Commun.*, 2577 (1994).
38. S. I. Yang, S. Prathapan, M. A. Miller, J. Seth, D. F. Bocian, J. S. Lindsey and D. Holten, *J. Phys. Chem. B.*, **105**, 8249 (2001).
39. S. I. Yang, R. K. Lammi, S. Prathapan, M. A. Miller, J. Seth, J. R. Diers, D. F. Bocian, J. S. Lindsey and D. Holten, *J. Mater. Chem.*, **11**, 2420 (2001).

40. S. Fukuzumi, K. Ohkubo, J. Ortiz, A. M. Gutierrez, F. Fernandez-Lazaro and A. Sastre-Santos, *J. Chem. Soc., Chem. Commun.*, 3814 (2005).
41. M. E. El-Khouly, Y. Araki, O. Ito, S. Gadde, A. L. McCarty, P. A. Karr, M. E. Zandler and F. D'Souza, *Phys. Chem. Chem. Phys.*, **7**, 3163 (2005).
42. F. D'Souza, M. E. El-Khouly, S. Gadde, A. L. McCarty, P. A. Karr, M. E. Zandler, Y. Araki and O. Ito, *J. Phys. Chem. B*, **109**, 10107 (2005).
43. F. Mitzel, S. FitzGerald, A. Beeby and R. Faust, *Chem. Eur. J.*, **9**, 1233 (2003).
44. T. Chandra, B. J. Kraft, J. C. Huffman and J. M. Zaleski, *Inorg. Chem.*, **42**, 5158 (2003).
45. W. J. Youngblood, D. T. Gryko, R. K. Lammi, D. F. Bocian, D. Holten and J. S. Lindsey, *J. Org. Chem.*, **67**, 2111 (2002).
46. F. Li, S. Gentemann, W. A. Kalsbeck, J. Seth, J. S. Lindsey, D. Holten and D. F. Bocian, *J. Mater. Chem.*, **7**, 1245 (1997).
47. J. Fajer, D. C. Borg, A. Forman, D. Dolphin and R. H. Felton, *J. Am. Chem. Soc.*, **92**, 3451 (1970).
48. J. F. Smalley, S. W. Feldberg and B. S. Brunschwig, *J. Phys. Chem.*, **87**, 1757 (1983).
49. R. Slota and G. Dyrda, *Inorg. Chem.*, **42**, 5743 (2003).
50. H. van Willigen and M. H. Ebersole, *J. Am. Chem. Soc.*, **109**, 2299 (1987).
51. A. P. Bobrovskii and V. E. Kholmogorov, *Russ. J. Phys. Chem.*, **47**, 983 (1973).
52. E. Ough, Z. Gasyna and M. J. Stillman, *Inorg. Chem.*, **30**, 2301 (1991).
53. D. Kim, *Bull. Korean Chem. Soc.*, **7**, 416 (1986).
54. H. Stiel, K. Teuchner, A. Paul, W. Freyer and D. Leupold, *J. Photochem. Photobiol. A: Chem.*, **80**, 289 (1994).
55. J. Feitelson and D. C. Mauzerall, *J. Phys. Chem.*, **97**, 8410 (1993).
56. K. Wynne, S. M. LeCours, C. Galli, M. J. Therien and R. M. Hochstrasser, *J. Am. Chem. Soc.*, **117**, 3749 (1995).
57. T. L. Netzel, M. A Bergkamp, C. K. Chang and J. Dalton, *J. Photochem.*, **17**, 451 (1981).
58. V. B. Evstigneev and V. A. Gavrilova, *Biofizika*, **14**, 43 (1969); *Chem. Abstr.*, **70**, 93263w (1969).
59. A. V. Barmasov, V. I. Korotkov and V. Ye. Kholmogorov, *Biofizika*, **39**, 263 (1994); *Chem. Abstr.*, **121**, 267490g (1994).
60. J. M. Furois, P. Brochette and M. P. Pileni, *J. Colloid Interface Sci.*, **97**, 552 (1984).
61. W. S. Szulbinski, *Inorg. Chim. Acta*, **228**, 243 (1995).
62. J. Zakrzewski and C. Giannotti, *Inorg. Chim. Acta*, **232**, 63 (1995).
63. A. Harriman, G. Poter and M.-C. Richoux, *J. Chem. Soc., Faraday Trans.*, **77**, 1175 (1981).
64. H. Ohtani, T. Kobayashi, T. Ono, S. Kato, T. Tanno and A. Yamada, *J. Phys. Chem.*, **88**, 4431 (1984).
65. G. I. Kobyshev, G. N. Lyalin and A. N. Terenin, *Dokl. Akad. Nauk SSSR*, **153**, 865 (1963); *Chem. Abstr.*, **60**, 14357e (1963).
66. C. Monnereau, J. Gomez, E. Blart and F. Odobel, *Inorg. Chem.*, **44**, 4806 (2005).
67. A. Gabrielsson, F. Hartl, H. Zhang, J. R. Lindsay Smith, M. Towrie, A. Vlcek Jr. and R. N. Perutz, *J. Am. Chem. Soc.*, **128**, 4253 (2006).
68. H. Imahori, *Org. Biomol. Chem.*, **2**, 1425 (2004).
69. H. Imahori and S. Fukuzumi, *Adv. Funct. Mater.*, **14**, 525 (2004).
70. H. Imahori, K. Tamaki, Y. Araki, T. Hasobe, O. Ito, A. Shimomura, S. Kundu, T. Okada, Y. Sakata and S. Fukuzumi, *J. Phys. Chem. A*, **106**, 2803 (2002).
71. H. Levanon, A. Regev, T. Galili, M. Hugerat, C. K. Chang and J. Fajert, *J. Phys. Chem.*, **97**, 13198 (1993).
72. H. Zhang, E. Schmidt, W. Wu, C. K. Chang and G. T. Babcock, *Chem. Phys. Lett.*, **234**, 133 (1995).
73. J. M. Zaleski, W. Wu, C. K. Chang, G. E. Leroi, R. I. Cukier and D. G. Nocera, *Chem. Phys.*, **176**, 483 (1993).
74. A. Osuka, S. Marumo, S. Taniguchi, T. Okada and N. Mataga, *Chem. Phys. Lett.*, **230**, 144 (1994).
75. A. Osuka, S. Marumo, K. Maruyama, N. Mataga, M. Ohkohchi, S. Taniguchi, T. Okada, I. Yamazaki and Y. Nishimura, *Chem. Phys. Lett.*, **225**, 140 (1994).
76. A. Osuka, F. Kobayashi, K. Maruyama, N. Mataga, T. Asahi, T. Okada, I. Yamazaki and Y. Nishimura, *Chem. Phys. Lett.*, **201**, 223 (1993).

77. S. Murov, I. Carmichael and G. Hug, *Handbook of Photochemistry*, Marcel Dekker, New York, 1993.
78. A. Berg, M. Rachamim, T. Galili and H. Levanon, *J. Phys. Chem.*, **100**, 8791 (1996).
79. S. Tunuli and J. H. Fendler, *J. Am. Chem. Soc.*, **103**, 2507 (1981).
80. A. Ilani, M. Woodle and D. Mauzerall, *Photochem Photobiol.*, **49**, 673 (1989).
81. Y. Kuwahara, M. Tamura and I. Yamazaki, *J. Biol. Chem.*, **257**, 11517 (1982).
82. J. Deguchi, M. Tamura and I. Yamazaki, *J. Biol. Chem.*, **260**, 15542 (1985).
83. E. Rabinowitch and J. Weiss, *Nature*, **138**, 1098 (1936).
84. E. Rabinowitch and J. Weiss, *Proc. Roy. Soc. London*, **A162**, 251 (1937).
85. M. Calvin and G. D. Dorough, *J. Am. Chem. Soc.*, **70**, 699 (1948).
86. F. M. Huennekens and M. Calvin, *J. Am. Chem. Soc.*, **71**, 4024 (1949).
87. F. M. Huennekens and M. Calvin, *J. Am. Chem. Soc.*, **71**, 4031 (1949).
88. J. Barrett, *Nature*, **215**, 733 (1967).
89. D. Mauzerall, *J. Am. Chem. Soc.*, **84**, 2437 (1962).
90. S. I. Beale and J. Cornejo, *J. Biol. Chem.*, **266**, 22341 (1991).
91. S. I. Beale and J. Cornejo, *J. Biol. Chem.*, **266**, 22333 (1991).
92. S. I. Beale and J. Cornejo, *J. Biol. Chem.*, **266**, 22328 (1991).
93. J.-H. Fuhrhop and D. Mauzerall, *Photochem. Photobiol.*, **13**, 453 (1971).
94. J.-H. Fuhrhop, *Angew. Chem., Int. Ed. Engl.*, **13**, 321 (1974).
95. J.-H. Fuhrhop, P. K. W. Wasser, J. Subramanian and U. Schrader, *Justus Liebigs Ann. Chem.*, 1450 (1974).
96. T. Matsuura, K. Inoue, C. R. Ranade and I. Saito, *Photochem. Photobiol.*, **31**, 23 (1980).
97. K. M. Smith, S. B. Brown, R. E. Troxler and J.-J. Lai, *Tetrahedron Lett.*, **21**, 2763 (1980).
98. J. C. Goedheer and J. P. J. Siero, *Photochem. Photobiol.*, **6**, 509 (1967).
99. J. C. Goedheer, *Photochem. Photobiol.*, **6**, 521 (1967).
100. G. A. Hendry, J. D. Houghton and S. B. Brown, *New Phytol.*, **107**, 25 (1987).
101. B. Kräutler and P. Matile, *Acc. Chem. Res.*, **32**, 35 (1999).
102. A. Gossauer and N. Engel, *J. Photochem. Photobiol. B: Biol.*, **32**, 141 (1996).
103. J. Iturraspe, N. Moyano and B. Frydman, *J. Org. Chem.*, **60**, 6664 (1995).
104. H. H. Inhoffen, *Pure Appl. Chem.*, **17**, 443 (1968).
105. H. Scheer and J. J. Katz, *Proc. Natl. Acad. Sci. USA*, **71**, 1626 (1974).
106. P. H. Hynninen, in *Chlorophylls* (Ed. H. Scheer), CRC Press, Boca Raton, 1991, pp. 145–209.
107. V. V. Gurinovich and M. P. Tsvirko, *J. Appl. Spectrosc.*, **68**, 110 (2001).
108. R. F. Troxler, K. M. Smith and S. B. Brown, *Tetrahedron Lett.*, **21**, 491 (1980).
109. C. A. Llewellyn, R. F. C. Mantoura and R. G. Brereton, *Photochem. Photobiol.*, **52**, 1037 (1990); *Photochem. Photobiol.*, **52**, 1043 (1990).
110. A. Peled, Y. Dror, I. Baal-Zedaka, A. Porat, N. Michrin and I. Lapsker, *Synth. Metals*, **115**, 167 (2000).
111. D. B. Min and J. M. Boff, *Compr. Rev. Food Sci. Food Saf.*, **1**, 58 (2002).
112. E. S. Nyman and P. H. Hynninen, *J. Photochem. Photobiol. B.: Biol.*, **73**, 1 (2004).
113. M. O. Senge, M. Fazekas, E. G. A. Notaras, W. J. Blau, M. Zawadzka, O. B. Locos and E. M. Ni Mhuircheartaigh, *Adv. Mater.*, **19**, 2737 (2007).
114. S. Yamada, K. Kuwata, H. Yonemura and T. Matsuo, *J. Photochem. Photobiol. A: Chem.*, **87**, 115 (1995).
115. C. Ingrosso, A. Petrella, P. Cosma, M. L. Curri, M. Striccoli and A. Agostiano, *J. Phys. Chem. B*, **110**, 24424 (2006).
116. Y. Osada, M. Mizumoto and H. Tsuruta, *J. Macromol. Sci., Chem.*, **A24**, 403 (1987).
117. Y. S. Shumov, V. I. Kromov, P. F. Sidorskikh and V. I. Mityaev, *Russ. J. Phys. Chem.*, **52**, 886 (1978).
118. Q. Luo, H. Tian, B. Chen and W. Huang, *Dyes Pigments*, **73**, 118 (2007).
119. M. Kobayashi, M. Yamamura, M. Akiyama, H. Kise, K. Inoue, M. Hara, S. Takaichi, N. Wakao, K. Yahara and T. Watanabe, *Anal. Sci.*, **14**, 1149 (1998).
120. N. A. Kuznetsova, N. S. Gretsova, V. M. Derkacheva, O. L. Kaliya and E. A. Lukyanets, *J. Porphyrins Phthalocyanines*, **7**, 147 (2003).
121. W. Maqanda, T. Nyokong and M. D. Maree, *J. Porphyrins Phthalocyanines*, **9**, 343 (2005).
122. N. A. Kuznetsova, V. V. Okunchikov, V. M. Derkacheva, O. L. Kaliya and E. A. Lukyanets, *J. Porphyrins Phthalocyanines*, **9**, 393 (2005).
123. H. Kropf and B. Kasper, *Justus Liebigs Ann. Chem.*, 2232 (1975).

124. H. J. Arpe, *Ullmann's Encyclopedia of Industrial Chemistry*, 5th Edition, Wiley-VCH, Weinheim, 1997.
125. J. C. Kephart, *Econ. Bot.*, **9**, 3 (1955).
126. M. Strell, A. Kalojanoff and F. Zuther, *Arzneimittel-Forsch.*, **5**, 640 (1955); *Chem. Abstr.*, **50**, 3712h (1955).
127. M. Strell, A. Kalojanoff and F. Zuther, *Arzneimittel-Forsch.*, **6**, 8 (1956); *Chem. Abstr.*, **50**, 5993f (1956).
128. J. W. Hein, *J. Am. Dent. Assoc.* **48**, 14 (1954).
129. D. L. Sudakin, *J. Toxicol. Clin. Toxicol.*, **41**, 195 (2003).
130. A. R. Waladkhani and M. R. Clemens, *Int. J. Mol. Med.*, **1**, 747 (1998).
131. S. Chernomorsky, A. Segelmann and R. D. Poretz, *Teratogen. Carcin. Mut.*, **19**, 313 (1999).
132. K. Morita, M. Ogata and T. Hasegawa, *Environ. Health Perspect.*, **109**, 289 (2001).
133. J. D. Spikes and J. C. Bommer, in *Chlorophylls* (Ed. H. Scheer), CRC Press, Boca Raton, 1991, pp. 1181–1204.
134. R. Bonnett, *Chemical Aspects of Photodynamic Therapy*; Gordon and Breach Sci. Publ., Amsterdam, 2000.
135. N. Himeshima and Y. Amao, *Energ. Fuel.*, 1641 (2003).
136. J. Woodward, M. Orr, K. Cordray and E. Greenbaum, *Nature*, **405**, 1014 (2000).
137. T. Inoue, S. N. Kumar, T. Karnachi and I. Okura, *Chem. Lett.*, 147 (1999).
138. B. D. Berezin, S. V. Rumyantseva, A. P. Moryganov and M. B. Berezin, *Russ. Chem. Rev.*, **73**, 185 (2004).
139. M. O. Senge and K. M. Smith, in *Anoxygenic Photosynthetic Bacteria* (Eds. R. E. Blankenship, M. T. Madigan and C. E. Bauer), Kluwer, Dordrecht, 1995, pp. 137–151.

CHAPTER **6**

Electrochemistry of organomagnesium compounds

JAN S. JAWORSKI

Faculty of Chemistry, University of Warsaw, 02-093 Warszawa, Poland
Fax: +48-22-822-5996; e-mail: jaworski@chem.uw.edu.pl

I. INTRODUCTION	219
II. ELECTROCHEMICAL SYNTHESIS OF ORGANOMAGNESIUM COMPOUNDS	221
III. CONDUCTIVITY OF SOLUTIONS OF ORGANOMAGNESIUM COMPOUNDS	224
IV. ANODIC OXIDATION OF ORGANOMAGNESIUM COMPOUNDS . . .	227
A. Simple Diorganomagnesium Compounds	227
B. Grignard Reagents	228
1. Oxidation at inert electrodes	229
2. Anodic addition to olefins	237
3. Oxidation on sacrificial anodes	237
4. Processes at semiconductor anodes	241
C. Other Compounds	243
V. CATHODIC REDUCTION OF ORGANOMAGNESIUM COMPOUNDS .	244
A. General Mechanism of the Reduction	244
B. Deposition of the Metallic Magnesium and the Reverse Process	245
VI. ORGANOMAGNESIUM COMPOUNDS AS INTERMEDIATES IN ELECTRODE REACTIONS	253
VII. CONCLUDING REMARKS: ELECTROCHEMICAL DATA IN THE ELUCIDATION OF REACTIVITY OF GRIGNARD REAGENTS	258
VIII. REFERENCES	260

I. INTRODUCTION

In the last 37 years a number of chapters and reviews have been published on electrochemistry of organoelemental and organometallic compounds[1–9] discussing electrode reactions of organomagnesium compounds, in particular Grignard reagents, the most important ones

The chemistry of organomagnesium compounds
Edited by Z. Rappoport and I. Marek © 2008 John Wiley & Sons, Ltd

in organic chemistry among the title compounds. However, so far no separate monograph has been devoted to that specific topic. Nevertheless, the investigations into electrochemical behavior of Grignard reagents have a rich and long history, going back to 1912 with the unsuccessful attempt of Jolibois[10, 11] to isolate the expected gaseous hydrocarbons during the electrolysis of their ethereal solutions, and the report of Nelson and Evans in 1917 on the conductivity of such solutions[12]. Kondyrew verified in 1925 that the loss of a magnesium anode in the electrolysis of these reagents fulfills Faraday laws and magnesium is deposited at a cathode[13]. Dimeric hydrocarbons as the main products of electrolysis were found by Gaddum and French in 1927[14]. The earliest research on the conductivity and electrolysis was continued and it helped to explain the nature of Grignard reagents in ethereal solutions[15]. A general mechanism of electrode processes in these solutions was established around 1940 by Evans and coworkers after a couple of years of investigations[15–23]. It was progressively found that the following reactions of alkyl or aryl radicals formed at electrodes are strongly dependent on the conditions of the electrolysis: the nature of the radical and the halide as well as on the electrode material. This behavior opened up a wide area of synthetic applications. It is one of the most characteristic and fascinating trends in the electrochemistry of organomagnesium compounds: most investigations were strongly directed to applications in industry and laboratory practice. A large amount of the results was patented. As a result, large-scale industrial production of tetraalkyl lead from Grignard reagents by the Nalco process started in 1964 (followed by the production of adiponitrile by the Monsanto process in the next year). This stimulated a rapid development of organic electrochemistry as a separate field with wide potential applications in industry. Although the next decades brought about a significant decrease in the production of R_4Pb because of environmental constraints, yet electroorganic methods are still thought to be particularly safe and valuable for 'green chemistry'. In this Chapter references to important, mostly US, patents are given; however, our attention is focused only on the reaction mechanisms and products distribution under the given conditions, which is the essential topic of interest for organic chemists. More details can be found in original documents easily available (using the given patent number) from websites, e.g. European Patent Office: ep.espacenet.com.

On the other hand, because of this strong interest in the practical use of organomagnesium compounds, as well as the beginnings of electrochemical studies in the early decades of the 20th century, many details of mechanisms of their heterogeneous reactions were not investigated later by modern and powerful electroanalytical and spectroscopic techniques. There is still a lack of such data, with the exception perhaps of some very recent studies which focus their interest on applications in rechargeable magnesium batteries (Section V.B) and a grafting of a silicon surface by anodic reactions of Grignard reagents for use in electronics (Section IV.B.4). It may also be interesting to note that these last investigations are strictly related to the 'modern face' of electrochemistry, which increasingly becomes the surface science investigating electrochemical reactions at well-defined solid surfaces, using different *in situ* spectroscopic techniques to determine the nature, structure and reactivity of the adsorbed species and open new directions, like material science and nanotechnology.

In this Chapter, first of all in Section II, the synthesis of diorganomagnesium compounds is reviewed, including the use of direct electrochemical reactions and combined methods with electrochemical and homogeneous steps. A brief review on the conductivity of solutions of organomagnesium compounds indicating a complex nature and dynamic behavior of ionic species present in solutions is given in Section III. The discussion of electrochemical behavior of the title compounds is divided into two parts. First, in Section IV, the anodic oxidation is given involving the reactions of organic radicals that are probably more interesting for the readers of this book. Then, in Section V the cathodic reduction is described, accompanied by the deposition of metallic magnesium.

Such a division should make the discussion clearer for readers, although it should be remembered that many investigations, in particular the research of Grignard reagents, were often performed in undivided electrochemical cells, where both kinds of processes occur at the same time. For anodic processes, consecutively the oxidation of simple diorganomagnesium compounds (Section IV.A) and Grignard reagents at various kinds of anodes (Section IV.B) are discussed. The example of the oxidation of other groups in organomagnesium compounds is mentioned in Section IV.C. For cathodic processes the general mechanism is presented in Section V.A, but the following reactions of organic radicals are the same as in anodic processes, discussed earlier. However, the deposition of metallic magnesium and the reverse process, important in recent years because of applications in rechargeable batteries, but also giving some interesting explanations of the nature of electroactive organomagnesium species, are discussed in Section V.B.

The use of sacrificial magnesium anodes in the electrochemical preparation of a number of organic compounds with high selectivity has been popular for decades. In most of the reported mechanisms magnesium cations produced from an anode form salts or complexes with organic anions. However, in a few cases the organomagnesium compounds are formed as intermediates and these processes are discussed in Section VI. Finally, the concluding remarks in Section VII focus on the use of some electrochemical data in order to elucidate the nature of Grignard reagents in solutions and to explain the most probable mechanism of homogeneous Grignard reactions.

In this Chapter the following common abbreviations are used, beside those used in this book: n_e, the number of electrons in a given reaction; CV, cycling voltammetry; OCP, open circuit potential; rds, rate-determining step; EDAX, elemental analysis by dispersive X-rays; SEM, scanning electron microscopy; STM, scanning tunneling microscopy; ATR, attenuated total reflection; XPS, X-ray photoelectron spectroscopy; EQCM, electrochemical quartz crystal microbalance; ECL, electrochemiluminescence; ACN, acetonitrile; TBAP, tetrabutylammonium perchlorate; TBAPF$_6$, tetrabutylammonium hexafluorophosphate; TBABF$_4$, tetrabutylammonium tetrafluoroborate; GC, glassy carbon; and M, mole dm^{-3}.

The quoted potentials are rarely expressed versus standard hydrogen electrode (SHE) but mainly versus an aqueous saturated calomel electrode (SCE) or versus Ag/Ag$^+$ couple in ACN or in a solvent used in particular experiments. However, in some cases the Mg/Mg^{2+} couple in THF or other solvents is used as the reference electrode.

II. ELECTROCHEMICAL SYNTHESIS OF ORGANOMAGNESIUM COMPOUNDS

In general, it is possible to obtain diorganomagnesium compounds R$_2$Mg (**1**) by the anodic oxidation of organoelemental complexes, such as Na[ZnR$_3$], Na[AlR$_4$] or Na[BR$_4$], using the sacrificial magnesium anode. For example, Et$_2$Mg (**1b**) can be obtained with 73% yield in the electrolysis at 150 °C of melted Na[BEt$_4$] using a Mg anode and a Hg cathode[24]; the second product Et$_3$B can be used to regenerate the electrolyte. The use of alkylaluminates (**2**) for this purpose was reviewed by Lehmkuhl[2]. R$_2$Mg•2AlR$_3$ formed at the magnesium anode and liquid alkali metal or its amalgam, depending on the metal of the cathode, are the products of electrolysis. In particular, from the electrolysis of a 1:1 mixture of Na[AlEt$_4$] (**2b**) and K[AlEt$_4$] using a mercury cathode it is possible to obtain Mg[AlEt$_4$]$_2$, which formally corresponds to Et$_2$Mg•2Et$_3$Al. In the method patented by Ziegler and Lehmkuhl[25] the electrolysis was performed in an inert gas atmosphere (e.g. nitrogen or argon) using melted NaF•2Et$_3$Al (m.p. 35 °C) as an electrolyte and the final product **1b** could be continuously extracted by Et$_3$Al (**3b**) with which it forms Et$_2$Mg•2Et$_3$Al. Volatile **3** is easily removed by heating at about 120 °C in vacuum and **1b** remains in the solid state. The anodic and cathodic spaces should be separated by a diaphragm, because otherwise magnesium is deposited at the iron cathode, instead of aluminium, which can be converted to **3b** and reused. Kobetz and Pinkerton patented a method[26] based on the electrolysis according to reaction 1 using a steel cathode and a magnesium anode in melted electrolytes, containing a mixture of two alkylaluminates (**2**) with methyl groups in at least one of them. The addition of a second component, with other alkyl or phenyl groups, results in lowering of the melting point and increasing the electrical conductivity of the mixture in comparison with the values characteristic of each component alone. However, then the product is the mixture of molecules of **1** with different R's, as is shown in Table 1 for the first entry. The yield of the main product is increased by the proper ratio of both components of the electrolyte and some results reported[26] are shown in Table 1. The original electrolysis product $R_2Mg \cdot 2R_3Al$ is floating on the electrolyte adjacent to the anode and the vacuum distillation of **3** (at 300 mm Hg) releases crystalline **1**. The distillation can be performed continuously during vacuum electrolysis or the electrolysis is carried out in the atmosphere of an inert gas, to avoid any contact with oxygen and moisture (**3** is flammable on air). Compound **3** is next used to regenerate the electrolyte in the reaction with $M[BR_4]$.

$$2\ M[AlR_4] \xrightarrow{\text{Mg anode}} 2M + 2R_3Al + R_2Mg$$

(**2**)   (**3**)   (**1**)

(1)

2	a	b	c	d	e	f	g	h
M	Na	Na	Na	Na	K	K	Li	Rb
R	Me	Et	n-Bu	Ph	Me	n-Pr	Me	Me

1 and **3**	a	b	c	d	e
R	Me	Et	n-Pr	n-Bu	Ph

The improved electrolytic production of magnesium dialkyls **1** with R containing from 2 to 6 carbon atoms using melted **2** as an electrolyte (with M = Na or a mixture of Na with up to 80% of K), a copper cathode and a magnesium anode, separated by a diaphragm in an originally designed apparatus was also patented by Ziegler and Lehmkuhl[27] and the example of their electrolysis is given in Table 1 in the last entry.

Versatile electrochemical generation of diorganomagnesium compounds **1** corresponding to unusual Grignard reagents, containing electrophilic groups, such as halogen, carbonyl and cyano, was proposed by Lund and coworkers[28]. Those substituents are reduced by magnesium and thus such reagents cannot be obtained by the classical reduction of

TABLE 1. Main products (after distillation of **3**) and temperatures of the electrolysis of **2** at the Mg anode[26,27]

Electrolyte		Temperature (°C)	Main product	Reference
components	mole ratio			
2a:2b	1:1	ca 100	**1b**[a]	26
2a:2b	1:3	ca 100	**1b**[b]	26
2e:2f	1:8	150	**1c**	26
2a:2d	2:3	175	**1a** + **1e**	26
2h:2a	1:5	140	**1a**	26
2g:2c	1:3	not specified	**1d**	26
2b	—	120	**1b**[c]	27

[a]The electrolysis at 7.4 V and 0.25 A cm^{-2}; other products are **1a** and MeEtMg.
[b]Ethyl groups in $R_2Mg \cdot 2R_3Al$ approach 90%.
[c]95% yield; the electrolysis at 5 V.

6. Electrochemistry of organomagnesium compounds

Pt cathode   Mg anode

KClO$_4$/DMSO

$2e \rightarrow 2K^+ \rightarrow 2K$

$2DMSO - H_2 \rightarrow 2\,dimsyl^-$

$Mg^{2+} \xleftarrow{-2e}$

dimsyl$_2$Mg

(4)

dimsyl$_2$Mg + 2RH $\longrightarrow$ R$_2$Mg + 2DMSO
(4)   (1)

SCHEME 1

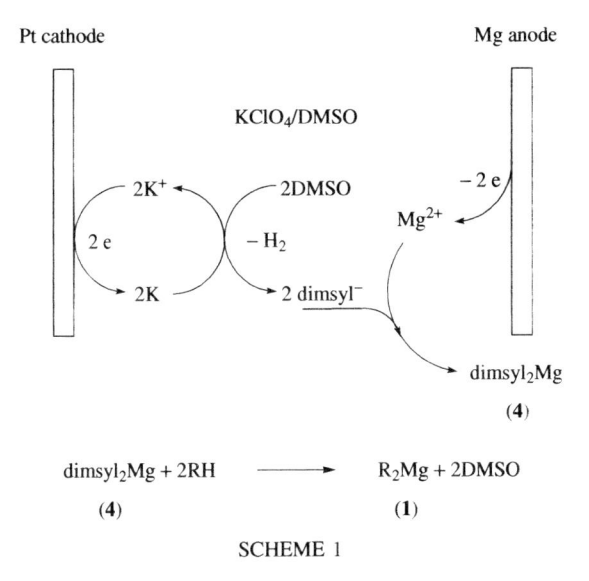

organic halide by Mg. The method proposed consists of the electrolysis of potassium perchlorate in dry and deaerated DMSO in an undivided cell with a platinum cathode and a sacrificial magnesium anode. The overall process[28], shown in Scheme 1, results in the formation of strong dimsyl base (i.e. the conjugate base of DMSO) in the reaction of potassium, formed at the cathode, with the solvent. Simultaneously, magnesium cations generated at an anode stabilize dimsyl anions through the interaction viewed as ion association (ion-pairs[28] or rather triple ion formation) in a magnesium salt dimsyl$_2$Mg (4). In a second nonelectrochemical step, the added weakly acidic substrate, RH, with pK_a < 26 (all pK_a values cited[28] refer to DMSO), is deprotonated by 4, resulting in the formation of 1.

RX + (bipyridine) $\xrightarrow[\text{anodic oxidation}]{\text{Mg}}$ RMgX · 6

(8)   (6)

R, R' = Alk, Ar

X = Cl, Br, I

XR'X + 2 6 $\xrightarrow[\text{anodic oxidation}]{\text{Mg}}$ R'Mg$_2$X$_2$ · 2 6

(9)

RX + R'$_4$NX $\xrightarrow[\text{anodic oxidation}]{\text{Mg, ACN}}$ R'$_4$N[RMgX$_2$ · ACN]

(7)

SCHEME 2

The effective deprotonation of fluorene (pK_a = 22.6), 2-bromofluorene (pK_a = 20.0), 2,7-dibromofluorene (p$K_a \leqslant$ 20.0), acetophenone (pK_a = 24.7) and phenylacetonitrile (pK_a = 21.9) was shown[28], but not for weaker acids such as 4-benzylpyridine (pK_a = 26.7). The usefulness of generated reagents **1** was illustrated[28] in reactions of nucleophilic addition to electrophiles, characteristic of the ordinary Grignard reagents (**5**, Tables 2 and 3), as will be reviewed in Section VI.

The direct electrochemical synthesis (Scheme 2) of the adducts of organomagnesium halides with 2,2'-bipyridine (**6**) and salts of organodihalogenomagnesium(II) anions (**7**) was reported by Hayes and coworkers[29]. Adducts of different stoichiometry and **7** were obtained in the electrochemical oxidation of magnesium in ACN solutions containing organic halides RX (**8**), α,ω-dihalides XR'X (**9**) and **8** with ammonium salts R$_4'$NX, respectively. All new products showed none of the typical reactions of Grignard reagents.

III. CONDUCTIVITY OF SOLUTIONS OF ORGANOMAGNESIUM COMPOUNDS

Ethereal solutions of organomagnesium compounds at room temperature show weak electric conductivity[15, 16, 30-35], as is evident from data collected in Tables 2 and 3. This behavior indicates the existence of ionic species at relatively low concentrations, c.

The conductivity (or the specific conductance in earlier literature), κ, of Grignard reagents EtMgBr (**5b**) and PhMgBr (**5e**) in Et$_2$O solutions of c = 0.5 M (Table 2) lies between those of MgBr$_2$ and the corresponding R$_2$Mg, **1b** and **1e**, respectively[15]. Thus, for some electrochemical measurements, in particular for **1**, the addition of a supporting electrolyte is necessary[36-38]. For higher concentration of **5** (c = 1 M) their conductivity in Et$_2$O (Table 3) is even higher than for MgBr$_2$. The values of κ for **5** in Et$_2$O solutions are not strongly dependent on the nature of R and in general they are higher for Et than for n-Bu, and higher for Bn than for Ph. However, for **5c**, κ is lower than for **5d** at room temperature (Table 3) but it is higher at lower temperatures[16]. Conductivities of n-PrMgBr (**5h**) and i-PrMgBr (**5i**) are similar for the same concentrations and temperatures[32] (κ is only 1.3 times higher for **5h**). On the other hand, the molar conductivity, $\lambda = \kappa/c$, of **5b** and **5e** in Et$_2$O decreases with dilution between 2 M and 0.5 M[16], but for **5b** it increases at much higher concentrations[30]. Similarly, the plot of molar conductivity of EtMgI (**5j**) solutions against c shows a maximum[15, 31]; it was observed[15] at c = 1.5 M

TABLE 2. Conductivity, κ, at 20 °C of 0.5 M solutions of organomagnesium compounds and MgBr$_2$ in Et$_2$O

Compound	R = Et $10^3 \kappa$ (Ω^{-1} m^{-1})	Reference	R = Ph $10^3 \kappa$ (Ω^{-1} m^{-1})	Reference
R$_2$Mg (**1**)	1	15	0.9	33
RMgBr (**5**)	1.6	15	1.2	16
MgBr$_2$	2	15	2	15

TABLE 3. Conductivity, κ, of 1.0 M solutions of RMgX (**5**) and MgBr$_2$ in Et$_2$O at 20 °C[16] and in THF at 22 °C[35]

Compound R X	**5a** Me Br	**5b** Et Br	**5c** n-Bu Br	**5d** Bn Br	**5e** Ph Br	**5f** Et Cl	**5g** n-Bu Cl	MgBr$_2$ — —
Et$_2$O $10^3 \kappa$ (Ω^{-1} m^{-1})	—	6.16	5.88	4.74	—	—	—	ca 1.9
THF $10^3 \kappa$ (Ω^{-1} m^{-1})	30.5	23.7	21.8	—	—	40.3	34.9	—

and was interpreted as a manifestation of the formation of ion associates higher than ion pairs. Moreover, the temperature coefficient of κ for solutions of **5** in Et_2O is often negative[16] but it depends on c and can change sign, as found for **5b**[30] and **5e**[16]. All the above observations show that there are complex and dynamic equilibria existing in various solutions.

The constitution of Grignard reagents in solutions depends first of all on the Schlenk equilibrium[39] (equation 2) including molecular association of R_2Mg, MgX_2 and RMgX. However, the association with solvent molecules is also important. A comprehensive view[40] on the Schlenk equilibrium is shown in Scheme 3. In general, the equilibria under consideration depend on the solvent, the R group and, to a lesser extent, on the halide, as well as on the temperature and concentration. Dimers are more favorable in Et_2O than THF (most probably because MgX_2 is solvated by four THF molecules but only by two Et_2O molecules) and more favorable for Alk, in particular Bu, than for Ar groups. Temperature effect on the composition of **5** in solutions can be either kinetic or thermodynamic in nature, and for the latter it should be remembered that the enthalpy changes for the Schlenk equilibrium in Et_2O and in THF have opposite signs[40].

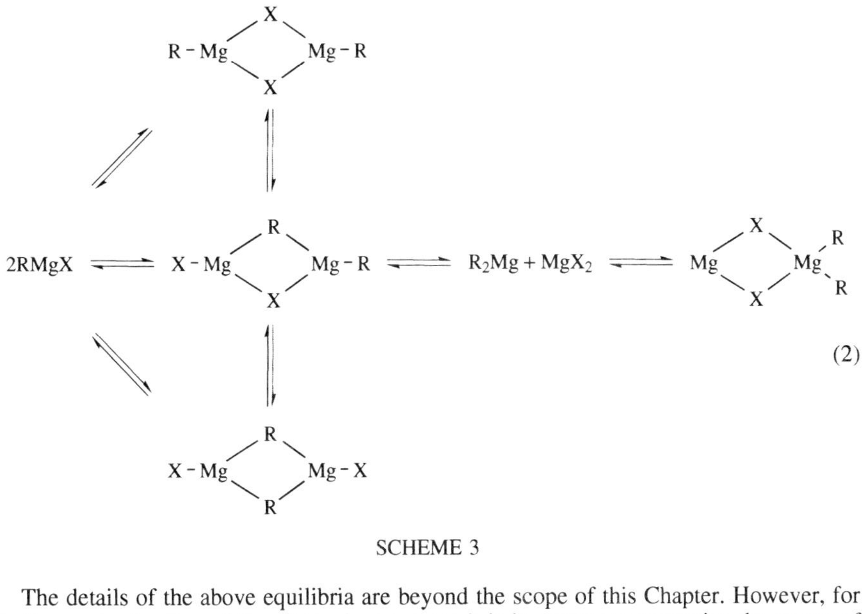

$$2RMgX \rightleftharpoons X-Mg\underset{X}{\overset{R}{\diamond}}Mg-R \rightleftharpoons R_2Mg + MgX_2 \rightleftharpoons Mg\underset{X}{\overset{X}{\diamond}}Mg\underset{R}{\overset{R}{}} \quad (2)$$

with additional equilibria to $R-Mg\underset{X}{\overset{X}{\diamond}}Mg-R$ (top) and $X-Mg\underset{R}{\overset{R}{\diamond}}Mg-X$ (bottom)

SCHEME 3

The details of the above equilibria are beyond the scope of this Chapter. However, for further understanding of the electrode processes it is important to recognize the nature of ions present in solutions. Thus, different ionization reactions (equations 3–6) postulated on the basis of conductivity measurements and other experimental data are listed below. Evans and coworkers considered[15, 19, 21] that the cations, RMg^+ and MgX^+, formed in simple ionization reactions 3a–3d, are coordinated with the Et_2O molecules and are relatively small, whereas the anions are large in size due to a coordination with **1**, **5** and $MgBr_2$. These processes can be summarized by a simplified equilibrium (equation 4) for Grignard reagents **5**[19, 21] and by the equilibrium in equation 5 for **1**[33, 34], but the participation of anions R_2MgX^- was also considered[17] as well as the equilibrium (equation 6) for **1e**[33]. Moreover, a nonlinear increase in the logarithm of the equivalent conductivity of **1e** solutions in 1,4-dioxane with log c found by Strohmeier[33] (in contrast to linear dependencies for Ph_2Cd and Ph_2Zn) supported the opinion that in solutions of **1e** there is no domination of a simple equilibrium. However, it should be added here that the molecules of 1,4-dioxane irreversibly coordinate to MgX_2, forming insoluble complexes; this ability is commonly used in the course of preparation of **1** from solutions of **5**.

$$RMgX \rightleftharpoons RMg^+ + X^- \tag{3a}$$

$$RMgX \rightleftharpoons MgX^+ + R^- \tag{3b}$$

$$R_2Mg \rightleftharpoons RMg^+ + R^- \tag{3c}$$

$$MgX_2 \rightleftharpoons MgX^+ + X^- \tag{3d}$$

$$2RMgX \rightleftharpoons RMgX_2^- + RMg^+ \tag{4}$$

$$2R_2Mg \rightleftharpoons R_3Mg^- + RMg^+ \tag{5}$$

$$Ph_2Mg + PhMg^+ \rightleftharpoons Ph_3Mg_2^+ \tag{6}$$

On the other hand, a strong effect of the nature of the solvent on the conductivity of **1b** and **1e** was reported[33,34]. It was explained only qualitatively in terms of two phenomena. One of them, previously suggested by Evans and Pearson[15], are donor–acceptor interactions between solvent molecules acting as donors and organomagnesium cations which have acceptor properties due to unoccupied orbitals. The other one is an ion association which increases with the decrease in the solvent electric permittivity, ε. Fortunately, nowadays Gutmann's donor number, DN[41], can be used as a quantitative measure of Lewis basicity for solvent molecules. A reasonable relationship between $\log \kappa$ and DN for solutions of **1e** is shown in Figure 1. The values of κ for **1b** solutions were measured[34] in solvents with greater variation of ε and thus a correlation (equation 7) with two explanatory parameters, DN and $1/\varepsilon$, must be applied. It holds with a correlation coefficient of $r = 0.9853$ and the addition of the second parameter is statistically significant with probability 78.7%; standard deviations are given in parentheses. The plot of the experimental $\log \kappa$ against the calculated value is shown in Figure 2.

$$\log \kappa = 0.11(\pm 0.03)DN - 10(\pm 2)/\varepsilon - 6(\pm 1) \tag{7}$$

Solvent effects on the conductivity of **5** also play a significant role in electrochemical applications. For example, the observation of a remarkable increase in the conductivity of **5f** solutions in $(n\text{-}BuOCH_2)_2$ caused by the addition of THF was patented[42] for

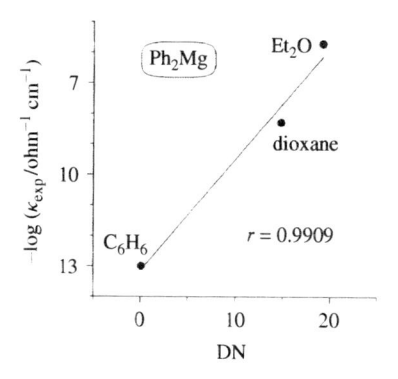

FIGURE 1. Dependence of the log of conductivity of 0.1 M solutions of **1e** measured[33] at 20 °C on the solvent donor number DN. The correlation coefficient is given

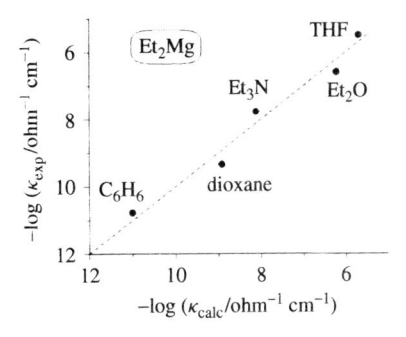

FIGURE 2. Relationship between experimental[34] $\log \kappa$ for 0.1 M solutions of **1b** at 20 °C and the calculated values from equation 7. The theoretical line with unit slope is shown

use in electrolytic preparation of organolead compounds. The above summary of complicated ionization phenomena in Grignard solutions can point to difficulties in detailed understanding and control of their electrochemical reactions.

IV. ANODIC OXIDATION OF ORGANOMAGNESIUM COMPOUNDS

A. Simple Diorganomagnesium Compounds

The polarographic behavior of simple diorganomagnesium compounds **1** in DME solutions (containing 1 mM of **1** and 0.1 M TBAP) was investigated by Psarras and Dessy[36]. For each compound the irreversible oxidation wave at a mercury electrode corresponding to the diffusion-controlled two-electron process was observed. The same half-wave potential for all the compounds was equal to $E_{1/2} = -1.2$ V vs. 1 mM AgClO$_4$/Ag electrode. However, this value is very uncertain because of pronounced maxima observed on the waves. Exhaustive controlled-potential oxidation of **1b** and **1e** confirmed that $n_e = 2$ and indicated two main products. One was the same for all the compounds under study and was identified as Mg(ClO$_4$)$_2$ by a comparison of its reduction potential $E_{1/2}$ with the potential found for the original compound ($E_{1/2} = -2.30$ V vs. Ag$^+$/Ag). The second product was assumed to be an organomercury compound R$_2$Hg (**10**) but only **10e**, the oxidation product of diphenylmagnesium (**1e**), could be identified by the reduction at the potential of $E_{1/2} = -3.34$ V vs. Ag$^+$/Ag. Reduction of other compounds **10** had to be beyond the discharge of the supporting electrolyte. However, it was also possible to identify HgBr$_2$ as the product formed in the oxidation reaction of MgBr$_2$ ($E_{1/2} = -0.6$ V vs. Ag$^+$/Ag) (equation 8). Then, for the other compounds a similar oxidation reaction (equation 9) was proposed[36].

$$MgBr_2 \xrightarrow[\text{Hg anode}]{-2e} Mg^{2+} + HgBr_2 \qquad (8)$$

$$R_2Mg \xrightarrow[\text{Hg anode}]{-2e} Mg^{2+} + R_2Hg \qquad (9)$$

(**1b**) R = Et   (**10b**) R = Et
(**1e**) R = Ph   (**10e**) R = Ph

228   Jan S. Jaworski

The oxidation of **1** in solutions is much easier at mercury and lead electrodes which form organometallic compounds[37] than at inert electrodes. For example, the oxidation of **1b** in THF containing 0.25 M TBAP at a lead electrode gives[37] a CV peak at a scan rate of 0.3 V s^{-1} with the half-peak potential equal to $E_{p/2} = -1.72$ V vs. 0.01 M Ag$^+$/Ag, whereas at a platinum electrode the process needs potentials over 1.5 V more positive. Thus, the formation of a carbon–lead bond during the electrode process was suggested[37]. Steady-state current/potential curves showed, after the first oxidation wave, a plateau with limiting current 1.0 mA cm^{-2} in 0.05 M **1b** solutions, i.e. much lower than expected for the diffusion-controlled process. This behavior indicates a slow chemical processe. Moreover, the Tafel slope, equal to 60 mV^{-1}, corresponds to the reversible electron transfer, contrary to the behavior found for Grignard reagent **5b**. However, the chemical reactions determining the overall rate of the oxidation of **1b** and **5b** are probably the same[37] and they are shown in Scheme 4.

Thus, the oxidation of **1**, as well as of **5**, using sacrificial anodes, yields the corresponding new organometallic compounds. For example, bis(indenyl) manganese was obtained with a good yield[43] by the electrolysis at 200 °C of bis(indenyl) magnesium in a saturated solution of Me$_2$O containing indene. A method of purifying organometallic complexes, in particular **1b** with NaF, by extraction with **11a** at 60 °C in order to remove EtMgOEt and (EtO)$_2$Mg contaminants, and further electrolysis of the above complex at 30 °C using a Pb anode and a Cu cathode, was patented[44] as a convenient procedure for the preparation of **11a**.

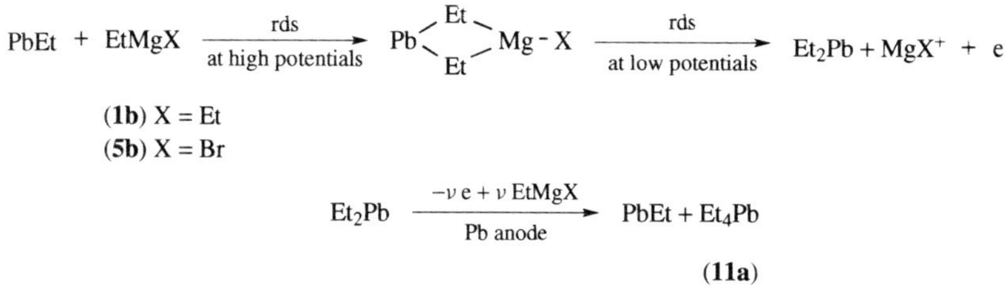

$$\text{PbEt} + \text{EtMgX} \xrightarrow[\text{at high potentials}]{\text{rds}} \text{Pb}\begin{smallmatrix}\text{Et}\\\text{Et}\end{smallmatrix}\text{Mg-X} \xrightarrow[\text{at low potentials}]{\text{rds}} \text{Et}_2\text{Pb} + \text{MgX}^+ + e$$

(**1b**) X = Et
(**5b**) X = Br

$$\text{Et}_2\text{Pb} \xrightarrow[\text{Pb anode}]{-\nu e + \nu \text{EtMgX}} \text{PbEt} + \text{Et}_4\text{Pb}$$

(**11a**)

SCHEME 4

B. Grignard Reagents

For the electrolysis of Grignard reagents **5**, it is well documented in numerous experiments[15–23] that the electroactive species at anodes contains the R group as well as magnesium. Thus, they can be represented by the anion RMgX$_2^-$ formed in equation 4. In general, its anodic oxidation involves an electron transfer and a bond cleavage with the formation of free radicals, R• (**12**), which follow a number of competitive chemical reactions depending on the nature of **12**, the solvent and the anode material. For inert anodes, made most often from platinum, the following reactions of **12**, shown in Scheme 5, can involve a hydrogen atom abstraction from solvent molecules (equation 10a) or an attack on another molecule of **5**, the disproportionation reaction between two radicals (equation 10b) yielding saturated, RH, and unsaturated, R(−H), hydrocarbons, the formation of unsaturated hydrocarbon accompanied by hydrogen evolution (equation 10c), or a coupling of two radicals (equation 10d). Moreover, an addition reaction (equation 10e) can occur with specially added reactants. On the other hand, metals from active anodes react with radicals **12**, in the reaction shown in equation 10f, called 'anodic transmetallation'. Examples discussed below show how the main products depend on the nature of the reactants and reaction conditions.

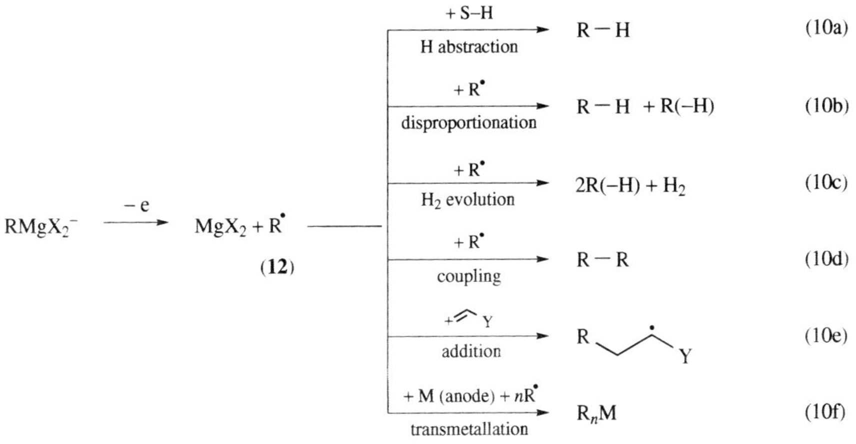

6. Electrochemistry of organomagnesium compounds

$$RMgX_2^- \xrightarrow{-e} MgX_2 + R^\bullet$$
(12)

$$R^\bullet \xrightarrow{+ S-H, \text{ H abstraction}} R-H \quad (10a)$$

$$R^\bullet \xrightarrow{+ R^\bullet, \text{ disproportionation}} R-H + R(-H) \quad (10b)$$

$$R^\bullet \xrightarrow{+ R^\bullet, \text{ H}_2 \text{ evolution}} 2R(-H) + H_2 \quad (10c)$$

$$R^\bullet \xrightarrow{+ R^\bullet, \text{ coupling}} R-R \quad (10d)$$

$$R^\bullet \xrightarrow{+ \diagup Y, \text{ addition}} R\diagdown\diagup{}^\bullet Y \quad (10e)$$

$$R^\bullet \xrightarrow{+ M \text{ (anode)} + nR^\bullet, \text{ transmetallation}} R_nM \quad (10f)$$

SCHEME 5

1. Oxidation at inert electrodes

Methane (**13**) and ethane (**14**) are the main organic products at Pt anodes of the electrolysis at a constant current density of methyl Grignard reagents MeMgX in Et$_2$O solutions (Table 4)[17]. However, the relative yields of these products depend strongly on the concentration as is shown in Figure 3 for **5a**. At a lower concentration **13** is mainly produced but also ethene (**15**), i-butene (**16**) and traces of n-butene (**17**) and n-propene (**18**). On the other hand, the yield of **14** increases with increase in the concentration of **5a** and finally **14** becomes the only product at $c = 3$ M. Thus, it is evident that the coupling reaction (equation 10d) dominates at higher concentrations. In that case the electrochemical yield of the electrolysis, given as moles of **14** per 1 Faraday, is equal to 43.8% (Table 4), close to the theoretical value of 50%. The formation of **13** at lower concentrations was explained[19] by the H atom abstraction from Et$_2$O molecules in equation 10a. The above reaction can also explain the formation of **15**, ethanol (**19**) and i-propanol (**20**), which were determined experimentally in small amounts, because on the basis of a pyrolysis of ethers, the formation of EtOMgX and Me$_2$CHOMgX were predicted[19]. The decomposition of ether was also supported by the formation of CO$_2$ (cf. Table 4). Moreover, Evans and Field found[19] that the fraction, Φ, of methyl radicals Me$^\bullet$ (**12a**), which couple to form **14** (equation 11),

$$\Phi = [2n_{\text{ethane}}/(2n_{\text{ethane}} + n_{\text{methane}})]100\% \quad (11)$$

is independent of the concentration of **5** but increases with increase in the current density during the electrolysis, as is shown in Figure 4 for MeMgI (**5k**). The relationships shown in Figure 4 can be explained taking into account that an increase in current density results[19] in an increase in the concentration of **12a** radicals at the electrode; hence their coupling (equation 10d) to **14** is more favored than the reaction with solvent molecules (equation 10a). The relationships shown in Figure 3 can be explained by a similar reasoning. Electrochemical efficiency, calculated as the number of methyl groups per Faraday, decreases in the order Cl > Br > I and also decreases linearly with increasing concentration (the formation of MeX at an anode followed by a regeneration of **5** at a cathode was suggested[19] to explain the last observation). A decrease in electrochemical efficiency

230   Jan S. Jaworski

TABLE 4. Distribution of anodic products after the electrolysis of MeMgX solutions in Et_2O and $n\text{-}Bu_2O$ on bright Pt electrodes[a]

5	X	C (M)[b]	I (A dm^{-2})[c]	Yield of gaseous products (%)				Y_{el} (%)[d]	Reference
Et_2O				13	14	15	16		
5a[e]	Br	1.09	1	79.3	0	3.5	17.2	56.6	17
		2.08	2	23.1	69.2	2.3	5.4	48.5	17
		2.83	2	0	100	0	0	43.8	17
5k	I	2.10	2	64.3	13.2	5.9	16.7	35.8	17
		4.11	2	56.6	27.8	6.9	8.5	46.3	17
				13	14	15	CO_2		
5k[f]	I	0.91	0.02	78.8	19.5	1.7	0	52	19
		0.91	0.545	30.7	68.2	0.5	0.6	43	19
		0.91	1.13	17.8	81.0	0.3	0.9	40	19
		0.91	2.62	14.9	84.2	0.3	0.6	38	19
$n\text{-}Bu_2O$				13	14	21	22		
5k[g,h]	I	0.95	0.2	75.4	23.2	0.45	0.95	34.3	20
5k[i]	I	0.95	0.04	88.5	5.8	1.9	3.8	34.3	20
5k[i,h]	I	0.95	1.60[j]	59.3	23.1	5.8	11.5	34.3	20

[a] In each case MgX_2 is also formed at the anode.
[b] Initial concentration of 5.
[c] Current density during electrolysis.
[d] Electrochemical efficiency equal to the number of moles of gaseous products per Faraday.
[e] Traces of 17 and 18 were also found.
[f] Electrolysis in refluxing solutions. 19 and traces of 20 were found but no 16 and H_2.
[g] Electrolysis at 90 °C. No 15 and CO_2 were found.
[h] Average yield from two measurements.
[i] Electrolysis at 143 °C. No 15 was found.
[j] 0.3% of CO_2 was found.

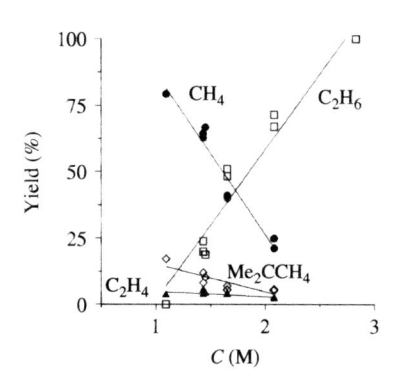

FIGURE 3. Dependence of product yields for the anodic oxidation of MeMgBr (5a) in Et_2O solutions on its concentration[17]. Constant current density 1 or 2 A dm^{-2}

at the same current density after increasing the effective anode area by the platinization was found[20], in agreement with the proposed explanation by the regeneration of 5.
  The reaction of radicals 12a with solvent molecules was supported by the electrolysis of 5k performed in $n\text{-}Bu_2O$ solutions. This process produced mainly[20] 13 and 14, but also

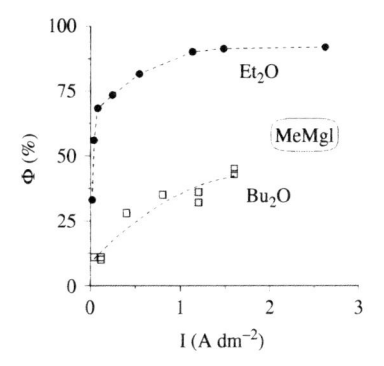

FIGURE 4. Effect of the current density on the fraction of Me• radicals coupled to ethane, Φ, for **5k** in refluxing Et_2O^{19} and in n-Bu_2O^{20} at 143 °C. Adapted with permission from Reference 20. Copyright 1936 American Chemical Society

butane (**21**) and butene-1 (**22**), whereas **15** was absent (cf. Table 4). It is evident from the data collected in Table 4 and Figure 4 that the hydrogen atom abstraction from n-Bu_2O molecules in reaction 10a is more favored than a similar reaction with Et_2O; moreover, a higher temperature favors reaction 10a as well. Small amounts of 1-butanol (**23**) and 2-pentanol (**24**) were also determined[20] after hydrolysis under nitrogen atmosphere of the solution remaining when the electrolysis was completed. The formation of all the products found was explained[20] by the reactions shown in Scheme 6; however, Evans and Field were not sure if the mechanism was radical or ionic. Gaseous products were also found[45] for the electrolysis of MeMgX solutions in pyridine.

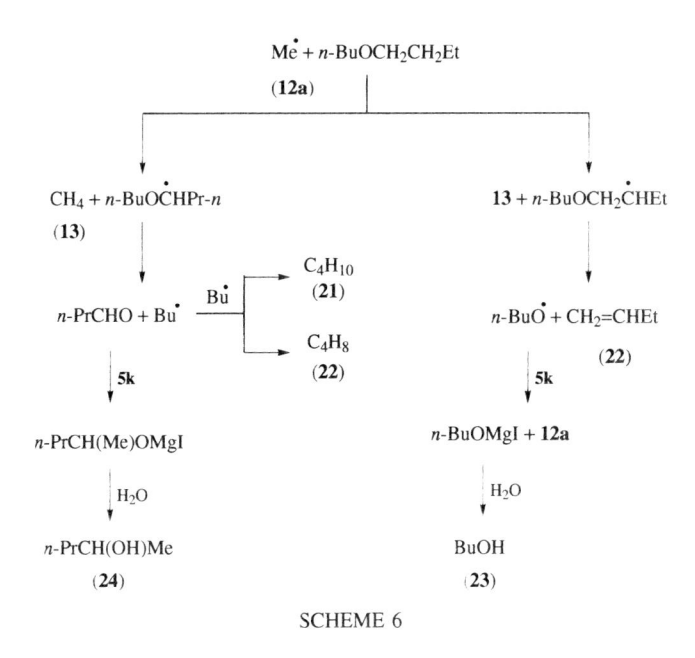

SCHEME 6

For higher alkyl radicals formed in the anodic oxidation of AlkMgX in Et_2O solutions, the participation of competitive reactions given in Scheme 5 is different and thus the products distribution is also different[17,21,22] than that found for MeMgX (cf. Tables 4 and 5). The corresponding alkane and alkene formed in an approximately equivalent amount are the main gaseous products, e.g. **14** and **15**, propane (**25**) and propene (**26**), **21** and **22** for Alk = Et, Pr and n-Bu, respectively. However, a small amount of hydrogen was also determined. The products distribution is independent of the concentration and current density. The above results indicate that the main route of the decay of Et• (**12b**), n-Pr• (**12c**) and i-Pr• (**12d**) radicals is the disproportionation in equation 10b but a secondary competitive reaction (equation 10c) also occurs. The high electrochemical efficiency for the above alkyls (>86%) supports the conclusion that reactions other than that in equation 10b take place to only a very small extent. The efficiency increases in the order of X of I < Br < Cl. The formation of small amounts of CO_2 and alcohols [**19** and **20** for **12d** and **19**, n-PrOH (**27**) and s-PenOH (**28**) for **12c**] was explained by reactions of alkyl radicals with Et_2O, similar to those proposed for **12a** in Scheme 6.

For higher alkyls the competition between disproportionation (equation 10b) and coupling (equation 10d) reactions is of particular interest. First of all, a tendency toward radical coupling increases for straight-chain radicals with their length, and for radicals with four or more carbon atoms the coupling approaches 100%. For example, for radicals **12b** the dimer **21** was not detected but its formation was suggested[17] on the basis of a determined number of carbon atoms in product molecules, equal to 2.15, i.e. higher than for pure **14**. **12c** has a 50% tendency to couple forming n-hexane (**29**)[21]. On the other hand, for n-Hex• (**12e**), n-Bu• (**12f**) (in experiments with a higher distance between the electrodes) and s-Bu• (**12g**) only the products of coupling were detected, although their isolation, in particular for the last radical, was poor. Second, the tendency toward coupling

TABLE 5. Products distribution after the anodic oxidation on Pt electrodes of solutions of AlkMgX (**5**) in Et_2O^a

5	Alk	X	Yield of gaseous products (%)^b			Y_{el} (%)^c	Reference
			14	15	H_2		
5j	Et	I	51.8	47.3	0.97	88.1	17
5b	Et	Br	48.7	50.3	1.1	89.8	17
5f	Et	Cl	50.9	48.0	1.1	95.1	17
			25	26	15	H_2	
5h	n-Pr	Br	50.5	48.5	—	1.0	96.3	17
5h	n-Pr	Br	46.8	46.7	1.4	1.5^d	~91^e	21
5i	i-Pr	Br	44.4	50.7	0.8	1.9^f	>90^g	21
			21	22	H_2		
5c^h	n-Bu	Br	52.3	41.7	0.3–2.5	~65^i	22

^{a}Current density in the range 0.4–2 A dm^{-2}. In each case MgX$_2$ is also formed at the anode.
^{b}Average values from two to seven measurements.
^{c}Electrochemical efficiency (moles of gaseous products per Faraday).
^{d}Other gaseous products: CO_2 0.6% and O_2 1.5% (probably formed at cathode).
^{e}Liquid products: **19**, **27**, **28** and **29**.
^{f}Other gaseous products: CO_2 0.3% and O_2 1.4% (probably formed at cathode).
^{g}Liquid products: **19** and **20**.
^{h}Gas was liberated only when the electrodes were nearly closed; other gaseous products: unsaturated hydrocarbons 1.2–2.1%, CO_2 0.8–2.0% and O_2 1.5–2.0%.
^{i}**26** is the main liquid product.

is reduced for branched-chain compounds. Finally, Evans and coworkers suggested[22] that the tendency to couple increases as the Et• radical (12b) becomes substituted with methyl groups, i.e. in the order of: Et < i-Pr < t-Bu < n-Pr < s-Bu < i-Bu. Similar trends were found later by Martinot[46] for the electrolysis under similar conditions and the results of both reports are given in Table 6. Radical dimerization in the electrolysis of **5** with various R's in Et$_2$O solutions at a Pt anode and a Hg cathode was also investigated by Morgat and Pallaud[47,48]. However, the yields of dimers were low, in the range of 35–60% even for long-chain radicals, e.g. for R = C$_{18}$H$_{37}$ it was 54%; a list of results was also reported in the review[2].

The behavior of BnMgBr (**5d**) is similar to that observed for compounds with higher alkyl groups, i.e. only the coupling product was detected[23] and the earlier report on the additional formation of benzyl alcohol[14] was not confirmed[23]. On the other hand, reactions of Ar• radicals formed in the anodic oxidation of aryl Grignard reagents are different from those established for Alk•, as is evident from the percent distribution of parent radicals in major products given in Table 7.

Reactions of Ph• radicals (**12h**) formed at anodes yield[23] styrene (**30**), biphenyl (**31**), p-terphenyl (**32**), insoluble hydrocarbon of high molecular weight and, in smaller amounts, benzene (**33**) as well as ethanol (**19**). **30** was the main product for substituted reagents **5q** and **5r** but for unsubstituted **5e** only if the current efficiency, Y_{el}, was low. For higher Y_{el} values **31** became the chief organic product. However, in contrast to aliphatic Grignard reagents, except methyl, the current efficiency was always much below 100%. In order to explain the above results the possibility of another route of anodic oxidation, different

TABLE 6. Percent participation of alkyl radicals which couple in the oxidation of AlkMgBr in Et$_2$O solutions at Pt anodes

AlkMgBr	5b	5h	5c	5l	5m	5n	5o	5p	Reference
Alk	Et	n-Pr	n-Bu	i-Bu	s-Bu	t-Bu	n-Hex	n-C$_7$H$_{15}$	
Alk$_2$ (%)	a	50	>85	96	43–49	b	82.5	c	22
	50	c	91	85	c	25	c	100	46

aOnly traces.
bA slight amount.
cNot investigated.

TABLE 7. Products distribution^a after the anodic oxidation^b of solutions of ArMgBr in Et$_2$O at Pt electrodes[23]

Reagent	Ar	Y_{el} (%)^c	Distribution of Ar• in major products (%)^d			
			30	Ar$_2$	32	Polymer^e
5e	Ph	14	49.7	5.5	11.0	5.5
		18	42.0	0	14.0	7.0
		41	18.0	29.9	3.5	2.0
		66	0	67.4	11.2	5.6
5q	p-Tol	31	72.5	0	—	f
5r	p-ClC$_6$H$_4$	20	27.5	0	—	f

aMgX$_2$ is also formed at the anode. Other minor products: **19** and **33**.
bCurrent density in the range 0.16–0.48 A dm^{-2}.
cElectrochemical efficiency given as the number of moles of **5** decomposed per Faraday.
dThe ratio of moles of Ar• radicals to moles of **5**.
eInsoluble polymeric hydrocarbon formed on the anode.
fA small amount.

from that discussed in Scheme 5, was suggested[15,23]. This route includes the formation of halogen atom (equations 12a, 12b or 12c) and a sequence of further reactions (equation 13) producing X_2 and ArX, which react with magnesium at the cathode yielding again **5**. The formation of bromobenzene (**34**) at the anode during the electrolysis conducted in a transference cell, where diffusion was avoided, supported[23] the last suggestion. It was also shown[15] that iodine, not aryl radicals, are formed during the electrolysis of **35** because crystalline iodine was collected upon the anode. The discharge of halogen instead of R is favored[15] by the high electronegativity of R and low electronegativity of X, as well as by high voltage and high current density. In full accordance with the above reasoning the current efficiency was lower for Ar than for Alk, and it was changing in the order of I < Br < Cl.

$$RMgX_2^- \xrightarrow{-e} X^\bullet + RMgX \qquad (12a)$$

$$RMgX^- \xrightarrow{-e} X^\bullet + RMg \qquad (12b)$$

$$MgX_3^- \xrightarrow{-e} X^\bullet + MgX_2 \qquad (12c)$$

$$2X^\bullet \longrightarrow X_2 \xrightarrow{RMgX} MgX_2 + RX \xrightarrow[\text{from cathode}]{Mg} RMgX \qquad (13)$$

$$PhC\equiv CMgI$$
(**35**)

Two series of reactions were considered by Evans and coworkers[23] in order to explain the products given in Table 7. The reactions in equations 14 of **12h** with solvent molecules (Scheme 7) produce progressively **33**, **30** and **19**. However, gaseous **14** and **15** shown in Scheme 7 were not detected. On the other hand, the radical coupling (equation 10d) yields **31**, which next gives **33** and **32** (equation 15), including the hydrogen atom abstraction from **31** by **12h** and the coupling of **12h** and **12i** radicals[23].

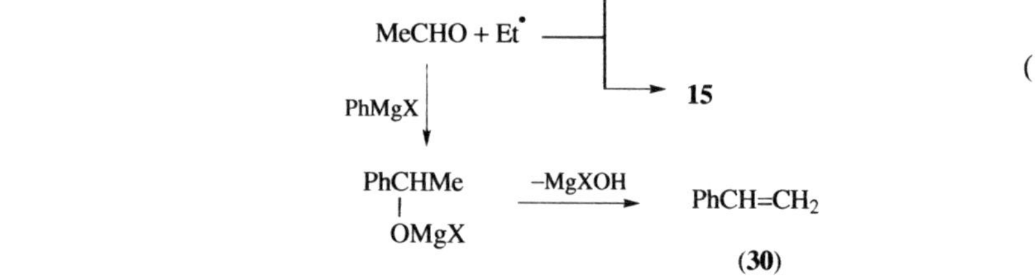

SCHEME 7

6. Electrochemistry of organomagnesium compounds

$$12h + Ph-Ph \longrightarrow 33 + Ph-C_6\overset{\bullet}{H_4} \xrightarrow{Ph^{\bullet}} Ph-C_6H_4-Ph \quad (15)$$
$$\quad\quad\quad\quad (31) \quad\quad\quad\quad (12i) \quad\quad\quad\quad (32)$$

The formation of dimers was also observed[47,48] in the electrolysis of RMgBr solutions in Et$_2$O with R = α-Naph (the yield of 1,1'-binaphthyl was 43%) and a number of Grignard reagents with R being the derivatives of terpenes.

It is interesting that a very marked anodic luminescence was observed during the electrolysis of ethereal solutions of **5e**[14] and thirteen other ArMgX, in particular those produced from p-MeC$_6$H$_4$I and 1,4-chlorobromonaphthalene[49]. For a number of compounds **5** a similar electron transfer step in an electrochemiluminescence (ECL) and a chemiluminescence caused by oxygen was suggested[14]. However, it was shown later[50] that the anodic emission of light during the electrolysis of **5e** could be caused by oxygen contaminations in solutions. The ECL mechanism of **5** still looked unclear in 1985[51]. A photovoltaic effect in the cell containing a solution of 1 M **5e** in Et$_2$O with gold and silver electrodes was also reported[49].

The effect of various R's on the anodic reactivity of **5** was investigated by Evans and coworkers[18] and Holm[52,53] for solutions in Et$_2$O and by Martinot[46] for solutions in THF. The decomposition potentials, E_d, corresponding to the beginning of the oxidation process in Et$_2$O solutions[18] are collected in Table 8. The back electromotoric forces determined for a Pt anode at the current density 0.06 A cm^{-2}, $\eta_{0.06}$, i.e. when the slopes of Tafel plots for each compound are identical[52] and the tentative standard oxidation potentials, E_o[53], recalculated from $\eta_{0.06}$ values, are also collected in Table 8. The corresponding bond dissociation energies, $D(R-MgBr)$, for a C—Mg bond obtained from thermochemical measurements[54] are given in Table 8 as well. A linear plot of E_o against $D(R-MgBr)$ was reported by Holm[55] and a plot of E_o against E_d was reported by Eberson[56], but they did not use these relationships to elucidate the electrochemical process.

However, standard oxidation potentials for the dissociative electron transfer, $E^o(RMgX/R^{\bullet}+MgX^+)$, described by the Savèant theory[57,58], can be expressed by the sum of the

TABLE 8. Decomposition potentials, E_d[18], anodic overvoltage for a current density 0.06 A cm^{-2}, $\eta_{0.06}$[52], standard oxidation potentials E_o[53] for the oxidation of RMgBr in Et$_2$O solutions and the bond dissociation energy, $D(R-MgBr)$, of the C—Mg bond[54] in RMgBr

RMgBr	R	E_d (V)^a	$\eta_{0.06}$ (V)^b	$-E_o$ (V vs. SHE)	$D(R-MgBr)$ (kJ mol^{-1})^c
5e	Ph	2.17	—	0.0^d	289
5a	Me	1.94	1.98	0.25	255
5c	n-Bu	1.32	1.70	0.53	213
5l	i-Bu	—	—	0.63	213^e
5h	n-Pr	1.42	—	—	209
5b	Et	1.28	1.57	0.66	205
5d	Bn	—	1.50	0.73	201
5s	All	0.86	1.07	1.16	201
5t	c-C$_5$H$_9$	—	1.35	0.88	201
5i	i-Pr	1.07	1.28	0.95	184
5m	s-Bu	1.24	1.36	0.87	184
5n	t-Bu	0.97	1.16	1.07	172

^{a}Measured at 22 °C for ca 1 M solutions.
^{b}Anodic overvoltage at a Pt anode relative to a Pt | Mg | MgBr$_2$ cathode at 20 °C for ca 0.8 M solutions.
^{c}Obtained for the reaction: RMgBr$_{(soln)}$ + HBr$_{(g)}$ → RH$_{(soln)}$ + MgBr$_{2(soln)}$ in Et$_2$O.
^{d}Value estimated in Reference 56 from the correlation between E_d and E_o.
^{e}Obtained for the reaction: RBr$_{(l)}$ + Mg$_{(s)}$ → RMgBr$_{(soln)}$ in Et$_2$O.

homolytic bond dissociation energy $D(R-MgBr)$ and the standard potential for the oxidation of $MgX^\bullet$ radicals (equation 16)

$$E^\circ(RMgX/R^\bullet + MgX^+) = D(R-MgBr) - T\Delta S^\circ + E^\circ(MgX^\bullet/MgX^+) \quad (16)$$

where ΔS° is the entropy change for the homolytic cleavage. For the series under consideration the last potential is constant and, if ΔS° is not strongly dependent on R, the linear correlation between standard potential and bond dissociation energy with a unit slope is expected. The expected relationship holds for E_o potentials originally determined by Holm[53] for 9 compounds (one point for R = All strongly deviates) and $D(R-MgBr)$ expressed in the same units, i.e. eV, with a correlation coefficient of $r = 0.963$ and the Snedecor F test 89.41 indicating statistical importance at the level of 99.997% (equation 17)

$$E_o = 0.96(\pm 0.12)D(R-MgBr) - 2.75(\pm 0.51) \quad (17)$$

where 95% errors are given in parentheses. In our opinion, equation 17 shows that the anodic process represents a concerted electron transfer and bond breaking. Thus, equation 17 also explains the order of electrochemical reactivity of compounds **5**: the lower the bond dissociation energy the easier the oxidation and the anodic potential become less positive.

The products of anodic oxidation of AlkMgX in THF and Et_2O solutions at bright Pt anodes are the same[46,59]. Using rotating electrodes in THF solutions containing 0.2–0.6 M RMgCl, Chevrot and coworkers found[60] that the anodic oxidation depends on the anode material (the easiest oxidation occurs at platinized Pt, next at Au and the most difficult one at bright Pt anodes) and it also decreases in the order t-Bu > Et > i-Pr, CH=CH$_2$ > n-Bu > Me > Ph. Thus, general trends for the reactivity of **5** with both halogens (Br and Cl) and in both solvents are similar and can be understood in terms of changes in the C—Mg bond dissociation energy.

On the other hand, details of the electrochemical steps are more complex. For the oxidation of AlkMgX (Alk = Et, n-Bu, i-Bu and t-Bu) Martinot found[46,59] that the potential of Pt electrodes depends linearly on log I (where I is the current density) according to the Tafel plot and the slope is equal to 0.2 V and 0.3 V in Et_2O and THF solutions, respectively. On the basis of polarization curves, reaction orders and capacity data, the ionic mechanism of the oxidation was proposed[59] with an initial electron transfer to anions $RMgX_2^-$ or R_3Mg^- as the rds, yielding the radical $R^\bullet$ (**12**), which is further oxidized at the electrode to the carbocation R^+. However, the electrochemical oxidation of **12** looks unlikely unless very positive potentials are applied; provisional standard potentials of the $R^+/R^\bullet$ couple in acetonitrile estimated by Eberson[61] are 1.91 V and 1.47 V more positive than E_o from Table 8 for R = Et and t-Bu, respectively. Moreover, for the electrolysis of a wider series of RMgBr in Et_2O solutions and using a wider range of I values (10^{-5} to 0.1 A cm^{-2}) Holm found[52,53] different slopes of Tafel plots. They were equal (after corrections for ohmic drops and concentrations) to 0.15 for **5d, 5n** and **5s**, but ca 0.30 for **5a**, whereas for **5b, 5c** and **5i** the slope was 0.30 at low I (ca 1 mA cm^{-2}), but it changes toward 0.15 at high I values. The reported behavior indicates beyond doubt that there exist two different mechanisms of the oxidation. Holm proposed[52] that surface-bonded radicals **12** act as catalysts for a further electron transfer at a 'radical saturated' platinum, decreasing the activation barrier and resulting in a lower Tafel slope. The surface saturation occurs for stable All$^\bullet$, Bn$^\bullet$ and t-Bu$^\bullet$ radicals at low I values, but for less stable radicals only at higher current densities, whereas for the least stable radical **12a** no saturation was reached at all. However, there is not enough experimental data to decide about the kinetics details.

2. Anodic addition to olefins

Anodic oxidation of Grignard reagents (**5**) in the presence of styrene (**30**), butadiene (**36**) or vinyl ethyl ether (**37**) was investigated by Schäfer and Küntzel[62] as an interesting (for preparative use) extension of other anodic reactions with olefins. The electrolysis was carried out at constant current density at Pt, Cu or graphite electrodes. It was found that the products obtained depend on the electrode material, as is seen from the data presented in Table 9.

The scheme of reactions proposed[62] to explain the products obtained is shown, after small modifications, in Scheme 8. Primary radicals **12** formed at the anodes produce with added **30** or **36** (equation 10e) the substituted benzyl or allyl radicals **38**, which can dimerize to **39** or can couple with the added olefin to form radicals **40** or **41**. For allyl radical (**38**) a 1,1'- or 1,3'-coupling is possible yielding **41** and **40**, respectively. Further couplings of **40** and **41** with the primary radical **12** produce **39** and head-to-tail dimer **42**, respectively. It was evident from the products obtained[62] that the coupling of **38** in the 1-position occurs 5 to 11 times faster than in the 3-position. However, for readily polymerizable olefins, rather polymerization occurs, in particular at graphite electrodes. At Pt electrodes both dimers **39** and **42** are formed, but for Cu electrodes exclusively dimers **39** were obtained with moderate yields. Thus, an indirect electrolysis including the oxidation of copper to Cu^+ ions and their further reaction with **5** yielding intermediate RCu was considered, but not proved[62].

On the other hand, the formation of unsaturated hydrocarbons **45**, **46** and **48** (Table 9) can be illustrated by the reactions shown in Scheme 9, with radicals **44** and **47** as intermediates.

3. Oxidation on sacrificial anodes

The electrooxidation of Grignard reagents (**5**) on reactive metal anodes produces the corresponding organometallic compounds, or more generally organoelemental compounds,

TABLE 9. Products of anodic oxidation of 0.2 M RMgBr in Et_2O solutions containing 0.1 M $LiClO_4$ in the presence of olefins[a]

RMgBr	Olefin	C (M)	Anode	Products	Yield (%)[b]
5c	30	0.7	Pt	6,8-diphenyldodecane **42a**	10
				6,7-diphenyldodecane **39a**	5
5c	30	2.0	Pt	polymer[c] **43**	2.6[d]
5c	30	0.7	Cu	6,7-diphenyldodecane **39a**	29
5n	30	0.7	Cu	4,5-diphenyl-2,2,7,7-tetramethyloctane **39b**	14
5c	36	2.0	Cu	6,7-divinyldodecane **39c**	3
				6-vinyl-8-tetradecaene **48a**	15
				6,10-hexadecadiene **46a**	15
				6-dodecene **45a**	7
5o	36	2.0	Cu	8-vinyl-10-octadecaene **48b**	8
				8,12-eicosadiene **46b**	11
				8-hexadecaene **45b**	6
5e	37	4.2	Pt	polymer[e] **43**	11[d]

[a] This Table was published in *Tetrahedron Letters*, H. Schäfer and H. Küntzel, 'Anodic addition of Grignard-reagents to olefins', 3333–3336, Copyright Elsevier, 1970.
[b] Current yield. Electrolysis at $I = 10$ mA cm^{-2} in a flow cell without diaphragm.
[c] Average molecular weight 2500.
[d] Yield in g Ah^{-1}.
[e] Average molecular weight 1040.

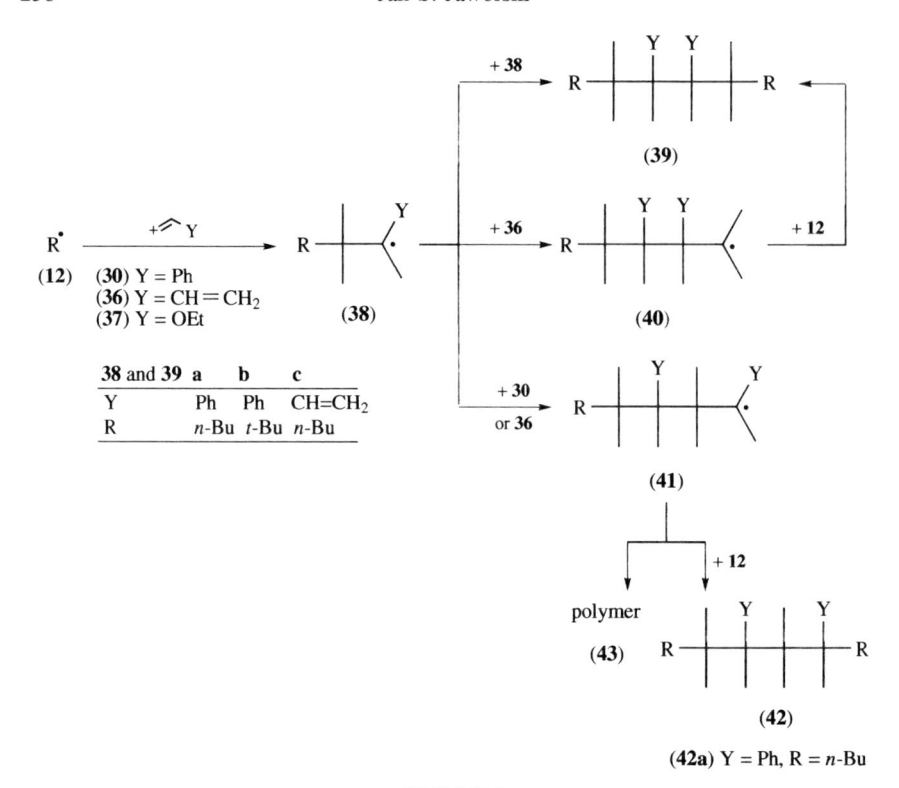

SCHEME 8

(30) Y = Ph
(36) Y = CH=CH$_2$
(37) Y = OEt

38 and 39	a	b	c
Y	Ph	Ph	CH=CH$_2$
R	n-Bu	t-Bu	n-Bu

(42a) Y = Ph, R = n-Bu

in the so-called 'anodic transmetallation' shown in Scheme 4 (equation 10f). The first observations on the dissolution of Mg, Al and Zn anodes, in amounts described by Faraday laws, during the electrolysis of **5b** solutions in Et$_2$O was reported in 1925 by Kondyrew[13]. French and Drane[63] supported the reactivity of Al, Zn and Cd anodes in the electrolysis of ethereal solutions of i-PenMgCl and suggested that 'metallic alkyls' are formed. The formation of Et$_3$Al was also suggested by Evans and Lee[17]. In further research the list of sacrificial anodes was extended to Pb, Bi, Mn, B, P and a number of industrial processes were patented[42,64–76]. Most of the investigations were devoted to the production of R$_4$Pb (**11**) and the yields of tetraalkyllead obtained in electrochemical processes increased from ca 73% in the first patent[64], when Et$_2$O was used as a solvent and the electrolysis needed the high voltage of 100 V, to 80–90% or even more in the Nalco process after finding better solvents and, above all, after the addition of extraneous organic halide (**8**) to the solutions of **5** in a molar ratio of ca 1:1 for **11a**[65] and up to 0.5:1 for Me$_4$Pb (**11b**)[67]. The added **8** reacts with the magnesium deposited on the cathode (equation 18b) recovering **5** and changing the overall process (Scheme 10) from the reactions in equations 18a to 18c. It should be added here that reaction 18a can occur in nonelectrochemical conditions[77] and this can explain[6] why current efficiencies of the electrolysis extend to 100%[66–69,75] (cf. data given in Table 10). The best media developed for the commercial production of **11** were anhydrous mixtures of organic solvents containing diethers of glycols [e.g. (MeOCH$_2$)$_2$ (**49**)[65], (n-BuOCH$_2$)$_2$ (**50**)[42,69], n-HexOC$_2$H$_4$OEt (**51**), Bz(OC$_2$H$_4$)$_3$OEt (**52**), (EtOC$_2$H$_4$)$_2$ (**53**) or others] with THF[42,66–68,75], which increases the conductivity (cf.

6. Electrochemistry of organomagnesium compounds

$$12 + 36 \longrightarrow$$

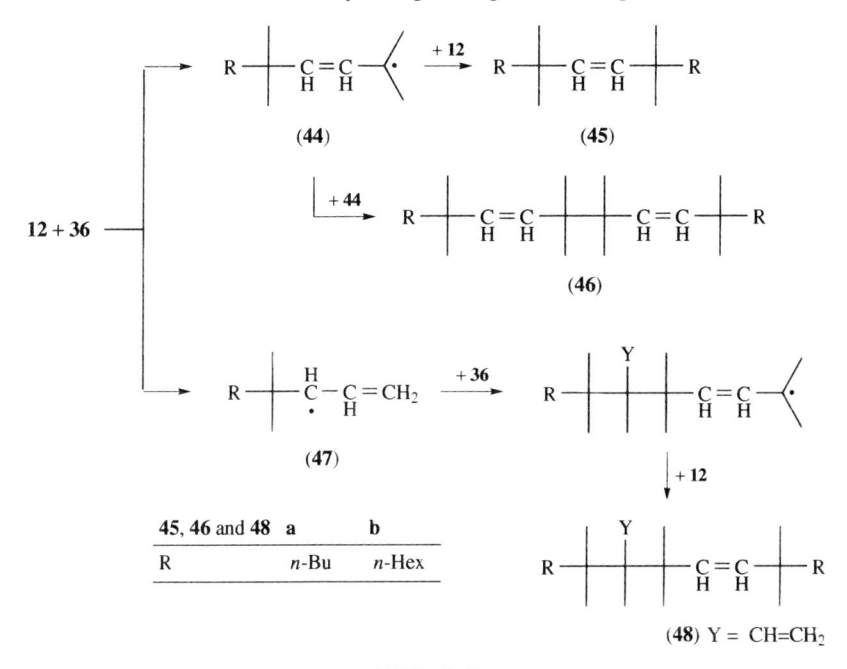

$$R-\underset{H}{\overset{}{C}}=\underset{H}{\overset{}{C}}\cdot \xrightarrow{+12} R-\underset{H}{\overset{}{C}}=\underset{H}{\overset{}{C}}-R$$
(44)   (45)

$$\xrightarrow{+44} R-\underset{H}{\overset{}{C}}=\underset{H}{\overset{}{C}}-\underset{H}{\overset{}{C}}=\underset{H}{\overset{}{C}}-R$$
(46)

$$R-\underset{\cdot}{\overset{H}{C}}-\underset{H}{\overset{}{C}}=CH_2 \xrightarrow{+36} R-\underset{}{\overset{Y}{|}}-\underset{H}{\overset{}{C}}=\underset{H}{\overset{}{C}}\cdot$$
(47)

$$\downarrow +12$$

$$R-\overset{Y}{|}-\underset{H}{\overset{}{C}}=\underset{H}{\overset{}{C}}-R$$
(48) Y = CH=CH₂

45, 46 and 48	a	b
R	n-Bu	n-Hex

SCHEME 9

Section III), thus increasing the efficiency of the electrolysis. Small amounts of aromatic hydrocarbons, like toluene or benzene (**33**)[66-70,75], were also added to the mixtures used. A number of typical examples described in patents are illustrated in Table 10, including the products and conditions of equation 10f.

$$Pb + 4RMgX \longrightarrow R_4Pb + 2Mg + 2MgX_2 \quad (18a)$$

$$2Mg + 2RX \longrightarrow 2RMgX \quad (18b)$$

$$Pb + 2RMgX + 2RX \longrightarrow R_4Pb + 2MgX_2 \quad (18c)$$
$$(5) \quad\quad (8) \quad\quad\quad\quad (11)$$

5	b	e	f	o	u	v
R	Et	Ph	Et	n-Hex	Me	CH=CH₂
X	Br	Br	Cl	Br	Cl	Cl

8	a	b	c	d	e	f	g
R	Ph	Ph	n-Hex	Me	Et	t-Bu	c-Hex
X	Br	Cl	Br	Cl	Cl	Cl	Cl

SCHEME 10

The industrial Nalco process for the production of **11a** and **11b** was conducted[78] in mixtures of **53** with THF, at 35–40 °C or 40–50 °C and about 2 kg cm^{-2} pressure, in a cell divided by porous diaphragms and with current densities of 1.5–3.0 A dm^{-2} at 15–30 V. However, production details are beyond the scope of this Chapter. Effective methods of recovery of **11** from mixtures after or during the electrolysis were elaborated[73,74]. Of

TABLE 10. Organoelemental compounds produced by electro-oxidation of Grignard reagents **5** on sacrificial anodes

Reactants (mole ratio)	Anode	Solvent	Conditions	Product (yield[a]/%) I_{eff}^b (%)	Reference
5b	Pb[c]	Et$_2$O	100 V	Et$_4$Pb (73)	64
5f + **8e** (1:1)	Pb[d]	**49**	50–65 °C, 12–26 V	Et$_4$Pb (81)	65
5u + **8d** (1:0.29)	Pb[d]	THF, **33**, **50**	29.9 °C, 27.2 V	Me$_4$Pb (92.1, 81.2[e]) (**9** + **10** 1.26%) 149	67
5u + **8d**	Pb[d]	**52**,THF	46 °C, 28 V	Me$_4$Pb (100, 82.5[e]) 174	68
5e + **8a**	Pb[d]	**50**	55 °C, 25.6–26.8 V	Ph$_4$Pb	69
5u + **8b** (2:1)	Pb[d]	**50**	38 °C, 30 V	R$_4$Pb[f]	69
5u + **8d**	Pb[d]	12% THF, 28% **33**, 60% **50**	30 °C, 27 V	Me$_4$Pb (89.6, 71[e]) 164	75
5u + **8d**	Pb[d]	10% THF, 45% **33**, 45% **51**	40 °C, 22 V	Me$_4$Pb (99.1, 94[e]) 161	75
5v	Pb[d]	THF	1.7–3.9 V	(CH$_2$=CH)$_4$Pb	76
5o + **8c**	Al[c]	**51**	35–45 °C, 26.5–27 V	(n-Hex)$_3$Al 134.2	69
5e + **8a**	Al[c]	**51**	55 °C	Ph$_3$Al	69
5o + **8c**	Zn, Cd Mn, Bi[c]	**51**	35–45 °C, 26.5–27 V	(n-Hex)$_x$M	69
5f	P$_{(black)}^g$	Et$_2$O		Et$_3$P	72
5f	B[d,h]	Et$_2$O		Et$_3$B	71

[a] Yield based on **5**.
[b] Current efficiency.
[c] Cathode from the same metal as the anode.
[d] Cathode: stainless steel.
[e] Yield based on Mg.
[f] A mixture of compounds with different R's.
[g] Pt cathode.
[h] Boron-coated tantalum.

course, equation 10f is general and products with various alkyl or aryl groups can be obtained; the list of R's in molecules **5** and **8** given in Scheme 10 includes only some compounds reported in patents[42,66–68,75] and mentioned in Table 10. For example, the production of R$_3$P with R being Ph, Bn, Tol and Alk from Me to C$_8$H$_{17}$ was described[72], as well as of R$_3$B with R being Ph and Alk groups from Me to Hex[71]. In equation 18c, the use of **5** with R from Me to n-Hex and c-Hex, as well as Ph and Bn, was suggested[66,69]. Moreover, by using different groups in **5** and **8** all possible molecules with mixed R's were produced.

The use of other sacrificial anodes, such as Ca, La, Hg, Tl, As, Te and Se, was also mentioned in patents[65,68,69], but no experimental evidence of their use was described.

As concerns the mechanism of anodic oxidation of **5** at sacrificial anodes, it can be noted that the process occurs at potentials close to those of the oxidation of the corresponding diorganomagnesium compounds (**1**). For example, half-peak potentials for the oxidation of **5b** and **1b** in THF containing 0.25 M TBAP at a lead electrode measured[37] at a scan rate of 0.3 V s^{-1} are equal to $E_{p/2} = -1.73$ and -1.72 V vs. 0.01 M Ag$^+$/Ag, respectively. However, the oxidation mechanism for both compounds is different, as shown

6. Electrochemistry of organomagnesium compounds 241

by different Tafel slopes: 0.12 and 0.06 V for **5b** and **1b**, respectively[37]. The high Tafel slope for **5b** means that the electron transfer is slow. Nevertheless, chemical reactions following the formation of the first Pb—Et bond and controlling the overall rate constants are the same for both kinds of compounds, as proposed by Fleischmann and coworkers[37] in Scheme 4 (Section IV.A). The second oxidation process (equation 19) observed at a potential of -1.2 V corresponds to the formation of the insoluble $PbBr_2$ layer on the electrode surface.

$$2MgBr^+ \xrightarrow[\text{Pb(anode)}]{-2e} PbBr_2 + 2Mg^{2+} \qquad (19)$$

Moreover, the oxidation process is strongly dependent on the state of the electrode surface[37]. At a freshly cleaned and polished lead the oxidation of **5b** occurs at $E_{p/2} = -1.52$ V on the first sweep, but on subsequent cycles the potential shifts in the cathodic direction approaching $E_{p/2} = -1.72$ V.

4. Processes at semiconductor anodes

In recent years there has been great interest in the derivatization of silicon surfaces and, beside other methods, electrochemical oxidations of **5** at silicon anodes were successfully used for this purpose[79–84]. Although a perfect electronic passivation toward electron–hole recombination of the (111)-oriented silicon surface can be obtained by hydrogen termination, yet its chemical stability toward oxidation is limited, in particular when it comes into contact with air or moisture. Molecular grafting of silicon surfaces by organic groups, first of all alkyl but in perspectives also biochemical, provides a promising approach to improve the stability of silicon interfaces and to develop silicon-based molecular electronic, biochip and sensing devices. In the electrochemical approach, the oxidation of **5** produces radicals **12** which form covalent C—Si bonds with anode atoms. Thus, there is some similarity to the oxidation of **5** on sacrificial metal anodes. However, the process is different because new bonds are formed only with surface atoms and there is no loss of the anode material. Moreover, oxidation processes at metal and semiconductor electrodes are different because of their different electronic properties.

Electrochemical grafting of methyl groups on the porous[79] as well as the atomically flat (111) Si surface[80] was reported. Fast methylation of the hydrogenated (111) surface of p-type silicon wafer, used as the anode, with the Cu counter electrode was performed[80] in 3 M solution of **5k** in Et_2O in a glove box under purified nitrogen by passing the anodic current from 0.1 to 5 mA cm^{-2} from 1 to 30 min. Differential attenuated total reflection (ATR) FTIR spectra, obtained with the electrode shaped as an ATR prism, allowing multiple reflections of the IR beam inside the plate and avoiding propagation across the electrolyte, supported substitution of the hydrogen atoms by methyl groups. Namely, a narrow single νSiH line at 2083 cm^{-1} in p polarization, characteristic of the stretching mode of Si—H bonds perpendicular to the surface, was not observed. On the other hand, lines of methyl groups appeared according to predictions[80]: the symmetric deformation δ_S mode at 1255 cm^{-1} in p polarization and the asymmetric δ_{AS} mode at 1410 cm^{-1} in s and p polarization. The presence of additional carbon on Si surfaces after grafting was evidently confirmed by high-resolution XPS spectra[83] which also confirmed the practical absence of surface oxidation. The *in situ* IR spectroscopy with a current-pulse method allowed researchers to investigate[81, 83] the kinetics of the electrochemical grafting. With the increase of the cumulated charge the integrated band intensities showed the loss of νSiH accompanied by a simultaneous gain of δ_{Me} (for methyls covalently bonded to Si), supporting the electrochemical character of the process. Independent *ex situ* IR measurements showed that the fraction of substituted hydrogens was of the order of 90%. There was no effect of concentration and solvent (Et_2O or THF) on the kinetics, but a larger current density caused faster variations in IR signals and increased the yield of surface modification. The overall grafting process (equation 20), including the transfer of positively charged holes, h^+, is fast and irreversible[81], with the participation of very short-lived intermediates, most probably radicals.

$$\mathord{>}SiH + 2\ \mathbf{5}\ + 2h^+ \longrightarrow \mathord{>}SiR + RH + 2MgX^+ \qquad (20)$$

The necessary breaking of SiH bonds may be realized either through direct potential activation, or more often through anodic generation of **12** at the first step of the oxidation of **5** at semiconductor electrodes[82]. However, for higher current densities a competition between reactions with silicon (equation 20) and other following reactions of **12**, similar to those accompanying the anodic oxidation of **5** at metal electrodes (cf. Scheme 5 in Section IV.B), was pointed out. Thus, a more detailed mechanism of grafting was proposed[83] as is shown in Scheme 11. **12** can abstract a hydrogen atom from the hydrogenated silicon surface (equation 21) and the dangling bond, then created at the Si surface, may react with **5** (equation 22a), or with another **12** (equation 22b) or may abstract a hydrogen atom from the solvent (equation 22c). The last competing step was confirmed by the observation of a weak reincrease in the SiH band after turning off the anodic current. The reactions in Scheme 11 as well as equation 20 correspond to two elementary charges per one attached R group, which is in best agreement with experimental data. Moreover, a detailed kinetic model was proposed[83] reproducing the shapes of kinetic curves and their dependence on experimental conditions. In conclusion, it was shown[83] that anodic alkylation of the Si surface by **5** is less favorable for attaining maximum coverage than chemical techniques, but it is much faster because the Faradaic efficiency may be close to unity, although the concentration of radicals **12** at the surface remains very low. Thus, grafting of a full monolayer requires only a charge of several hundred $\mu C\ cm^{-2}$, which can be completed in one second[82, 83].

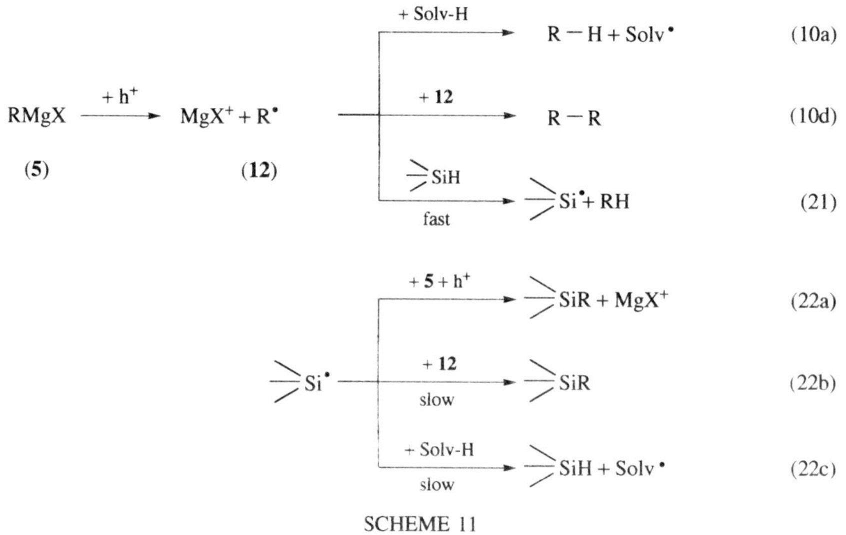

$$RMgX \xrightarrow{+h^+} MgX^+ + R^\bullet$$
$$(\mathbf{5}) \qquad\qquad (\mathbf{12})$$

$R^\bullet \xrightarrow{+\text{Solv-H}} R-H + Solv^\bullet \qquad (10a)$

$R^\bullet \xrightarrow{+\mathbf{12}} R-R \qquad (10d)$

$R^\bullet \xrightarrow[\text{fast}]{\mathord{>}SiH} \mathord{>}Si^\bullet + RH \qquad (21)$

$\mathord{>}Si^\bullet \xrightarrow{+\mathbf{5}+h^+} \mathord{>}SiR + MgX^+ \qquad (22a)$

$\mathord{>}Si^\bullet \xrightarrow[\text{slow}]{+\mathbf{12}} \mathord{>}SiR \qquad (22b)$

$\mathord{>}Si^\bullet \xrightarrow[\text{slow}]{+\text{Solv-H}} \mathord{>}SiH + Solv^\bullet \qquad (22c)$

SCHEME 11

6. Electrochemistry of organomagnesium compounds

Investigations of **5** with different R groups showed[82] that fast grafting can be obtained for the most inert radicals, R = Alk (from Me to $C_{18}H_{37}$) and ethynyl, whereas for more reactive **12**, e.g. Ar•, side reactions were observed[82], in particular electropolymerization on the silicon surface. For example, such behavior was found for **5r** and the first steps of the formation of a polymeric layer can be described by equations 23.

$$\geq SiC_6H_4Cl + \dot{C}_6H_4Cl \rightleftharpoons \geq Si\dot{C}_6H_4 + ClC_6H_4Cl$$
$$\geq Si\dot{C}_6H_4 + \dot{C}_6H_4Cl \rightleftharpoons \geq SiC_6H_4C_6H_4Cl$$
(23)

However, if the molecules of **5** had R alkyl chains longer than Me, the steric hindrance prevented 100% substitution[83] and IR examinations indicated a 50% less derivatization. Moreover, XPS analysis showed that the surface is partly modified by substitution of hydrogen by halogen[83]. In the case of **5** with X = I and to some extent X = Br, the formation of X• radicals (besides **12**) in a secondary reaction was reported[81, 83]. They participate in reactions analogous to equations 21 and 22b, but with X• instead of **12**, and attach to the Si surface improving the electronic passivation of the surface at defect sites, sterically inaccessible to **12**. A possibility that surface dangling bonds may also appear in the charged states was discussed as well[83].

On the other hand, alkylation of silicon surfaces using **5** can be achieved by chemical methods: chlorination with PCl_5 or Cl_2 followed by alkylation with **5**. A comparison of the electrical properties and chemical stability of (111) silicon surfaces alkylated by different chemical and electrochemical methods was reported by Webb and Lewis[84]. They found that the surfaces prepared by anodization of Si in 3 M solutions of **5k** in Et_2O displayed extensive oxidation in air and higher initial charge-carrier surface recombination velocities than those observed for the samples prepared by chemical methods. However, it should be added here that even in the thermal grafting of hydrogenated silicon surfaces using **5**, some electrochemistry is hidden[85]. Namely, a zero-current electrochemical step was proposed[85] in order to explain the following experimental results: (i) the addition of **8** to **5** significantly increased the grafting efficiency of alkyl chains, (ii) the grafting is also possible in solutions containing only **8** and, moreover, (iii) the process in 1 M $C_{10}H_{21}MgBr$ solution in Et_2O is much faster on n-type than on p-type silicon. The last result indicates that the rds is of electrochemical nature. A reaction model containing simultaneous oxidation of **5** and reduction of **8** at the silicon surface, with the second step acting as rds, was proposed[83] on the basis of electrochemical thermodynamic considerations.

C. Other Compounds

Basically, it is possible to obtain organomagnesium compounds with electroactive groups oxidized without the cleavage of the Mg−C bond. They are formally beyond the scope of this Chapter, and thus only one example is mentioned. **54** (written as the *S, S* diastereomer), having a dimethylaminomethylferrocenyl unit, which is (*C,N*)-bidentate ligand with the α-carbon atom from the substituted Cp ring and the amine nitrogen atom as donors, was investigated[86] at a platinum electrode in CH_2Cl_2 containing 0.2 M $TBAPF_6$ electrolyte. Reversible oxidation of both ferrocene moieties was found with a two-electron CV peak at $E_{pa} = 0.41$ V vs. SCE at any scan rate, which indicates no electronic communication between the two ferrocene units. A yellow-to-blue color change, typical of the formation of the ferrocenium cation, corroborated the nature of the electrochemical process.

(54)

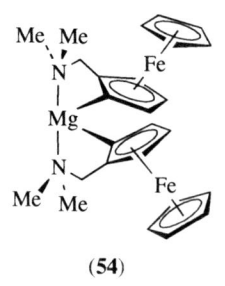

V. CATHODIC REDUCTION OF ORGANOMAGNESIUM COMPOUNDS
A. General Mechanism of the Reduction

The earliest investigations on the electrode reduction of organomagnesium compounds already indicated the magnesium deposition on the cathode. It was found in 1927 for solutions of **5** in Et_2O using a Pt cathode[14] and later evidently supported[87] by the formation of a magnesium amalgam on a Hg electrode. Evans and coworkers excluded[19, 21] the existence of magnesium(II) ions in Et_2O solutions (considered earlier[14]) and suggested[15], on the basis of conductivity measurements and transference study of solutions of **5c**, the formation of RMg^+ and MgX^+ ions (equations 3a–3c in Section III). The overall equations 24 proposed[15] indicated the final products of the reduction, but details were unclear, although the participation of free radicals $RMg^{\bullet}$ and $MgX^{\bullet}$, as intermediates formed in the one-electron transfer from the electrode, was postulated.

$$2RMg^+ + 2e \longrightarrow Mg + R_2Mg \qquad (24a)$$

$$2MgX^+ + 2e \longrightarrow Mg + MgX_2 \qquad (24b)$$

Dessy and Handler supported[88] earlier findings of Evans and Pearson[15] on the mass balance during the electrolysis of **5b** solutions in Et_2O, which indicated the magnesium migration to both Pt electrodes, the loss of Et in preference to Br in the cathode compartment (separated by a stopcock bore) and the existence of large aggregates of ions. The most interesting result was obtained[88] from a study of radioactivity balance in the cell containing the Grignard reagent prepared by mixing **1b** and labeled $^{28}MgBr_2$. After the exchange two different types of magnesium were found in the solution, indicating not only that **5b** should be represented by $Et_2Mg \cdot MgBr_2$, in accordance with the Schlenk equilibrium (equation 2), but also that the magnesium deposited at the cathode has its origin in **1b**, whereas $MgBr_2$ migrates to the anode compartment. Thus, it was postulated that the cathodic reaction (equation 24b) is not involved in the electrochemical process and the RMg^+ ion plays the main role in the reduction. It should be noted that **1b** was reduced at high voltage equal to 160 V^{88}. However, the nature of the intermediates and the detailed mechanism was not explained until recent years (see next Section).

On the other hand, the polarographic behavior of organomagnesium compounds in DME solutions containing 0.1 M TBAP, reported by Psarras and Dessy[36], was quite different. Compounds **1** with R = Alk and Ph were not reducible before a supporting electrolyte discharge. However, solutions of **1f**, **1g** and **1h**, with R = Bn, All and C_5H_5, respectively, i.e. containing groups capable of forming in DME fairly stable carbanions (**55**) (they are formed if the pK values of the parent hydrocarbons are lower than 44)[89], were easily reduced (equation 25). The polarographic waves observed were irreversible, diffusion-controlled and corresponded to $n_e = 1$, as proved by the controlled-potential electrolysis at −3 V. Their half-wave potentials are given in Table 11. Radicals $RMg^{\bullet}$, formed in the first step, decomposed with the deposition of Mg on the cathode and formation of radicals **12**. Their further reactions were the same as for **12** formed during the oxidation at anodes, including mainly equations 10a, 10b and 10d[36], as was discussed in Section IV.B.

$$\begin{array}{c|ccc}
\text{1 and 55} & \text{f} & \text{g} & \text{h} \\
\hline
\text{R} & \text{Bn} & \text{All} & \text{C}_5\text{H}_5
\end{array} \qquad R_2Mg \xrightarrow{+e} R^- + RMg^\bullet \longrightarrow R^\bullet + Mg \qquad (25)$$
$$\qquad\qquad\qquad\qquad\qquad\quad (1) \qquad\qquad\quad (55) \qquad\qquad (12)$$

Furthermore, the polarographic reduction of RMgBr under the same conditions gave two waves[36] (with the exception of **5a**, when only the first wave was observed). The first one had half-wave potentials, $E^I_{1/2}$, given in Table 11, close to the value $E_{1/2} = -2.47$ V characteristic of the reduction of MgBr$_2$ (which is different than $E_{1/2} = -2.3$ V for the reduction of Mg^{2+} ions). However, the reduction of MgBr$_2$ was the two-electron process yielding Mg and 2Br$^-$, but the coulombic analysis of the reduction of **5** showed the total number of electrons for both waves of $n_e = 1$. This behavior was explained by taking into account the Schlenk equilibrium (equation 2)[36]. In full accordance with the equilibrium 2, the addition of MgBr$_2$ to a solution of **5i** caused an increase in the height of both waves and the addition of i-Pr$_2$Mg (**1i**) caused an increase in the second wave at the expense of the first one. It should be remembered that compounds **1**, participating in the Schlenk equilibrium, were not reducible[36]. Moreover, $E_{1/2}$ values found for **5** and for mixtures of MgBr$_2$ and the corresponding **1** are practically the same (Table 11).

On the other hand, the values for $E^{II}_{1/2}$ for the second reduction wave of **5** (Table 11) are dependent on the nature of R. It is evident that the reduction becomes more difficult with the increase in the size of R in the order of Et < i-Pr < i-Bu < Ph. All the observations above indicated[36] that the second polarographic waves correspond to the reduction of **5** themselves (equation 26). Thus, for both types of compounds, **1** and **5**, the unstable radicals RMg$^\bullet$ formed in the reduction of parent, neutral molecules are responsible for the magnesium deposition[36] and radicals **12** are involved in further reactions, already discussed in Section IV.B.

$$RMgBr + e \longrightarrow Br^- + RMg^\bullet \longrightarrow Mg + R^\bullet \qquad (26)$$
$$\qquad (5) \qquad\qquad\qquad\qquad (12)$$

B. Deposition of the Metallic Magnesium and the Reverse Process

Although the electrochemical deposition of magnesium from Grignard reagents is not interesting for organic chemists, yet the investigations into such processes undertaken

TABLE 11. Half-wave potentials for the polarographic reduction of 2 mM solutions of **1** and **5** in DME + 0.1 M TBAP[a]

Compound	$-E_{1/2}$ (V)[b]	Compound	$-E^I_{1/2}$ (V)[b]	$-E^{II}_{1/2}$ (V)[b]	Mixture[c] of MgBr$_2$ + **1**	$-E^I_{1/2}$ (V)[b]	$-E^{II}_{1/2}$ (V)[b]
1f	2.74	**5a**	2.49	—	**1a**	2.49	—
1g	2.65	**5b**	2.44	2.70	**1b**	2.43	2.66
1h	2.54	**5i**	2.44	2.75	**1i**	2.46	2.74
		5l	2.46	2.75	**1j**	2.50	2.78
MgBr$_2$	2.47	**5e**	2.46	2.80	**1e**	2.50	2.83

[a] Adapted with permission from Reference 36. Copyright 1966 American Chemical Society.
[b] Expressed vs. Ag/1 mM AgClO$_4$. Precision of $E_{1/2}$ values is ±0.04 V.
[c] Corresponding to **5** given in 3rd column.

in recent years should be mentioned here because they shed new light on the nature of organomagnesium cations active at electrodes and on the details of electrochemical processes, in particular the reasons of their reversibility, observed only in specific electrolytes. It also resulted in the synthesis and electrochemical investigations of organomagnesium complexes with organoaluminum compounds.

Preparation of a relatively thick, uniform and coherent microcrystalline layer of magnesium is important for some galvanotechnique applications, like the production of parabolic reflectors used for solar collectors and antennas. Aqueous solutions as well as those of the most common organic solvents cannot be used for this purpose. On the other hand, exceptionally good results were obtained using ethereal solutions of **5**, following evidence of magnesium deposition from these solutions reported in early studies[13, 14, 30, 90]. Patented methods[91-93] preferably used THF solutions of **5b**, if **8g** was added after starting electrodeposition at a rate sufficient to dissolve sponge-like magnesium deposits but low enough to avoid corrosion of the magnesium compact layer[91]. Other methods used electrolytes containing among others **1** and **3** in, e.g., toluene[93, 94].

Even more important in recent years were continuous efforts[95] to develop rechargeable batteries with magnesium anodes, which would be cheap, environment-friendly and safe to handle, and would substitute for the commonly used lead–acid and nickel–cadmium systems for heavy load applications[96]. In such devices the metallic magnesium could be applied as a rechargeable negative electrode if reversible cycling of the magnesium occurred in organic electrolytes, i.e. the Mg^{2+} ions were released from the electrode during discharge and a redeposition of Mg^0 would occur during recharge. The electrodissolution of magnesium in most organic solutions is also difficult, similarly to the electrodeposition, due to the dense passivation layer formed on the magnesium surface by reduction products of the solution species. The electrochemical Mg dissolution can only occur via breakdown of the surface films at relatively high overpotentials, whereas the electrochemical Mg deposition on electrodes covered by a passivating layer is impossible. In the review given below our attention is focused on the nature of species existing in solutions and the mechanism of processes at electrodes, but many electrochemical problems discussed in the cited references are omitted.

Cathodic deposition and anodic dissolution of magnesium from ethereal solutions of **5** have been repeatedly reported[46, 59, 97, 98]. For example, in 0.5 M solutions of **5b** in THF at 293 K using a Cu electrode both processes were reversible[98], with an exchange current density of 1 mA cm^{-2}. Moreover, the addition of magnesium bromide etherate in refluxing THF at 338 K enhances the rate of Mg deposition; however, no deposition was observed from solutions of $MgBr_2$ alone. A good magnesium deposition from organoboron complexes with **5** in ethers was found[99]. However, better crystalline deposits with purities of 99.99+% can be obtained[100] from mixtures of less expensive and toxic $AlCl_3$ (0.1 M solution in THF) with 0.8–1.5 M solutions of **5**, in particular organomagnesium chlorides, but not bromides or iodides. In the cathodic deposition (equation 24a) mainly a participation of RMg^+ ions, formed in reactions 4 and 27, were considered[100]. For the dissolution of the Mg anode equations 28 and 29 were proposed[100] followed by further dissociation of $Mg(AlX_4)_2$.

$$RMgX + AlX_3 \longrightarrow AlX_4^- + RMg^+ \quad (27)$$

$$Mg + 2RMgX_2^- \longrightarrow R_2Mg + 2MgX_2 + 2e \quad (28)$$

$$Mg + 2AlX_4^- \longrightarrow Mg(AlX_4)_2 + 2e \quad (29)$$

However, the solutions of **5** were found to be unsuitable as electrolytes for rechargeable batteries. The deposition of Mg at high current efficiency is possible[100] from solutions in

THF of aminomagnesium halides, like (N-methylaniline)MgCl, but much better was the solution of magnesium dibutyldiphenylborate, $Mg(BBu_2Ph_2)_2$ (**56**) in THF, even without the addition of other organomagnesium compounds. By using a solution of **56** in a mixture of THF and DME it was possible to construct[100] and patent[101] the first cell with a magnesium anode which could be discharged and recharged, but only in four cycles. It was noted[100] that a few magnesium compounds which appear to be capable of Mg deposition from organic solvents have a relatively covalent nature of bonds to the Mg atom.

The rate of the reoxidation of Mg deposits is controlled by their morphology[102], which in turn depends on the substrate material. Smooth and compact deposits were obtained using silver or gold, but not nickel or copper. It was also established[103] that the open circuit potential (OCP) of magnesium electrode (a fresh deposit on a Pt) in concentrated solutions of **5** depends strongly on the solvent used. In THF solutions with c around 1 M at 22 °C under argon atmosphere, the values of OCP for **5a**, **5b** and **5f** were equal to -2.8, -2.73 and -2.77 V vs. Ag^+/Ag, respectively[35].

The electrodeposition of magnesium from THF solutions containing **5**, amidomagnesium halides or magnesium organoborates was compared by Liebenow and coworkers[35]. Conductivity of solutions of **5** in THF (cf. Table 3 in Section II) is lower than that for a solution of bis(trimethylsilyl) magnesium chloride (**57**), but the highest value (0.31 Ω^{-1} cm^{-1} at a concentration of 0.4 M) was found for solutions of $Mg(B(s-Bu)(n-Bu)_3)_2$ (**58**). The reversibility of the magnesium deposition in solutions of **5** was high and the reoxidation efficiency of magnesium deposit on silver in 10 cycles was 100%, much better than for solutions of **58**. However, the last electrolyte was found to be the most stable for irreversible oxidation, which is favorable in energy storage devices. The best result at that time for a complete cell was obtained[35, 104] with a solution of the magnesium salt **56**. Conducting polymer electrolytes were also proposed[104–106] in order to construct all solid-state magnesium rechargeable batteries.

Significant progress in understanding the mechanism of the processes discussed was recently reached by the group of Aurbach[96, 107–117]. The results obtained, little known by organic chemists, can also be interesting for understanding the processes of classical preparation of **5** by the reaction with magnesium, most often used in organic chemistry. A comparison of Mg electrochemistry in solutions of **5b**, **5f** and **5g** in THF and different magnesium salts in dipolar aprotic solvents was performed[107] under a very pure argon atmosphere at room temperature using a number of modern techniques: CV, impedance spectroscopy, surface sensitive FTIR spectroscopy, elemental analysis by dispersive X-rays (EDAX), scanning and tunneling electron microscopy (SEM and STM) and electrochemical quartz crystal microbalance (EQCM). FTIR spectra showed evidently that the surface chemistry of magnesium electrode in THF solutions (without **5**) is dominated primarily by possible reactions of the salt anions and atmospheric contaminants, such as trace water. The electrochemical examinations supported the conclusion that in solutions of all the solvents used, magnesium electrodes are covered by surface films formed spontaneously even if the surface is freshly prepared in solutions. Organic or inorganic magnesium salts, MgO and $Mg(OH)_2$, as well as some hydrated forms of the two last compounds, formed at the surface do not conduct Mg ions and this passivation of the electrode makes the Mg deposition impossible. It was believed[96, 114] that magnesium ions are located at a specific site in the lattice and their mobility is close to zero. On the other hand, those layers also prevent a spontaneous reaction of the metal with the solvent and/or solution components. In THF, which was the least polar solvent under examination, the pristine surface films are very stable but thick passivation layers are not formed. Those films must be confined to a monolayer scale as was supported[115] by X-ray photoelectron spectroscopy (XPS). Moreover, THF and other ethers are not reactive with Mg.

Thus, the Mg electrode in a solution of the supporting electrolyte [0.1 M $(CH_3COO)_2Mg$ and 1 M TBAP] in THF was blocked and did not show any voltammetric activity in the potential range from -3.5 to $+0.5$ V vs. Ag/Ag^+. However, after the addition of **8g** to that solution, the Mg dissolution and deposition was obtained at relatively low overpotentials and the oxidation and reduction peaks appear in voltammetric curves. This means that the presence of alkyl halides leads to a breakdown of 'native' films (some electron tunneling through the surface films was assumed[107]) and the formation of soluble **5** with unreactive anions toward magnesium metal. Thus, a reversible electrochemical behavior was observed only in passivating surface-film free conditions. It was also emphasized[117] that the molecules of **1** as Lewis bases are effective scavengers for trace atmospheric contaminants and react with them in the bulk of solution, preventing the magnesium surface from the passivation. The CV-EQCM experiments supported the conclusion that stable passivating surface films were not formed on Mg electrodes in THF solutions of **5** and that the magnesium deposited does not react with THF but remains electrochemically active. Moreover, the anodic process around the 1.5 V vs. Mg/Mg^{2+} reference electrode was considered[107] as related to the oxidation of Alk groups. On the other hand, there were indications of some adsorption/desorption processes and it was evidently shown that the electroreduction corresponds to equation 24a as well as the reduction of Alk_2Mg but not Mg^{2+} ion.

Most important for practical application as electrolytes for rechargeable batteries were the solutions in THF or polyethers of some of the complexes synthesized[96] by Aurbach and coworkers. These complexes are based on Mg organohaloaluminate salts, such as $Mg(AlCl_3R)_2$ (**59**) and $Mg(AlCl_2RR')_2$ (**60**), preferably **59a, 59b** and **60a**. The electrochemical deposition–dissolution of magnesium in these solutions is reversible with almost 100% efficiency, as is illustrated in Figure 5 for **60a** in THF. Moreover, it is evident from Figure 5 that the electrolyte decomposition of **60a** solution occurs at the most negative potential in comparison with other electrolytes, i.e. the anodic stability is the best, giving a potential range of more than 2.5 V in which the solution is inactive. Using these electrolytes, a magnesium anode and a $Mg_xMo_3S_4$ cathode with intercalated Mg ions, it was possible to develop the Mg battery system with promise for applications[96, 113].

In subsequent papers[109–111] Aurbach and coworkers extended an examination of the mechanism to the ethereal solutions of complexes of general formula $Mg(AX_{4-n}R_{n'}R'_{n''})_2$ (A = Al or B, X = halide, R, R′ = Alk or Ar and $n' + n'' = n$), which can be considered as products of the interaction between R′RMg Lewis bases and $AX_{3-n}R_{n'}R'_{n''}$ Lewis acids. The order of anodic stability found[110, 114] is **59a** > **60a** > **56** > **5g**.

$Mg(AlCl_3R)_2$ $Mg(AlCl_2RR')_2$

(**59**)

59	a	b
R	n-Bu	Et

(**60**)   (**60a**) R = n-Bu, R′ = Et

The morphology of Mg deposition on a gold electrode is strongly dependent on the composition of solutions, as indicated by *in situ* imaging using the STM method and by the high impedance measured. That behavior is caused by the adsorption of different species present in each solution in the course of deposition. Surface residuals of carbon, aluminium and chlorine were detected[115] by the XPS technique in THF solutions containing **60a**, but they are restricted to the outermost part of the surface, as physically adsorbed species. The mass balance of the Mg deposition–dissolution process is zero, whereas the cycling efficiency, describing the charge balance, depends on the solution composition and is

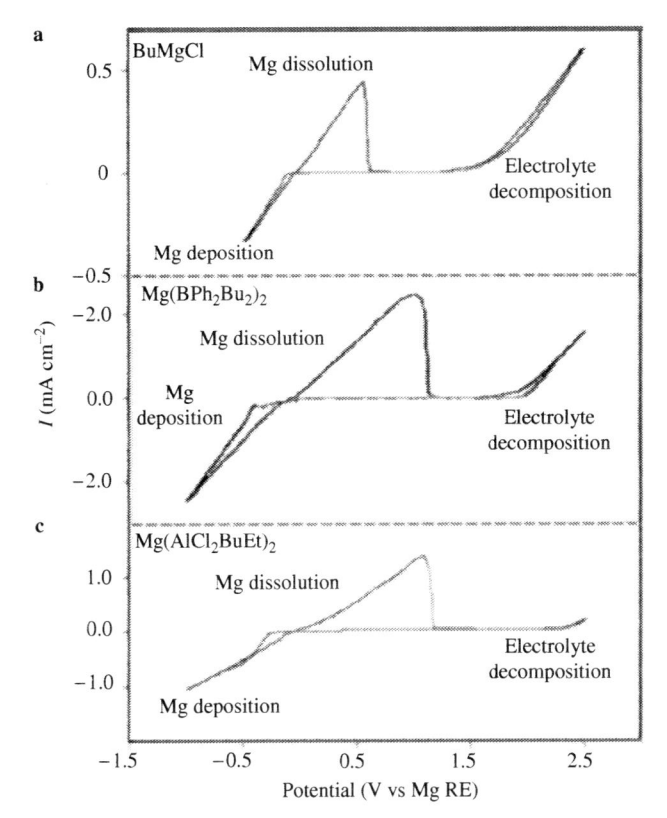

FIGURE 5. Cycling voltammograms recorded at 5 mV s^{-1} using a Pt working and Mg counter and reference electrodes in THF solutions containing: (a) 2 M n-BuMgCl (**5g**), (b) 0.25 M Mg(BPh$_2$Bu$_2$)$_2$ (**56**), (c) 0.25 M Mg(AlCl$_2$BuEt)$_2$ (**60a**). Reprinted by permission from Macmillan Publishers Ltd: *Nature*, **407**, 724, copyright 2000

higher in THF solutions of **5g** and **60a**. Taking into account the complicated structures of **5** as well as organomagnesium chloroaluminate complexes in THF solutions, it was assumed[110] that the electrochemical processes of **5** occur via electron transfer to cations such as RMg$^+$ or MgX$^+$ adsorbed at the electrode surface, as shown in Scheme 12 for the case of RMg$^+$ ions. For chloroaluminate complexes the equilibria (equations 30 and 31) were considered[110], but also the adsorption of other species, like Mg$_2$Cl$_3$(nTHF)$^+$ and Mg$_x$Cl$_y$R$_z$(nTHF)$^+$.

$$Mg(AlCl_{3-n}R_{n+1})_2 \rightleftharpoons MgR_2 + 2AlCl_{3-n}R_n \tag{30}$$

$$Mg(AlCl_{3-n}R_{n+1})_2 \rightleftharpoons (AlCl_{3-n}R_{n+1})^- + (AlCl_{3-n}R_{n+1})Mg^+ \tag{31}$$

In order to explain the results of EQCM and microscopy, it was postulated that in all the solutions studied the deposition occurs as a two-stage process. Initially, when the amount of charge involved is small (< 0.4 C cm^{-2}), a porous and irregular-in-shape magnesium deposit is formed. EDAX spectra of the electrode in a solution of **5g** revealed that this initial deposit contains, in addition to magnesium, also carbon and chlorine. This means

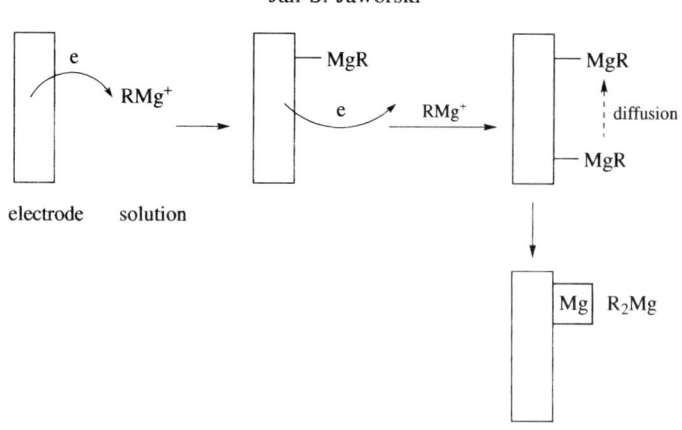

SCHEME 12

that traces of the electrolyte are trapped in the porous structure of the initial deposit. However, while the Mg deposition proceeds further from all kinds of solutions[110], **5g**, **60a** or **56**, the Mg layer becomes compact and crystalline, mostly composed of distorted, pyramid-shaped magnesium crystals. The pure magnesium deposition from the THF solution of **60a** was supported by XPS measurements[115]. The mass per mole of the electrons for this step is close to 12 g mol^{-1}, reflecting the deposition of pure magnesium from solutions of **5**[110]. The nature of the species adsorbed at electrodes was investigated *in situ* by FTIR spectroscopy, using an internal reflectance mode[111]. In those experiments the working electrode was prepared by plating first a platinum layer and next a gold layer on the spectroelectrochemical KBr window. FTIR spectra measured at open circuit voltage for gold electrodes on which magnesium was earlier deposited at -0.5 V vs. Mg/Mg^{2+}, in solutions containing **5g**, **5u** or BnMgCl (**5v**), showed new peaks around 550 cm^{-1} and 600–650 cm^{-1} which were attributed to Mg–C and Mg–Cl bonds, respectively. This was supported by a comparison with FTIR spectra obtained in transmittance mode for **5** and MgX$_2$ salts pelletized with KBr. Similar spectra obtained after deposition from **60a** and **59a** solutions also revealed bands in that range attributed to Mg–Cl and Mg–C bonds; for the last solution an additional peak at 466 cm^{-1} was attributed to Al–Cl bonds. For the solution of **56**, new peaks at 3000 cm^{-1} found after a deposition have higher wave numbers than those attributed to the υ_{C-H} mode of the phenyl groups. In conclusion, the adsorption of the following species on the Mg electrode surface was suggested[111]: RMg$^+$ or RMg$^{\bullet}$ for solutions of **5**, Mg$_x$Cl$_y^+$, e.g. Mg$_2$Cl$_3^+$ (nTHF), as well as the species with Al–Cl bonds [e.g. (AlCl$_{4-n}$R$_{n'}$R$'_{n''}$)Mg$^+$ (nTHF) etc.] and Mg–C bonds for complex solutions of Mg(Al$_{4-n}$R$_n$) and probably PhMg$^+$ and B(Ph$_2$Bu$_2$)Mg$^+$ for the solutions of **56**. All of the above species are stabilized by THF molecules coordinated to the Mg ions [i.e. RMg$^+$ (nTHF) exists instead of RMg$^+$], as was supported by bipolar IR peaks, similar but shifted to a lower wave number as compared with the peaks of the bulk THF molecules. Thus, a general mechanism of Mg deposition involves an adsorption of the cationic species to the electrode surface during the polarization to low potentials and then charge transfer to them followed by disproportionation of the adsorbed species in the radical state to form Mg deposits and solution species. The proposed mechanism for solutions of **5** is shown[110] in Scheme 12 (similar processes for RX$^+$ cations also take place[110]) and is described[111] by equations 32 or 33. On the other hand, the magnesium dissolution is represented[111] by equations 34. However, for some species in solutions the Schlenk equilibrium (equation 2) and the ionization equilibrium (equation 4) should be taken into account in the overall mechanism.

$$2RMg^+ + 2e \rightleftharpoons 2RMg^{\bullet}_{(ad)} \quad (32a)$$

$$2RMg^{\bullet}_{(ad)} \rightleftharpoons Mg + R_2Mg_{(sol)} \quad (32b)$$

$$2R_2Mg + 2e \rightleftharpoons 2RMg^{\bullet}_{(ad)} + 2R^- \quad (33a)$$

$$R^- + RMg^+ \longrightarrow R_2Mg_{(sol)} \quad (33b)$$

$$R^- + MgX_2 \longrightarrow RMgX_2^-{}_{(sol)} \quad (33c)$$

$$R^- + 2RMgX \longrightarrow R_2Mg_{(sol)} + RMgX_2^-{}_{(sol)} \quad (33d)$$

$$Mg + R_2Mg \rightleftharpoons 2e + 2RMg^+ \quad (34a)$$

$$Mg + 2RMgX_2^- \rightleftharpoons 2e + 2RMgX + MgX_2 \quad (34b)$$

The adsorbed cationic species are next transformed to the adsorbed radical species by the electron transfer. Finally, the Mg deposition may occur via the adsorbed radicals that disproportionate laterally on the electrode surface to form Mg metal and solution species. as represented, for example, by equations 35[111].

$$2(M \cdots MgR) \xrightarrow{\text{surface}} M \cdots Mg + R_2Mg_{(sol)} \quad (35a)$$

$$2(M \cdots MgAlCl_{4-n}R_{n'}'R_{n''}') \xrightarrow{\text{surface}} M \cdots Mg + Mg(AlCl_{4-n}R_{n'}'R_{n''}')_{2(sol)} \quad (35b)$$

Very high impedance (>10 kΩ cm^2) of reversible Mg electrode systems studied also indicated[112] adsorption processes. However, overpotentials of only several tens of millivolts are sufficient to break down the adsorbed layers, resulting in a much lower impedance (< 100 Ω cm^2) during the electrochemical processes.

The electrochemical behavior of a number of complexes with formal formulae Mg $(AX_{4-n}R_n)_2$ (where A = Al, B, Sb, P, As, Fe and Ta; X = Cl, Br and F; R = Bu, Et, Ph and Bn) in several solvents, including THF, Et$_2$O, diglyme and tetraglyme, was also investigated[114]. It was found that a highest decomposition potential >2.1 V vs. Mg/Mg^{2+} couple and a cycling efficiency close to 100% can be obtained for $(Bu_2Mg)_x(EtAlCl_2)_y$ complexes (considered as products of the reaction between a **1d** base and the EtAlCl$_2$ (**61**) Lewis acid) in THF or tetraglyme solutions. It was clearly demonstrated that electrochemical processes are influenced by both the nature of the Lewis acid–base systems and the solvent molecules. A series of experiments with precipitation of crystals and their redissolution in THF indicated that the structures of the solution species and of the crystals are different. The structure of the solution species is undefinable and should be described as a series of complicated equilibria, depending on the acid–base ratio, the nature of the solvent and other molecules in the solution, and temperature, in the same manner as in the solutions of **5**. The most interesting conclusion of Aurbach and coworkers[114] for this Chapter is that the electrochemically active cation includes more than one Mg ion and may have a general structure of $Mg_2R_{3-n}Cl_n^{+\bullet}$ solv, while the anion probably has the structure of $AlCl_{4-n}R_n$. In any case, the presence of the R group in the cation is crucial for reversible magnesium deposition. However, the concentration of **61** cannot be too high because a sufficient content of R groups is necessary for a reversible behavior. Thus, the optimal ratio of **61** to **1d** in solutions of interest was found[114] to be close to 2:1. Moreover, the determination of the anodic stability in THF solutions without Mg^{2+} ions, but containing 0.25 M AlEt$_3$Cl$^-$, AlEt$_2$Cl$_2^-$, AlEtCl$_3^-$ and AlCl$_4^-$, showed oxidation potentials equal to 2.0, 2.2, 2.5 and 2.6 V vs. Mg/Mg^{2+}, respectively[116]. The above result means that the anodic stability is mainly determined by the weakest Al−C bond, and not, as thought previously[114], by C−Mg bonds, as in **5**. The existence of different species in solutions, depending on the components ratio, was also suggested by nonmonotonous changes in the conductivity, κ, of 0.25 M solutions in THF against the acid-to-base ratio. The highest κ (1.6 mS cm^{-1} for R = Bu and 0.8 mS cm^{-1} for R = Et) was obtained for ratios of **1:61** equal to 0.5:1 and 2:1, whereas for a ratio of 1:1 a lowest $\kappa = 0.4$ mS cm^{-1} was observed[116] for both R groups. On the basis of ^{1}H, ^{13}C, ^{27}Al and ^{25}Mg NMR measurements (for **1d**) and in agreement with the above-mentioned conductivity data, the existence of major components in each solution was identified[116], as shown by equations 36a–c for complexes with acid-to-base ratio of 2:1, 1:1 and 1:2, respectively, hexacoordination of magnesium species was assumed in equations 36. However, it was not clear if the same species are responsible for magnesium electrodeposition and the authors rather suggested[116] that other, possibly organomagnesium, species undetected by NMR play a major role in the Mg deposition and dissolution from 2:1 complex solutions[116]. The last suggestion is in line with the observed increase in the rates of magnesium electrochemical processes after the addition of **1d** to 2:1 complex solutions.

$$2EtAlCl_2\bullet THF + Et_2Mg\bullet 4THF \longrightarrow Et_2ClAl\text{-}Cl\text{-}AlClEt_2^-$$
$$+ MgCl^+\bullet 5THF + THF \quad (36a)$$

$$EtAlCl_2\bullet THF + Et_2Mg\bullet 4THF \longrightarrow Et_3Al\bullet THF + MgCl_2\bullet 4THF \quad (36b)$$

$$EtAlCl_2\bullet THF + 2Et_2Mg\bullet THF \longrightarrow Et_4Al + EtMg^+\bullet 5THF + MgCl_2\bullet 4THF \quad (36c)$$

The kinetics of the Mg deposition on Pt microelectrodes in THF solutions of different compositions, starting from a complex containing only Cl ligands (MgCl$_2$ + AlCl$_3$) to the all-organic electrolyte **1d**, was analyzed[117] in detail on the basis of Tafel and Allen & Hickling plots. The exchange current density increased considerably as the ratio of organic ligand to Cl ligand increased and this acceleration of the Mg deposition was attributed to the change in proportions of the electroactive BuMg$^+$ and MgCl$^+$ species in solutions. However, the Faradaic efficiency is much higher for electrolytes containing chloride anions. For all the solution compositions studied, the transfer coefficients for the cathodic and anodic reactions at overpotentials close to 0 were 0.5 and 1.5, respectively, indicating two one-electron steps with the first one being the rds. On the other hand, Tafel slopes increase remarkably at higher overvoltages, giving finally the cathodic transfer coefficient of 0.09, which indicates a more complex reaction mechanism. The results obtained were interpreted in terms of a three-stage electrocrystallization mechanism, which describes the growth of the metallic deposits on the faces, steps and kinks of the substrate metal. The first stage, which is the formation of ad-atoms on the metallic faces, is the most interesting for the present Chapter. For the case of BuMg$^+$ as electroactive species in a solution, this stage involves fast diffusion to the metal face (equation 37a), slow adsorption (equation 37b), followed by a slow first electron transfer (equation 37c) yielding a radical BuMg$^\bullet$ on the metal surface and a fast electron transfer to it with the formation of magnesium ad-atom on the metal face (equation 37d). Note the difference in the second electron transfer from the electrode instead of the disproportionation between the adsorbed radicals (equation 32b), as considered previously[110,111]. However, at higher overvoltage the rate of electron transfer (equation 37c) increases and the adsorption of BuMg$^+$ becomes the rds, resulting in unusually high Tafel slopes. Finally, the reaction of Bu$^-$ with BuMg$^+$ yields **1d**, which is highly soluble in THF, contrary to insoluble MgCl$_2$. This may explain the acceleration of Mg deposition in all-organic electrolyte in comparison with the process in solutions containing only MgCl$_2$ and AlCl$_3$. In the last

6. Electrochemistry of organomagnesium compounds

solutions, the overall mechanism is similar but with $MgCl^+$ ions instead of $BuMg^+$ as the electroactive species. On the other hand, for the magnesium dissolution at small positive overpotentials the same steps in the opposite direction, from the process in equation 37d to 37a, was suggested[117] and the same electron transfer as the rds. However, at higher positive overpotentials the nonelectrochemical step in which the magnesium atom from the metallic lattice is placed near the kink, before the formation of Mg ad-atom, becomes the rds.

$$BuMg^+_{(sol)} \xrightarrow[\text{fast}]{\text{diffussion}} BuMg^+_{\text{(near Mg metal face)}} \quad (37a)$$

$$BuMg^+_{\text{(near Mg metal face)}} \xrightarrow[\text{slow}]{\text{adsorption}} BuMg^+_{\text{(surface)}} \quad (37b)$$

$$BuMg^+_{\text{(surface)}} \xrightarrow[\text{slow}]{+e} BuMg^\bullet_{\text{(surface)}} \quad (37c)$$

$$BuMg^\bullet_{\text{(surface)}} \xrightarrow[\text{fast}]{+e} Mg_{\text{(surface)}} + Bu^- \quad (37d)$$

It should be added that nonaqueous electrolytes for high-energy rechargeable electrochemical cells developed by Aurbach and coworkers were patented for use as solutions in organic solvents[118] or as gel-type solids[119].

Moreover, the magnesium deposition on silver substrate from 0.25 M solution of **60a** in THF is accompanied[120] by the formation of silver–magnesium alloy, which decreases the overpotential of deposition–dissolution processes and promotes the cycling efficiency.

VI. ORGANOMAGNESIUM COMPOUNDS AS INTERMEDIATES IN ELECTRODE REACTIONS

A number of electroorganic syntheses based on the use of sacrificial magnesium anodes have been described. However, in only a few reactions, listed below, was the formation of intermediate organomagnesium compounds postulated or documented.

Shono and coworkers[121] reported that magnesium electrodes promote cyclocoupling reactions of 1,3-dienes (**62**) with aliphatic carboxylic esters (**63**) resulting (Scheme 13) in the formation of cyclic products **64** with a five-membered ring. The electrode process was carried out in a single-compartment cell with a magnesium rod cathode and the same anode, which were alternated at an interval of 15 s. The electrolysis was carried out under nitrogen atmosphere in dry THF containing $LiClO_4$ and 5 Å molecular sieves. The electrodes from other metals (Pt, Al, Zn, Cu, Ni or Pb) did not yield **64**. The role of a sacrificial magnesium anode in the synthesis based on electrogenerated reagents was well recognized as the source of Mg^{2+} ions which stabilize intermediate organic anions (often called electrogenerated bases) forming with them coordination complexes or magnesium organic salts, like carboxylates or enolates; references can be found in reviews[122,123] and in a more recent paper[124]. However, Shono and coworkers[121] found the same product **64** if a solution of a diene (**62**) was reduced electrochemically at the Mg electrode and an ester (**63**) then added, after the current was terminated. The above result led the authors to suggest that a magnesium–diene compound (**65**) is formed during the electroreduction and not the Mg complex with an ester. It is known that organomagnesium compounds can be obtained in the reaction of 1,3-dienes with metallic magnesium, in particular, when highly reactive magnesium, prepared by the reduction of $MgCl_2$ in THF,

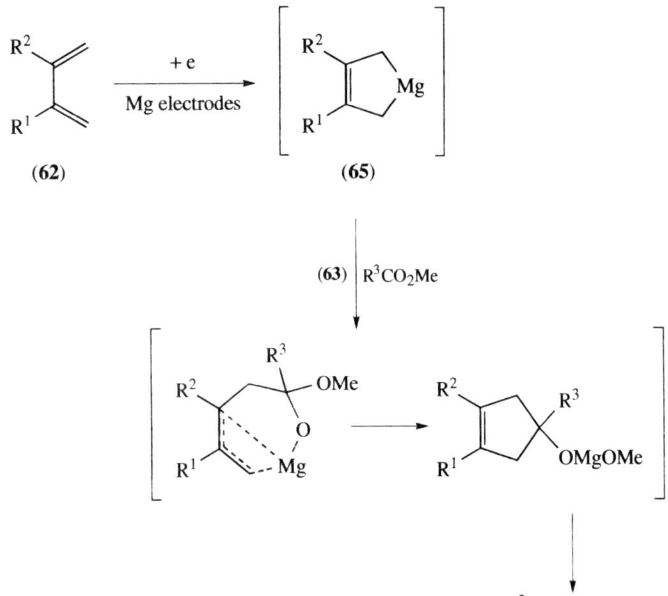

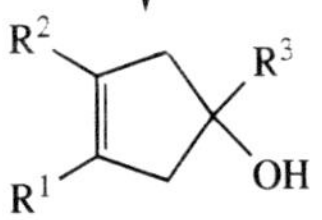

SCHEME 13

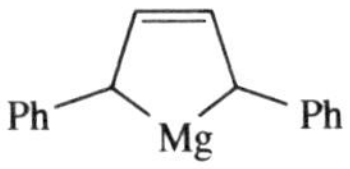

62 and 65		63	64
R^1	R^2	R^3	Yield
H	Me	n-Bu	76%
H	Me	i-Pr	71%
H	Me	$PhCH_2CH_2$	56%
H	$Me_2C=CHCH_2CH_2$	Et	63%
Me	Me	i-Pr	88%

is used[125]. A cyclic structure for **65** was proposed[121] by analogy to (1,4-diphenyl-2-butene-1,4-diyl)magnesium (**66**), the crystal structure of which was shown[126] to be a five-membered ring.

The cathodic cyclocoupling reaction (equation 38) with a diene under similar conditions was also observed[121] for 1-vinylcyclohexene (**67a**) and 1-vinylcycloheptene (**67b**), giving **68** as the final product with satisfactory yields. Thus, similar intermediate metallocycles can be expected for this reaction. On the other hand, for the cathodic coupling of styrenes the formation of a Mg compound was not proved[121], although a magnesium electrode was necessary in order to obtain a 2-phenylcyclopropanol-type product.

During electrochemical reduction of bicyclic 1,1-dibromocyclopropanes (**69a–69c**) in DMF or THF in the presence of chlorotrimethylsilane (**70**), using a stainless steel cathode and a sacrificial magnesium anode in an undivided cell, the replacement of one or both bromine atoms (depending on the amount of current passed) by a trimethylsilyl group was observed[127]. The reaction (equation 39) was stereoselective giving **71**, the *exo*-silyl bromide with the trialkylsilyl group in the *cis* position to hydrogen atoms at the ring junction, as the major product (and the *endo*-silyl compound **72** as the second product), in contrast to the reaction of **69** with n-BuLi and **70** which yield **72** as the major product. The by-product **73** was formed by a hydrogen atom abstraction from a solvent molecule.

A detailed mechanism was proposed for **69a** as an example[127]. The overall reaction (equation 40) finally produced the disilyl compound **74a** with a yield not much higher than 50%. However, unexpected results were obtained using ultrasonic irradiation during the electrolysis. The apparent current efficiency increased up to 200% giving, for example, 71% of disilane **74b** (and by-products) but no monosilane **71b** or **72b** after passage of 2.5 Faradays per mole of **69b**.

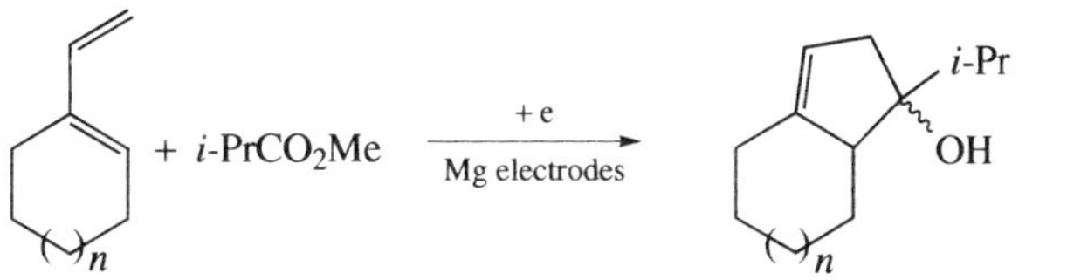

$$(67) + i\text{-PrCO}_2\text{Me} \xrightarrow[\text{Mg electrodes}]{+e} (68) \tag{38}$$

67 and 68	Yield
(a) $n=1$	62%
(b) $n=2$	72%

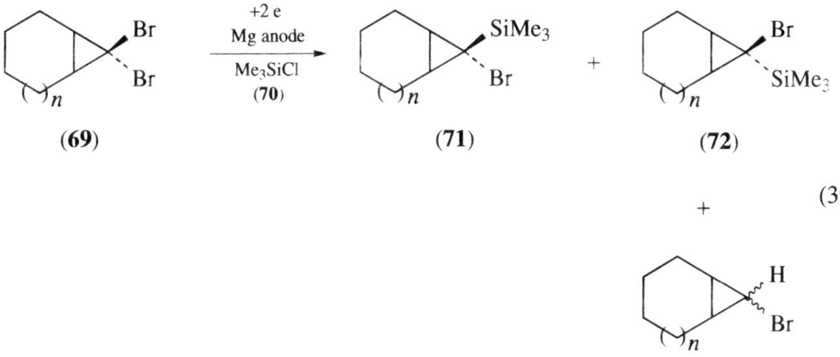

$$(69) \xrightarrow[\text{Me}_3\text{SiCl (70)}]{+2e, \text{ Mg anode}} (71) + (72) + (73) \tag{39}$$

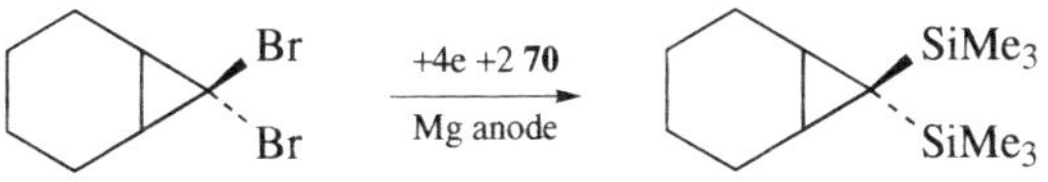

$$(69a) \xrightarrow[\text{Mg anode}]{+4e +2\ 70} (74a)$$

69, 70, 71, 72, 73 and 74

(a) $n=1$
(b) $n=2$ (40)
(c) $n=3$

Moreover, in the divided cell the *exo:endo* ratio of bromosilanes was 91:9 in the anode compartment but only 52:48 in the cathode compartment. Thus, the nature of the ultrasonic effect was explained assuming that beside the electrochemical silylation at the cathode, a parallel silylation process occurs at a magnesium anode, namely the silylation by **70** of an intermediate Grignard reagent produced from dibromide **69**. It appears as a rare example of the 'anodic reduction'[127]. However, the increase in the current density during electrolysis caused a decrease in the apparent current efficiency. This observation indicates a chemical nature of the anodic process. Of course, the ultrasonic irradiation facilitates the formation of the organomagnesium intermediate at the sacrificial anode[127] and the authors reported[128] a similar ultrasonic effect for the nonelectrochemical but purely sonochemical reaction in THF in the presence of a bulk magnesium rod; the latter reaction had given even higher yields and stereoselectivity.

Nucleophilic addition of diorganomagnesium compounds **1** with unusual substituents, generated by an indirect electrochemical method according to Scheme 1 in Section II, to the electrophiles benzaldehyde and its 4-substituted derivatives (**75**) or 2,2,2-trifluoroacetophenone (**76**) was reported by Lund and coworkers[28]. We recall that the electrochemical step involves only the formation of the magnesium salt, dimsyl$_2$Mg (**4**), and **1** is formed in the nonelectrochemical reaction of **4** with fluorene or its derivatives (**77**), acetophenone (**78**) and phenylacetonitrile (**79**). Nevertheless, overall reactions are carried out without isolation of intermediate compounds from the electrochemical cell. The overall processes[28] and isolated yields of the main products are shown in equations 41 and 42 and in Scheme 14. It should be emphasized that, in general, these products cannot be obtained in classical reactions. For example, the Grignard reagent with a cyano group, α to acidic hydrogens, necessary for the preparation of (Z)-2,3-diphenylacrylonitrile (**80**), cannot be generated in a direct reaction with magnesium. However, **80** was obtained with an isolated yield of 92% (equation 41). Similarly, 9-benzylidenefluorene (**81a**) was obtained via magnesium fluorenide (equation 42) with 89% yield, whereas the Grignard reagent, 9-fluorenylmagnesium bromide (**82**), cannot be obtained directly from Mg and 9-bromofluorene. If **82** was obtained indirectly, it gave **81a** as a by-product only[28].

75 and 81	a	b	c	d	e	f
Z	H	CN	OMe	NMe$_2$	NO$_2$	Br
Yield (%)	89	87	61	42	51	61

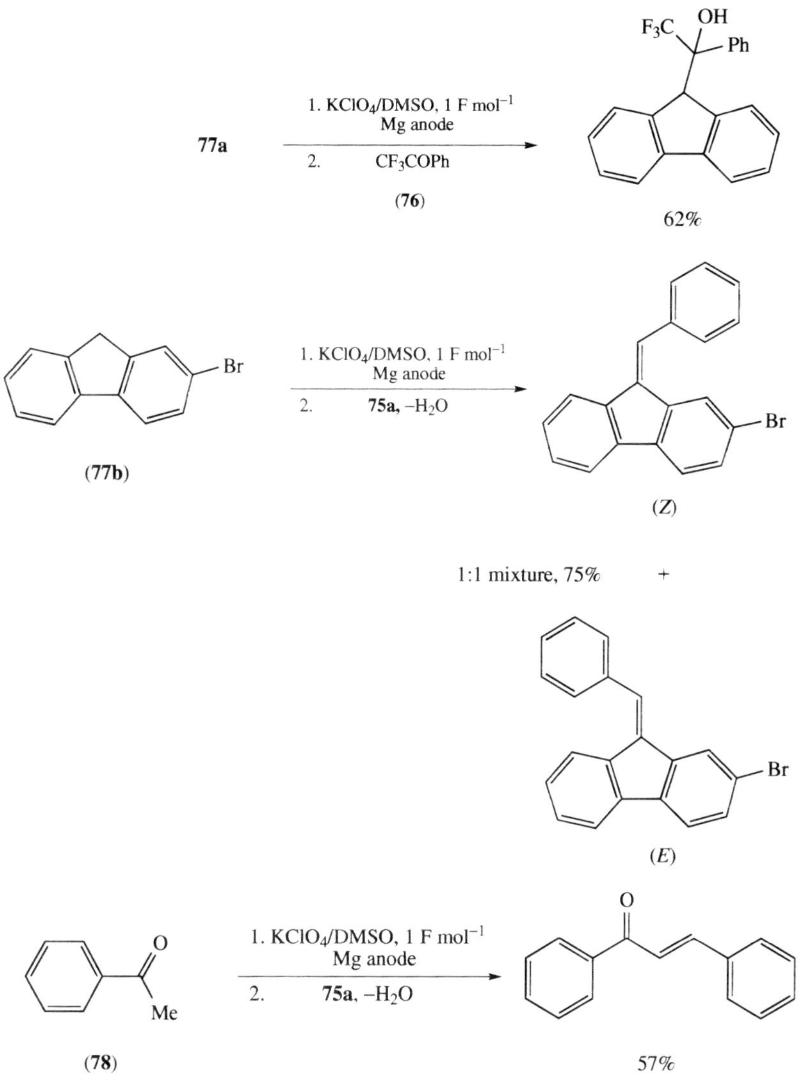

6. Electrochemistry of organomagnesium compounds

SCHEME 14

The electrochemical method of conversion *in situ* halogenated organic contaminants RX_n in wet soil formations or groundwaters to intermediate Grignard reagents, followed by their hydrolysis to RH_n and HOMgX [or $Mg(OH)_2$ produced in a competing reaction] was patented[129]. The electrical potential applied between magnesium and counter electrodes allows enhanced reactions between RX_n and magnesium metal and also allows one to clean magnesium surfaces from oxidation products after periodical reversal of the polarity of the potential.

VII. CONCLUDING REMARKS: ELECTROCHEMICAL DATA IN THE ELUCIDATION OF REACTIVITY OF GRIGNARD REAGENTS

It is evident from the review presented above that the most recent electrochemical investigations of organomagnesium compounds, like magnesium deposition or grafting on silicon surfaces, are far removed from the main topics of organic chemistry. On the other hand, progress in contemporary synthetic use of Grignard reagents does not include electrochemical methods. Nevertheless, for a detailed elucidation of the mechanisms of Grignard reactions and the formation of Grignard reagents by classical methods, the electrochemical approach can be helpful. A brief review of some of the problems in which electrochemical data were used in the chemistry of organomagnesium compounds is given below.

As was already mentioned, the constitution of reagents **5** in ethereal solutions and of the complexes formed by them with a number of Lewis acids was investigated by electrochemical methods[15, 88, 109–111] and the nature of existing ions and electroactive species was elucidated. It was also possible to evaluate the equilibrium constant for the Schlenk equilibrium for **5b**, **5e** and **5i** in DME solutions from polarographic measurements[36].

The basicity of a number of reagents **1**, **5** and related compounds in THF solutions was determined voltammetrically by Chevrot and coworkers[38, 130]. Using platinized platinum and hydrogen electrodes, they measured the cathodic–anodic current as a function of potential for overall electrode reactions (equations 43a and 43b)

$$\mathbf{5} + {}^1/_2 H_2 \underset{+e}{\overset{-e}{\rightleftarrows}} RH + MgX^+ \qquad (43a)$$

$$\mathbf{1} + {}^1/_2 H_2 \underset{+e}{\overset{-e}{\rightleftarrows}} RH + RMg^+ \qquad (43b)$$

and found that zero current potentials of the hydrogen electrode are equilibrium potentials, and thus the corresponding pH is a measure of the pK_a of organomagnesium compounds. A summary of pK_a values reported by Chevrot and coworkers in a number of papers and typical trends for different compounds were recently reviewed[131].

The reaction mechanisms of Grignard reactions, in particular those with ketones, were the subject of a long debate as a result of which a reactivity spectrum was ascertained between the radical mechanism with the single electron transfer (SET) from **5** to ketone, the classic polar mechanism and the concerted mechanism. A history of these findings was reviewed in 1996 by Blomberg[132]. Linear correlations between logarithms of homogeneous rate constants, $\log k$, and oxidation potentials of **5**, obtained from electrochemical measurements, were used as a strong support of the SET step or at least of 'a significant amount of radical character of the transition state'[52]. Such correlations were found, for example, for pseudo-first-order rate constants for the reaction of **5a**, **5b**, **5e**, **5i** and **5n** with di-*tert*-butyl peroxide in Et$_2$O[133] and for the reaction of AlkMgBr (cf. Table 8) with benzophenone (**83**)[52, 53] and azobenzene (**84**)[53] in Et$_2$O. Moreover, the kinetic data for Grignard reactions were also analyzed[53] in terms of the Marcus theory of electron transfer kinetics, assuming that SET is the rds and using the oxidation potentials of **5**, E_{ox}, in order to calculate the thermodynamic driving force. Marcus nonlinear plots of the activation barrier against the reaction free-energy change were reported[53] for reactions of a series of **5** with **83** and **84**. It was also found on the basis of the Marcus approach that the SET steps are feasible for the reaction of 5-hexenylmagnesium bromide with 3-phenylimido-2-phenyl-3*H*-indole and with 2-methoxy-1-nitronaphthalene in THF but are not feasible for the reaction of **5b** with pyrazine in ethers[56]. However, later investigations[134] on kinetic isotope effects in reactions of **83** with a series of reagents **5** showed different rds steps for different reagents **5**, although all the reactions followed the SET mechanism (a different mechanism for **5s** was invoked recently[135]). In conclusion, the correlations of $\log k$ and E_{ox} reported earlier were explained[134] rather as an indication of an electron transfer preequilibrium before the rds.

Nevertheless, redox potentials (evaluated in part by electrochemical methods) can be useful for support of the SET step and for predicting the formation of radical products. For example, it was reported[136,137] that when the oxidation potentials of carbanions ($E°_{R•/R^-}$, obtained in DMF+TBABF$_4$ and expressed vs. SCE) are compared with the reduction potentials of ketones, they may be helpful in predicting the reaction path of the Grignard reaction. For example, in reactions of **83** ($E° = -1.72$ V) with t-BuMgCl ($E°_{R•/R^-} = -1.77$ V) and s-BuMgCl ($E°_{R•/R^-} = -1.72$ V), radical products were obtained, whereas with **5k** ($E°_{R•/R^-} = -1.19$ V) and **5v** ($E°_{R•/R^-} = -1.40$ V) no radical products were found. On the other hand, in the reaction with fluorenone ($E° = -1.19$ V) all of the **5** mentioned above give the same products ratio as was found for reactions of the corresponding **8** with the electrogenerated radical anion of fluorenone.

In general, it is evident[61] that a SET process is governed by the difference in the oxidation potential of a nucleophile and the reduction potential of a partner reactant. The difference between these two potentials, $\Delta E = E_{ox} - E_{red}$, was used by Okubo and coworkers[138–142] to estimate the relative efficiency of SET (the so-called 'ΔE approach'). The distribution of products obtained in polar and radical routes for reactions between magnesium compounds, including ArMgBr, as well as ArSMgBr, ArNHMgBr, ArN(MgBr)$_2$ and ArOMgBr, with a number of carbonyl, nitro and cyano compounds, was correlated with ΔE even for multistep reactions.

Irreversible oxidation peaks of **5u**, PhMgCl and n-HexMgCl recorded at a Pt electrode in THF containing 0.1 M LiBr were found[143] in the same potential range (-2.5 to -1.5 V vs. Ag$^+$/Ag) as for the irreversible reduction peak of the Cl-terminated Si(111) surface (-2.5 V). This result evidently shows that **5** can reduce chlorine bonds on the surface, and thus a SET step participates in the alkylation, as included in the mechanism proposed[143].

For the unusual reactivity of ferrocenylsilanes toward **5u** in THF, affording ketones instead of the expected tertiary alcohols, a mechanism was proposed[144] including the inner-sphere electron transfer from **5u** within a reactant complex. The proposition was based on an electrochemical CV examination, which indicated that the outer-sphere process is thermodynamically unfavorable.

Finally, it may be noted that the formation of Grignard reagents from organic halides (**8**) and metallic magnesium is a heterogeneous reaction and starts by a SET from magnesium to **8**, as is now commonly accepted[145]. However, many aspects of this reaction are not clear, in particular those connected with its surface nature[146,147]. Similarities to electrochemical reduction of **8** were considered in order to explain the reaction mechanism. Logarithms of relative rate constants for reactions of Mg with a series of substituted bromobenzenes in Et$_2$O and Bu$_2$O–C$_6$H$_{12}$ mixture[148] and with a series of alkyl chlorides in Et$_2$O[149] were correlated with electrochemical $E_{1/2}$ values obtained in DMF for the reduction of the corresponding **8** at a Hg or a glassy carbon (GC) electrode. However, the real sense of these correlations is not straightforward, taking into account the different electrode mechanisms for both series under examination, as is now well documented[58]. Moreover, the absolute rate constants for the formation of **5** reported recently[146] showed much smaller variations with the nature of **8** and the lack of correlations with $E_{1/2}$ values, indicating that the SET step is not the rds. The last conclusion was also supported[146] by free energies of activation, determined for the same reactions, which were substantially smaller than the literature intrinsic activation barriers for the dissociative electron transfer to **8**.

The electrochemical reduction at GC electrodes in ACN of a number of organic bromides (mainly with cyclopropyl systems), used as radical clocks in reactions of the formation of **5**, was applied to support the concerted electron transfer and cleavage of a carbon–halogen bond in both types of processes[150] and this conclusion looks quite justified.

On the other hand, the results concerning the heterogeneous nature of the formation of **5** should be interpreted with special care, in particular, in comparison with rearranged products obtained from the radical clock reactions under homogeneous conditions. Recent experiments[147] comparing the behavior of potassium and magnesium in THF and Et_2O solutions containing the precursor of the aryl radical clock, 1-bromo-2-(3-butenyl)benzene, evidently indicate that the reactions with both metals are comparable to the heterogeneous electron transfer occurring at a cathode, whereas a similar reaction of potassium in the presence of crown ethers corresponds to homogeneous SET processes observed in redox catalysis. In conclusion, it was emphasized[147] that certain unclear problems in the formation of **5**, like a hypothesis of the participation of dianions, can be most probably resolved by treating the dissolution of a metal in these reactions in a similar approach to the one recently developed for elementary steps occurring at electrodes.

VIII. REFERENCES

1. C. K. Mann and K. Barnes, *Electrochemical Reactions in Nonaqueous Systems*, Chap. 13.1, Dekker, New York, 1970.
2. H. Lehmkuhl, in *Organic Electrochemistry. An Introduction and a Guide* (Ed. M. M. Baizer), Chap. 18, Dekker, New York, 1973, pp. 621–678.
3. M. D. Morris, in *Electroanalytical Chemistry* (Ed. A. J. Bard), Vol. VII, Dekker, New York, 1974, pp. 79–160.
4. S. G. Mairanovskii, *Russ. Chem. Rev.*, **45**, 298 (1976).
5. M. D. Morris and G. L. Kok, in *Encyclopedia of Electrochemistry of the Elements* (Eds. A. J. Bard and H. Lund), Vol. XIII, Chap. XIII-1, Dekker, New York, 1979, pp. 41–76.
6. D. A. White, in *Organic Electrochemistry. An Introduction and a Guide* (Eds. M. M. Baizer and H. Lund), 2nd ed., Chap. 19, Dekker, New York, 1983, pp. 591–636.
7. A. P. Tomilov, I. N. Chernyh and Yu. M. Kargin, *Elektrokhimiya elementoorganicheskih soiedinienij*, Nauka, Moscow, 1985.
8. L. Walder, in *Organic Electrochemistry. An Introduction and a Guide* (Eds. H. Lund and M. M. Baizer), 3rd ed., Dekker, New York, 1991, pp. 809–875.
9. J. Yoshida and S. Suga, in *Organic Electrochemistry* (Eds. H. Lund and O. Hammerich), 4th ed., Chap. 20, Dekker, New York, 2002, pp. 765–794.
10. P. Jolibois, *Compt. Rend.*, **155**, 353 (1912).
11. P. Jolibois, *Compt. Rend.*, **156**, 712 (1913).
12. J. M. Nelson and W. V. Evans, *J. Am. Chem. Soc.*, **39**, 82 (1917).
13. N. W. Kondyrew, *Chem. Ber.*, **58**, 459 (1925).
14. L. W. Gaddum and H. E. French, *J. Am. Chem. Soc.*, **49**, 1295 (1927).
15. W. V. Evans and R. Pearson, *J. Am. Chem. Soc.*, **64**, 2865 (1942).
16. W. V. Evans and F. H. Lee, *J. Am. Chem. Soc.*, **55**, 1474 (1933).
17. W. V. Evans and F. H. Lee, *J. Am. Chem. Soc.*, **56**, 654 (1934).
18. W. V. Evans, F. H. Lee and C. H. Lee, *J. Am. Chem. Soc.*, **57**, 489 (1935).
19. W. V. Evans and E. Field, *J. Am. Chem. Soc.*, **58**, 720 (1936).
20. W. V. Evans and E. Field, *J. Am. Chem. Soc.*, **58**, 2284 (1936).
21. W. V. Evans and D. Braithwaite, *J. Am. Chem. Soc.*, **61**, 898 (1939).
22. W. V. Evans, D. Braithwaite and E. Field, *J. Am. Chem. Soc.*, **62**, 534 (1940).
23. W. V. Evans, R. Pearson and D. Braithwaite, *J. Am. Chem. Soc.*, **63**, 2574 (1941).
24. K. Ziegler and H. Lehmkuhl, German Patent to Ziegler, DE1212085 (1966).
25. K. Ziegler and H. Lehmkuhl, US Patent to Ziegler, US2985568 (1961).
26. P. Kobetz and R. C. Pinkerton, US Patent to Ethyl Corp, US3028319 (1962).
27. K. Ziegler and H. Lehmkuhl, US Patent to Ziegler, US3306836 (1967).
28. H. Lund, H. Svith, S. U. Pedersen and K. Daasbjerg, *Electrochim. Acta*, **51**, 655 (2005).
29. P. C. Hayes, A. Osman, N. Seudeal and D. G. Tuck, *J. Organomet. Chem.*, **291**, 1 (1985).

30. N. W. Kondyrew and D. P. Manojew, *Chem. Ber.*, **58**, 464 (1925).
31. N. W. Kondyrew and A. K. Ssusi, *Chem. Ber.*, **62B**, 1856 (1929).
32. N. W. Kondyrew and A. I. Zhel'vis, *J. Gen. Chem. USSR.*, **4**, 203 (1934); *Chem. Abstr.*, **29**, 25 (1935).
33. W. Strohmeier, *Z. Elektrochem.*, **60**, 396 (1956).
34. W. Strohmeier and F. Seifert, *Z. Elektrochem.*, **63**, 683 (1959).
35. C. Liebenow, Z. Yang and P. Lobitz, *Electrochem. Comm.*, **2**, 641 (2000).
36. T. Psarras and R. E. Dessy, *J. Am. Chem. Soc.*, **88**, 5132 (1966).
37. M. Fleischmann, D. Pletcher and C. J. Vance, *J. Organomet. Chem.*, **40**, 1 (1972).
38. C. Chevrot, J. C. Folest, M. Troupel and J. Périchon, *J. Electroanal. Chem.*, **54**, 135 (1974).
39. W. Schlenk and W. Schlenk, Jr., *Chem. Ber.*, **62B**, 920 (1929).
40. K. C. Cannon and G. R. Krow, in *Handbook of Grignard Reagents* (Eds. G. S. Silverman and P. E. Rakita), Chap. 13, Dekker, New York, 1996, pp. 271–290.
41. V. Gutmann, *The Donor–Acceptor Approach to Molecular Interactions*, Plenum Press, New York, 1978.
42. J. Linsk and E. A. Mayerle, US Patent to Standard Oil Co, US3155602 (1964).
43. A. P. Giraitis, T. H. Pearson and R. C. Pinkerton, US Patent to Ethyl Corp, US2960450 (1960).
44. A. P. Giraitis, US Patent to Ethyl Corp, US2944948 (1960).
45. C. E. Thurston and K. A. Kobe, *Philippine J. Sci.*, **65**, 139 (1938); *Chem. Abstr.*, **32**, 6956 (1938).
46. L. Martinot, *Bull. Soc. Chim. Belg.*, **75**, 711 (1966).
47. J. L. Morgat and R. Pallaud, *C. R. Acad. Sci. C (Paris)*, **260**, 574 (1965).
48. J. L. Morgat and R. Pallaud, *C. R. Acad. Sci. C (Paris)*, **260**, 5579 (1965).
49. R. T. Dufford, D. Nightingale and L. W. Gaddum, *J. Am. Chem. Soc.*, **49**, 1858 (1927).
50. T. H. Bremer and H. Friedman, *Bull. Soc. Chim. Belg.*, **63**, 415 (1954).
51. G. A. Tolstikov, R. G. Bulgakov and V. P. Kazakov, *Russ. Chem. Rev.*, **54**, 1058 (1985).
52. T. Holm, *Acta Chem. Scand.*, **B28**, 809 (1974).
53. T. Holm, *Acta Chem. Scand.*, **B37**, 567 (1983).
54. T. Holm, *J. Chem. Soc., Perkin Trans. 2*, 464 (1981).
55. T. Holm, *Acta Chem. Scand.*, **B42**, 685 (1988).
56. L. Eberson, *Acta Chem. Scand.*, **B38**, 439 (1984).
57. J. M. Savèant, *J. Am. Chem. Soc.*, **109**, 6788 (1987).
58. J. M. Savèant, *Elements of Molecular and Biomolecular Electrochemistry*, Chap. 3, Wiley-Interscience, Hoboken, 2006, pp. 182–250.
59. L. Martinot, *Bull. Soc. Chim. Belg.*, **76**, 617 (1967).
60. C. Chevrot, M. Troupel, J. C. Folest and J. Périchon, *C. R. Acad. Sci. C (Paris)*, **273**, 493 (1971).
61. L. Eberson, *Electron Transfer Reactions in Organic Chemistry*, Chap. IV, Springer-Verlag, Berlin, 1987, pp. 39–66.
62. H. Schäfer and H. Küntzel, *Tetrahedron Lett.*, 3333 (1970).
63. H. E. French and M. Drane, *J. Am. Chem. Soc.*, **52**, 4904 (1930).
64. D. G. Braithwaite, US Patent to Nalco Chemical Co, US3007857 (1961).
65. D. G. Braithwaite, US Patent to Nalco Chemical Co, US3007858 (1961).
66. D. G. Braithwaite, US Patent to Nalco Chemical Co, US3234112 (1966).
67. D. G. Braithwaite, US Patent to Nalco Chemical Co, US3256161 (1966).
68. D. G. Braithwaite, US Patent to Nalco Chemical Co, US3312605 (1967).
69. D. G. Braithwaite, US Patent to Nalco Chemical Co, US3391066 (1968).
70. D. G. Braithwaite, US Patent to Nalco Chemical Co, US3391067 (1968).
71. J. W. Ryznar and J. G. Premo, US Patent to Nalco Chemical Co, US3100181 (1963).
72. W. P. Hettinger, Jr., US Patent to Nalco Chemical Co, US3079311 (1963).
73. J. Linsk, US Patent to Standard Oil Co, US3118825 (1964).
74. J. Linsk, W. C. Ralph and E. Field, US Patent to Standard Oil Co, US3164537 (1965).
75. J. Linsk, US Patent to Standard Oil Co, US3298939 (1967).
76. G. C. Robinson, US Patent to Ethyl Corp, US3522156 (1970).
77. G. Calingaert and H. Shapiro, US Patent to Ethyl Corp, US2535193 (1950).
78. D. E. Danly, in *Organic Electrochemistry. An Introduction and a Guide* (Eds. M. M. Baizer and H. Lund), 2nd ed., Chap. 30, Dekker, New York, 1983, pp. 959–994.

79. T. Dubois, F. Ozanam and J.-N. Chazalviel, *Electrochem. Soc. Proc.*, **97-7**, 296 (1997).
80. A. Fidélis, F. Ozanam and J.-N. Chazalviel, *Surf. Sci.*, **444**, L7 (2000).
81. J.-N. Chazalviel, S. Fellah and F. Ozanam, *J. Electroanal. Chem.*, **524-525**, 137 (2002).
82. A. Teyssot, A. Fidélis, S. Fellah, F. Ozanam and J.-N. Chazalviel, *Electrochim. Acta*, **47**, 2565 (2002).
83. S. Fellah, A. Teyssot, F. Ozanam, J.-N. Chazalviel, J. Vigneron and A. Etcheberry, *Langmuir*, **18**, 5851 (2002).
84. L. J. Webb and N. S. Lewis, *J. Phys. Chem. B*, **107**, 5404 (2003).
85. S. Fellah, R. Boukherroub, F. Ozanam and J.-N. Chazalviel, *Langmuir*, **20**, 6359 (2004).
86. N. Seidel, K. Jacob, A. K. Fischer, C. Pietzsch, P. Zanello and M. Fontani, *Eur. J. Inorg. Chem.*, 145 (2001).
87. N. W. Kondyrew, *J. Russ. Phys. Chem. Soc.*, **60**, 545 (1928); *Chem. Abstr.*, **23**, 1321 (1929).
88. R. E. Dessy and G. S. Handler, *J. Am. Chem. Soc.*, **80**, 5824 (1958).
89. R. E. Dessy, W. Kitching, T. Psarras, R. Salinger, A. Chen and T. Chivers, *J. Am. Chem. Soc.*, **88**, 460 (1966).
90. D. M. Overcash and F. C. Mathers, *Trans. Electrochem. Soc.*, **64**, 305 (1933).
91. E. Findl, M. A. Ahmadi and K. Lui, US Patent to Xerox Corp., US3520780 (1970).
92. J. Eckert and K. Gneupel, GDR Patent to Technische Hochschule 'C. Schorlemmer', DD 243722 (1987).
93. A. Mayer, US Patent to US Dept. of Energy, US4778575 (1988).
94. A. Mayer, *J. Electrochem. Soc.*, **137**, 2806 (1990).
95. P. Novák, R. Imhof and O. Haas, *Electrochim. Acta*, **45**, 351 (1999).
96. D. Aurbach, Z. Lu, A. Schechter, Y. Gofer, H. Gizbar, R. Turgeman, Y. Cohen, M. Moshkovich and E. Levi, *Nature*, **407**, 724 (2000).
97. H. Göhr and A. Seiler, *Chem.-Ing.-Tech.*, **42**, 196 (1970).
98. J. D. Genders and D. Pletcher, *J. Electroanal. Chem.*, **199**, 93 (1986).
99. A. Brenner, *J. Electrochem. Soc.*, **118**, 99 (1971).
100. T. D. Gregory, R. J. Hoffman and R. C. Winterton, *J. Electrochem. Soc.*, **137**, 775 (1990).
101. R. J. Hoffman, R. C. Winterton and T. D. Gregory, US Patent to Dow Chemical Company, US4894302 (1990).
102. C. Liebenow, *J. Appl. Electrochem.*, **27**, 221 (1997).
103. C. Liebenow, Z. Yang and P. Lobitz, *Bull. Electrochem.*, **15**, 424 (1999).
104. C. Liebenow, *Electrochim. Acta*, **43**, 1253 (1998).
105. C. Liebenow, *Solid State Ionics*, **136-137**, 1211 (2000).
106. O. Chusid, Y. Gofer, H. Gizbar, Y. Vestfrid, E. Levi, D. Aurbach and I. Riech, *Adv. Mater.*, **15**, 627 (2003).
107. Z. Lu, A. Schechter, M. Moshkovich and D. Aurbach, *J. Electroanal. Chem.*, **466**, 203 (1999).
108. D. Aurbach, M. Moshkovich, A. Schechter and R. Turgeman, *Electrochem. Solid State Lett.*, **3**, 31 (2000).
109. D. Aurbach, Y. Cohen and M. Moshkovich, *Electrochem. Solid State Lett.*, **4**, A113 (2001).
110. D. Aurbach, A. Schechter, M. Moshkovich and Y. Cohen, *J. Electrochem. Soc.*, **148**, A1004 (2001).
111. D. Aurbach, R. Turgeman, O. Chusid and Y. Gofer, *Electrochem. Comm.*, **3**, 252 (2001).
112. D. Aurbach, Y. Gofer, A. Schechter, O. Chusid, H. Gizbar, Y. Cohen, M. Moshkovich and R. Turgeman, *J. Power Sources*, **97-98**, 269 (2001).
113. D. Aurbach, Y. Gofer, Z. Lu, A. Schechter, O. Chusid, H. Gizbar, Y. Cohen, V. Ashkenazi, M. Moshkovich, R. Turgeman and E. Levi, *J. Power Sources*, **97-98**, 28 (2001).
114. D. Aurbach, H. Gizbar, A. Schechter, O. Chusid, H. E. Gottlieb, Y. Gofer and I. Goldberg, *J. Electrochem. Soc.*, **149**, A115 (2002).
115. Y. Gofer, R. Turgeman, H. Cohen and D. Aurbach, *Langmuir*, **19**, 2344 (2003).
116. H. Gizbar, Y. Vestfrid, O. Chusid, Y. Gofer, H. E. Gottlieb, V. Marks and D. Aurbach, *Organometallics*, **23**, 3826 (2004).
117. Yu. Viestfrid, M. D. Levi, Y. Gofer and D. Aurbach, *J. Electroanal. Chem.*, **576**, 183 (2005).
118. D. Aurbach, Y. Gofer, A. Schechter, L. Zhonghua and C. Gizbar, US Patent to Bar Ilan University, US6316141 (2001).
119. D. Aurbach, O. Chasid, Y. Gofer and C. Gizbar, US Patent to Bar Ilan University, US6713212 (2004).
120. Z. Feng, Y. NuLi, J. Wang and J. Yang, *J. Electrochem. Soc.*, **153**, C689 (2006).

121. T. Shono, M. Ishifune, H. Kinugasa and S. Kashimura, *J. Org. Chem.*, **57**, 5561 (1992).
122. J. Simonet and J.-F. Pilard, in *Organic Electrochemistry* (Eds. H. Lund and O. Hammerich), 4th ed., Chap. 29, Dekker, New York, 2002, pp. 1163–1225.
123. J. H. P. Utley and M. F. Nielsen, in *Organic Electrochemistry* (Eds. H. Lund and O. Hammerich), 4th ed., Chap. 30, Dekker, New York, 2002, pp. 1227–1257.
124. R. Yee, J. Mallory, J. D. Parrish, G. L. Carroll and R. D. Little, *J. Electroanal. Chem.*, **593**, 69 (2006).
125. R. D. Rieke and H. Xiong, *J. Org. Chem.*, **56**, 3109 (1991).
126. Y. Kai, N. Kanehisa, K. Miki, N. Kasai, K. Mashima, H. Yasuda and A. Nakamura, *Chem. Lett.*, 1277 (1982).
127. A. J. Fry and J. Touster, *Electrochim. Acta*, **42**, 2057 (1997).
128. J. Touster and A. J. Fry, *Tetrahedron Lett.*, **38**, 6553 (1997).
129. P. A. Rock, W. H. Casey and R. B. Miller, US Patent to Regents of the University of California, US6004451 (1999).
130. C. Chevrot, K. Kham, J. Périchon and J. F. Fauvarque, *J. Organomet. Chem.*, **161**, 139 (1978).
131. W. Kosar, in *Handbook of Grignard Reagents* (Eds. G. S. Silverman and P. E. Rakita), Chap. 23, Dekker, New York, 1996, pp. 441–453.
132. C. Blomberg, in *Handbook of Grignard Reagents* (Eds. G. S. Silverman and P. E. Rakita), Chap. 11, Dekker, New York, 1996, pp. 219–248.
133. W. A. Nugent, F. Bertini and J. K. Kochi, *J. Am. Chem. Soc.*, **96**, 4945 (1974).
134. H. Yamataka, T. Matsuyama and T. Hanafusa, *J. Am. Chem. Soc.*, **111**, 4912 (1989).
135. T. Holm, *J. Org. Chem.*, **65**, 1188 (2000).
136. H. Lund, K. Daasbjerg, T. Lund, D. Occhialini and S. U. Pedersen, *Acta Chem. Scand.*, **51**, 135 (1997).
137. H. Lund, K. Skov, S. U. Pedersen, T. Lund and K. Daasbjerg, *Collect. Czech. Chem. Commun.*, **65**, 829 (2000).
138. M. Okubo, T. Tsutsumi, A. Ichimura and T. Kitagawa, *Bull. Chem. Soc. Jpn.*, **57**, 2679 (1984).
139. M. Okubo, T. Tsutsumi and K. Matsuo, *Bull. Chem. Soc. Jpn.*, **60**, 2085 (1987).
140. M. Okubo, K. Matsuo, N. Tsurusaki, K. Niwaki and M. Tanaka, *J. Phys. Org. Chem.*, **6**, 509 (1993).
141. K. Matsuo, R. Shiraki and M. Okubo, *J. Phys. Org. Chem.*, **7**, 567 (1994).
142. M. Okubo and K. Matsuo, *Nippon Kagaku Kaishi*, 1 (2000); *Chem. Abstr.*, **132**, 200217 (2000).
143. E. J. Nemanick, P. T. Hurley, B. S. Brunschwig and N. S. Lewis, *J. Phys. Chem. B*, **110**, 14800 (2006).
144. A. Alberti, M. Benaglia, B. F. Bonini, M. Fochi, D. Macciantelli, M. Marcaccio, F. Paolucci and S. Roffia, *J. Phys. Org. Chem.*, **17**, 1084 (2004).
145. C. Hamdouchi and H. M. Walborsky, in *Handbook of Grignard Reagents* (Eds. G. S. Silverman and P. E. Rakita), Chap. 10, Dekker, New York, 1996, pp. 145–218.
146. B. J. Beals, Z. I. Bello, K. P. Cuddihy, E. M. Healy, S. E. Koon-Church, J. M. Owens, C. E. Teerlinck and W. J. Bowyer, *J. Phys. Chem. A*, **106**, 498 (2002).
147. H. Hazimeh, J.-M. Mattalia, C. Marchi-Delapierre, R. Barone, N. S. Nudelman and M. Chanon, *J. Phys. Org. Chem.*, **18**, 1145 (2005).
148. H. R. Rogers, R. J. Rogers, H. L. Mitchell and G. M. Whitesides, *J. Am. Chem. Soc.*, **102**, 231 (1980).
149. J. J. Barber and G. M. Whitesides, *J. Am. Chem. Soc.*, **102**, 239 (1980).
150. H. M. Walborsky and C. Hamdouchi, *J. Am. Chem. Soc.*, **115**, 6406 (1993).

CHAPTER 7

Analytical aspects of organomagnesium compounds

JACOB ZABICKY

Department of Chemical Engineering, Ben-Gurion University of the Negev, P. O. Box 653, Beer-Sheva 84105, Israel
Fax: +972 8 6472969; e-mail: zabicky@bgu.ac.il

I. ACRONYMS	266
II. INTRODUCTION AND SCOPE OF THE CHAPTER	267
III. ELEMENTAL ANALYSIS OF MAGNESIUM	269
A. Introduction	269
B. Sample Preparation	271
1. Matrix obliteration	271
2. Preconcentration	272
C. Column Separation Methods	273
1. Ion chromatography	273
2. High-performance liquid chromatography	274
3. Electrophoresis	274
D. Electrochemical Methods	275
1. Ion-selective electrodes	275
2. Electroanalytical determination	276
E. Spectral Methods	277
1. Atomic absorption spectrometry	277
2. Atomic emission spectrometry	278
3. Ultraviolet-visible spectrophotometry and colorimetry	279
4. Ultraviolet-visible fluorometry	283
5. Chromatic chemosensors	285
6. Nuclear magnetic resonance spectroscopy	286
7. Mass spectrometry	287
IV. SPECIATION ANALYSIS OF ORGANOMAGNESIUM COMPOUNDS	288
A. Titration Methods	288
1. Visual endpoint	288
2. Potentiometric and other instrumental titrations	291
B. Chromatographic Methods	292

The chemistry of organomagnesium compounds
Edited by Z. Rappoport and I. Marek © 2008 John Wiley & Sons, Ltd

1. Liquid chromatography	292
2. Gas chromatography	293
C. Spectral Methods	295
1. Infrared spectroscopy	295
2. Gilman's color tests	295
3. Nuclear magnetic resonance spectroscopy	296
4. Electron spin resonance spectroscopy	299
D. Cryoscopy	299
V. GRIGNARD REAGENTS AS ANALYTICAL REAGENTS AND AIDS	299
A. Active Hydrogen	299
B. Derivatization Reagents	300
1. Gas chromatography of organometallic compounds	300
2. Analysis of glycerides and waxes	301
C. Ancillary Applications	301
1. Surface conditioning	301
2. Electrochemical behavior	305
VI. REFERENCES	306

I. ACRONYMS

AAS	atomic absorption spectroscopy/spectrometry
AED	atomic emission detection/detector
AES	atomic emission spectroscopy/spectrometry
ANN	artificial neural network
AOAC	Association of Official Analytical Chemists
ASTM	American Society for Testing and Materials
CE	capillary electrophoresis
CPE	carbon paste electrode
CRM	certified reference material
CZE	capillary zone electrophoresis
DA	diode array
DCTA	1,2-diaminocyclohexylidenetetraacetic acid
DIN	direct injection nebulizer
DRIFTS	diffuse reflectance infrared Fourier transform spectroscopy
EDTA	ethylenediaminetetraacetic acid
EGTA	ethylene glycol bis(β-aminoethyl ether)-N,N,N',N'-tetraacetic acid
EI	electron impact
ESR	electron spin resonance
ETAAS	electrothermal AAS
EXSY	2D ^{1}H–^{1}H exchange spectroscopy
FAAS	flame AAS
FAES	flame AES
FFGD	fast flow glow discharge
FIA	flow injection analysis
FLD	fluorescence detection/detector
FPD	flame photometric detection/detector
GCE	glassy carbon electrode
GFAAS	graphite furnace AAS
GPC	gel permeation chromatography
HMPT	hexamethylphosphoric triamide
IC	ion chromatography

ICP	inductively coupled plasma	
ISE	ion-selective electrode(s)	
LLE	liquid–liquid extraction	
LOD	limit(s) of detection	
LOQ	limit(s) of quantitation	
LSCSV	linear sweep cathodic strip voltametry	
LTA	low temperature ashing	
MIP	microwave-induced plasma	
MSD	mass spectrometric detection/detector	
NCI	negative-ion chemical ionization	
NMR	nuclear magnetic resonance	
PCI	positive-ion chemical ionization	
PDVB	polystyrene-divinylbenzene	
PEBBLE	probe encapsulated by biologically localized embedding	
PEO	poly(ethylene oxide)	
PLS	partial least squares	
PQCD	piezoelectric quartz crystal detection/detector	
RP	reversed phase	
RSD	relative standard deviation	
SDS	sodium dodecylsulfate	
SEFT	spin-echo Fourier transform	
SFE	supercritical fluid extraction	
SIA	sequential injection analysis	
SIM	selected ion monitoring	
SNR	signal to noise ratio	
SSCE	silver–silver chloride electrode	
SWAdSV	square wave adsorptive stripping voltametry	
TAG	triacylglycerol	
TCD	thermal conductivity detection/detector	
THF	tetrahydrofuran	
TPPI	time proportional phase increment	
UVD	UVV detection/detector	
UVV	ultraviolet-visible	

II. INTRODUCTION AND SCOPE OF THE CHAPTER

Research related to magnesium belongs to various groups of substances: (i) The metal, its alloys and its intermetallic compounds, all outside the scope of this chapter; (ii) combinations of Mg(II) cations with inorganic and organic anions, without or with additional ligands, to which reference will be made in Section III; and (iii) the organomagnesium compounds, containing at least one C–Mg σ-bond, dealt with in Sections IV and V. Minor types of Mg compounds may be incorporated into one of these main groups.

The earlier research period of the inorganic and metallorganic compounds of group (ii), extending from the 18th century to the first quarter of the 20th century, consists mainly of the development of chemical and pharmacological knowledge. The most relevant research efforts on Mg compounds in the modern period, extending from 1926 to the present, relate to biological and biomedical subjects; they began with recognition of the essential character of this element and were followed with development of our knowledge of the physiological, epidemiological and clinical aspects[1]. The bibliography mentioned in Table 1 will help to appreciate the enormous research effort invested in this field; it is a small selection of reviews published in the last quarter of a century, dealing with biological and biomedical subjects related to magnesium. As for the organomagnesium compounds

TABLE 1. Selection of reviews on biological and biomedical subjects related to magnesium[a]

General subjects	Specialized areas and bibliography
Analytical issues	(i) X-ray elemental microanalysis[12,13]. (ii) Ion-selective electrodes for clinical use[14–21]. (iii) Electron probe and electron energy loss analysis[22]. (iv) Intracellular measurements[21,23–27]. (v) Determination of Mg in human tissues and fluids[21,25,28–34]. (vi) Trace elements in hair[35]. (vii) Determination of Ca and Mg in wines[36].
Biological issues	(i) Mg bioavailability, metabolism and physiology[21,37–57]. (ii) Cell proliferation and differentiation[21,58]. (iii) Animal husbandry[59]. (iv) Magnesium in blood[60–63]. (v) Genetic regulation[61,64–67]. (vi) Mineral phase composition of bone and teeth[68–70]. (vii) Brain and nervous system[21,71–73]. (viii) Renal handling of magnesium[21,74,75].
Biomedical issues	(i) General clinical analysis[b 3,20,21,31,38,39,41,49,54,76–90]. (ii) Blood conditions[c 21,63,91–96]. (iii) Cardiovascular diseases[21,62,84,97–111]. (iv) Kidney diseases[b 13,64–66,112–117]. (v) Lung diseases[118]. (vi) Mental diseases[119]. (vii) Nutrition[b 21,57,70,112,120–130]. (viii) Gynecology and obstetrics[131,132]. (ix) Pediatrics[42,67,70,116,124,133–135]. (x) Geriatrics[136–139].
Pharmacological issues	(i) Renal handling of magnesium[74,140,141]. (ii) Metabolic effects of diuretics[142]. (iii) Myocardial infraction[143]. (iv) Hypomagnesemia[144]. (v) Central nervous system injury[145].

[a] References were picked up among more than 700 reviews and belong to the period from 1980 to 2006.
[b] References 85, 112 and 126 belong to veterinary medicine.
[c] Reference 93 is about short-term space flights in Shuttle and Skylab.

of group (iii), their study began in the second half of the 19th century. However, the most significant research started after 1900, when Victor Grignard synthesized the alkylmagnesium halides (the Grignard reagents), of utmost importance in synthetic organic chemistry[2]. Chemical research of organomagnesium compounds continues to the present in the direction of multinuclear or functionalized Grignard reagents, mainly to extend their capabilities as synthons, as shown in many reviews[3–11].

Elemental analysis is an important feature of organic analysis. In the case of organometallics, determination of Mg usually involves a mineralization step, by which an inorganic salt of Mg(II) is obtained before proceeding to the end analysis. Methods for determination of Mg(II) have long been established. However, advancements in analytical science of Mg are still made for determination of Mg(II) related to the subjects listed in Table 1, because of the low LOD required and the difficulties of speciation of this ion in complex biological matrices. Methods for Mg(II) analysis are presented in Section III.

The classical methods for the analysis of organomagnesium compounds in general and of Grignard reagents in particular were developed in the first half of the 20th century[146]. However, some advances took place more recently, with the appearance of new instrumentation, especially for the various chromatographic modalities. The analytical speciation and quantitative analysis of organomagnesium compounds are discussed in Section IV. The compounds addressed in that section are mostly uninuclear. Although analytical speciation of multinuclear compounds such as the dinuclear MeMgMgF and MeMgFMg, the trinuclear MeMgFMg$_2$[147] and **1** or the tetranuclear **2**[148] is usually outside the scope of the chapter, brief consideration is given to some multinuclear compounds to illustrate the application of various analytical techniques.

(1)   (2)

Grignard reagents may be used in the determination of other analytes and as ancillary agents for various analytical applications. The use of Grignard reagents in analysis is presented in Section V.

III. ELEMENTAL ANALYSIS OF MAGNESIUM

A. Introduction

The most characteristic element of organomagnesium compounds usually is magnesium and its analysis may afford important information about the identity and the quality of the sampled material. In contrast to speciation analysis, which may require delicate handling and involves sample preparation procedures to preserve the analyte or its identity features, elemental analysis of Mg frequently requires destructive processes leading to mineralization of the element to form a salt. Although Mg elemental analysis for quality control usually requires sample dilution before the end analysis, the present section will also deal with methods for trace and ultratrace analysis, as sometimes required for samples of biological and environmental origin. The sample preparation processes will preferably refer to this type of materials, as they are more akin to those required for organometallic compounds. Reviews appeared on the various steps of determination of Mg in biological materials, from sampling to end analysis[149, 150].

Speciation of Mg in certain complex matrices refers to distinguishing among Mg bound to different fractions of the matrix, as opposed to identification of definite magnesium compounds; according to IUPAC's recommendations, the latter distinction should be referred to as speciation analysis[151]. An important speciation case is that of Mg in plasma, for which various values may be distinguished: (i) total Mg, (ii) Mg strongly bound in metalloproteins, (iii) Mg weakly bound to proteins, (iv) Mg bound to low molecular mass anions, such as amino acids, carboxylates, carbonate, ascorbate, salicylate, etc., and (v) free hydrated Mg ions. The set of these values is controlled by the pH, temperature, ionic strength and concentration of other metal ions. All or part of these values are of clinical relevance and pertinent analytical methods will be discussed below. It was proposed to save the labor and expense involved in such speciation analyses, making instead estimations by applying the artificial neural network (ANN) methodology. During the ANN training phase pH, $[Mg]_{total}$ and $[Mg(II)]_{free}$ data were used, which were determined for the plasma of patients. Cross-validation of the ANN method for a given set of pH and $[Mg(II)]_{total}$ values showed an average error of 8% for estimated $[Mg(II)]_{free}$

values[152]. A method was proposed for determination of classes (ii) to (v) in plasma and serum, combining ISE determination of $[Mg(II)]_{free}$ (Section III.D.1), AAS determination of $[Mg(II)]_{total}$ (Section III.E.1) and, after contolled ultrafiltration, AAS determinations of $[Mg(II)]_{protein}$ retained on the filter and of $([Mg(II)]_{free} + [Mg(II)]_{complexed})$ in the filtrate[153]. Reviews appeared on the determination of the total contents and activity of Ca(II) and Mg(II) in serum[154] and on the physiological and clinical aspects of magnesium in human beings, paying attention to the principal analytical methods applied for determining the various magnesium species[153, 155]. See also pertinent reviews listed among the analytical issues in Table 1.

Approximately 1 mol of Mg is found in the adult human body, equally distributed among bones and soft tissues. Only about 0.3% of the total body Mg is present in serum, yet most of the analytical results are for this body fluid. Speciation of Mg for an individual is difficult, for the lack of fast and accurate assays for intracellular magnesium, but determination of total and free magnesium in tissues and physiological tests may give helpful information[29, 77]. Mg(II) is the most abundant divalent cation within cells, followed by Ca(II) by far; it is the fourth most abundant cation in the body, after Na(I), K(I) and Ca(II), and is the second most common cation in the intracellular free fluid, after Na(I). Magnesium alone or bound to proteins is essential for many cellular functions, for example, acting as cofactor for hundreds of enzymatic reactions, and being required for protein and nucleic acid synthesis, signal transduction, energy metabolism, maintenance of cytoskeletal and mitochondrial integrity, and the modulation of various ion transport pumps, carriers and channels[156]. In plasma Mg(II), as well as Ca(II), is found in three forms: (i) an ultrafiltrable fraction consisting of free Mg(II) (70 to 80%), (ii) complex-bound Mg(II) (1 to 2%) and (iii) a protein-bound non-ultrafiltrable fraction (20 to 30%). Free Mg(II) is an important parameter in clinical analysis. The reference range for total Mg concentration in adult blood plasma is 0.65 to 1.05 mM and 0.55 to 0.75 mM for free Mg (usually determined with ISE, Section III.D.1); the range for total Mg in erythrocytes is 1.65 to 2.65 mM[155]. The Mg content in the food intake affects the level of certain trace (e.g. Pb, Rb, Sr) and ultratrace (e.g. As, Au, Ba, Ir, Mo, Se, Ta) elements in plasma, of which As, Mo, Rb, Pb and Se have already been shown to be essential. The concentration of As, Au, Ir, Rb, Sr and Ta was significantly higher in the plasma of rats fed with a low-Mg diet than in the control group fed with normal Mg levels[157].

Methods for determining magnesium in body fluids fall into several major categories: (i) complexometric titration, (ii) atomic absorption spectrometry, (iii) atomic emission spectrometry, (iv) fluorometry and (v) various spectrophotometric techniques, which include enzymatic and dye binding methods[158]. The presence of heparin solution in sampling syringes, used to avoid coagulation of blood, may introduce a significant negative bias in the determination of Ca(II) and Mg(II)[159].

Another important speciation refers to Mg bound to RNA, which is essential to the folding and function of this macromolecule. A computational approach to this analysis was presented for site-bound and diffusively bound Mg(II) ions in RNA. This method confirmed the locations of experimentally determined sites and pointed to potentially important sites not currently annotated as Mg binding sites but deserving experimental follow-up in that direction[160].

Mg speciation can also be applied to olives and olive oil: (i) Total Mg can be determined by AAS, after mineralization of a sample by nitric acid digestion; (ii) covalently bound Mg is extracted with CCl_4 from homogenized olives and is determined after evaporating the solvent and dissolving the residue in AcBu-i; (iii) extraction of homogenized olives with $CHCl_3$ leads to chlorophyll-bound Mg; and (iv) extraction of homogenized olives with water is also performed. Four fractions can be defined for Mg in the water extract, which are obtained following a definite experimental procedure: (iv-a) Mg in particulate matter, passing a paper filter but not a 0.45 μm polymeric filter; (iv-b) polyphenol-bound Mg; (iv-c) polysaccharide-bound Mg; and (iv-d) free cationic Mg. End analysis of Mg in all these fractions is carried out by FAAS or ETAAS, after dissolving in an adequate solvent[161].

B. Sample Preparation

1. Matrix obliteration

An easy method for eliminating the organic and volatile components of the matrix is dry ashing. The sample should be initially dried in air or in an oven to avoid losses by sputtering, and be placed in a furnace at temperatures sufficiently high to burn the organic matter and sufficiently low to avoid volatilizing of analytes. Heating to the final calcination temperature should be gradual to avoid losses by kindling the sample. The ashes are dissolved in dilute HCl or HNO_3 and submitted to end analysis. Sometimes the ashes contain remnants of the organic matrix, appearing as carbonate, sulfate, phosphate or silicate anions, which may interfere with the analysis of Mg by certain methods. For example, milk fermentation samples were dried, calcined in a furnace at 600 °C, the ash was dissolved in 0.03 M HCl, the solution was centrifuged and the supernatant was analyzed[162]. Determination of Ca, Cu, Fe, K, Mg, Mn, Na and Zn in foodstuffs involves placing the oven-dried sample in a furnace at 200 °C and raising the temperature in 50 °C steps every time the sample stops fuming, and finally it is left for 16 h at 450 °C. End analysis of Ca and Mg was by FAAS, after dissolving the ashes in dilute HCl and adding $LaCl_3$[163].

Dry ashing is unfit for volatile analytes or when it is necessary to preserve the mineral matter structure. In such cases the organic parts of the matrix can be eliminated by low temperature ashing (LTA), the main variant of which consists of exposing the sample to a MIP in a low-pressure atmosphere of pure oxygen. The slow combustion of the organic matter develops temperatures below 200 °C and possibly below 150 °C. For example, LTA followed by microwave-aided extraction of the analytes with aqua regia was applied for determination of As and Hg in coal, coal fly ash and slag[164]. Lower ashing temperatures (60 to 70 °C) may be achieved on developing the MIP in oxygen diluted with helium[165].

Wet mineralization is also frequently applied for elimination of organic matter present in the matrix and is recommended in many official methods[166]. It is mainly based on the oxidizing action of HNO_3 alone or with additives aiming at reinforcing the oxidative action, such as H_2O_2, increasing the temperature of the process, such as H_2SO_4 and $HClO_4$, or achieving special effects, such as HF for volatilization of Si. Various examples follow of wet mineralization prior to end analysis.

The details for the mineralization of infant milk formulas with HNO_3–$HClO_4$ for the analysis of major and trace elements are given in AOAC Official Method 984.27[166a]. A household microwave oven on-line with a flow system was proposed for mineralization of urine in the presence of 1 M HNO_3. The off-line end analysis of Ca and Mg was by FAAS with good results[167]. Mineralization of milk samples for Mg analysis by FAAS was carried out by three methods: (i) wet digestion of a sample after adding thrice the volume of a 4:1 mixture of concentrated HNO_3 and $HClO_4$, heating for 65 min at 120 °C, cooling and adding water to a 25-fold dilution of the sample; (ii) placing in an acid digestion bomb 0.2 mL of sample and 2.5 mL of 1 N HNO_3 and digesting for 90 s in a microwave oven at full power, cooling to room temperature for 1 h and adding water to 125-fold dilution of the sample; and (iii) drying of the sample at 80 to 100 °C for 4 h, calcinating at 550 °C for 5 h, dissolving the white ash in some 1 N HNO_3 and adding water to a 25-fold dilution of the sample and end analysis by FAAS[168]. Total hair digestion with HNO_3–H_2O_2 is hard to accomplish. When reaching a pale brown stage any fatty residue can be emulsified on addition of a surfactant. After adding La(III) in agar solution the end analysis of Ca and Mg by FAAS was consistent with that obtained by the dry ashing method[169].

A comparison study was made of ultrasonic extraction and microwave digestion of plant material for FAAS determination of Mg, Mn and Zn. The optimal conditions for recovery of these elements by ultrasound assisted extraction were: 3 min sonication, at 30% ultrasonic amplitude, of 0.1 g of material milled to <50 μm particle size, dispersed in 5 mL of 0.3% (m/v) HCl solution; after 4 min of centrifugation at 4500 rpm, the supernatant was used for FAAS end analysis. For the microwave digestion, 0.1 g of sample was placed in 5 mL of 69.5% (m/m) HNO_3 and 0.5 mL of 48% (m/m) HF; the mixture was heated in the microwave oven for 1 min at 40 psi, 1 min at 80 psi and 5 min at 120 psi; after cooling the digestion vessel with ice, the solution was evaporated to dryness, the residue was dissolved in 37% (m/m) HCl and diluted with water to 5 mL for end analysis. Analytical results for both methods were similar. The advantages of ultrasound-assisted extraction over microwave-assisted digestion are: (i) much shorter processing time, (ii) simpler proceedings, (iii) lower consumption of reagents and (iv) better safety as no harsh acids, heating or high pressure operations are involved. Some disadvantages are: (i) lower amounts of sample can be processed, (ii) milling to <50 μm particle size is required and (iii) aging of the ultrasonic probe surface may reduce the extraction efficiency[170].

Determination of Mg(II) in a cellulose matrix requires dissolution of the matrix in 70% H_2SO_4, dilution, neutralization with NaOH, and addition of $CaCl_2$ and K_2CO_3 solutions in that order. A quatitative coprecipitation of Mg(II) takes place with Ca(II) and CO_3^{2-} ions. The double salt is filtered, washed with water and dissolved in dilute HCl for end analysis by FAAS[171]. A comparison was made of methods for mineralization of plant material. Dry ashing in a furnace at 470 °C followed by ash dissolution in 5 M HCl is preferable to digestion in a 9:2:1 by volume mixture of HNO_3–H_2SO_4–$HClO_4$, followed by dilution. The wet method seems to be problematic because of $CaSO_4$ formation; however, this may be corrected by leaving the dilute digest overnight, when slow dissolution takes place. End analysis of the metallic elements is by AAS[172]. If N and P also are to be determined from the mineralized sample, then digestion in the Kjeldahl fashion is called for, using concentrated H_2SO_4 containing ca 1% w/v Se[173].

Sometimes the nature of the solid matrix allows sample preparation without mineralization. In the determination of Cu and Mg in polyethylene, the sample was milled to a fine powder and suspended in a solution of a detergent (Triton X-100) in EtOH. A reference solution of the same viscosity was prepared by adding ethylene glycol instead of polymer to the supporting solution. The end analysis by FAAS was carried out using Sr(II) as releasing agent and n-BuOH as enhancement reagent[174].

2. Preconcentration

Problems attaining the GFAAS determination of ppb levels of Ca, Mg, Sr and Si in saturated NaCl brines for electrolytic chloroalkali cells were discussed; Mg is the most problematic among these analytes because the boiling points of $MgCl_2$ and NaCl are very near to each other and special handling procedures are needed for improving the analytical quality. A general problem concerning analysis at ppb levels is avoiding contamination of the sample and instrumentation and some recommendations were given in this regard[175]. Traces of Ca(II) and Mg(II) present in salt were concentrated by passing a solution of the sample through an ion-exchange resin containing chelating groups. End analysis after elution was by FAAS[176,177].

An interesting method for preconcentration of ultratrace amounts of Mg(II) in water consists of treating a PTFE tube with 2 M NaOH solution, at 70 °C for 3 h. On passing the weakly alkaline sample of water through the tube, Mg(II) is adsorbed on the wall. The analyte is recovered with dilute HCl for end analysis, e.g. by fluorometry (Section III.E.4)[178].

C. Column Separation Methods

1. Ion chromatography

Modern techniques for ion chromatography (IC) allow simultaneous determination of anions and cations in the same run and with a unique detector. One possibility for attaining such results is introducing into the eluting solution chelating agents such as EDTA or DCTA, which are totally ionized at the high pH of the solution (e.g. $NaHCO_3$–Na_2CO_3 buffer), and form negatively charged chelated ions with the cationic analytes, such as **3a**, **3b**, **4a** and **4b**, respectively. The concentration of the chelates may be determined by UVD. A frequently used detection method in IC is conductivity measurement (CND); however, certain eluting solutions are not appropriate for this. Such solutions can be simplified by the use of suppression membranes, allowing easier determination by CND. If the suppression membrane is a cation exchanger in acidic form, it collects Na(I) ions and transfers them to a regenerating solution supplying H^+ ions to the analysis solution[179]. Some technical problems and their solutions were discussed regarding the use of suppression membranes in the determination of the four major cations found in human plasma[180]. Also, the **3a** and **3b** chelates are decomposed and the conductivity of the free EDTA is measured together with that of other anions present in the analytical sample. LOD in the ppb range were achieved for Ca(II) and Mg(II), with precisions better than 1%[181]. This approach varied from another separation technique where gradient elution was applied to accelerate the analysis[182]. An alternative to CND is piezoelectric detection with a PQCD, which is responsive to conductivity and permittivity of a solution; however, one of its advantages is being free from errors stemming from a double electric layer or Faraday impedance. IC-PQCD was applied for determination of Ca(II) and Mg(II) in saliva and urine, which were passed through a 0.45 μm filter membrane. The chromatographic column was a cation exchange resin on PDVB copolymer support and the mobile phase was an aqueous solution at pH 4.0 of 4.0 mM tartaric acid and 2.0 mM ethylenediamine. The LOD (SNR = 3) were 0.4 and 0.2 ppm, with linear behavior in the ranges from 0.8 to 500 ppm and from 1.0 to 500 ppm for Ca(II) and Mg(II), respectively[183].

(3) (a) M = Mg
(b) M = Ca

(4) (a) M = Mg
(b) M = Ca

Magnesium speciation (Section III.A) in serum was carried out using an anion exchange column for protein separation, with mobile phase at pH 7.4; the effluent was collected in an automatic fraction collector. On-line quantitation of the protein fractions was carried out by DA-UVD, and Mg determination was carried out from the automatic sampler in a GFAAS apparatus, measuring at 202.8 nm[184].

2. High-performance liquid chromatography

Oxine (**5**) forms complexes of analytical applicability with various metal ions[185]. A RP-HPLC-FLD method ($\lambda_{ex} = 370$ nm, $\lambda_{fl} = 516$ nm) was proposed for simultaneous determination of Al(III) and Mg(II), using a C_{18} column. Various details of the method are noteworthy: Optimization of the method showed that for both ions it is best to have also precolumn and in-column complex formation, caused by the presence of **5** in the injection loop and in the carrier solution; FLD detection is preferable to simple UVD because it avoids the background of **5** and interference of various ions forming nonfluorescent chromogenic complexes, e.g. Ca(II) and Zn(II); the intensity of the fluorescence can be increased by micelle formation on addition of SDS and neutralized N,N-bis(2-hydroxyethyl)-2-aminoethanesulfonic acid (**6**). The LOD (SNR = 3) were 0.74 μM (18 ppb) Mg(II) and 0.60 μM (16 ppb) Al(III); the latter was attributed in part to residual impurities in the purified water[186, 187].

OH (HOCH$_2$CH$_2$)$_2$NCH$_2$CH$_2$SO$_3$H

(**5**)   (**6**)

3. Electrophoresis

After denaturation of the protein in plasma with trichloroacetic acid and centrifugation, the concentration of Na(I), K(I), Mg(II) and Ca(II) in the supernatant was determined by CZE with DA-UVD. The background electrolyte was an aqueous solution containing 20 mM imidazole (**7**), 0.5 mM oxalic acid (**8**) and 5% (v/v) MeOH, brought to pH 2.8 with 0.1 M HCl; Cd(II) served as internal standard. Separations were carried out by hydrodynamic injection at the anodic side, on a 50 μm capillary coated with polyvinyl alcohol, in positive mode, applying a constant 30 kV potential. The cations were detected at 214 nm (λ_{max} of **7**). The LOD (SNR = 3), LOQ (SNR = 10) and linearity range, in ppm, for the ions in their order of emergence were: K(I) 0.25, 0.75, 0.75–50, Ca(II) 0.50, 0.90, 0.9–50, Na(I) 1.00, 4.00, 4–400 and Mg(II) 0.20, 0.50, 0.5–50[188].

(**7**)   (**8**)

At slightly acidic pH values weak dibasic acids H$_2$L give on dissociation anions HL$^-$, forming ion pairs MHL with metal ions. These ion pairs are neutral for M(I), which is the case of Na(I), K(I) and ammonium ions, and electrophoretically mobile for M(II), such as Ca(II) and Mg(II). A chromophore BH/B consisting of a weak base B, which at slightly acidic pH values is in equilibrium with its conjugate acid BH, also has electrophoretic mobility due to the latter ion and may serve for indirect UVD of the M(II) ions. These principles have been applied as a CE method for determination of trace concentrations of Ca(II) and Mg(II) in aqueous solutions containing more than 5000-fold concentrations of Na(I). Two systems of weak acid ligand/weak base chromophore proved especially efficient under the particular conditions of a developed CE assay: 2 mM oxalic acid (**8**)/10 mM creatinine (**9**) at pH 4.6 and 8 mM tartaric acid (**10**)/14.4 mM benzylamine at pH 4.8; in both cases UVD was carried out at 214 nm. A LOD of 4 µM was achieved for a simulated matrix containing 500 ppm Na(I)[189].

(9)   (10)

D. Electrochemical Methods

1. Ion-selective electrodes

Electrically neutral magnesium ionophores should fulfill the following requirements for their use in ISE: (i) Ionophores should be lipophilic to assure their longevity and stable response on the electrode. (ii) The complex formed with the main analyte cation should be stable, however, not too stable to avoid emulation of classical anion exchangers; a lipophilic anion should be present on the membrane to induce cation permeability and to reduce anion interference. (iii) The electromotive force developed by the ion-selective electrode depends on the selectivity coefficients of the analyte *vs* the other cations present and the activity of all the cations present in the sample. Most Mg-selective electrodes are designed for analysis of biological fluids, where the most abundant cations are Na(I), K(I), Ca(II) and Mg(II) (Section III.A). Selectivity toward the univalent cations is usually high, however Ca(II) may interfere in the determination of Mg(II). The following were found among ionophores with a selectivity for Mg(II) over Ca(II) of at least one order of magnitude: ETH 5220 (**11a**), ETH 4030 (**11b**), ETH 7025 (**12**) and ETH 3832 (**13**)[190]. Ion-selective electrodes incorporating in the membrane **11a**[191, 192] or **12**[193] can be used for potentiometric determination of Mg(II); these electrodes are used in clinical ionic analyzers, for determination of free Mg(II) in blood and its derived fractions.

$$RR'NCOCH_2CONH(CH_2)_8NHCOCH_2CONRR'$$
(**11**) (a) R = H, R′ = C$_8$17-*n*
(b) R = Me, R′ = C$_7$H$_{15}$-*n*

n-C$_7$H$_{15}$N(CH$_3$)COCH$_2$CON[(CH$_2$)$_8$NHCOCH$_2$CON(CH$_3$)C$_7$H$_{15}$-*n*]$_2$
(**12**)

1,3,5-[(CH$_2$)$_5$NHCOCH$_2$CON(CH$_3$)C$_7$H$_{15}$-*n*]$_3$C$_6$H$_3$
(**13**)

Instruments are offered in the market for clinical determination of electrolytes in blood, plasma or serum. One of them, for example, carries out simultaneous determinations of Na, K, Ca, Mg, hematocrit and pH. The cations are of the free type (see Section III.A) and are measured with specific ion-selective electrodes. In complex matrices such as blood or its derived fractions the concentration of free Ca and Mg is affected by the pH of the solution, for example, a slight change of pH will produce or neutralize anionic sites in the proteins, binding or releasing these cations; furthermore, the response of the Mg-selective electrode is also affected by the concentration of free Ca(II). The correction for the concentration of cation M determined at pH x, is made for pH 7.4, a standard value for blood, applying equation 1, and the correction for the interference of Ca(II) on the Mg-selective electrode is made using the selectivity constant K_{MgCa}, based on calibration measurements[194-196]. Application of Mg ISE in clinical practice and research has been reviewed[9, 14-18].

$$\log[M]_{pH\ 7.4} = \log[M]_{pH\ x} - 0.24(7.4 - x) \tag{1}$$

2. Electroanalytical determination

Direct electroanalytical determination of Mg(II) ions is of little practical value because of the interference by hydrogen, aluminum and alkali earth metal ions[197]. To avoid these and other difficulties, an indirect method was proposed based on the voltametric determination of sodium pentothal (**14**). The voltametric (LSCSV) determination of **14** is carried out in two steps in a phosphate buffer solution at pH ca 10.5, using a hanging mercury drop electrode and an auxiliary Pt wire electrode. In the preconcentration (deposition) step at -0.1 V the thiolate ion of **14** is attached to mercury cations generated $in\ situ$ according to equation 2, whereas in the stripping step ending at -0.8 V the Hg(II) ions are reduced to Hg^0, according to equation 3. The LOD is about 10 ppb for 180 s deposition time, and no interference is observed for equimolar concentrations of Ca(II), Cr(VI), Cu(II), Fe(III), Ni(II), Pb(II) or Zn(II). On the other hand, addition of 2 μM of Mg(II) to the same concentration of **14** caused about a fourfold peak current increase[198]. The latter behavior of Mg(II) was the basis for another method for determination of this cation, after introducing some modifications in the method used for **14**. Thus, instead of a hanging Hg drop, a mercury film was developed on a CPE, and the electroanalytical technique was SWAdSV, using sodium phosphate buffer at pH 10.75. A calibration is necessary for the increment in the peak cathodic current, ΔI_p, measured on adding Mg(II) to the solution of **14**, which is a function of the cation concentration and the deposition time. The LOD is 0.14 ppb Mg(II) for 60 s deposition time, with RSD 0.5% ($n = 5$). The method was applied to analysis of urine and tap water, and the results were in good agreement with FAAS determinations[199].

(RSNa, **14**)

$$\left.\begin{array}{l} Hg \longrightarrow Hg^{2+} + 2e^- \\ Hg^{2+} + 2RS^- \longrightarrow Hg(SR)_2 \end{array}\right\} \tag{2}$$

$$Hg(SR)_2 + 2e^- \longrightarrow Hg + 2RS^- \tag{3}$$

A setup proposed for simultaneous determination of Al(III) and Mg(II) included a working GCE, a reference SSCE and a Pt auxiliary electrode, operating under N_2 atmosphere, at pH 5.0, in the presence of 0.01M KNO_3 and 0.02 M Me_4NCl, according to the Osteryoung square wave stripping voltametric technique. After 120 s deposition time at -0.8 V the scan proceeds in the positive direction, with the peaks of Mg(II) and Al(III) appearing at ca -0.42 and $+0.20$ V, respectively. The LOD are as low as 0.4 nM Mg(II)

7. Analytical aspects of organomagnesium compounds

and 0.05 nM Al(III); no interference is observed for Ba(II), Ca(II), Cd(II), Co(II), Fe(III), K(I), Mn(II), Na(I), Pb(II), Sr(II), Ti(IV), Zn(II) or UO_2^{2+}; however, Cu(II) can be determined as it shows a peak at ca -0.05 V. The method was applied to the analysis of a Portland cement CRM after HCl digestion[200]. A catalytic polarographic method for determination of water hardness (Ca + Mg) was proposed. The method is based on reduction of Mg(II) which has been displaced from its complex with EDTA (**3a**) by an added metal ion, such as Cu(II), which forms a much stronger complex with EDTA. The catalytic signal current shows about 100-fold amplification relative to the diffusion signal. The method was claimed to be of the same precision as the EDTA complexometric titration (Section III.E.3) but less cumbersome[201].

An alkaline solution of the complex formed by Mg(II) and Bromopyrogallol Red (**15**) shows a polarographic wave at -1.30 V, the intensity of which is linear with the Mg(II) concentration in the 0.05 to 2 ppm range. The LOD is 0.01 ppm Mg(II). The method is sensitive and selective; it was applied to determination of Mg(II) in food and the results corresponded to those obtained by AAS[202].

(**15**)

E. Spectral Methods

1. Atomic absorption spectrometry

For various mineralization methods prior to FAAS determination of Mg in many materials, see Section III.B.1. Analysis of the four major elements (Na, K, Ca and Mg) in drinking and other types of water by the FAAS method is well established and is the subject of several national and international standards. Nevertheless, there is a continuous discussion about the improvement of analytical quality and efficiency. A comparison was made of certain details of Polish standards with other national or international ones (ASTM, USEPA, ISO) and published research; of special concern were the presence of interfering ions and the effect of avoiding sample dilution on the analytical results. The latter consideration affords considerable savings in time and solvent expenses[203, 204]. Both Ca and Mg can be determined in solutions by FAAS with air-acetylene, measuring at 422.8 and 285.2 nm, respectively. When applying the method to analysis of urine[205], minor nutrients in fertilizers (AOAC Official Method 965.09[166b]), in water (AOAC Official Method 974.27[166c]) or cheese after dry ashing (AOAC Official Method 991.25[166d]), $LaCl_3$ can be used as releasing reagent, to avoid possible interference by phosphate, sulfate or silicate ions present in the matrix. Analysis of some toxic and essential elements in eggs of various origins was carried out after drying and mineralizing with a HNO_3–$HClO_4$–H_2O_2 mixture. Ca, Fe and Mg were determined by FAAS (LOD 20 to 70 ppb), whereas Cu, Pb and Zn by GFAAS, after adding a modifier containing palladium nitrate, yttrium nitrate and citric acid (LOD 1 to 0.03 ppb)[206]. The serum of patients receiving total parental nutrition was analyzed for trace elements after 10-fold dilution with Triton X-100 solution. Mg was determined by FAAS using an air-acetylene flame and measuring at 258.2 nm from a deuterium lamp; Cu, Mn, Pb and Zn were determined by GFAAS[207]. A procedure for FAAS analysis of Mg in a water-soluble multivitamin pharmaceutical preparation was validated and found adequate for the purpose[208].

Mg concentration in plasma was determined by FAAS, after centrifuging whole blood samples, acidifying and centrifuging again. A rise to 130% of the baseline levels took place during cerebral ischemia (stopping blood supply) in gerbils, which gradually returned to normal after reperfusion[209]. Blood fractions were analyzed for Cu, Fe, Mg, Se and Zn by FAAS, after suitable sample preparation. Blood with added EDTA was separated by centrifugation into plasma and erythrocytes; the latter were hemolyzed by freezing and towing and further treated with HCl, before determination of Cu, Fe and Zn. In each of these fractions the protein was precipitated by CF_3CO_2H and centrifugation, taking the supernatant for analysis; determination of selenium required digestion with $HNO_3-HClO_4-H_2SO_4$ mixture and reduction of Se(VI) to Se(IV)[210]. A device for determining Mg in the extracellular fluids of the gerbil brain consisted of an on-line microdialysis unit implanted in the organ; the collected fluids are passed by a microinjection pump together with a diluent to a sample collector, from which an automatic sampler injects the solution into a GFAAS device. The mean Mg after on-line dilution of the basal dialysate was 1.50 $\mu g\,L^{-1}$, and it significantly decreased to about 40% during cerebral ischemia, gradually returning to the basal value on reperfusion[209, 211]. Implanting more than one microdialysis unit for sample collection and ETAAS may afford important information on changes taking place simultaneously in an organism[212].

A multiparametric flow system was devised for the automated determination of Na(I) and K(I) by FAES and Ca(II) and Mg(II) by FAAS, to be applied in the quality control of large-volume parental solutions and concentrated hemodialysis solutions. The latter are rather concentrated pharmaceutical solutions whereas the determination methods operate at trace concentrations, thus requiring dilution and addition of reagents such as a La salt as releasing agent. The automated system allows a sampling frequency of nearly 60 h^{-1} for Mg and 70 h^{-1} for the other ions[213]. The same spectrophotometric methods were applied for determination of these elements in surgically excised cataracteous lenses after $HNO_3-H_2O_2$ digestion[214].

Research is carried out to find noninvasive pathogenesis indicators for cancer. Trace elements in scalp hair have been investigated. As the morphology and other characteristics of this material drastically change with age, sex and ethnic group, proper healthy control groups are needed to evaluate the results. The hair samples of a group of stage III breast cancer patients and a corresponding healthy control group were digested in hot concentrated HNO_3, properly diluted and Cu, Mg and Zn were determined by FAAS with Zeeman-effect background correction. Significant differences were found in the patients for Cu (nearly doubled) and Zn (nearly halved), whereas a 4% average decrease in Mg was not considered significantly different (Student's test). Other reported behavior of these trace elements is as follows: Zn was higher with malignant breast tumors as compared with benign tumors or healthy persons; it was lower in cases of prostatic carcinoma, nasopharyngeal and lung cancers. No significant difference was found for Cu in hair for lung cancer (it increased in plasma). Trace Mg in hair was significantly lower for esophageal cancer, acute lymphoblastic leukemia and malignant lymphoma[215].

2. Atomic emission spectrometry

Tracking local variations of trace element concentration in body fluids requires sensitive methods capable of returning sufficiently accurate and precise results with small samples.

7. Analytical aspects of organomagnesium compounds 279

The analytical quality in such cases can be better assessed when CRMs of similar nature are available. For example, samples of arterial blood weighing 5 to 15 mg were withdrawn from different parts of a rabbit and were subjected to mineralization with hot 50% (v/v) HNO_3 in a closed vessel microwave device. After adequate dilution end analysis of Ca, Mg and Fe was carried out by ICP/AES[216].

A method for determination of various elements in infant milk formulas consists of mineralization by wet digestion with HNO_3-HClO_4, and end analysis by ICP-AES. In the AOAC Official Method 984.27 the following measuring wavelengths in nm units are recommended, where * denotes the need for background correction: Ca (317.9), Cu (324.7*), Fe (259.9), K (766.5), Mg (383.2), Mn (257.6), Na (589.0), P (214.9) and Zn (213.8*)[166a]. The ICP-AES determination of Na, K, Ca and Mg in urine shows quantitative recovery for the alkali elements; however, the analysis of Ca and Mg is affected by the presence of the other cations and the anions in the matrix. The problem was solved by 10-fold dilution of the sample with water[217].

Determination of Mg in the hard tissues (shell and pearl) of shellfish by the ICP-AES method involves dissolution of the sample by hot concentrated nitric acid, hydrochloric acid and perchloric acid. However, the large excess of Ca in the matrix strongly interferes with the end analysis and causes damage to the torch. After adjusting the pH to 4.5, the Mg ions were extracted by a 0.01 M solution of 3-methyl-1-phenyl-4-trifluoroacetylpyrazol-5-one (**16**) in dibutyl ether and the ICP-AES analysis was carried out by direct injection of the organic solution[218].

(**16**)

A method was described for simultaneous determination of Ca, Mg, Fe, Cu, Zn and P in blood serum by spark AES using DA-UVD. For this modality of AES, a few milligrams of dry sample are placed in the hollow graphite anode and a spark is produced between the anode and the tapered cathode for a few seconds. The method development took into account various sources of systematic error and means of correcting them. Preconcentration and mineralization was carried out by dry ashing for 2 h in an oven at 450 °C; this avoids evaporative losses of analytes and reduces organic matter to a level where it ceases to interfere. The elements in serum can be classified for their abundance into macroelements (Na, Ca, Mg, K), which are easily ionized elements, and microelements. By far the most abundant metallic element in serum is Na, which at low concentrations causes a significant intensity enhancement of the emission lines of certain analytes; however, at higher concentration (4% was used) this effect is minor. The presence of Ca, Mg and K in the matrix causes underestimation of the microelements, and this can be corrected with a concentration-dependent factor, if the Na concentration is kept constant[219].

3. Ultraviolet-visible spectrophotometry and colorimetry

The formation of complexes of various metal ions with oxine (**5**) and some of its derivatives, their extraction into $CHCl_3$ and their λ_{max} and ε values have been reviewed[185]. Purpurin (LH_3, **17**) forms with Mg(II) colored complexes of varied composition. However, at pH 9.5 the complex MgLH is formed and can be measured at 540 nm ($\varepsilon =$

9200 L mol^{-1} cm^{-1}); first derivative spectra are of advantage over direct absorption spectra. For a 0.1 mM Mg(II) solution no interference was observed from alkali metal cations, about 25-fold excess of Al(III), Cu(II), Fe(II), Hg(II), Mn(II), Mo(VI), Ti(IV) and V(V), about 50-fold excess of Ba(II), Ca(II), Cd(II), Ni(II), Pb(II), Sr(II) and Zn(II) or about 100-fold excess of Br$^-$, Cl$^-$, ClO$_4^-$, NO$_3^-$, PO$_4^{3-}$ and SO$_4^{2-}$. The strong interference of Fe(III) can be avoided on adding ascorbic acid to the sample solution which reduces the ion to Fe(II). The LOD for normal spectrophotometry is 75 ppb Mg(II) and 34 ppb for first derivative measurements. The method was applied for analysis of Mg in cement clinker[220]. A spot test for Mg(II) is based on formation of a blue lake with quinalizarin (**18**) in alkaline solution, in contrast to a violet color obtained for a blank test. The LOD is 0.25 µg Mg, with interference by Be, Ce, La, Nd, Pt, Th and Zr[221].

(**17**) (**18**)

FIA systems have the disadvantage of employing one manifold per determination; if more than one determination has to be carried out on the same manifold, the reagents have to be changed. This limitation is avoided in SIA systems, where the sample and reagents are sequentially introduced into a holding coil by means of a selection valve; on reversing the flow the stacked zones mix on their way to the detector. Additional manifolds can be easily added to the computer-controlled SIA systems[222]. This subject has been reviewed[223]. A SIA method was proposed for the digestion of food samples and subsequent colorimetric determination of Ca(II), Mg(II) and Fe(III), including an in-line microwave digestion unit, from which the analytical samples are withdrawn. The sequential operation for end analysis is as follows: (1) For Mg(II) determination, an aliquot of the digested sample is mixed with aliquots of solutions of EGTA (**19**), serving as masking reagent for Ca(II), and *o*-cresolphthalein (**20**), the chromogenic reagent for Mg(II); in the detector a transient signal is measured at 535 nm, the intensity of which is proportional to the analyte concentration. (2) An aliquot of the digested sample is mixed with one of **20**, serving as chromogenic reagent for both Ca(II) and Mg(II) and the intensity of the color is similarly measured. (3) An aliquot of digested sample is mixed with one of reagent solution of *o*-phenanthroline (**21**), forming a colored complex with Fe(II), which is obtained by reduction of Fe(III) with ascorbic acid present in the reagent solution, and further measured in the detector. Each reagent carries its own buffer and other additives[224]. A SIA procedure was used for determination of hardnes (Ca + Mg) and alkalinity of water. A common complexing agent was used for Ca and Mg, Cresolphthalein Complexone (**22**), measuring at 572 nm. However, to distinguish between the two ions EGTA (**19**) served to mask Ca(II) in the Mg(II) determination and oxine (**5**) to mask Mg(II) in the Ca(II) determination. Bromocresol Green (**23**) was used for determination of alkalinity, measuring at 611 nm[225]. An analogous SIA protocol was applied for determination of Mg in pharmaceutical preparations, based on complexation with **22**, masking with EGTA (**19**) and UVD at 570 nm[226]. A FIA system was designed for the simultaneous determination of Ca(II) and Mg(II) ions, by which the complexes of both ions with Methylthymol Blue (**24**) were measured simultaneously at λ_{max} = 610 nm. On application of the zone sampling technique, part of the solution in the first loop was mixed with a solution of oxine

(5), to mask the Mg(II) ions, and a second measurement was made for Ca(II) alone. The method was applied for analysis of white, rose and red wines[227].

3-(2-Carboxy-4-nitrophenylazo)-4,5-dihydroxy-2,7-naphthalenedisulfonic acid (**25**) at pH 10.4 forms a colored complex with Mg(II) ions with high selectivity in the presence of Ca(II) and minor elements such as Al(III), Cd(II), Co(II), Cr(III), Cu(II), Fe(II), Hg(II), In(III), Mn(II), Mo(II), Ni(II), Sn(II), Ti(II) and Zn(II). Application of the spectrophotometric method for determination of Mg with this reagent requires measurement at 582 nm, because at the maximum for the complex ($\lambda_{max} = 560$ nm) there is considerable interference of **25** itself. To correct for this interference a second measurement is taken at

505 nm. The complex has $Mg_2(\mathbf{25})$ composition, with stability constant $K = 1.92 \times 10^4$ and $\varepsilon = 2.25 \times 10^4$ L mol^{-1} cm^{-1} at 582 nm[228]. A kinetic method was proposed for determation of ultratrace concentrations of Mg(II), based on the inhibition that this ion causes to the Mn(II)-catalyzed decoloration of Acid Chome Blue K (**26**) by KIO_4, in Britton–Robinson buffer at pH 11.9. The LOD was 7.6 ppb, with linearity up to 0.48 ppm. The method was applied for analysis of soybean and human serum[229].

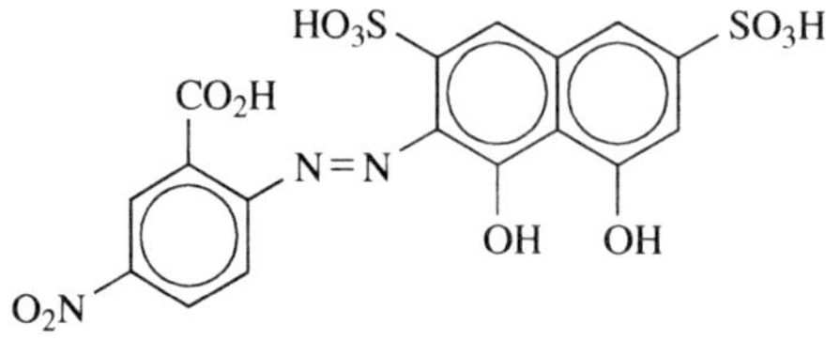

(**25**)

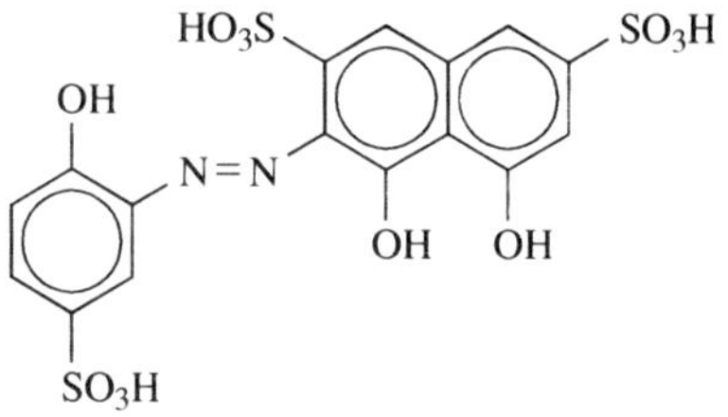

(**26**)

A commercial kit consists of two solutions, one containing calmagite (**27**) serving as chromogenic complexant for Mg(II) and the second one containing EDTA, as chelating agent for Ca(II). Measurements are carried out at 500 nm. The kit is recommended for serum, urine, water and soil analysis[230]. Another kit, recommended for biological fluids, is based on the action of 1,5-bis(3,5-dichloro-2-hydroxyphenyl)-3-cyanoformazan (**28**) as chromogenic complexant for Mg(II), with 1,2-bis(2-aminophenoxy)ethane-N,N,N',N'-tetraacetic acid (**29**) as Ca(II) masking agent, both on a dry slide. After contact with the sample, quantitation is carried out by colorimetric measurements[231]. Magon (**30**) forms a red complex with Mg(II), which is measured spectrophotometrically at 520 and 600 nm, with no significant interference from Ca(II), phosphate, albumin or billirubin. The method was applied in clinical analysis for Mg(II) determination in serum[158, 232]. Aqueous Mg(II) at pH 9.5 can be extracted by a solution of N-p-tolyl-2-thenohydroxamic acid (**31**) in chloroform. A colored complex is formed on addition of quinalizarin (**18**) to the extract ($\lambda_{max} = 590$ nm, $\varepsilon = 2800$ L mol^{-1} cm^{-1}). No interference was observed by most common ions[233].

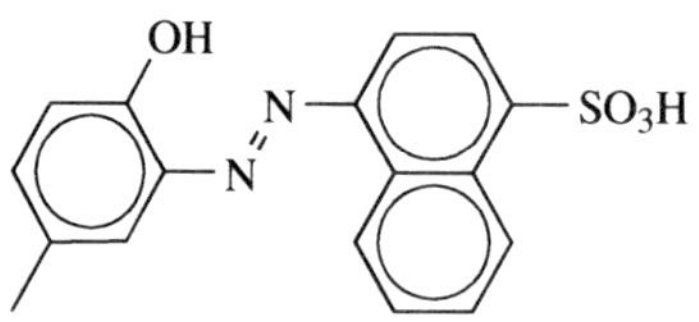

(**27**)

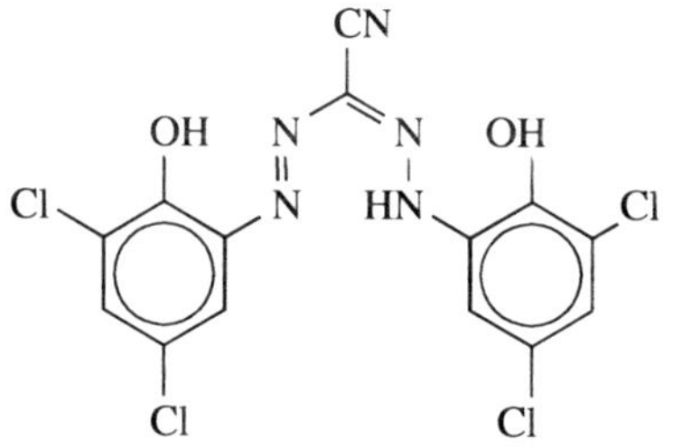

(**28**)

The complexometric method for determination of Ca(II) and Mg(II) is based on two titrations with EDTA in alkaline solution, one where both ions are determined together and the second after one of them has been masked with a specific complexing agent. The effect of interfering heavy metals such as Cu, Fe, Mn or Zn can be avoided by adding cyanide. The AOAC Official Method 964.01 for determination of acid-soluble

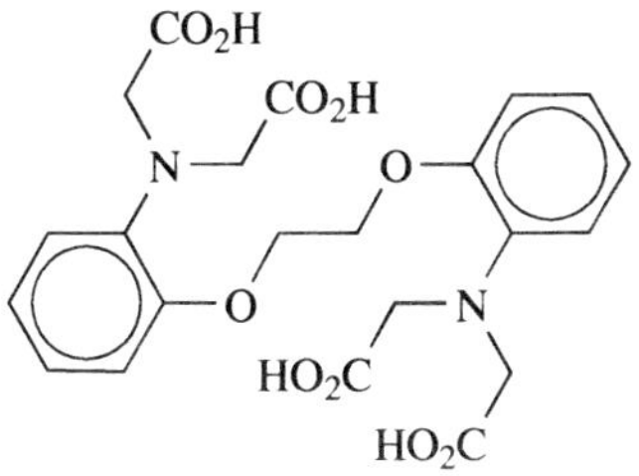

(29) 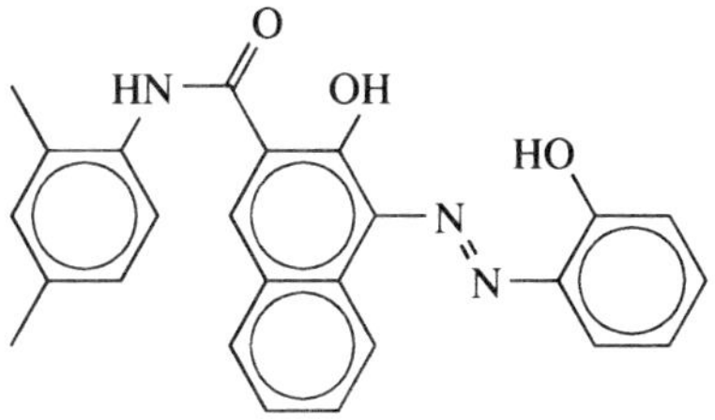   (30)

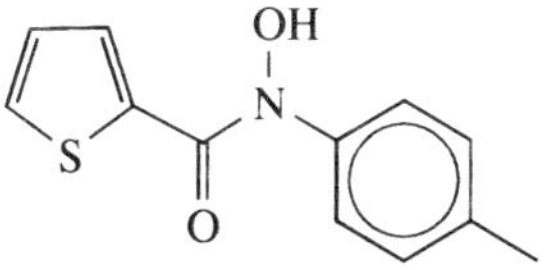

(31)

Mg in fertilizers is based on such proceedings[166e]. This standard method or variation thereof has been applied on multiple occasions. In a recent publication it was used for milk fermentation, where the samples were dried, calcined in a furnace at 600 °C, the ash was dissolved in 0.03 M HCl, the solution was centrifuged and the supernatant was thus analyzed[162]. The complexometric method for determination of Ca(II) and Mg(II) can be carried out in a single titration with EDTA in alkaline solution, using a Ca-ISE for potentiometric determination of two endpoints. This is accomplished on digitally plotting pCa values measured by the ISE as a function of the volume V of titrant added to the aliquot of analyte; the first and second inflection points of the curve mark the Ca(II) and Mg(II) equivalences, respectively[234].

4. Ultraviolet-visible fluorometry

A spectrophotometric and photochemical study was carried out on Mg(II)-selective fluorophores in their free and complexed form. The lifetimes of the excited state of Mag-quin-1 (32a), Mag-quin-2 (32b), Magnesium Green (33a) and Magnesium Orange (33b) increased two- to ten-fold on Mg(II) binding, whereas the presence of this cation did not affect those of Mag-fura-2 (34a), Mag-fura-5 (34b), Mag-indo-1 (35) and Mag-fura Red (36). On applying phase modulation fluorometry, it was found for 32b and 33a that a much wider Mg(II) sensitivity range is available than from intensity measurements. These two dyes undergo significant photochemical change under intense and prolonged illumination[235].

Mg(II) forms a complex with 8-hydroxyquinoline-5-sulfonic acid (37) at pH 9.0 with Tris-HCl buffer, which can be determined by FLD ($\lambda_{ex} = 388$ nm, $\lambda_{fl} = 495$ nm) with micellar enhancement by cetyltrimethylammonium chloride (38). Masking of Ca(II) is achieved by EGTA (19). The method was applied in a SIA system for analysis of natural waters[236]. After elution of the Mg(II) ions adsorbed on an alkali-activated PTFE tube with 0.1 M HCl and addition of N,N'-bis(salicylidene)-2,3-diaminobenzofuran (39), the end analysis was by fluorometric determination of the Mg(II) complex ($\lambda_{ex} = 475$ nm, $\lambda_{fl} = 545$ nm). Possible interference of Ca(II) is masked on addition of the chelating agent

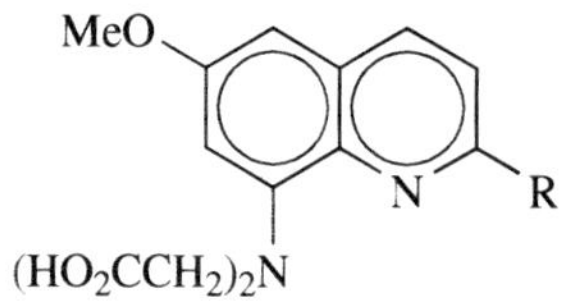

(32) (a) R = Me
(b) R = H

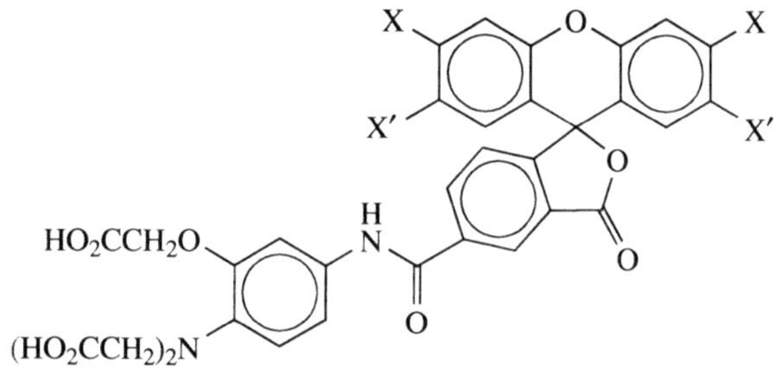

(33) (a) X = OH, X' = Cl
(b) X = NMe$_2$, X' = H

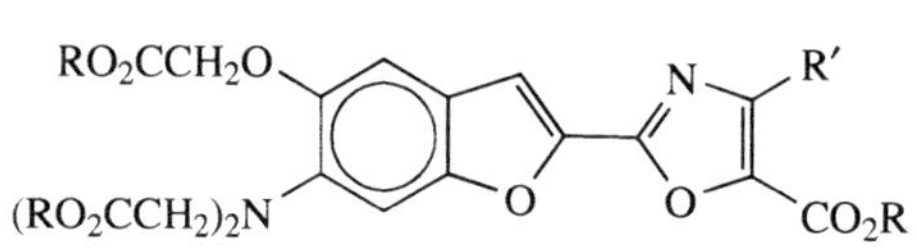

(34) (a) R = CH$_2$OAc, R' = H
(b) R = H, R' = Me

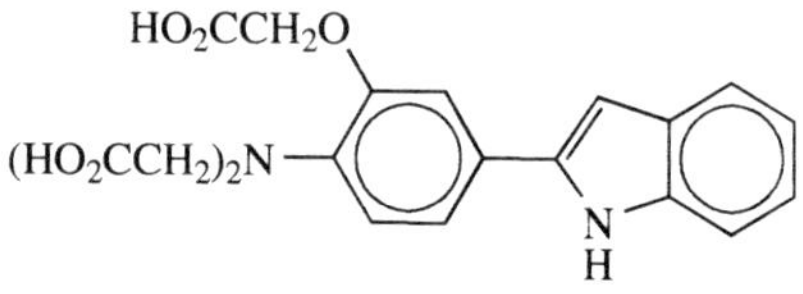

(35)

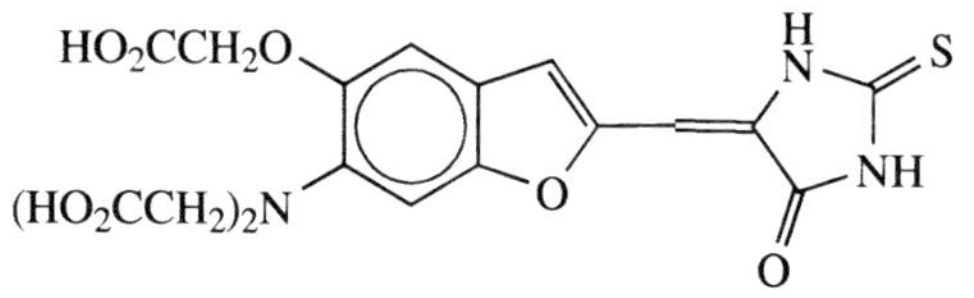

(36)

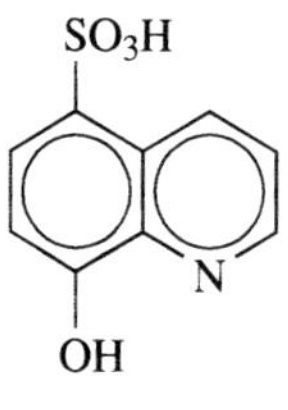

(37) (38) n-C$_{18}$H$_{37}$NMe$_3$Cl (39)

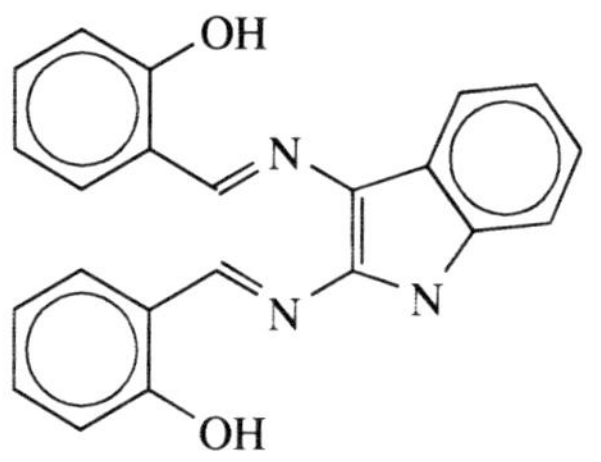

29. LOD is 82 ppt. A sample of distilled water showed 3.1 ppb Mg(II), in good agreement with the result obtained by ICP-AES[178].

Fluorescent probes for microscopic evaluation of free intracellular Mg(II) should fulfil requirements such as adequate photochemical properties (excitation with laser-based instrumentation, high extinction coefficient and quantum yield, reduced interference from autofluorescence), low toxicity and low photochemical damage. Several Mg(II) microfluorescent probes are in the market, for example, Mag-fura-2 (**34a**), Mag-indo-1 (**35**) and

7. Analytical aspects of organomagnesium compounds

Magnesium Green (**33a**). Although the formation constant, K_{Ca}, of the Ca(II) compex with **33a**, **34a** and **35** is about two orders of magnitude larger than the corresponding K_{Mg}, the method takes advantage of the much reduced intracellular concentration of free Ca(II) as compared to that of free Mg(II). Nevertheless, it would be of advantage to have probes with higher Mg(II) selectivity, such as Coumarin 343 (**40a**), which, however, does not penetrate the cell membrane. This is overcome by esterification with an acetyloxymethyl group which yields fluorophores such as KMG-20-AM (**40b**) and KMG-27-AM (**40c**), with K_{Mg}/K_{Ca} ca 3.0, by which the presence of intracellular free Ca(II) cannot interfere with free Mg(II)[237]. An alternative to this approach is the PEBBLE technique, by which a Mg(II) complexant, such as **40a** with K_{Mg}/K_{Ca} of ca 2.0, and an unreactive reference fluorophore, such as Texas Red–dextran (**41**), are encapsulated in biocompatible polyacrylamide nanospheres. The PEBBLEs are introduced into the cell by gene gun injection and do not interfere with the cell normal functions. Fluorometric measurements are carried out by exciting the sample at 445 nm and recording the resulting emission from 460 (for the Mg(II) complex) to 640 nm (for the reference). The LOD was 340 μM for a dynamic range of 0 to 30 mM[156].

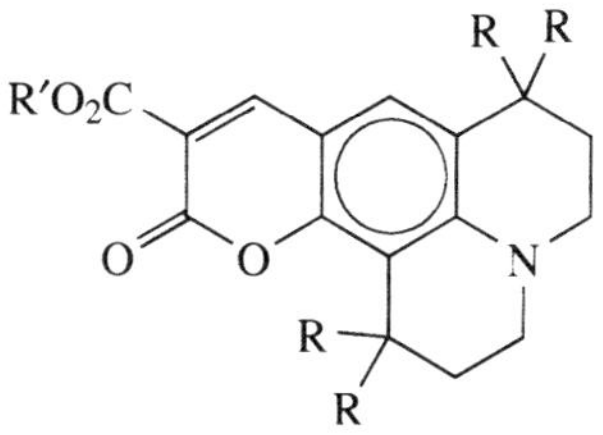

(**40**) (a) R = R′ = H
(b) R = H, R′ = CH$_2$OAc
(c) R = Me, R′ = CH$_2$OAc

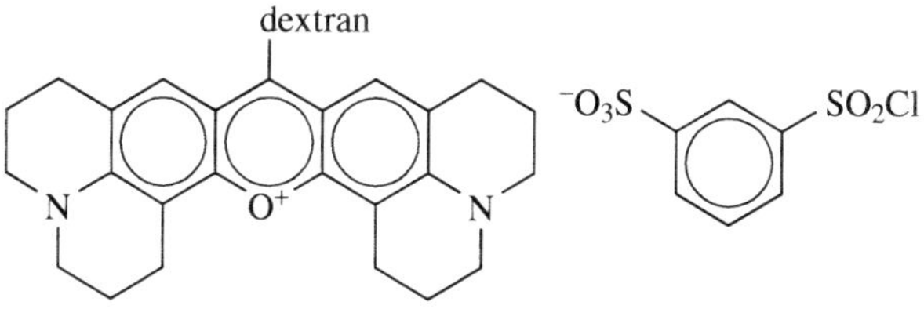

(**41**)

5. Chromatic chemosensors

A potentially useful group of polymers changes coloration mainly due to their state of aggregation and may serve as sensors in solid state or solution for various ions. An example for this functionality is afforded by regioregular head-to-tail poly(thiophene-3-alkanoic acid)s (**42**), which are electrically conducting with low band gaps due to their ability for self-assembling into planar π-stacked aggregates. Polythiophene derivatives are also known for chromatic response to various stimuli, showing properties such as affinity chromism[238], biochromism[239,240], electrochromism[241], ionochromism[242], photochromism[239,243], piezochromism and thermochromism[239]. Polymers **42a** and **42b** are not very soluble in ordinary organic solvents and appear as violet crystals. On the other hand, they are soluble in water yielding the corresponding polycarboxylate ions **43a** and **43b**, on addition of an equivalent amount of a base, according to equation 4. The solutions show ionochromism from violet to yellow, depending on the length of the pending carboxylic acid chain and the size of the univalent base cation (ammonium and alkali metal) used to produce the carboxylate. The mechanism responsible for the ionochromism in solution seems to be self-assembly of π-stacked regions with color shift to violet; however, if the cations are too large (e.g. Bu_4N^+ or Cs^+), ion pair formation with the carboxylate groups of **43a** causes unzipping of the chains and shift to yellow color; in the case of **43b**, with longer pending chains, color is only slightly affected by cation size. Polymers such as **42a** may serve for detection of divalent cations. Addition of divalent cations to a red solution of **43a** with Et_4N^+ counterions causes color changes and precipitate formation, probably by interchain ion pairing taking place with the divalent cations. Thus, solutions turn purple and purple precipitates form on adding small amounts of Fe(II), Mg(II) or Mn(II) salts; more stable red to purple solutions can be attained with Cd(II), Co(II), Cu(II), Hg(II), Ni(II) or Zn(II), however, magenta to brown solids are formed much before the equivalent concentration is reached[244].

$$\underset{\textbf{(42)}\ \ (\textbf{a})\ m = 2\ \ (\textbf{b})\ m = 7}{\text{thiophene-}(CH_2)_m CO_2H} \underset{H^+}{\overset{OH^-}{\rightleftharpoons}} \underset{\textbf{(43)}\ \ (\textbf{a})\ m = 2\ \ (\textbf{b})\ m = 7}{\text{thiophene-}(CH_2)_m CO_2^-} \quad (4)$$

6. Nuclear magnetic resonance spectroscopy

EDTA forms stable chelate complexes with Mg(II) and Ca(II) ions (**3**) at high pH, when the dissociation of the organic ligand is total[245]. This allows application of SEFT[246] or single pulse[247] ^1H NMR techniques for determination of these ions in mM concentrations. For quantitative analysis of magnesium and calcium, a reusable sealed capillary containing a solution of sodium salt of 3-(trimethylsilyl)propionic acid-d_4 (**44**) in D_2O is inserted coaxially; the **44** signal serves as a chemical shift and quantitation reference while deuterium oxide provides the field-frequency deuterium lock. Two types of proton can be distinguished, an AB multiplet for acetate and a singlet for ethylene which is easier to handle in quantitative analysis. As the chemical shifts of both proton types are slightly different for **3a** and **3b**, simultaneous analysis of Mg and Ca is enabled. Application of the EXSY and TPPI techniques demonstrated that a slow exchange takes place between the free tetravalent anion and the complexed anions **3a** and **3b**. The method was proposed for clinical determination of these ions and various organic analytes in plasma and erythrocytes, as it requires no separation or mineralization steps[247].

$$Me_3SiCD_2CD_2CO_2Na$$
(**44**)

Mg(II) forms a complex with adenosine triphosphate (ATP, **45**), which at pH 7.2 and 37 °C has dissociation constant $K_d = 3.8 \times 10^{-5}$ mol L^{-1}. The fraction of the total ATP which did not undergo complexation present in a cell, φ, can be estimated on the basis of ^{31}P NMR spectra by means of equation 5, where the subscripts $\alpha\beta$ denote the chemical shift of P_β relative to that of P_α and the superscripts denote the value measured for the cell, pure ATP and the Mg complex under similar pH and temperature conditions. The concentration of free Mg(II) in the cell is calculated by equation 6. This method was applied to determine the concentration of free Mg(II) in erythrocytes and was in accord with values determined by other methods[192, 248, 249]. Although ATP hydrolysis *in vivo* releases Mg(II), certain muscle conditions may minimize this[250].

(45)

$$\varphi = \frac{\delta_{\alpha\beta}^{cell} - \delta_{\alpha\beta}^{MgATP}}{\delta_{\alpha\beta}^{ATP} - \delta_{\alpha\beta}^{MgATP}} \tag{5}$$

$$[Mg(II)]_{free} = K_d[(1/\varphi) - 1] \tag{6}$$

7. Mass spectrometry

Short reviews appeared on the various MS techniques for quantitation of stable isotopes and long-lived radioisotopes[251] and the application of Mg stable isotopes as tracers in biology and medicine[252]. The radioactive isotope ^{28}Mg is not usually available and has a short half-life (21.3 h), hence its limited usefulness as a tracer[252]. The sensitivities and interference problems encountered in activation analysis for Al, Mg, Mo, P, Si and Zr were discussed. Much higher sensitivities were found for cyclotron-produced than for reactor-produced fast neutrons or 14 MeV neutrons[253].

The reverse isotope dilution technique can be applied for accurate determination of the Mg contents in a sample, Q_{sample}, on applying equation 7, by measuring the isotope ratio of a selected pair of stable isotopes, R_{mix}, in a weighed mixture of the sample with an isotopically enriched CRM. The average atomic masses m and the isotopic ratios R of Mg in the enriched CRM and in nature are known. The method was applied for determination of Mg in plant material using a CRM isotopically enriched with ^{26}Mg, measuring with an ICP/MS instrument[251].

$$Q_{sample} = Q_{CRM} \left(\frac{R_{CRM} - R_{mix}}{R_{mix} - R_{nature}} \right) \left(\frac{m_{nature}}{m_{CRM}} \right) \tag{7}$$

More than 300 enzyme systems of the human being are dependent on the presence of magnesium, hence the clinical importance of determining the input–output balance and homeostatic levels of this element. For estimation of Mg absorption two possible input avenues are considered, oral and intravenous injection, two output avenues, feces and urine, and the general pool in plasma. The input amounts are known by design;

determination of the output and plasma amounts requires digestion of the samples and final analysis by FAAS after dilution with 0.5% $LaCl_3$ to a Mg concentration of 0.1 to 0.4 ppm. The Mg absorption calculated from the oral input and feces output is undervalued because of the fecal endogenous excretion (FEE), by which Mg is transferred from the blood to the digestive tract. The FEE cannot be estimated from the FAAS analyses. Additional information is obtained on labeling the inputs with different stable Mg isotopes, ^{26}Mg for the oral intake and ^{25}Mg for the intravenous injection. The end analysis of the digested plasma, urine and feces samples is carried out by the ICP-MS method, measuring the abundance of the ^{24}Mg, ^{25}Mg and ^{26}Mg isotopes. Following the development of the $^{25}Mg:^{24}Mg$ and $^{26}Mg:^{24}Mg$ isotope ratios in the samples, it is possible to evaluate the FEE and to correct the fecal Mg balance and the Mg absorption. In fact, following the double isotope labeling of the Mg inputs, one can save the labor and expense involved in the feces analysis, leaving only plasma and urine analyses for clinical purposes[254,255]. The fractional absorption of Mg in the intestine of rats was studied on applying a dietary intake enriched with ^{25}Mg. Solid samples (feces and bone) were mineralized by calcination in a furnace at 500 °C and dissolution of the ashes in HNO_3/H_2O_2; liquid samples (urine, plasma and red blood cells) were diluted to the appropriate Mg concentration before end analysis by FAAS for total Mg and ICP/MS for isotope ratios[256].

A study of the exchange taking place between Mg pool masses was based on measurement of stable isotope ratios after administration of ^{25}Mg or ^{26}Mg doses, and analysis of plasma by ICP-MS[257,258]. In a study of Mg transport in epithelial cells the isotope ratios were determined by MS for intracellular free Mg(II) after cell dissolution, as obtained from a feed enriched with ^{25}Mg, whereas the total free Mg(II) in the cell was determined by a microfluorimetric method using the complex of Mag-fura-2 (**34a**) with Mg(II) (Section III.E.4)[252].

IV. SPECIATION ANALYSIS OF ORGANOMAGNESIUM COMPOUNDS
A. Titration Methods

Many of the titration methods for organomagnesium compounds are similar to those for organolithium compounds, which have been reviewed elsewhere[146,259,260].

1. Visual endpoint

Titrations of organomagnesium or organolithium compounds take advantage of two characteristic properties of these compounds: they may act as strong Lewis bases forming salts with acids, e.g. butanol, as in equations 8 and 9, or forming charge-transfer complexes, usually with aromatic bases. The latter property affords good indicators for titration; for example, 1,10-phenanthroline (**21**) or 2,2′-biquinoline (**46**) form deeply colored complexes with alkyl Grignard reagents or with dialkylmagnesium compounds, which persist until the titrant is in excess[261,262]; the same titration method may be applied to assess the quality of $LiAlH_4$[262]. 1-Pyreneacetic acid (**47a**) and 1-pyrenemethanol (**47b**) were proposed as titrant/indicator for Grignard reagents. On dropwise addition of the analyte to an aliquot of **47a**, the endpoint is marked by a turn to intense red due to formation of a dianion. The titrant can be recovered[263]. Not all indicators suitable for titration of organolithium compounds are fit for titration of organomagnesium compounds, probably due to the lesser reactivity of the latter. Thus, 9-methylfluorene (**48**) and N-phenyl-1-naphthylamine (**49**) were successfully used for titration of various organolithium compounds with a solution of s-BuOH in THF according to equation 8, by virtue of the red discoloration obtained on losing their active proton in the presence of the organometallic analyte. Neither MeMgCl nor t-BuMgCl could be titrated with either indicator, PhMgCl responded only to **49**, and only with vinylmagnesium chloride could both indicators be used[264]. The requirement for compounds such as **49** to serve as indicator in this type of titration is that their basicity be lower than that of the Grignard reagent[265]. However, the analytical quality of the titration of this compound according to equation 8 with o-phenanthroline (**21**) as indicator is uncertain when the vinylmagnesium halide has deteriorated (see Section IV.B.1)[266]. Arylmagnesium halides were dissolved in xylene and titrated with s-BuOH using **49** as endpoint indicator, turning to a colorless solution. These titrations served for validation of two chromatographic methods for speciation analysis of the Grignard reagents and their impurities (Section IV.B)[267]. 4-(Phenylazo)diphenylamine (**50**) was used as indicator for the titration of MeMgI with PrOH, the color turning from pink to yellow[268].

$$RMgX + s\text{-BuOH} \longrightarrow RH + s\text{-BuOMgX} \tag{8}$$

$$R_2Mg + 2s\text{-BuOH} \longrightarrow 2RH + (s\text{-BuO})_2Mg \tag{9}$$

(**46**)

(**47**) (a) $X = CO_2H$
(b) $X = OH$

(**48**)   (**49**)   (**50**)

Gilman and coworkers[269, 270] pointed out that titration of the alkalinity produced by hydrolysis of the sample according to equations 10 and 11 can yield high values. Organomagnesium compounds may undergo deterioration on being exposed to oxygen, moisture or carbon dioxide in the environment; also, long storage at room temperature may cause condensation, elimination or rearrangement reactions. Various titration methods have been proposed for the quality assessment of organomagnesium compounds in general and Grignard reagents in particular. The double titration method has been often applied for organolithium compounds. One aliquot of the compound is hydrolyzed and the alkalinity produced by RLi, ROLi and LiOH present in the sample is determined by titration with acid. A second aliquot is treated with a specific reagent for the organolithium compound, e.g. BnCl, and titrated for the alkalinity of ROLi and LiOH. The difference between these values is the content of organometallic compound. The ASTM E233-90 standard method for assay of n-butyllithium solutions is an example of such proceedings[271]. The subject was briefly reviewed[260a] and is of applicability to other organometallics, such as Grignard reagents, for which CCl_4 has been proposed as reagent for the second titration[272].

Titrations with water, based on equations 10 and 11, have been of wide application, and are appropriate for concentrations as low as about 2 mM[273].

$$RMgX + H_2O \longrightarrow RH + \tfrac{1}{2}MgX_2 + \tfrac{1}{2}Mg(OH)_2 \qquad (10)$$

$$R_2Mg + H_2O \longrightarrow 2RH + Mg(OH)_2 \qquad (11)$$

Diphenyl ditelluride (**51**) serves both as titrant and indicator for organometallic compounds as shown in equation 12. Reagent **51** imparts a deep red discoloration when dissolved in THF or other solvents used in carbene chemistry; on reaching the endpoint the color turns to yellow. The chemical process is more complicated than in equation 12 when dealing with analytes of basicity stronger than that of Grignard reagents or alkynyllithium compounds; however, the stoichiometry of the overall titration is not disturbed. No interference was observed by alkoxide impurities present in organometallic compounds[274].

$$\underset{(\mathbf{51})}{PhTeTePh} + RM \xrightarrow[(M = Li,\ MgX)]{} PhTeR + PhTeM \qquad (12)$$

Salicylaldehyde phenylhydrazone (**52**) is an easily prepared and inexpensive titrant/indicator for organometallics such as Grignard reagents (equation 13) and organolithium compounds and for hydride species such as lithium aluminum hydride and sodium bis(2-methoxyethoxy)aluminum hydride. A solution of the analyte is added to THF containing a weighed amount of **52**; a yellow color appears due to the presence of the phenolate ion **53**. On reaching the endpoint the solution turns bright orange due to formation of the **54** dianion[275].

(**52**) →[RMgX] (**53**, yellow) →[RMgX] (**54**, red)   (13)

A titration method for organomagnesium, organozinc or organolanthanides is based on the reaction of these compounds with iodine dissolved in a saturated solution of LiCl in THF. The analyte solution is added to an aliquot of I_2 solution until the brown color disappears. The chemical process for organozinc or organomagnesium is as shown in equations 14 and 15[276]. In the case of the organolanthanides, conveniently prepared

7. Analytical aspects of organomagnesium compounds

from a Grignard reagent[277], the titration proceeds as shown in equation 16[276]. Note in equations 14 to 16 that the equivalence is one mol of I_2 per mol of organo group; the X on the right hand side are halide ions present in the solution.

$$RMX + I_2(LiCl/THF) \xrightarrow[(M = Mg, Zn)]{} RI + MX_2 \cdot LiCl \tag{14}$$

$$R_2M + 2I_2(LiCl/THF) \xrightarrow[(M = Mg, Zn)]{} 2RI + MX_2 \cdot LiCl \tag{15}$$

$$RLnCl_2 + I_2(LiCl/THF) \xrightarrow[(Ln = La, Ce, Nd)]{} RI + LnX_3 \cdot 2LiCl \tag{16}$$

In a typical titration of Grignard reagents with substituted diimidosulfur compounds (**55**) the analyte is added dropwise to the red-orange solution of **55** until it becomes colorless, yielding monomeric (**56**) or dimeric adducts (**57**), as shown in equation 17. Dialkyl magnesium and alkyllithium compounds undergo a similar reaction. No interference occurs by alkoxides, hydroxides and other products stemming from degradation of the analytes on storing[278].

(17)

(**56**)   (**55**)   (**57**)

Ar = 4-MeC$_6$H$_4$

2. Potentiometric and other instrumental titrations

Determination of organomagnesium compounds with various titrants can be carried out with a potentiometric endpoint, using a Ag working electrode and a Ag/AgClO$_4$/Bu$_4$NClO$_4$/THF reference electrode. Thus, MeMgCl and EtMgCl could be titrated with THF solutions of AgClO$_4$, BuOH or PhNH$_2$, and PhMgCl with the latter two titrants[279]. A combined Pt–Ag/AgCl electrode for use in nonaqueous titrations has to be prepared from an ordinary electrode of this type, by changing the KCl electrolyte solution with a saturated solution of LiCl in THF, to which a few drops of AgNO$_3$ solution were added. The endpoint is chosen at the inflexion point of the potential *vs* titrant curve, or more clearly from the first derivative of this curve. This method is suitable for automatic titration of Grignard reagents with BuOH, according to equation 8. An alternative method for determining the titration endpoint is based on in-line FTIR. Spectra of the titration solution are collected at a rate of 2 min^{-1}, over the 4000 to 600 cm^{-1} range, with a 4 cm^{-1} resolution. For example, in the case of MeMgCl, calculated difference spectra, where the spectrum of the solvent alone is subtracted from the collected spectra, show peaks at 1070 and 911 cm^{-1}, due to C–O–C stretching and ring breathing in complex **58**; the corresponding strong peaks of the THF solvent are at 1037 and 884 cm^{-1}. Disappearance of the peaks of **58** in the difference spectra marks the titration endpoint. The potentiometric and FTIR methods for determining the endpoint were in good agreement[280].

Potentiometric titration with KI$_3$ dissolved in HMPT, according to equations 18 and 19, was proposed. The method is limited by the relatively low shelf stability of the titrant

(58)

solution; for example, the oxidation shown in equation 20 can take place in the presence of moisture[281].

$$RMgX + KI_3 \longrightarrow RI + MgXI + KI \quad (18)$$

$$R_2Mg + 2KI_3 \longrightarrow 2RI + MgI_2 + 2KI \quad (19)$$

$$\underset{\substack{\diagup N \diagdown \\ OP-N \\ \diagup N \diagdown}}{\overset{CH_3}{\diagdown}} \xrightarrow{3KI_3 + H_2O} \underset{\substack{\diagup N \diagdown \\ OP-N \\ \diagup N \diagdown}}{\overset{H}{\diagdown}} + CH_2O + 2HI_3 + 3KI \quad (20)$$

Various methods have been proposed for avoiding interference by the usual degradation products of Grignard reagents. Thus, a sample of the analyte on treatment with excess of an aromatic ketone followed by dilution with an alcohol is converted to a tertiary alcohol, as shown in equation 21. The excess of benzophenone (**59a**) is measured at 333 nm; however, the method is limited to primary alkyl groups R [282]. The method is of more general applicability when the titrating reagent is acetophenone (**59b**), the excess of which is determined at 243 nm[283].

$$RMgX + \underset{Ph}{\overset{R'}{\diagdown}}=O \longrightarrow R\underset{Ph}{\overset{R'}{-\!\!\!\mid\!\!\!-}}OH \quad (21)$$

(**59**) (a) R' = Ph
(b) R' = Me

Thermometric titrations were applied for determination of organometallic compounds and, in particular, of Grignard reagents. The method takes advantage of the negative enthalpy of reaction with an alcohol (e.g. equation 8), which causes heating of the reaction mixtures as the titrant solution is added. A typical run consists of measuring the sample temperature while adding the titrant solution at a constant rate. A gradual temperature rise (reaction in progress) is observed until the curve starts bending down (reaction approaching the equivalence point) and then decreasing (reaction finished and mixture cooling down). The endpoint is determined by the intersection of the tangents to the increasing and decreasing parts of the curve. This titration was applied for determination of MeMgBr, MeMgI, EtMgI and PhMgBr with i-PrOH[284].

B. Chromatographic Methods

1. Liquid chromatography

The titration methods of Section IV.A do not allow detection and quantitation of the impurities accompanying the organomagnesium compound analytes. A method was proposed for fast derivatization (less than 1 min) of aromatic organomagnesium compounds

7. Analytical aspects of organomagnesium compounds   293

with carbon dioxide, according to equation 22, followed by dilution with aqueous MeOH. End analysis was by RP-HPLC on a C_8 column, with a mobile phase of aqueous acetonitrile containing 0.05% formic acid to ensure that the analyte derivative is an arenecarboxylic acid. An alternative to CO_2 derivatization was quenching with water to get the arene, according to equation 23. Detection was by EI-MS, in the full scan negative ion mode. For example, titer determination of a *ca* 1 M 4-fluorophenylmagnesium bromide sample with CO_2 showed in the order of emergence from the column 4-fluorophenol, 4-fluorobenzoic acid, fluorobenzene, 4'-fluorobiphenyl-3-carboxylic acid, 4'-fluorobiphenyl-4-carboxylic acid, 4-fluorobiphenyl and 4,4'-difluorobiphenyl; except for 4-fluorobenzoic acid all were in trace amounts. The same sample did not show 4-fluorobenzoic acid by the water quenching method and fluorobenzene was a strong peak[267].

$$\text{ArMgX} + \text{CO}_2 \longrightarrow \text{ArCO}_2\text{MgX} \xrightarrow{\text{H}^+/\text{H}_2\text{O}} \text{ArCO}_2\text{H} + \text{Mg}^{2+} \quad (22)$$

$$\text{ArMgX} + \text{H}_2\text{O} \longrightarrow \text{ArH} + \text{Mg(X)OH} \xrightarrow{\text{H}^+/\text{H}_2\text{O}} \text{ArH} + \text{Mg}^{2+} \quad (23)$$

A precolumn derivative recommended for vinylmagnesium halide and alkyl Grignard reagents is a secondary alcohol (**61**) derived from 9-anthraldehyde (**60**), as shown in equation 24. End analysis after appropriate dilution is by RP-HPLC-UVD, measuring at 227 nm and using PhAc as internal standard. Although the chromatographic method is more involved than direct titration with *s*-butanol, according to equation 8, wrong results were obtained by titration for the titer of a reagent that partially decomposed on strorage[266].

(24)

(**60**)   (**61**)

Separation and quantitation of mixtures of analytes in complex matrices is possible with a combination of sample preparation, column, eluent and detection method. Thus, magnesium lithospermate B (**62**), lithospermic acid (**63**) and rosmarinic acid (**64**), present in *Salvia miltiorrhiza Bge (Labiatae),* used in traditional Chinese medicine, were determined in dog serum by LC/tandem MSD. The serum samples, either spiked with the analytes or obtained after intravenous infusion, were spiked with silibinin (**65**), serving as internal standard, treated with formic acid, subjected to LLE with AcOEt and centrifuged to separate the protein. The supernatant extract was evaporated to dryness, dissolved in aqueous 25% acetone, centrifuged and 10 µL of the supernatant were injected into the LC/MS/MS instrument, equipped with a C_{18} column, in isocratic regime, using as mobile phase water–MeCN (6:4), containing 0.5% formic acid. The highest sensitivity was attained for negative ion operation, optimized for $[M - H]^-$ ions. Analytical runs of about 3 min were carried out. Analytical figures of merit for **62**, **63** and **64** were, respectively, as follows: LOD 1.0, 1.5 and 1.0 ng mL^{-1}; LOQ 8, 4 and 4 ng mL^{-1}; linearity ranges 8 to 2048, 4 to 1024 and 4 to 1024 ng mL^{-1} [285].

2. Gas chromatography

Dialkylmagnesium compounds R_2Mg and Grignard reagents RMgX can be derivatized to yield volatile compounds containing the R group, and the products may be subjected to

(62)

(63)

(64)

(65)

end analysis by GC. Although reaction with active hydrogen compounds such as butanol and water (equations 8 to 11 and 23) is fast and quantitative, the lower members of the R series may produce hydrocarbons that are too volatile and require stringent analytical conditions to avoid losses. Derivatization with carbonyl compounds such as CO_2 (equation 22), esters, ketones (equation 21) or aldehydes (equation 24) requires a hydrolysis step of the R'O–MgX intermediate to obtain the volatile derivatives; furthermore, the R'OH compounds may need further derivatization to assure good chromatographic behavior.

Coupling of Grignard reagents RMgX with reactive halides R'X has the advantage of directly yielding volatile compounds RR' which may be analyzed by GC. This method was applied for determination of vinyl chloride by coupling with Bu_3SnCl, using dodecane as internal standard[286], and of methyllithium and methylmagnesium compounds by coupling with $Me_2PhSiCl$, using cumene as internal standard[287]. In Section V.B analytical methods are discussed for various organometallic and inorganic compounds based on the coupling reaction with Grignard reagents acting as derivatization agents.

C. Spectral Methods

1. Infrared spectroscopy

The course of development of a process in the solid state usually poses difficult analytical problems. A nondestructive, noninvasive method for determination of multicomponent mixtures uses an IR reflectance technique such as DRIFTS, with data processing by procedures such as the multivariate PLS regression and the RMSEP parameter. A practical advantage of the DRIFTS method is the simplicity of the analytical procedure, with no sample preparation other than proper placement in the instrument, and no post-analytical procedures, as no reagents are used. The method requires previous knowledge of the spectra of the individual components of interest, and it was applied to follow the composition of rocks containing calcium carbonate, magnesium carbonate and magnesium oxide in a process aiming at separation of these components[288].

2. Gilman's color tests

Gilman and coworkers developed color tests for identifying classes of organometallic compounds, including those containing magnesium and lithium. The methods for organomagnesium compounds are easy to perform and may be helpful to assess the development of processes involving Grignard reagents. The first such test consists of treating the sample solution with a 1% solution of Michler's ketone (**66**) in C_6H_6, followed by addition of H_2O and subsequent acidifying with dilute AcOH. A characteristic greenish blue color develops for compounds containing the C–MgX moiety. A deep blue or purple color develops in the presence of metallic Mg, therefore the solution where a Grignard reagent was synthesized should be filtered before carrying out the test[289,290]. The mechanism can be explained as formation of a tertiary alcohol derived from **66**, which forms a colored carbonium ion on acidification[291]. The color test was used for assessing the rate of formation of Grignard reagents in Et_2O for various alkyl and aryl halides in the presence of Mg turnings, with or without addition of I_2 as catalyst[292], and for monitoring the progress of the reaction of various Grignard reagents with azobenzene[293]. However, when tracking reaction mixtures interference with the color test was observed for certain halogen compounds, such as BzBr, Cl_2CO, ClCOCOCl and PCl_5 and amino compounds such as pyrrole (**67**), PhNHMe, Bu_2NH, All_2NH, $PhNAlkyl_2$, but none was observed for Et_2NH, Ph_2NH, BnNHPh or piperidine (**68**)[294-296]. Two color tests using less sensitive

(66)   (67)   (68)

reagents were developed for alkyllithium and aryllithium compounds are ineffective for Grignard reagents[297]. Solutions of various amino compounds develop a discoloration in the presence of alkyllithium compounds but not of Grignard reagents[298].

3. Nuclear magnetic resonance spectroscopy

The titer of organolithium compounds, Grignard reagents and lithium diisopropylamide solutions in ordinary (nondeuteriated) organic solvents can be easily determined by the so-called No-D NMR spectroscopy, based on ^{1}H NMR spectra, taking advantage of the power, signal stability and operational capabilities of modern instrumentation. Typical concentrations of commercially acquired reagents are in the 1 to 2 M range while those of the solvents are about 10 M, thus the solutes can be clearly seen in the presence of the solvent; furthermore, the analytical quality of the measurements can be improved with various provisions, allowing good quantitation of the solutes. A precisely measured amount of an adequate standard should be added to the solution; cyclooctene (**69**), with three multiplets at δ5.615, 2.14 and 1.49 ppm, in 1:2:4 ratio[299a], or 1,5-cyclooctadiene (**70**), with two peaks at δ5.558 and 2.36 ppm in 1:2 ratio[299b] (data for CDCl$_3$ solution), were used in the following examples. A commercial sample of allylmagnesium chloride (**71**) in THF using **69** as reference compound showed the peaks of solvent and reference, a quintuplet at δ6.0 and a doublet at δ2.1. The latter peaks correspond to H2 and H1,3 of **71**, respectively; in this compound the protons at positions 1 and 3 are equivalent by virtue of a fast 1 $\rightleftharpoons$ 3 rearrangement of MgCl taking place at 25 °C. Integration of the peaks of **69** and **71** pointed to a 1.2 M concentration of the acquired product. The No-D NMR method may also be helpful to assess the quality of an organometallic sample. Thus, a solution of vinylmagnesium bromide (**72**) in THF, using **70** as reference, showed that it contained substantial amounts of vinyl bromide (CH$_2$=CHBr) and 1,3-butadiene (CH$_2$=CHCH=CH$_2$), as all the protons of these compounds appeared in the spectrum, besides the quadruplets of H$_a$, H$_b$ and H$_c$ of **72**[300]. An earlier version of the No-D NMR method was proposed for titer determination of Grignard reagents and other alkylmetal solutions in Et$_2$O, THF or hexane, using CH$_2$Cl$_2$ or C$_6$H$_6$ as internal standard. Most frequently the peaks used for integration were those of the α-C protons, as they stood apart from those of solvent or standard. No coupling was observed for these protons with the ^{25}Mg nucleide (10% natural abundance and $I = \frac{5}{2}$)[301].

(69)   (70)   (71)   (72)

^{1}H and ^{13}C NMR spectroscopies afford a means of following the development of complex processes in solution. Equation 25 illustrates the power of this instrumental method, which allows identification and relative quantitation of the various species present. Triisopropylsilylamine reacts with dimethylmagnesium to yield methane by reaction of Me$_2$Mg with an active hydrogen compound (see Section V.A) and a dimeric heteroleptic organomagnesium compound (**73**), which is further stabilized by coordination with the THF solvent. No equilibrium involving **73** with a homoleptic compound was observed. Furthermore, two sets of chemical shifts (relative to TMS) were obtained for the protons at the secondary and primary positions of the isopropyl groups, pointing to the existence of two diastereoisomers of **73**, with the *N*-triisopropylsilyl groups being at the same or opposite sides of the plane determined by the N$_2$Mg$_2$ cycle[302].

$$2\ i\text{-Pr}_3\text{SiNH}_2 + 2\text{MgMe}_2 \xrightarrow{\text{THF}} \text{(73)} + 2\text{CH}_4 \qquad (25)$$

(**73**)

The dimeric complex **74** reacts with phenylacetylene or ferrocenylacetylene to yield the tetrameric complexes **75a** and **75b**, respectively, according to equation 26. These complexes are stable in CDCl$_3$ solution in the absence of air and can be characterized by ^{1}H and ^{13}C NMR spectroscopies. The low solubility of **75a** in unreactive organic solvents precludes detailed studies of the solution structure; in reactive solvents it decomposes to a dimeric complex, **76**, according to equation 27[303]. The association behavior of these complexes resembles that of analogous organolithium compounds[260b, 303].

$$2\ \text{(74)} + R\text{—}\equiv\text{—}H \xrightarrow{\text{RT, 18 h OEt}_2} \text{(75)} + 4\text{CH}_4 \qquad (26)$$

(**74**)   (**75**) (**a**) R = Ph
              (**b**) R = C$_5$H$_4$FeC$_5$H$_5$

The degradation of a compound in THF-d_8 solution could be followed by ^{31}P{^{1}H} NMR spectroscopy. Thus, for example, the heteroleptic dimeric phosphanide Grignard analogue **77** does not dissolve well in nondonor solvents; the fresh solution in THF shows two main peaks A and B, at $\delta -107.0$ and -103.1 ppm, respectively; peak B is tentatively attributed to the homoleptic compound **78**, obtained on loss of MgBu$_2$. After 1 h, peak A is markedly decreased and peaks C and D appear at $\delta -104.4$ and -31.8 ppm, respectively. Peak C possibly corresponds to an oligomeric alkoxophosphanide complex such as **79** while D belongs to the tertiary phosphane **80**. After 36 h, peaks A and B have almost disappeared, leaving fully developed peaks C and D. The fate of the Mg compounds lost on forming **80** is unknown[304].

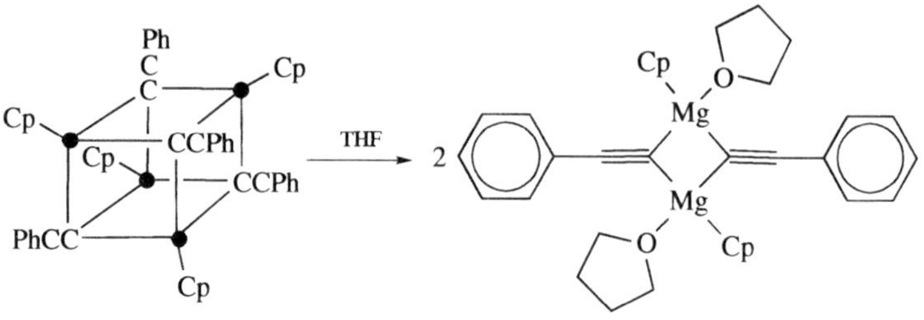

(75a) → (76) (27)

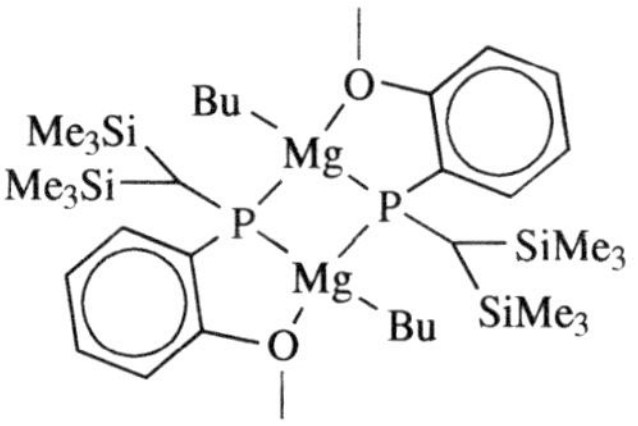

(77)

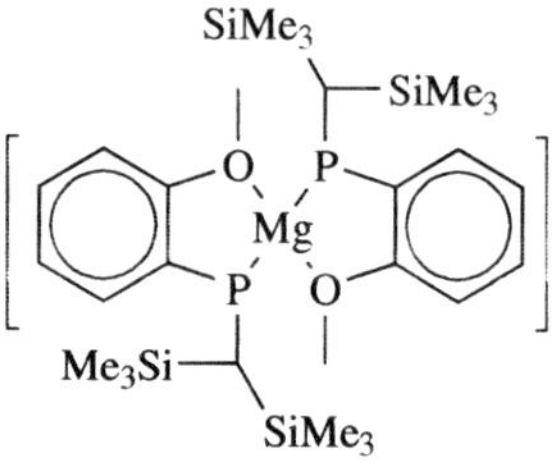

(78)

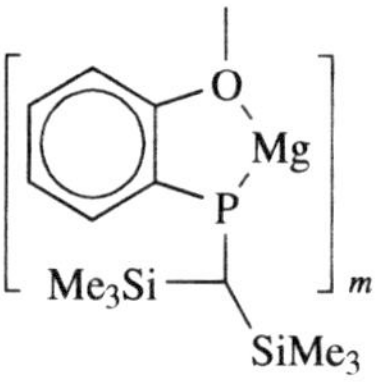

(79)

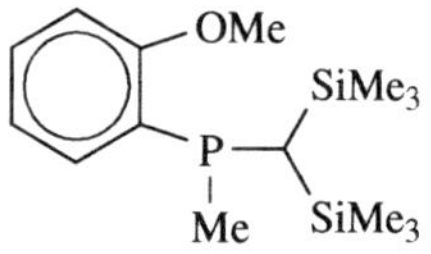

(80)

The transformation of a silyl Grignard reagent (**81**, δ 0.457 ppm) into a disilylmagnesium compound (**82**, δ 0.465 ppm) by the action of dioxan in C_6D_6 solution, according to equation 28, could be followed in time by ^1H NMR spectroscopy. It should be noted that the THF ligands are strongly coordinated to the Mg atom[305].

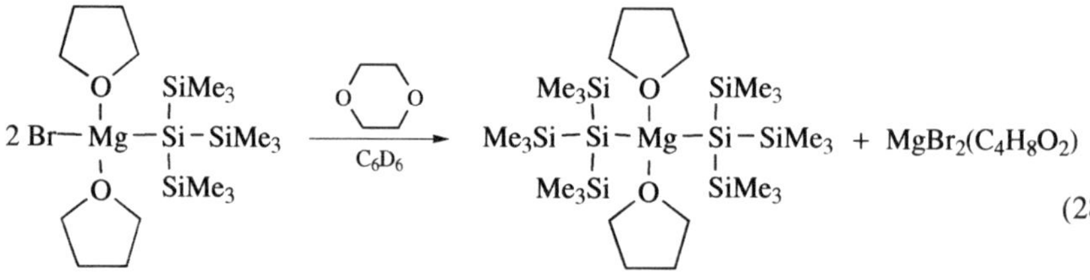

$$2 \text{ (81)} \xrightarrow{C_6D_6, \text{ dioxan}} \text{(82)} + MgBr_2(C_4H_8O_2) \quad (28)$$

4. Electron spin resonance spectroscopy

The ESR spectrum of the 2-thenyl free radical can be observed at 77 K when preparing a Grignard reagent (**85**), according to equation 29. When thenyl bromide (**83**, X = Br) is in the presence of metallic Mg, a quadruplet with full width of *ca* 55 G is shown, due to the spin interaction of the protons of the methylene group and at positions 3, 4 and 5 of the ring; the ESR spectrum of the free radical observed for 2-thenyl iodide (**83**, X = I) is poorly resolved, however its width corresponds to that of X = Br. In the particular case of **83** (X = Cl), that multiplet is superimposed on a singlet with half width 9 ± 2 G, attributed to the free radical pair **84**, including a Mg cluster. The difference in spectral behavior was correlated with the dissociation energy of the C−X bond in **83**[306].

$$\underset{(83)}{\text{thenyl-X}} \xrightarrow[X=Cl]{Mg} \left[\underset{(84)}{\text{thenyl-}\overset{\bullet}{\text{CH}}_2 \overset{\bullet}{\text{Mg}}_n \text{Cl}^-} \right] \longrightarrow \underset{(85)}{\text{thenyl-MgCl}} \quad (29)$$

X = Cl, Br, I

D. Cryoscopy

Cryoscopic determination of the molecular weight of $t\text{-Bu}_2\text{Mg}$ in benzene solution pointed to a dimer (**86**), the structure of which was confirmed by ^{1}H and ^{13}C NMR spectroscopy and ultimately determined by XRD crystallographic analysis[307].

$$t\text{-Bu}\text{---Mg} \overset{\times}{\underset{\times}{}} \text{Mg}\text{---Bu-}t$$

(**86**)

V. GRIGNARD REAGENTS AS ANALYTICAL REAGENTS AND AIDS
A. Active Hydrogen

Since Zerevitinov's publication in 1913[308], his method underwent analytical refinements and has been applied on multiple occasions. Alcohols (e.g. equation 10, R = Me) and amines evolve at room temperature 1 mol of methane per mol of functional group. Certain compounds containing 'active' CH groups α to aromatic rings are less reactive and require warming. Compounds achieving only partial evolution of methane after warming were considered to be tautomeric. An apparatus was described for processes involving MeMgI, by which both the volume of CH_4 evolved and the excess Grignard reagent can be determined. Such processes need not necessarily be reaction with active hydrogen[309,310]. Volumetric determination of MeMgI was proposed instead of the Karl Fischer reagent, for determination of water in metal powders, such as Pb, Sn, Ag and Mg, by volumetric measurement of the evolved methane[311]. A method for selective determination of active hydrogen consists of analyzing the sample by GC-TCD and collecting the emerging fractions in a MeMgI/Bu$_2$O solution, which is analyzed for the formation of CH_4 by GC on a short silica gel column, using a second TCD. The method was verified with a mixture of EtOH, AcOEt, AcEt, PhH and BuOH[312].

Care should be taken when developing analytical processes using Grignard reagents, choosing the right solvents and conditions of reaction, lest the analytical results will be affected by spurious factors. An example is the simultaneous determination of active hydrogen and carbonyl group in certain aromatic aldehydes, as deduced from methane evolution and total consumption of MeMgI. Vanillin (**87a**) and isovanillin (**87b**) show incomplete reaction in dioxan and the presence of one active hydrogen and one carbonyl group in pyridine solution. Veratraldehyde (**87c**), on the other hand, shows one carbonyl group in xylene, whereas in dioxan or pyridine the analysis shows less than one carbonyl group and the presence of active hydrogen. Indeed, it happens that **87c** undergoes an equilibrium dimerization in the latter solvents to yield **88**, which explains the results[313].

(**87**) (**a**) R = H, R′ = Me
(**b**) R = Me, R′ = H
(**c**) R = R′ = Me

The concentration of silanol groups in porous borosilicate glass could be determined by the Grignard titration method. The value found (1.945 mmol g^{-1}) for a sample with mesopores of 10.0 nm diameter and 0.938 mL g^{-1} volume was significantly larger than that (0.681 mmol g^{-1}) found for a sample with micropores of 2.0 nm diameter and 0.263 mL g^{-1} volume. On the other hand, the values obtained for these samples by thermogravimetric analysis, due to weight loss by condensation of the silanol groups, were similar to each other. The analytical discrepancy was attributed to restricted access of the Grignard reagent to the silanol groups in the sample with the micropores[314].

The detection and determination of active hydrogen have been reviewed[315-317].

B. Derivatization Reagents

1. Gas chromatography of organometallic compounds

Cost-effective methods of good analytical quality are needed for detection and speciation of organometallic compounds, for research, quality control, forensic analysis and environmental pollution control. Of special interest are organotin compounds, for their widespread commercial application as additives in many materials. Frequently, sample preparation includes extraction of organotin compounds with tropolone (**89**) solutions, of organolead compounds with dithizone (**90**) or sodium N,N-diethyldithiocarbamate (**91**) solutions[318] and of organoantimony with ammonium pyrrolidinedithiocarbamate (**92**) solutions[319]. A widespread procedure for the analysis of organometallics bearing halogen atoms on the metallic atom consists of treating the sample with a Grignard compound, destroying the excess of reagent, extracting the products and carrying out the end analysis by GC using various detection methods. The synthesis of volatile derivatives of organometallic compounds for GC speciation anlysis has been reviewed[320,321]. The partially alkylated analytes exchange the halogen atoms with the group of the Grignard compound, therefore the latter should bear an organic group different from those of the analyte, and the nature of this group should vary according to the effect to be achieved. For example, added volatility is attained with small alkyl groups, diminished volatility with large alkyl groups or fingerprinting with perfluoroalkyl groups. Although MSD is probably the most informative detection method, other less expensive detectors can give satisfactory results, especially in routine analysis. Applications of GC in speciation analysis of organometallic compounds after derivatization with Grignard reagents are summarized in Table 2.

(89)   PhNHNHCON=NPh (90) $Et_2NCS_2^-Na^+$ (91) $N-CS_2^-NH_4^+$ (92)

2. Analysis of glycerides and waxes

The abundance of fatty acids in triacylglycerols (TAG) can be determined by conversion of the acyl groups into tertiary alcohols on reaction with an alkyl Grignard reagent, followed by chromatographic separation. The method was found to be of advantage over saponification and conversion to a methyl ester, especially in the determination of short-chain fatty acids, which suffer losses by volatilization[355].

A standard procedure for identifying the fatty acid at the β-position of TAG is carrying out a pancreatic lipase hydrolysis. Analysis of the acyl groups at α-positions, however, is affected by isomerization taking place in the diacylglycerols[356]. An alternative approach was proposed based on partial conversion of the acyl groups to tertiary alcohols with MeMgBr or EtMgBr. The process lasts a few seconds, resulting in a mixture of α- and β-monoacylglycerols, α,α'- and α,β-diacylglycerols, unreacted TAG and tertiary alcohols[357,358]. After separation by preparative TLC on silicic acid impregnated with boric acid to prevent isomerization, each fraction can be further analyzed for identification (besides that indicated by the R_f values) and quantitation. For example, the glycerol esters can be converted to methyl esters by direct addition on the TLC spot of MeOH containing BF_3, followed by extraction and GC. The method was demonstrated by the regiodistribution analysis of tuna oil and milk fat[358]. Derivatization of TAG with EtMgBr for determination of the distribution of fatty acids in TAG was part of extensive analytical investigations of the oil extracted from *Rhodococcus opacus* strain PD630[359] or from evening primrose oil[360].

The fatty acid/fatty alcohol distribution of jojoba wax was determined after derivatization with EtMgBr, leaving a mixture of tertiary alcohols stemming from the fatty acid and the primary alcohols which were present as esters in the wax. The progress of wax disappearance due to the Grignard reaction could be assessed by TLC, with only traces left after 10 min. MeMgBr was even faster; however, reaction with *n*-BuMgBr or BnMgBr was complete only after 2 h. End analysis was by GC-FID[361].

C. Ancillary Applications

1. Surface conditioning

Fused silica capillary tubes were variously coated for capillary electrophoresis. The chemical process involving a Grignard reaction is shown in equation 30. The silanol groups on the silica surface are treated with alkali, dried, converted to chlorosilanes with thionyl chloride and vinylmagnesium bromide replaces the chlorine atoms with vinyl

TABLE 2. Derivatization of organometallic compounds with Grignard reagents for GC analysis

Analytes and comments	Grignard reagent	Detection method
Organotin compounds		
Bu$_3$SnCl. Study of SPE effectiveness in environmental samples. After dipping the SPE extraction tube in the sample extract, it was eluted with AcOEt, the eluate evaporated and the residue derivatized, extracted and subjected to end analysis[322]. Better results are obtained for derivatization with n-PenMgBr[323].	MeMgBr	MSD[322] FPD[323]
BuSnCl$_3$, Bu$_2$SnCl$_2$, Bu$_3$SnCl, PhSnCl$_3$, Ph$_2$SnCl$_2$, Ph$_3$SnCl, cyhexatin (**93**), fenbutatin oxide (**94**). Determination of butyltin and phenyltin pollutants in mussels, after tissue destruction with NMe$_4$OH, extraction with tropolone (**89**) solution and derivatization. Almost similar results were obtained for *in situ* ethylation with NaBEt$_4$ in acetate buffer at pH 4[323, 324]. Elution of pesticide residues **93** and **94** in food with AcOH/Et$_2$O, derivatization, cleaning by GPC. LOD were 0.02 ppm **93** and 0.05 ppm **94**. Better results are obtained for derivatization with n-PenMgBr[323].	EtMgBr	AED[324] FPD[323, 325, 326]
Et$_3$SnBr, Bu$_3$SnBr, Me$_4$Sn, Et$_4$Sn, cyhexatin (**93**), fenbutatin oxide (**94**). General procedure[327, 328]. Elution of pesticide residues **93** and **94** in crops with HCl/Me$_2$CO, transfer to CH$_2$Cl$_2$, cleaning by GPC[329]. Better results are obtained for derivatization with n-PenMgBr[323].	n-PrMgCl	FFGD-MSD[328] FPD[323, 329]
MeSnCl$_3$, Me$_2$SnCl$_2$, Me$_3$SnCl, BuSnCl$_3$, Bu$_2$SnCl$_2$, Bu$_3$SnCl, PhSnCl$_3$, Ph$_2$SnCl$_2$, Ph$_3$SnCl. General review[330]. Simultaneous determination of Sn, Pb and Hg organometallics[331]. Contamination analysis of various items: (i) Antarctic bivalve *Adamussium colbecki*[332]; (ii) water[333]; (iii) fish and marine sediments, digestion with concentrated HCl, LLE with **89**/PenH[334]; (iv) mussels and marine sediments; higher sensitivity for the n-pentyl than for the n-propyl derivatives; better recoveries for the n-pentyl than for the methyl or n-propyl derivatives[335]; (v) lard, MeSn(Pr-n)$_3$ as internal standard for FPD[336]; (vi) sediments after LLE with **91**[318]; (vii) sediments after SFE[337]; (viii) occupational exposure to organotin compounds in particulates or as vapors[338]; (ix) water and sediments. No difference was observed for the FPD[339], AAS or AES[340] results after pentylation of the alkyltin compounds with n-PenMgBr or ethylation with NaBEt$_4$; however, slightly better results were reported for the Grignard reagent[323].	n-PenMgBr	EI-MSD[330, 332, 333] PCI-MSD[330] EI-MS-MSD[335] FPD[323, 330, 336, 339] AED[318, 331, 337, 338, 340, 341] ETAAS[340] FAAS[334]
MeSnCl$_3$, Me$_2$SnCl$_2$, Me$_3$SnCl. (i) Determination in wastewaters includes LLE with CH$_2$Cl$_2$, and	n-HexMgBr	FPD

TABLE 2. (continued)

Analytes and comments	Grignard reagent	Detection method
evaluation of an efficiency factor for the extraction[342]; (ii) ppb levels in seawater, tissues and marine sediments, homogenization in acid, LLE with CH_2Cl_2; LOD in tissues and sediments were 0.29, 0.12 and 0.1 ng for mono-, di- and tributyltin, respectively[343].		
Arsenic compounds		
As(III) and As(V) compounds. Comparison with derivatization with N-t-butyldimethylsilyl-N-methyltrifluoroacetamide with 1% t-butyldimethylchlorosilane[344].	4-FC_6H_4MgBr	EI-MSD PCI-MSD NCI-MSD
Lead compounds		
Me_2PbCl_2, Et_2PbCl_2, Me_3PbCl, Et_3PbCl, Me_4Pb, Et_4Pb. LLE from water with **91**/HexH, derivatization with n-Pr groups gave better resolution than with n-Bu groups[345].	n-PrMgCl	AAS[345]
Pb(II), Me_2PbCl_2, Me_3PbCl, Et_2PbCl_2, Et_3PbCl, Et_4Pb, Bu_4Pb. Determination of organolead compounds in airborne particulates; Et_4Pb as internal standard[346]. Derivatization was performed after preconcentration of water pollutants on polymer beads functionalized with dithizone (**90**) groups[347]. LLE from water with **91**/HexH, derivatization with n-Pr groups gave better resolution than with n-Bu groups[345]. Determination of Pb(II) and organolead compounds in human urine, after LLE of the complexes with **91** and derivatization; Et_4Pb as internal standard[348]. The aqueous solution is evaporated to dryness, the solid residue is derivatized and the alkylated Pb was evaporated into an ETAAS instrument; the LOD (SNR 2) was 7 ppb[349].	n-BuMgCl	ICP-MSD[346] SIM-MSD[348] AAS[345,347] ETAAS [a] [349]
Me_3PbCl, Me_3PbAc, Et_3PbAc, Pr_3PbAc, Me_2PbAc_2, Et_2PbAc_2, Pr_2PbAc_2. Simultaneous determination of Sn, Pb and Hg organometallics[331]. Determination of organotin and organolead pollutants in environmental sediments after LLE with **91** solution and derivatization[318].	n-PenMgBr	AED[318,331]
Pb. Determination in urine and whole blood by isotope dilution. An aliquot of ^{204}Pb is added to a urine or blood sample and digested with HNO_3/H_2O_2, the residue is dissolved with lithium bis(trifluoroethyl)dithiocarbamate (**95**) to give chelate **96a**, which yields the volatile **97a** after derivatization with the Grignard reagent. Total Pb by AAS; ^{204}Pb/^{206}Pb, ^{204}Pb/^{207}Pb and ^{204}Pb/^{208}Pb isotope ratios by MSD (see equation 7R718)[350].	4-FC_6H_4MgBr	EI-SIM-MSD TE-AAS

(*continued overleaf*)

TABLE 2. *(continued)*

Analytes and comments	Grignard reagent	Detection method
Antimony compounds		
Sb(III) compounds, Ph$_3$Sb. After extraction with **92** solution, evaporation to dryness and derivatization.	PhMgBr	TE-AAS[319]
Mercury compounds		
MeHgCl, inorganic Hg(II). Extraction of MeHgCl from sediments by SFE or steam distillation, derivatization and end analysis[351].	*n*-BuMgCl	DIN-ICP-MSD MIP-AED
MeHgCl, EtHgCl, PhHgOAc, C$_7$H$_{15}$HgCl, (Me$_3$SiCH$_2$)$_2$Hg, Ph$_2$Hg. Simultaneous determination of Sn, Pb and Hg organometallics[331].	*n*-PenMgBr	AED
Cadmium compounds		
Cd(II). The aqueous solution is evaporated to dryness, the solid residue is derivatized and the alkylated Cd is evaporated and carried into an ICP-AES instrument; the LOD was 11 pg[352].	EtMgCl	ICP-AES[a]
Gallium compounds		
Ga(III). The aqueous solution is evaporated to dryness, the solid residue is derivatized and the alkylated Ga is evaporated and carried into an ICP-AES instrument; the LOD was 1.9 ng[353].	EtMgCl	ICP-AES[a]
Tellurium compounds		
Te(IV) and other forms. Determination of Te in urine by isotope dilution after intake of the antitumor drug AS 101 (**98**). An aliquot of ^{120}Te is added to a urine sample and digested with HNO$_3$/H$_2$O$_2$, the residue is dissolved with a solution of **95** to give chelate **96b**, which yields the volatile **97b** after derivatization with the Grignard reagent. End analysis of total Te by AAS and ^{120}Te/^{130}Te isotope ratio by MSD (see equation 7R718)[354].	4-FC$_6$H$_4$MgBr	EI-SIM-MSD TE-AAS

[a] End analysis without GC separation.

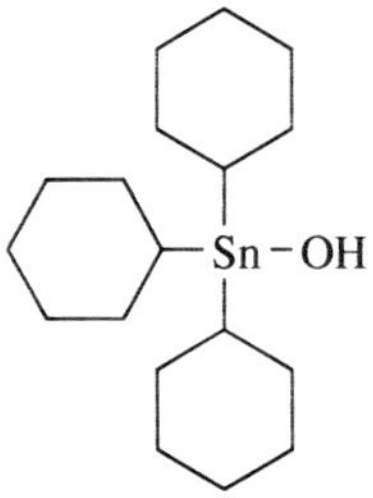

(93)

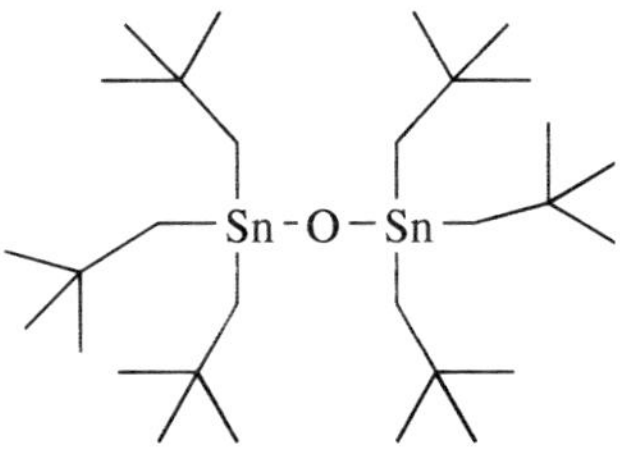

(94)

(F$_3$CCH$_2$)$_2$NCS$_2$Li   [(F$_3$CCH$_2$)$_2$NCS$_2$]$_2$M   (4-FC$_6$H$_4$)$_2$M $\left[\begin{array}{c}\text{O}\\\text{O}\end{array}\text{TeCl}_3\right]^-$ NH$_4^+$

(**95**)   (**96**) (a) M = Pb   (**97**) (a) M = Pb   (**98**)
                  (b) M = Te   (b) M = Te

groups. These groups participate in a polymerization reaction in the presence of vinyl acetate, and the product is hydrolyzed by alkali to yield a silica capillary with a coating of polyvinyl alcohol (**99**) which is very resistant in operation at high pH. These capillaries were used for CE of dsDNA fragments with excellent results[362,363].

$$\text{Si}-\text{OH} \xrightarrow[\text{(alkalinization)}]{\text{NaOH}} \text{Si}-\text{ONa} \xrightarrow[\text{(chlorination)}]{\text{SOCl}_2} \text{Si}-\text{Cl} \xrightarrow[\text{(vinylation)}]{\text{CH}_2=\text{CHMgBr}} \text{Si}-\text{CH}=\text{CH}_2$$

$$\xrightarrow[\text{(copolymerization)}]{\text{AcOCH}=\text{CH}_2}$$

(Si–[OAc,OAc,OAc / OAc,OAc,OAc] branched structure) $\xrightarrow[\text{(saponification)}]{\text{NaOH}}$ (Si–[OH,OH,OH / OH,OH,OH] branched structure)   (30)

(**99**)

The HPLC separation capabilities of poly(chlorotrifluoroethylene) can be modified by treatment with Grignard reagents, by which a Wurtz-type reaction takes place with the Cl moieties of the polymer. Thus, Kel-F 6300 or Kel-F 6061 altered with long alkyl groups (C$_8$ or C$_{18}$) had a poor lifetime; when the alkyl group was short (C$_1$) separation was ineffective. Insertion of aryl groups (Ph, Naph), and especially Ph, conferred the column a good lifetime and the RP-HPLC performance was similar or better when compared to silica-C$_{18}$ and PRP-1, for the separation of benzene compounds bearing various functional groups[364].

2. Electrochemical behavior

EtMgBr solutions in poly(ethylene oxide) containing a small amount of THF or Et$_2$O are electrically conducting. Best conductivity is achieved for an ethylene oxide–Mg ratio of 4, e.g. 0.1 mS cm^{-1} at 40 °C was found. In contrast, PEO solutions of MgCl$_2$, Mg(ClO$_4$)$_2$ or Mg(SCN)$_2$ show only low electrical conduction below 100 °C. Furthermore, in the presence of EtMgBr solutions Mg can be deposited by cathodic reduction or dissolved by anodic oxidation. Practical application of these solutions are limited by their low thermal and electrochemical stability[365].

VI. REFERENCES

1. J. Durlach, N. Pages, P. Bac, M. Bara and A. Guiet-Bara, *Magnesium Res.*, **17**, 163 (2004).
2. V. Grignard, *Ann. Chim.*, **24**, 433 (1901); V. Grignard, *Notice sur les Titres et Travaux Scientifiques de M. V. Grignard*, Imprimerie J. Marlhens, Lyon, 1926; (http://gallica.bnf.fr/ark:/12148/bpt6k90654d); J. J. Eisch, *Organometallics*, **21**, 5439 (2002).
3. F. Bickelhaupt, *Chem. Soc. Rev.*, **28**, 17 (1999).
4. F. Bickelhaupt, in *Grignard Reagents* (Ed. H. G. Richey, Jr.), Wiley, Chichester, 2000, pp. 367–393.
5. V. V. Smirnov, L. A. Tjurina and I. P. Beletskaya, in *Grignard Reagents* (Ed. H. G. Richey, Jr.), Wiley, Chichester, 2000, pp. 395–410.
6. O. G. Kulinkovich, *Pure Appl. Chem.*, **72**, 1715 (2000).
7. P. Knochel, E. Hupe and H. Houte, *Actualite Chim.*, 12 (2003).
8. P. Knochel, W. Dohle, N. Gommermann, F. F. Kneisel, F. Kopp, T. Korn, I. Sapountzis and V. A. Vu, *Angew. Chem. Int. Ed.*, **42**, 4302 (2003).
9. H. Shinokubo and K. Oshima, *Eur. J. Org. Chem.*, 2081 (2004).
10. H. Ila, O. Baron, A. J. Wagner and P. Knochel, *Chem. Lett.*, **35**, 2 (2006).
11. H. Ila, O. Baron, A. J. Wagner and P. Knochel, *Chem. Commun.*, 583 (2006).
12. I. L. Cameron and N. K. Smith, *Scanning Electron Microsc. Pt. 2*, 463 (1980).
13. K. M. Kim, H. B. Alpaugh and F. B. Johnson, *Scanning Electron Microsc. Pt. 3*, 1239 (1985).
14. U. Oesch, D. Ammann and W. Simon, *Clin. Chem.*, **32**, 1448 (1986).
15. G. J. Kost, *Crit. Rev. Clin. Lab. Sci.*, **30**, 153 (1993).
16. N. Fogh-Andersen and O. Siggaard-Andersen, *Scand. J. Clin. Lab. Invest. Suppl.*, **217**, 89 (1994).
17. M. A. Olerich and R. K. Rude, *New Horiz.*, **2**, 186 (1994).
18. B. T. Altura, *Scand. J. Clin. Lab. Invest. Suppl.*, **217**, 5 (1994).
19. B. T. Altura, J. L. Burack, R. Q. Cracco, L. Galland, S. M. Handwerker, M. S. Markell, A. Mauskop, Z. S. Memon, L. M. Resnick, Z. Zisbrod et al., *Scand. J. Clin. Lab. Invest. Suppl.*, **217**, 53 (1994).
20. B. M. Altura and B. T. Altura, *Scand. J. Clin. Lab. Invest. Suppl.*, **224**, 211 (1996).
21. N. E. Saris, E. Mervaala, H. Karppanen, J. A. Khawaja and A. Lewenstam, *Clin. Chim. Acta*, **294**, 1 (2000).
22. A. P. Somlyo and H. Shuman, *Ultramicroscopy*, **8**, 219 (1982).
23. R. Y. Tsien, *Ann. Rev. Biophys. Bioeng.*, **12**, 91 (1983).
24. R. K. Gupta, P. Gupta and R. D. Moore, *Ann. Rev. Biophys. Bioeng.*, **13**, 221 (1984).
25. R. E. London, *Ann. Rev. Physiol.*, **53**, 241 (1991).
26. H. Koppel, R. Gasser and U. Spichiger, *Wien. Med. Wochenschr.*, **150**, 321 (2000).
27. B. B. Silver, *J. Am. Coll. Nutr.*, **23**, 732S (2004).
28. W. E. Wacker, *Magnesium*, **6**, 61 (1987).
29. R. J. Elin, *Magnesium Trace Elem.*, **10**, 172 (1991–1992).
30. W. R. Kulpmann, *Wien. Klin. Wochenschr., Suppl.*, **192**, 37 (1992).
31. G. A. Quamme, *Clin. Lab. Med.*, **13**, 209 (1993).
32. E. Murphy, *Miner. Electrolyte Metab.*, **19**, 250 (1993).
33. H. Millart, V. Durlach and J. Durlach, *Magnesium Res.*, **8**, 65 (1995).
34. C. Ritter, M. Ghahramani and H. J. Marsoner, *Scand. J. Clin. Lab. Invest. Suppl.*, **224**, 275 (1996).
35. G. Chittleborough, *Sci. Total Environ.*, **14**, 53 (1980).
36. C. Baluja-Santos, A. Gonzalez-Portal and F. Bermejo-Martinez, *Analyst*, **109**, 797 (1984).
37. W. Merlevede, J. R. Vandenheede, J. Goris and S. D. Yang, *Curr. Top. Cell. Regul.*, **23**, 177 (1984).
38. R. A. Reinhart, *Arch. Intern. Med.*, **148**, 2415 (1988).
39. G. Baltzer, *Med. Klin. (Munich)*, **83**, 370 (1988).
40. O. Ferment and Y. Touitou, *Presse Med.*, **17**, 584 (1988).
41. R. J. Elin, *Dis.-Mon.*, **34**, 161 (1988).
42. L. Paunier, *Monatsschr. Kinderheilkd.*, **140** (Suppl. 1), S17 (1992).
43. H. Benech and J. M. Grognet, *Magnesium Res.*, **8**, 277 (1995).
44. R. Mehrotra, K. D. Nolph, P. Kathuria and L. Dotson, *Am. J. Kidney Dis.*, **29**, 106 (1997).
45. R. Whang, *Comp. Ther.*, **23**, 168 (1997).

7. Analytical aspects of organomagnesium compounds 307

46. W. Vierling, *Herz*, **22** (Suppl 1), 3 (1997).
47. A. M. Romani and A. Scarpa, *Front. Biosci.*, **5**, D720 (2000).
48. M. D. Yago, M. Manas and J. Singh, *Front. Biosci.*, **5**, D602 (2000).
49. S. Iannello and F. Belfiore, *Panminerva Med.*, **43**, 177 (2001).
50. J. Durlach, N. Pages, P. Bac, M. Bara and A. Guiet-Bara, *Magnesium Res.*, **15**, 49 (2002).
51. M. E. Maguire and J. A. Cowan, *Biometals*, **15**, 203 (2002).
52. R. D. Grubbs, *Biometals*, **15**, 251 (2002).
53. A. M. P. Romani and M. E. Maguire, *Biometals*, **15**, 271 (2002).
54. J. Durlach, N. Pages, P. Bac, M. Bara and A. Guiet-Bara, *Magnesium Res.*, **15**, 203 (2002).
55. K. P. Schlingmann and T. Gudermann, *J. Physiol.*, **566** (Pt. 2), 301 (2005).
56. V. Chubanov, T. Gudermann and K. P. Schlingmann, *Eur. J. Physiol.*, **451**, 228 (2005).
57. F. H. Nielsen and H. C. Lukaski, *Magnesium Res.*, **19**, 180 (2006).
58. F. I. Wolf and A. Cittadini, *Front. Biosci.*, **4**, D607 (1999).
59. T. L. Barton, *Poultry Sci.*, **75**, 854 (1996).
60. P. W. Flatman, *Magnesium Res.*, **1**, 5 (1988).
61. J. G. Henrotte, *Magnesium*, **7**, 306 (1988).
62. M. Gawaz, *Fortschr. Med.*, **114**, 329 (1996).
63. T. Gunther, *Magnesium Res.*, **19**, 190 (2006).
64. S. J. Scheinman, L. M. Guay-Woodford, R. V. Thakker and D. G. Warnock, *New Eng. J. Med.*, **340**, 1177 (1999).
65. I. C. Meij, L. P. van den Heuvel and N. V. Knoers, *Adv. Nephrol. Necker Hosp.*, **30**, 163 (2000).
66. N. V. A. M. Knoers, J. C. de Jong. I. C. Meij, L. P. W. J. van den Heuvel and R. J. M. Bindels, *J. Nephrol.*, **16**, 293 (2003).
67. K. P. Schlingmann, M. Konrad and H. W. Seyberth, *Pediat. Nephrol.*, **19**, 13 (2004).
68. F. C. Driessens, *Z. Naturforsch. C*, **35**, 357 (1980).
69. R. A. Terpstra and F. C. Driessens, *Calcif. Tissue Int.*, **39**, 348 (1986).
70. J. J. Steichen and W. W. Koo, *Monatsschr. Kinderheilkd.*, **140** (Suppl. 1), S21 (1992).
71. M. E. Morris, *Magnesium Res.*, **5**, 303 (1992).
72. Y. Liu and J. Zhang, *Chin. Med. J.*, **113**, 948 (2000).
73. M. J. Cevette, J. Vormann and K. Franz, *J. Am. Acad. Audiol.*, **14**, 202 (2003).
74. G. A. Quamme, *Magnesium*, **5**, 248 (1986).
75. G. A. Quamme, *Kidney Int.*, **52**, 1180 (1997).
76. D. Juan, *Surgery*, **91**, 510 (1982),
77. R. J. Elin, *Clin. Chem.*, **33**, 1965 (1987).
78. E. Ryzen, K. L. Servis and R. K. Rude, *J. Am. Coll. Nutr.*, **9**, 114 (1990).
79. R. Whang, D. D. Whang, K. W. Ryde and T. O. Oei, *Magnesium Res.*, **3**, 267 (1990).
80. J. W. Van Hook, *Crit. Care Clin.*, **7**, 215 (1991).
81. M. Salem, R. Munoz and B. Chernow, *Crit. Care Clin.*, **7**, 225 (1991).
82. M. F. Ryan, *Ann. Clin. Biochem.*, **28** (Pt. 1), 19 (1991).
83. E. L. Tso and R. A. Barish, *J. Emergency Med.*, **10**, 735 (1992).
84. R. Whang, E. M. Hampton and D. D. Whang, *Ann. Pharmacother.*, **28**, 220 (1994).
85. N. Dhupa and J. Proulx, *Vet. Clin. North Am.*, **28**, 587 (1998).
86. M. F. Ryan and H. Barbour, *Ann. Clin. Biochem.*, **35** (Pt. 4), 449 (1998).
87. Z. S. Agus, *J. Am. Soc. Nephrol.*, **10**, 1616 (1999).
88. G. T. Sanders, H. J. Huijgen and R. Sanders, *Clin. Chem. Lab. Med.*, **37**, 1011 (1999).
89. J. M. Topf and P. T. Murray, *Rev. Endocr. Metab. Disorders*, **4**, 195 (2003).
90. D.-H. Liebscher and D.-E. Liebscher, *J. Am. Coll. Nutr.*, **23**, 730S (2004).
91. M. Gonella and G. Calabrese, *Magnesium Res.*, **2**, 259 (1989).
92. H. O. Garland, *Magnesium Res.*, **5**, 193 (1992).
93. C. S. Leach, *Microgravity Q.*, **2**, 69 (1992).
94. L. Tosiello, *Arch. Intern. Med.*, **156**, 1143 (1996).
95. H. W. de Valk, *Neth. J. Med.*, **5**, 139 (1999).
96. C. Hermes Sales and L. de F. Campos Pedrosa, *Clin. Nutr.*, **25**, 554 (2006).
97. B. M. Altura and B. T. Altura, *Magnesium*, **4**, 226 (1985).
98. B. M. Altura and B. T. Altura, *Magnesium*, **4**, 245 (1985).
99. R. Whang, *Magnesium*, **5**, 127 (1986).
100. L. T. Iseri, *Magnesium*, **5**, 111 (1986).

101. L. M. Resnick, *Am. J. Med.*, **93** (A), 11S (1992).
102. J. Durlach, V. Durlach, Y. Rayssiguier, M. Bara and A. Guiet-Bara, *Magnesium Res.*, **5**, 147 (1992).
103. M. A. Arsenian, *Prog. Cardiovasc. Dis.*, **35**, 271 (1993).
104. H. S. Rasmussen, *Dan. Med. Bull.*, **40**, 84 (1993).
105. C. V. Leier, L. Dei Cas and M. Metra, *Am. Heart J.*, **128**, 564 (1994).
106. J. McCord and S. Borzak, *Hosp. Pract.*, **29**, 47, 53, 57 (1994).
107. C. G. Osborne, R. B. McTyre, J. Dudek, K. E. Roche, R. Scheuplein, B. Silverstein, M. S. Weinberg and A. A. Salkeld, *Nutr. Rev.*, **54**, 365 (1996).
108. T. L. Shirey, *J. Anesth.*, **18**, 118 (2004).
109. A. Hordyjewska and K. Pasternak, *Ann. Univ. Mariae Curie-Sklodowska, Sect. D*, **59**, 108 (2004).
110. W. Bobkowski, A. Nowak and J. Durlach, *Magnesium Res.*, **18**, 35 (2005).
111. R. Rylander, *Clin. Calcium*, **15**, 11 (2005).
112. L. D. Lewis and M. L. Morris, Jr., *Vet. Clin. North Am.*, **14**, 513 (1984).
113. J. J. Pahira, *Urol. Radiol.*, **6**, 74 (1984).
114. M. Labeeuw and N. Pozet, *Magnesium Res.*, **1**, 187 (1988).
115. T. D. Mountokalakis, *Magnesium Res.*, **3**, 121 (1990).
116. O. Richard and M. T. Freycon, *Pediatrie*, **47**, 557 (1992).
117. S. Ekane, T. Wildschutz, J. Simon and C. C. Schulman, *Acta Urol. Belg.*, **65**, 1 (1997).
118. G. B. Fedoseev, A. B. Emel'ianov, V. A. Goncharova, K. K. Malakauskas, V. L. Emanuel' and T. M. Sinitsyna, *Klin. Med. (Moscow)*, **72**, 13 (1994).
119. A. Castillo and L. A. Ordóñez, *Acta Cient. Venez.*, **32**, 123 (1981).
120. R. Masironi and A. G. Shaper., *Ann. Rev. Nutr.*, **1**, 375 (1981).
121. T. Hazell, *World Rev. Nutr. Diet.*, **46**, 1 (1985).
122. J. R. Marier, *Magnesium*, **5**, 1 (1986).
123. I. E. Dreosti. *Nutr. Rev.*, **53** (Pt. 2), S23 (1995).
124. B. Lonnerdal, *Physiol. Rev.*, **77**, 643 (1997).
125. T. Clausen and I. Dorup, *Bibl. Nutr. Dieta*, [54], 84 (1998).
126. J. P. Goff, *Food Animal Pract.*, **15**, 619 (1999).
127. J. G. Dorea, *J. Am. Coll. Nutr.*, **19**, 210 (2000).
128. L. K. Massey, *J. Nutr.*, **131**, 1875 (2001).
129. F. Gaucheron, *Reprod. Nutr. Dev.*, **45**, 473 (2005).
130. S. Monarca, F. Donato, I. Zerbini, R. L. Calderon and F. F. Craun, *Eur. J. Cardiovasc. Prev. Rehabil.*, **13**, 495 (2006).
131. M. S. Seelig, *Magnesium Res.*, **3**, 197 (1990).
132. L. Spatling, *Gynakol.-geburtshilfliche Rundschau*, **33**, 85 (1993).
133. F. Mimouni and R. C. Tsang, *Magnesium Res.*, **4**, 109 (1991).
134. G. L. Klein and D. N. Herndon, *Magnesium Res.*, **11**, 103 (1998).
135. R. Mittendorf, P. G. Pryde, R. J. Elin, J. G. Gianopoulos and K.-S. Lee, *Magnesium Res.*, **15**, 253 (2002).
136. R. B. Costello and P. B. Moser-Veillon, *Magnesium Res.*, **5**, 61 (1992).
137. M. P. Vaquero, *J. Nutr. Health Aging*, **6**, 147 (2002).
138. X. Boman, T. Guillaume and J. M. Krzesinski, *Rev. Med. Liege*, **58**, 104 (2003).
139. S. Onishi and S. Yoshino, *Intern. Med.*, **45**, 207 (2006).
140. M. P. Ryan, *Magnesium*, **5**, 282 (1986).
141. H. Lajer and G. Daugaard, *Cancer Treat. Rev.*, **25**, 47 (1999).
142. L. E. Ramsay, W. W. Yeo and P. R. Jackson, *Cardiology*, **84** (Suppl. 2), 48 (1994).
143. K. L. Woods, *Br. J. Clin. Pharmacol.*, **32**, 3 (1991).
144. J. Atsmon and E. Dolev, *Drug Safety*, **28**, 763 (2005).
145. R. Vink and I. Cernak, *Front. Biosci.*, **5**, D656 (2000).
146. M. S. Kharasch and O. Reimuth, *Grignard Reactions of Nonmetallic Substances*, Prentice Hall, New York, 1954, p. 19.
147. A. V. Nemukhin, I. A. Topol and F. Weinhold, *Inorg. Chem.*, **34**, 2980 (1995).
148. U. Casellato and F. Ossola, *Organometallics*, **13**, 4105 (1994).
149. R. G. Martinek, *J. Am. Med. Technol.*, **36**, 241 (1974).
150. H. G. Seiler, *Met. Ions Biol. Syst.*, **26**, 611 (1990).

151. D. M. Templeton, F. Ariese, R. Cornelis, L.-G. Danielsson, H. Muntau, H. P. van Leeuwen and R. Łobiński, *Pure Appl. Chem.*, **72**, 1453 (2000).
152. A. Liparini, S. Carvalho and J. C. Belchior, *Clin. Chem. Lab. Med.*, **43**, 939 (2005).
153. B. T. Altura and B. M. Altura, *Scand. J. Clin. Lab. Invest. Suppl.*, **217**, 83 (1994).
154. A. Jensen and E. Riber, *Metal Ions Biol. Syst.*, **16**, 151 (1983).
155. N.-E. L. Saris, E. Mervaala, H. Karppanen, J. A. Khawaja and A. Lewenstam, *Clin. Chim. Acta*, **294**, 1 (2000).
156. E. J. Park, M. Brasuel, C. Behrend, M. A. Philbert and R. Kopelman, *Anal. Chem.*, **75**, 3784 (2003).
157. M. Kimura, K. Honda, A. Takeda, M. Imanishi and T. Takeda, *J. Am. Coll. Nutr.*, **23**, 748S (2004).
158. H. M. Barbour and W. Davidson, *Clin. Chem.*, **34**, 2103 (1988).
159. C. S. Shin, C. H. Chang and J.-H. Kim, *Yonsei Med. J.*, **47**, 191 (2006).
160. D. R. Banatao, R. B. Altman and T. E. Klein, *Nucleic Acids Res.*, **31**, 4450 (2003).
161. S. B. Yasar and Ş. Güçer, *Anal. Chim. Acta*, **505**, 43 (2004).
162. C. Jiwoua Ngounou, R. Ndjouenkeu, C. M. F. Mbofung and L. Noubi, *J. Food Eng.*, **57**, 301 (2003).
163. *Compendium of Methods for Chemical Analysis of Foods*, Bureau of Chemical Safety, Health Products and Food Branch, Health Canada, Ottawa, Ont., Canada, Method LPFC-137 (1985). (http://www.hc-sc.gc.ca/fn-an/res-rech/analy-meth/chem/reg_prep_dry_ashing-reg_prep_ echantillons_calcination_e.html).
164. J. Moreda-Piñeiro, E. Beceiro-González, E. Alonso-Rodríguez, E. González-Soto, P. López-Mahía, S. Muniategui-Lorenzo and D. Prada-Rodríguez, *At. Spectrosc.*, **22**, 422 (2001).
165. A. R. Shirazi and O. Lindqvist, *Fuel*, **72**, 125 (1993).
166. P. Cunniff (Ed.), *Official Methods of Analysis of AOAC International*, AOAC International, Gaithersburg, MD, USA, 1996, (a) Chap. 50, pp. 15–16; (b) Chap. 2, pp. 25–26; (c) Chap. 11, pp. 15–16; (d) Chap. 33, pp. 60–61; (e) Chap. 2, pp. 30–31.
167. L. M. Coelho, E. R. Pereira-Filho and M. A. Z. Arruda, *Quim. Anal.*, **20**, 243 (2002); *Chem. Abstr.*, **137**, 43744 (2002).
168. R. Moreno-Torres, M. Navarro, M. D. Ruiz-López, R. Artacho and C. López, *Lebens. Wiss. Technol.*, **33**, 397 (2000).
169. L. Liu, P. Li, Y. Li and L. Sun, *Guangpuxue Yu Guangpu Fenxi*, **21**, 560 (2001); *Chem. Abstr.*, **135**, 315474 (2001).
170. A. V. Filgueiras, J. L. Capelo, I. Lavilla and C. Bendicho, *Talanta*, **53**, 433 (2000).
171. F. Sugimoto, *Sen'i Gakkaishi*, **59**, 304 (2003); *Chem. Abstr.*, **140**, 340647 (2004).
172. K. L. Sahrawat, G. Ravi Kumar and J. K. Rao, *Commun. Soil Sci. Plant Anal.*, **33**, 95 (2002).
173. K. L. Sahrawat, G. Ravi Kumar and K. V. S. Murthy, *Commun. Soil Sci. Plant Anal.*, **33**, 3757 (2002).
174. L. Lin and L. Zhu, *Fenxi Huaxue*, **30**, 819 (2002); *Chem. Abstr.*, **137**, 311443 (2002).
175. L. A. Powell and R. L. Tease, *Anal. Chem.*, **54**, 2154 (1982).
176. X. Ou, T. Lin, P. Wang, L. Jiang and L. Li, *Guang pu xue yu guang pu fen xi*, **20**, 79 (2000); *PubMed ID*, 12953457.
177. J. Chang, *Yejin Fenxi*, **21**, 65 (2001); *Chem. Abstr.*, **136**, 111798 (2002).
178. K. Watanabe, T. Okada and M. Itagaki, *Bunseki Kagaku*, **52**, 55 (2003); *Chem. Abstr.*, **138**, 146813 (2003).
179. Dionex, *Anion Micromembrane Suppressor® III—Cation Micromembrane Suppressor® III*, Product Manual, Document No. 031727, Dionex Corporation, 2004. (http://www1.dionex.com/en-us/webdocs/4366_31727-03_MMS_Combined_V21.pdf).
180. J. E. Van Nuwenborg, D. Stöckl and L. M. Thienpont, *J. Chromatogr. A*, **770**, 137 (1997).
181. R. García-Fernández, J. I. García-Alonso and A. Sanz-Medel, *J. Chromatogr. A*, **1033**, 127 (2004).
182. M. C. Bruzzoniti, E. Mentasti and C. Sarzanini, *Anal. Chim. Acta*, **382**, 291 (1999).
183. B.-S. Yu, Q.-G. Yuan, L.-H. Nie and S.-Z. Yao, *J. Pharm. Biomed. Anal.*, **25**, 1027 (2001); erratum, *J. Pharm. Biomed. Anal.*, **29**, 969 (2002).
184. B. Godlewska-Żyłkiewicz, B. Leśniewska and A. Hulanicki, *Anal. Chim. Acta*, **358**, 185 (1998).
185. F. Umland, *Fresenius Z. Anal. Chem.*, **190**, 186 (1962).
186. T. Takeuchi, S. Inoue and T. Miwa, *J. Microcolumn Sep.*, **12**, 450 (2000).

187. T. Takeuchi, S. Inoue, M. Yamamoto, M. Tsuji and T. Miwa, *J. Chromatogr. A*, **910**, 373 (2001).
188. E. Nemutlu and N. Özaltin, *Anal. Bioanal. Chem.*, **383**, 833 (2005).
189. S. Motellier, S. Petit and P. Decambox, *Anal. Chim. Acta*, **410**, 11 (2000).
190. M. Maj-Żurawska, *Chem. Anal. (Warsaw)*, **42**, 187 (1997).
191. H. E. van Ingen, H. J. Hutjgen, W. T. Kok and G. T. B. Sanders, *Clin. Chem.*, **40**, 52 (1994).
192. A. Malon, B. Wagner, E. Bulska and M. Maj-Żurawska, *Anal. Biochem.*, **302**, 220 (2002).
193. Z. Cao, C. Tongate and R. J. Elin, *Scand. J. Clin. Lab. Invest.*, **61**, 389 (2001).
194. Nova Biomedical, *Nova Electrolyte/Chemistry Analyzers*, Nova Biomedical Corporation, Waltham, MA, USA. (http://www.novabiomedical.com/clinical/electrolyte.html#top).
195. S. Unterer, H. Lutz, B. Gerber, T. M. Glaus, M. Hässig and C. E. Reusch, *Am. J. Vet. Res.*, **65**, 183 (2004).
196. S. Unterer, B. Gerber, T. M. Glaus, M. Hässig and C. E. Reusch, *Vet. Res. Commun.*, **29**, 647 (2005).
197. J. Wang, P. A. M. Farias and J. S. Mahmoud, *Anal. Chim. Acta*, **172**, 57 (1985).
198. A. M. M. Ali, O. A. Farghaly and M. A. Ghandour, *Anal. Chim. Acta*, **412**, 99 (2000).
199. O. A. Farghaly, *Talanta*, **63**, 497 (2004).
200. N. Abo El-Maali, D. Abd El-Hady, M. Abd El-Hamid and M. M. Seliem, *Anal. Chim. Acta*, **417**, 67 (2000).
201. M. C. Cheney, D. J. Curran and K. S. Fletcher III, *Anal. Chem.*, **52**, 942 (1980).
202. X. Zhou, L. Wang, Y. Zeng, G. Lu and P. Zhen, *J. AOAC Int.*, **89**, 782 (2006).
203. Z. Jońca and W. Lewandowski, *Pol. J. Environ. Stud.*, **13**, 275 (2004).
204. Z. Jońca and W. Lewandowski, *Pol. J. Environ. Stud.*, **13**, 281 (2004).
205. Y. Bai, J. M. Ouyang, Y. Bai and M. L. Chen, *Guangpuxue Yu Guangpu Fenxi*, **24**, 1016 (2004); *Chem. Abstr.*, **142**, 193656 (2005).
206. Z. Kiliç, O. Acar, M. Ulaşan and M. Ilim, *Food Chem.*, **76**, 107 (2002).
207. T. Papageorgiou, D. Zacharoulis, D. Xenos and G. Androulakis, *Nutrition*, **18**, 32 (2002).
208. A. Abarca, E. Canfranc, I. Sierra and M. L. Marina, *J. Pharm. Biomed. Anal.*, **25**, 941 (2001).
209. M.-C. Lin, Y.-L. Huang, H.-W. Liu, D.-Y. Yang, J.-B. Lee and F.-C. Cheng, *J. Am. Coll. Nutr.*, **23**, 556S (2004).
210. L. E. Walther, K. Winnefeld and O. Sölch, *J. Trace Elements Med. Biol.*, **14**, 92 (2000).
211. M.-C. Lin, Y.-L. Huang, H.-W. Liu, D.-Y. Yang, C.-P. Lee, L.-L. Yang and F.-C. Cheng, *J. Am. Coll. Nutr.*, **23**, 561S (2004).
212. D.-Y. Yang, J.-B. Lee, M.-C. Lin, Y.-L. Huang, H.-W. Liu, Y.-J. Liang and F.-C. Cheng, *J. Am. Coll. Nutr.*, **23**, 552S (2004).
213. M. Pistón, I. Dol and M. Knochen, *J. Autom. Methods Manage. Chem.*, Article ID 47627 (2006). (http://www.hindawi.com/GetArticle.aspx?doi=10.1155/JAMMC/2006/47627).
214. N. Dilsiz, A. Olcucu and M. Atas, *Cell Biochem. Funct.*, **18**, 259 (2000).
215. E. Kilic, A. Demiroglu, R. Saraymen and E. Ok, *J. Trace Elem. Exp. Med.*, **17**, 175 (2004).
216. Z. Yang, X. Hou, B. T. Jones, D. C. Sane, M. J. Thomas and D. C. Schwenke, *Microchem. J.*, **72**, 49 (2002).
217. A. Krejčová, T. Černohorský and E. Čurdová, *J. Anal. At. Spectrom.*, **16**, 1002 (2001).
218. H. Fukui, O. Fujino and S. Umetani, *Bunseki Kagaku*, **53**, 1329 (2004); *Chem. Abstr.*, **142**, 351615 (2005).
219. E. V. Polyakova and O. V. Shuvaeva, *J. Anal. Chem.*, **60**, 937 (2005).
220. K. A. Idriss, H. Sedaira and H. M. Ahmed, *Talanta*, **54**, 369 (2001).
221. E. Jungreis, in *Encyclopedia of Analytical Chemistry* (Ed. R. A. Meyers), Vol. 15, Wiley, Chichester, 2000, p. 13609.
222. J. Ruzicka and G. D. Marshall, *Anal. Chim. Acta*, **237**, 329 (1990).
223. G. D. Christian, *Analyst*, **119**, 2309 (1994).
224. C. C. Oliveira, R. P. Sartini and E. A. G. Zagatto, *Anal. Chim. Acta*, **413**, 41 (2000).
225. R. B. R. Mesquita and A. O. S. S. Rangel, *Anal. Sci.*, **20**, 1205 (2004).
226. Z. O. Tesfaldet, J. F. van Staden and R. I. Stefan, *Talanta*, **64**, 981 (2004).
227. D. G. Themelis, P. D. Tzanavaras, A. V. Trellopoulos and M. C. Sofoniou, *J. Agric. Food Chem.*, **49**, 5152 (2001).
228. H.-W. Gao, J.-X. Yang, Z.-Z. Zhou and J.-F. Zhao, *Phytochem. Anal.*, **14**, 91 (2003).
229. C. Liu and C. Han, *Fenxi Huaxue*, **28**, 594 (2000); *Chem. Abstr.*, **133**, 37409 (2000).

7. Analytical aspects of organomagnesium compounds

230. BioAssay Systems, *QuantiChrom Magnesium Assay Kit (DIMG-250)*, BioAssay Systems, Hayward, CA, USA. (http://www.bioassaysys.com/DIMG.pdf).
231. Ortho-Clinical Diagnostics, *Vitros Mg Slide*, Johnson & Johnson Co, Rochester, NY, USA. (http://www.fda.gov/cdrh/pdf2/k023876.pdf).
232. I. M. Papazachariou, A. Martinez-Isla, E. Efthimiou, R. C. N. Williamson and S. I. Girgis, *Clin. Chim. Acta*, **302**, 145 (2000).
233. D. Nasser and Y. K. Agrawal, *Iran. J. Chem. Chem. Eng., Int. Eng. Ed.*, **23**, 65 (2004).
234. T. F. Christiansen, J. E. Bush and S. C. Krogh, *Anal. Chem.*, **48**, 1051 (1976).
235. H. Szmacinski and J. R. Lakowicz, *J. Fluoresc.*, **6**, 83 (1996).
236. G. de Armas, A. Cladera, E. Becerra, J. M. Estela and V. Cerdà, *Talanta*, **52**, 77 (2000).
237. Y. Suzuki, H. Komatsu, T. Ikeda, N. Saito, S. Araki, D. Citterio, H. Hisamoto, Y. Kitamura, T. Kubota, J. Nakagawa, K. Oka and K. Suzuki, *Anal. Chem.*, **74**, 1423 (2002).
238. K. Faid and M. Leclerc, *J. Am. Chem. Soc.*, **120**, 5274 (1998).
239. M. Leclerc, *Adv. Mater.*, **11**, 1491 (1999).
240. K. Peter, R. Nilsson and O. Inganäs, *Nature Mater.*, **2**, 419 (2003).
241. K. Faid, R. Cloutier and M. Leclerc, *Macromolecules*, **26**, 2501 (1993).
242. A. Boldea, I. Levesque and M. Leclerc, *J. Mater. Chem.*, **9**, 2133 (1999).
243. H. Mochizuki, Y. Nabeshima, T. Kitsunai, A. Kanazawa, T. Shiono, T. Ikeda, T. Hiyama, T. Maruyama, T. Yamamoto and N. Koide, *J. Mater. Chem.*, **9**, 2215 (1999).
244. P. C. Ewbank, R. S. Loewe, L. Zhai, J. Reddinger, G. Sauvé and R. D. McCullough, *Tetrahedron*, **60**, 11269 (2004).
245. L. G. Sillen (Ed.), *Stability Constants of Metal-Ion Complexes*, **25**, Suppl. 1, The Chemical Society, London, 1971.
246. J. K. Nicholson, M. J. Buckingham and P. J. Sadler, *Biochem. J.*, **211**, 605 (1983).
247. B. S. Somashekar, O. B. Ijare, G. A. Nagana Gowda, V. Ramesh, S. Gupta and C. L. Khetrapala, *Spectrochim. Acta Part A*, **65**, 254 (2006).
248. W. B. Geven, G. M. Vogels-Mentink, J. L Willems, C. W v. Os, C. W. Hilbers, J. J. M. Joordens, G. Rijksen and L. A. H. Monnens, *Clin. Chem.*, **37**, 2076 (1991).
249. A. Malon and M. Maj-Zurawska, *Anal. Chim. Acta*, **448**, 251 (2001).
250. R. F. Burton, *Comp. Biochem. Phys.*, **65A**, 1 (1980).
251. J. Dombovári, J. S. Becker and H.-J. Dietze, *Int. J. Mass Spectrom.*, **202**, 231 (2000).
252. W. Stegmann and G. A. Quamme, *J. Pharmacol. Toxicol. Meth.*, **43**, 177 (2000).
253. W. Bäuerle, V. Krivan and H. Münzel, *Anal. Chem.*, **48**, 1434 (1976).
254. M. Sabatier, W. R. Keyes, F. Pont, M. J. Arnaud and J. R. Turnlund, *Am. J. Clin. Nutr.*, **77**, 1206 (2003).
255. M. Sabatier, F. Pont, M. J. Arnaud and J. R. Turnlund, *Am. J. Physiol., Regulatory Integrative Comp. Physiol.*, **285**, R656 (2003).
256. C. Coudray, C. Feillet-Coudray, D. Grizard, J. C. Tressol, E. Gueux and Y. Rayssiguier, *J. Nutr.*, **132**, 2043 (2002).
257. C. Feillet-Coudray, C. Coudray, E. Gueux, V. Ducros, A. Mazur, S. Abrams and Y. Rayssiguier, *Metabolism*, **40**, 1326 (2000).
258. C. Feillet-Coudray, C. Coudray, E. Gueux, A. Mazur and Y. Rayssiguier, *Magnesium Res.*, **15**, 191 (2002).
259. T. R. Compton, *Chemical Analysis of Organometallic Compounds*, Vol. 1, Academic Press, New York, 1973.
260. J. Zabicky, in *The Chemistry of Organolithium Compounds* (Eds. Z. Rappoport and I. Marek), Wiley, Chichester, 2004; (a) p. 336; (b) pp. 355 ff.
261. S. C. Watson and J. F. Eastham, *J. Organomet. Chem.*, **9**, 165 (1967).
262. J. Villieras, M. Rambaud and B. Kirschleger, *J. Organomet. Chem.*, **249**, 315 (1983).
263. H. Kiljunen and T. A. Hase, *J. Org. Chem.*, **56**, 6950 (1991).
264. M. E. Bowen, B. R. Aavula and E. A. Mash, *J. Org. Chem.*, **67**, 9087 (2002).
265. D. E. Bergbreiter and E. Pendergrass, *J. Org. Chem.*, **46**, 219 (1981).
266. J. O. Egekeze, H. J. Perpall, C. W. Moeder, G. R. Bicker, J. D. Carroll, A. O. King and R. D. Larsen, *Analyst*, **122**, 1353 (1997).
267. J. Kelly, L. Wright, T. Novak, M. Huffman and V. Antonucci. *J. Liq. Chromatogr. Relat. Technol.*, **24**, 15 (2001).
268. B. J. Magerlein and W. P. Schneider, *J. Org. Chem.*, **34**, 1179 (1969).

269. H. Gilman, P. D. Wilkinson, W. P. Fishel and C. H. Meyers, *J. Am. Chem. Soc.*, **45**, 150 (1923).
270. H. Gilman, E. A. Zoellner and J. B. Dickey, *J. Am. Chem. Soc.*, **51**, 1576 (1929).
271. L. Bernhard, N. C. Furcola, E. L. Gutman, S. L. Kauffman, J. G. Kramer, C. M. Leinweber and V. A. Mayer (Eds.), *1995 Annual Book of ASTM Standards*, Vol. 15.01, American Society for Standards and Materials, Philadelphia, PA, USA, 1995, pp. 311–313.
272. T. Vlismas and R. D. Parker, *J. Organomet. Chem.*, **10**, 193 (1967).
273. J.-Y. Gal, O. Perrier and T. Yvernault, *C. R. Acad. Sci. Paris Ser. C*, **271**, 1561 (1970).
274. Y. Aso, H. Yamashita, T. Otsubo and F. Ogura, *J. Org. Chem.*, **54**, 5627 (1989).
275. B. E. Love and E. G. Jones, *J. Org. Chem.*, **64**, 3755 (1999).
276. A. Krasovskiy and P. Knochel, *Synthesis*, 890 (2006).
277. A. Krasovskiy, F. Kopp and P. Knochel, *Angew. Chem. Int. Ed.*, **45**, 497 (2006).
278. J. Kuyper and K. Vrieze, *Chem. Commun.*, 64 (1976).
279. K. Kham, C. Chevrot, J. C. Folest, M. Troupel and J. Perichon, *Bull. Soc. Chim. Fr. (Pt. 1)*, 243 (1977).
280. Y. Chen, T. Wang, R. Helmy, G. X. Zhou and R. LoBrutto, *J. Pharm. Biomed. Anal.*, **29**, 393 (2002).
281. J.-Y. Gal, O. Mertinat-Perrier and T. Yvernault, *C. R. Acad. Sci. Paris Ser. C*, **277**, 1343 (1973).
282. R. D'Hollander and M. Anteunis, *Bull. Soc. Chim. Belg.*, **72**, 77 (1963).
283. S. Görög and G. Szepesi, *Analyst*, **95**, 727 (1970).
284. R. D. Parker and T. Vlismas, *Analyst*, **93**, 330 (1968).
285. X. Li, C. Yu, W. Sun, G. Liu, J. Jia and Y. Wang, *Rapid Commun. Mass Spectrom.*, **18**, 2878 (2004).
286. A. Wowk and S. Digiovanni, *Anal. Chem.*, **38**, 742 (1966).
287. H. O. House and W. L. Respess, *J. Organomet. Chem.*, **4**, 95 (1965).
288. L. Marder, P. Tomedi, M. F. Ferrão, A. Jablonski and C. U. Davanzo, *J. Braz. Chem. Soc.*, **17**, 594 (2006).
289. H. Gilman and F. Schulze, *J. Am. Chem. Soc.*, **47**, 2002 (1925).
290. H. Gilman and F. Schulze, *Bull. Soc. Chim. Fr.*, **41**, 1479 (1927).
291. H. Gilman and R. G. Jones, *J. Am. Chem. Soc.*, **62**, 1243 (1940).
292. H. Gilman and R. J. Vander Wal, *Bull. Soc. Chim. Fr.*, **45**, 344 (1929).
293. H. Gilman, L. L. Heck and N. B. St. John, *Recl. Trav. Chim. Pays-Bas*, **49**, 212 (1930).
294. H. Gilman and L. L. Heck, *J. Am. Chem. Soc.*, **52**, 4949 (1930).
295. H. Gilman and L. L. Heck, *Recl. Trav. Chim. Pays-Bas*, **49**, 218 (1930).
296. H. Gilman, O. R. Sweeney and L. L. Heck, *J. Am. Chem. Soc.*, **52**, 1604 (1930).
297. H. Gilman and J. Swiss, *J. Am. Chem. Soc.*, **62**, 1847 (1940).
298. H. Gilman and L. A. Woods, *J. Am. Chem. Soc.*, **65**, 33 (1943).
299. AIST, *Spectral Database for Organic Compounds (SDBS)*, National Institute of Advanced Industrial Science and Technology (AIST), Japan, (a) SDBS No. 2051, (b) SDBS No. 2054. (http://www.aist.go.jp/RIODB/SDBS/cgi-bin/display_frame_disp.cgi?sdbsno=2051).
300. T. R. Hoye, B. M. Eklov and M. Voloshin, *Org. Lett.*, **6**, 2567 (2004).
301. R. Jones, *J. Organomet. Chem.*, **18**, 15 (1969).
302. M. Westerhausen, T. Bollwein, N. Makropoulos and H. Piotrowski, *Inorg. Chem.*, **44**, 6439 (2005).
303. A. Xia, M. J. Heeg and C. H. Winter, *Organometallics*, **22**, 1793 (2003).
304. S. Blair, K. Izod, W. Clegg and R. W. Harrington, *Eur. J. Inorg. Chem.*, 3319 (2003).
305. J. D. Farwell, M. F. Lappert, C. Marschner, C. Strissel and T. D. Tilley, *J. Organomet. Chem.*, **603**, 185 (2000).
306. A. M. Egorov, S. V. Kuznetsova and A. V. Anisimov, *Russ. Chem. Bull., Int. Ed.*, **49**, 1544 (2000).
307. K. B. Starowieyski, J. Lewinski, R. Wozniak, J. Lipkowski and A. Chrost, *Organometallics*, **22**, 2458 (2003).
308. T. Zerevitinov, *Ber. Dtsch. Chem. Ges.*, **45**, 2384 (1913).
309. E. P. Kohler, J. F. Stone, Jr. and R. C. Fuson, *J. Am. Chem. Soc.*, **49**, 3181 (1927).
310. E. P. Kohler and N. K. Richtmyer, *J. Am. Chem. Soc.*, **52**, 3736 (1930).
311. K. Banas and M. Rachtan, *Rudy Met. Niezelaz.*, **26**, 557 (1981); *Chem. Abstr.*, **96**, 173541 (1982).

312. D. Ishii, T. Tsuda and N. Tokoro, *Bunseki Kagaku*, **21**, 367 (1972); *Chem. Abstr.*, **77**, 42920 (1972).
313. M. Lieff, G. F. Wright and H. Hibbert, *J. Am. Chem. Soc.*, **61**, 865 (1939).
314. T. Yazawa, H. Tanaka and K. Eguchi, *Nippon Kagaku Kaishi*, 1338 (1984); *Chem. Abstr.*, **101**, 135658 (1984).
315. F. T. Weiss, in *Treatise on Analytical Chemistry* (Eds. I. M. Kolthoff and P. J. Elving), Part 2, Vol. 13, Interscience, New York, 1966, pp. 33–94.
316. G. R. Leader, *Appl. Spectrosc. Rev.*, **11**, 287 (1976).
317. M. A. Williams and J. E. Ladbury, 'Protein-ligand interactions from molecular recognition to drug design', in *Methods and Principles in Medicinal Chemistry* (Eds. H.-J. Böhm and G. Schneider), Vol. 19, Wiley-VCH, Weinheim, 2003, pp. 137–161.
318. Y. K. Chau and F. Yang, *Appl. Organomet. Chem.*, **11**, 851 (1997).
319. M. B. de la Calle-Guntiñas and F. C. Adams, *J. Chromatogr. A*, **764**, 169 (1997).
320. Y. K. Chau and P. T. S. Wong, in *Analysis of Trace Organics in the Aquatic Environment* (Eds. B. K. Afghan and A. S. Y. Chau), CRC Press, Boca Raton, 1989, pp. 283–312.
321. Y. K. Chau, *Analyst*, **117**, 571 (1992).
322. J. M. F. Nogueira, P. Teixeira and M. H. Florencio, *J. Microcolumn Sep.*, **13**, 48 (2001).
323. M. B. de la Calle-Guntiñas, R. Scerbo, S. Chiavarini, P. Quevauviller and R. Morabito, *Appl. Organomet. Chem.*, **11**, 693 (1997).
324. Y. K. Chau, F. Yang and M. Brown, *Anal. Chim. Acta*, **338**, 51 (1997).
325. M. D. Müller, *Anal. Chem.*, **59**, 617 (1987).
326. C. Yamamoto, M. Nakamura, N. Kibune and Y. Maekawa, *Shokuhin Eiseigaku Zasshi*, **37**, 288 (1996); *Chem. Abstr.*, **126**, 59025 (1997).
327. Q. Zhou, G. Jiang and D. Qi, *Fenxi Huaxue*, **27**, 1197 (1999); *Chem. Abstr.*, **132**, 30085 (2000).
328. K. Newman and R. S. Mason, *J. Anal. At. Spectrom.*, **20**, 830 (2005).
329. K. Tonami, C. Sasaki and F. Sakamoto, *Ishikawa-ken Hoken Kankyo Senta Nenpo*, **33**, 95 (1995, publ. 1997); *Chem. Abstr.*, **127**, 46396 (1997).
330. M. A. Unger, J. Greaves and R. J. Huggett, in *Organotin* (Eds. M. A. Champ and P. F. Seligman), Chapman & Hall, London, 1996, pp. 124–134.
331. Y. Liu, V. Lopez-Avila and M. Alcaraz, *J. High Resolut. Chromatogr.*, **17**, 527 (1994).
332. E. Magi, M. Di Carro and P. Rivaro, *Appl. Organometal. Chem.*, **18**, 646 (2004).
333. S. Tsunoi, T. Matoba, H. Shioji, L. T. Huong Giang, H. Harino and M. Tanaka, *J. Chromatogr. A*, **962**, 196 (2002).
334. I. Martin-Landa, F. de Pablos and I. L. Marr, *Anal. Proc.*, **26**, 16 (1989).
335. J. L. Martínez Vidal, A. Belmonte Vega, F. J. Arrebola, M. J. González-Rodríguez, M. C. Morales Sánchez and A. Garrido Frenich, *Rapid Commun. Mass Spectrom.*, **17**, 2099 (2003).
336. G.-b. Jiang and Q.-f. Zhou, *J. Chromatogr. A*, **886**, 197 (2000).
337. Y. Liu, V. Lopez-Avila and M. Alcaraz, *Anal. Chem.*, **66**, 3788 (1994).
338. L. M. Allan, D. K. Verma, F. Yang, Y. K. Chau and R. J. Maguire, *Am. Ind. Hyg. Assoc. J.*, **61**, 820 (2000).
339. L. A. Uzal Barbeito and J. L. Wardell, *Quim. Anal.*, **14**, 158 (1995).
340. M. Ceulemans and F. C. Adams, *Anal. Chim. Acta*, **317**, 161 (1995).
341. R. Łobiński, W. M. R. Dirkx, M. Ceulemans and F. C. Adams, *Anal. Chem.*, **64**, 159 (1992).
342. F. Rodigari, *Proc. Moving toward the 21st Century*, Philadelphia, Aug. 3–6, 1997; *Chem. Abstr.*, **129**, 305920 (1998).
343. M. O. Stallard, S. Y. Cola and C. A. Dooley, *Appl. Organomet. Chem.*, **3**, 105 (1989).
344. S. K. Aggarwal, R. Fitzgerald and D. A. Herold, *BARC News.*, **201**, 95 (2000); *Chem. Abstr.*, **134**, 157030 (2001).
345. M. Radojević, A. Allen, S. Rapsomanikis and R. M. Harrison, *Anal. Chem.*, **58**, 658 (1986).
346. I. A. Leal-Granadillo, J. I. García Alonso and A. Sanz-Medel, *Anal. Chim. Acta*, **423**, 21 (2000).
347. B. Salih, *Spectrochim. Acta, Part B*, **55**, 1117 (2000).
348. B. Pons, A. Carrera and C. Nerín, *J. Chromatogr. B*, **716**, 139 (1998).
349. K. Fujiwara, Y. Okamoto, M. Ohno and T. Kumamaru, *Anal. Sci.*, **11**, 829 (1995).
350. S. K. Aggarwal, M. Kinter and D. A. Herold, *Clin. Chem.*, **40**, 1494 (1994).
351. H. Emteborg, E. Björklund, F. Ödman, L. Karlson, L. Mathiasson, W. Frech and D. C. Baxter, *Analyst*, **121**, 19 (1996).

352. S. Tao and T. Kumamaru, *Anal. Chim. Acta*, **310**, 369 (1995).
353. T. Kumamaru, S. Tao, M. Uchida and Y. Okamoto, *Anal. Lett.*, **27**, 2331 (1994).
354. S. K. Aggarwal, M. Kinter, J. Nicholson and D. A. Herold, *Anal. Chem.*, **66**, 1316 (1994).
355. M. Pina, D. Montet, J. Graille, C. Ozenne and G. Lamberet, *Rev. Fr. Corps Gras*, **38**, 213 (1991).
356. M. Yurkowski and H. Brockerhoff, *Biochim. Biophys. Acta*, **125**, 55 (1966).
357. C. Franzke, E. Hollstein, J. Kroll and H. J. Noske, *Fette, Seifen, Anstrichmittel*, **75**, 365 (1973).
358. F. Turon, P. Bachain, Y. Caro, M. Pina and J. Graille, *Lipids*, **37**, 817 (2002).
359. M. Wältermann, H. Luftmann, D. Baumeister, R. Kalscheuer and A. Steinbüchel, *Microbiology*, **146**, 1143 (2000).
360. P. R. Redden, X. Lin and D. F. Horrobin, *Chem. Phys. Lipids*, **79**, 9 (1996).
361. M. Pina, D. Ploch and J. Graille, *Lipids*, **22**, 358 (1987).
362. T. Moritani, K. Yoon, M. Rafailovich and B. Chu, *Electrophoresis*, **24**, 2764 (2003).
363. T. Moritani, K. Yoon and B. Chu, *Electrophoresis*, **24**, 2772 (2003).
364. N. D. Danielson, S. Ahmed, J. A. Huth and M. A. Targrove, *J. Liq. Chromatogr.*, **9**, 727 (1986).
365. C. Liebenow, *Solid State Ionics*, **136–137**, 1211 (2000).

CHAPTER 8

Biochemistry of magnesium

JAMES WESTON

Institut für Organische Chemie und Makromolekulare Chemie, Friedrich-Schiller-Universität, Humboldtstraße 10, D-07743 Jena, Germany
Fax: +49(0)3641-9-48212;   e-mail: c9weje@uni-jena.de

I. MAGNESIUM IN BIOLOGICAL SYSTEMS	316
II. MAGNESIUM AS AN ESSENTIAL COFACTOR	316
III. BASIC COORDINATION SPHERE	318
IV. FUNDAMENTAL BINDING MODES	320
A. Carboxylate Ligands	321
B. Phosphate Functionalities	322
C. Nitrogen Ligands	324
V. MAGNESIUM TRANSPORT SYSTEMS	324
VI. UNIVERSAL ENERGY CURRENCY OF LIFE	327
A. MgATP^{2-}	328
B. Biosynthesis of ATP	329
C. Hydrolysis of ATP	331
VII. INTERACTIONS WITH DNA AND RNA	333
A. Magnesium and Structural Stability	334
B. Ribozymes	335
1. Self-splicing in group I introns	337
2. Hammerhead ribozymes	339
VIII. PROTEIN-BASED ENZYMES	342
A. General Enzymatic Modes of Action	342
1. Template enzymes	342
2. Sequential systems	343
3. Allosteric systems	343
B. Metabolic Enzymes	345
1. Protein kinases	345
2. Enolases	348
C. DNA Replication and Repair	350
1. DNA replicases	351
2. Base excision repair	353
3. Generic two-ion mechanism	354

The chemistry of organomagnesium compounds
Edited by Z. Rappoport and I. Marek   © 2008 John Wiley & Sons, Ltd

        D. Magnesium and Photosynthesis 355
            1. Chlorophyll 356
            2. Rubisco—a photosynthetic CO_2 fixing enzyme 357
    IX. REFERENCES 359

I. MAGNESIUM IN BIOLOGICAL SYSTEMS

Although most of the magnesium found on earth is tied up in mineral deposits, many magnesium salts are highly soluble in water. Continual leaching processes lead to a constant and relatively high concentration of Mg^{2+} both in the soil and in the hydrosphere (for example, the concentration of Mg^{2+} in the ocean is ca 55 mM[1]) which guarantees a high bioavailability. This together with its unique physico-chemical properties has ensured that Mg^{2+} is an indispensable cofactor in countless life processes. One of the distinctive characteristics of Mg^{2+} is that it is typically an *intra*cellular metal cation. Similar to K^+, the concentration of Mg^{2+} is generally an order of magnitude higher inside cells than in extracellular milieu (in contrast to this, Ca^{2+} and Na^+ are extracellular cations and tend to accumulate outside the cells)[1]. Magnesium is present in every cell type in every known organism and is the fourth most common metal cation found in biological systems (Ca^{2+} > K^+ > Na^+ > Mg^{2+})[2]. For example, the human body contains a steady-state concentration of ca 25 g of magnesium which has to be replenished at a rate of about 1/2 g per day[2].

For quite some time, the biochemistry of magnesium has been neglected in favor of studying transition metal ions such as iron or copper—mostly because the latter are much easier to study. Magnesium is spectroscopically silent and notoriously difficult to detect and/or monitor[1]. Even solid-state X-ray analysis proves rather difficult due to the extremely low electron density and small size of the Mg^{2+} cation. As progress in methodological and spectroscopic approaches has been realized and now that theoretical approaches (especially DFT theory and MM/MD modeling techniques) can be applied to larger biomolecules, more and more mechanistic details for magnesium-containing biomolecules are becoming available. However, we are far from being able to summarize the biochemistry of magnesium in its entirety. The intent of this chapter is thus not that of a comprehensive review but rather an illustrative article in which selected systems are discussed in order that a reader with a general chemical background may gain an overview of the versatility and fascinatingly complex mechanistic biochemistry of magnesium.

II. MAGNESIUM AS AN ESSENTIAL COFACTOR

Magnesium is perhaps the most versatile metal cation found in living systems. It can and does interact with an extremely wide variety of biomolecules, thus giving rise to multiple biological roles of fundamental importance in life processes. A comprehensive discussion of all biosystems that have an absolute dependency on Mg^{2+} as a cofactor would currently have to include hundreds of unrelated examples and more are being discovered all the time.

The biochemistry of Mg^{2+} begins with cell membranes where metal cations, especially Na^+ and Mg^{2+}, are needed to help reduce the strong repulsions between negatively charged phosphates in the densely packed lipids that make up cellular membranes[2]. Since the concentration of Mg^{2+} is higher inside cells than outside, nature has to have a way to move it across cell membranes and then keep it there—a task accomplished by magnesium-specific transport proteins[3]. All cationic membrane transport systems (Na^+, K^+, Mg^{2+} and Ca^{2+}) face certain challenges. They must be able to specifically recognize and interact with a hydrated cation. Next, they must strip away most, if not all, of the water ligands and then move the cation across the membrane. However, Mg^{2+} is the most challenging of all to transport due to its very small ionic radius and the fact that its hydration radius and transport number is larger than those of the others[1,4]. In addition, it has the slowest exchange rate (3 orders of magnitude lower) of solvent waters[1]. As a result, Mg^{2+} transporters are generally quite unusual members of the transport family[3].

Once in the cell, the most common physiological role of a Mg^{2+} cation is to bind ATP or other nucleoside triphosphates (NTP). Differing estimates of the total amount of intracellular Mg^{2+} that directly interacts with ATP are available in the literature. These vary from 50%[1] to 75%[5]. Since ATP is the fundamental biochemical unit of energy in life processes, the importance of this function cannot be underestimated. It is believed that one of the purposes of Mg^{2+} binding to ATP is to activate it towards specific phosphate hydrolysis. Of significance in this task is the high Lewis acidity of Mg^{2+}, i.e. its general ability to polarize functional groups (such as a carbonyl group in a peptide backbone), stabilize anions (carboxylates or phosphates, for example) and polarize water molecules so that they can be more easily deprotonated to provide active nucleophiles for general phosphate or peptide bond hydrolysis.

Magnesium is an intrinsic component of cellular signaling processes in higher organisms[6]. As such, Mg^{2+}-dependent enzymes are found in virtually every known metabolic pathway where they often function as key mediators. The well-known glycolytic pathway is no exception to this. In addition to metabolic pathways, magnesium is often an essential cofactor in enzymatic and ribozymatic DNA and RNA replication, repair and transcription. As a consequence, numerous unrelated enzymatic families, both protein and ribozyme based that, in addition, may or may not depend on ATP, have an absolute dependency upon magnesium[7,8]. In the active site of enzymes, magnesium can play several different mechanistic roles, the more important probably being the following[2]:

Stabilization of an intermediate, I:

$$Mg^{2+} + S \longrightarrow Mg^{2+}\text{-}I \longrightarrow Mg^{2+} + P$$

Stabilization of a leaving group, LG (or product P):

$$Mg^{2+} + S\text{-}LG \longrightarrow S + Mg^{2+}\text{-}LG$$

Bring two substrates S together for reaction:

$$Mg^{2+} + S_1 + S_2 \longrightarrow S_1\text{-}Mg^{2+}\text{-}S_2 \longrightarrow S_1\text{-}S_2$$

Provide an activated, water-based nucleophile for hydrolysis:

$$Mg^{2+} + \text{'}H_2O\text{'} + S_1\text{-}S_2 \longrightarrow S_1\text{-}H + S_2\text{-}OH + Mg^{2+}$$

Moving away from enzymatic activity, it is a well-known fact that both the structure (conformation as well as topology) and the function of DNA and RNA depend strongly on specific interactions with divalent cations, especially Ca^{2+} and Mg^{2+} [9,10]. These cations stabilize base pairing and stacking by relieving electrostatic repulsion between phosphates[11]. Among other things, Mg^{2+} stimulates the formation of, as well as stabilizes DNA/RNA helices and other structural motifs[11,12].

Finally, a very unusual binding situation of magnesium, namely its interaction with the porphyrin ring in chlorophyll, plays an essential role in the most thermodynamically demanding reaction to be found in biology—the synthesis of carbohydrates from CO_2 in plants and cyanobacteria using light energy from the sun. This extremely complex process of photosynthesis is considered to be fundamental for sustaining essentially all life on our planet[13].

III. BASIC COORDINATION SPHERE

Before one can study the biochemistry of magnesium in detail, one first needs to understand its structure and behavior under physiological conditions—which basically means understanding the interaction of a single Mg^{2+} cation first with water and then with a few selected organic functionalities. It is a well-known fact that, in aqueous solutions, Mg^{2+} binds six water molecules in an octahedral arrangement to generate a hexaaquomagnesium $[Mg(H_2O)_6]^{2+}$ ion—a species which has been the subject of numerous experimental and theoretical studies[14]. The existence of $[Mg(H_2O)_6]^{2+}$ as an independent structural unit is also corroborated by countless solid-state structures of inorganic and bioinorganic compounds in structural data banks worldwide[15]. As compared to other metal cations, the low electronic density of Mg^{2+} makes resolving the solid-state structures of large biomolecules in the region of the magnesium ion quite difficult. Although there is continual progress in this field, there are still relatively few solid-state structures of magnesium-containing biomolecules available. Of the known structures, most, but (significantly) not all, possess an octahedral coordination sphere.

The interaction of a single Mg^{2+} cation with water is fundamentally electrostatic in nature and is traditionally considered to be quite 'rigid'. An older (1984) analysis of solid-state structures, for example, reported that a hexacoordinated Ca^{2+} ion has a much greater angular flexibility with deviations of up to $40°$ from the ideal $90°$ whereas the variation in comparable Mg^{2+} complexes is no more than $5–10°$[16]. The octahedral arrangement allows for ligand exchange; the water molecules in the first coordination sphere of Mg^{2+} are in dynamic equilibrium with individual waters in the looser second coordination sphere (transition region between bulk water and the 'ionic cavity' generated by $[Mg(H_2O)_6]^{2+}$). However, this exchange is significantly slower—3 to 4 orders of magnitude—than the other common biologically relevant metal cations (Na^+, K^+ and Ca^{2+}) and approaches that found for transition metal ions[1]. It is generally accepted that a maximum of 12 water molecules can be accommodated in the second coordination sphere of Mg^{2+}[4,17,18]. The size of a hydrated Mg^{2+} ion is surprisingly large and its volume is ca 400-fold larger than its dehydrated ionic form[1]. For comparison, the hydrated volume of the intrinsically much larger Ca^{2+} is only 25-fold larger than its ionic volume[1].

Experiments on isolated $[Mg(H_2O)_6]^{2+}$ ions in the gas phase indicate that the coordination number of the central Mg^{2+} is principally somewhat more flexible than is generally assumed. Two different isomers are present—the expected octahedral $[Mg(H_2O)_6]^{2+}$ ion and a previously unknown pentacoordinated $[Mg(H_2O)_5]^{2+} \cdot H_2O$ species with the sixth water in the second coordination sphere[19,20]. The metal binding site of magnesium-containing biomolecules which contain a variable number of organic ligands presents a problematic situation with an environment intermediate between that of bulk water where a hexacoordinated magnesium ion is clearly preferred and the gas phase where the coordination number of Mg^{2+} is unusually flexible. Although quite a few theoretical studies of ligand exchange processes (H_2O against HCO_2^-, HCO_2H, formamide, methanol etc.) have been published, all of these studies have *assumed* an octahedral geometry at the central Mg^{2+} ion[14]. Only very recently has the possibility of alternative coordination modes begun to be considered. The fact that this is necessary is illustrated by a solid-state structure of a Mg^{2+}–GDP complex[21]. This biomolecule crystallizes with two different central coordination geometries for Mg^{2+}—the expected hexacoordination observed in all other known members of the GTPase family and an 'unusual' pentacoordinated structure (Figure 1)[21]. Moving away from simple aqueous solutions, organomagnesium compounds, especially those containing aprotic organic ligands such as THF, often exhibit a coordination number of four (tetrahedral geometry)[22,23].

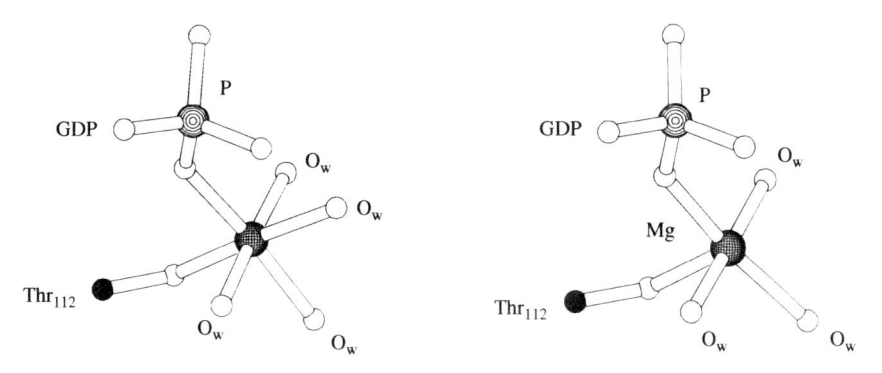

FIGURE 1. The two different coordination geometries of the Mg^{2+} cation in the solid-state structure of a Mg^{2+}–GDP complex

Quite a few enzymes with specific Mg^{2+} binding sites catalyze either the hydrolysis of a phosphate ester or the hydrolytic transfer of a phosphate group[24]. It is often postulated that hydrolysis proceeds over the initial deprotonation of a water ligand to generate a nucleophilic metal-bound hydroxide that is then capable of attacking the substrate. However, very few investigations on species containing a Mg–OH functionality have been performed with most of these being theoretical studies in the gas phase on small $[Mg(OH)_n]^{2-n}$ ($n = 1-3$) complexes[25]. Only one recent DFT study explicitly considered the deprotonation of $[Mg(OH_2)_6]^{2+}$ and discovered that the expected hexacoordinated $[Mg(OH)(OH_2)_5]^+$ species is intrinsically unstable[14]. Upon deprotonation of a water ligand, $[Mg(OH_2)_6]^{2+}$ spontaneously lowers its coordination number to five with accompanying migration of one water ligand to the second coordination sphere (Figure 2). It is quite interesting that this study (performed at a relatively high level of theory where thermodynamic accuracy can be expected) predicts that deprotonation is a slightly *exothermic* ($\Delta G = -4.1$ kcal mol^{-1}) process and thus probably highly relevant in biological processes[14]. Fluoride ligands (isoelectronic with hydroxide) behave quite similarly; DFT calculations predict a pentacoordinate geometry for $[MgF(H_2O)_5]^+$ [14], a finding which is

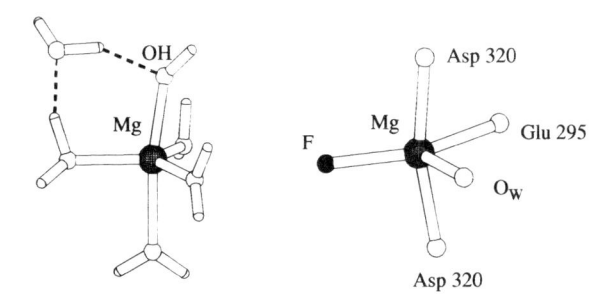

FIGURE 2. Left: deprotonation of $[Mg(OH_2)_6]^{2+}$ lowers the coordination number of magnesium. Reprinted with permission from Reference 14. Copyright 2005 American Chemical Society. Right: pentacoordinate geometry of Mg^{2+} in the solid-state structure of the Mg^{2+}–F^-–P_i complex of yeast enolase

supported by a solid-state structure of the fluoride-inhibited $Mg^{2+}-F^--P_i$ complex of yeast enolase (Figure 2)[26].

IV. FUNDAMENTAL BINDING MODES

Magnesium differs from all other alkaline earth and transition metal ions in that two fundamentally different binding mechanisms ('outer sphere' and 'inner sphere') occur that are capable of competing with each other (Figure 3)[7,27]. The small ionic radius of Mg^{2+} together with its oxophilicity (high affinity for oxygen ligands) results in unusually strong water–metal interactions. As a result, ligand exchange reactions are quite slow, which makes it relatively difficult to replace a water ligand with a bulky organic ligand. As a consequence, $[Mg(H_2O)_6]^{2+}$ has a certain tendency to act as an independent entity when it interacts with biomolecules. Binding interactions in this 'outer sphere' case occur indirectly via strong hydrogen bond interactions between a substrate S and one or more of the water ligands. Alternatively, an organic functionality in a biomolecule may displace one or more of the water ligands in $[Mg(H_2O)_6]^{2+}$, thus effectively binding the (now partially dehydrated) Mg^{2+} cation in an 'inner sphere' coordination mode. Theoretical investigations have indicated that the degree of local solvation is probably one of the major factors in determining the binding mode of a magnesium ion. The tendency for an inner-sphere binding mode increases as the dielectric constant of the local medium decreases (increasing hydrophobicity of an active site binding pocket that is only partially solvent accessible, for example)[28].

For reasons not yet entirely understood, the oligonucleotides in DNA and RNA often preferably interact with $[Mg(H_2O)_6]^{2+}$ via an outer-sphere binding mode[8]. Extensive hydrogen bonding of magnesium-bound water molecules to heteroatoms in the bases and to the phosphate backbone are the predominant interactions[7]. Guanine, for example, appears to actively promote outer sphere binding of magnesium in RNA; an analysis of a solid-state structure of an RNA strand containing a total of 27 hydrated Mg^{2+} ions revealed that 21 of them undergo outer-sphere contacts with guanine; only three inner-sphere contacts were observed[29]. A typical outer-sphere interaction of $[Mg(H_2O)_6]^{2+}$ with two GC base pairs in the major groove of DNA is illustrated in Figure 4[30,31]. A theoretical study (RHF) of this bonding interaction attributed this preference to a cooperative enhancement of charge transfer from guanine to magnesium mediated by the hydration sphere of the ion[32].

If a magnesium cation is incorporated into the active site of a protein-based enzyme, it usually binds via an inner-sphere mechanism in which one or more water ligands in $[Mg(OH_2)_6]^{2+}$ are exchanged for organic ligands L_x originating from side chains. The

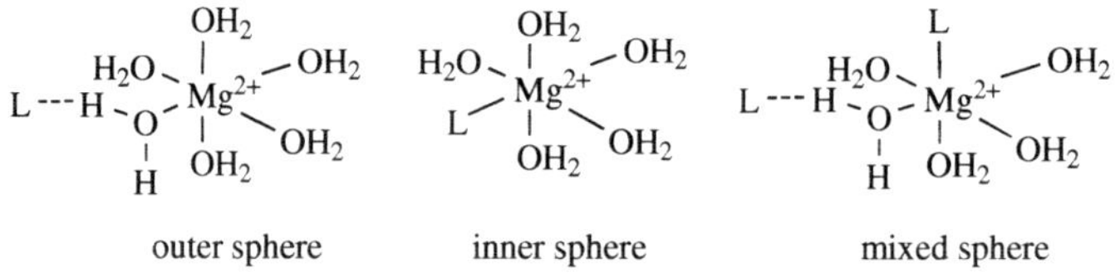

FIGURE 3. Difference between 'outer' and 'inner' sphere coordination modes of magnesium (Reproduced by permission of Elsevier from Reference 7) as well as the possibility of mixed sphere coordination. L is an organic ligand

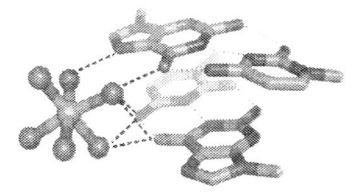

FIGURE 4. Typical outer-sphere interaction of $[Mg(H_2O)_6]^{2+}$ with two GC base pairs in the major groove of a DNA double helix

number of inner-sphere binding contacts is quite variable and values of x from 1–5 have been observed. Sometimes, a mixed coordination occurs in which ligands (L_x) from protein side chains fix Mg^{2+} in the active site. At the same time, a substrate S interacts with the magnesium via an outer-sphere binding mechanism (Figure 3). Again, the number of inner-sphere binding contacts is quite variable.

Statistical studies of the inner-sphere binding mode of Mg^{2+} in the solid-state structures of metalloproteins have revealed that oxygen, as expected, is the preferred bonding partner[33,34]. Approximately 77% of all Mg–X bonds are Mg–O bonding situations in which either water or negatively charged oxygen functionalities such as carboxylates (Asp, Glu) are the preferred ligands[35]. The second most commonly occurring situation is a Mg–N interaction which can occur either in the form of a porphyrin ring (chlorophyll) or a nitrogen ligand originating from the side chain of lysine or the imidazole ring in histidine. Magnesium–sulfur bonds are extremely seldom in natural systems, having only been observed in a single chlorophyll chromophore[36].

A. Carboxylate Ligands

After water, the second most common biological ligand for magnesium is a carboxylate which usually originates from a Glu or Asp side chain. Quite a few theoretical studies of the first coordination sphere of Mg^{2+} have therefore included carboxylates[14]. With their negative charge, they are significantly better ligands than water for a hexacoordinated Mg^{2+} cation. The equilibrium position is quite favorable for successive exchanges of up to three carboxylates bound in a monodentate manner[14]. However, it is clear that Mg^{2+} will not exchange all of its first-shell water molecules[37]. Current calculations indicate that the maximum number of monodentately bound carboxylates will likely not exceed four[38]. In accord with this, a PDB data bank analysis revealed that of 82 solid-state structures available in 2006 for magnesium-containing protein binding sites, 52 contain one, 25 have two, 3 have three and only two structures contain four carboxylate ligands[38]. In the case of two (or more) carboxylate ligands, the resulting $[Mg(RCO_2)_x(OH_2)_{6-x}]^{2-x}$ complexes are asymmetric and several isomers are possible due to differentiation between axial and equatorial positions. The energy difference between these different possibilities is usually rather small[14,37,38] and examples of both orientations can be found in the PDB data bank. Monodentately bound carboxylates can and do stabilize themselves further via two basic interactions—hydrogen bonding in the first coordination sphere to Mg^{2+}-bound water molecules and, as observed in almost all solid-state structures available, additional hydrogen bonding to second-sphere functionalities such as backbone peptide groups, Asn, Gln, Lys or Arg side chains[39]. The situation is further complicated

monodentate   bidentate   bridge bidentate

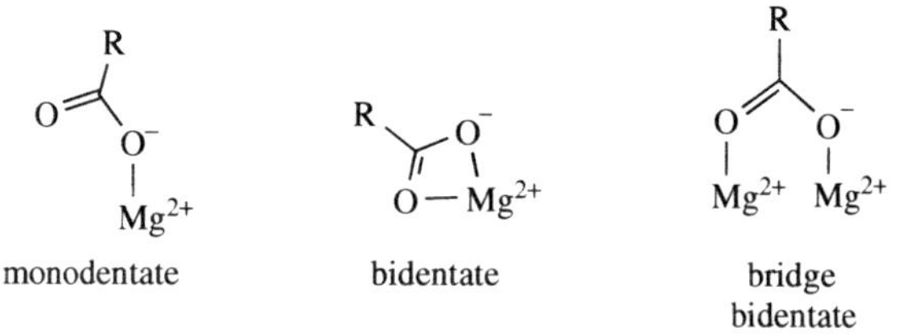

FIGURE 5. Coordination modes available to a carboxylate ligand when interacting with Mg^{2+}

due to the fact that carboxylates possess three fundamentally different binding modes: mono-, bidentate and bridge bidentate (Figure 5). Until now, only the monodentate mode has been discussed. However, all three have been observed in the solid-state structures of magnesium-containing biomolecules.

It has been suggested that a carboxylate ligand 'is quite indifferent to its coordination mode' which is postulated to be mainly determined by the presence/absence of second-sphere hydrogen bonding[40]. It is believed that an equilibrium between both modes (the 'carboxylate shift'[41]) could be important in enzymatic modes of action[42]. This seems to be especially true for biomolecules that must discriminate between Ca^{2+} and Mg^{2+}. A survey (2004) of the PDB data bank revealed that Mg^{2+} clearly favors the monodentate mode with only 4% of the carboxylate ligands binding in a bidentate manner whereas 29% of the Ca^{2+} structures were bidentate[43]. A change in the binding mode (monodentate for Mg^{2+}, bidentate for Ca^{2+}) has been postulated to be the deciding factor in discriminating between these ions in *E. coli* ribonuclease H1[44]. Systematic theoretical studies of this equilibrium[43,45] suggest that the binding mode is determined by a fine balance of several factors—only one of them being the identity of the metal ion. The total charge in the region of the metal ion, electrostatic properties of the interacting substrate, the presence/absence of stabilizing H-bond donors in the first and second coordination sphere and the general dielectric medium all play a critical role. The effect of ionic charge density has recently been investigated spectroscopically on droplets of aqueous $Mg(OAc)_2$ solutions[46,47]. Significant changes in the coordination mode were observed upon increasing salt concentration—isolated contact ion pairs (outer sphere solvation) are in equilibrium with monodentate coordination at lower salt concentrations. As the solution becomes increasingly saturated, bidentate contacts and then bridge bidentate modes begin to be detected. The bridging bidentate mode is usually only observed in protein-based enzymes where it is a very common motif employed in binuclear active sites in order to position two metal ions in near proximity (3–5 Å) to each other[48,49].

B. Phosphate Functionalities

After a carboxylate, the next most significant biological ligand for Mg^{2+} is a phosphate. Phosphates, in addition to being present as inorganic phosphate (P_i), also happily polymerize to form not only diphosphates (for example, PP_i and ADP) but also triphosphates (nucleoside triphosphates such as ATP). Many of these polyphosphates have fundamental biochemical importance as evidenced by the ATP $\rightleftharpoons$ ADP + P_i reaction.

Each of the species mentioned above can be expected to exhibit different interactions with Mg^{2+} and the binding situation is even more complex than with carboxylates. In addition to outer-sphere interactions, phosphates, like carboxylates, are capable of monodentate, bidentate and bridge bidentate inner-sphere binding modes. Due to the complexity of naturally occurring phosphorus chemistry, our understanding of the specific interactions of phosphate functionalities with metal ions is still rather limited. Comprehensive statistical analyses of available solid-state structures supported by detailed computational investigations are still missing in the literature. There is a definitive need for such studies since the molecular mechanisms of many biochemical reactions depend upon specific phosphate–metal ion interactions.

A common bonding situation for phosphates is a monodentate inner-sphere mode. An older (1998) study of available solid-state structures, which did not consider outer-sphere binding, concluded that a negatively charged (-1) phosphate group (phosphate diester) clearly favors the monodentate over a bidentate binding situation which had, at that time, not yet been observed for Mg^{2+} in a biological molecule[50]. One of the few systems which has been studied in detail is the interaction of a dimethylphosphate monoanion [$(MeO)_2PO_2$]$^-$ with [$Mg(OH_2)_6$]$^{2+}$. Recent DFT calculations which explicitly considered solvent effects concluded that an outer-sphere binding situation is preferred over a monodentate inner-sphere mode[51,52]. Most likely, the choice between an outer-sphere and inner-sphere situation is a fine balance between many effects, among the most likely candidates being solvent exposure, identity and number of the organic fragments bound to the phosphate, the charge on the phosphate and the electrostatic potential in the immediate vicinity. A recent QM/MM study on a DNA fragment containing a GC base pair investigated the question of outer- vs. inner-sphere interactions of [$Mg(OH_2)_6$]$^{2+}$ with a diphosphate linker[53]. Three binding modes (two inner and one outer sphere) are fundamentally possible (Figure 6). Solvent effects were not considered in this study.

Upon dimerization of two phosphates to form pyrophosphate, PP_i (with the distinct possibility of various anionic forms depending on the pH), the situation becomes even more complex. Magnesium can now interact with just one of the phosphate groups—or with both simultaneously. In addition, a considerable conformational mobility (rotation about the P-O-P anhydride and R-O-P ester linkages) is present and the number of stable conformers grows exponentially. The dimethyl diphosphate dianion [$(MeO)_2P_2O_7^{2-}$], for example, has been reported to exist in at least nine different stable conformations—before interacting with magnesium[54]. Unfortunately, most computational studies have investigated the interaction of a naked Mg^{2+} ion with various forms of PP_i in the gas phase—results which can scarcely be applied to physiological conditions where [$Mg(OH_2)_6$]$^{2+}$ is present[54–56]. One theoretical study did, however, partially investigate the interaction of [$Mg(OH_2)_6$]$^{2+}$ with a simple glycosidic linkage (glucose-PP_i) in the gas phase (investigation of the complete manifold of conformational possibilities was not possible)[57]. Even so, eight representative conformations—all with comparable stabilities—were reported.

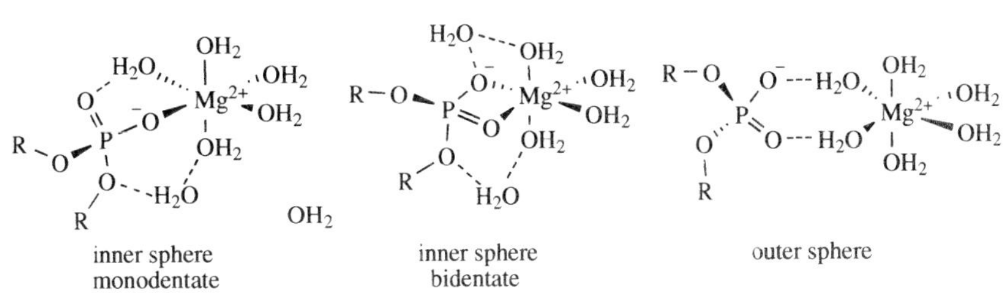

FIGURE 6. Possible binding interactions of a phosphate linker in DNA with [$Mg(H_2O)_6$]$^{2+}$. Adapted with permission from Reference 53. Copyright 2006 American Chemical Society

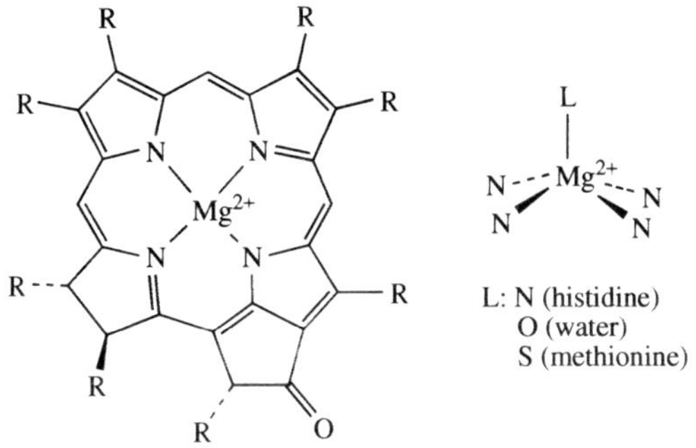

FIGURE 7. Porphyrin ring structure and unusual binding situation of Mg^{2+} in chlorophylls and bacteriochlorophylls

In over half of these conformations, Mg^{2+} interacted in an inner-sphere binding mode with both phosphate groups (on average with three oxygen ligands). Due to the biological significance of ATP, the interaction of a triphosphate linkage with Mg^{2+} has been extensively studied and will be discussed later.

C. Nitrogen Ligands

Due to the oxophilicity of Mg^{2+}, the majority (ca 77%) of all bonding interactions are with oxygen ligands[35]. However, magnesium can and does occasionally interact with nitrogen. Usually, the amine group in the side chain of Lys or a nitrogen atom in the imidazole side chain of His replaces a water ligand to form a monodentate Mg−N bond. It is quite characteristic of magnesium that more than a single Mg−N interaction in a biomolecule is seldom observed; the other ligands involved are oxygen based (H_2O, carboxylate etc.)[35].

However, a discussion of the binding modes of magnesium is not complete without mentioning the extremely unusual situation found in chlorophylls and bacteriochlorophylls (Figure 7) which is the only case where magnesium interacts with *four* (and sometimes five) nitrogen atoms in a biomolecule. The central porphyrin ring in chlorophyll binds Mg^{2+} irreversibly to the four ring nitrogens in complete disregard for its preferred hexacoordinate geometry and high oxophilicity. The magnesium is pentacoordinated in a distorted square-pyramidal geometry with the fifth ligand being located above the ring in an apical position. This fifth ligand is usually either an imidazole nitrogen of a histidine side chain or a water molecule[58]. In one remarkable case (a primary electron acceptor in photosystem I), this fifth ligand is a sulfur atom from a methionine residue. This is the only authenticated Mg−S bond found to date in nature[59].

V. MAGNESIUM TRANSPORT SYSTEMS

Cation transporters generally function by either partially or completely dehydrating the cation before it is delivered across the membrane[60]. The immense volume change between hydrated and dehydrated Mg^{2+} together with the unusually strong Mg^{2+}−water ligation energy has the consequence that the demands on a magnesium transporter are much larger than for other cations. It had therefore been postulated, long before specific information was available, that magnesium transporters would lack homology to all other known transport systems[61]—a hypothesis which has subsequently been supported by all experimental evidence available to date[60].

It has been recognized for a very long time that cellular Mg^{2+} levels depend on a regulated magnesium transmembranal transport. Since the concentration of Mg^{2+} is generally an order of magnitude higher inside cells than outside them, magnesium transporters must be capable of working uphill against a considerable thermodynamic gradient[62]. In the case of higher vertebrates, this goes as far as to recover Mg^{2+} from bodily secretions such as urine[63]. Although the physiology of Mg^{2+} transport in higher organisms (specifically mammals) has been extensively studied, detailed information at a molecular level is still lacking[64, 65]. However, three different Mg^{2+} transport proteins (CorA, MgtA/B and MgtE) have been identified in bacteria and archaea[3]. Although their precise classification as 'ion channels' or 'transporters' is still being debated[66, 67], these systems do not depend upon ATP for energy but instead utilize membrane potentials to unidirectionally drive Mg^{2+} uptake into the cell[66, 68]. Only under conditions where the intracellular concentration is dangerously high, is the fluxional direction reversed and Mg^{2+} is effluxed[60].

Of these three transport systems, only CorA has been studied in depth at a molecular level. It does not share a sequence homology with any other known ion transport system and is considered to be the primary Mg^{2+} transporter in both bacteria and archaea[3]. In addition to Mg^{2+}, CorA is also capable of transporting Co^{2+} and Ni^{2+} [69]. This 'leakage' is postulated to provide some or even all of the cell's requirements for these trace elements[60]. Most recently, several very similar solid-state structures of a CorA transporter (*Thermatoga maritima*) have become available[67, 70–72]. At first considered to be a homotetramer[73, 74], CorA is now known to possess a homopentameric quaternary structure with each subunit containing an extremely long α-helix. With a total length of *ca* 100 Å, this is the longest α-helix observed in a protein to date[72]. Five of these helices come together to provide a central ion-conduction channel or pore that transects the membrane and then extends a considerable distance into the cytoplasm (Color Plate 1). This central core of five helices, termed the 'stalk', provides most of the inner surface of the transport channel and has a narrow mouth (*ca* 6 Å) at the membrane surface and a much wider, funnel-like opening into the cell. An unusual feature of CorA is that the inside of the ion channel formed by the five stalk helices is negatively polarized along its entire length but does not contain a single charged residue. CorA obviously mediates the influx of a highly charged cation without the help of specific electrostatic interactions[60]. It should be mentioned that all available structures of CorA represent a 'closed' transporter configuration.

At the entrance of the closed transporter, a ring of highly conserved Asn residues provides a gate. When this gate opens, the initial interaction is believed to be with a fully hydrated $[Mg(H_2O)_6]^{2+}$ cation since CorA transports all hexacoordinated metal ions whose radii are comparable to Mg^{2+} (Ni^{2+} and Co^{2+})[69]. Discrimination between cations probably occurs at a later stage since $[Co(NH_3)_6]^{3+}$, very nearly the same size as $[Mg(H_2O)_6]^{2+}$, is a very potent inhibitor[75]. The incoming Mg^{2+} is suspected to be at least partially or even fully dehydrated during passage since ions with larger radii (Mn^{2+}, for example) are not transported[75]. In one solid-state structure crystallized in the presence of Co^{2+}, a single Co^{2+} is trapped just behind the closed Asn gate[71]. Once past the gate, the channel narrows and a passage with a diameter of *ca* 3.3 Å composed of a circular arrangement of conserved residues (a Thr and a Met per stalk), termed the hydrophobic girdle (HG), is encountered. At this point, dehydration is suspected to begin[71]. The real obstacle, however, is encountered somewhat further along at the membrane–cytosol interface. Here, pairs of conserved Leu and Met residues (one pair per subunit) form a hydrophobic belt (HB) that further narrows the passage down to 2.5 Å in the closed conformation. This region is believed to be the primary bottleneck for ion movement[72].

For the duration of its passage through the membrane, a shorter ring of five additional transmembranal α-helices provides an outside collar for the stalk. The ends of these helices are arranged so that an unusual ring of 20 positively charged lysine residues—called the sphincter—encircles the stalk at the level of the membrane–cytoplasm interface. At exactly this level, two additional lysine side chains in each helix point outside and away from the central channel which increases the total number of lysine residues in the sphincter region to 30. This sphincter tightly surrounds the hydrophobic girdle. In the closed conformation, the concentrated electrostatic potential generated by the sphincter will strongly repel positively charged Mg^{2+} cations, especially if they have already been partially dehydrated. After the hydrophobic girdle, the pore gradually widens into the funnel region. At this point, another metal binding position has been identified[71]. The region from the hydrophobic girdle until the funnel has opened considerably is characterized by successive rings of negative polarity due to side chains bearing negatively polarized hydroxyl and carbonyl ligands[71].

At the wide cell interface opening, a second array of α-helices (2 per subunit) also surround the stalk and hang 'down' similar to the branches of a weeping willow tree. Surrounding these is an additional arrangement of β-sheets. The tips of these willow helices and β-sheets contain an extraordinary number (a total of 50) of aspartic and glutamic acids which build a ring of negative charge, which undoubtedly helps to counterbalance the positively charged sphincter region. Taken together, the willow helices and β-sheets form an $\alpha\beta\alpha$ sandwich domain with a novel type of 'funnel' fold containing two Mg^{2+} binding sites (occupied by either Ca^{2+} or Co^{2+} ions in some solid-state structures[71]) on the outside of the funnel.

The current model for Mg^{2+} transport is illustrated in Figure 8 and is believed to be controlled by a 'magnesium sensor' consisting of two specific Mg^{2+} binding sites located on the outside of the funnel in the $\alpha\beta\alpha$ sandwich domain that spans the willow helices and β-sheets[60,71]. Studies of similar binding sites in other proteins[6] indicates that these sites will probably have an affinity for Mg^{2+} that is slightly less than the average free concentration of Mg^{2+} in cells[60]. When the concentration of Mg^{2+} in the cell drops below

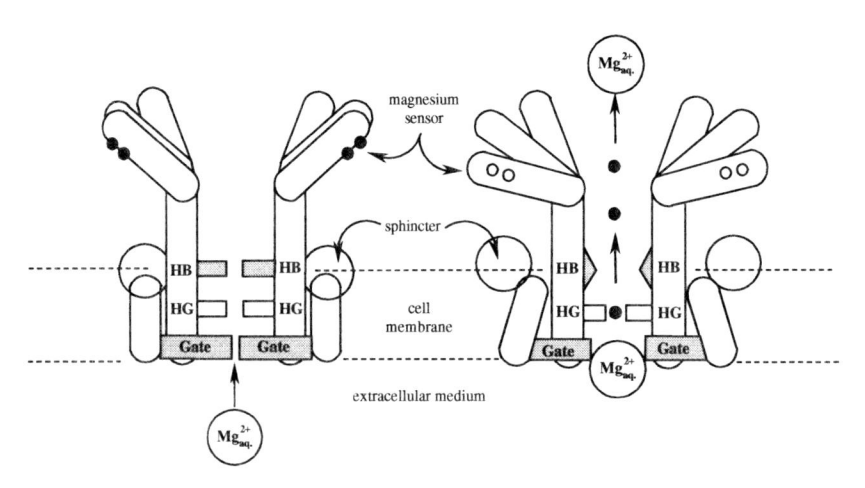

FIGURE 8. Current model for the gating mechanism in the CorA transporter. Left: closed conformation with the magnesium binding sites in the sensor occupied. Right: postulated open conformation allowing ion transport into the cell. Adapted by permission of Macmillan Publishers Ltd. from Reference 71

8. Biochemistry of magnesium   327

this critical level, these binding sites will lose their metal cations. It has been suggested that this is the pivot point of a lever system which allows binding/debinding events to drive gate closing/opening[67]. Loss of Mg^{2+} in one or more of these binding sites is postulated to trigger rotation or other movement which would move the willow helices away from the stalk, pulling the positively charged sphincter with them and thus opening the Asn gate and the Leu/Met HB bottleneck[71]. Due to the length of the stalk, even a small movement at the binding site could create a large leverage movement[72]. Once Mg^{2+} (partially or even fully dehydrated) has passed the bottleneck, it will be drawn along the pore by the successive negatively polarized rings, probably much like a bead on a chain[71]. As the pore widens into the funnel, the Mg^{2+} will be spontaneously rehydrated.

VI. UNIVERSAL ENERGY CURRENCY OF LIFE

Living organisms depend on the continual availability of free energy in an immediately useable bioequivalent form. This fuel of life is used to sustain countless biological functions. For example, mechanical processes (cellular motions, muscle contractions etc.), the active transport of small molecules and ions as well as many enzymatic activities and the synthesis of large biomolecules all require an energy source. The requirements for this biofuel are quite strict. It must be a small organic molecule that contains a large amount of energy stored in its molecular structure that, in addition, is stable under general physiological conditions. However, it must also be able to instantaneously undergo a simple chemical reaction which releases the stored energy upon demand. One of the few organic functionalities that meets these stringent requirements is a phosphorus anhydride bond in a polyphosphate. A P-O-P linkage stores a good deal of chemical energy, is generally immune towards general base or acid catalyzed hydrolysis in aqueous solutions at pH ca 7 and can be specifically hydrolyzed upon demand.

It is believed that the energy carrier at the beginning of life was triphosphate ($P_3O_{10}^{4-}$), which is often called polyphosphate[76]. A vestige of this activity can still be found in some bacterial enzymes [poly(P)ATP-NAD kinases] which are capable of utilizing triphosphate as an energy source[77,78]. In the meantime, evolutionary adaptation has resulted in the development of nucleoside triphosphates (NTP and dNTP), one of which (adenosine triphosphate, ATP) now represents the general biofuel of life (Figure 9)[79]. Hydrolytic removal of the γ-phosphate to form adenosine diphosphate (ADP) releases a considerable amount of free energy ($\Delta G \approx -30.5$ kJ mol^{-1} [80]) which can be employed in further chemical reactions. Most biological processes which require energy are now driven by an ATP $\rightleftharpoons$ ADP cycle. As a consequence, the bioturnover of ATP in organisms is extremely high. For example, the daily turnover of ATP in a human approximates half of his/her body weight[81]. In addition to the ATP-ADP cycle, some biosyntheses (synthesis of DNA

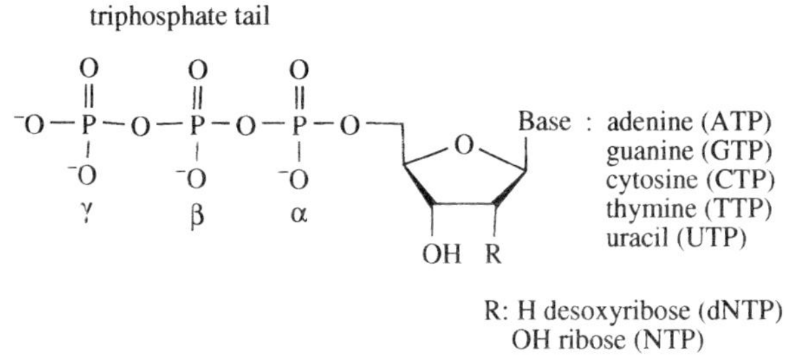

FIGURE 9. The chemical structure of nucleoside triphosphates (NTP and dNTP)

and RNA sequences, for example) are driven by ATP analogues in which adenosine has been substituted for another nucleobase.

A. MgATP^{2-}

Almost all processes involving ATP depend on magnesium as an essential cofactor. Although this has been recognized since the 1940s, only in the past few years has a detailed understanding of the interaction of magnesium with ATP begun to become available. ATP binds quite strongly to Mg^{2+} in aqueous solutions. It has been estimated that 75% or more of all free ATP in cells is actually present in a MgATP^{2-} complex which, in addition, is generally accepted as being the biologically active form of ATP[1,5]. ATP has numerous oxygen atoms in the phosphate tail, all of which are capable of binding to Mg^{2+}. The sugar and nucleobase components also contain several heteroatoms (oxygen and nitrogen) which could also conceivably interact with Mg^{2+}. Solid-state structures of MgATP^{2-} are unfortunately not available due to spontaneous nonenzymatic hydrolysis of the triphosphate tail during crystallization[82]. However, one structure of MgATP^{2-} was obtained upon cocrystallization with bis(2-pyridyl)amine[83,84]. In this structure, two distinct coordination modes for magnesium are present (Figure 10). An [(ATP)$_2$Mg]$^{6-}$ complex resulted in which all of the water ligands were replaced by monodentate oxygen ligands originating from each of the phosphates in ATP. The second magnesium in the form of [Mg(OH$_2$)$_6$]$^{2+}$ stabilizes the complex via outer-sphere coordination. The biological relevance of such a structure is probably minimal.

Although ATP predominantly exists in the '*anti*' configuration (the nucleobase points away from the sugar ring) in aqueous solutions, it is an extremely flexible molecule[85]. As has already been discussed for the much simpler pyrophosphate (PP$_i$), a complex mixture of conformational isomers can be expected. NMR studies on ATP/Mg^{2+} solutions have shown that the triphosphate tail is completely deprotonated at pH 7 and a MgATP^{2-} complex is the dominant form[86]. At or around pH 7, magnesium only interacts with ATP via the phosphate chain and the nucleotide part of ATP does not participate in chelation[87]. NMR experiments have furthermore sought to determine exactly how the phosphate tail coordinates with magnesium in solution, with varying results. Some experiments suggest that magnesium coordinates with one oxygen from each of the three phosphates (Figure 10)[88]. Other studies indicate that only one oxygen from the γ- and the β-phosphate coordinate

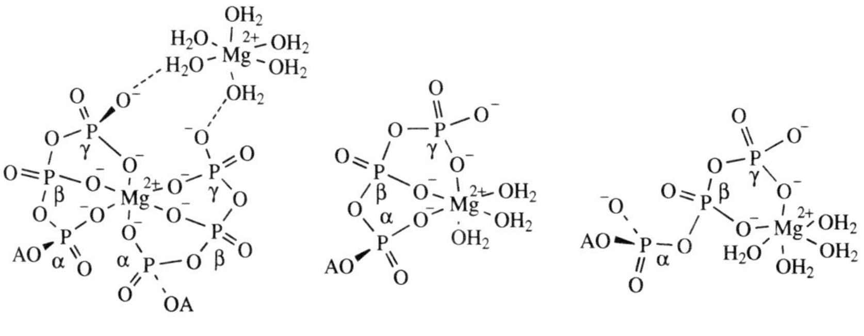

FIGURE 10. Left: solid-state structure of MgATP^{2-} (cocrystallized with *bis*(2-pyridyl)amine). Adapted by permission of the Royal Society of Chemistry from Reference 83, and with permission from Reference 84. Copyright 2000 American Chemical Society. Right: the two predominant conformations of MgATP^{2-} in aqueous solutions. Adapted by permission of Springer Science and Business Media from Reference 90

and that the α-phosphate is not involved[87, 89]. A more recent molecular dynamics study of $MgATP^{2-}$ in water has found a possible explanation for these disparate findings. According to this study, magnesium is equally likely to coordinate an oxygen atom of either the two end phosphates or of all three phosphates[90]. In addition, a relatively high barrier exists between the two conformations; conformational switching is thus rather unlikely to occur[90].

B. Biosynthesis of ATP

The biosynthesis of ATP is quite challenging due to the fact that a considerable thermodynamic gradient must be overcome in order to drive this extremely endothermic reaction. This synthesis is achieved in the cell membranes of bacteria and mitochondria through a very effective combination of a complex biomechanical motor with chemical processes[91]. ATP synthases consist of two basic components—a catalytic F_1 unit which is responsible for the chemical reaction (binding of ADP and P_i with subsequent oxidative phosphorylation) and the F_0 transmembranal unit which delivers the necessary energy in the form of an electrochemical proton gradient (Figure 11)[92]. Solid-state structures of the F_1 unit have revealed that it consists of a hexagonal array of α and β subunits bound on a shaft comprised of three further subunits (γ, δ and ε)[93, 94] which is driven by F_0 and rotates in discrete 120° steps during catalysis[95, 96]. The synthesis of ATP is tightly coupled to and driven by the rotating shaft[97]—quite analogous to a drive shaft in a mechanical motor[98]. The ATPase motor has a further remarkable property. Reversing the rotary motion switches the enzymatic mode from ATP synthesis to ATP hydrolysis[99].

Energy transmitted over the drive shaft induces the release of newly synthesized ATP on one of the β-subunits, simultaneously promotes ATP synthesis on the next β-subunit and concomitantly binds ADP, Mg^{2+} and P_i to the third[100]. It is suspected that ADP is bound before P_i[101]. A binding-change mechanism is often used to illustrate this rotational behavior. One of the three β-subunits is believed to be in an 'open' O conformation. Upon rotation, this changes to a 'loose' L conformation with a high binding affinity for ADP and P_i. After these substrates are bound, rotation results in a tightly closed conformation T in which ATP is synthesized. Further rotation opens the conformation (now O) and ATP is released[102]. Upon a rotation of 360°, each β-subunit has successively bound ADP and P_i, synthesized ATP and released it.

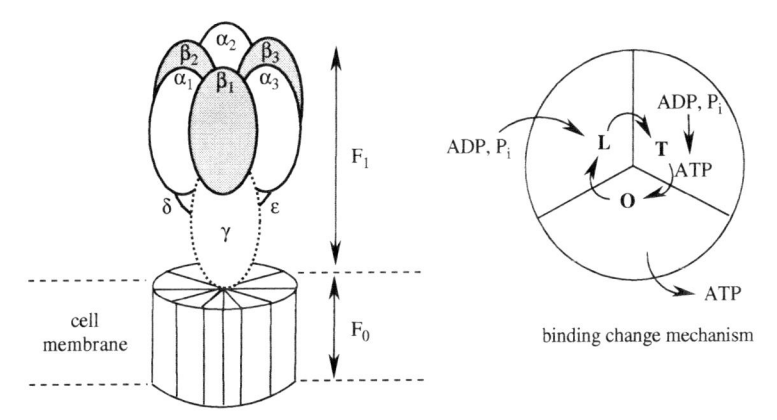

FIGURE 11. Left: simplified schematic diagram of a mitochondrial ATP-synthase. Adapted from Reference 97 with permission from AAAS. Right: binding change mechanism for the synthesis of ATP

Although our knowledge of the general motoric details of ATPase is considerable, critical chemical events taking place in the active sites in the β-subunits, especially the role of the essential Mg^{2+} ion, are just now beginning to be unraveled. For quite some time, traditional organic mechanisms (linear pentacoordinated transition structures/intermediates, for example) were postulated as being the critical chemical step in the formation of the β,γ-phosphorus anhydride bond in ATP[103]. In these mechanisms, the role of Mg^{2+} was believed to be limited to binding, orientating and activating ADP and P_i[104]. However, such traditional mechanisms are not capable of explaining exactly how mechanical motion is effectively translated into a single phosphorus anhydride bond[104] and the chemical mechanism of phosphorylation has remained an enigma[105].

Valuable insights have been gained by a remarkable phenomenon recently discovered: ATPase shows a large magnetic magnesium isotope effect[104]. The rate of ATP synthesis in which $^{25}Mg^{2+}$ (nuclear spin 5/2; magnetic moment -0.855 Bohr magneton) is present is twice to three times higher than in the presence of the spinless, nonmagnetic nuclei $^{24}Mg^{2+}$ and/or $^{26}Mg^{2+}$ [106,107]. The magnesium isotope effect is a nonclassical phenomenon discovered in 1976[108] and it has since then been shown that *any* reaction involving this isotopic effect is spin-selective; paramagnetic intermediates, e.g. radicals, radical ions and/or radical ion pairs, *must* be involved[109,110].

One of the interesting properties of ATPase is that, during or directly after ADP, P_i and Mg^{2+} are bound, a structural change (mechanical motion) compresses the active site and 'squeezes' water out of the active pocket (change from the 'loose' to the 'closed' conformation). As a result, the primary hydration shell of Mg^{2+} is partially removed[104]. This concentrates the positive charge on the ion itself and drastically increases its ability to accept a single electron. The mechanics of ATPase thus directly transforms Mg^{2+} into a very reactive electron acceptor (at this point, mechanical energy becomes chemical energy). It is now postulated that the first step in ATP synthesis is a one-electron transfer from the terminal phosphate in ADP to Mg^{2+} [104]. A radical-ion pair $[Mg^+-ADP^\bullet]$ is generated which, according to the large isotope effect observed, is also the rate-limiting step in the phosphorylation (Figure 12). This primary radical-ion pair is in the singlet spin state, as are all thermally generated radical-ion pairs, and the radical electron is localized on one of the β-oxygen atoms in the phosphate tail of ADP. This extremely reactive oxyradical now attacks the P=O bond in P_i. On the way to ATP, a species results in which the γ-phosphate is in a pentacoordinated geometry. This undergoes a fast β-decomposition which releases P_i and ATP. A hydroxyl radical in the immediate vicinity of Mg^+ remains behind. Fast electron spin relaxation in $OH^\bullet$ (10^{-11} s or even shorter)

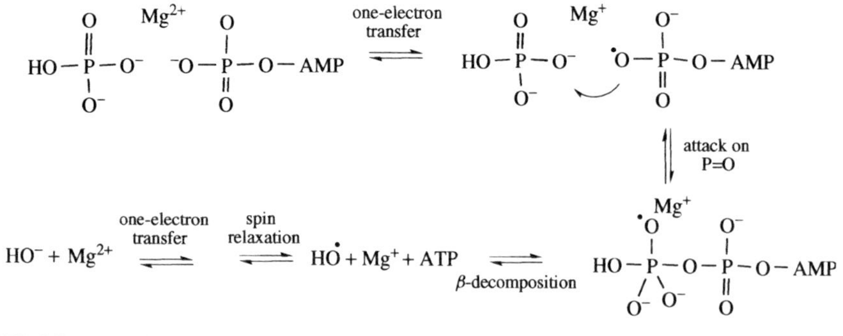

FIGURE 12. Radical-ion pair mechanism for the biosynthesis of ATP. Reproduced by permission of Springer Science and Business Media from Reference 104

makes the reverse electron transfer possible and Mg^{2+} is regenerated[104]. The calculated isotope effect for this mechanism is in good agreement with experimental results[111]. Such a radical-ion pair mechanism allows for the possibility of isotopic label (^{18}O) transfer from P_i to water. It has been recognized since as early as 1953 that such cross-transfer reactions accompany oxidative phosphorylation[112]; a mechanistic explanation has not, until now, been offered for this finding.

C. Hydrolysis of ATP

All of the countless biochemical functions of ATP involve specific hydrolysis of either the $P_\gamma-O-P_\beta$ or $P_\beta-O-P_\alpha$ phosphorus anhydride bond. The underlying organic mechanism of phosphorus ester bond hydrolysis has, of course, been the subject of numerous experimental[113] and theoretical[114-116] studies. In the absence of a metal cofactor (Mg^{2+}), two general mechanistic pathways are fundamentally possible: an associative and a disassociative reaction (Figure 13). The associative pathway resembles a general S_N2-type reaction; in-line attack of water initiates hydrolysis which proceeds over a pentacoordinated transition structure or hypervalent intermediate. In the dissociative pathway, P—OR bond breakage precedes the water attack. A more or less 'free' metaphosphate (PO_3^-) species is generated which then accepts the water[117,118]. In the dissociative pathway, a linear approach of the water is not absolutely necessary.

The influence of a single hydrated Mg^{2+} ion on the general mechanism of hydrolysis has only been considered in two independent studies using $MeP_2O_7^{3-}$ as a model for ATP[119,120]. The presence of Mg^{2+} does not change the two possible basic mechanisms (associative vs. dissociative). Both studies concluded that a hydrated Mg^{2+} cation helps to promote hydrolysis (as compared to the nonmetallated, gas-phase case) primarily by stabilizing the transition structures/hypervalent intermediates and end products. However, this stabilization is definitely not enough to overcome the considerable kinetic barrier that hinders spontaneous hydrolysis in aqueous solutions. This is in accord with the fact that free $MgATP^{2-}$ is quite stable in water at pH 7. Extensive kinetic studies on ATP (and other NTPs) have conclusively demonstrated that *two* metal ions have to be coordinated to the triphosphate chain before hydrolysis can take place[121,122]. These studies clearly indicate that binding of two Mg^{2+} ions to the triphosphate tail in a $M_{\alpha,\beta}M_\gamma$ mode (Figure 14) clearly promotes the hydrolysis of the $O_\beta-P-O_\gamma$ bond[123,124].

Although comprehensive structural analyses of ATP binding sites are not yet available, it is clear that the primary binding interactions usually originate from hydrophobic

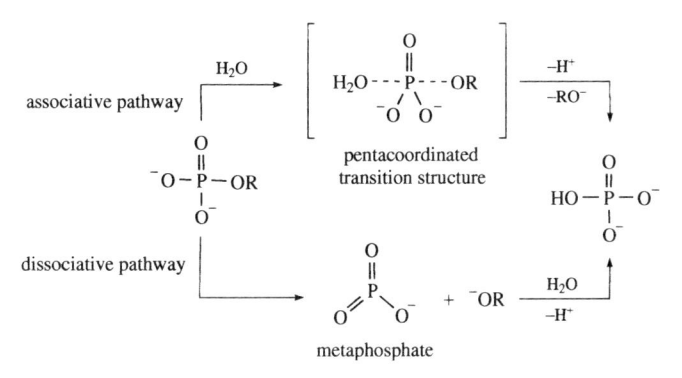

FIGURE 13. Fundamental pathways for the hydrolysis of a phosphorus ester bond

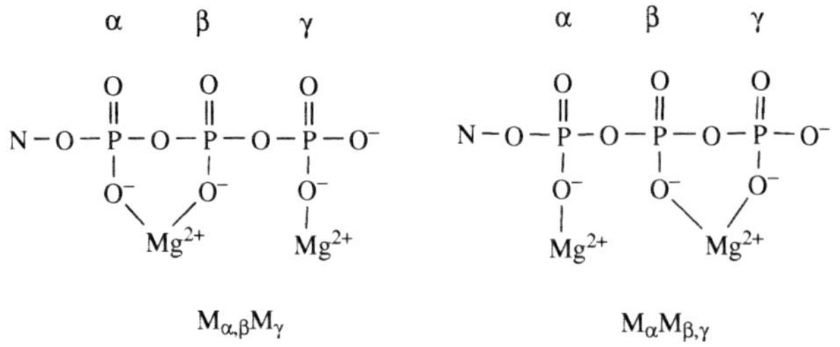

FIGURE 14. General binding modes of two Mg^{2+} ions with nucleoside (N) triphosphates. Reproduced by permission of The Royal Society of Chemistry from Reference 123

interactions of adenosine (or other nucleotides) with the enzyme pocket. Interactions between Mg^{2+} and protein residues are usually minimal, although a magnesium ion sometimes coordinates in a monodentate fashion with one or maximum two oxygen ligands from protein side chains. This leaves the phosphate tail relatively free and a $M_{\alpha,\beta}M_\gamma$ binding situation is often observed[123]. In the meantime, it is clear that this motif is extremely relevant in the mode of action of enzymes that promote transphosphorylations, i.e. enzymes that transfer the γ-phosphate of an NTP molecule to a substrate[122, 125]. If the active site is unusually hydrophobic, sometimes one of the two metal ions can be replaced by an ionic interaction (e.g. Arg)[126].

Several other major classes of enzymes, among them the nucleic acid polymerases, activate ATP (and other NTPs) in a completely different manner[127]. Similar to transphosphorylation enzymes, they utilize two metal ions for catalysis[128]. However, steric interactions are purposely employed in order to reverse the preferred binding situation. A $M_\alpha M_{\beta,\gamma}$ motif is generated which weakens the $P_\alpha-O-P_\beta$ linkage[123, 124]. This allows a nucleoside monophosphate group to be transferred (under liberation of PP_i), a process which is essential in the biosynthesis of DNA and RNA sequences.

A current molecular dynamics study using a DFT-based method looked at the magnesium-catalyzed hydrolysis of the $P_\alpha-O-P_\beta$ linkage in guanosine triphosphate (GTP) and discovered that weakening this linkage may result in a fundamental change in the hydrolytic mechanism[129, 130]. This study explicitly considered solvation effects by performing all computations in a boundary box containing 180 water molecules. As $[Mg(OH_2)_6]^{2+}$ approaches the negatively charged oxygen atoms in the P_γ and P_β phosphate tail, it begins to bind to them in a bridge bidentate manner (Figure 15). This causes a spontaneous lowering of the coordination number on magnesium from six to four, as four water ligands spontaneously leave. A similar reduction in the coordination number has been previously discussed for hydroxide ligands[14]. This initial tetrahedral complex is unstable; as the simulation time proceeds, it spontaneously embarks on a remarkable reaction path in which $MgGTP^{2-}$ is completely decomposed. Four extremely unusual, very unstable species result: two equivalents of metaphosphate (PO_3^-), the radical cation $[Mg(OH_2)_2]^+$, *molecular* oxygen O and a $GMP^{-\bullet}$ radical anion[129].

Water immediately transforms the PO_3^- into HPO_4^{3-} which presents no further problems. $[Mg(OH_2)_2]^+$ also spontaneously reacts with water in the presence of a proton in a series of complex steps to finally yield $[Mg(OH_2)_6]^{2+}$ [129]. High-level quantum-chemical calculations revealed that the first and most important step in this procedure is a single electron transfer from Mg^+ to H^+ which occurs in the conical intersection between the singlet and triplet hypersurface of the following reaction[131]:

$$[Mg(OH_2)_2]^+ + H_3O^+ \longrightarrow [Mg(OH_2)_3]^{2+} + H$$

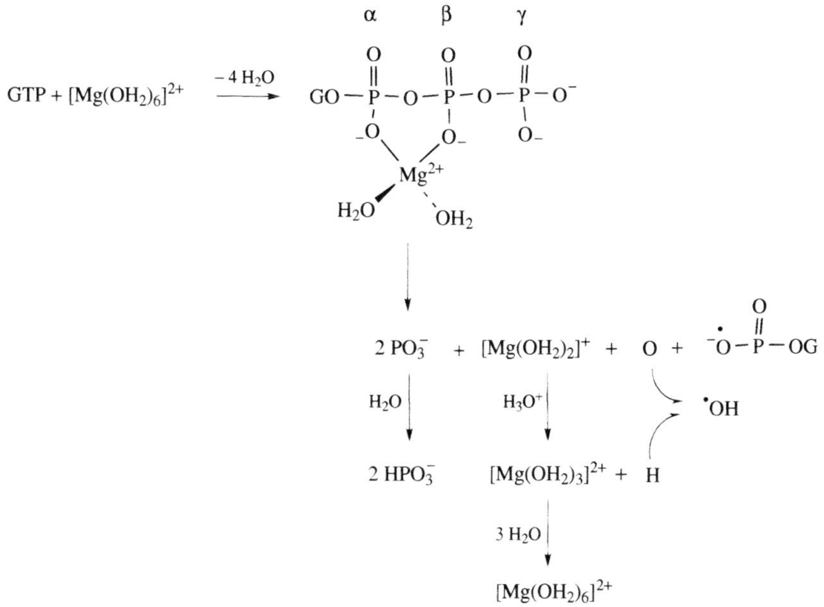

8. Biochemistry of magnesium

$$GTP + [Mg(OH_2)_6]^{2+} \xrightarrow{-4 H_2O}$$

FIGURE 15. Spontaneous decomposition of MgGTP^{2-} upon coordination of Mg^{2+} in a M$_{\gamma,\beta}$ motif. Reproduced by permission of the PCCP Owner Societies from Reference 129

As the proton is converted into molecular hydrogen, it is immediately expelled from the complex at a high velocity[129]. The [Mg(OH$_2$)$_3$]$^{2+}$ species left behind is in its ground state and can now simply accept further water ligands until Mg^{2+} has restored its optimal octahedral coordination geometry[132]. The proton ejected is assumed to collide with the oxygen atom in the immediate vicinity and a hydroxyl radical (•OH) is produced. If the GMP$^{-•}$ radical anion is generated in the heart of an active site with the hydroxyl group of a substrate R already lined up and positioned, it will attack it and form a new P–O–R bond under ejection of molecular hydrogen from the hydroxyl group. This is immediately trapped by the hydroxyl radical (•OH) and both nasty species end up as water.

This mechanism is still quite controversial[133, 134]. However, CIDNP (chemically induced dynamic nuclear polarization) experiments on a magnesium-dependent DNA-I polymerase (which hydrolyzes the P$_\alpha$–O–P$_\beta$ linkage in a NTP in order to insert a new nucleoside in a growing DNA chain) being fed MgGTP^{2-} revealed that GMP$^{-•}$ radical anions may indeed be involved[129].

VII. INTERACTIONS WITH DNA AND RNA

Only four nucleobases encode sequence-specific information in both DNA and RNA. In comparison to proteins, DNA/RNA obviously provides for much less chemical diversity (Figure 16). Nevertheless, nature manages to encode genetic information in its entirety using this simple strategy. Although a fact often forgotten or neglected in textbooks, both DNA and RNA are polyelectrolytes and contain a negatively charged phosphate group per nucleotide. As such, they have an absolute requirement for charge equalization. Usually, hydrated metal cations (Na$^+$, K$^+$, Mg^{2+} and Ca^{2+}) in the neighborhood of DNA and

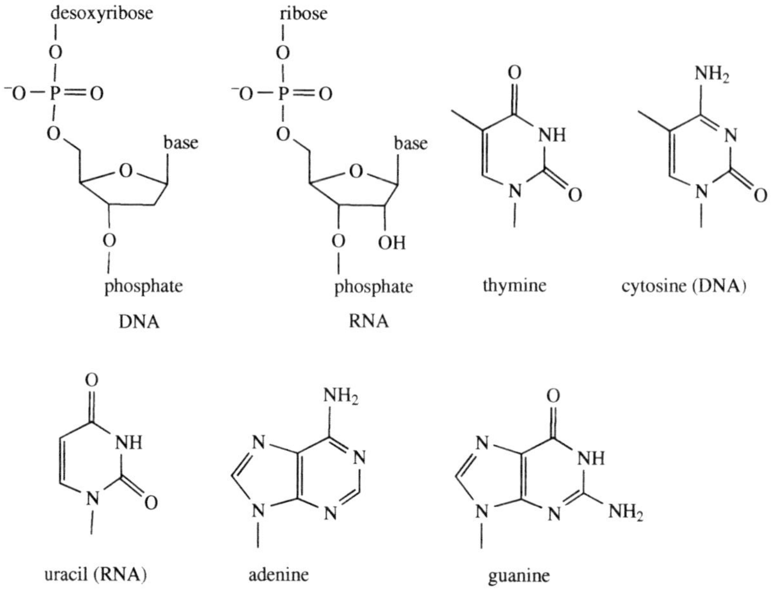

FIGURE 16. Fundamental subunit of the basic polymeric structure of DNA and RNA as well as the individual nucleobases involved

RNA strands take care of this job. However, on occasion, positively charged polyamines or even side chains of proteins can neutralize the backbone charge[135].

Metal ions interact with DNA or RNA in at least four different ways[136]. The most general mode is that of position unspecific, diffuse binding in which the presence of the (hydrated) metal cation simply provides charge screening to overcome electrostatic repulsions between RNA backbone fragments[10]. These diffuse interactions can be modeled by mathematical approximations such as the Poisson–Boltzman distribution[137, 138]. When DNA or RNA folds, it sometimes creates well-defined sites, 'holes', with a high concentration of negative charge. These sites often contain electrostatically localized metal ions that do not further interact with the substructure. Binding is controlled by simple electrostatics and the size of the 'hole'[139]. Many different ion types, including protonated polyamines, can occupy these sites. Specific binding interactions start with metal ions (often Mg^{2+}) bound through outer-sphere coordination modes[140]. In many cases, an extensive network of specific hydrogen-bonding interactions is built up between the DNA or RNA and the water ligands of the metal ion[136]. Finally, the metal ions can interact directly via an inner-sphere coordination. As in the case of protein interactions, this can occur with a variable number (1–3) of ligands, the most common being the phosphoryl oxygens in the backbone, purine N7, base carbonyl groups and the 2′–OH in ribose (RNA)[141]. Finally, a mixed-sphere mode which combines both inner- and outer-sphere binding is sometimes observed[142].

A. Magnesium and Structural Stability

The high negative charge in the phosphate backbone of both DNA and RNA is detrimental towards folding into compact structures such as the well-known double-helix motif.

Due to its high charge, small size and propensity for promoting the formation of extensive stabilizing networks of hydrogen bonds, Mg^{2+} is often the most effective ion for reducing electrostatic repulsions and promoting folding. As an example, it was recognized very early on (1972) that millimolar concentrations of Mg^{2+} stabilize RNA tertiary structures that are only marginally stable in the presence of monovalent ions[143]. In the meantime, a massive amount of data has unequivocally demonstrated the essential role of metal ions in determining the folding kinetics and maintaining the thermal stability of DNA and RNA helices and other secondary structures[10, 11]. It is clear that metal ion interactions play a significant role in determining local conformation and topology. However, our understanding of just how important they really are is still rather limited[9].

Perhaps the most common site-specific magnesium binding sites in both DNA and RNA are localized in the major and minor grooves of double-helix strands. Backbone folding in these regions forms cavities that are lined with lone electron pairs from the nucleobases (typically the N7 purine nitrogen and the O6 carbonyl group) that provide excellently positioned electrostatic contacts for a hydrated Mg^{2+} ion[9, 144]. An example of this has already been illustrated for a DNA double helix in Figure 4, where crystallization of the Dickerson–Drew dodecamer (sequence CGCCAATTCGCG) revealed one bound Mg^{2+} located in the major groove per duplex (helix dimer)[30, 31]. Raised Mg^{2+} concentrations lead to an improved crystal quality and the use of synchrotron radiation allowed a much better resolution (1.1 Å). Shortly after the first structure was published, five different specific magnesium binding sites were located in this dodecamer (see Color Plate 2)[145]. Four of them bind in an outer-sphere mode. In the major groove, Mg1 provides a bridge linking two guanosines from opposite strands via an outer-sphere coordination. Mg1 is the only ion present in crystals grown with lower Mg^{2+} concentrations, which indicates that this binding site has the highest affinity and is the first one to be occupied[145]. This magnesium affects the conformation of the duplex in that it introduces a slight 'kink' (11°) in the helix axis. In the minor groove, Mg_2 bridges phosphates from opposite strands. This clearly affects the width of the groove by effectively 'zipping' it up, thus allowing the negatively charged backbones to closely approach each other. For example, the width of this groove is 2 Å less than the same dodecamer when crystallized with larger Ca^{2+} ions[146]. Mg3 is the only ion that interacts via an inner-sphere binding mode and binds in a monodentate manner to the phosphate backbone. Mg4 and Mg5 are located on the outside of the phosphate backbone and provide charge equalization via outer-sphere interactions.

Many RNA structural motifs also specifically bind Mg^{2+} with a striking example being the Loop E motif illustrated in Color Plate 2[147]. In this motif, four magnesium ions provide a zipper for the minor groove which results in a stable, compact secondary structure. Mg1 binds monodentately to the backbone and Mg_2 shows a typical outer-sphere binding pattern. The most interesting feature, however, is a very unusual Mg_2 cluster in which three water ligands bridge two magnesium ions. This cluster is bound directly to the backbone via a monodentate bridging motif. Such Mg_2 clusters are occasionally found to stabilize specific RNA folding patterns, a further example being a fragment of a 5S rRNA domain (Color Plate 2)[147]. This fragment adopts its characteristic structure only in the presence of Mg^{2+}. Twelve magnesium ions bind in a specific manner[148]. Two binuclear clusters are present: one is buried deep in a major groove and the other is located on the outside of a major groove that has been twisted and compacted by the presence of a twisted Loop E motif in which four further magnesium ions participate. A partial Loop E motif with two more Mg^{2+} ions glues the fragment ends together.

B. Ribozymes

Traditionally seen as a passive carrier of genetic information, it came as a shock when the catalytic properties of RNA were discovered in early 1980s[149, 150]. In the meantime, it is clear that RNA catalysis is quite ancient; it may have played an important role in the early stages of evolution and it is quite possible that RNA was used to support a primitive metabolism[151]. Although some RNA activities may be a 'fossil' remnant of an earlier world, the modern chemical function of RNA is extremely complex and highly varied. RNA is deeply involved in almost every aspect of cellular metabolism. In spite of its basic chemical simplicity, RNA sequences serve in a surprising number of multifunctional roles that vary from structural scaffolds[152], conformational riboswitches[153], regulatory signaling[151] and catalytic systems (ribozymes[12]). RNA catalysis is considerably more widespread than was originally believed[154].

Three different families of naturally occurring ribozymes are known today—the large and small phosphoryl transfer ribozymes and the aminoacyl esterase ribozymes. The large phosphoryl transfer enzymes are a widely varied collection of huge molecular machines which include the group I[155] and group II[156] introns, ribonuclease P[157] and, perhaps, the eukaryotic spliceosome[12]. They all cleave or synthesize phosphodiester linkers in the backbones of RNA (and often DNA), usually via a classical S_N2 mechanism involving the in-line attack of a nucleophile (typically alcohol or water) at phosphorus. The small phosphoryl transfer enzymes are typically found in the genome of primitive viruses. These are small self-cleaving RNA motifs which are responsible for cutting long RNA strands into individual genes[158]. Among others, they include the hammerhead family[159] and the hepatitis delta[160] ribozymes. At the interface between the RNA and protein worlds, the ribosomal aminoacyl esterases catalyze the making and breaking of amide bonds. Since 2000, it has become increasingly clear that ribosomal RNA actually catalyzes the synthesis of proteins[161].

In contrast to this, catalytic DNA has not yet been observed in nature, although synthetic 'DNAzymes' have been successfully designed[162]. This gives rise to the question of exactly why there are no natural DNAzymes. Is this due to a natural 'fluke' in evolution or does the exchange of one single nucleobase (cytosine for uracil) and the presence of the additional hydroxyl group (ribose instead of desoxyribose) in the sugar have something to do with promoting the catalytic properties of RNA? No satisfactory explanation has yet been offered.

Most ribozymes are folded into compact, stable tertiary structures that possess a unique conformation that is responsible for the chemical activity observed. In addition, most of these systems exhibit a clear requirement for Mg^{2+} ions[12]. It is currently accepted that the dynamics of RNA folding follows a typical two-step procedure (Figure 17)[12, 163]. In the first step, the secondary structure with regions of single and double strands is formed. It has been recognized for quite some time that this secondary structure formation can be stimulated from almost anything which screens charges and brings polyanionic backbones in close proximity to each other (mono- and divalent metal cations, protonated polyamines and even proteins)[164]. In contrast to this, adoption of the tertiary structure under collapse into a specific 3D conformation (second step) is usually controlled by stringent metal ion requirements (often Mg^{2+})[165, 166]. The hypersurface of RNA folding is quite complex and stable misfolded intermediates easily occur. The overall rate of folding is therefore often determined by the time it takes for a misfolded structure to unfold again (kinetic trap)[167]. Metal cations may be solely responsible for directing folding processes and stabilizing the active tertiary structure (structural function). However, they also provide a means of overcoming the deficit of chemical diversity in RNA in a striking manner[12]. One of the greatest challenges in current RNA research is that of separating structural functions from chemical (catalytic) processes[168].

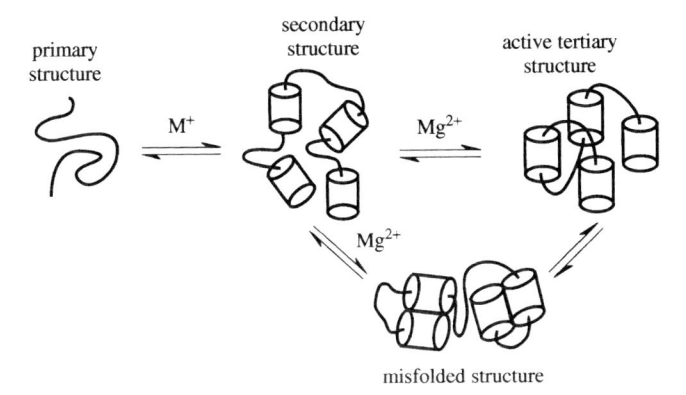

FIGURE 17. Simplified view of two-step folding processes in RNA. Adapted by permission of Springer Science and Business Media from Reference 12

1. Self-splicing in group I introns

Group I introns are the most abundant category of ribozymes known to date and more than 2000 sequences have been identified so far[169]. All of these ribozymes self-splice using a common strategy of two consecutive transesterifications using an external guanosine nucleotide as a cofactor[170]. They are distinguished from the group II introns which employ an internal adenosine to initiate self-splicing[154]. It is quite interesting that, apart from a few critical nucleotides, the sequence conservation in the active site is quite poor. However, the 3D core structure of the active site is very well conserved[171], a fact indicating that the topology of the active site is extremely important. Group I introns have a specific requirement for Mg^{2+} ions both to ensure proper folding[172] and to promote catalysis[12].

Although extensively studied for well over two decades (the group I intron from *Tetrahymena thermophila* was the very first ribozyme to be discovered), structural information has been quite elusive. Nevertheless, with the help of 'metal ion rescue experiments'[173], the general mechanism of self-splicing illustrated in Figure 18 could be elucidated[12, 159]. In the presence of Mg^{2+} ions, the active site binds an external guanosine nucleotide (G) which then functions as a cofactor. Splicing is initiated by nucleophilic attack of the O3' hydroxyl group of G on the phosphate linker (P) at the 5'-splice site. The attack occurs via an 'in line' S_N2 mechanism over a pentacoordinated intermediate/transition structure[152]. A conformational change then removes the G cofactor from the active site and brings the terminal nucleotide of the hydrolyzed 5'-exon into the proximity of the 3'-splice site and positions it for the second phosphoryl transfer. This time, the nucleophile is believed to be the 5'-hydroxyl group (5'-exon) which attacks the phosphate linker at the 3'-splice site, again via an 'in line' S_N2 mechanism.

Mechanistic studies based mainly on metal ion rescue experiments have identified six oxygen atoms involved in metal ion coordination in the active site (the oxygens in bold font in Figure 19)[174]. Metal ion rescue experiments substitute a potential oxygen ligand with a 'soft' atom, usually sulfur, that is much less inclined to coordinate a 'hard' Mg^{2+} ion. If the addition of a 'soft' cation such as Cd^{2+} restores activity, the oxygen

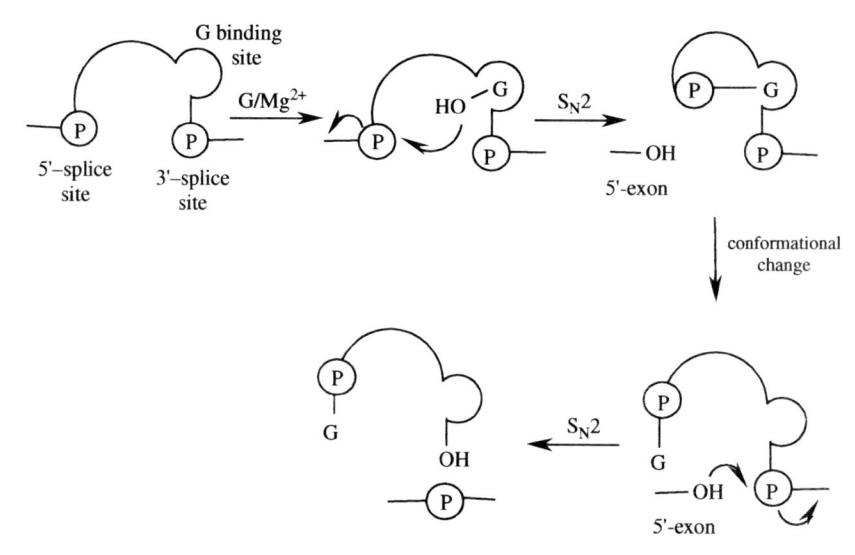

FIGURE 18. Schematic representation of the self-splicing reaction catalyzed by group I introns. G is the guanosine nucleotide cofactor, P a phosphate linker

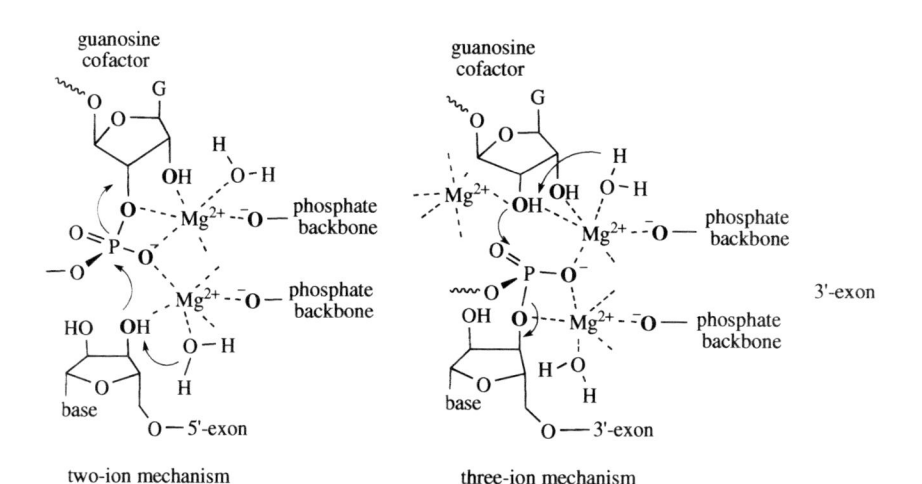

FIGURE 19. The two- and three-metal-ion mechanisms for phosphoryl transfer in the group I introns. Oxygens in bold have been identified as metal ligands by ion rescue experiments. From Reference 177. Reprinted with permission from AAAS

is considered to be a metal ligand[174]. These results have led to possible mechanisms for the phosphoryl transfer steps which involve either two[175] or three[176] metal ions. In the three-ion mechanism, illustrated for the first phosphoryl transfer step, the guanosine cofactor interacts with two Mg^{2+} ions via both hydroxyl groups. One of the ions activates the nucleophilic hydroxyl group while the other two ions stabilize the developing pentacoordinated transition structure. As negative charge builds up in the leaving group, it is stabilized by the third metal ion, most probably by providing a proton originating from a water ligand. The two-metal-ion mechanism, illustrated for the second phosphoryl transfer step, is quite similar: one of the magnesium ions activates the nucleophile, both stabilize the pentacoordinated transition structure and the second Mg^{2+} stabilizes the developing charge on the leaving group. Only recently have the first solid-state structures for group I introns become available. A total of four are available to date and, in each case, chemical modifications or deletions were made in order to capture one of the reaction intermediates indicated in Figure 19[174]. All four solid-state structures support the two-metal-ion mechanism but do not rule out the possibility of a three-ion mechanism[177].

2. Hammerhead ribozymes

First identified in 1986 as the catalytic active element in the replication cycle of certain viruses, the hammerhead ribozymes (HHRz) are the smallest known, naturally occurring RNA endonucleases[178, 179]. They consist of a single RNA motif which catalyzes a reversible, site-specific cleavage of one of its own phosphodiester bonds[180]. Truncation of this motif allowed a minimal HHRz to be constructed which was the very first ribozyme to be crystallized[181]. HHRz minimal motifs are characterized by a core of eleven conserved nucleotides (bold font in Figure 20) from which three helices of variable length radiate. Selective mutation of any of these conserved residues results in a substantial loss of activity[182]. In the absence of metal ions the structure is relaxed ('extended'), but upon addition of Mg^{2+}, hammerhead ribozymes spontaneously fold into a Y-shaped conformation (Figure 20; Color Plate 3)[183].

It has been postulated that, in this Y-conformation, the 2'-OH functionality of the ribose at C17 initiates a nucleophilic attack on the phosphodiester linkage over a pentacoordinated transition structure (classical S_N2-type reaction; Figure 21)[183]. The leaving group is believed to be protonated by a general acid in the vicinity. Quite a bit of experimental support for such an S_N2 mechanism is available, with perhaps the most important fact being that inversion at phosphorus invariably occurs[184, 185]. However, solid-state structures of minimal HHRz constructs indicate that achievement of the in-line orientation required for an S_N2 reaction is practically impossible, the angle between the three atoms being

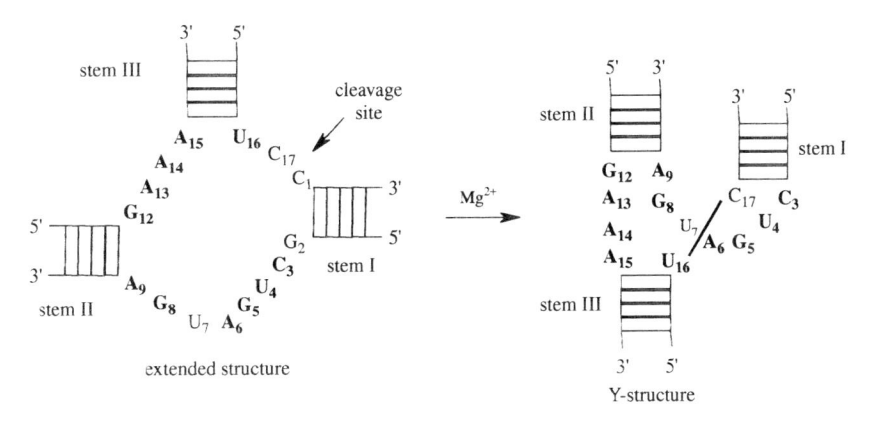

FIGURE 20. The catalytically active RNA sequence of a typical 'minimal' model of hammerhead ribozymes. Reproduced by permission of Wiley-VCH Verlag GmbH & Co. KGaA from Reference 190

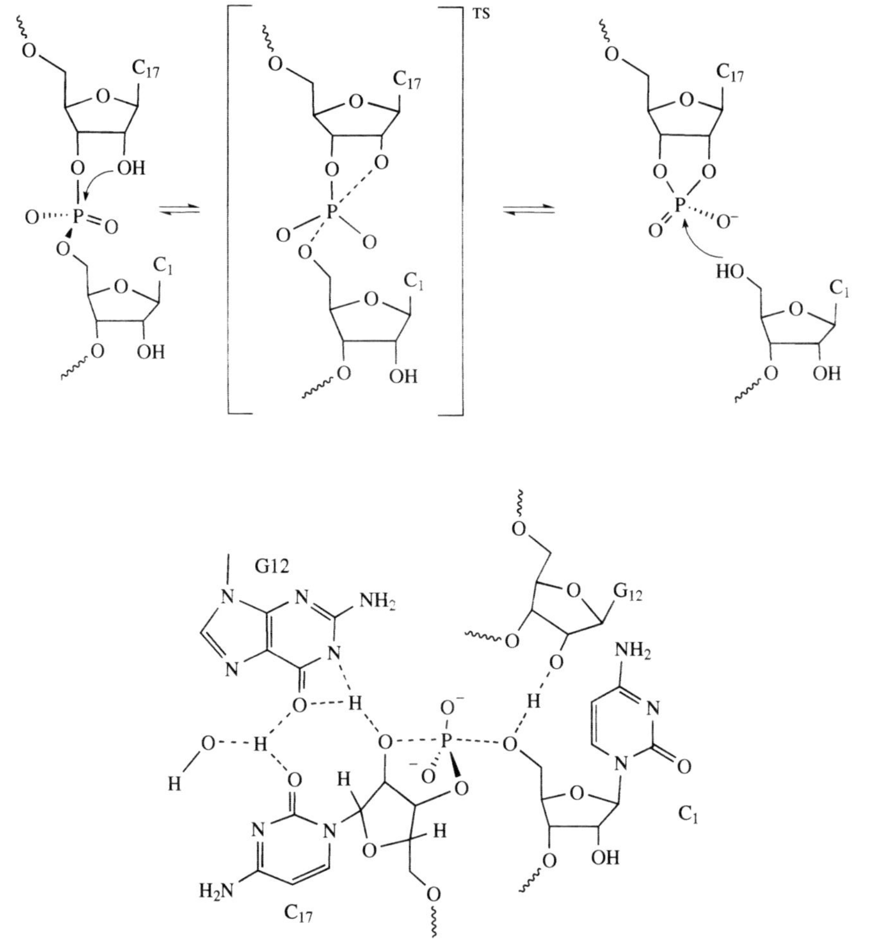

FIGURE 21. Above: initially postulated mechanism of phosphordiester bond cleavage. Below: a modified model of the transition structure in natural hammerhead ribozymes. Reproduced by permission of Wiley-VCH Verlag GmbH & Co. KGaA from Reference 190

nearly 90° instead of the required linear orientation[186, 181]. This problem was further compounded by the fact that cleavage could be achieved by soaking the crystals with Mg^{2+} ions, which indicated that major structural rearrangements are not necessary to reach the transition structure[187]. Structures of minimal HHRz constructs in the solid state were also consistent with solution data[183]. These findings are incompatible with results obtained from metal ion rescue experiments which clearly indicate that, in the active conformation, a single Mg^{2+} ion bridges the phosphate groups of residue A9 and the phosphate being cleaved[188, 189]. In the solid-state structures of minimal constructs, these phosphates are over 20 Å apart. Taken together, these disparate findings gave rise to a long-standing debate—the 'structure-function dilemma'[180]—for the hammerhead ribozymes which is just now beginning to be resolved[190].

8. Biochemistry of magnesium 341

Perhaps the first step in this direction consisted of the realization that the 'minimal' models usually employed for mechanistic investigations (which fold into the Y-structure) are suboptimal[191, 192]. These truncated motifs require high (millimolar) concentrations of Mg^{2+} ions for proper folding and subsequent catalytic activity (typical rates of ca 1 min^{-1})[180]. In contrast to this, natural hammerhead ribozymes are active at submillimolar (physiological) concentrations and exhibit much higher cleavage rates (ca 870 min^{-1} in the case of HHRzs isolated from schistosomes)[193].

In a complete hammerhead ribozyme, additional loops and bulges in stems I and II permit tertiary interactions which considerably alter the folding dynamics and allow for activity under physiological conditions[194]. Recent photocross-linking experiments have identified some of these tertiary interactions[195] and a solid-state structure of a new conformation has been determined[196]. This structure is characterized by a much more compact arrangement of the active site with additional tertiary base interactions which are illustrated in Color Plate 3. More important is the fact that a perfect in-line geometry of the 2'-oxygen in C17 with the phosphate and the 5'-oxygen in C1 is now present. Nucleobase A9 is now in the immediate vicinity of the phosphate (not illustrated) and most of the discrepancies which lead to the 'structure-function dilemma' have now been explained. Nucleobase substitution experiments[197] together with kinetic studies[193] have given rise to a modified cleavage mechanism in which N1 of G12 probably functions as the general base to deprotonate the 2'-OH group (Figure 21)[190]. The general acid is postulated to be the hydroxyl group in the sugar functionality of G8. The exact role of Mg^{2+} in the mode of action of natural hammerhead ribozymes is still unclear. The bridging Mg^{2+} ion indicated by mechanistic experiments[188, 189] was not observed in this solid-state structure—probably due to a high concentration of monovalent cations during crystallization[196]. However, the presence of this specific Mg^{2+} binding site was recently confirmed spectroscopically[198]. According to this study, Mg^{2+} retains 4 water ligands and binds in a monodentate fashion to the N7 atom of guanosine in G10 and to the pro-R phosphate atom in the A9 linkage—which places it in the immediate vicinity of the catalytic activity in the active conformation. However, another study indicated that the A9/G10 site can interact with divalent metal ions via both an inner- and an outer-sphere manner[199].

Is Mg^{2+} solely responsible for achieving proper folding into the transient active conformation (structural role) or does it also actively participate in the catalytic reaction? Again, conflicting information is available in the chemical literature. Originally thought to be absolutely dependent upon Mg^{2+}, many studies have now demonstrated that the mode of action is much more complex than previously assumed. Under certain conditions, hammerheads function sluggishly in the presence of (very) high concentrations of monovalent cations such as NH_4^+ or Li^+. In the meantime, there is evidence for the possibility of at least three reaction pathways—a monovalent, a divalent and a cooperative pathway that involves both mono- and divalent metal ions[200, 201].

Although quite some time and effort have been expended to develop computationally based models for direct Mg^{2+}-ion participation in the catalysis (both one-[202] and two-ion[203, 204] mechanisms have been suggested), it is clear that more information is needed before the relevance of these studies can be judged. It is also quite possible that the mode of action of minimal HHRz constructs could fundamentally differ from the natural ribozymes—and this may be responsible for the differential catalytic rates observed. It is generally accepted that inversion of configuration is conclusive proof for a $S_N 2$-type inline attack. However, a recent computational study argued that, due to the possibility of a facile pseudorotation at phosphorus, an adjacent (90°) attack with simultaneous inversion of configuration cannot principally be ruled out[205]. An adjacent attack mechanism requires that a normally unstable apical oxyanion must be stabilized in the transition structure in order for inversion of configuration to occur. This theoretical study demonstrated that a Mg^{2+} ion is capable of such a stabilization[205].

VIII. PROTEIN-BASED ENZYMES

In 2000, more than 350 protein-based enzymes (not including the metabolic cycles) with a specific requirement for Mg^{2+} had been described in the chemical literature[62] and many more have been discovered since then. In the past three or four years, our knowledge of structural and mechanistic details for many of these enzymes has increased exponentially and currently more than 3000 solid-state structures of protein-based biomolecules containing magnesium are available in the data banks[206]. There is a real need for comprehensive studies, as a systematic description of the common structural characteristics and mode of action of magnesium-based enzymes is not yet available. In an attempt to provide an initial approach to this wide field of current research, this review first provides an overview of general enzymatic modes of action followed by a more detailed discussion of several common types of magnesium-based systems. Selected examples of basic metabolic processes are presented. Magnesium involvement in DNA/RNA replication and repair is illustrated and, finally, the role of magnesium in the important process of photosynthesis is discussed.

A. General Enzymatic Modes of Action

Quite some progress has been made in understanding general enzymatic catalysis since the first simple 'key and lock' model first proposed by Emil Fischer, still to be found in many biochemistry textbooks. In the meantime, it is clear that such a simple model usually does not represent biochemical reality in the slightest. Enzymes are fascinatingly complex systems—and their individual modes of action can vary quite widely. Perhaps the most convenient categorization for enzymatic reactions to date is to sort them into three different mechanistic classes—the template, sequential and allosteric systems. Of course, some enzymes will show borderline behavior with characteristics belonging to more than one of these categories.

1. Template enzymes

As applied to metalloenzymes in general, the most simple mechanistic behavior—and that corresponding most closely to the earlier key and lock model of enzymatic catalysis—is that of a template system. In these enzymes, the metal cation is irreversibly bound in a well-defined active site on or near the surface of the enzyme with a channel or opening that allows substrate approach and product departure. Template enzymes function primarily via the 'coordination template effect'[207], also known as the 'scaffold effect'[208], in which the role of the metal cation is to specifically recognize the substrate(s), bring it/them together and activate it/them (usually through direct metal–substrate interactions) and finally to catalyze the desired chemical reaction, sometimes over one or more metal-bound intermediates, after which the product departs from the active site. Perhaps the best definition of a template metalloenzyme is that the metal ion functions as a 'true catalyst' and, although absolutely necessary in order to catalyze the chemical reaction, it remains bound to the active site and does not directly participate in the catalytic turnover [does not enter or leave the active site with the substrate(s)/product(s)]. The active site of template enzymes is rather small and the chemical reaction is strictly *localized* in the immediate vicinity of the metal ion. Due to this characteristic feature, small organometallo complexes that model the immediate structural and electronic features of the active site sometimes make good biomimetica for template enzymes. As

8. Biochemistry of magnesium   343

a consequence, most (probably more than 95%) of the biomimetical work—of both experimental and computational nature—reported in the literature to date is based on template models. Although the majority of template enzymes contain a single metal ion in their active site, quite a few, especially hydrolases, contain two[48, 209, 210]. Some even contain three metal ions, although the fundamental mode of action should probably be considered borderline, as the third metal ion (often magnesium) usually exhibits sequential or even allosteric behavior. Interestingly enough, very few magnesium-based enzymes exhibit 'template' behavior.

2. Sequential systems

The next stage of mechanistic complexity occurs in the sequential enzymes. In this case, the metal ion is an intrinsic part of the catalytic turnover. It enters the catalytic circle, interacts with the active site, substrate(s) and/or product and leaves again at some point in the turnover. Due to the ability of water to effectively compete with general organic ligands for magnesium, this metal ion is predestined to display sequential behavior. The resting state of these enzymes does not necessarily contain magnesium in the active site and the enzymatic activity obviously depends on the immediate bioavailability of free $[Mg(OH_2)_6]^{2+}$. The intracellular Mg^{2+} concentration is, however, usually high enough to guarantee enzymatic activity. Many magnesium-based enzymes, especially those involved in metabolic pathways, are suspected to be regulated by controlling the amount of free magnesium present in the immediate vicinity.

The role of magnesium in a sequential enzyme can fall into one of three general categories—two of which involve direct binding interactions with the active site. In the course of a single catalytic turnover, Mg^{2+} may enter the active site, bind to it and fulfill a critical catalytic role—one that may even be attributed to 'template' behavior, at least for one or more critical steps in the turnover—and then leave again. Alternatively, magnesium binding can trigger a structural change in and around the active site, thus regulating some important aspect of the catalytic turnover. Such binding sites are often termed 'allosteric' regulatory sites. Once this function has been fulfilled, it departs again and the structure of the active site returns to its previous state. The third category are enzymes in which Mg^{2+} has little or no direct interaction with the active site itself but interacts principally with the substrate(s) and/or product(s). In this case, the magnesium cofactor usually functions as a Lewis acid and helps to activate a bound nucleophile (deprotonation of water, for example), stabilize a critical intermediate and/or transition structure—often via outer-sphere interactions or helps to stabilize a leaving group such as pyrophosphate (PP_i) or inorganic phosphate (P_i).

The active site in a sequential enzyme, which must now accommodate at least one (partially or fully solvated) metal ion as well as substrate/product, is generally a bit larger than those observed for template systems—especially when a regulatory binding site is present. When a metal ion binds to such a site, it usually induces structural changes. In many magnesium-based enzymes, these regulatory conformational changes occur in a relatively limited region (10–50 Å in diameter) in and around the active site, which can still be considered to be localized. Examples of such localized motion include switching between active and inactive conformations, inducing a loop movement that clamps a 'lid' on the active site during a critical chemical reaction etc. Sometimes the borderline to true allosteric behavior is quite fluxional.

3. Allosteric systems

The most complex mechanistic behavior is displayed by an allosteric system in which metal binding/debinding events trigger structural changes not just in and around the active site (or sites) but in the *entire* enzyme. Indeed, it is often difficult to speak of an active site as such. The metal ion may bind to a regulatory site far away from where the reaction of interest is taking place. Although the region where the substrate is being transformed into product can often be identified, it is impossible to consider this region as being independent of the rest of the enzyme. As a consequence, it is very difficult to perform mechanistic studies on allosteric systems—and even more difficult to develop models for studying their behavior[211]. Even employing the most simple of molecular modeling techniques, it is still beyond the capacity of modern computers to perform calculations on these enzymes which usually tend to be quite large, often possessing several interacting subunits[212]. Their complexity, together with the fact that Mg^{2+} is spectroscopically silent, very small and difficult to detect experimentally, currently limits available knowledge of the mechanistic details occurring in allosteric systems involving magnesium.

One of the few magnesium-based enzymes which exhibits allosteric behavior that has been studied in more detail is alkaline phosphatase (AP), which hydrolyzes phosphate monoesters nonspecifically under both acidic and alkaline conditions[213]. AP is a relatively small enzyme consisting of two subunits. In each subunit is a trinuclear active site which contains two Zn^{2+} and one Mg^{2+} ion (Figure 22). The two Zn^{2+} ions exhibit 'template' behavior and are responsible for the catalytic behavior. They are essential for the activity whereas magnesium alone is not active. Reference 214 contains a detailed discussion of the actual hydrolysis; this chapter concentrates on illustrating the ancillary role of Mg^{2+} in modifying the behavior of AP. In the solid state, AP consists of two symmetric subunits (homodimer)[215]. In the absence of Mg^{2+}, solutions of AP are also homodimeric[216]. However, Mg^{2+} binding/debinding triggers reversible dynamic refolding and AP undergoes continual structural rearrangements in solution[217]. Each subunit can assume one of two distinct, inherently nonequivalent conformations which are illustrated by squares and circles in Figure 22[218, 219]. In addition, Mg^{2+} binds to AP with negative

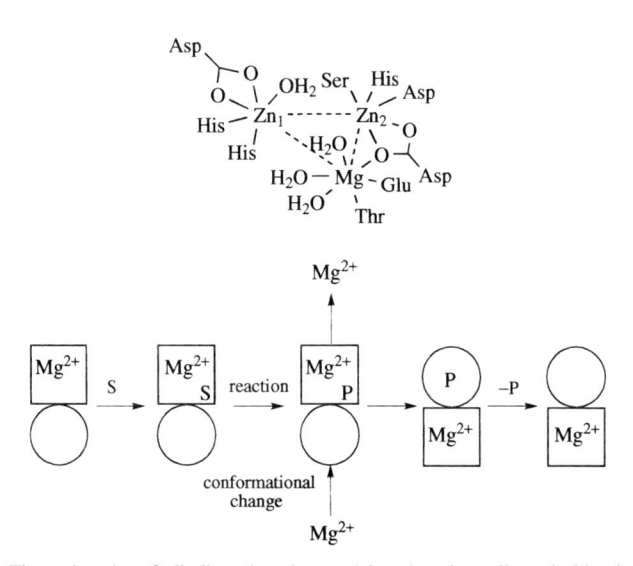

FIGURE 22. The active site of alkaline phosphatase (above) and an allosteric kinetic switch mechanism (below) for the regulatory function of the Mg^{2+} ions in controlling the conformation of the nonequivalent subunits (square and circle). Reprinted with permission from Reference 214. Copyright 2005 American Chemical Society

cooperativity and the dimer prefers to have only one Mg^{2+} present (square)[220]. The subunit with bound magnesium exhibits a higher binding affinity for both the substrate S (which preferentially docks on this subunit) and the product P. After hydrolysis, the product is not easily displaced from this subunit. However, binding of Mg^{2+} to the second subunit (circle) triggers an allosteric conformational change in both subunits. The first subunit now becomes a circle with a drastically reduced affinity for both Mg^{2+} and product, both of which are now easily ejected from the active site. Allosteric regulation of AP through reversible binding/debinding of Mg^{2+} results in a 'kinetic switch' which accelerates the overall rate via conformationally controlled accelerated dissociation of the product P^{218}.

B. Metabolic Enzymes

Many of the metabolic pathways in higher organisms are mediated by magnesium-dependent enzymes. The well-known glycolytic cycle (found in all biochemistry textbooks and illustrated in Figure 23) is a typical example; of the 10 enzymes involved, half of them have a specific requirement for magnesium. In glycolysis, glucose is oxidized and split in half to create two equivalents of pyruvate. The overall reaction [glucose + 2 ADP + 2 P_i + 2 NAD^+ → 2 pyruvate + 2 ATP + 2 NADH + 2 H^+ + 2 H_2O] is exothermic and, along the way, the energy released is converted into ATP. In addition, the NADH generated during glycolysis is used to fuel ATP synthases which produce further equivalents of ATP. The net yield of ATP per glucose molecule is either 6 or 8, depending on which shuttle mechanism (glycerol phosphate or malate-aspartate) is employed to transport the electrons from NADH into the mitochondria. The key in regulating the glycolytic pathway is the rate-limiting step catalyzed by the magnesium-dependent enzyme phosphofructokinase. Among other regulatory mechanisms, phosphofructokinase is inhibited by ATP. When ATP is abundant, the turnover of phosphofructokinase slows down, which prevents wasting glucose on making energy when it is not needed.

In spite of the wide chemical variance and high structural and kinetic diversity encountered in magnesium-based metabolic enzymes, they generally follow a common mechanistic theme[6]. They usually possess at least two (and sometimes three) magnesium binding sites and exhibit a typical sequential (and sometimes allosteric) behavior which is illustrated in Figure 24. The catalytic turnover is initiated when Mg^{2+} binds to the apoenzyme which triggers local (or even allosteric) conformational changes, thus activating the enzyme towards substrate binding. The second Mg^{2+} ion either enters the active site with the substrate ($MgATP^{2-}$, for example) or binds either before or after the substrate (often in a kinetically ordered fashion) and helps to activate the substrate. Reaction then takes place and the product and the second Mg^{2+} ion leave (sometimes together as a magnesium complex). The initially generated E-Mg^{2+} intermediate remains behind and can then either accept another substrate or lose the first Mg^{2+} (this step often underlies external regulatory control mechanisms).

1. Protein kinases

Protein kinases catalyze the reversible transfer of the γ-phosphate group in ATP to a hydroxyl group in a substrate. All kinases are dependent upon at least one divalent metal ion, usually Mg^{2+}, which they need to assist in the binding of ATP and to facilitate phosphoryl transfer[221]. Such phosphorylations are perhaps one of the most important regulatory reactions occurring in the cell and kinases, as a group, represent one of the fundamental building blocks of complex signal transduction processes. They are 'traffic cops' that help to regulate and/or are intrinsically involved in countless, extremely varied processes that include entire metabolic cycles (glycolysis, for example), DNA transcription/replication,

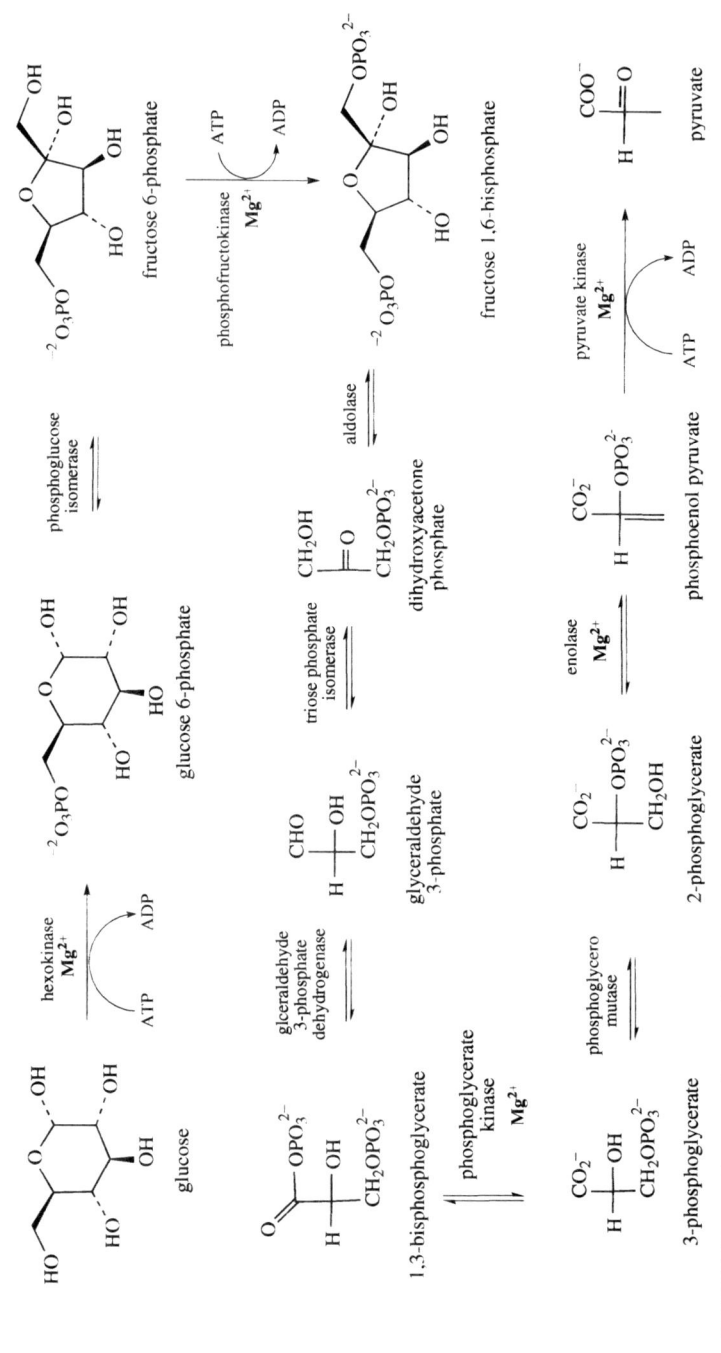

FIGURE 23. Involvement of magnesium in the glycolytic pathway

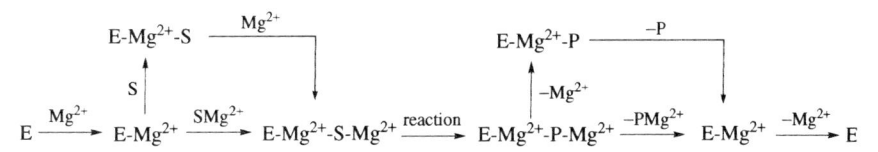

$$E \xrightarrow{Mg^{2+}} E\text{-}Mg^{2+} \xrightarrow{S} E\text{-}Mg^{2+}\text{-}S \xrightarrow{Mg^{2+}} E\text{-}Mg^{2+}\text{-}S\text{-}Mg^{2+} \xrightarrow{\text{reaction}} E\text{-}Mg^{2+}\text{-}P\text{-}Mg^{2+} \xrightarrow{-PMg^{2+}} E\text{-}Mg^{2+} \xrightarrow{-Mg^{2+}} E$$

(with upper branches: $E\text{-}Mg^{2+}\text{-}S$ via SMg^{2+}, and $E\text{-}Mg^{2+}\text{-}P$ via $-Mg^{2+}$ and $-P$)

FIGURE 24. General mode of action of magnesium-dependent metabolic enzymes. Reproduced [© 2002 Kluwer Academic Publishers] by permission of Springer Science and Business Media from Reference 6

the biosynthesis of neurotransmitters in the brain—right up to and including events that can be directly observed at the microscopic (cellular differentiation) or macroscopic level (muscle contractions). It is therefore not surprising that higher vertebrates are estimated to have more than 2000 different protein kinases[222]. For example, it has been estimated that *ca* 2% of the human genome (as understood in 2001) contains protein kinase domains[223]. A relatively simple core structure (*ca* 250–300 amino acids) containing the active site is conserved over the entire family[224, 225]; however, a kinase frequently has additional auxiliary components[226], and/or is bound to or otherwise directly interacts with other regulatory proteins[227] or complex domains that either enhance or repress the kinase activity of the core subunit via complex allosteric interactions[228, 229]. This gives rise to an incredible degree of familial variance—from both a structural and mechanistic viewpoint.

One kinase, the cAMP-dependent protein kinase (known as PKA), has been the focus of mechanistic research for several decades and is perhaps the best understood member of the kinases to date[221]. PKA catalyzes the phosphorylation of a serine or threonine hydroxyl contained in an Arg–Arg–X–Ser/Thr–Y sequence where X is a small amino acid (Ala, for example) and Y is a hydrophobic residue[230]. In order to maintain cellular homeostasis, hormones are released from organs. These are detected by the G-proteins which then activate adenylate cyclase, which produces a cyclic form of adenosine monophosphate, cAMP. This is a hormonal second messenger and is the regulatory signal controlling the activity of PKA. In the absence of an activating signal (cAMP), PKA is a heterotetramer containing two core catalytic subunits (C) and, as an auxiliary component, a regulatory dimer (R_2). When the concentration of cAMP in the immediate vicinity rises, the regulatory unit binds four cAMP molecules, an event which instantly causes the heterotetramer to fall apart. A $R_2(cAMP)_4$ complex remains behind and two relatively small, active C subunits are released[231].

The catalytic subunit C of PKA consists of two domains, one composed mostly of α-helices and one of β-strands, which are connected by a small linker region[232]. The ATP binding site is located deep in the active site between the two domains; the binding site of the larger substrate is at the mouth of the pocket (Color Plate 4). A flexible 'activation' loop is postulated to function as a door for the active site and is believed to be directly involved in regulating PKA[221]. PKA has a 'disordered' or random binding mechanism. When the 'door' is open, both the substrate and ATP have unhindered access to the active site and the binding of one does not influence the other[233].

Asp184 provides the primary binding site for the catalytic Mg^{2+} which probably enters as MgATP^{2-} (Figure 25)[234]. This residue is absolutely essential, since mutation to Ala completely destroys the catalytic activity[235]. This catalytic Mg^{2+} chelates oxygen atoms from the β- and γ-phosphates and clearly properly positions the phosphate tail and helps to activate the γ-phosphate towards phosphorylation. A second Mg^{2+} ion also helps to fix the conformation. This ion is not absolutely required; however, its presence clearly accelerates catalysis[236]. Lys72 is essential for activating ATP, since mutation of this residue destroys the catalytic activity but does not change the ATP binding affinity[237]. Lys168

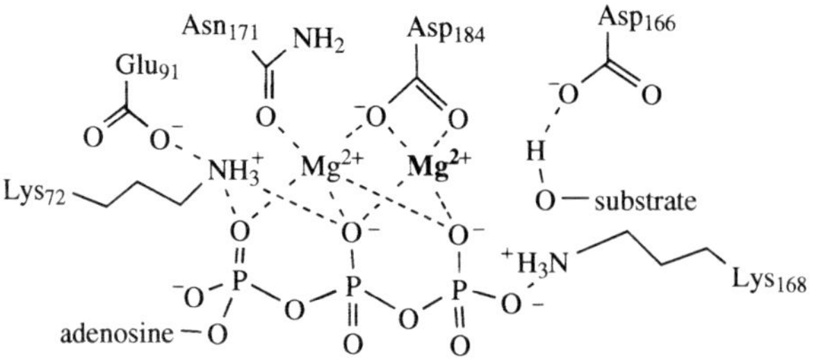

FIGURE 25. Key interactions in the active site of the cAMP-dependent protein kinase (PKA); the additional H_2O ligands on both hexacoordinated Mg^{2+} ions have been omitted for clarity and the essential Mg^{2+} is bold. Modified with permission from reference 234 with permission from cold spring habor laboratory press

is also involved in positioning the γ-phosphate and, in addition, may be involved in substrate binding[221]. The exact role of Asp166 in the phosphorylation has been a topic of considerable discussion[221,238]. This residue is not absolutely necessary; however, it considerably enhances the catalytic rate[221]. In the meantime, it is clear that its primary role is that of a general base with only a minor contribution, if any, to substrate binding[239]. It undergoes a direct hydrogen-bonding interaction with the hydroxyl group of the substrate.

The experimental evidence is somewhat contradictory as to whether the phosphate transfer occurs over an associative (S_N2-type) or dissociative (metaphosphate) pathway (Figure 13). Kinetic experiments demonstrate that the activity is not pH dependent and that a solvent deuterium isotope effect is clearly missing; both facts indicating that the transferred proton is still bound to the hydroxyl in the rate-determining step, as would be expected in the metaphosphate pathway[240]. On the other hand, a solid-state structure of a transition state analogue (PKA was crystallized with a substrate peptide, ADP and AlF_3) indicated a strong $O_{Asp166}-H-O_{hydroxyl}$ interaction in a possible S_N2-type transition structure[241]. DFT calculations on large models of the active site which included Asp166 (as well as the functional groups of all other conserved residues in the immediate vicinity of both ATP and the substrate) have shown that the two-step dissociative mechanism is probably favored[242,243]. In the first step, the phosphorus anhydride bond is broken to form a trigonal planar metaphosphate that is stabilized by a hydrogen-bonding interaction with the hydroxyl group in the substrate. In the second step, the hydroxyl group attacks the metaphosphate. It was concluded that Asp166 first helps in substrate binding and then functions as a general base mediator to transfer the hydroxyl proton from serine to the phosphate during the second phosphorylation step. A third computational study using DFT QM/MM methods confirmed that Asp166 primarily functions as a proton trap[244].

2. Enolases

Enolases, also known as 2-phospho-D-glycerate hydrolases, catalyze the reversible dehydration of 2-phosphoglycerate (2-PGA) to phosphoenol pyruvate (PEP) in complex metabolic systems such as the glycolytic pathway illustrated in Figure 23[245]. Eukaryotic enolases generally have a high degree of family resemblance and those isolated from widely varied sources such as yeast, lobster and human usually possess a sequence homology greater than 60%[246]. One member of this very interesting family, yeast enolase, has been studied in quite some detail due to its ease in isolation and propensity towards crystallization. It was one of the very first enzymes to be successfully crystallized (1941)[247].

Since then, a considerable amount of structural and mechanistic information has been collected and yeast enolase is probably the best understood sequential enzyme to date. It is a homodimer[248] and requires two Mg^{2+} ions per active site for catalytic activity under physiological conditions, although magnesium can be replaced with a variety of divalent metal ions *in vitro*[249]. During a catalytic turnover, the metal ions bind to the active site in a kinetically ordered, sequential manner with differential binding affinities[250]. The mode of action of yeast enolase is illustrated in Figure 26 and is unusually well understood since several solid-state structures for each intermediate identified with kinetic methods have been determined.

It is quite interesting that the enolase subunit can exist in four major conformational states—an inactive *apo*-form, as well as open, semiclosed and closed conformations[251, 252]. The resting state of yeast enolase is the *apo*-form. Binding of the first Mg^{2+} ion (highest affinity) to the apoenzyme[253] induces a conformational change in the active site which activates it towards substrate binding[249]. This Mg^{2+} ion is thus often called the conformational ion. The [E−Mg^{2+}] intermediate is now in the open conformation which is observed in the absence of substrates or inhibitory analogues[254]. The activated [E−Mg^{2+}] complex now binds the substrate (2-PGA) to generate an [E−Mg^{2+}−S] intermediate[255]. The third step in the catalytic turnover is the binding of the second 'catalytic' Mg^{2+} ion[250, 256]. This causes a concerted movement of three short, flexible loops in the region of the substrate canal which partially closes the active site—similar to placing a lid on a pot (illustrated in Color Plate 5)[253, 257]. Solid-state structures of the [E−Mg^{2+}−S−Mg^{2+}] complex[255, 258] exhibit this closed conformation in which the chemical reaction (dehydration) takes place after which a [E−Mg^{2+}−P−Mg^{2+}] intermediate results[259]. After the reaction, the 'lid' has relaxed somewhat and the product complex exhibits the semiclosed conformation.

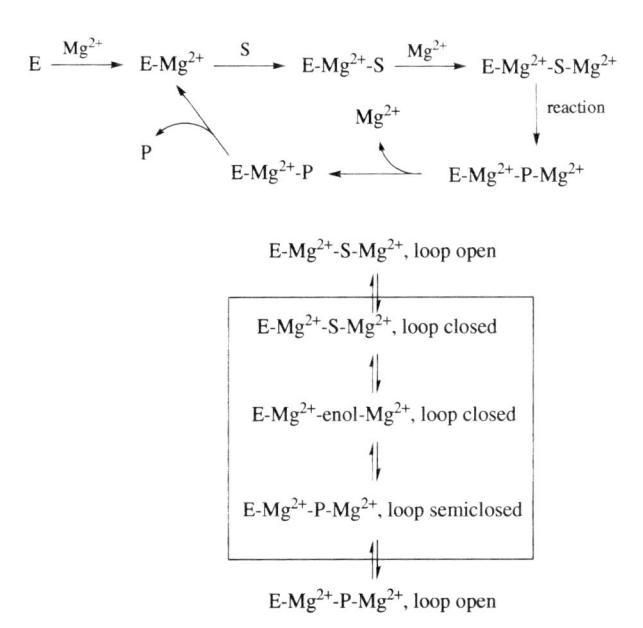

$$E \xrightarrow{Mg^{2+}} E\text{-}Mg^{2+} \xrightarrow{S} E\text{-}Mg^{2+}\text{-}S \xrightarrow{Mg^{2+}} E\text{-}Mg^{2+}\text{-}S\text{-}Mg^{2+}$$

$$\downarrow P \qquad \qquad \uparrow Mg^{2+} \qquad \downarrow \text{reaction}$$

$$E\text{-}Mg^{2+}\text{-}P \longleftarrow E\text{-}Mg^{2+}\text{-}P\text{-}Mg^{2+}$$

E-Mg^{2+}-S-Mg^{2+}, loop open
⇅
E-Mg^{2+}-S-Mg^{2+}, loop closed
⇅
E-Mg^{2+}-enol-Mg^{2+}, loop closed
⇅
E-Mg^{2+}-P-Mg^{2+}, loop semiclosed
⇅
E-Mg^{2+}-P-Mg^{2+}, loop open

FIGURE 26. Above: the kinetically ordered sequential mode of action of yeast enolase. Reprinted with permission from Reference 250. Copyright 2001 with permission of the American Chemical Society. Below: conformational changes in the chemical step

After a conformational change has opened up the active site, dissociation is kinetically ordered with the catalytic Mg^{2+} ion leaving first, followed by the product and then the conformational Mg^{2+} ion. It has long been suspected that dehydration occurs via a two-step mechanism over a metastable enol intermediate[260, 261]. This is the only intermediate in the mode of action of yeast enolase that has not yet been either crystallized or observed spectroscopically. However, QM/MM calculations indicate that such an enol intermediate is indeed a viable metastable intermediate along the reaction pathway between the ternary substrate and product complexes[262, 263].

To date, most quantum-chemical studies on sequential enzymes have been performed using small 'cut outs' of the active site which strongly resemble organometallic template complexes. The electrostatic interactions between the metal ion(s) and organic ligands hold the quantum-chemical model together. While this template strategy may suffice for the proper description of selected intermediates or individual reaction steps, it is fundamentally incapable of being employed for calculating the entire mode of action. Due to the fact that metal ion movement is an intrinsic part of catalysis, the model used needs to be stable towards metal ion exchange as well as ionic movement relative to the amino acid residues that make up the bulk of the active site. Metal-ion-centered template models are incapable of this; they simply fall apart *in silicio* when the metal ions are removed or even slightly displaced from their optimal positions. The only remedy to date is to employ a QM/MM method[264, 265] with a large enough 'cut out' to permit limited ionic movement. QM/MM approaches are based on a method gradient with a small region in the center of the model being calculated with a higher-level, more accurate method and the larger periphery region with a lower-level (and thus faster) method. Due to technical difficulties in implementing the overlap region which is always accompanied by method overlap errors, the results of such calculations are unsystematic and can be quite inaccurate.

Very recently, a new strategy has been developed for systematic quantum-chemical investigations on sequential enzymes. Instead of a method gradient, a structure gradient is employed which generates a 'soccer ball' model[266]. In this approach, all known solid-state structures of the enzyme are overlaid. As long as the mode of action is quasi-localized, the backbone residues will begin to overlap at some point in space moving out and away from the active site. At this point, a sphere containing the active site is cut out. All open valencies of residues on the surface of this sphere are completed with hydrogen atoms and their positions are frozen in space. This creates a hard (fixed) outer shell. Using a structure gradient approach (first dihedral then bond angles and finally bond lengths are freed), the fixed outer shell is connected to a freely optimizable inside[267]. In this manner, not only the limitations of metal-ion-centered models can be completely overcome, but it is possible to employ a single computational method (DFT, for example) with approximately the same size/time advantages as QM/MM methods but without the additional overlap error[268]. The coordinated movement and/or chemical reactions of metal ions, substrate, product as well as flexible side-chain residues (general acids and bases involved in catalysis) and specific solvation waters can now be explicitly studied. This strategy has recently been used to develop an initial model for the active site of yeast enolase (illustrated in Color Plate 5)[269] which is now being enlarged with the goal of studying the localized loop movements in the mode of action.

C. DNA Replication and Repair

Genetic information necessary for the propagation of all life forms is stored in compressed form in genomes, the data storage compartments of nature, which are basically composed of two long DNA strands wound together in a double helix. Several large, quite diverse classes of enzymes are responsible for manipulating this genetic code. Helicases, for example, unwind double-stranded DNA helices and thus prepare them for further manipulations[270, 271]. These single (primed) strands are then worked on by the DNA polymerases which are responsible for replicating and maintaining the DNA[272]. During DNA replication, the new DNA is synthesized in a template-dependent process that faithfully copies the original DNA molecule. Replicative DNA polymerases synthesize very long DNA molecules with an incredible accuracy[273, 274]. Usually, these replicases consist of a macromolecular assembly of several proteins that function together as a single unit[275]. As early as 1976, it was realized that the accuracy of this process depends on the presence of Mg^{2+} ions[276]. Replacing Mg^{2+} in a DNA polymerase by Co^{2+} or Mn^{2+}, for example, usually results in a considerable loss of replication fidelity[277]. Ever since then, almost all newly discovered enzymatic systems for DNA processing also involve Mg^{2+} as an essential cofactor[278].

DNA polymerases are now broadly classified into two groups—the DNA replicases that are responsible for copying DNA and the repair polymerases that fix damaged DNA. Far from being the incredibly stable molecule originally believed, it is now known that DNA is a dynamic system that is constantly being damaged by a horde of potential mutagens[279]. These include, among many others, reactive oxygen species formed during metabolic processes, chemical mutagens absorbed over the skin, eaten or breathed, as well as sun light (UV radiation). Damage can also occur as a result of replication or recombination mistakes. Without an ability to repair genomic information, life encoded by DNA would be altered so fast that an organism could not thrive[280]. Cells have therefore developed several different repair mechanisms designed to repair localized damage[272]. Base excision, for example, replaces a damaged nucleobase[280] and nucleotide excision replaces the entire nucleotide[281]. There are also mechanisms for correcting mismatches (replication errors)[282] as well as for repairing breaks in DNA single[283] and double[284] strands. In light of the incredible specificity of enzymes that work on DNA, it is quite remarkable that all of these repair pathways are characterized by the ability to perform broad-band repair with most of the enzymes involved recognizing multiple types of DNA damage[285].

1. DNA replicases

DNA replicases are quite unusual enzymes in that they employ a DNA substrate as a template in order to guide the synthesis of the product (DNA replicant). This is an extremely complex task. The enzyme must first recognize and bind the primed DNA strand. Here the first challenge is encountered. The strand must be bound with a high affinity, but it must be capable of being released—without complete dissociation—to be repositioned at the end of each catalytic turnover so that the next nucleotide can be incorporated. After primer binding, the enzyme must then recognize and bind the proper nucleoside (desoxynucleoside 5'-triphosphate; dNTP). All four dNTPs must be specifically recognized, but how does the enzyme decide that it has the proper one? The next step consists in matching the nucleoside to the template in order to form the proper base pair before it is incorporated, via a chemical reaction into the growing DNA replicant[272].

The first DNA replicase to be crystallized was the Klenow fragment of the *E. coli* polymerase I[286]. Full length polymerase I is a single polypeptide that contains three functional domains—a polymerization domain, a 3'-exonuclease and a 5'-nuclease. Removal of the 5'-nuclease (a proofreading fragment which reduces the error rate of 1 in 10^4 base pairs to 1 in $10^{8\,272}$) yields the Klenow fragment which, in itself, is a fully functional replicase[287]. Subsequent solid-state structures of various other polymerases have revealed that, in spite of considerable sequence inhomogeneity, all DNA replicases share several features that are critical for activity[272]. Perhaps the most important feature common to almost all polymerases is the shape of the catalytic domain. This resembles a half-open hand with the 'palm' forming a deep cavity with the 'thumb' to the right and the 'fingers' to the left (Color Plate 6). The 'fingers' hold the DNA primer template and interact with the incoming dNTP. The 'thumb' positions and fixes the newly synthesized DNA replicant and the chemistry takes place in the 'palm'. The growing replicant leaves the active site at a 90° angle to the template[288].

Extensive kinetic experiments on the Klenow fragment (and other replicases) have resulted in the general mechanism illustrated in Figure 27[289,290]. Except for variations in the individual rate constants, this mechanism seems to be valid for a series of DNA polymerases[127,291]. The first step is to fetch a nucleoside N (MgdNTP^{2-}). At this point, differentiation between nucleosides occurs; the binding constant of the 'correct' one to form an E-D$_n$-N complex is an order of magnitude higher than for an 'incorrect' one[292]. It is believed that nucleoside selection is achieved through geometrical constraints that allow the formation of correct Watson–Crick base pairing to the template and rejects nucleosides that do not have the proper shape[293]. A slow, Mg^{2+}-dependent, rate-limiting conformational change which only occurs when the 'correct' nucleoside is present ('induced fit mechanism') now closes the active site[292]. This rearrangement is thought to deliver and bind the nucleoside to the active site. After the chemical reaction (phosphorus anhydride bond formation) which incorporates the nucleoside into the replicant, the conformationally active Mg^{2+} ion dissociates. A takes place second conformational change which returns the active site to an open state with concomitant pyrophosphate release (most likely in the form of MgPP$_i$). Dissociation of either the template or the replicant from the active site is rather slow; replicases tend to function in a repetitive manner[275] and it is not unusual when thousands of nucleosides are processed per binding/debinding event[272].

A considerable amount of evidence gathered on several replicases strongly indicates that all of them possess the same general phosphorylation mechanism (Figure 28)[272,288]. The active site contains two absolutely conserved Asp residues that play an essential role in the phosphoryl transfer step[294,295]. Two Mg^{2+} ions are essential[296] and they are fixed in the active site via bridge bidentate coordination with the two Asp residues[297,298]. This bimetallic arrangement binds and positions the dNTP in a manner in which the usual M$_\gamma$ M$_{\beta\gamma}$ binding motif has been reversed in favor of a M$_\alpha$M$_{\gamma,\beta}$ motif. One of the metal ions (M$_\alpha$) is ideally positioned to lower the pK_a of the hydroxyl group of the last nucleotide in the growing replicant and is believed to facilitate its deprotonation by a general base in the vicinity[299]. The resulting hydroxide attacks the α-phosphate of the dNTP[300]. The other Mg^{2+} binds to the phosphate tail, holds it in a position favorable for a S$_N$2-type reaction and stabilizes charge separation in the pentacoordinated transition structure (associative mechanism)[301]. Most likely this second Mg^{2+} also stabilizes the pyrophosphate (which is

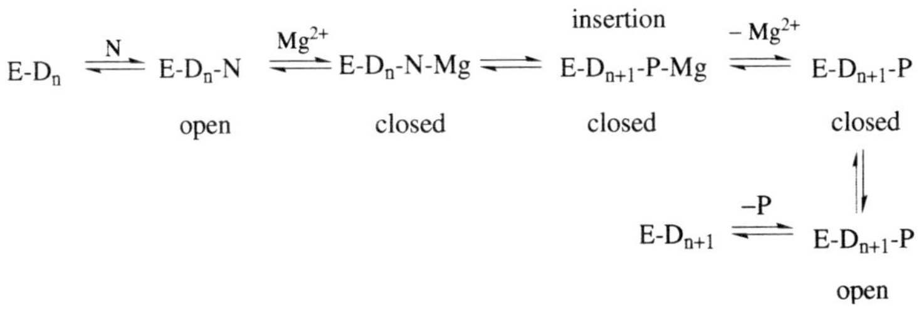

$$E\text{-}D_n \xrightleftharpoons{N} E\text{-}D_n\text{-}N \xrightleftharpoons{Mg^{2+}} E\text{-}D_n\text{-}N\text{-}Mg \xrightleftharpoons{\text{insertion}} E\text{-}D_{n+1}\text{-}P\text{-}Mg \xrightleftharpoons{-Mg^{2+}} E\text{-}D_{n+1}\text{-}P$$

open   closed   closed   closed

$$E\text{-}D_{n+1} \xrightleftharpoons{-P} E\text{-}D_{n+1}\text{-}P$$

open

FIGURE 27. General kinetic mechanism of DNA replication. E is the replicase (polymerase), D the growing DNA replicant, N is MgdNTP^{2-} and P represents MgPP$_i$. Adapted with permission from Reference 290. Copyright 2006 American Chemical Society

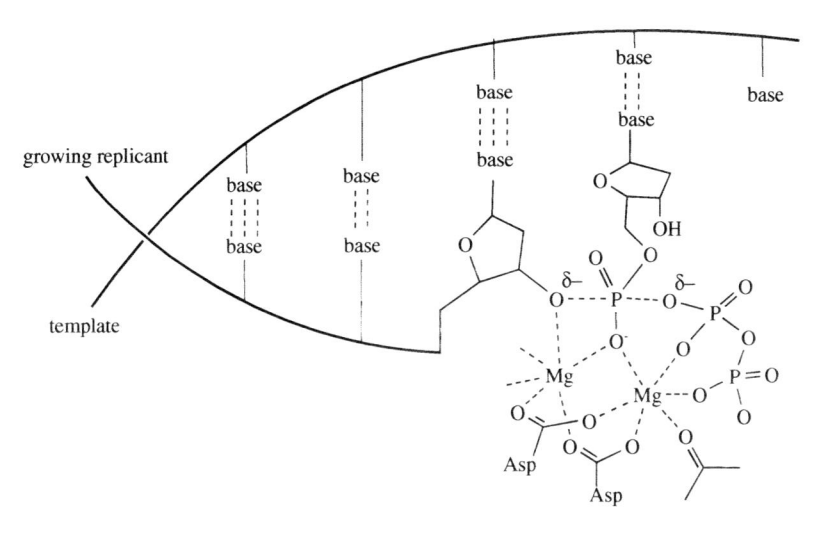

FIGURE 28. Proposed transition structure for the transphosphorylation in DNA replication. Reproduced by permission of Bentham Science Publishers from Reference 272

believed to be protonated by a general acid) as it dissociates from the active site after the nucleoside transfer is complete.

2. Base excision repair

Damage to an individual nucleobase is the most common type of DNA damage that occurs[285]. A multitude of environmental factors leads to spontaneous depurinations, depyrimidations, deaminations, oxidations and alkylations of the heterocyclic nucleobase[302]. For example, thousands of damaged nucleobases must be repaired in a single human cell every day of its life in order to maintain genomic integrity[302]. Higher organisms have therefore evolved a common base excision repair (BER) strategy[303], which is perhaps the major cellular pathway for dealing with most DNA damage[304]. If this damage control system is not working properly, consequences such as early aging, cancer and neurodegenerative diseases result[305].

Base excision repair is carried out with a 'cut and paste' strategy[306, 307]. First, the damaged nucleobase is identified and then excised. This is usually carried out by glycolylases that target distinct nucleobase lesions. They work by flipping the damaged nucleotide out of the helical structure and then cleaving the N-glycosidic bond to remove the nucleobase[308]. Monofunctional glycolyases use water as the nucleophile; bifunctional glycolases employ an amine residue in their active site to first generate a Schiff base (covalently bound intermediate) which then, depending on the glycolase, undergoes either a β- or a β,δ-elimination (Figure 29). Further processing is pathway-dependent and requires, depending on the intermediate generated by the glycolase, an AP endonuclease, a phosphodiesterase or a phosphatase. After this, a polymerase (usually polymerase β) pastes in a new nucleotide after which a ligase tucks it back into its proper place in the DNA strand[285].

The different enzymes involved in base excision repair (and other DNA repair pathways) are quite diverse. Even among the same family, sequence homology is usually

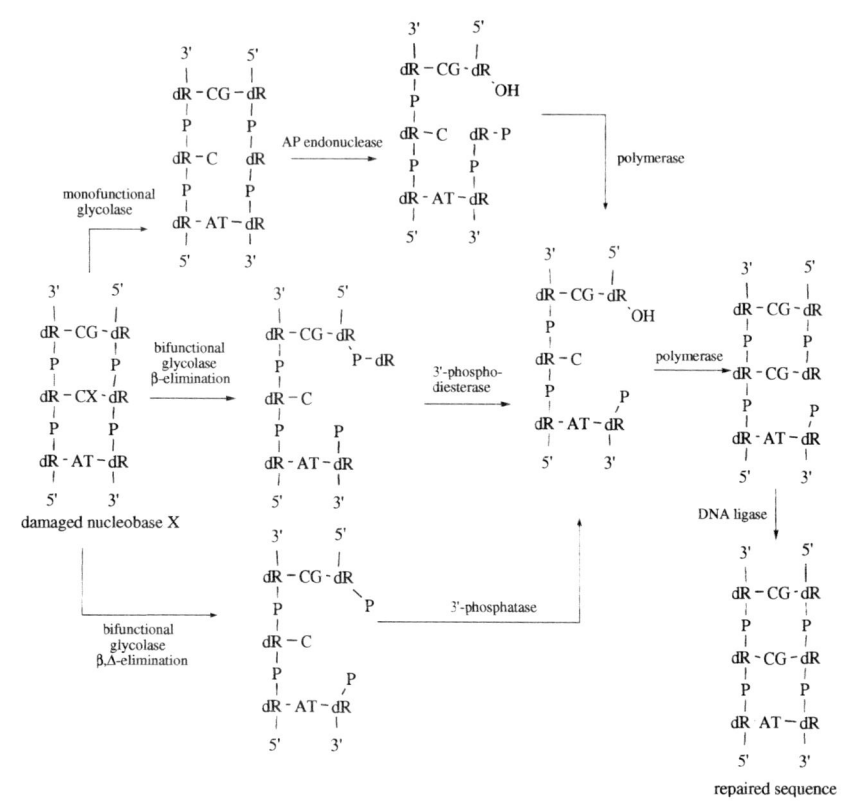

FIGURE 29. Minimal biochemical pathway for base excision repair. Adapted with permission from Reference 285. Copyright 2006 American Chemical Society

severely limited from organism to organism and from task to task. Although more is being learned about these systems on a daily basis, a detailed structure-functional understanding of their individual modes of action is still in the very early stages of being developed. However, recent research efforts have discovered one single feature that all of these enzymes have in common. They normally require divalent metal ions, usually Mg^{2+}, for activation[27].

3. Generic two-ion mechanism

DNA and many RNA molecules have helical duplex structures with identical, unvarying phosphate backbones that surround uniformly stacked base pairs that have a very high degree of chemical and topological similarity. This poses the question of exactly how do the countless protein-based enzymes that act upon DNA and RNA (and, in addition, how do ribozymes) manage to pick out their substrate, one particular phosphate linker or nucleotide among literally thousands of others, with such a high degree of selectivity?

A detailed analysis of solid-state structures of several unrelated systems that catalyze the hydrolysis of diphosphate esters (alkaline phosphatase[309] and the Klenow fragment[310], among others) revealed that their active sites invariably contain conserved carboxylate residues that are capable of binding two divalent metal ions in a bridge bidentate manner[175]. The metal ions are ca 4 Å apart and enzyme–substrate complexes indicate that they are ideally positioned with respect to both the phosphate backbone and the substrate so as to enable a phosphoryl transfer reaction over a linear $S_N 2$-type transition structure (Figure 19)[175]. This gave rise in 1993 to the postulate of a general 'two-ion mechanism' for phosphate ester hydrolysis or transphosphorylation in which one metal ion reduces the pK_a of the hydroxyl nucleophile (or of H_2O when a simple hydrolysis is being performed); both support a $S_N 2$-type associative transition structure and the second ion stabilizes the oxyanion in the leaving group—which is then protonated by a general acid (or water) in the immediate vicinity[175]. Since then, mechanistic investigations on many protein-based enzymes involving various phosphoryl transfer reactions in DNA and RNA have provided concrete evidence for a $S_N 2$-type reaction involving a pentacoordinated intermediate/transition structure with accompanying inversion of configuration at phosphorus[311,312]. In addition, two metal ions have consistently been found in every DNA and RNA replicase identified to date[313,314].

A unique characteristic of nucleic acid phosphoryl transfer is an extremely high substrate specificity. For example, the error rate of a replicase inserting a wrong nucleotide is ca 10^{-3} to 10^{-4}, even without a proofreading element present[315]. However, the free energy difference between a perfect Watson–Crick base-pair match and a mismatch is ca 2 kcal mol^{-1}, a value which would lead to an error rate of only 10^{-1} to 10^{-2}[316]. This discrepancy is generally believed to be resolved by the induced-fit mechanism discussed above; however, newer findings indicate that this may not be the sole answer[274]. It is now postulated that the metal ions play an important role in helping to determine substrate specificity[312].

The situation is not as clear for many other classes of enzymes that act upon DNA and RNA with typical examples being the exo- and endonucleases (exonucleases remove nucleotides from the end of a strand; endonucleases incise internal sites[317]). In order to crystallize nuclease–substrate complexes, this cleavage must be artificially inhibited. Successful strategies include using inert substrate analogues, chemical modification of specific residues or employment of a divalent cation that does not promote catalysis (Mn^{2+}, for example). This inevitably perturbs the active site and can lead to changes in the positions (and number) of the metal ions and catalytic residues involved[312]. Controversial findings are present in the literature for many systems, thus making it difficult to decide exactly how many metal ions are involved[318]. For example, three different mechanisms involving one, two and even three metal ions have been proposed for the phosphate diester bond cleavage catalyzed by type II restriction endonucleases[319,320]. It is clear that more information is needed before mechanistic similarities for these widely varied enzymes can be recognized.

D. Magnesium and Photosynthesis

Magnesium is directly involved in one of the most thermodynamically demanding reactions to be found in biology—the synthesis of carbohydrates from CO_2 in plants and cyanobacteria using light energy from the sun. This extremely complex process is considered to be the 'engine of life' and is fundamental for sustaining essentially all life on our planet[13]. The first step in this process is initiated by two very large, incredibly complex, coupled biomolecules—the photosystems I (PS-I) and II (PS-II)[321,322]. Oversimply stated, both photosystems work together to convert light energy into electrons, which are then transported across the cell membrane in photosynthesizing organisms where they are used to drive the synthesis of ATP. In addition, photosystem I provides the electrons necessary to reduce $NADP^+$ to NADPH and photosystem II oxidizes water to O_2 (a major source of

1. Chlorophyll

Sunlight is captured in both photosystems by large antenna systems, also known as light harvesting complexes, which basically consist of a complex 3D array of magnesium-based chlorophyll pigments (Figure 7) and carotenoids. The number of chlorophyll cofactors in such antenna systems varies quite widely. For example, six chlorophylls are located in the antenna of both photosystems in purple bacteria[323, 324], 27 in the light harvesting complex (LHC-II) of *Rhodopseudomonas acidophila*[325], 48 in the LHC-II of spinach[326], 96 in the PS-I of *Synechococcus elongates*[36], 200–300 in the PS-I and PS-II of plants and algae and, finally, a huge number (*ca* 200,000) in the chlorosomes of green bacteria[327]. A subunit of one of the more simple systems is illustrated in Color Plate 7[328]. In this complex process of light harvesting, specific pigment–protein and pigment–pigment interactions are used by nature to finetune the absorption properties of the individual light gathering processes according to environmental demands.

Light energy initially absorbed by the antenna pigments is transferred via complex photochemical processes which are not yet well understood to a primary reaction center located at the base of the antenna, where it is then transformed into electrical energy. An example for such a reaction center is the chlorophyll dimer known as P700 (illustrated for the PS-I of *Synechococcus elongates*; Color Plate 7)[36]. The incoming photoenergy excites the P700 core. A singlet excited P700* state results which promptly ejects an electron to generate a P700$^{+\bullet}$ radical cation. The electron is immediately transferred across the thylakoid membrane by a complex chain of electron carriers[329]. The P700$^{+\bullet}$ species remaining behind is then reduced by either plastocyanine (plants) or cytochrome c6 (cyanobacteria), which returns it to its resting state[322].

The unusual binding situation of the central Mg^{2+} ion (illustrated in Figure 7) together with the photophysical properties of the chlorophyll ring system is obviously the key element underlying the transformation of photochemical into electrical energy. However, photoabsorption processes in chlorophylls are extremely complex processes which are still not fully understood—and chemical modifications in the ring periphery as well as very small perturbations in protein–chlorophyll interactions can have large consequences for the photophysical properties[330]. Magnesium coordination to the porphyrin ring results in a quasi-planar system which has two additional axial positions (above and below) available for additional coordination to the central Mg^{2+}. Most of the chlorophylls in the naturally occurring antenna systems are pentacoordinated to the N^{τ}-nitrogen of the imidazole ring in a histidine side chain of the protein backbone[331]. The presence of a fifth ligand pulls the central Mg^{2+} ion slightly out of the porphyrin plane. Biomimetic studies have shown that hexacoordinated species are only formed when the ligand concentration is extremely high[332]. The identity of the fifth ligand on magnesium (usually N, but it can be O or even S) modifies the absorption spectra and it is believed that this ligand is involved in stabilization of the charge separation process in photosynthesis. Not only this, the very presence of a fifth ligand introduces a diastereotopic environment (both *syn* and *anti* diastereomers are possible) which causes small but nontrivial changes in the energetic levels of both ground and excited states[330].

Unraveling the individual molecular interactions in these very complex photosystems is a challenging focus of current research, particularly since the demand for alternative energy sources has become critical in the past few years. Progress in this area is hampered, however, due to the extreme complexity of the natural systems coupled with the fact that the synthesis of biomimetic porphyrin models is quite challenging. Experimental methods to directly study the interaction of magnesium with porphyrin systems are still being developed; a new technique is, for example, solid-state ^{25}Mg NMR spectroscopy[333]. In addition, quantum-chemical studies are just now becoming feasible; calculations of the spectroscopic properties of these extensively conjugated systems have to be performed using higher multireference methodology (which, for these large molecules, is often beyond the limit of available computational capacities) if they are to be at all accurate[334]. Current calculations are often limited to semiempirical methods[335]. However, recent progress in calculating excited-state electron dynamics[336] combined with progress in time-resolved femtosecond spectroscopy[337] is quite promising for future studies on the ultrafast photodynamics in these fascinating systems.

2. Rubisco — a photosynthetic CO_2 fixing enzyme

Most of the carbon in us, the food we eat and, in general, the biosphere which surrounds us has been extracted from CO_2 at some time or another by the world's most abundant enzyme — D-ribulose 1,5-bisphosphate carboxylase/oxygenase (rubisco)[338]. This enzyme catalyzes the initial step of carbon metabolism, the fixation of CO_2, in all organisms that rely upon photosynthesis. It has been estimated that the yearly turnover of CO_2 processed by rubisco is well over 10^{11} metric tons[339]. The overall reaction is the addition of H_2O and CO_2 to D-ribulose 1,5-bisphosphate, a multistep process which ends up splitting the C2–C3 bond to yield two molecules of 3-phosphoglycerate (Figure 30). One of the interesting things about rubisco is, although it performs an essential biochemical role in sustaining life, it is actually quite a 'bad' enzyme with a very poor performance. It has an extremely low catalytic rate and, among a multitude of other side reactions[340], a high tendency to confuse CO_2 with O_2, a phenomenon which leads to photorespiration in plants[341].

The minimal functional unit (quite well conserved among all rubiscos) is a homodimer in which the active sites are located at the subunit interface. Residues from both subunits contribute to each active site, which is illustrated in Color Plate 8. All known forms (at present, four different types) consist of these basic dimeric units which are arranged into various larger multimer arrays — dimers, tetramers and even pentamers[342]. The different forms of rubisco all have a common evolutionary origin and existing solid-state structures of the active sites are nearly superimposable[343].

Before substrate binding can take place, rubisco must first be activated. This occurs via carbamylation (reaction with CO_2) of an essential Lys residue[344]. This promotes the binding of an essential Mg^{2+} ion after which the active site is complete. Rubisco can now recognize and bind the first substrate which is ribulose-P_2 (D-ribulose 1,5-bisphosphate)[345]. The substrate is bound to the Mg^{2+} ion via an inner-sphere coordination of the C2-carbonyl and C3-hydroxyl groups which appropriately positions and activates the ribulose-P_2 for subsequent reaction. Substrate binding causes a flexible loop to close over the active site which buries the active site deep within the protein and restricts access to a small channel just large enough for CO_2 (and O_2)[346].

The catalytic circle begins when the substrate is converted into a reactive enediol or enediolate[345]. The presence of CO_2 is not required for enolization. This is a reversible process, facilitated by the prepositioning of the Mg^{2+}-bound substrate[347]. Isotope exchange experiments indicate that a general base, most likely the carbamate bound to Lys201, could abstract the C3-proton[339]. This carbamate appears to be part of a possible proton relay for transporting H^+ out and away from the reaction center[339]. Alternatively, a theoretical study indicates that a direct transfer of the hydrogen from the C3 center to the C2 carbonyl group with subsequent proton exchange of the resulting hydroxyl group with the medium is possible[347].

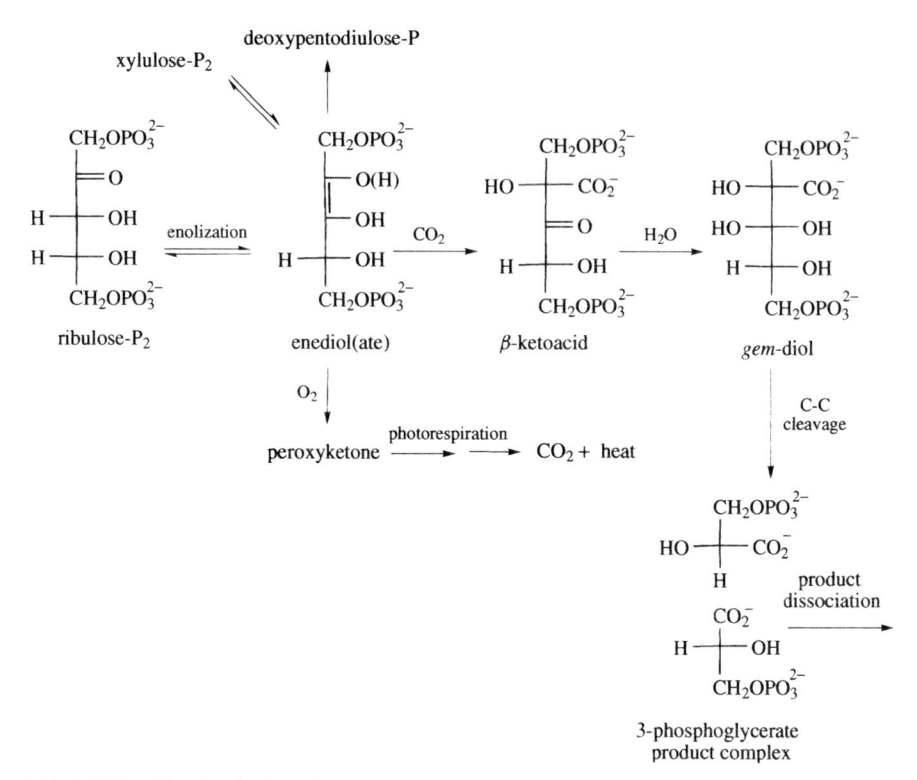

FIGURE 30. The chemical reaction carried out by rubisco

The enediol(ate) has a number of possible fates—it can tautomerize via H-transfer over the 'wrong' face of the double bond and xylulose-P_2 results, which is a tightly bound inhibitor[348]. This side reaction is held partially responsible for the low catalytic rate of rubisco[349]. Another nonproductive pathway is elimination of the C1-phosphate which generates deoxypentodiulose-P[350]. If the enediol(ate) survives these processes, a carboxylase/oxygenase bifunctionality becomes possible. The enediol(ate) bound in the active site is not capable of efficiently discriminating between CO_2 and O_2. If O_2 reacts with the enediol, the secondary pathway of photorespiration is opened up. The products of this reaction are metabolized in the photorespiration pathway which eventually produces CO_2 and dissipates energy as heat—thus wasting important resources. It is estimated that a typical rubisco loses, depending on the relative atmospheric concentrations of CO_2 and O_2, ca 25%–50% of its turnover to photorespiration[341]. This single concurrence reaction dictates the overall efficiency in which plants use their light, water and nitrogen resources—and as such, is currently a target of intense biotechnological efforts aimed at improving the catalytic properties of rubisco and engineering such improvements into crop plants[341].

DFT calculations indicate that an incoming CO_2 displaces a water ligand at magnesium[351]. Coordination of one of the oxygen atoms in CO_2 to Mg^{2+} bends, and thus polarizes, the central carbon atom which becomes sufficiently electrophilic to attack the *Si* face (observed experimentally[352]) of the enediolate C2 atom, which leads to the formation of the β-ketoacid intermediate over a product-like transition structure. It has recently been suggested that, due to the chemical inertness of CO_2, the specificity of rubisco is determined in a late transition structure in which CO_2 closely resembles a carboxylate group. This would maximize the structural difference between the competing transition structures for carboxylation and oxygenation. However, if the transition structure is too 'close' to the β-ketoacid, this would cause it to bind so tightly that subsequent reactions would be slowed down or even stopped. Rubisco is thus forced to make a compromise between CO_2/O_2 selection and the maximum rate of catalytic turnover[353].

In the subsequent hydration step, a water molecule, which has been positioned by the carbamylated Lys residue (which acts as a general base to accept H^{+} [354]), now adds to the now positive polarized C3 atom to form a *gem*-diol (experimentally verified[355])[351]. The C−C bond cleavage step is the rate-determining step and is known to proceed with inversion of configuration at C2 of the first and the addition of a proton to the *Si* face of the *aci*-carboxylate form of the second 3-phosphoglycerate being formed[339]. Computational studies (HF, DFT, MP2) indicate that two completely different mechanisms for C−C bond rupture are theoretically possible—an intramolecular, pericyclic reaction over a five-membered-ring transition structure with a great deal of radical character (homolytic case)[356, 357] and a heterolytic bond rupture which is initiated when a general base in the immediate vicinity (possibly Lys201) deprotonates the C3−OH group[351]. However, a much higher level of theory is needed before a differentiation between these two alternatives can be made.

IX. REFERENCES

1. M. E. Maguire and J. A. Cowan, *Biometals*, **15**, 203 (2002).
2. F. I. Wolf and A. Cittadini, *Mol. Asp. Med.*, **24**, 3 (2003).
3. D. G. Kehres and M. E. Maguire, *Biometals*, **15**, 261 (2002).
4. J. A. Cowan, *Inorg. Chim. Acta*, **275–276**, 24 (1998).
5. R. D. Grubbs, *Biometals*, **15**, 251 (2002).
6. J. A. Cowan, *Biometals*, **15**, 225 (2002).
7. C. B. Black, H. W. Huang and J. A. Cowan, *Coord. Chem. Rev.*, **135–136**, 165 (1994).
8. A. Sreedhara and J. A. Cowan, *Biometals*, **15**, 211 (2002).
9. M. Egli, *Chem. & Biol.*, **9**, 277 (2002).
10. D. E. Draper, *RNA*, **10**, 335 (2004).
11. Z. J. Tan and S. J. Chen, *Biophys. J.*, **90**, 1175 (2006).
12. A. M. Pyle, *J. Biol. Inorg. Chem.*, **7**, 679 (2002).
13. J. F. Kasting and J. L. Siefert, *Science*, **296**, 1066 (2002).
14. S. Kluge and J. Weston, *Biochemistry*, **44**, 4877 (2005).
15. See, for example, the Cambridge Structural Database (CSD): www.ccdc.cam.ac.uk.
16. H. Einspahr and C. E. Bugg, *Met. Ions Biol. Syst.*, **17**, 51 (1984).
17. R. Caminiti, G. Licheri, G. Piccaluga and G. Pinna, *Chem. Phys. Lett.*, **61**, 45 (1979).
18. H. M. Diebler, G. Eigen, G. Ilgenfritz, G. Maass and R. Winkler, *Pure Appl. Chem.*, **20**, 93 (1969).
19. S. E. Rodriguez-Cruz, R. A. Jockusch and E. R. Williams, *J. Am. Chem. Soc.*, **121**, 1986 (1999).
20. S. E. Rodriguez-Cruz, R. A. Jockusch and E. R. Williams, *J. Am. Chem. Soc.*, **121**, 8898 (1999).
21. P. J. Focia, H. Alam, T. Lu, U. D. Ramirez and D. M. Freymann, *Proteins: Struct. Funct. Bioinf.*, **54**, 222 (2004).
22. P. R. Markies, O. S. Akkerman, F. Bickelhaupt, W. J. J. Smeets and A. L. Spek, *Adv. Organomet. Chem.*, **32**, 147 (1991).
23. N. Walker, M. P. Dobson, R. R. Wright, P. E. Barran, J. N. Murrell and A. J. Stace, *J. Am. Chem. Soc.*, **122**, 11138 (2000).
24. W. W. Cleland and A. C. Hengge, *Chem. Rev.*, **106**, 3252 (2006).

25. M. Trachtman, G. D. Markham, J. P. Glusker, P. George and C. W. Bock, *Inorg. Chem.*, **40**, 4230 (2001).
26. L. Leiboda, E. Zhang, K. Lewinski and J. M. Brewer, *Proteins: Struct. Funct. Gen.*, **16**, 219 (1993).
27. J. A. Cowan, *Chem. Rev.*, **98**, 1067 (1998).
28. T. Dudev and C. Lim, *Chem. Rev.*, **103**, 773 (2003).
29. K. Juneau, E. Podell, D. J. Harrington and T. R. Cech, *Structure*, **9**, 221 (2001).
30. X. Shui, L. McFail-Isom, G. G. Hu and L. D. Williams, *Biochemistry*, **37**, 8341 (1998).
31. I. Berger, V. Tereshko, H. Ikeda, V. E. Marquez and M. Egli, *Nucleic Acids Res.*, **26**, 2473 (1998).
32. A. S. Petrov, G. Lamm and G. R. Pack, *J. Phys. Chem. B*, **106**, 3294 (2002).
33. M. M. Harding, *Acta Crystallogr.*, **D62**, 678 (2006).
34. M. M. Harding, *Acta Crystallogr.*, **D57**, 401 (2001) and literature contained therein.
35. C. W. Bock, A. K. Katz, G. D. Markham and J. P. Glusker, *J. Am. Chem. Soc.*, **121**, 7360 (1999).
36. P. Jordan, P. Fromme, H. T. Witt, O. Klukas, W. Saenger and N. Krauß, *Nature*, **411**, 909 (2001).
37. T. Dudev, J. A. Cowan and C. Lim, *J. Am. Chem. Soc.*, **121**, 7665 (1999).
38. T. Dudev and C. Lim, *J. Am. Chem. Soc.*, **128**, 1553 (2006).
39. T. Dudev, Y. L. Lin, M. Dudev and C. Lim, *J. Am. Chem. Soc.*, **125**, 3168 (2003).
40. U. Ryde, *Biophys. J.*, **77**, 2777 (1999).
41. R. L. Rardin, W. B. Tolmann and S. J. Lippard, *New J. Chem.*, **15**, 417 (1991).
42. G. C. Dismukes, *Chem. Rev.*, **96**, 2909 (1996).
43. T. Dudev and C. Lim, *J. Phys. Chem. B.*, **108**, 4546 (2004).
44. C. S. Babu, T. Dudev, R. Casareno, J. A. Cowan and C. Lim, *J. Am. Chem. Soc.*, **125**, 9318 (2003).
45. E. Rezabal, J. M. Mercero, X. Lopez and J. M. Ugalde, *J. Inorg. Biochem.*, **100**, 374 (2006).
46. L. Y. Wang, Y. H. Zhang and L. J. Zhao, *J. Phys. Chem. A.*, **109**, 609 (2005).
47. A. Wahab, S. Mahiuddin, G. Hefter, W. Kunz, B. Minofar and P. Jungwirth, *J. Phys. Chem. B*, **109**, 24108 (2005).
48. D. E. Wilcox, *Chem. Rev.*, **96**, 2435 (1996).
49. N. Sträter, W. N. Lipscomb, T. Klabunde and B. Krebs *Angew. Chem., Int. Ed. Engl.*, **35**, 2024 (1996).
50. B. Schneider and M. Kabeláč, *J. Am. Chem. Soc.*, **120**, 161 (1998).
51. A. S. Petrov, J. Funseth-Smotzer and G. R. Pack, *Int. J. Quantum Chem.*, **102**, 645 (2005).
52. A. S. Petrov, G. R. Pack and G. Lamm, *J. Phys. Chem. B*, **108**, 6072 (2004).
53. N. Sundaresan, C. K. S. Pillai and C. H. Suresh, *J. Phys. Chem. A*, **110**, 8826 (2006).
54. I. Tvaroska, I. André and J. P. Carver, *J. Mol. Struct. (Theochem.)*, **469**, 103 (1999).
55. W. J. McCarthy, D. M. A. Smith, L. Adamowicz, H. Saint-Martin and I. Ortega-Blake, *J. Am. Chem. Soc.*, **120**, 6113 (1998).
56. H. Saint-Martin, L. E. Ruiz-Vicent, A. Ramírez-Solís and I. Ortega-Blake, *J. Am. Chem. Soc.*, **118**, 12167 (1996).
57. I. André, I. Tvaroska and J. P. Carter, *J. Phys. Chem. A*, **104**, 4609 (2000).
58. S. M. Prince, M. Z. Papiz, A. A. Freer, G. McDermott, A. M. Hawthornthwaite-Lawless, R. J. Cogdell and N. W. Isaacs, *J. Mol. Biol.*, **268**, 412 (1997).
59. A. S. Pedrares, W. Teng and K. Ruhlandt-Senge, *Chem. Eur. J.*, **9**, 2019 (2003).
60. M. E. Maguire, *Front. Biosci.*, **11**, 3149 (2006).
61. R. D. Grubbs and M. E. Maguire, *Magnesium*, **6**, 113 (1987).
62. A. M. P. Romani and A. Scarpa, *Front. Biosci.*, **5**, 720 (2000).
63. J. Sahni, B. Nelson and A. M. Scharenberg, *Biochem. J.*, **401**, 505 (2007).
64. A. M. P. Romani and M. E. Maguire, *Biometals*, **15**, 271 (2002).
65. A. M. P. Romani, *Front. Biosci.*, **12**, 308 (2007).
66. M. Kolisek, G. Zsurka, J. Samaj, J. Weghuber, R. J. Schweyen and M. Schweigel, *EMBO J*, **22**, 1235 (2003).
67. V. V. Lunin, E. Dobrovetsky, G. Khutoreskaya, R. Zhang, A. Joachimiak, D. A. Doyle, A. Bochkarev, M. E. Maguire, A. M. Edwards and C. M. Koth, *Nature*, **440**, 833 (2006).
68. E. M. Froschauer, M. Kolisek, F. Dieterich, M. Schweigel and R. J. Schweyen, *FEMS Microbiol. Lett.*, **237**, 49 (2004).

8. Biochemistry of magnesium

69. M. D. Snavely, J. B. Florer, C. G. Miller and M. E. Maguire, *J. Bacteriol.*, **171**, 4761 (1989).
70. S. Eshaghi, D. Niegowski, A. Kohl, D. M. Molina, S. A. Lesley and P. Nordlund, *Science*, **313**, 354 (2006).
71. J. Payandeh and E. F. Pai, *EMBO J.*, **25**, 3762 (2006).
72. M. E. Maguire, *Cur. Opin. Struct. Biol.*, **16**, 432 (2006).
73. S. Z. Wang, Y. Chen, Z. H. Sun, Q. Zhou and S. F. Sui, *J. Biol. Chem.*, **281**, 26813 (2006).
74. M. A. Warren, L. M. Kucharski, A. Veenstra, L. Shi, P. F. Grulich and M. E. Maguire, *J. Bacteriol.*, **186**, 4605 (2004).
75. L. M. Kucharski, W. J. Lubbe and M. E. Maguire, *J. Biol. Chem.*, **275**, 16767 (2000).
76. A. Kornberg, *J. Bacteriol.*, **177**, 491 (1995).
77. S. Garavaglia, A. Galizzi and M. Rizzi, *J. Bacteriol.*, **185**, 4844 (2003).
78. S. Kawai, S. Mori, T. Mukai, S. Suzuki, T. Yamada, W. Hashimoto and K. Murata, *Biochem. Biophys. Res. Commun.*, **276**, 57 (2000).
79. H. Sigel, *Inorg. Chim. Acta*, **198–200**, 1 (1992).
80. L. Stryer, *Biochemie*, 4th Edition, Spectrum Akademischer Verlag, Heidelberg, 1996.
81. G. Oster, *Nature*, **417**, 25 (2002).
82. M. Souhassou, C. Lecomte and R. H. Blessing, *Acta Crystallogr.*, **B48**, 370 (1992).
83. R. Cini, M. C. Burla, A. Nunzi, G. P. Polidori and P. F. Zanazzi, *J. Chem. Soc., Dalton Trans.*, 2467 (1984).
84. C. V. Grant, V. Frydman and L. Frydman, *J. Am. Chem. Soc.*, **122**, 11743 (2000).
85. P. Wang, R. M. Izatt, J. L. Oscarson and S. E. Gillespie, *J. Phys. Chem.*, **100**, 9556 (1996).
86. P. Wang, J. L. Oscarson, R. M. Izatt, G. D. Watt and C. D. Larsen, *J. Solution Chem.*, **24**, 989 (1995).
87. L. Jiang and X. A. Mao, *Spectrochim. Acta*, **A57**, 1711 (2001).
88. A. S. Mildvan, *Magnesium*, **6**, 28 (1987).
89. M. Cohn and T. R. Hughes Jr., *J. Biol. Chem.*, **237**, 176 (1962).
90. J. C. Liao, S. Sun, D. Chandler and G. Oster, *Eur. Biophys. J.*, **33**, 29 (2004).
91. P. D. Boyer, *Annu. Rev. Biochem.*, **66**, 717 (1997).
92. T. M. Duncan, V. V. Bulygin, Y. Zhou, M. L. Hutcheon and R. L. Cross, *Proc. Natl. Acad. Sci. U.S.A.*, **92**, 10964 (1995).
93. M. A. Bianchet, J. Hullihen, P. L. Pedersen and L. M. Amzel, *Proc. Natl. Acad. Sci. U.S.A.*, **95**, 11065 (1998).
94. J. P. Abrahams, A. G. W. Leslie, R. Lutter and J. E. Walker, *Nature*, **370**, 621 (1994).
95. D. Sabbert, S. Engelbrecht and W. Junge, *Nature*, **381**, 623 (1996).
96. H. Noji, R. Yasuda, M. Yoshida and K. Kinosita Jr., *Nature*, **386**, 299 (1997).
97. Y. Sambongi, Y. Iko, M. Tanabe, H. Omote, A. Iwamoto-Kihara, I. Ueda, T. Yanagida, Y. Wada and M. Futai, *Science*, **286**, 1722 (1999).
98. H. Noji and M. Yoshida, *J. Biol. Chem.*, **276**, 1665 (2001).
99. Y. Rondelez, G. Tresset, T. Nakashima, Y. Kato-Yamada, H. Fujita, S. Takeuchi and H. Noji, *Nature*, **433**, 773 (2005).
100. R. Yasuda, H. Noji, M. Yoshida, K. Kinosita Jr. and H. Itoh, *Nature*, **410**, 898 (2001).
101. V. N. Kasho, M. Stengelin, I. N. Smirnova and L. D. Faller, *Biochemistry*, **36**, 8045 (1997).
102. J. P. Abrahams, A. G. W. Leslie, R. Lutter and J. E. Walker, *Nature*, **370**, 621 (1994).
103. Y. H. Ko, S. Hong and P. L. Pedersen, *J. Biol. Chem.*, **274**, 28853 (1999).
104. A. L. Buchachenko and D. A. Kuznetsov, *Mol. Biol.*, **40**, 9 (2006).
105. J. Weber and A. E. Senior, *FEBS Lett.*, **545**, 61 (2003).
106. A. L. Buchachenko, D. A. Kouznetsov, S. E. Arkhangelsky, M. A. Orlova and A. A. Markarian, *Mitochondrion*, **5**, 67 (2005).
107. A. L. Buchachenko, D. A. Kouznetsov, S. E. Arkhangelsky, M. A. Orlorva and A. A. Markarian, *Cell. Biochem. Biophys.*, **43**, 243 (2005).
108. A. L. Buchachenko, E. M. Galimov, V. H. Ershov, G. A. Nikiforov and A. P. Pershin, *Proc. USSR Acad. Sci.*, **228**, 379 (1976).
109. A. L. Buchachenko and V. L. Berdinsky, *Chem. Rev.*, **102**, 603 (2002).
110. A. L. Buchachenko, *J. Phys. Chem. A*, **105**, 9995 (2001).
111. A. L. Buchachenko, N. N. Lukzen and J. B. Pedersen, *Chem. Phys. Lett.*, **434**, 139 (2007).
112. M. Cohn, *J. Biol. Chem.*, **201**, 735 (1953).
113. G. R. J. Thatcher and R. Kluger, *Adv. Phys. Org. Chem.*, **25**, 99 (1989).
114. M. Bianciotto, J. C. Barthelat and A. Vigroux, *J. Am. Chem. Soc.*, **124**, 7573 (2002).

115. N. Iché-Tarrat, J. C. Barthelat, R. Rinaldi and A. Vigroux *J. Phys. Chem. B.*, **109**, 22570 (2005).
116. N. Iché-Tarrat, M. Ruiz-Lopez, J. C. Barthelat and A. Vigroux, *Chem. Eur. J.*, **13**, 3617 (2007).
117. B. L. Grigorenko, A. V. Nemukhin, M. S. Shadrina, I. A. Topol and S. K. Burt, *Proteins; Struct. Funct. Bioinf.*, **66**, 456 (2007).
118. B. L. Grigorenko, A. V. Nemukhin, I. A. Topol, R. E. Cachau and S. K. Burt, *Proteins; Struct. Funct. Bioinf.*, **60**, 495 (2005).
119. E. Franzini, P. Fantucci and L. De Gioia, *J. Mol. Catal. A*, **204–205**, 409 (2003).
120. J. Akola and R. O. Jones, *J. Phys. Chem. B*, **107**, 11774 (2003).
121. H. Sigel, *Pure Appl. Chem.*, **70**, 969 (1998).
122. H. Sigel, *Coord. Chem. Rev.*, **100**, 453 (1990).
123. H. Sigel and R. Griesser, *Chem. Soc. Rev.*, **34**, 875 (2005).
124. H. Sigel, *Chem. Soc. Rev.*, **33**, 191 (2004).
125. L. W. Tari, A. Matte, H. Goldie and L. T. J. Delbaere, *Nat. Struct. Biol.*, **4**, 990 (1997).
126. H. Sigel and R. Tribolet, *J. Inorg. Biochem.*, **40**, 163 (1990).
127. T. A. Steitz, *J. Biol. Chem.*, **274**, 17395 (1999).
128. P. H. Patel and L. A. Loeb, *Nat. Struct. Biol.*, **8**, 656 (2001).
129. A. A. Tulub, *Phys. Chem. Chem. Phys.*, **8**, 2187 (2006).
130. A. A. Tulub, *Russ. J. Inorg. Chem.*, **50**, 1884 (2005).
131. W. Domcke, D. Yarkony and H. Koppel, *Conical Intersections: Electronic Structure, Dynamics and Spectroscopy*, World Scientific Press, New York, 2004.
132. S. A. Pavlenko and A. A. Voitjuk, *J. Struct. Chem.*, **32**, 590 (1991).
133. J. N. Harvey, J. Zureck, U. Pentikäinen and A. J. Mulholland, *Phys. Chem. Chem. Phys.*, **8**, 5366 (2006).
134. A. A. Tulub, *Phys. Chem. Chem. Phys.*, **8**, 5368 (2006).
135. J. A. Subirana and M. Soler-López, *Annu. Rev. Biophys. Biomol. Struct.*, **32**, 27 (2003).
136. R. K. O. Sigel and A. M. Pyle, *Chem. Rev.*, **107**, 97 (2007).
137. G. S. Manning, *Biopolymers*, **69**, 137 (2003).
138. I. A. Shkel, O. V. Tsodikov and M. R. Record Jr., *Proc. Natl. Acad. Sci. U.S.A.*, **99**, 2597 (2002).
139. V. K. Misra and D. E. Draper, *Proc. Natl. Acad. Sci. U.S.A.*, **98**, 12456 (2001).
140. R. T. Batey, R. P. Rambo, L. Lucast, R. Rha and J. A. Doudna, *Science*, **287**, 1232 (2000).
141. D. J. Klein, P. B. Moore and T. A. Steitz, *RNA*, **10**, 1366 (2004).
142. J. Anastassopoulou, *J. Mol. Struct.*, **651–653**, 19 (2003).
143. P. E. Cole, S. K. Yang and D. M. Crothers, *Biochemistry*, **11**, 4358 (1972).
144. T. K. Chiu and R. E. Dickerson, *J. Mol. Biol.*, **301**, 915 (2000).
145. V. Tereshko, G. Minasov and M. Egli, *J. Am. Chem. Soc.*, **121**, 470 (1999).
146. G. Minasov, V. Tereshko and M. Egli, *J. Mol. Biol.*, **291**, 83 (1999).
147. C. C. Correll, B. Freeborn, P. B. Moore and T. A. Steitz, *Cell*, **91**, 705 (1997).
148. N. B. Leontis, P. Gosh and P. B. Moore, *Biochemistry*, **25**, 7386 (1986).
149. K. Kruger, P. J. Grabowski, A. J. Zaug, J. Sands, D. E. Gottschling and T. R. Cech, *Cell*, **31**, 147 (1982).
150. C. Guerrier-Takada, K. Gardiner, T. Marsh, N. Pace and S. Altman, *Cell*, **35**, 849 (1983).
151. G. F. Joyce, *Nature*, **418**, 214 (2002).
152. D. M. Zhou and K. Taira, *Chem. Rev.*, **98**, 991 (1998).
153. B. J. Tucker and R. R. Breaker, *Cur. Opin. Struct. Biol.*, **15**, 342 (2005).
154. J. A. Doudna and T. R. Cech, *Nature*, **418**, 222 (2002).
155. S. A. Woodson, *Cur. Opin. Struct. Biol.*, **15**, 324 (2005).
156. R. K. O. Sigel, *Eur. J. Inorg. Chem.*, 2281 (2005).
157. A. V. Kazantsev and N. R. Pace, *Nat. Rev. Microbiol.*, **4**, 729 (2006).
158. R. H. Symons, *Nucl. Acid Res.*, **25**, 2683 (1997).
159. R. G. Kuimelis and L. W. McLaughlin, *Chem. Rev.*, **98**, 1027 (1998).
160. H. Liu, J. J. Robinet, S. Ananvoranich and J. W. Gauld, *J. Phys. Chem. B*, **111**, 439 (2007).
161. N. Ban, P. Nissen, J. Hansen, P. B. Moore and T. A. Steitz, *Science*, **289**, 905 (2000).
162. R. Fiammengo and A. Jäschke, *Cur. Opin. Biotech.*, **16**, 614 (2005).
163. I. Tinoco Jr. and C. Bustamante, *J. Mol. Biol.*, **293**, 271 (1999).
164. D. E. Draper, D. Grilley and A. M. Soto, *Annu. Rev. Biophys. Biomol Struct.*, **34**, 221 (2005).

165. S. A. Woodson, *Cur. Opin. Chem. Biol.*, **9**, 104 (2005).
166. V. K. Misra and D. E. Draper, *J. Mol. Biol.*, **317**, 507 (2002).
167. T. R. Sosnick and T. Pan, *Cur. Opin. Struct. Biol.*, **13**, 309 (2003).
168. V. J. DeRose, *Cur. Opin. Struct. Biol.*, **13**, 317 (2003).
169. J. J. Cannone, S. Subramanian, M. N. Schnare, J. R. Collett, L. M. D'Souza, Y. Du, B. Feng. N. Lin, L. V. Madabusi, K. M. Müller, N. Pande, Z. Shang, N. Yu and R. R. Gutell, *BMC Bioinformatics*, **3**, 2 (2002).
170. T. R. Cech, *Annu. Rev. Biochem.*, **59**, 543 (1990).
171. Q. Vicens and T. R. Cech, *Trends Biochem. Sci.*, **31**, 41 (2006).
172. J. Pan, D. Thirumlalai and S. A. Woodson, *Proc. Natl. Acad. Sci. U.S.A.*, **96**, 6149 (1999).
173. E. L. Christian and M. Yarus, *Biochemistry*, **32**, 4475 (1993).
174. M. R. Stahley and S. A. Strobel, *Cur. Opin. Struct. Biol.*, **16**, 319 (2006).
175. T. A. Steitz and J. A. Steitz, *Proc. Natl. Acad. Sci. U.S.A.*, **90**, 6498 (1993).
176. S. Shan, A. V. Kravchuk, J. A. Piccirilli and D. Herschlag, *Biochemistry*, **40**, 5161 (2001).
177. M. R. Stahley and S. A. Strobel, *Science*, **309**, 1587 (2005).
178. G. A. Prody, J. T. Bakos, J. M. Buzayan, I. R. Schneider and G. Bruening, *Science*, **231**, 1577 (1986).
179. C. J. Hutchins, P. D. Rathjen, A. C. Forster and R. H. Symons, *Nucl. Acids Res.*, **14**, 3627 (1986).
180. K. F. Blount and O. C. Uhlenbeck, *Annu. Rev. Biophys. Biomol. Struct.*, **34**, 415 (2005).
181. H. W. Pley, K. M. Flaherty and D. B. McKay, *Nature*, **372**, 68 (1994).
182. D. E. Ruffner, G. D. Stormo and O. C. Uhlenbeck, *Biochemistry*, **29**, 10695 (1990).
183. C. Hamman and D. M. Lilley, *ChemBioChem*, **3**, 690 (2002).
184. G. Slim and M. J. Gait, *Nucl. Acids Res.*, **19**, 1183 (1991).
185. M. Koizumi and E. Ohtsuka, *Biochemistry*, **30**, 5145 (1991).
186. W. G. Scott, J. T. Finch and A. Klug, *Cell*, **81**, 991 (1995).
187. J. B. Murray, D. P. Terwey, L. Maloney, A. Karpeisky, N. Usman, L. Beigelman and W. G. Scott, *Cell*, **92**, 665 (1998).
188. S. Wang, K. Karbstein, A. Peracchi, L. Beigelman and D. Herschlag, *Biochemistry*, **38**, 14363 (1999).
189. K. Suzmura, M. Warashina, K. Yoshinari, Y. Tanaka, T. Kuwabara, M. Orita and K. Taira, *FEBS Lett.*, **473**, 106 (2000).
190. R. Przybilski and C. Hammann, *ChemBioChem*, **7**, 1641 (2006).
191. M. de la Peña, S. Gago and R. Flores, *EMBO J.*, **22**, 5561 (2003).
192. A. Khvorova, A. Lescoute, E. Westhof and S. D. Jayasena, *Nat. Struct. Biol.*, **10**, 708 (2003).
193. M. Roychowdhury-Saha and D. H. Burke, *RNA*, **12**, 1846 (2006).
194. J. C. Penedo, T. J. Wilson, S. D. Jayasena, A. Khvorova and D. M. J. Lilley, *RNA*, **10**, 880 (2004).
195. D. Lambert, J. E. Heckman and J. M. Burke, *Biochemistry*, **45**, 7140 (2006).
196. M. Martick and W. G. Scott, *Cell*, **126**, 309 (2006).
197. J. Han and J. M. Burke, *Biochemistry*, **44**, 7864 (2005).
198. M. Vogt, S. Lahiri, C. G. Hoogstraten, R. D. Britt and V. J. DeRose, *J. Am. Chem. Soc.*, **128**, 16764 (2006).
199. Y. Tanaka, Y. Kasai, S. Mochizuki, A. Wakisaka, E. H. Morita, C. Kojima, A. Toyozawa, Y. Kondo, M. Taki, Y. Takagi, A. Inoue, K. Yamasaki and K. Taira, *J. Am. Chem. Soc.*, **126**, 744 (2004).
200. Y. Takagi, A. Inoue and K. Taira, *J. Am. Chem. Soc.*, **126**, 12856 (2004).
201. J. M. Zhou, D. M. Zhou, Y. Takagi, Y. Kasai, A. Inoue, T. Baba and K. Taira, *Nucl. Acids Res.*, **30**, 2374 (2002).
202. R. A. Torres, F. Himo, T. C. Bruice, L. Noodleman and T. Lovell, *J. Am. Chem. Soc.*, **125**, 9861 (2003).
203. F. Leclerc and M. Karplus, *J. Phys. Chem. B*, **110**, 3395 (2006).
204. M. Boero, M. Tateno, K. Terakura and A. Oshiyama, *J. Chem. Theory Comput.*, **1**, 925 (2005).
205. R. Stowasser and D. A. Usher, *Bioorg. Chem.*, **30**, 420 (2002).
206. See, for example, the Jena library of biological macromolecules at www.fli-leibniz.de/IMAGE.html.
207. D. H. Busch and N. A. Stephenson, *Coord. Chem. Rev.*, **100**, 119 (1990).
208. S. W. Ragsdale, *Chem. Rev.*, **106**, 3317 (2006).

209. W. T. Lowther and B. W. Matthews, *Chem. Rev.*, **102**, 4581 (2002).
210. N. Miti, S. J. Smith, A. Neves, L. W. Guddat, L. R. Gahan and G. Schenk, *Chem. Rev.*, **106**, 3338 (2006).
211. J. A. Hardy and J. A. Wells, *Cur. Opin. Struct. Biol.*, **14**, 706 (2004).
212. D. Kern and E. R. P. Zuiderweg, *Cur. Opin. Struct. Biol.*, **13**, 748 (2003).
213. G. Parkin, *Chem. Rev.*, **104**, 699 (2004).
214. J. Weston, *Chem. Rev.*, **105**, 2151 (2005).
215. M. H. Le Du, C. Lamoure, B. H. Muller, O. V. Bulgakov, E. Lajeunesse, A. Ménez and J. C. Boulain, *J. Mol. Biol.*, **316**, 941 (2002).
216. E. Dirnbach, D. G. Steel and A. Gafni, *Biochemistry*, **40**, 11219 (2001).
217. V. Subramaniam, N. C. H. Bergenhem, A. Gafni and D. G. Steel, *Biochemistry*, **34**, 1133 (1995).
218. V. Bučević-Popović, M. Pavela-Vrančić and R. Dieckmann, *Biochimie*, **86**, 403 (2004).
219. S. Orhanović and M. Pavela-Vrani, *Eur. J. Biochem.*, **270**, 4356 (2003).
220. M. F. Hoylaerts, T. Manes and J. L. Millán, *J. Biol. Chem.*, **272**, 22781 (1997).
221. J. A. Adams, *Chem. Rev.*, **101**, 2271 (2001).
222. T. Hunter, *Semin. Cell Biol.*, **5**, 367 (1994).
223. E. S. Lander et al., *Nature*, **409**, 860 (2001).
224. S. K. Hanks and T. Hunter, *FASB J.*, **9**, 576 (1995).
225. S. S. Taylor and E. Radzio-Andezelm, *Structure*, **2**, 345 (1994).
226. M. G. Gold, D. Barford and D. Komander, *Curr. Opin. Struct. Biol.*, **16**, 693 (2006).
227. P. Pellicena and J. Kuriyan, *Curr. Opin. Struct. Biol.*, **16**, 702 (2006).
228. Z. Shi, K. A. Resing and N. G. Ahn, *Curr. Opin. Struct. Biol.*, **16**, 686 (2006).
229. A. Reményi, M. C. Good and W. A. Lim, *Curr. Opin. Struct. Biol.*, **16**, 676 (2006).
230. B. E. Kemp, D. J. Graves, E. Benjamini and E. G. Krebs, *J. Biol. Chem.*, **252**, 4888 (1977).
231. S. S. Taylor, J. A. Buechler and W. Yonemoto, *Annu. Rev. Biochem.*, **59**, 971 (1990).
232. J. Zheng, D. R. Knighton, L. F. Ten Eyck, R. Karlsson, N. H. Xuong, S. S. Taylor and J. M. Sowadski, *Biochemistry*, **32**, 2154 (1993).
233. B. Grant and J. A. Adams, *Biochemistry*, **35**, 2022 (1996).
234. Madhusudan, E. A. Trafny, N. H. Xuong, J. A. Adams, L. F. Ten Eyck, S. S. Taylor and J. M. Sowadski, *Prot. Sci.*, **3**, 176 (1994).
235. C. S. Gibbs and M. J. Zoller, *J. Biol. Chem.*, **266**, 8923 (1991).
236. P. F. Cook, M. E. Neville Jr., K. E. Vrana, F. T. Hartl and J. R. Roskoski Jr., *Biochemistry*, **21**, 5794 (1982).
237. M. J. Robinson, P. C. Harkins, J. Zhang, R. Baer, J. W. Haycock, M. H. Cobb and E. J. Goldsmith, *Biochemistry*, **35**, 5641 (1996).
238. V. E. Anderson, M. W. Ruszczycky and M. E. Harris, *Chem. Rev.*, **106**, 3236 (2006).
239. V. T. Skamnaki, D. J. Owen, M. E. M. Noble, E. D. Lowe, G. Lowe, N. G. Oikonomakos and L. N. Johnson *Biochemistry*, **38**, 14718 (1999).
240. J. Zhou and J. A. Adams, *Biochemistry*, **36**, 2977 (1997).
241. Madhusudan, P. Akamine, N. H. Xuong and S. S. Taylor, *Nat. Struct. Biol.*, **9**, 273 (2002).
242. N. Diaz and M. J. Field, *J. Am. Chem. Soc.*, **126**, 529 (2004).
243. M. Valiev, R. Kawai, J. A. Adams and J. H. Weare, *J. Am. Chem. Soc.*, **125**, 9926 (2003).
244. Y. Cheng, Y. Zhang and J. A. McCammon, *J. Am. Chem. Soc.*, **127**, 1553 (2005).
245. J. Stubbe and R. H. Abeles, *Biochemistry*, **19**, 5505 (1980).
246. S. Duqueroy, C. Camus and J. Janin, *Biochemistry*, **34**, 12513 (1995).
247. O. Warburg and W. Christian, *Biochem. Z.*, **310**, 384 (1941).
248. B. Stec and L. Leiboda, *J. Mol. Biol.*, **211**, 235 (1990).
249. J. M. Brewer and G. Weber, *J. Biol. Chem.*, **241**, 2550 (1966).
250. R. R. Poyner, W. W. Cleland and G. H. Reed, *Biochemistry*, **40**, 8009 (2001).
251. J. Qin, G. Chai, J. M. Brewer, L. L. Loveland and L. Lebioda, *Biochemistry*, **45**, 793 (2006).
252. P. A. Sims, A. L. Menefee, T. M. Larsen, S. O. Mansoorabadi and G. H. Reed, *J. Mol. Biol.*, **355**, 422 (2006).
253. J. E. Wedekind, R. R. Poyner, G. H. Reed and I. Rayment, *Biochemistry*, **33**, 9333 (1994).
254. L. Leiboda, B. Stec and J. M. Brewer, *J. Biol. Chem.*, **264**, 3685 (1989).
255. T. M. Larsen, J. E. Wedekind, I. Rayment and G. H. Reed, *Biochemistry*, **35**, 4349 (1996).
256. E. Zhang, M. Hatada, J. M. Brewer and L. Lebioda, *Biochemistry*, **33**, 6295 (1994).
257. J. M. Brewer, C. V. C. Glover, M. J. Holland and L. Lebioda, *J. Prot. Chem.*, **22**, 353 (2003).

8. Biochemistry of magnesium

258. E. Zhang, J. M. Brewer, W. Minor, L. A. Carreira and L. Lebioda, *Biochemistry*, **36**, 12526 (1997).
259. G. H. Reed, R. R. Poyner, T. M. Larsen, J. E. Wedekind and I. Rayment, *Cur. Opin. Struct. Biol.*, **6**, 736 (1996).
260. D. A. Vinarov and T. Nowak, *Biochemistry*, **38**, 12138 (1999).
261. S. Duquerroy, C. Camus and J. Janin, *Biochemistry*, **34**, 12513 (1995).
262. H. Liu, Y. Zhang and W. Yang, *J. Am. Chem. Soc.*, **122**, 6560 (2000).
263. C. Alhambra, J. Gao, J. C. Corchado, J. Villà and D. G. Truhlar, *J. Am. Chem. Soc.*, **121**, 2253 (1999).
264. H. M. Senn and W. Thiel, *Curr. Opin. Chem. Biol.*, **11**, 182 (2007).
265. R. A. Friesner and V. Guallar, *Annu. Rev. Phys. Chem.*, **56**, 389 (2005).
266. J. Weston, unpublished results (2007).
267. I. Wagner, Diplomarbeit (Master's thesis), Friedrich-Schiller-Universität, Jena, Germany (2006).
268. S. Kluge, PhD Dissertation, Friedrich-Schiller-Universität, Jena, Germany (2007).
269. D. Mollenhauer, Diplomarbeit (Master's thesis), Friedrich-Schiller-Universität, Jena, Germany (2007).
270. E. Delagoutte and P. H. von Hippel, *Q. Rev. Biophys.*, **36**, 1 (2003).
271. E. Delagoutte and P. H. von Hippel, *Q. Rev. Biophys.*, **35**, 431 (2002).
272. M. M. Hingorani and M. O'Donnel, *Cur. Org. Chem.*, **4**, 887 (2001).
273. T. A. Kunkel, *J. Biol. Chem.*, **279**, 16895 (2004).
274. C. M. Joyce and S. J. Benkovic, *Biochemistry*, **43**, 14317 (2004).
275. T. A. Baker and S. P. Bell, *Cell*, **92**, 295 (1998).
276. M. A. Sirover and L. A. Loeb, *Biochem. Biophys. Res. Commun.*, **70**, 812 (1976).
277. M. A. Sirover and L. A. Loeb, *J. Biol. Chem.*, **252**, 3605 (1977).
278. A. Hartwig, *Mutat. Res.*, **475**, 113 (2001).
279. J. H. J. Hoeijmakers, *Nature*, **411**, 366 (2001).
280. J. J. Truglio, D. L. Croteau, B. Van Houten and C. Kisker, *Chem. Rev.*, **106**, 233 (2006).
281. A. Sancar, *Annu. Rev. Biochem.*, **65**, 43 (1996).
282. R. D. Kolodner and G. T. Marsischky, *Cur. Opin. Gen. Devel.*, **9**, 89 (1999).
283. K. W. Caldecott, *DNA Repair*, **6**, 443 (2007).
284. X. Xing and C. E. Bell, *Biochemistry*, **43**, 16142 (2004).
285. P. J. O'Brian, *Chem. Rev.*, **106**, 720 (2006).
286. D. L. Ollis, P. Brick, R. Hamlin, N. G. Xuong and T. A. Steitz, *Nature*, **313**, 762 (1985).
287. H. Klenow and I. Henningsen, *Proc. Natl. Acad. Sci. U.S.A.*, **65**, 168 (1970).
288. W. A. Beard and S. H. Wilson, *Chem. Rev.*, **106**, 361 (2006).
289. S. S. Patel, I. Wong and K. A. Johnson, *Biochemistry*, **30**, 511 (1991).
290. A. K. Showalter, B. J. Lamarche, M. Bakhtina, M. I. Su, K. H. Tang and M. D. Tsai, *Chem. Rev.*, **106**, 340 (2006).
291. K. A. Johnson, *Annu. Rev. Biochem.*, **62**, 685 (1993).
292. I. Wong, S. S. Patel and K. A. Johnson, *Biochemistry*, **30**, 526 (1991).
293. S. Lone and L. J. Romano, *Biochemistry*, **46**, 2599 (2007).
294. J. Wang, A. K. M. A. Satar, C. C. Wang, J. D. Karam, W. H. Konigsberg and T. J. Steitz, *Cell*, **89**, 1087 (1997).
295. C. A. Brautigam and T. J. Steitz, *Cur. Opin. Struct. Biol.*, **8**, 54 (1998).
296. X. Zhong, S. S. Patel and M. D. Tsai, *J. Am. Chem. Soc.*, **120**, 235 (1998).
297. C. M. Joyce and T. A. Steitz, *Annu. Rev. Biochem.*, **63**, 777 (1994).
298. S. Doublié, S. Tabor, A. M. Long, C. C. Richardson and T. Ellenberger, *Nature*, **391**, 251 (1998).
299. P. Lin, L. C. Pedersen, V. K. Batra, W. A. Beard, S. H. Wilson and L. G. Pedersen, *Proc. Natl. Acad. Sci. U.S.A.*, **103**, 13294 (2006).
300. J. Flórián, M. F. Goodman and A. Warshel, *J. Am. Chem. Soc.*, **125**, 8163 (2003).
301. C. Castro, E. Smidansky, K. R. Maksimchuk, J. J. Arnold, V. S. Korneeva, M. Götte, W. Konigsberg and C. E. Cameron, *Proc. Natl. Acad. Sci. U.S.A.*, **104**, 4267 (2007).
302. T. Lindahl, *Nature*, **362**, 709 (1993).
303. J. L. Huffman, O. Sundheim and J. A. Tainer, *Mutat. Res. Fund. Mol. Mech. Mut.*, **577**, 55 (2005).
304. R. D. Wood, *Annu. Rev. Biochem.*, **65**, 135 (1996).

305. D. M. Wilson III and V. A. Bohr, *DNA Repair*, **6**, 544 (2007).
306. K. Hitomi, S. Iwai and J. A. Tainer, *DNA Repair*, **6**, 410 (2007).
307. T. Izumi, L. R. Wiederhold, G. Roy, R. Roy, A. Jaiswal, K. K. Bhakat, S. Mitra and T. K. Hazra, *Toxicology*, **193**, 43 (2003).
308. J. T. Stivers and Y. L. Yang, *Chem. Rev.*, **103**, 2729 (2003).
309. E. E. Kim and H. W. Wyckoff, *J. Mol. Biol.*, **218**, 449 (1991).
310. L. S. Beese and T. A. Steitz, *EMBO J.*, **10**, 25 (1991).
311. K. Mizuuchi, T. J. Nobbs, S. E. Halford, K. Adzuma and J. Qin, *Biochemistry*, **38**, 4640 (1999).
312. W. Yang, J. Y. Lee and M. Nowotny, *Mol. Cell*, **22**, 5 (2006).
313. K. D. Westover, D. A. Bushnell and R. D. Kornberg, *Cell*, **119**, 481 (2004).
314. Y. W. Yin and T. A. Steitz, *Cell*, **116**, 393 (2004).
315. T. A. Kunkel and K. Bebenek, *Annu. Rev. Biochem.*, **69**, 497 (2000).
316. E. C. Friedberg, G. C. Walker, W. Siede, R. D. Wood, R. A. Schultz and T. Ellenberger, *DNA Repair and Mutagenesis*, 2nd edition, ASM Press, Washington DC, 2005.
317. T. M. Marti and O. Fleck, *Cell Mol. Life Sci.*, **61**, 336 (2004).
318. A. Pingoud and A. Jeltsch, *Nucl. Acids Res.*, **29**, 3705 (2001).
319. A. Pingoud, M. Fuxreiter and W. Wende, *Cell. Mol. Life Sci.*, **62**, 685 (2005).
320. A. Jeltsch, J. Alves, H. Wolfes, G. Maass and A. Pingoud, *Proc. Natl. Acad. Sci. U.S.A.*, **90**, 8499 (1993).
321. J. Barber, *Q. Rev. Biophys.*, **36**, 71 (2003).
322. P. Fromme, P. Jordan and N. Krauß, *Biochim. Biophys. Acta*, **1507**, 5 (2001).
323. N. Kamiya and J. R. Shen, *Proc. Natl. Acad. Sci. U.S.A.*, **100**, 98 (2003).
324. A. Zouni, H. T. Witt, J. Kern, P. Fromme, N. Krauß, W. Saenger and P. Orth, *Nature*, **409**, 739 (2001).
325. G. McDermott, S. M. Price, A. A. Freer, A. M. Hawthornthwaite-Lawless, M. Z. Papiz, R. J. Cogdell and N. W. Isaacs, *Nature*, **374**, 517 (1995).
326. Z. Liu, H. Yan, K. Wang, T. Kuang, J. Zhang, L. Gui, X. An and W. Chang, *Nature*, **428**, 287 (2004).
327. G. A. Montaño, B. P. Bowen, J. T. LaBelle, N. W. Woodbury, V. B. Pizziconi and R. E. Blankenship, *Biophys. J.*, **85**, 2560 (2003).
328. W. Kühlbrandt, D. N. Wang and Y. Fujiyoshi, *Nature*, **367**, 614 (1994).
329. S. Santabarbara, P. Heathcote and M. C. W. Evans, *Biochim. Biophys. Acta*, **1708**, 283 (2005).
330. T. S. Balaban, P. Fromme, A. R. Holzwarth, N. Krauß and V. I. Prokhorenko, *Biochim. Biophys. Acta*, **1556**, 197 (2002).
331. S. M. Prince, M. Z. Papiz, A. A. Freer, G. McDermott, A. M. Hawthornthwaite-Lawless, R. J. Cogdell and N. W. Isaacs, *J. Mol. Biol.*, **268**, 412 (1997).
332. A. J. van Gammeren, F. B. Hulsbergen, C. Erkelens and H. J. M. de Groot, *J. Biol. Inorg. Chem.*, **9**, 109 (2004).
333. A. Wong, R. Ida, X. Mo, Z. Gan, J. Poh and G. Wu, *J. Phys. Chem. A*, **110**, 10084 (2006).
334. J. Linnanto and J. Korppi-Tommola, *Phys. Chem. Chem. Phys.*, **8**, 663 (2006).
335. M. Nsango, A. B Fredj, N. Jaidane, M. G. K. Njock and Z. B. Lakhdar, *J. Mol. Struct. (Theochem.)*, **681**, 213 (2004).
336. I. Barth, J. Manz, Y. Shigeta and K. Yagi, *J. Am. Chem. Soc.*, **128**, 7043 (2006).
337. B. Dietzek, R. Maksimenka, W. Kiefer, G. Hermann, J. Popp and M. Schmidt, *Chem. Phys. Lett.*, **415**, 94 (2005).
338. S. Gutteridge and J. Pierce, *Proc. Natl. Acad. Sci. U.S.A.*, **103**, 7203 (2006).
339. W. W. Cleland, T. J. Andrews, S. Gutteridge, F. C. Hartman and G. H. Lorimer, *Chem. Rev.*, **98**, 549 (1998).
340. F. G. Pearce, *Biochem. J.*, **399**, 525 (2006).
341. R. J. Spreitzer and M. E. Salvucci, *Annu. Rev. Plant Biol.*, **53**, 449 (2002).
342. H. J. Imker, A. A. Fedorov, E. V. Fedorov, S. C. Almo and J. A. Gerlt, *Biochemistry*, **46**, 4077 (2007).
343. I. Andersson and T. C. Taylor, *Arch. Biochem. Biophys*, **414**, 130 (2003).
344. G. H. Lorimer, *Biochemistry*, **20**, 1236 (1981).
345. J. Pierce, G. H. Lorimer and G. S. Reddy, *Biochemistry*, **25**, 1636 (1986).
346. J. Newman and S. Gutteridge, *J. Biol. Chem.*, **268**, 25876 (1993).
347. O. Tapia, M. Oliva, V. S. Safont and J. Andrés, *Chem. Phys. Lett.*, **323**, 29 (2000).

348. D. L. Edmondson, H. J. Kane and T. J. Andrews, *FEBS Lett.*, **260**, 62 (1990).
349. F. G. Pearce and T. J. Andrews, *Plant Physiol.*, **117**, 1059 (1998).
350. F. G. Pearce and T. J. Andrews, *J. Biol. Chem.*, **278**, 32526 (2003).
351. H. Mauser, W. A. King, J. E. Gready and T. J. Andrews, *J. Am. Chem. Soc.*, **123**, 10821 (2001).
352. J. V. Schloss and G. H. Lorimer, *J. Biol. Chem.*, **257**, 4691 (1982).
353. G. G. B. Tcherkez, G. D. Farquhar and T. J. Andrews, *Proc. Natl. Acad. Sci. U.S.A.*, **103**, 7246 (2006).
354. T. C. Taylor and I. Andersson, *J. Mol. Biol.*, **265**, 432 (1997).
355. G. Schneider, Y. Lindqvist and C. I. Brändén, *Annu. Rev. Biophys. Biomol. Struct.*, **21**, 119 (1992).
356. M. Oliva, V. S. Safont, J. Andrés and O. Tapia, *Chem. Phys. Lett.*, **340**, 391 (2001).
357. M. Oliva, V. S. Safont, J. Andrés and O. Tapia, *J. Phys. Chem. A*, **105**, 9243 (2001).

CHAPTER 9

Theoretical studies of the addition of RMgX to carbonyl compounds

SHINICHI YAMABE and SHOKO YAMAZAKI

Department of Chemistry, Nara University of Education, Takabatake-cho, Nara 630-8528, Japan
Fax: +81-742-27-9208; e-mail: yamabes@nara-edu.ac.jp

I. INTRODUCTION	369
II. EXPERIMENTAL BACKGROUND	370
III. CALCULATIONS OF GRIGNARD REAGENTS	374
IV. CALCULATIONS OF MODEL GRIGNARD REACTIONS	380
V. A COMPREHENSIVE COMPUTATIONAL STUDY OF GRIGNARD REACTIONS	384
A. Computational Methods	384
B. Structures of Grignard Reagents	384
C. Additions of Grignard Reagents to Carbonyl Compounds without Solvent Molecules	387
D. Additions with Solvent Molecules	389
E. Chelation-controlled Addition Models	391
F. A Model for SET	396
G. Concluding Remarks	399
VI. REFERENCES	401

I. INTRODUCTION

In this chapter, theoretical studies of geometries of Grignard reagents and paths of Grignard reactions are presented. Since numerous theoretical studies have been reported, we confine ourselves here to *ab initio* studies which are thought to give reliable results. Molecular orbital calculations began to be practical when examining reactions of small model systems with the software GAUSSIAN of computational organic chemistry. It had the function of geometry optimizations (GAUSSIAN 80[1]). However, around the year 1980, only small model systems were used to simulate reactions. With the version GAUSSIAN 92/DFT[2], the density functional theory calculations became available. DFT calculations

The chemistry of organomagnesium compounds
Edited by Z. Rappoport and I. Marek   © 2008 John Wiley & Sons, Ltd

display a good compromise between the accuracy of calculated results and performance. Then, from 1993, reliable reacting systems began to be studied to trace the paths. At present, the computational study is thought to be an indispensable tool to precisely understand the reaction mechanism. As for the Grignard reactions, however, the mechanism has been veiled for a long time. One difficulty in dealing with the reaction is a well known problem: there are both polar and single-electron-transfer (SET) mechanisms. One question is why the closed-shell system, ketone plus Grignard reagent, is converted to singlet biradical species. In other words, what is the driving force for forming the biradical species? Those questions have been recently solved and will be discussed in Section V of this chapter. However, we will first present the experimental background and earlier theoretical studies.

II. EXPERIMENTAL BACKGROUND

The Grignard reaction has a 100-year history and is one of the most important organic reactions for C−C bond formation[3], and is still extensively utilized in organic syntheses nowadays. The structure of Grignard reagents has been gradually revealed by X-ray analyses and other spectroscopic methods[3b,c]. However, the detailed mechanism (in particular, C−C bond formation) of carbonyl addition of Grignard reagents is still unclear. The mechanism is considered to be complex and varies depending on alkyl groups, halogens, solvent, concentration and temperature. The two mechanistic possibilities, polar vs SET (single-electron transfer) shown in the process, **1** → **2**, of Scheme 1 have been discussed for many years[3b,c,f−k].

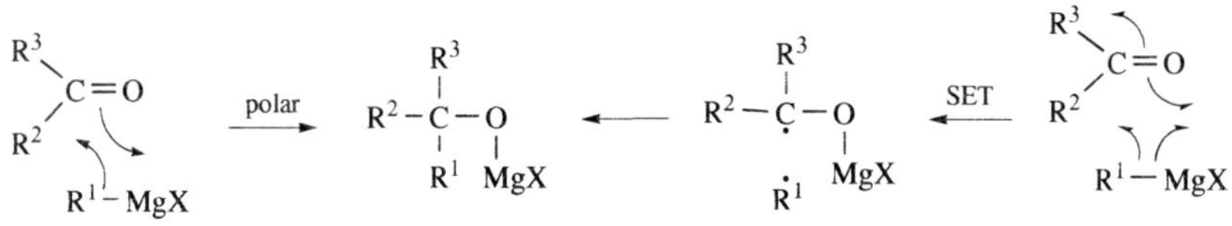

$$
\begin{array}{c}
R^3 \\
\diagdown \\
=O \\
R^2 \diagup \\

\end{array}
\quad \xrightarrow{\text{polar}} \quad
\begin{array}{c}
R^3 \\
| \\
R^2-C-O \\
| \quad | \\
R^1 \quad MgX
\end{array}
\quad \longleftarrow \quad
\begin{array}{c}
R^3 \\
| \\
R^2-\overset{\cdot}{C}-O \\
| \quad | \\
\overset{\cdot}{R^1} \quad MgX
\end{array}
\quad \xleftarrow{\text{SET}} \quad
\begin{array}{c}
R^3 \diagdown \\
C=O \\
R^2 \diagup \\
R^1-MgX
\end{array}
$$

(**1**)   (**2**)   (**1**)

SCHEME 1. Two traditional mechanisms for C−R^1 bond formation

Numerous stereoselective carbonyl additions of Grignard reagents, including enantioselective examples, have been developed recently. Such stereoselective additions have been considered through the polar mechanism, and Cram's selectivity involving chelation control is used in order to explain the high diastereoselectivity[4,5]. Since the detailed C−C bond formation steps including transition states has not been clear, further development of higher efficiency (improving yields), chemoselectivity (minimizing side-reactions) and stereoselectivity in the addition steps is difficult.

The actual composition/structure of Grignard reagents—commonly written as RMgX—has been a matter of some dispute[6]. It appears to depend on the nature of R and also on the solvent. Thus, the ^{1}H NMR spectrum or MeMgBr in Et$_2$O indicates that it is present largely as MgMe$_2$ + MgBr$_2$[3k]. On the other hand, X-ray measurements on crystals of PhMgBr, isolated from Et$_2$O solution, indicate that it has the composition PhMgBr·2Et$_2$O, with the four ligands arranged tetrahedrally around the Mg atom[3b,c]. In any event, Grignard reagents may be regarded as acting as sources of negatively polarized carbon, i.e. as $^{\delta-}$R(MgX)$^{\delta+}$.

There is evidence of complexing the Mg atom of the Grignard reagent with the carbonyl oxygen atom (**3** in Scheme 2), and it is found that two molecules of R^{1}MgX are involved

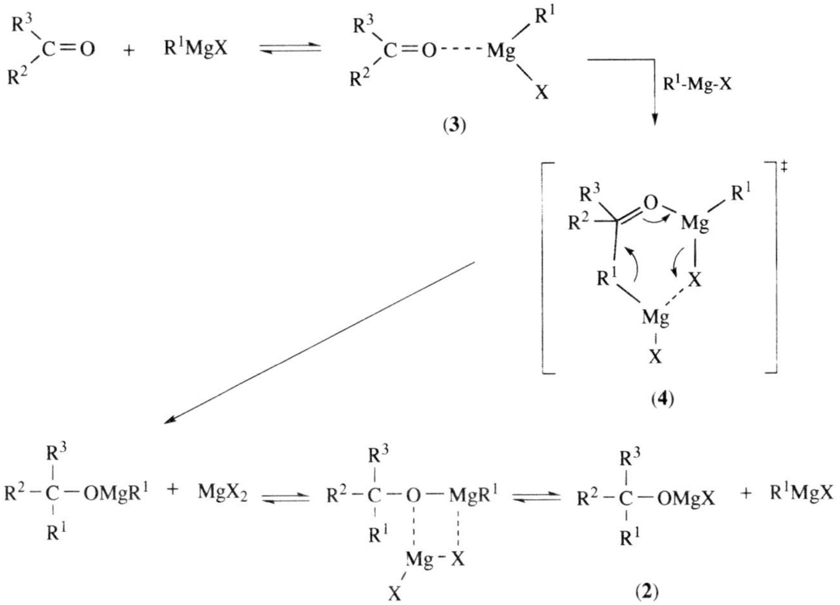

SCHEME 2. A termolecular mechanism for Grignard reagent addition to carbonyl compounds via the cyclic transition state **4**

in the termolecular addition reaction, in some cases at least, possibly via a cyclic transition state such as **4** (Scheme 2)[7,8].

In this termolecular mechanism, the second molecule of R^1MgX could be regarded as a Lewis acid catalyst, increasing the positive polarization of the carbonyl carbon atom through complexing with oxygen. In practice, it is found that the addition of Lewis acids such as $MgBr_2$[3f] enhances the rate of Grignard additions. The details of the mechanism of Grignard reagent addition to C=O have been scarcely studied for such a well-known reaction; however, pathways closely analogous to that shown above (i.e. via **4** in Scheme 2) can be invoked to explain the following two further important observations.

The first is that Grignard reagents bearing hydrogen atoms on the β-carbon atom of the alkyl group ($R^4CH_2CH_2MgX$ in **3a**) tend to reduce the extent of the transformation, carbonyl group → an alcohol, while being converted to alkenes ($R^4CH=CH_2$ in **5**). In this process, transfer of H rather than $R^4CH_2CH_2$ takes place via **4a** (Scheme 3).

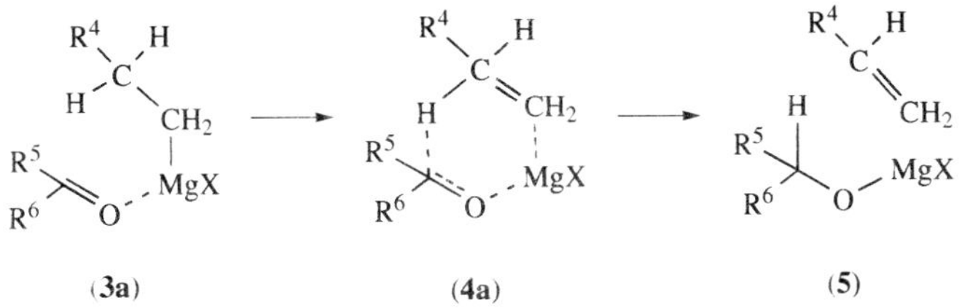

SCHEME 3. A mechanism for alkene formation via a Grignard reaction

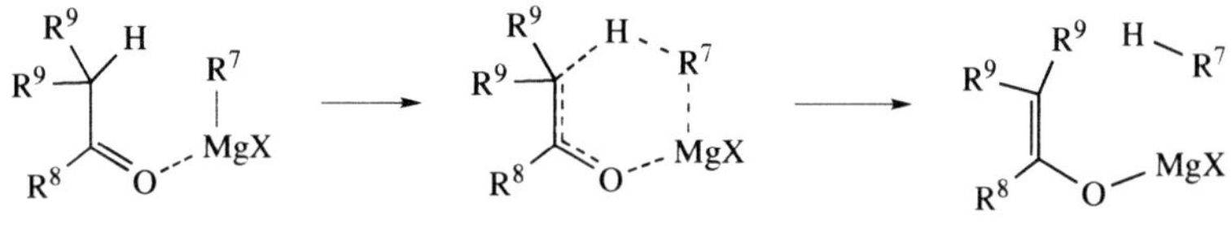

(3b)   (4b)   (6)

SCHEME 4. The mechanism for alkane formation in a Grignard reaction

The second is that sterically hindered ketones bearing hydrogen atoms on their α-carbons, $R_2^9CH(CO)R^8$ (cf. **3b**), tend to be converted to their enolates (**6**), where the Grignard reagent, R^7MgX, is lost as R^7-H in the process via **4b** (Scheme 4).

Grignard reagents act as strong nucleophiles and the addition reaction is almost always substantially irreversible (cf. conjugate addition to C=C—C=O). The initial products are alcohols, but it is important to emphasize that the utility of Grignard and similar additions to C=O is a general method of connecting different carbon atoms together, i.e. the original products can then be further modified in a wide variety of reactions. In the past organozinc compounds were used in a similar way, but they are largely displaced by Grignard reagents. In turn, Grignard reagents tend to be displaced gradually by lithium alkyls RLi and aryls ArLi, respectively. These latter reagents (RLi and ArLi) tend to give more of the normal addition product with sterically hindered ketones than Grignard reagents, as well as more of the 1,2-product and less 1,4-additions with C=C—C=O than Grignard reagents.

Holm and Crossland reported the product distribution (Scheme 5 and Table 1)[9] in a reaction between t-BuMgCl and benzophenone. While the 1,2-addition affords the normal product, 1,6- and 1,4-additions should involve the *tert*-butyl radical for *ortho*- and *para*-additions, and the mechanism involves a single-electron transfer (SET). The product distribution indicates that the more sterically crowded benzophenones give more of the SET products. Ashby and Smith[10a] obtained relative rates for reactions of acetone and benzophenone (Table 2)[10b]. Noteworthy is that those ketones have opposite reactivity orders toward R^1MgCl for the R^1 variation.

There are now two types of Grignard reagents[11]; one gives a large kinetic isotope effect (KIE), a large Hammett ρ value for the substituted benzophenones and large steric rate retardation. Examples are MeMgX, ArMgBr and PhCH$_2$MgBr, and R transfer is regarded as rate-determining. The other (e.g. allylic MgBr) gives a near-unity KIE, a small ρ value and negligible steric rate retardation, and SET is regarded as rate-determining. However, as shown in Scheme 6, t-BuMgCl shows a different pattern, i.e. a small KIE, a large ρ value and no steric rate retardation. The two last reactivity features reported by Holm were interpreted in terms of the rate-determining SET mechanism[9]. Yamataka and coworkers

TABLE 1. Product distributions in the reaction of substituted benzophenones and t-BuMgCl

Benzophenone	Pinacol (%)	1,2-Adduct (%)	1,4-Adduct (%)	1,6-Adduct (%)
Parent	6	44	0	50
4,4'-Dimethyl	12	55	0	33
4,4'-Di-t-butyl	21	40	39	0
4,4'-Dichloro	0	50	21	29
2,4,6-Trimethyl	0	0	0	100
2,4,6,4'-Tetramethyl	0	0	0	100
2,3,5,6-Tetramethyl	0	0	0	100

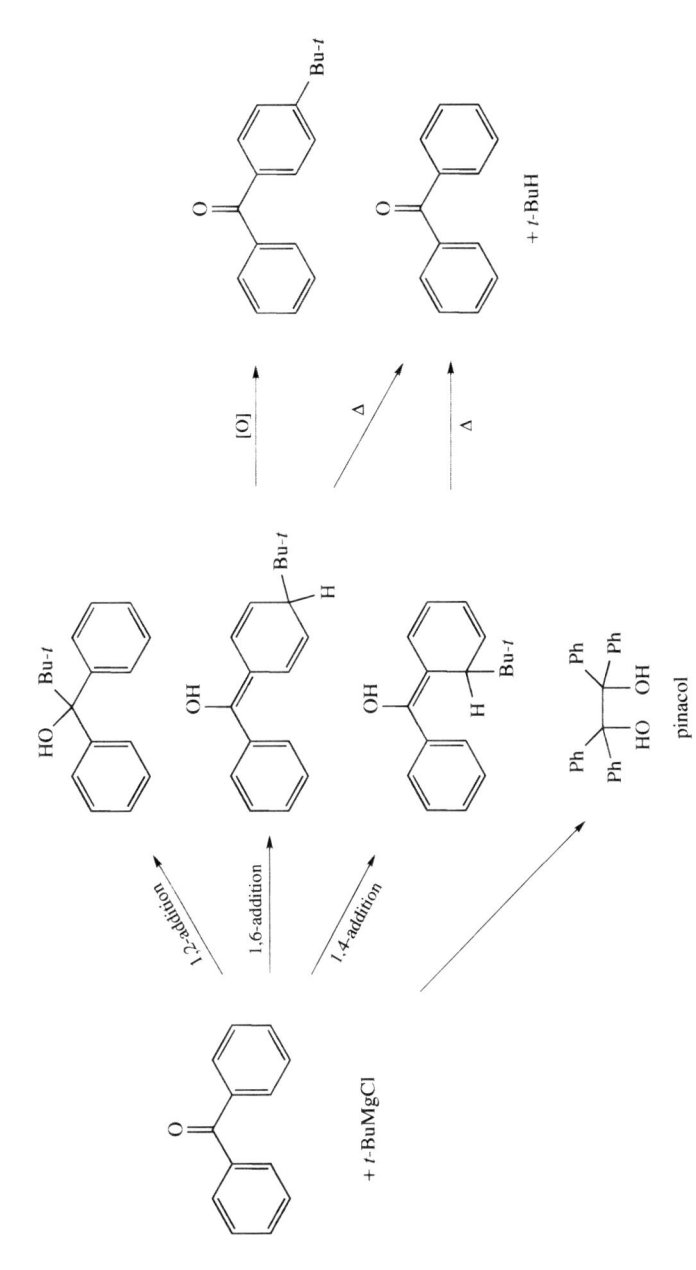

SCHEME 5. The product distribution in the 1,2-, 1,4- and 1,6-additions of *t*-BuMgCl to benzophenone

TABLE 2. Relative reaction rates of typical ketones and Grignard reagents R^1MgCl

R^1	Acetone	Benzophenone
CH_3	1114	30
CH_3CH_2	2324	408
$(CH_3)_2CH$	272	4027
$(CH_3)_3C$	9	5363

suggested[11], however, that the Hammet ρ value for the SET step is small as observed in the reactions of allylic Grignard reagents. The large ρ value of 3.0 reported for the reaction of t-BuMgCl with benzophenones is rather indicative of the presence of electron-transfer equilibrium prior to the rate-determining step. It is assumed that in the t-BuMgCl reaction the product formation from the first intermediate, I, is retarded compared to the rate with MeMgCl due to the steric bulk of t-Bu, and another route via the second intermediate, II, becomes important. The rate-determining step of the reaction is then the isomerization of I to II. This interpretation is consistent with the large ρ value as well as a small KIE observed for this reaction.

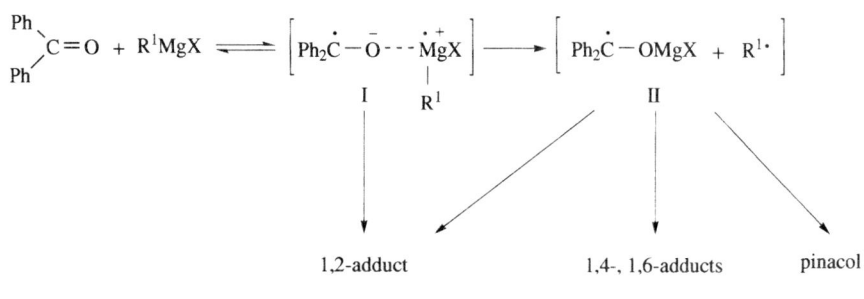

$$\underset{Ph}{\overset{Ph}{\diagdown}}C=O + R^1MgX \rightleftharpoons \left[Ph_2\overset{\cdot}{C}-\overset{-}{O}\cdots\overset{+}{MgX} \atop {|} \atop R^1 \right] \longrightarrow \left[Ph_2\overset{\cdot}{C}-OMgX + R^{1\cdot} \right]$$

I → 1,2-adduct
II → 1,4-, 1,6-adducts
→ pinacol

R^1MgX /solvent	$^{12}k/^{14}k$ at $0.0 \pm 0.1°C$	ρ value
$CH_3CH=CHCH_2MgBr/Et_2O$	0.999 ± 0.002	0.01 ± 0.03
t-BuMgCl/Et_2O	1.010 ± 0.007	3.0

SCHEME 6. Kinetic isotope effect (KIE) and Hammett ρ value in the SET mechanism

III. CALCULATIONS OF GRIGNARD REAGENTS

First, the geometry, stability and harmonic frequency of the CH_3MgCl monomer are reported. The calculations confirm the C_{3v}-symmetric, CH_3MgCl structure for the Grignard reagent and indicate that the $Mg + CH_3Cl \rightarrow CH_3MgCl$ reaction is quite exothermic, with the heat of reaction being 58.8 and 47.5 $kcal\,mol^{-1}$ in the 3-21G and 6-31G* basis sets, respectively[12]. The calculated energies, dipole moments, and geometrical parameters for CH_3Cl, and CH_3MgCl are listed in Table 3, together with the experimental values[13] of CH_3Cl. For CH_3Cl, the geometry calculated from the 6-31G* basis set is found to be in good agreement with the experimental data. The 3-21G geometry is similar except that, due to the neglect of d functions on the heavy atoms, the C—Cl bond turns out to be much too long. It is of interest to note that the 6-31G* C—Mg bond length (2.09 Å) is in fair agreement with the value of 2.16 Å found experimentally for the C—Mg bond of

9. Theoretical studies of the addition of RMgX to carbonyl compounds 375

TABLE 3. Energies, dipole moments and geometries of CH_3Cl and CH_3MgCl

Parameter	CH_3Cl			CH_3MgCl	
	3-21G	6-31G*	exptl	3-21G	6-31G*
r(C-Cl), Å	1.892	1.785	1.778		
r(C-H), Å	1.074	1.078	1.084	1.088	1.088
r(C-Mg), Å				2.090	2.090
r(Mg-Cl), Å				2.278	2.211
∠(HCCl or HCMg), deg	106.3	108.6	108.4	111.3	111.7
μ, Debye	2.87	2.44	1.94	3.52	2.42

Reprinted with permission from Reference 12. Copyright 1982 American Chemical Society.

TABLE 4. Normal-mode vibrational frequencies (cm^{-1}) of CH_3Cl and CH_3MgCl

		CH_3Cl				CH_3MgCl		
		theory		experimental[a]		theory		
	Mode	3-21G	6-31G*	normal modes	measured	3-21G	6-31G*	exptl[b]
a_1	sym C-H st[c]	3282	3280	3074.4	2967.8	3156	3179	2805
	sym C-H d[d]	1501	1538	1382.6	1354.9	1371	1327	1306
	C-Cl st	663	782	740.2	732.8			
	C-Mg st					656	647	
	Mg-Cl st					370	376	
e	asym C-H st	3401	3376	3165.9	3039.2	3224	3235	
	asym C-H d	1639	1641	1481.8	1452.1	1642	1607	
	rocking	1096	1138	1038.0	1017.3	703	637	530
	Me-Mg-Cl bend					119	123	

[a] From Reference 16. For CH_3Cl 'normal modes' are frequencies corrected by the anharmonicity effects for measured frequencies.
[b] From Reference 17.
[c] Stretch.
[d] Deformation.

Reprinted with permission from Reference 12. Copyright 1982 American Chemical Society.

$MgC_2H_5BrEt_2O^{14}$, and the 6-31G* Mg—Cl bond length is only 0.04 Å larger than that determined for $MgCl_2$[15].

The calculated normal-mode vibrational frequencies are compared with the available experimental values[16] in Table 4. For CH_3Cl, the 3-21G and 6-31G* basis sets yield similar results except that the former basis underestimates the C—Cl stretching frequency. With the 6-31G* basis utilized in assignments of the CH_3MgCl modes, the harmonic frequencies of CH_3Cl are overestimated on an average by 7%, and the assignment of the modes of CH_3Cl is straightforward.

Second, formation of CH_3MgX is reported. Theoretical calculations using self-consistent field (SCF) and Møller–Plesset perturbation theory, up to the fourth order (MP4), have been carried out on the gas-phase Mg + CH_3X → CH_3MgX reaction surface for X = F and Cl[18]. The transition-state energies, geometries and vibrational frequencies for both reactions are presented and compared to those of the Mg + HX → HMgX reaction. The transition states for both X = F and X = Cl are found to possess Cs symmetry and to be almost identical in structure. The activation energy for the Mg + fluoromethane reaction is found to be 31.2 kcal mol^{-1}, while that for the chloromethane reaction is

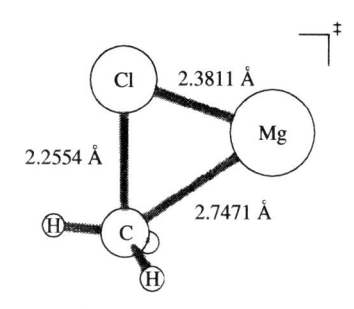

FIGURE 1. Transition-state structure for Mg + CH$_3$Cl → CH$_3$MgCl optimized at the MP2/6-311G (d,p) level. Reprinted with permission from Reference 18. Copyright 1991 American Chemical Society

substantially higher, at 39.4 kcal mol^{-1}, calculated at the MP4SDTQ level by using the 6-311G(d,p) basis set. The intrinsic reaction coordinate has been followed down from the transition state toward both reactants and product for the Mg + CH$_3$F → CH$_3$MgF reaction, confirming the connection of these points on the potential energy surface. The structure for the transition state of Mg + CH$_3$Cl → CH$_3$MgCl is shown in Figure 1. The H−C−Cl−Mg dihedral angle of 180.0° in the transition state is found to be the same at both the SCF and MP2 levels, in contrast to that found for the CH$_3$FMg‡. Therefore, both the CH$_3$FMg‡ and CH$_3$ClMg‡ transition states have analogous structures at the MP2 level, belonging to the Cs point group. The following comparisons between the two TS structures will be for those calculated at the MP2 level. The C−Mg bond length is slightly longer by 0.11 Å in the CH$_3$ClMg‡ transition state than that for CH$_3$FMg‡, while the C−H bond lengths are the same to within a few thousandths of an angstrom. The C−Cl and Mg−Cl bonds are necessarily longer due to the larger size of the Cl atom. The H−C−Cl angle is within only 0.4° of that in CH$_3$FMg‡, and the H'−C−Cl angles differ by only 0.56°. Here, H' refers to the hydrogen atom in the Cs symmetry plane. The H'−C−X−H dihedral angles are also about the same (only a 1.5° difference). The Mg−Cl bond distance in the transition state is fairly close to its value in the CH$_3$MgCl Grignard structure, indicative of a strong interaction between these two atoms.

Theoretical calculations using self-consistent field and Møller–Plesset perturbation theory through second order (MP2) have been carried out on the gas-phase Mg + C$_2$H$_3$X → C$_2$H$_3$MgX reaction for X = F, Cl[19]. Optimized geometries for the reactants, transition states (TSs) and products have been determined along with relative energies and vibrational frequencies. The intrinsic reaction coordinate has been followed from the TS to reactants and products, confirming that the located structures all lie on the reaction potential energy surface (Figure 2). The transition state is found to possess C_1 symmetry, while the product belongs to point group Cs (Figure 3). The activation energies are calculated to be 22.8 kcal mol^{-1} for the Mg + C$_2$H$_3$F reaction and 29.7 kcal mol^{-1} for the Mg + C$_2$H$_3$Cl reaction. The overall exothermicity for both reactions is 54.3 kcal mol^{-1} at the MP2/6-31G** level. The geometry of Cl−Mg−C$_2$H$_3$ is shown in Figure 4.

The mechanism of the Grignard reagent formation was studied.[20] The results of density functional calculations are reported for CH$_3$Mg$_2$, CH$_3$Mg$_{4(T)}$ and for CH$_3$Mg$_{5(TB)}$Cl model clusters, with T = tetrahedral and TB = trigonal bipyramid. These calculations aim at a simulation of the migration of a methyl group in the proposed intermediates RMg$_n^{(I)}$ ($n = 2$ and 4) and of the succession of steps from the substrate to RMgX. The mono-coordination of the methyl group in the clusters CH$_3$Mg$_n$ ($n = 2$ or 4) represents the most stable structure. The CH$_3$Mg$_5$Cl geometries and energies are shown in Figure 5. The energy

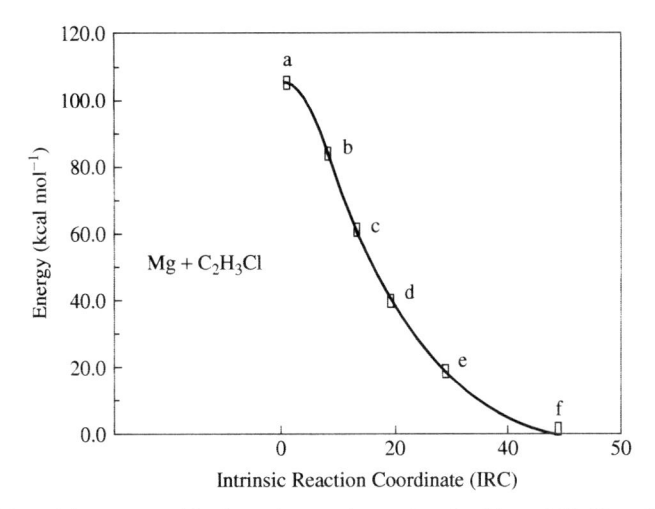

9. Theoretical studies of the addition of RMgX to carbonyl compounds 377

FIGURE 2. Potential energy profile along the reaction surface for $Mg + C_2H_3Cl \rightarrow C_2H_3MgCl$ at the RHF/3-21G* level. The zero point on the abscissa along the IRC is the TS and the product is toward the positive direction. Reprinted with permission from Reference 19. Copyright 1991 American Chemical Society

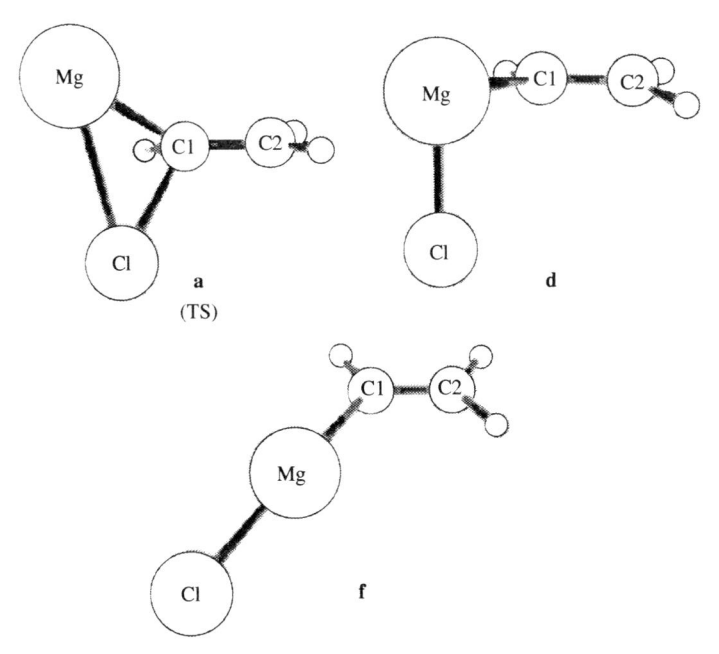

FIGURE 3. Optimized geometries for selected points along the IRC for the $Mg + C_2H_3Cl \rightarrow C_2H_3MgCl$ reaction calculated at the RHF/3-21G* level. See Figure 2 for the points selected. Reprinted with permission from Reference 19. Copyright 1991 American Chemical Society.

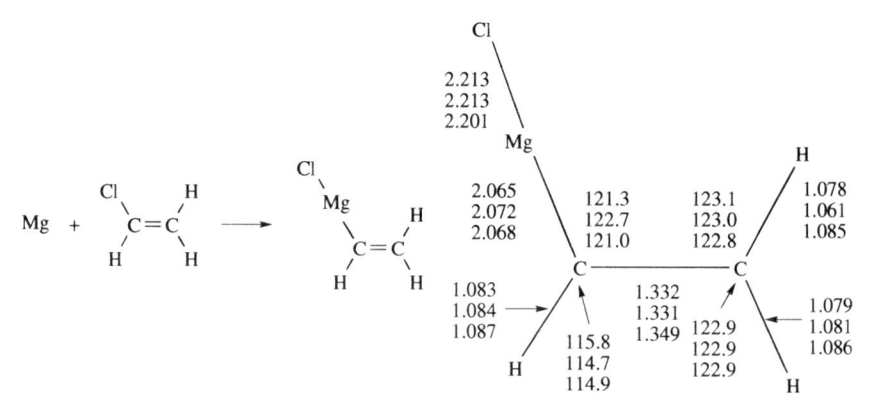

FIGURE 4. Optimized geometries for the C_2H_3MgCl Grignard molecule. Units are in angstroms and degrees. Values from top to bottom are at the RHF/3-21G*, RHF/6-31G** and MP2/6-31G** levels, respectively

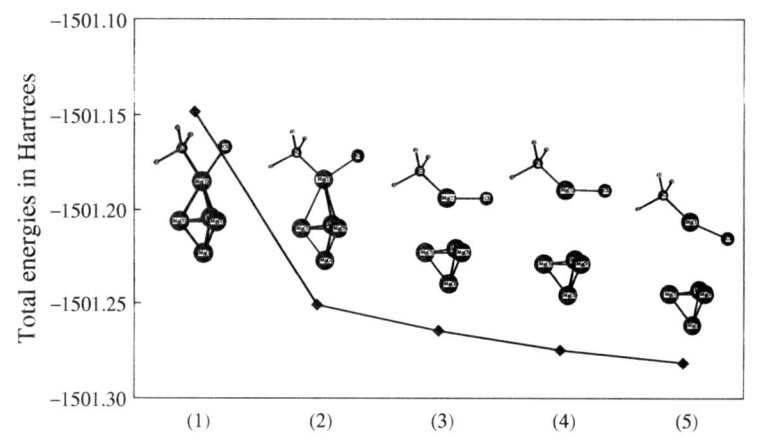

FIGURE 5. Formation of the Grignard reagent CH_3MgCl[20]. 1 Hartree = 627.51 kcal mol^{-1}. Along the abscissa, sequential changes of geometries, (1) → (2) → (3) → (4) → (5), are shown

barrier to pass from poly-coordination structure to mono-coordination structure is low (1.213 eV = 27.9 kcal mol^{-1}). It is sufficient, however, to prevent the methyl migration from a magnesium atom to another one and the migration motion is probably frozen at lower temperatures. The reaction would evolve then in an irreversible pathway toward the Grignard reagent RMgX and a magnesium cluster with $n - 1$ magnesium atoms.

Ab initio molecular orbital calculations were used to study the modified Schlenk equilibrium[21]: $2R^1MgCl$ (**7**) $\rightleftarrows (R^1MgCl)_2 \rightleftarrows MgR_2^1 + MgCl_2 \rightleftarrows Mg(Cl_2)MgR_2^1$ with $R^1 =$ H and CH_3 (Scheme 7) by Axten and coworkers.[22] In the absence of a solvent, calculations indicate that the formation of the various possible bridged dimers $(R^1MgCl)_2$ (**9, 10** and **11**) is substantially exothermic (Figure 6). When the dimer **10** is decomposed nonequivalently, **8** is obtained. It is very unstable (Figure 6) and formation of its dimeric form (**12**) is further unlikely. With dimethyl ether as a model solvent, only the formation

9. Theoretical studies of the addition of RMgX to carbonyl compounds

SCHEME 7. Schlenk equilibrium which describes the composition of a wide range of Grignard solutions, $R^1 = Me$

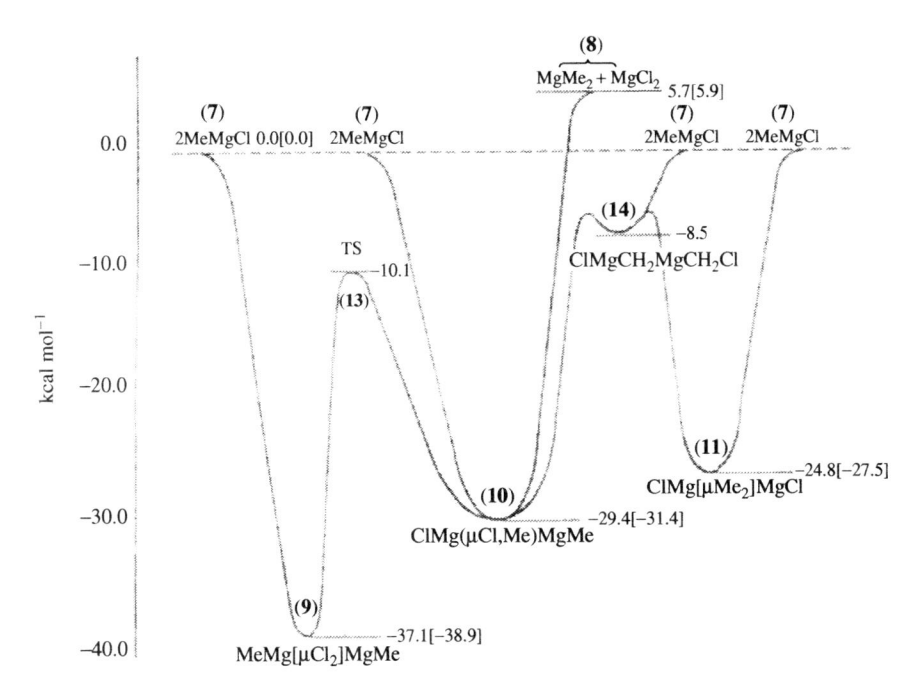

FIGURE 6. Energies (kcal mol^{-1}) of the various reaction channels for CH$_3$MgCl. Unbracketed values are at the MP4SDTQ/6-31G*//HF/6-31G* level and bracketed values are at the MP2/6-31G*//MP2/6-31G* level. Reproduced by permission of Springer Science + Business Media from Reference 22

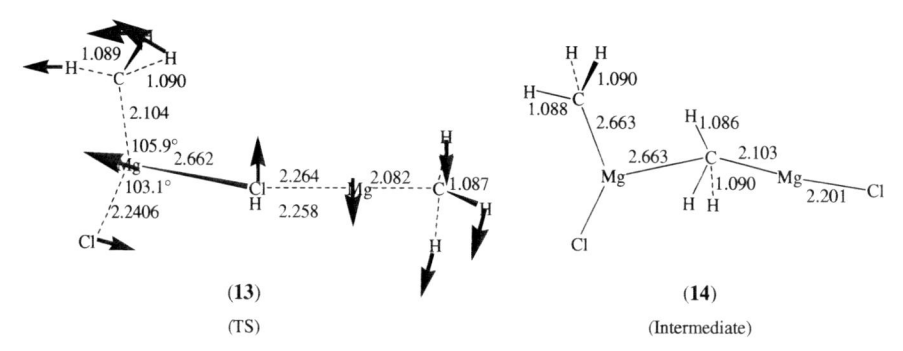

FIGURE 7. Optimized geometries of structures **13** and **14**. Distances are in angstrom. For **13**, reaction-coordinate vectors are also shown. Reproduced by permission of Springer Science + Business Media from Reference 22

of the dimer $(Me_2O)(CH_3)Mg(\mu Cl_2)Mg(CH_3)(OMe_2)$ is exothemic when entropic effects are included (i.e. in Gibbs free energies). Geometries of the isomerization TS (**13**) and the transient intermediate (**14**) are shown in Figure 7.

IV. CALCULATIONS OF MODEL GRIGNARD REACTIONS

The mechanism of the Grignard reaction was investigated for the first time by *ab initio* SCF MO theory by Nagase and Uchibori[23] for a model reaction composed of formaldehyde and MgH_2 molecules. A reactant complex **3c** is formed, which is stabilized by 23.4 kcal mol^{-1} (Figure 8) relative to the isolated molecules. Complex **3c** is isomerized to a four-centered transition state (TS) with a small activation energy of 12.7 kcal mol^{-1}. The product, H_3COMgH, is afforded with a large exothermic energy, 63.4 kcal mol^{-1}. The geometries of the complex **3c** and TS are shown in Figure 9. Although the model, $H_2C=O + MgH_2$, is far from real reacting systems, this pioneering work prompted further computational studies of reaction paths.

Electronic and conformational effects on π-facial stereoselectivity in nucleophilic additions to carbonyl compounds have been studied by the use of RHF/3-21G and RHF/6-31G* methods[24]. Figure 10 shows a comparison of predicted and experimental selectivities for methyl Grignard additions. Satisfactory agreement of the ratios of *anti* and equatorial attacks of MeMgX on the carbonyl carbon atoms was reported.

A theoretical study on the addition of organomagnesium reagents (CH_3Mg^+, CH_3MgCl, $2CH_3MgCl$) to the carbonyl group of chiral α-alkoxy carbonyl compounds (2-hydroxypropanal, 3-hydroxybutanone, and 3,4-di-*O*-methyl-1-*O*-(trimethylsilyl)-L-erythrulose) was carried out[25,26]. Analytical gradients SCF MO and second derivatives at the *ab initio* method at the HF/3-21G basis set level were applied to identify the stationary points on potential energy surfaces. The geometry, harmonic vibrational frequencies, transition vectors and electronic structures of the transition structures were obtained. The dependence of the results obtained upon the computation method and the model system is analyzed, discussed and compared with available experimental data (Scheme 8). The first step corresponds to the exothermic formation of a chelate complex **17** without energy barrier. This stationary point corresponds to a puckered five-membered ring, determining the stereochemistry of the global process, which is retained throughout the reaction pathway. The second and rate-limiting step is associated with the C—C bond formation via 1,3-migration of the nucleophilic methyl group (R in M—R) from the organomagnesium compound to the carbonyl carbon. For an intramolecular mechanism, the transition

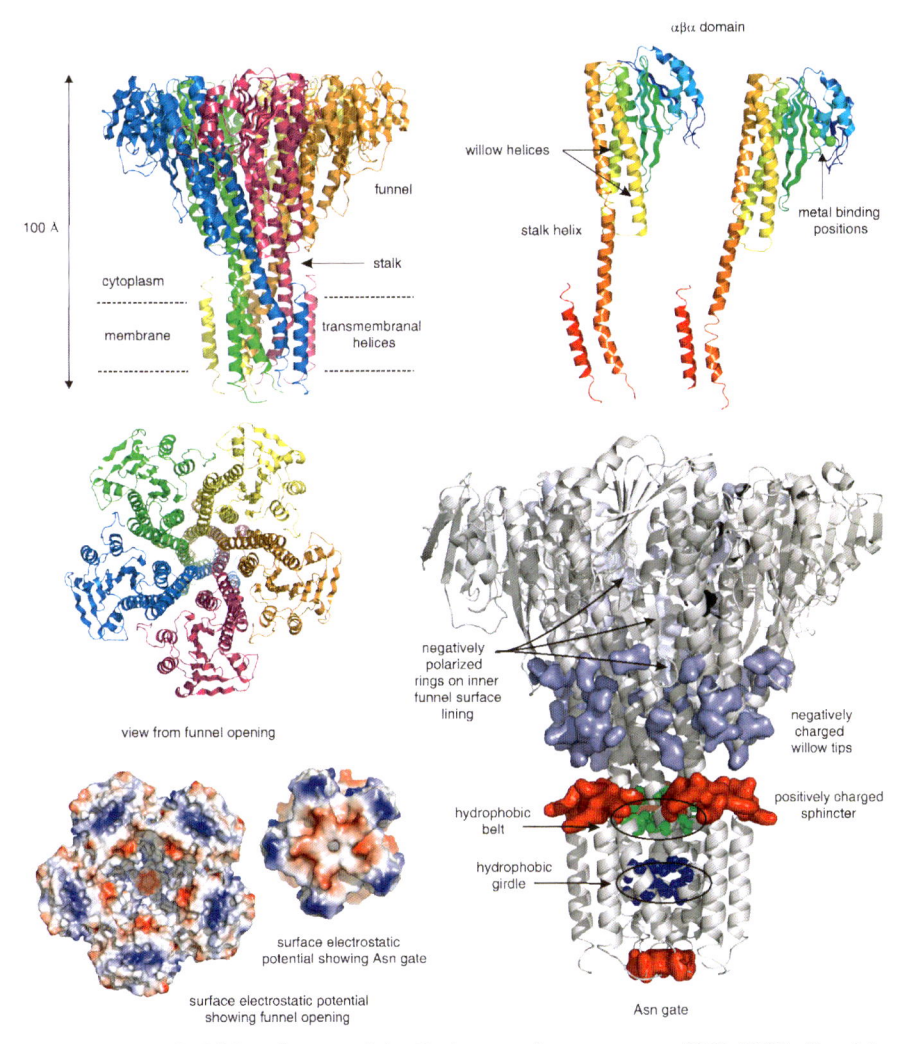

PLATE 1 Top left: Ribbon diagram of the CorA magnesium transporter (PDB 2BBJ). Top right: Monomeric subunit. Middle and bottom left: Various views of the funnel and membrane openings. Bottom right: Illustration of critical structural features

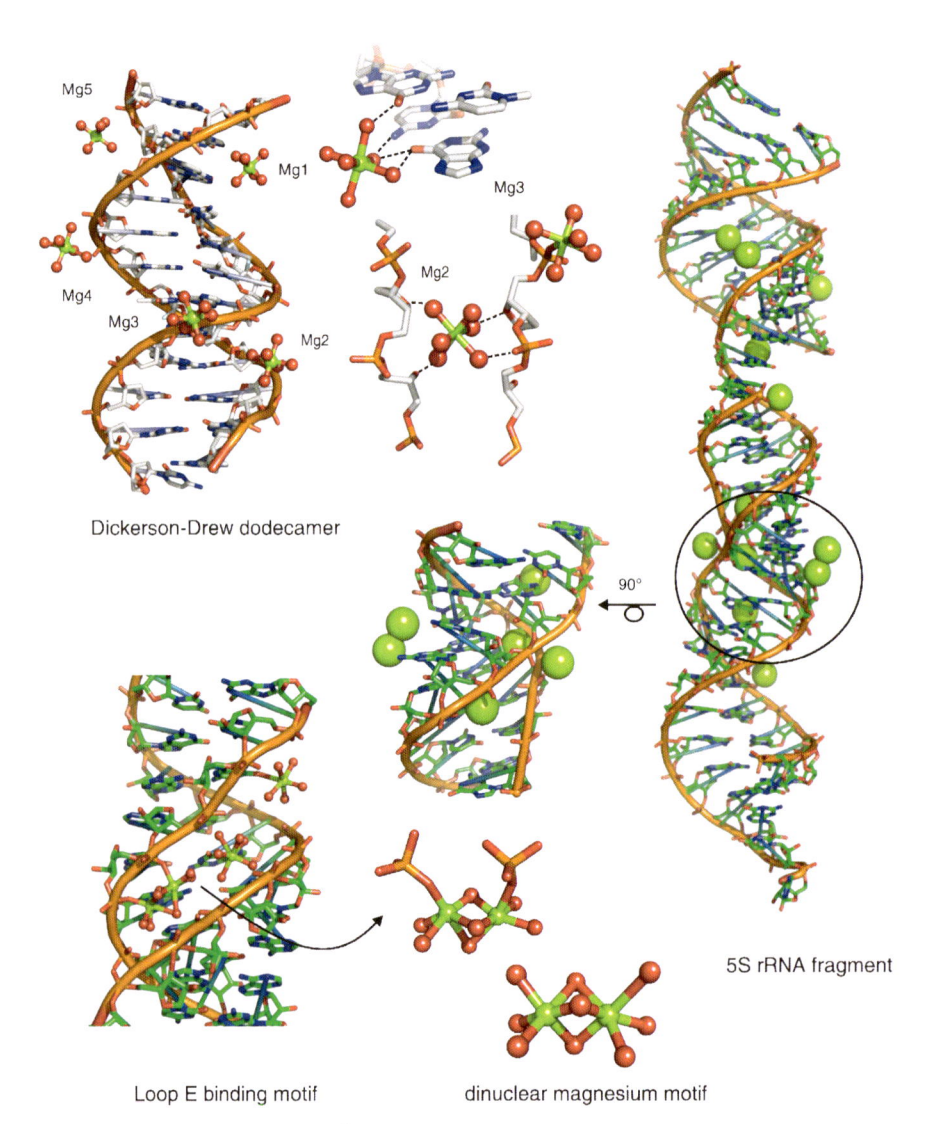

PLATE 2 Examples of specific Mg^{2+} interactions with DNA and RNA. Upper left: the Dickerson-Drew DNA fragment CGCCAATTCGCG (NDB BD0007). Lower left: the RNA loop E backbone zipper motif containing a dinuclear magnesium cluster (NDB URL064). Right: the 5S rRNA fragment from ribosomal *E. coli* (NDB file URL065) containing two dinuclear magnesium clusters and a twisted loop E motif (blow-up)

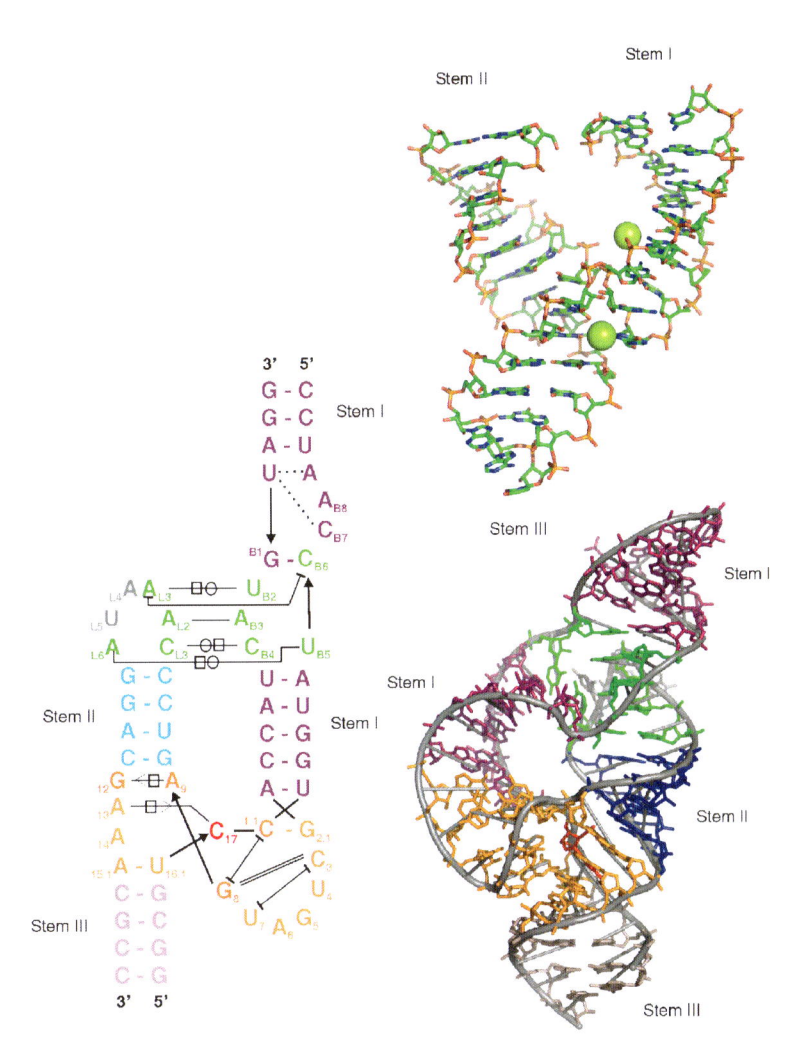

PLATE 3 Top right: The Y-structure of a minimal hammerhead construct (PDB 301D). Left: Sequence, secondary structure and tertiary interactions of the *Schistosoma mansoni* ribozyme. Stems I, II, and II are purple, blue and lilac, respectively. Nucleotides involved in tertiary interactions are green. The catalytic core is orange and the cleavage site is red. Thick black arrowed lines denote backbone continuity and thin lines show tertiary interactions; T-termini represent stacking interactions and ○□/□○ denotes a Watson-Crick/Hoogsteen interaction and □▷/◁□ is a Hoogsteen/sugar edge interaction. Reproduced with permission from reference 190 © Wiley-VCH Verlag GmbH & Co. KGaA. Bottom right: Solid state structure drawn with the color notation indicated above (PDB 2goz)

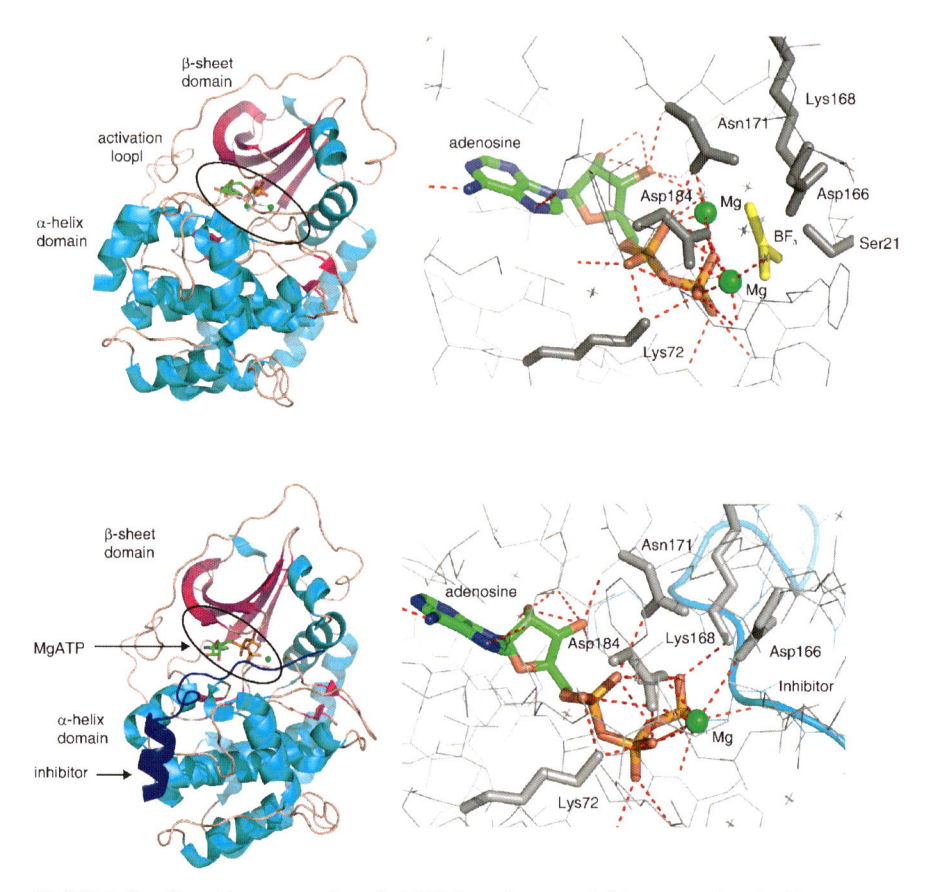

PLATE 4 Top: Transition state analog of cAMP dependent protein kinase complexed with ADP, two Mg^{2+} ions and BF_3 (left) as well as a blowup of the active center (right) showing polar interactions with essential side chains (PDB 1L3R). Bottom: An engineered variant of cAMP dependent protein kinase complexed with $MgATP^{2-}$ and the inhibitor peptide fragment 5–24 (left) and a blowup of the active site (right). Drawn from the PDB file 1Q24

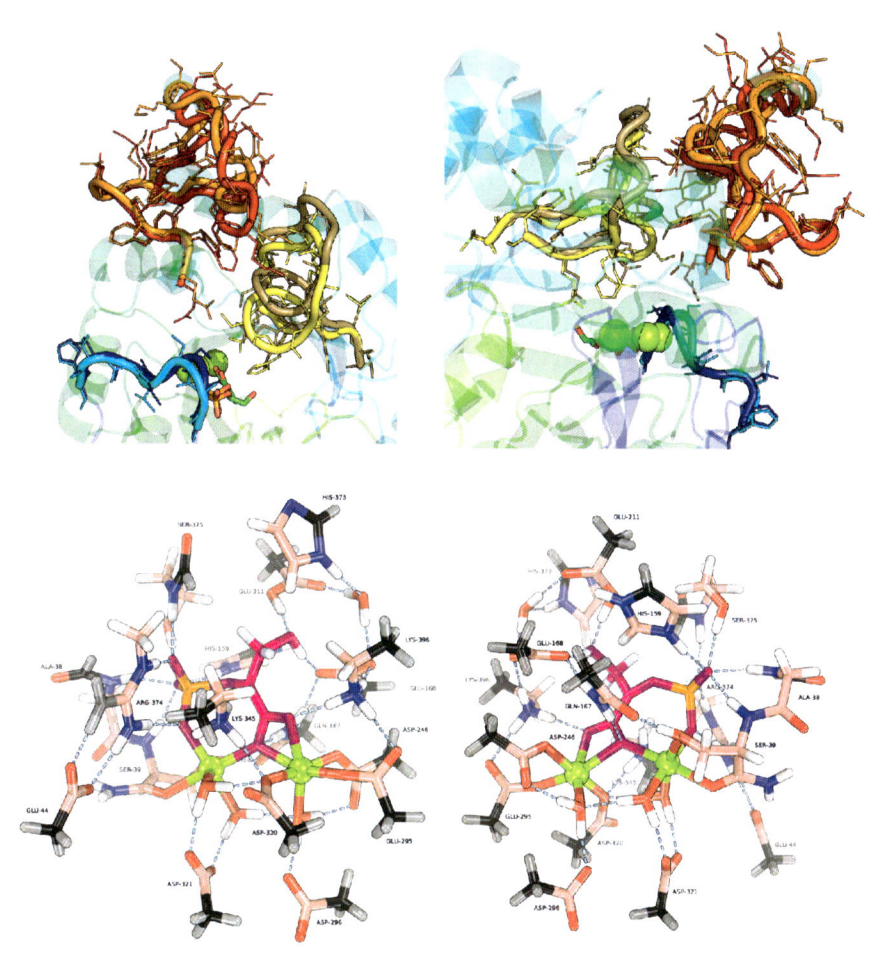

PLATE 5 Loop movement in the active site of yeast enolase. Upper left: 'closed' conformation (PDB 2AL1) superimposed upon the 'open' conformation (PDB 1P43). Upper right: view from the back. Lower left: A quantum chemical 'soccer ball' model for yeast enolase illustrated on the enol-intermediate and calculated at the TPSS(MARI-J;COSMO)/SV(P) level of theory. Lower right: view from the back

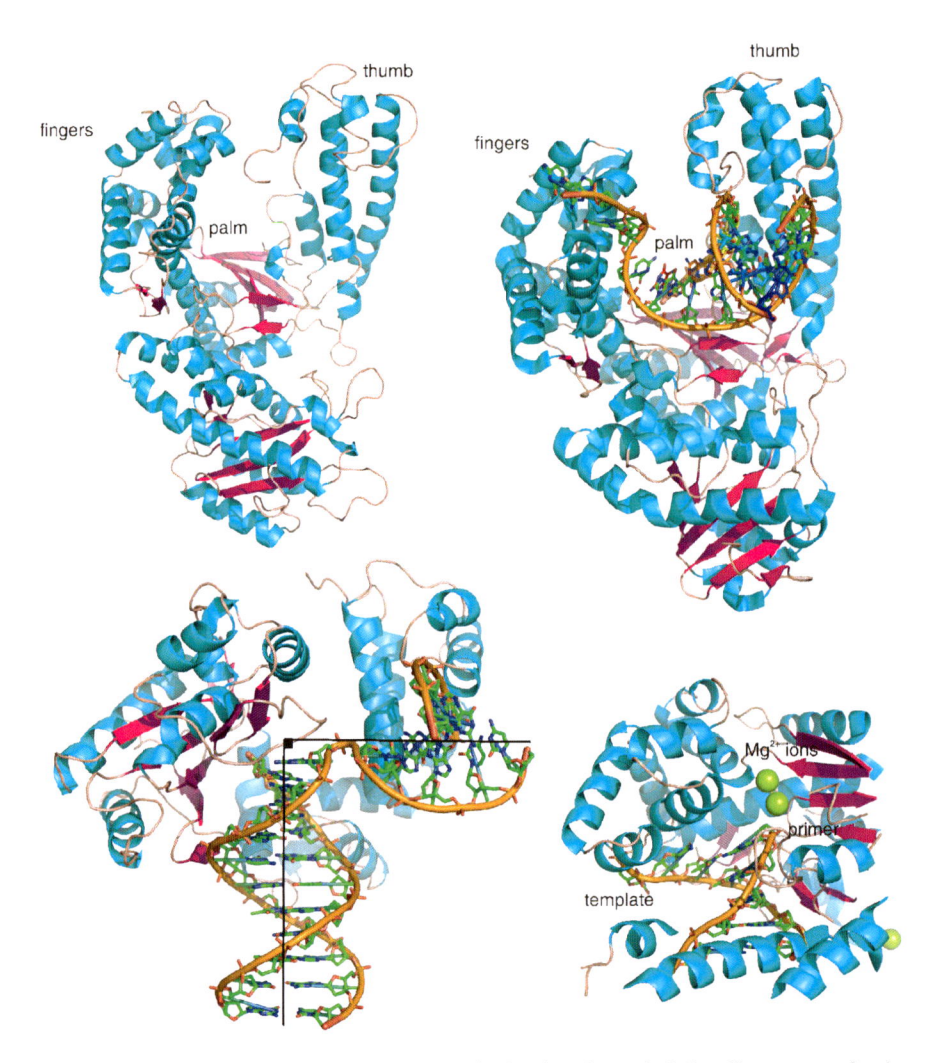

PLATE 6 Top left: Klenow fragment (engineered) showing the typical "hand" structure of polymerases (PDB 2KZZ). Top right: Klenow fragment of *E. coli* polymerase I (*Bacillus stearothermophilus*) complexed with 9 base pairs of duplex DNA (PDB 1L3S). Bottom left: Human DNA polymerase β complexed with a gapped DNA inhibitor showing the typical 90° orientation of the template to the growing replicant (PDB 1BPX). Bottom right: Active site of T7 DNA demonstrating the position of the metal ions with respect to the primer and template (PDB 1T7P)

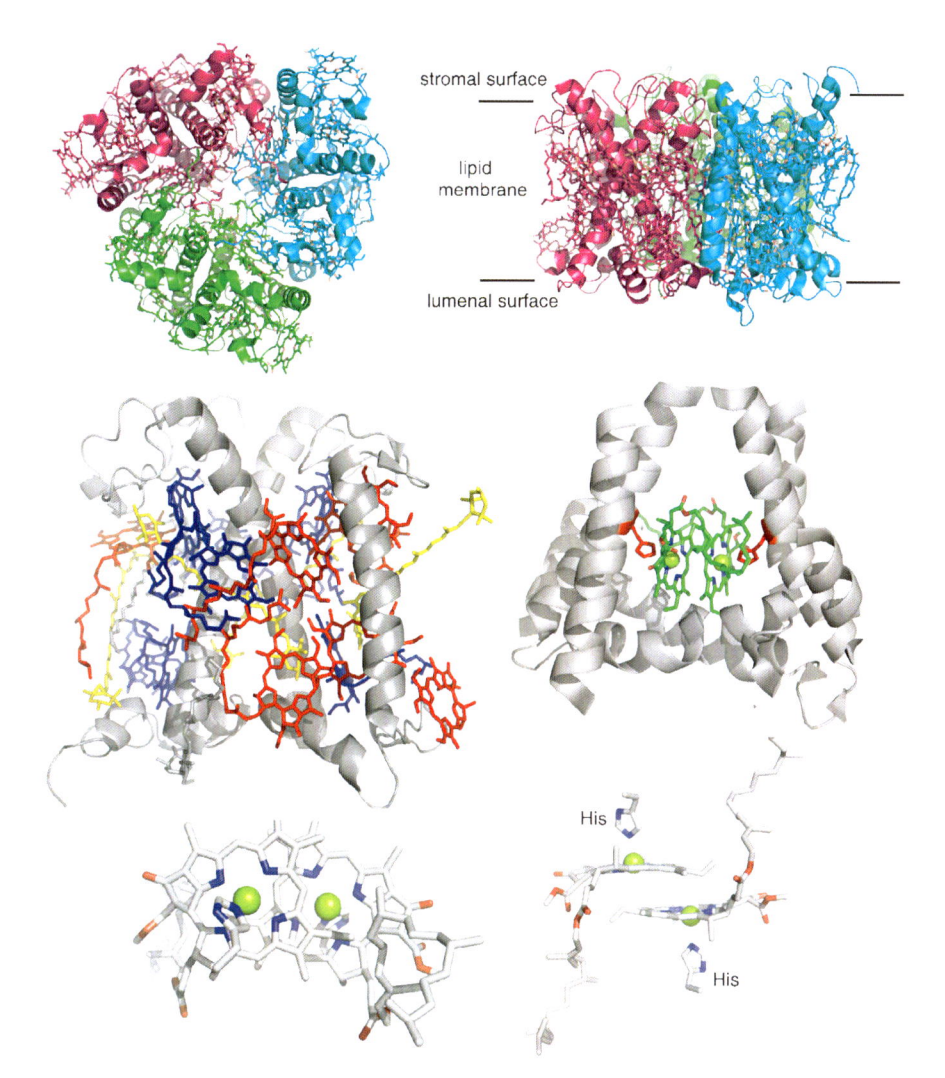

PLATE 7 Top left: Plant (pea) light harvesting complex LHC-II—view from the lumenal membrane surface (PDB 2BHW). Top right: View from the side. Middle left: One sub unit of LHC-II showing the placement of chlorophyll a (blue), chlorophyll b (red) and carotinoid cofactors (yellow). Middle right: The primary electron donor P700 and its protein environment in the photosystem I of *Synechococcus* elongates (PDB 1JB0). Bottom left: A closer view (from the top) of P700. Bottom right: View from the side

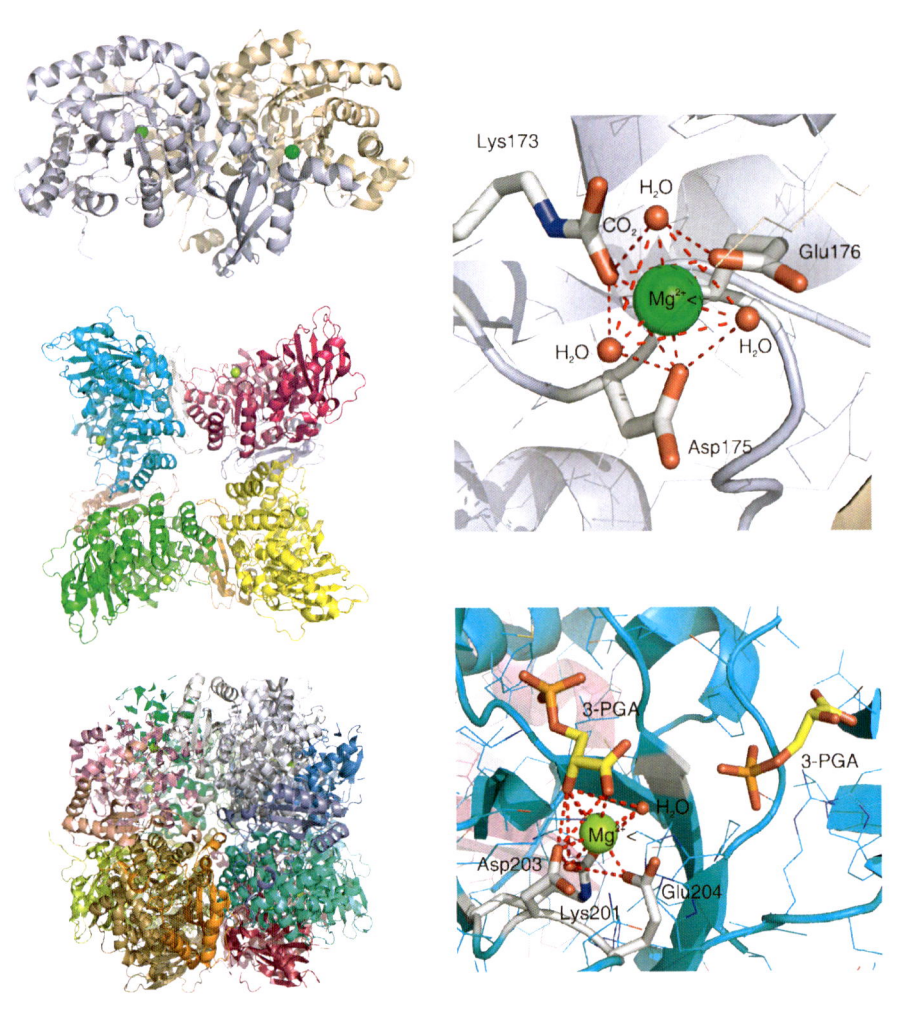

PLATE 8 Top left: Structure of a rubisco-like protein showing the basic dimeric unit (PDB 2OEK). Top right: Active site of this rubisco-like protein showing conserved residues and the coordination sphere of Mg^{2+}. Middle left: Structure (dimer of dimers) of activated spinach rubisco complexed with product (PDB 1AA1). Bottom right: Active site of this spinach rubisco product complex. Bottom left: Structure of activated green algae rubisco (tetramer of dimers)

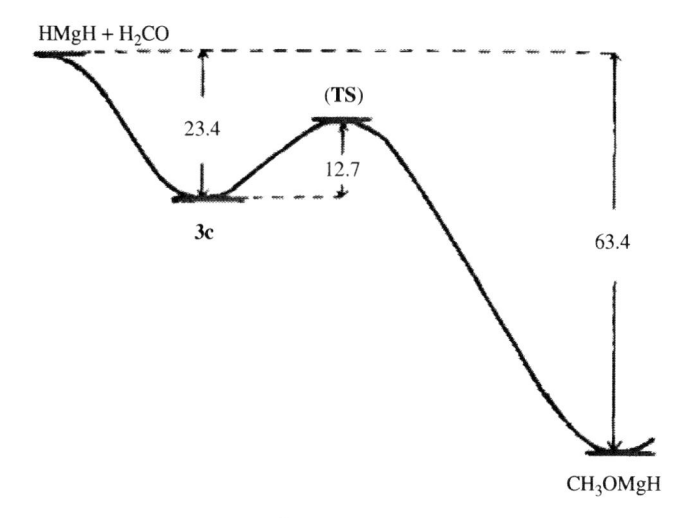

FIGURE 8. The energy profile (kcal mol^{-1}) for the HMgH + H$_2$CO reaction along a polar pathway. Reprinted from Reference 23, copyright 1982, with permission from Elsevier

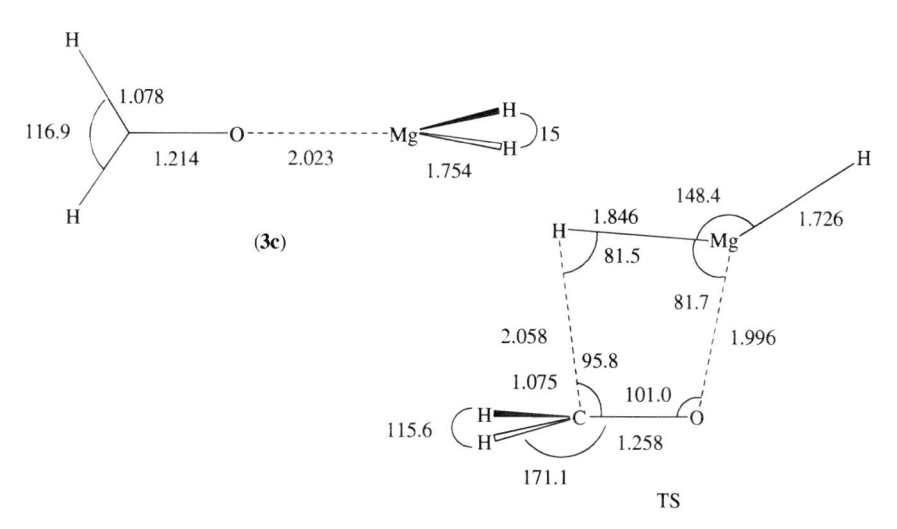

FIGURE 9. Optimized geometries in Å and deg. for an intermediate (**3c**) and a transition state (TS). Reprinted from Reference 23, copyright 1982, with permission from Elsevier

structure can be described as a four-membered ring (TS in Scheme 8). The inclusion of a second equivalent of CH$_3$MgCl, corresponding to an intermolecular mechanism, decreases the barrier height, and the process can be considered as an assisted intermolecular mechanism: the first equivalent forms the chelate structure and the second CH$_3$MgCl carries out the nucleophilic addition to the carbonyl group. The most favorable pathway corresponds to an intermolecular mechanism via an *anti* attack. Analysis of the results reveals that

FIGURE 10. Comparisons of experimental and calculated *anti* to equatorial isomer ratios

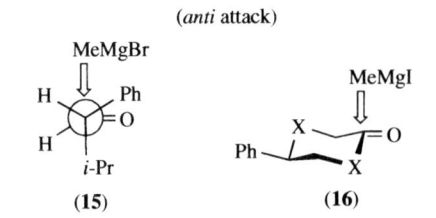

Substrate		Stereochemistry	
		Exptl.	Calc.
15		45:55	36:64
16a	X=CH$_2$	45:55	68:32
16b	X=O	98:2	94:6
16c	X=S	7:93	3:97

SCHEME 8. Schematic representation for the chelate-controlled addition of an organometallic reagent (M—R) to the carbonyl group of a chiral α-alkoxy carbonyl compound (17). Two diastereomers 18 with different orientation of R with respect to CH$_2$R^3 can be obtained. The *syn* diastereomer is obtained when the nucleophilic attack of R takes place on the same face of the plane, defined by the carbonyl group and the R-substituted carbon atom, where CH$_2$R^3 is located in the chelate complex

the nature of transition structures for the intramolecular and intermolecular mechanisms is a rather robust entity. There is a minimal molecular model with a transition structure which describes the essentials of the chemical addition process, and the corresponding transition vector is an invariant feature.

The following results were derived from those calculations for the paths shown in Figure 11 (see also Scheme 9).

SCHEME 9. 2-Hydroxypropanal (19), 3-hydroxy-2-butanone 20 and 3,4-di-O-methyl-1-O-(trimethylsilyl)-L-erythrulose (21), used as the carbonyl substrate in 17

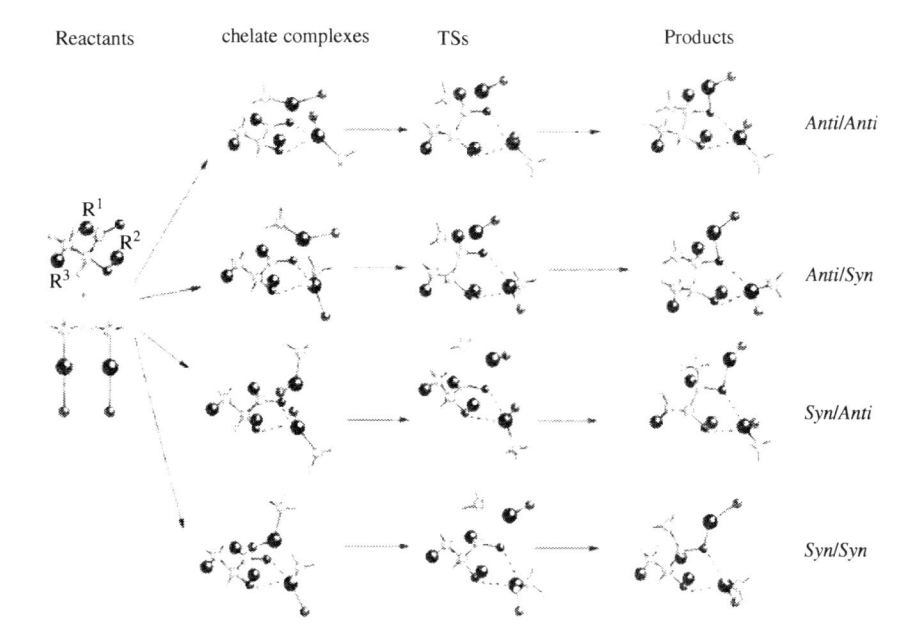

FIGURE 11. Reactants, chelate complexes, TSs and products for the *anti* (two upper paths) and *syn* (two lower paths) addition of CH_3MgCl to 19, 20 and 21 with the participation of 2 equivalents of CH_3MgCl. The relative orientation of the methyl group of the chelating CH_3MgCl can be *anti* (first and third paths) or *syn* (second and fourth paths) with respect to the CH_2R^3 group of the model. Reprinted with permission from Reference 25. Copyright 1996 American Chemical Society

(i) The formation of *syn* and *anti* chelate complexes is the first step in the addition process and takes place without an energy barrier.

(ii) The magnesium is coordinated to the lone pair of the carbonyl oxygen and to the methoxy oxygen. The chelate complexes can be described as puckered five-membered rings.

(iii) The chelate conformation is maintained throughout the reaction path, being the thermodynamic controls for the *syn* and *anti* pathways dominated by the relative stability between the corresponding chelate complexes and products. Cram's model based on chelation-controlled carbonyl addition can explain the energetic results.

(iv) The C—C bond-forming stage is the second and rate-limiting step for the addition process. The TSs are four-membered rings, corresponding to the 1,3-intramolecular migration from the chelate complex to products. The calculations[25] adopted models of α-hydroxy aldehyde **19** and ketones (**20** and **21**) along with the CH_3MgCl dimer. The dimer-participating reactions were computed to be favorable energetically. However, one CH_3MgCl molecule is retained as a chelate complexed catalyst. The dimer formed in the Schlenk equilibrium seems to be more active in general Grignard reactions.

V. A COMPREHENSIVE COMPUTATIONAL STUDY OF GRIGNARD REACTIONS

In spite of various theoretical studies of Grignard reactions shown in the previous section, their mechanisms seem to be as yet unsettled. In particular, the connection between the R^1MgX dimer in the Schlenk equilibrium and its reactivity toward carbonyl groups is still unclear. In this section, a systematic computational study of several Grignard reactions is presented[27] in order to reveal their unclear points. The polar vs. SET problem will be explained in a forthcoming new mechanism.

A. Computational Methods

Geometries were fully optimized by the B3LYP density functional theory (DFT) method[28] together with the SCRF[29] solvent effect (dimethyl ether Me_2O, dielectric constant $= 5.02$). The basis set used is 6-31G*. Vibrational frequency calculations gave a sole imaginary frequency for all transition structures, which verifies that the geometries obtained are correctly of the saddle point. From the reactant precursor, partial geometry optimizations were repeatedly carried out with fixed Mg–O and C–C distances. Through the partial optimizations, an approximate transition state (TS) structure could be obtained. Next, by use of the approximate structure and the force constants (second derivatives of total energies), TS geometries were determined. In this case, the negative values (ca -0.05 hartree au^{-2}) of the Hessian diagonal force constants of the bond-forming Mg–O and C–C distances should be included in the input line. All calculations were performed using GAUSSIAN 98[30].

B. Structures of Grignard Reagents

Grignard reagents (R^1MgX) in ether solution form aggregates[3b, c, 21]. The degree of aggregation depends on the halogen (X), the concentration, the alkyl group R^1 and the solvent. For simple alkyl or aryl magnesium chlorides in diethyl ether, the predominant species is considered to be a solvated halogen-bridged dimer[3f, g]. As described in Section III, Axten and coworkers revealed by *ab initio* calculations that the dimer of MeMgCl is much more stable than the monomer in calculations which do not take the solvent into account[22].

The dimers of the Schlenk equilibrium (Scheme 10) have been investigated by B3LYP/6-31G* calculations. The coordination of the dimethyl ether solvent was included along with the SCRF solvent effect. In Figure 12, three constitutional isomers of (Me-Mg-Cl)$_2$(Me$_2$O)$_n$ ($n = 2$ and 4) are compared. For both $n = 2$ and 4, **9a** and **9b** are most stable and have two bridged Cl atoms. This result is in accord with Axten's results[22], although our present model with ether molecules is more realistic. The structure shows good resemblance to the crystal structures of halogen-bridged dimers $(MgX)_2$[31].

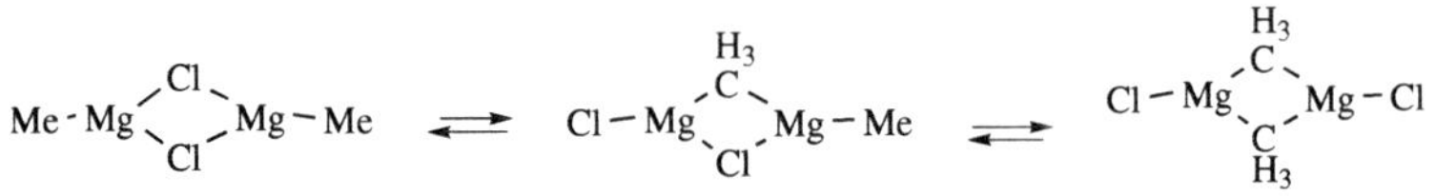

SCHEME 10. Schlenk equilibrium for methylmagnesium chloride

(a) (Me-Mg-Cl)$_2$·(Me$_2$O)$_2$

(9a) 0 kcal mol^{-1} (most stable)

(10a) +4.4 kcal mol^{-1}

(11a) +8.4 kcal mol^{-1}

(b) (Me-Mg-Cl)$_2$·(Me$_2$O)$_4$

(9b) 0 kcal mol^{-1} (most stable)

(10b) +4.7 kcal mol^{-1}

(11b) +6.9 kcal mol^{-1}

FIGURE 12. Three geometric isomers of Schlenk equilibrium (Grignard reagents) with two solvent molecules (a) are shown in the upper row. Those with four solvent molecules (b) are shown in the lower row. Relative energies to **9a** and **9b** are also shown, where the positive values correspond to less stable systems. The stability of the dimer **9a** relative to that of two monomers, MeMgCl·Me$_2$O + MeMgCl·Me$_2$O → (MeMgCl)$_2$·(Me$_2$O)$_2$ (**9a**), is also calculated. The energy change ΔE_{rel} for dimerization is -23.1 kcal mol^{-1}

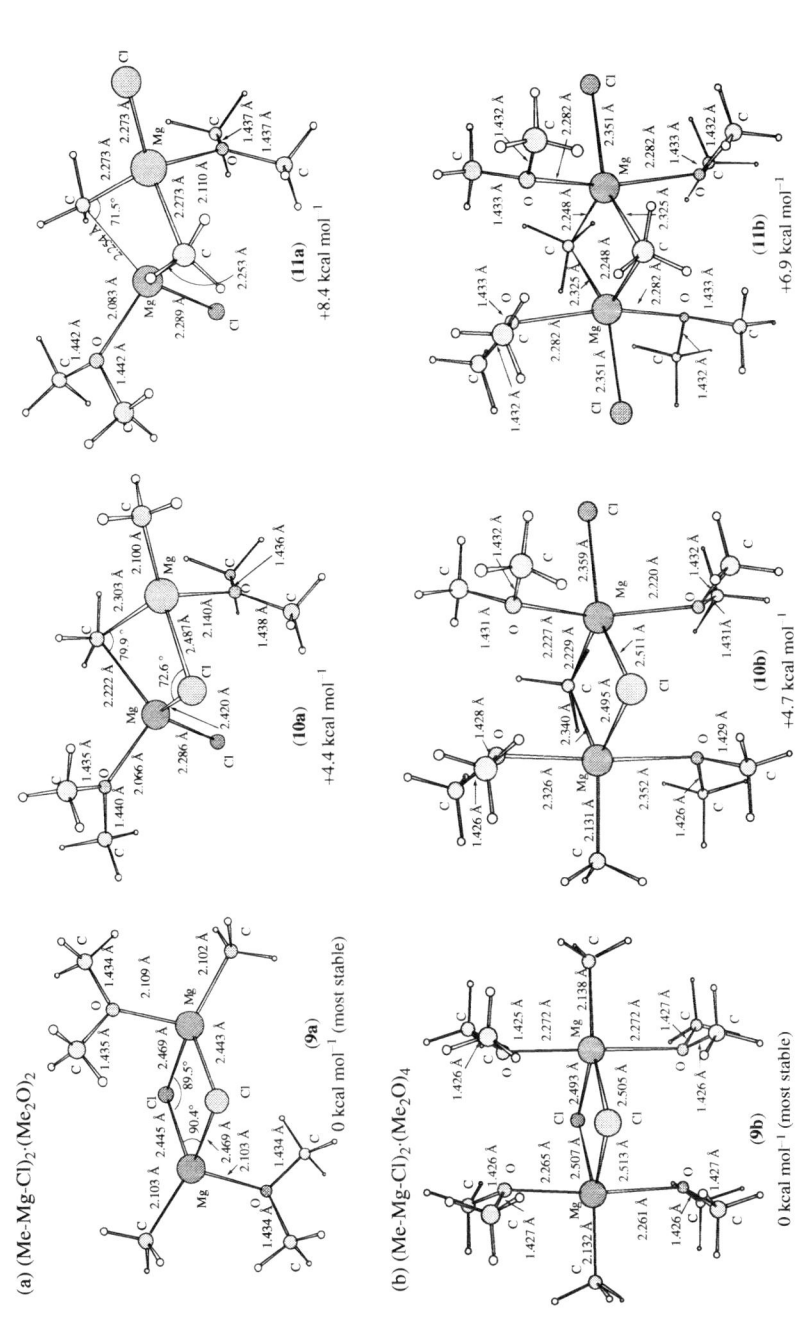

In order to determine how many ether molecules are favored by the Schlenk dimer, MeMgCl$_2$MgMe in Scheme 10, the geometry of **9b** (four ethers) is compared to that of **9a** (two ethers). In **9a**, two Mg–O distances (2.103 Å and 2.109 Å) are close to that (2.104 Å) of the MgO ionic crysral. In **9b**, they are 2.265 Å, 2.261 Å, 2.272 Å and 2.272 Å and are larger than those in **9a**. In spite of the large Mg to O affinity, two Mg atoms do not favor the coordination of four ether molecules. Thus, **9a** is a saturated complex, although there seems to be room on the two Mg atoms for further nucleophilic coordination. Mg atoms seem to persist in tetra-coordination. Ether solvation of the Schlenk equilibrium species does not block reaction channels completely.

In order to confirm the saturation in **9a**, free-energy changes for the following stepwise clustering reactions were calculated:

$$(\text{Me-Mg-Cl})_2(\text{Me}_2\text{O})_{n-1} + \text{Me}_2\text{O} \rightarrow (\text{Me-Mg-Cl})_2(\text{Me}_2\text{O})_n$$

When the addition of the n-th ether molecule gives a negative $\Delta G_{n-1,n}$ value, the addition is favored. In contrast, a positive $\Delta G_{n-1,n}$ value means that the n-th molecule is not bound to the $(n-1)$ cluster. The calculated values are $\Delta G_{0,1} = -8.2$, $\Delta G_{1,2} = -4.8$, $\Delta G_{2,3} = +7.1$ and $\Delta G_{3,4} = +10.9$ kcal mol^{-1} as shown in Scheme 11; $\Delta H_{n-1,n}$ values are also given in Scheme 11. The addition of the third and fourth ether molecules are less exothermic ($\Delta H_{2,3} = -5.6$ and $\Delta H_{3,4} = -0.9$ kcal mol^{-1}), which are overcome by entropy changes leading to the positive $\Delta G_{2,3}$ and $\Delta G_{3,4}$ values.

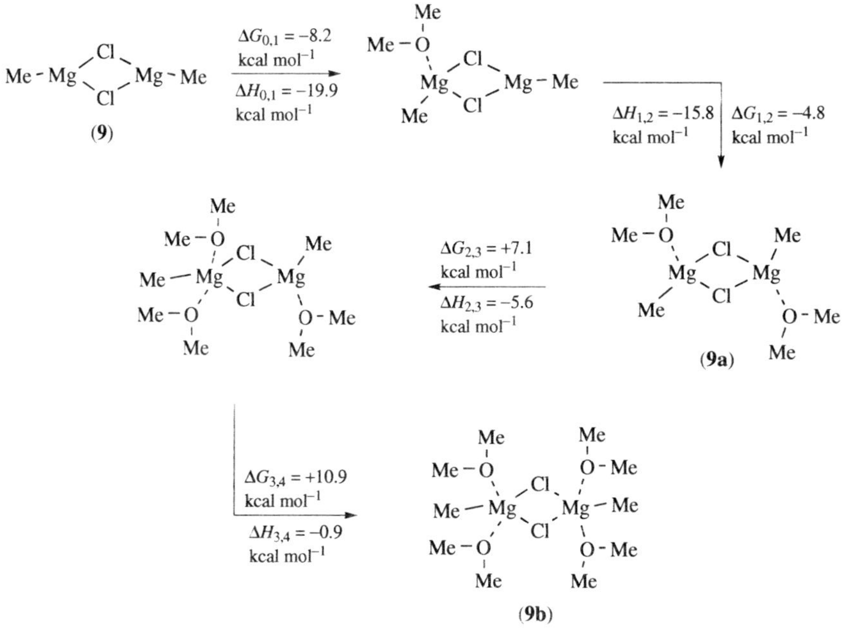

SCHEME 11. Free-energy differences $\Delta G_{n-1,n}$ and enthalpy differences $\Delta H_{n-1,n}$ for the successive addition (see above) of Me$_2$O molecules to (Me-Mg-Cl)$_2$.

Clearly, the third and fourth molecules cannot be bound to the (Me-Mg-Cl)$_2$(Me$_2$O)$_2$ cluster, **9a**. Therefore, **9a** is a saturated shell. Since the double trigonal–bipyramidal geometry of **9b** ($n = 4$) in Figure 12 does not involve significant steric congestion, expulsion of $n = 3$ and 4 ether molecules arises from the poor ability of Mg atoms for the fifth coordination. Among Schlenk equilibrium dimers in Figure 12, **9a** is a most likely reactant for Grignard reactions.

C. Additions of Grignard Reagents to Carbonyl Compounds without Solvent Molecules

The SET mechanism is known to be operative for reactions of Grignard reagents and aromatic ketones such as benzophenone[3f, h, k, 11]. In reactions of Grignard reagents and aliphatic ketones and aldehydes, the polar mechanism seems to be major[3b, c]. The reactions of aliphatic ketones and aldehydes are more widely utilized in organic syntheses. Thus, at first, the polar addition mechanism was examined.

First, a mechanism without solvent molecules was examined for simplification and initial formation of a carbonyl–Mg atom complex was expected. The previously proposed polar addition mechanism is shown in Scheme 12 [7, 8], which has been quoted widely[32].

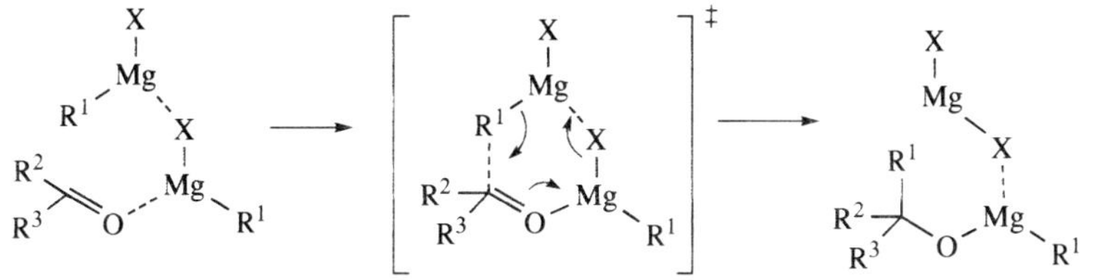

SCHEME 12. Termolecular polar mechanism proposed by Swain and Boyles[7] and Ashby and coworkers[8]. The first reaction in Scheme 2

The potential intermediacy of precursor **22** (a model of **3**) was investigated. However, the initial geometry of the model **22** converged by geometry optimizations to that of MeMgCl$_2$MgMe−O=CH$_2$ (**23**).

The convergence arises from the stability of the two chlorine-bridged structures in the Schlenk equilibrium (i.e. **9** in Scheme 10). Thus, the proposed concerted mechanism containing a cyclic transition state in Scheme 12 is unlikely for Grignard reactions. This result is in contrast to the carbonyl additions with organolithium reagents[33]. The formation of a 1:1 complex was reported for the reactions of Grignard reagents and ketones[3i, j]. Also, stoichiometric amounts of Grignard reagent to carbonyl compounds are generally enough. Therefore, an intermediate model **24** consisting of MeMgCl$_2$MgMe and two formaldehyde molecules was calculated. Reactions between the Schlenk dimer, (MeMgCl)$_2$, and two formaldehyde molecules are shown in Scheme 13. The (**24** → **25** → **26**) process is shown in Figure 13 and the (**26** → **27** → **28** → **29** → **30**) process in Figure 14. Geometries of

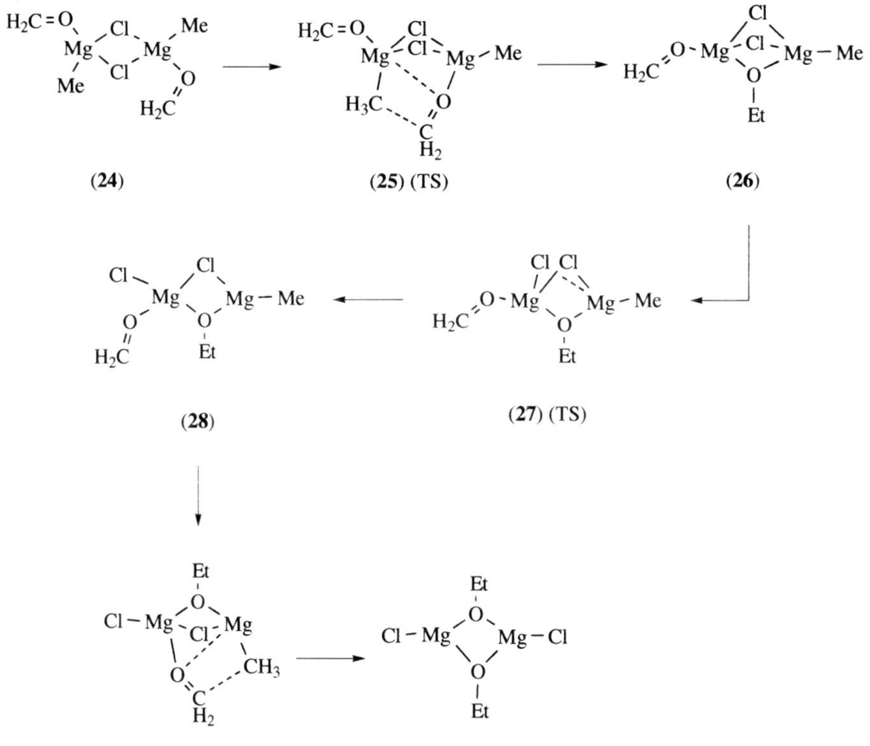

SCHEME 13. Reactions between the Schlenk dimer, (MeMgCl)$_2$, and two formaldehyde molecules

the bromide analogue of Figure 13 will be shown in Figure 15. Solvent ether molecules will be taken into account in Figure 16.

In the reactant-like complex **24**, H$_2$C=O molecules are bound to Mg atoms. One H$_2$C=O is shifted leftward to be linked with the left methyl group. At the same time, the carbonyl oxygen of H$_2$C=O is directed to the left Mg atom. Thus, a concerted C–C and O–Mg bond formation is shown in TS **25**. Reaction-coordinate vectors in Figure 17 indicate the formation clearly. A C–C covalent bond (1.525 Å in **26**) is established, and the original dichlorine bridge **24** in the square is replaced by the dichlorine and oxygen bridges. The two strong Mg–O bonds (1.943 Å and 2.008 Å) in the bridge of **26** are reflected by a large exothermic energy (−49.3 kcal mol^{-1}). The triply-bridged structure of **26** is presumably caused by preference of tetra-coordination of magnesium atoms. The reaction pathway **24** → **25**(TS) → **26** would be very important among the polar Grignard reactions; the minimum essential is a four-center reaction (Scheme 14).

For the addition transition state structure **25**, a remarkable feature has been found: the bond-forming carbonyl carbon and the Me group do not reside on the same magnesium atom but on the vicinal magnesium atom of the bridged dimer. The di-Cl- and O-bridged product **26** has newly formed Mg–O and C–C bonds. The energy barrier for this process (**24** → **25**(TS)) is very small (+2.4 kcal mol^{-1}).

The first addition product **26** can proceed to the second addition stage (Scheme 13 and Figure 14). The intermediate **26** transforms to **28** through bridged-Cl opening TS **27**. **28**

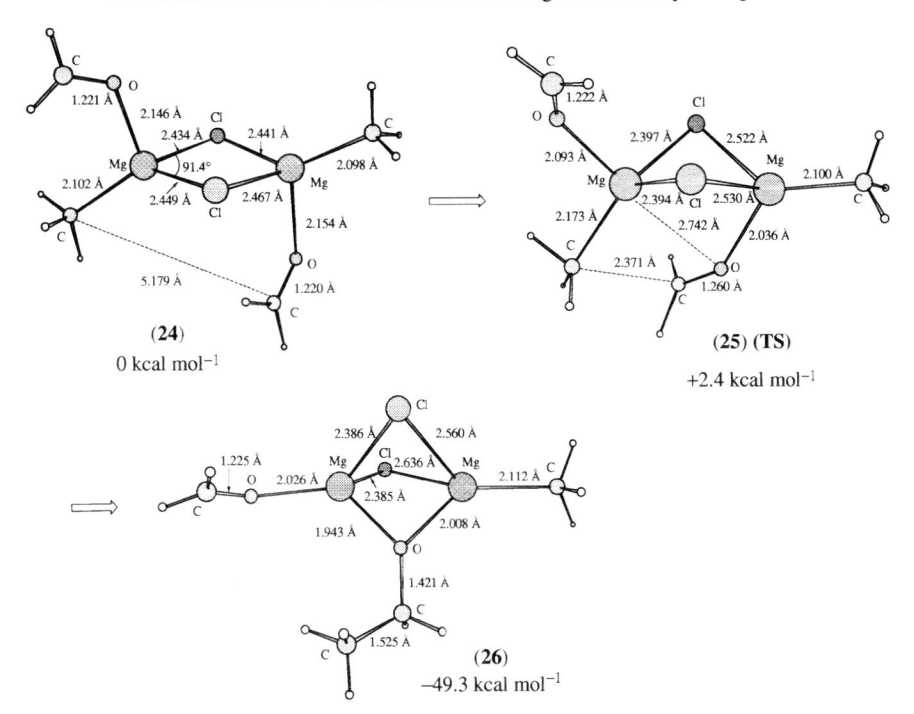

FIGURE 13. Carbonyl addition 1:1 complex (no solvent molecule) (the first stage) corresponding to that of Scheme 13. The geometry of **24** was determined in the intrinsic reaction coordinate to **25** (TS). A symmetric reactant geometry (C_{2h}) is 1.8 kcal mol^{-1} less stable than **24**

undergoes the second addition step (**28** → **29**(TS) → **30**), similar to the first addition step (**24** → **25**(TS) → **26**). The final product **30** is highly stable due to the four Mg−O bonds. Although the energy barrier for the second addition step (+12.9 kcal mol^{-1}) is larger than the first addition step (+2.4 kcal mol^{-1}), transformation of **28** to **30** is a highly exothermic process ($-97.5 - (-52.6) = -44.9$ kcal mol^{-1}).

The calculation of the bromide analog for the first critical addition step **24-Br** → **25-Br** (TS) → **26-Br** was also carried out for comparison with that in (Me-Mg-Cl)$_2$(H$_2$C=O)$_2$ of Figure 13 (Figure 15). The structure of the intermediates, and transition state and energy barriers, are similar to those of the chloride models. This similarity suggests that the addition of bromo-Grignard reagent requires dimeric species. Although Grignard reagents are known to be in a different aggregate state depending on halogen atoms[3f, g], this result shows that the reactivity toward carbonyl compounds does not depend on whether the halogen is Br or Cl. The high reactivity of Grignard reagents toward carbonyl compounds can be understood by the model of Scheme 14 starting from the coordination of the magnesium to the carbonyl oxygen and transforming the C−Mg to strong C−C and O−Mg bonds.

D. Additions with Solvent Molecules

The polar addition process was investigated also by calculation with dimethyl ether molecules as a more realistic system. Two H$_2$C=O molecules were added to the sole

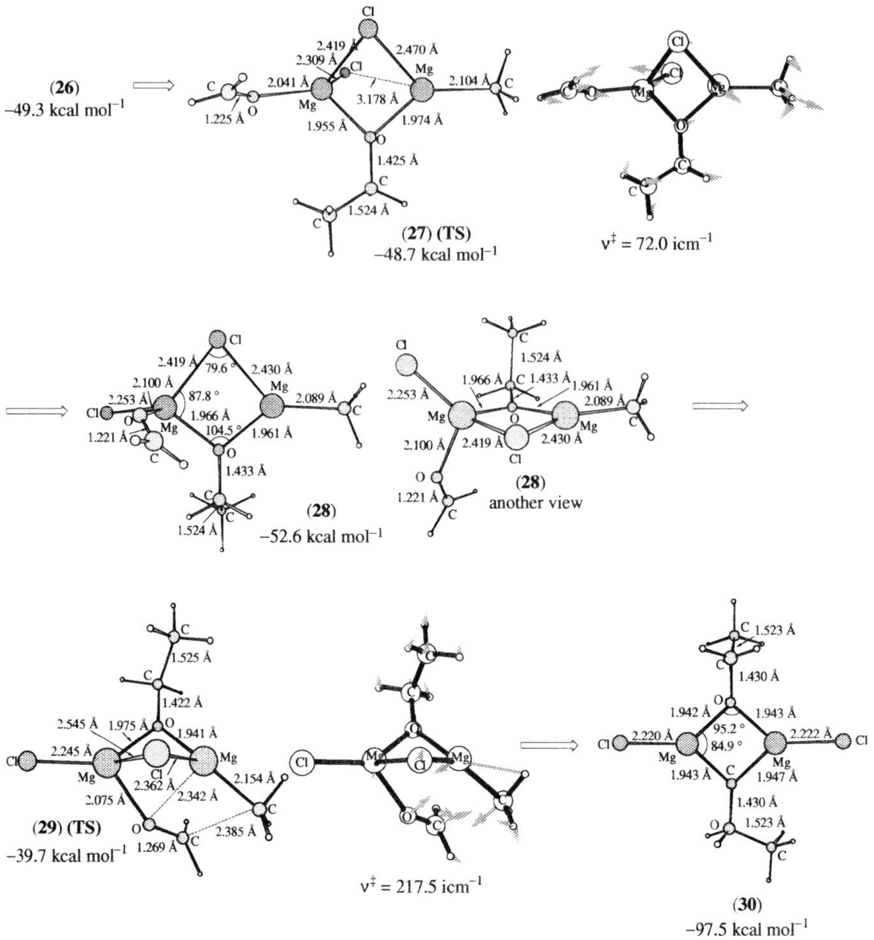

FIGURE 14. Carbonyl addition 1:1 complex (no solvent molecules) (the second stage) corresponding to that of Scheme 13

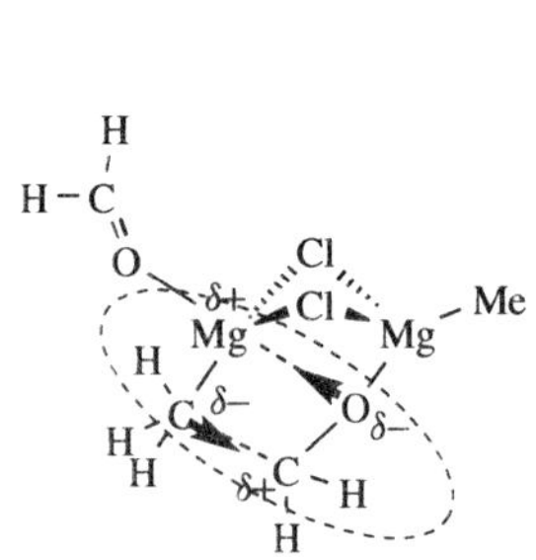

SCHEME 14. A four-center reaction in **25**(TS)

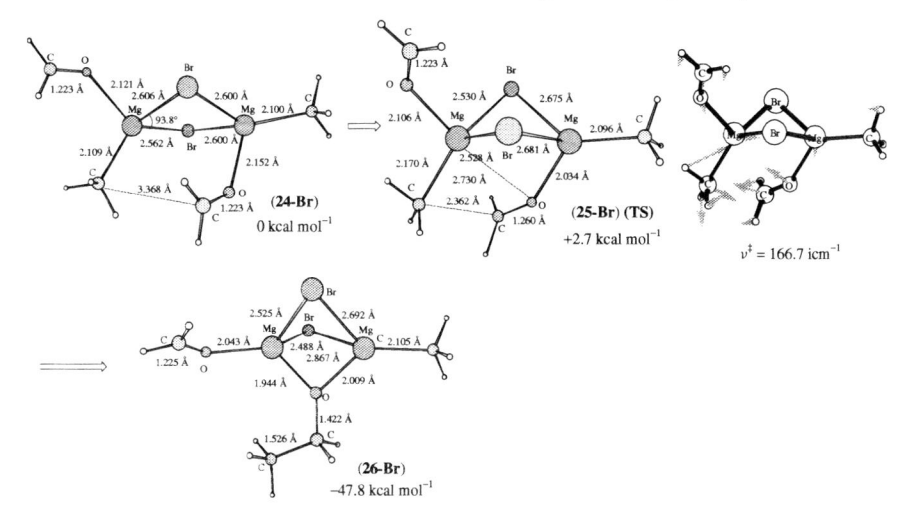

FIGURE 15. The first reaction channel in $(Me-Mg-Br)_2(H_2C=O)_2$, 24-Br → 25-Br (TS) → 26-Br

reactant **9a** in Scheme 11 and Figure 12. The dimer **31**, the transition state **32** and a product **33** were calculated (Scheme 15). Their geometries are shown in Figure 16. The structure obtained for the precursor **31** has dimeric five-coordinated magnesiums. The structure is similar to the reported X-ray structure having dimeric five-coordinated magnesium atoms[34]. As in the non-solvated model system (**25**(TS) in Figure 13), the bond-forming carbonyl carbon and Me group reside on vicinal magnesium atoms of the bridged dimer in the addition transition state structure **32**(TS) (see the reaction-coordinate vectors in Figure 17). The energy barrier for this process is very small ($+0.3$ kcal mol^{-1}) and shows that the process is very facile. Conversion from **31** to the product **33** takes place exothermally (-69.3 kcal mol^{-1}). Thus, the dichlorine-bridged four-membered structure is retained in the transition state.

In Figure 16, the Me$_2$O molecules do not affect the polar reaction path significantly in comparison with the path in Figure 13. However, some bond distances are appreciably different between the two precursors **24** (Figure 13) and **31** (Figure 16). A Mg–Cl distance is 2.449 Å in **24**, while it is 2.739 Å in **31**. When the Mg–OMe$_2$ distances of **9a** (Figure 12) and **31** are compared, the distance of 2.368 Å in **31** is larger than that (2.109 Å) in **9a**. These results indicate that the tetravalent Mg atoms form tight covalent and coordination bonds and the pentavalent Mg atoms form somewhat loose chemical bonds.

E. Chelation-controlled Addition Models

The high diastereoselectivity for the addition of Grignard reagents to carbonyl compounds is explained by the proposed four-centered process (Scheme 14). Reactions of chiral carbonyl compounds (Scheme 16a) and chiral Grignard reagents (Scheme 16b) are examined. The examples in Scheme 16a are stereoselective addition reactions of Grignard reagents to chiral α-alkoxy ketones[4a]. Some other examples of stereoselective Grignard reagent addition to various α-alkoxy carbonyl derivatives[4b-d, 26] and a related chiral sulfinyl imine[35] were reported. The reaction examples of Scheme 16b are additions of chiral γ-alkoxy magnesium halides to ketones[36]. Some other examples of chelated

392   Shinichi Yamabe and Shoko Yamazaki

SCHEME 15. The first C—C bond-forming reaction of Figure 16

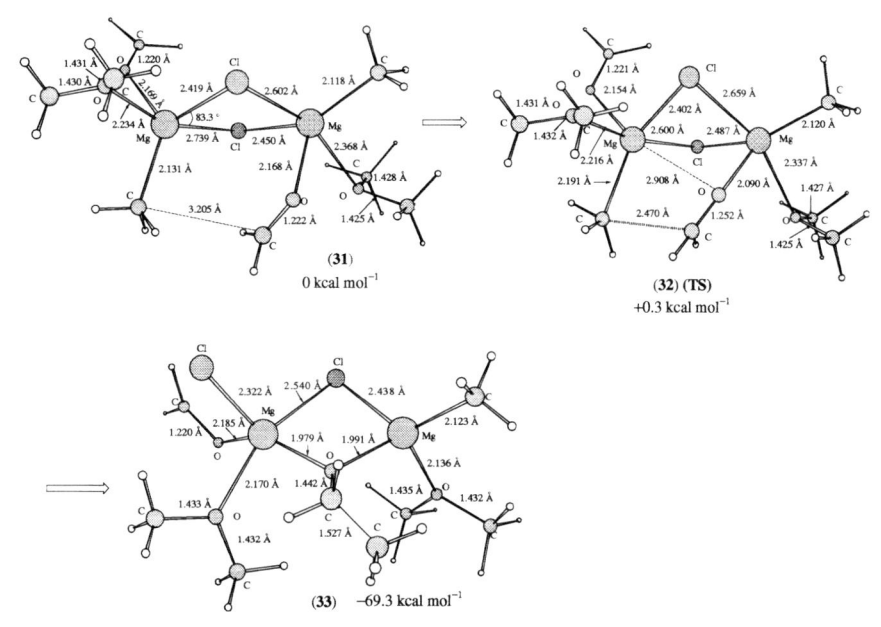

FIGURE 16. Carbonyl addition process (with two solvent molecules) corresponding to that of Scheme 15, $(MeMgCl)_2(H_2CO)_2(Me_2O)_2$

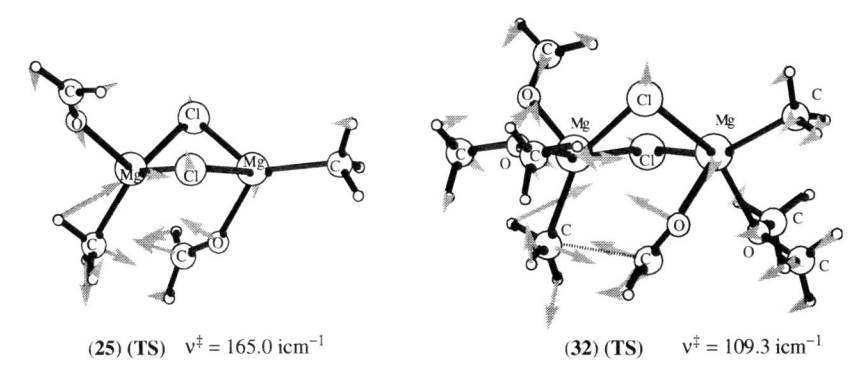

(25) (TS) $v^{\ddagger} = 165.0 \text{ icm}^{-1}$ **(32) (TS)** $v^{\ddagger} = 109.3 \text{ icm}^{-1}$

FIGURE 17. Reaction-coordinate vectors corresponding to the sole imaginary frequencies $v^{\ddagger}$ for **25**(TS) in Figure 13 and **32**(TS) in Figure 16

Grignard reagents, such as tetrahydroisoquinoline Grignard species discovered by Seebach and coworkers[37] and α-alkoxy Grignard reagents[38], were also reported. Because a stable coordination of an ether solvent to magnesium is shown in the model system of Figure 16, the coordination of the ether oxygen in substrates to magnesium may work effectively to form chelation. The proposed mechanism of chelation-controlled addition could be determined by a theoretical study.

SCHEME 16. The observed stereochemistry of the reported reactions by the use of (a) chiral ketones[4a] and (b) Grignard reagents[36] in Grignard reactions. Asymmetric carbon atoms are denoted by R or S

The addition paths of two separated MeMgCl molecules to the carbonyl group of chiral α-alkoxy carbonyl compounds[25], which are shown in Schemes 8 and 9 and in Figure 11 in Section IV, were traced. However, in these models, one MeMgCl molecule bridges two oxygen atoms and acts merely as a catalyst. Instead, the Schlenck dimer (MeMgCl)$_2$ should be considered for the carbonyl reactant coordination.

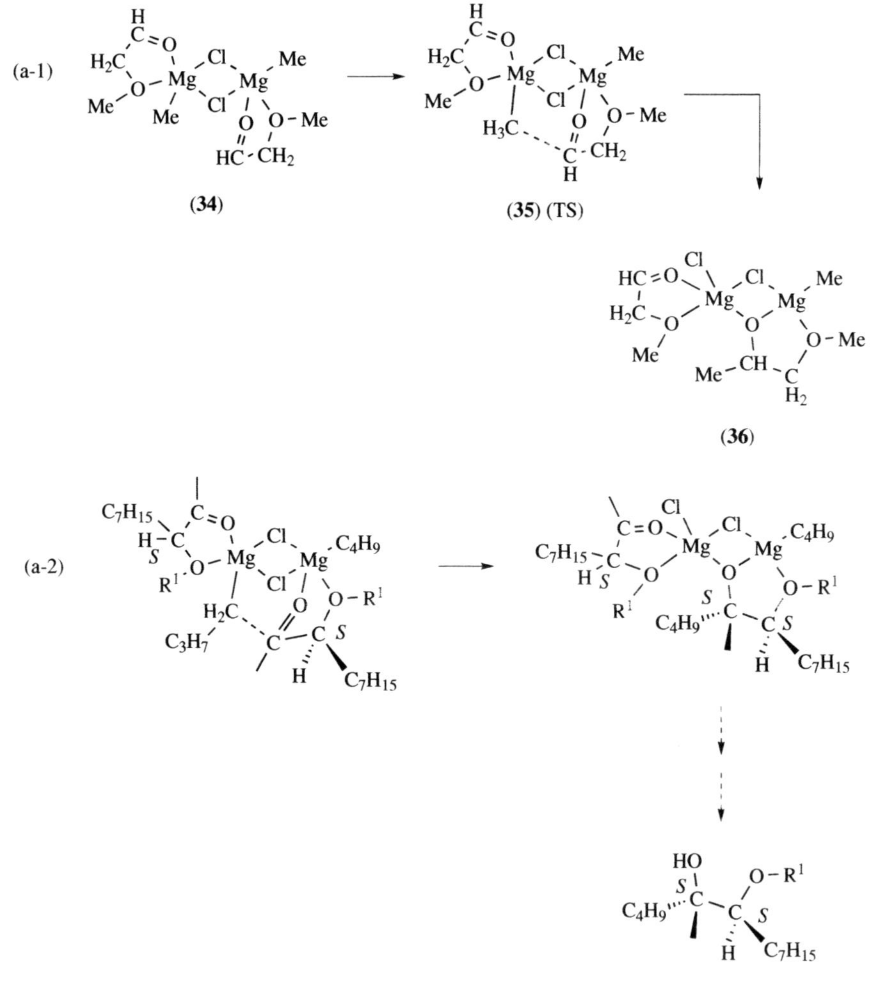

SCHEME 17a. Reaction of methylmagnesium chloride and 2-methoxyacetaldehyde (a-1) and its extension to the chiral reaction (a-2)

As an example of a reaction in Scheme 16a, the reaction between MeMgCl and 2-methoxyacetaldehyde as a model for chelated carbonyl compounds was investigated. By taking the dimeric forms in Scheme 11 and Figure 12, the dimeric addition process of (MeMgCl/OHCCH$_2$OMe) was examined. In Scheme 17a-1 and Figure 18a, a reaction pathway with effective chelation at Mg was obtained. Geometries of **34**, **35** and **36** are shown in Figure 18a. The polar addition precursor as a dimer **34**, transition state **35** and product **36** have features similar to that of the solvent-attached MeMgCl/O=CH$_2$ system in Figure 16. In the first addition transition-state structure **35**, the bond-forming carbonyl carbon and Me group reside on vicinal magnesium atoms of the bridged dimer. The dichlorine-bridged four-membered structure is retained in the transition state as well.

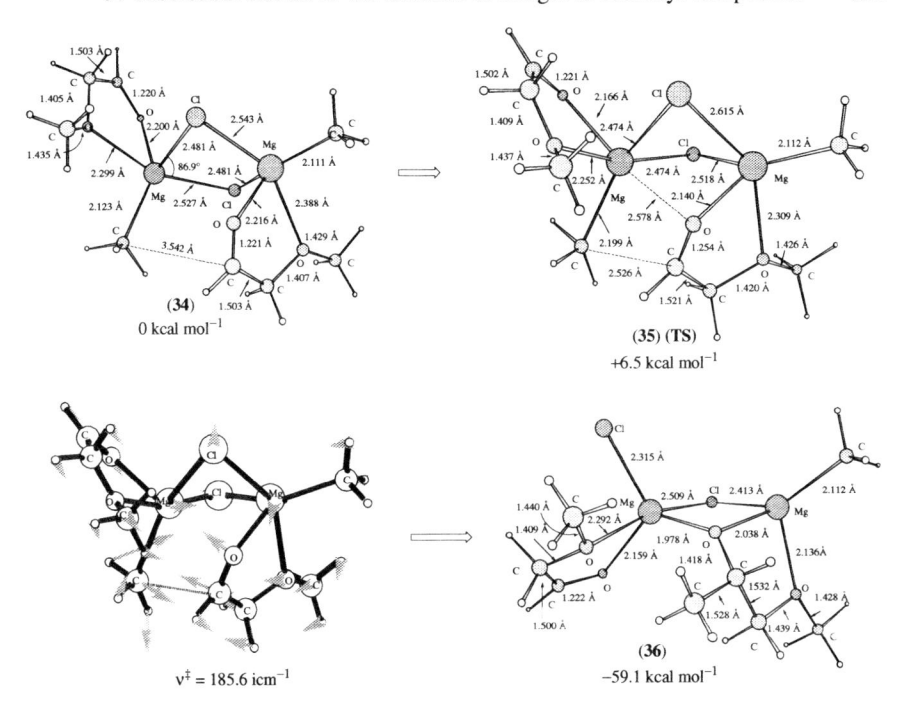

FIGURE 18a. Carbonyl addition process for the alkoxy carbonyl compound model corresponding to that of Scheme 17a-1

Owing to chelation of the OMe group, the aldehyde was fixed configurationally. Thus, the present model calculation has revealed that the chelation control results in the highly stereoselective addition processes observed. The reaction occurs via a dimeric form which has not been considered so far. This model uses achiral substrates and does not create asymmetric carbons. However, through replacement of the α-H of an aldehyde by a heptyl group, in addition to some other substitutions (Scheme 17a-2), the stereoselective pathway can be explained as follows. The C_7H_{15} group would be located on the outside of the dimer. The configuration is fixed by chelation. The nucleophile (n-C_4H_9 group) would attack from the less hindered H side. The stereochemical result is the same as in Cram's original concept[5]. The result from the extension of the model study is in accord with the major product stereochemistry in Scheme 16a.

As an example of Scheme 16b, the reaction between 3-methoxypropylmagnesium chloride which is a model for chelated Grignard reagents and formaldehyde was adopted. The first addition process of $(MeOCH_2CH_2CH_2MgCl/O=CH_2)_2$ was examined (Scheme 17b-1). A pathway with an effective chelation was also obtained in this model reaction and geometries of **37**, **38** (TS) and **39** are shown in Figure 18b. Thus, the example of Scheme 16b also shows that the effective chelation control works in a dimeric form as well as that of Scheme 16a. When the magnesium-substituted carbon configuration is fixed by chelation, stereoselection by steric effects could occur. In the example of Scheme 17b-2, the stereoselectivity is determined by the step of the formation of the conformationally stable Grignard reagent. Various remote steric effects based on the chelation control to create diastereomeric carbons have been suggested so far[4a,34–38]. The present

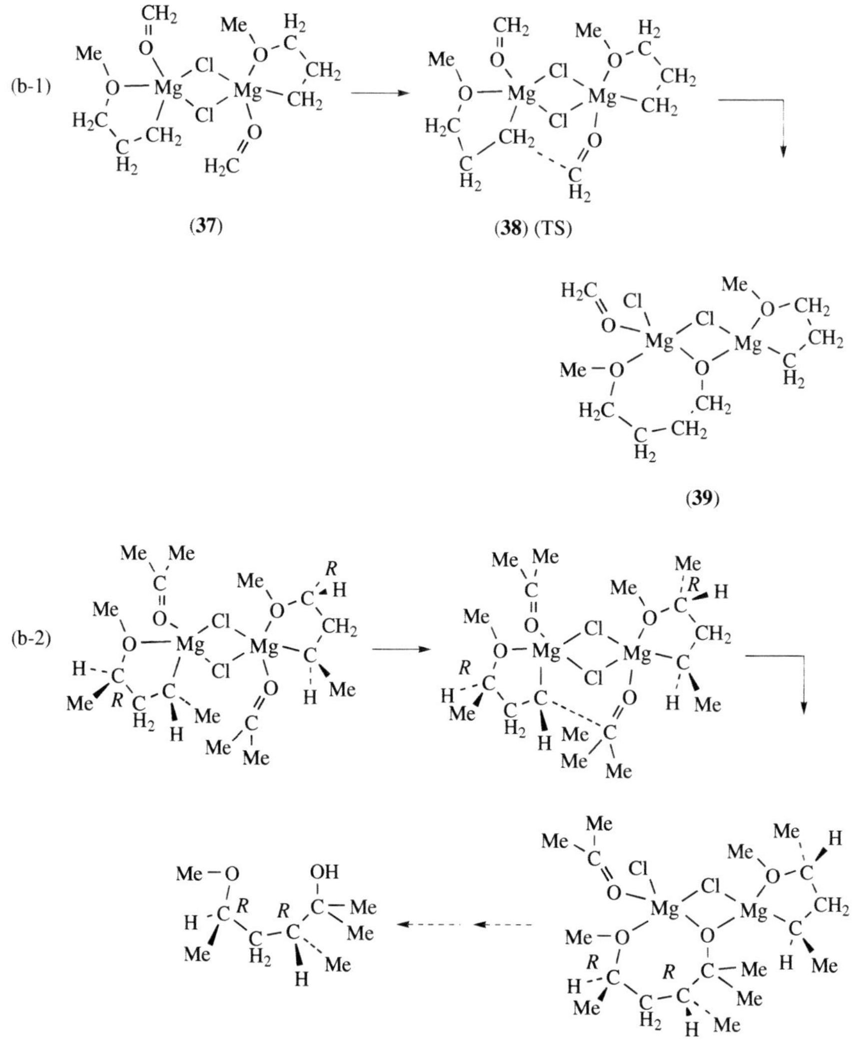

SCHEME 17b. An example of Scheme 16b

dimeric intermediate model rather than the monomeric one may reasonably describe the stereochemistry of Grignard reactions in the framework of the polar mechanism.

F. A Model for SET

In order to facilitate the four-center reaction in Scheme 14, steric congestion between the alkyl group R^1 and the carbonyl carbon needs to be avoided. In Scheme 18, a reaction using the bulky $(t\text{-BuMgCl})_2(\text{acrolein})_2$ system was examined.

Here a model conjugate ketone, i.e. cis-acrolein, was adopted. From the dimeric precursor **40**, the four-center reaction path was traced but could not be accomplished. The

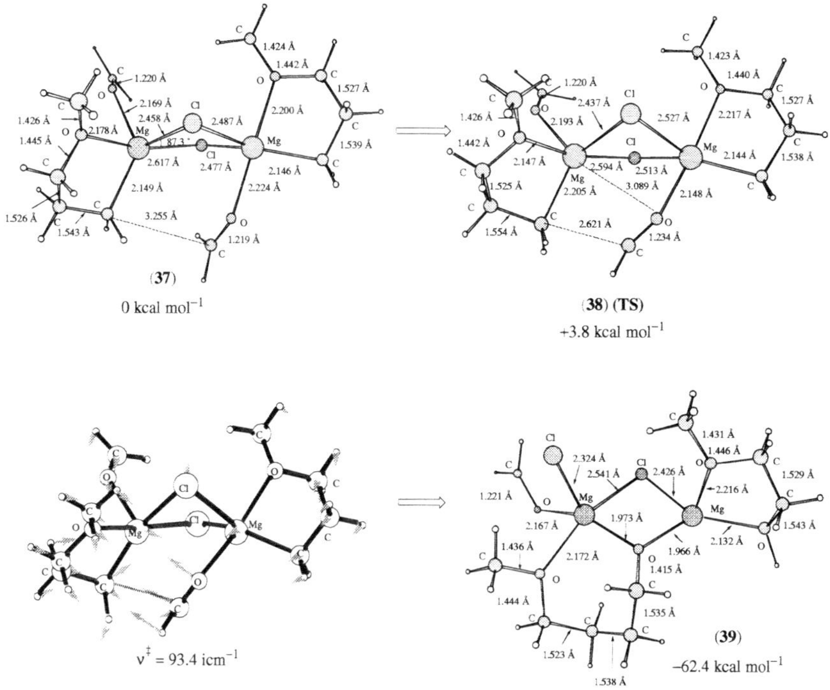

FIGURE 18b. Carbonyl addition process for the alkoxy Grignard model corresponding to that of Scheme 17b-1

nucleophilic center is blocked by the bulky *t*-Bu group. Other paths were sought starting from **40**. As a result, a singlet biradical forming path was found uniquely and is shown in Scheme 18 and Figure 19. The carbonyl oxygen approaches the left-side Mg, in a process similar to the four-center reaction in Scheme 14. However, instead of the simultaneous C−C bond formation, the *t*-Bu group is pushed away from Mg. The motion is described in TS **41** (see the reaction-coordinate vectors in Figure 19). With the decrease in the Mg−O distance, the Mg−C distance is gradually enlarged. Partial optimizations with a fixed Mg−O distance were repeated, and a complex potential surface with various extremely shallow energy minima was found. A local minimum with smaller relative energy than that (=8.5 kcal mol^{-1}) of TS **41** was sought. However, the attempted geometry converged to that of **42**, probably due to the complex potential curve and spin contamination, and **42** was calculated to be slightly less stable than TS**41**. A singlet biradical **42** is obtained, where the spin densities are localized on the tertiary carbon of *t*-Bu and on three carbon atoms of the reaction-center acrolein. Noteworthy is a very small spin density on the left-side Mg atom despite the Mg−C homolytic dissociated product. The left-side Mg persists in its tetravalency, and the *t*-Bu group has been pushed away. In **42**, two singlet radical centers (*t*-Bu$^{\bullet}$ and O-CH=CH-CH$_2{}^{\bullet}$) are distant; the corresponding triplet state has similar geometry and spin density (except signs) distributions. In **42**, an allyl radical moiety is formed. Radical−radical recombinations leading to the normal Grignard addition and the conjugate addition are possible. In fact, the reactions of α,β-unsaturated carbonyl compounds (e.g. 3-hexen-2-one, 4-methyl-3-penten-2-one, 2-cyclohexenone, *t*-butyl crotonate and crotonaldehyde) give 1,2-adducts and 1,4-adducts[39]. As explained for Scheme 5 and

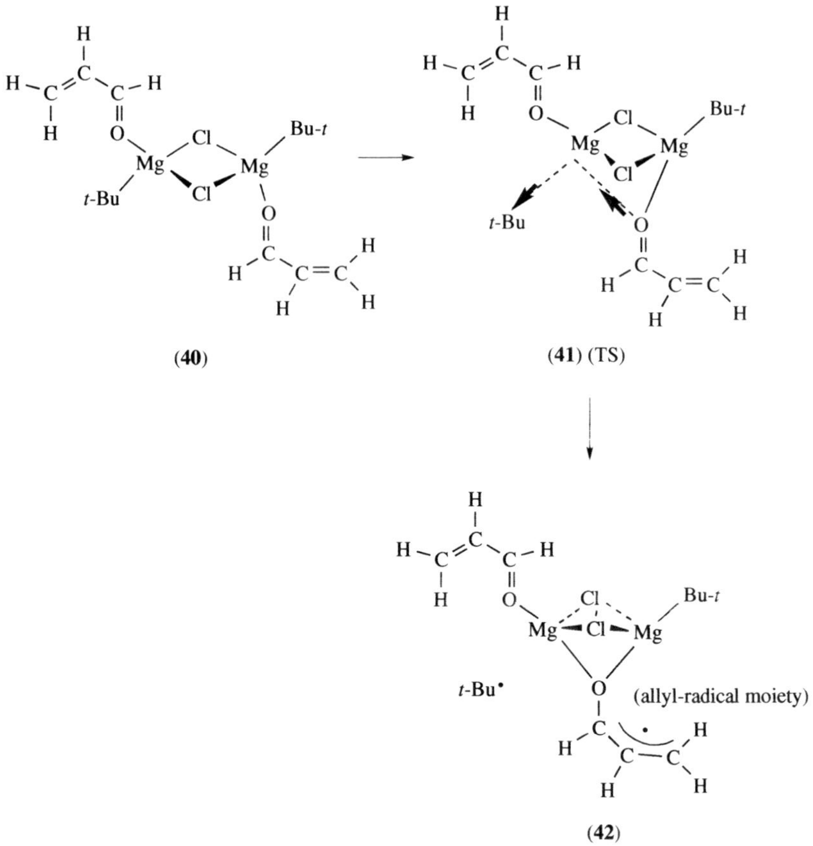

SCHEME 18. A singlet-biradical forming process. When the Mg−O bond formation proceeds, the left-side *t*-Bu group is pushed away as a *t*-Bu• via homolytic C−Mg cleavage

Table 1, a kinetic study suggested both polar and homolytic mechanisms for conjugate addition, depending on the substrate and the substrate conformation[39a].

The singlet biradical intermediate **42** is less stable ($\Delta E_{el} = +8.7$ kcal mol^{-1}, the difference of total electronic energies) than precursor **40**. When the *tert*-butyl group was separated infinitely, the instability, $\Delta E_{el} = +13.7$ kcal mol^{-1}, was calculated. However, the instability is cancelled out in Gibbs free energies. This Gibbs free energy ($T = 300$ K) of *t*-Bu• and the residual radical ($= (\mathbf{42}-t\text{-Bu})^{\bullet}$) is 1.0 kcal mol^{-1} smaller (i.e. the species is more stable) than that of **40**. As the temperature is raised to > 300 K, the biradical separated state is even more stable than precursor **40**. This stability ($\Delta G = -1.0$ kcal mol^{-1}, $T = 300$ K) of the homolytically dissociated products (two doublet radicals, *t*-Bu• and $(\mathbf{42}-t\text{-Bu})^{\bullet}$) compared to **40** is in sharp contrast to the significant instability ($\Delta G = +24.9$ kcal mol^{-1}) of homolytically dissociated products from **24** (Figure 13), Me• and the residual radical, $(\mathbf{24}-\text{Me})^{\bullet}$. The precursor **24** cannot cause such a reaction as in Figure 19. SET is *entropy-driven* in the reaction between *t*-BuMgCl and acrolein.

In Scheme 5 and Table 1, the product distributions in reactions between the substituted benzophenones and *t*-BuMgCl have been shown. While the parent benzophenone

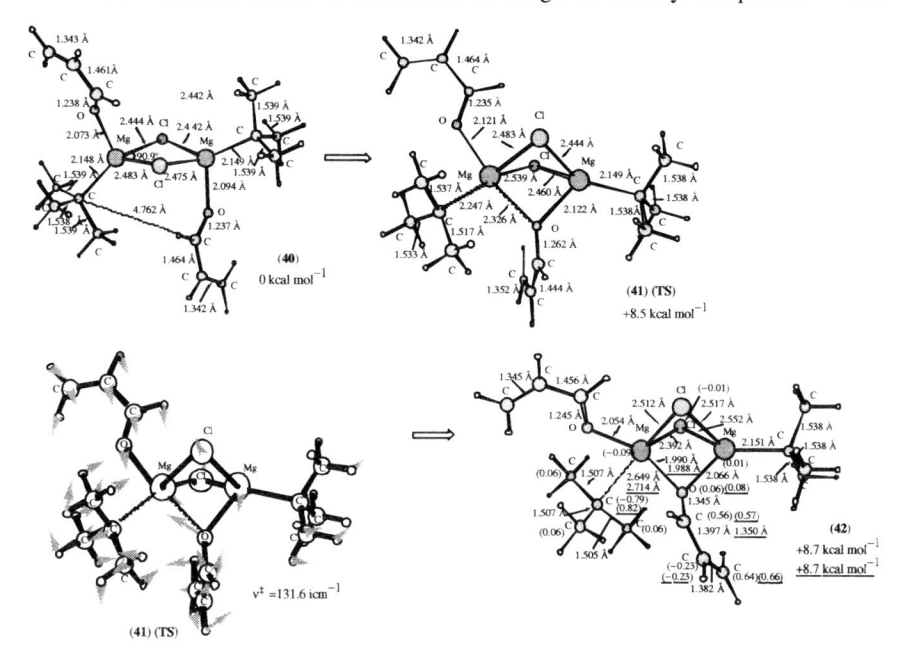

FIGURE 19. Geometries of $(t\text{-BuMgCl})_2(\text{acrolein})_2$ optimized by (U)B3LYP/6-31G*SCRF and reaction-coordinate vectors corresponding to the sole imaginary frequencies $v^{\ddagger}$ for **41**(TS). Those models are explained in Scheme 18. For **41** and **42**, the singlet biradical state was calculated with the symmetry-broken (iop(4/13 = 1)) initial orbitals. In **42**, the triplet-spin geometry optimized data are shown by the underlined numbers. In **42**, spin densities are exhibited in parentheses

affords 1,2-(normal) and 1,6-adducts almost equally, 2,4,6-trimethylbenzophenone form the 1,6-(abnormal) adduct exclusively. The geometries of two singlet biradicals **43** and **44** (Scheme 19) were determined and are shown in Figure 20. In the biradical intermediate **43**, the evolved t-Bu radical may recombine with the bridged coordinated benzophenone which lacks steric crowding. That is, the carbonyl carbon C may undergo addition of t-Bu leading to the normal 1,2-adduct.

In contrast, in the radical **44** there is steric congestion (particularly by the *ortho* methyl groups) around the carbonyl carbon, C$^{\bullet}$. The t-Bu radical cannot be bound to the carbonyl carbon C$^{\bullet}$ anymore, and normal Grignard adduct formation is prohibited. The steric hindrance in **44** is consistent with the experimental evidence (no normal adduct).

In view of the present calculated results, the SET mechanism would be described as follows. Basically, the polar four-center reaction in Scheme 14 leads to C—C bond formation. However, when the alkyl group is bulky, only the two-center (Mg—O) reaction takes place. The alkyl—Mg bond is cleaved homolytically owing to the persistent Mg tetravalency and the stability of the resultant radical species. Hence, biradical intermediates are formed not by a single electron transfer but by the C—Mg homolytic scission.

G. Concluding Remarks

Section V has revealed a new mechanism of addition of Grignard reagents to carbonyl compounds. The mechanism was thought to be very complex due to aggregation and a competing SET mechanism. No attempts to elucidate the correlation between the

SCHEME 19. The singlet biradical **43** is formed in the reaction between the two parent benzophenones and $(t\text{-BuMgCl})_2$. The radical **44** is formed from the two 2,4,6-trimethylbenzophenones and $(t\text{-BuMgCl})_2$. A, B, C and D denote for the respective phenyl and mesityl rings

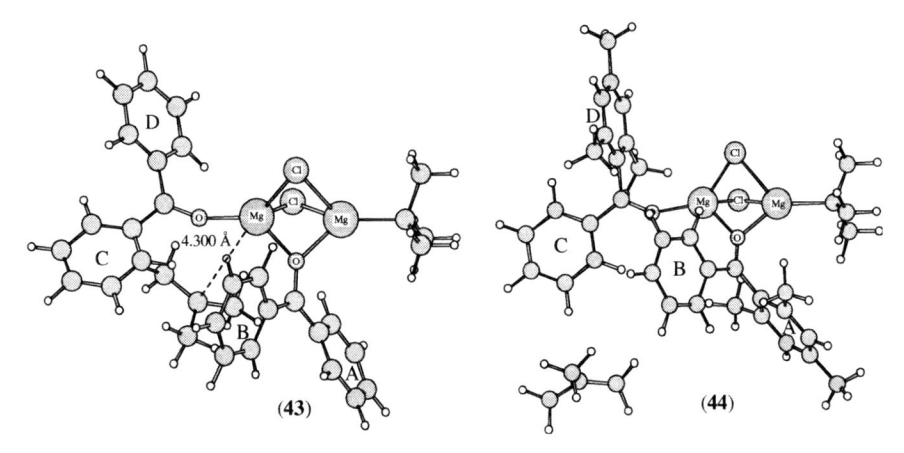

FIGURE 20. Geometries of the two singlet-biradical intermediates, **43** and **44**, corresponding to those in Scheme 19. In **43**, the t-Bu group is adjacent to the carbonyl carbon of the bridged benzophenone molecule. In **44**, the group is far away from it and cannot be bound with the carbon atom owing to the steric crowding

Schlenk intermediates and the addition path have been made so far. However, our results seem to provide simple and realistic pathways. The Grignard addition occurs in a dimeric dichlorine-bridged form. A vicinal-magnesium bonding alkyl and C=O interaction causes C—C bond formation via a four-center interaction as shown in Scheme 20 (polar mechanism). When the interaction is improbable owing to the steric effect, the Mg—O=C bond formation precedes the C—C bond formation and the Mg—C bond is ruptured (SET).

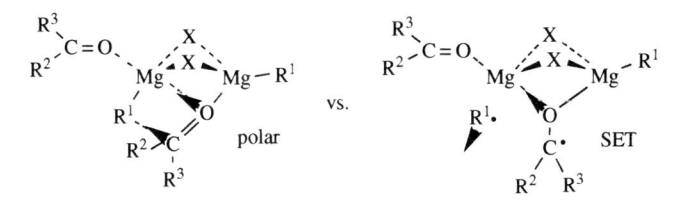

SCHEME 20. Slightly different geometric changes lead to two mechanisms

The rupture is caused by the preference of the tetravalency of the Mg atom, whereas the preference is ambiguous (e.g., elongation and weakening of the bridged Mg−Cl bonds) in the polar mechanism.

Solvent ether molecules may be bound effectively and flexibly to Mg atoms in retaining their tetravalency. When only the reactants $(R^1MgX)_2$ and $(R^2R^3C=O)_2$ are taken into account (e.g. in Scheme 13), trivalent Mg states such as **28** and **30** are inevitably formed. The solvent molecules compensate for the lack of chemical bonds to the Mg atoms, through formation of appropriate Mg−O coordination bonds. Even in this case they do not interfere with intrinsic reaction channels.

VI. REFERENCES

1. J. S. Binkley, R. A. Whiteside, R. Krishnan, R. Seeger, D. J. Defrees, H. B. Schlegel, S. Topiol, L. R. Kahn and J. A. Pople, *Gaussian 80*, Carnegie-Mellon Quantum Chemistry Publishing Unit, Pittsburgh, PA, 1980.
2. M. J. Frisch, G. W. Trucks, H. B. Schlegel, P. M. W. Gill, B. G. Johnson, M. W. Wong, J. B. Foresman, M. A. Robb, M. Head-Gordon, E. S. Replogle, R. Gomperts, J. L. Andres, K. Raghavachari, J. S. Binkley, C. Gonzalez, R. L. Martin, D. J. Fox, D. J. Defrees, J. Baker, J. J. P. Stewart and J. A. Pople, *Gaussian 92/DFT*, Gaussian, Inc., Pittsburgh, PA, 1993.
3. (a) V. Grignard, *Compt. Rend.*, **130**, 1322 (1900).
   (b) G. S. Silvermann and P. E. Rakita (Eds.), *Handbook of Grignard Reagents*, Marcel Dekker, New York, 1996.
   (c) H. G. Richey Jr. (Ed.), *Grignard Reagents: New Developments*, Wiley, New York, 2000.
   (d) B. J. Wakefield, *Organomagnesium Methods in Organic Synthesis*, Academic Press, London, 1995.
   (e) M. S. Kharasch and O. Reinmuth, *Grignard Reactions of Nonmetallic Substances*, Prentice-Hall, NJ, 1954.
   (f) E. C. Ashby, *Pure Appl. Chem.*, **52**, 545 (1980).
   (g) E. C. Ashby and W. E. Becker, *J. Am. Chem. Soc.*, **85**, 118 (1963).
   (h) K. Maruyama and T. Katagiri, *J. Am. Chem. Soc.*, **108**, 6263 (1986).
   (i) T. Holm, *Acta Chem. Scand.*, **20**, 2821 (1966).
   (j) S. G. Smith and G. Su, *J. Am. Chem. Soc.*, **86**, 2750 (1964).
   (k) E. C. Ashby, J. Laemmle and H. M. Neumann, *Acc. Chem. Res.*, **7**, 272 (1974).
4. (a) W. C. Still and J. H. McDonald, *Tetrahedron Lett.*, **21**, 1031 (1980).
   (b) K-Y. Ko and E. L. Eliel, *J. Org. Chem.*, **51**, 5353 (1986).
   (c) S. V. Frye and E. L. Eliel, *J. Am. Chem. Soc.*, **110**, 484 (1988).
   (d) W. F. Bailey, D. P. Reed, D. R. Clark and G. N. Kapur, *Org. Lett.*, **3**, 1865 (2001).
5. D. J. Cram and K. R. Kopecky, *J. Am. Chem. Soc.*, **81**, 2748 (1959).
6. P. Sykes, *A Guidebook to Mechanism in Organic Chemistry*, 5th edn., Chap. 8, Longman Group Ltd, London & New York, 1981.
7. C. G. Swain and H. B. Boyles, *J. Am. Chem. Soc.*, **73**, 870 (1951).
8. E. C. Ashby, R. B. Duke and H. M. Neumann, *J. Am. Chem. Soc.*, **89**, 1964 (1967).
9. T. Holm and I. Crossland, *Acta Chem. Scand.*, **25**, 59 (1971).
10. (a) E. C. Ashby and R. S. Smith, unpublished results.
    (b) M. Okubo and K. Maruyama, *Kagaku*, **35**, 338 (1980); *Chem. Abstr.* **94**, 14633 (1981).

11. H. Yamataka, T. Matsuyama and T. Hanafusa, *J. Am. Chem. Soc.*, **111**, 4912 (1989).
12. S. Sakai and K. D. Jordan, *J. Am. Chem. Soc.*, **104**, 4019 (1982).
13. *Tables of Interatomic Distances*; *Chem. Soc. Spec. Publ.*, No. 18, Suppl., 1965.
14. L. J. Guggenberger and R. E. Rundle, *J. Am. Chem. Soc.*, **86**, 5344 (1964).
15. P. A. Akishin and V. P. Spiridnov, *Sov. Phys. Crystallogr. (Engl. Transl.)*, **2**, 472 (1957).
16. J. L. Duncan, A. Allan and D. C. McKean, *Mol. Phys.*, **18**, 289 (1970).
17. B. S. Ault, *J. Am. Chem. Soc.*, **102**, 3480 (1980).
18. S. R. Davis, *J. Am. Chem. Soc.*, **113**, 4145 (1991).
19. L. Liu and S. R. Davis, *J. Phys. Chem.*, **95**, 8619 (1991).
20. E. Péralez, J-C. Négrel, A. Goursot and M. Chanon, *Main Group Metal Chem.*, **22**, 201 (1999).
21. W. Schlenk and W. Schlenk, Jr. *Ber. Dtsch. Chem. Ges.*, **62**, 920 (1929).
22. J. Axten, J. Troy, P. Jiang, M. Trachtman and C. W. Bock, *Struct. Chem.*, **5**, 99 (1994).
23. S. Nagase and Y. Uchibori, *Tetrahedron Lett.*, **23**, 2585 (1982).
24. Y-D. Wu and K. N. Houk, *J. Am. Chem. Soc.*, **109**, 908 (1987).
25. V. S. Safont, V. Moliner, M. Oliva, R. Castillo, J. Andrés, F. González and M. Carda, *J. Org. Chem.*, **61**, 3467 (1996).
26. (a) M. Carda, F. González, S. Rodríguez and J. A. Marco, *Tetrahedron: Asymmetry*, **3**, 1511 (1992).
    (b) M. Carda, F. González, S. Rodríguez and J. A. Marco, *Tetrahedron: Asymmetry*, **4**, 1799 (1993).
27. S. Yamazaki and S. Yamabe, *J. Org. Chem.*, **67**, 9346 (2002).
28. (a) A. D. Becke, *J. Chem. Phys.*, **98**, 5648 (1993).
    (b) C. Lee, W. Yang and R. G. Parr, *Phys. Rev. B*, **37**, 785 (1998).
29. L. Onsager, *J. Am. Chem. Soc.*, **58**, 1486 (1936).
30. M. J. Frisch, G. W. Trucks, H. B. Schlegel, G. E. Scuseria, M. A. Robb, J. R. Cheeseman, V. G. Zakrzewski, J. A. Montgomery, R. E. Stratmann, J. C. Burant, S. Dapprich, J. M. Millam, A. D. Daniels, K. N. Kudin, M. C. Strain, O. Farkas, J. Tomasi, V. Barone, M. Cossi, R. Cammi, B. Mennucci, C. Pomelli, C. Adamo, S. Clifford, J. Ochterski, G. A. Petersson, P. Y. Ayala, Q. Cui, K. Morokuma, D. K. Malick, A. D. Rabuck, K. Raghavachari, J. B. Foresman, J. Cioslowski, J. V. Ortiz, B. B. Stefanov, G. Liu, A. Liashenko, P. Piskorz, I. Komaromi, R. Gomperts, R. L. Martin, D. J. Fox,, T. Keith, M. A. Al-Laham, C. Y. Peng, A. Nanayakkara, C. Gonzalez, M. Challacombe, P. M. W. Gill, B. G. Johnson, W. Chen, M. W.Wong, J. L. Andres, M. Head-Gordon, E. S. Replogle, and J. A. Pople, *Gaussian 98*, Revision A.7, Gaussian, Inc., Pittsburgh, PA, 1998.
31. (a) A. L. Spek, P. Voorbergen, G. Schat, C. Blomberg and F. Bickelhaupt, *J. Organomet. Chem.*, **77**, 147 (1974).
    (b) J. Toney and G. D. Stucky, *J. Chem. Soc., Chem. Commun.*, 1168 (1967).
32. For example, see: F. A. Carey and R. J. Sundberg, *Advanced Organic Chemistry Part B: Reactions and Synthesis*, Plenum Press, New York, 1990, p. 376.
33. M. Nakamura, E. Nakamura, N. Koga and K. Morokuma, *J. Am. Chem. Soc.*, **115**, 11016 (1993).
34. M. Marsch, K. Harms, W. Massa and G. Boche, *Angew. Chem., Int. Ed. Engl.*, **26**, 696 (1987).
35. B. Z. Lu, C. Senanayake, N. Li, Z. Han, R. P. Bakale and S. A. Wald, *Org. Lett.*, **7**, 2599 (2005).
36. W. H. Miles, S. L. Rivera and J. D. del Rosario, *Tetrahedron Lett.*, **33**, 305 (1992).
37. (a) D. Seebach and M. A. Syfrig, *Angew. Chem., Int. Ed. Engl.*, **23**, 248 (1984).
    (b) D. Seebach, J. Hansen, P. Seiler and J. M. Gromek, *J. Organomet. Chem.*, **285**, 1 (1985).
    (c) P. Zhang and R. E. Gawley, *Tetrahedron Lett.*, **33**, 2945 (1992).
38. G. J. McGarvey and M. Kimura, *J. Org. Chem.*, **47**, 5422 (1982).
39. (a) T. Holm, *Acta Chem. Scand.*, **45**, 925 (1991).
    (b) E. R. Coburn, *Organic Synthesis*, Wiley, New York, 1955; Collect. Vol. No. III, p. 696.
    (c) J. Munch-Petersen, *Organic Synthesis*, Wiley, New York, 1973; Collect. Vol. No. V, p. 762.
    (d) F. S. Prout, R. J. Hartman, P.-Y. Huang, C. J. Korpics and G. R. Tichelaar, *Organic Synthesis*, Wiley, New York, 1963; Collect. Vol. No. IV, p. 93.

CHAPTER 10

Organomagnesium-group 15- and Organomagnesium-group 16-bonded complexes

KATHERINE L. HULL and KENNETH W. HENDERSON

Department of Chemistry and Biochemistry, 251 Nieuwland Science Hall, University of Notre Dame, Notre Dame, IN 46556, USA
Fax: +57 46 31 66 52; e-mail: khenders@nd.edu

I. INTRODUCTION	403
II. ORGANOMAGNESIUM-GROUP 15-BONDED COMPLEXES	404
A. Organomagnesium Amides	404
1. Synthesis	404
2. Structural characterization	412
3. Reactivity studies	419
a. Metalation reactions	419
b. Disproportionation reactions	422
c. Reactions with oxygen	422
d. Reactions with aldehydes and ketones	423
B. Organomagnesium Heavy Pnictogenides	427
III. ORGANOMAGNESIUM-GROUP 16-BONDED COMPLEXES	427
A. Organomagnesium Alkoxides and Aryloxides	428
1. Synthesis	428
2. Structural characterization	428
3. Reactivity studies	432
B. Organomagnesium Heavy Chalcogenides	433
IV. REFERENCES	434

I. INTRODUCTION

This chapter is devoted to the chemistry of organomagnesium-group 15- and group 16-bonded complexes, with emphasis on their synthesis, structural characterization and utility. In particular, organomagnesium amides will be the central focus of the first section, as

The chemistry of organomagnesium compounds
Edited by Z. Rappoport and I. Marek © 2008 John Wiley & Sons, Ltd

these complexes have received by far the most attention. For the purposes of this particular review the compounds of discussion are limited to those that contain both magnesium to carbon and magnesium to pnictogen or chalcogen bonds. Metal species containing dative interactions to the group 15 or group 16 elements will not be considered and charge-separated species containing these elements are also excluded. Also, only homometallic complexes will be considered. It should, however, be noted that a set of heterodimetallic reagents containing magnesium has recently received a good deal of interest due to their ability to act as highly selective Brönsted bases. Selections of these reagents contain alkyl units, amide units, an alkali metal and a divalent or a trivalent metal (Mg, Zn, Mn or Al). These are highly interesting compounds from both a structural and a synthetic perspective and have been recently reviewed extensively elsewhere[1,2]. A good deal of work has been carried out on organomagnesium β-diketiminates and tris(pyrazolyl)-hydroborates, and these complexes are included as they illustrate general reactivity patterns for this class of complexes.

II. ORGANOMAGNESIUM-GROUP 15-BONDED COMPLEXES
A. Organomagnesium Amides
1. Synthesis

A summary of the general routes for the synthesis of organomagnesium amides is given in Scheme 1. The most commonly used synthesis of organomagnesium amides is by the reaction of a diorganomagnesium, R_2Mg, with one molar equivalent of a protic amine. This simple procedure is over a century old and is still the most convenient method for preparing a wide variety of compounds within this class[3]. In general, dialkylmagnesium bases have been most commonly employed in these reactions, although a few examples of diaryl or mixed alkyl/aryl magnesium reagents have also been utilized. Also, the commercial availability of the reagent dibutylmagnesium has made this route attractive for many researchers.

Alkane elimination:

$$R_2Mg + R^1_2NH \longrightarrow R^1_2NMgR + RH$$

Ligand Redistribution:

$$R_2Mg + Mg(NR^1_2)_2 \longrightarrow 2\,R^1_2NMgR$$

$$R^1_2NMgR + R^2Li \longrightarrow R^1_2NMgR^2 + RLi$$

Metathesis:

$$R_2Mg + M(NR^1_2) \longrightarrow R^1_2NMgR + MR$$

$$RMgX + M(NR^1_2) \longrightarrow R^1_2NMgR + MX$$

SCHEME 1

Table 1 gives a complete list of the crystallographically characterized compounds synthesized from the reaction of diorganomagnesium compounds with protic amines or metal amides. A variety of dialkylmagnesium precursors have been used in these reactions. In general, one alkyl group acts as the base and extracts the amine's proton, while the second alkyl group remains attached to the magnesium center. It should be noted that the identity of the alkyl and amide groups present is important for the successful completion

TABLE 1. Structurally characterized organomagnesium amides synthesized from the reaction of amines or metal amides with diorganomagnesium precursors

Amine/Amide	R_2Mg	Solvent	Product number	Product	Reference
$Me_2NCH_2CH_2NHMe$	Me_2Mg	Et_2O	1		4
$(Me_3Si)_2NH$	$n\text{-}Bu(s\text{-}Bu)Mg$	Hexane	2		5
$Tl\{HB(3\text{-}C_3N_2\text{-}t\text{-}BuH_2)_3\}$	R_2Mg $R=Me, i\text{-}Pr$	THF	3a,b		6–8
$K\{HB(3,5\text{-}C_3N_2(CH_3)_2H)_3\}$	R_2Mg $R=CH_2SiMe_3$	THF	4		8, 9
$Tl\{PhB(3\text{-}C_3N_2\text{-}t\text{-}BuH_2)_3\}$	R_2Mg $R=Me, Et$	Benzene	5a,b		10
$Tl\{HB(3\text{-}C_3N_2PhH_2)_3\}$	Et_2Mg	THF	6		11
(carbazole with t-Bu groups)	Et_2Mg	THF	7		12

(continued overleaf)

TABLE 1. (continued)

Amine/Amide	R_2Mg	Solvent	Product number	Product	Reference
t-BuNH$_2$	t-Bu$_2$Mg	Toluene	8	(dimeric Mg amide with thf, t-Bu and N-Bu-t groups)	13
RNH$_2$ R=dipp dipp = 2,6-i-Pr$_2$-C$_6$H$_3$	Et$_2$Mg	Toluene	9	(cluster structure; ○ = Et, ● = dippNH)	13
(β-diketiminate, dipp/H/dipp)	Me$_2$Mg	Et$_2$O or THF	10a,b	(Mg complex with L = Et$_2$O, THF)	14, 15
(β-diketiminate, dipp/H/dipp)	R_2Mg R=Me, n-Bu	Toluene	11a,b	(dimeric Mg complex with bridging R groups)	14–16
(β-diketiminate, dipp/H/dipp)	t-Bu$_2$Mg	Toluene	12	(Mg–Bu-t complex)	14
(β-diketiminate, dipp/H/dipp)	[(PhCH$_2$)$_2$Mg(thf)$_2$]	THF	13	(Mg–CH$_2$Ph complex with thf)	17
(β-diketiminate, dipp/H/dipp)	[La(η^3-C$_3$H$_5$)$_3$(μ-C$_4$H$_8$O$_2$)·Mg(η^1-C$_3$H$_5$)$_2$(μ-C$_4$H$_8$O$_2$)$_{1.5}$]$_n$	THF	14	(Mg–allyl complex with thf)	18
(β-diketiminate, t-Bu/dipp/H/t-Bu/dipp)	Me$_2$Mg	THF	15	(Mg–Me complex with thf)	19

TABLE 1. (continued)

Amine/Amide	R$_2$Mg	Solvent	Product number	Product	Reference
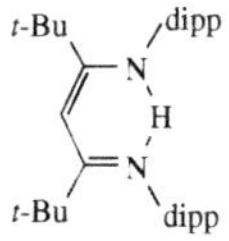 t-Bu, dipp, H, N, N, t-Bu, dipp (β-diketiminate)	[(C$_3$H$_5$)$_2$Mg(thf)$_n$]	Toluene	16	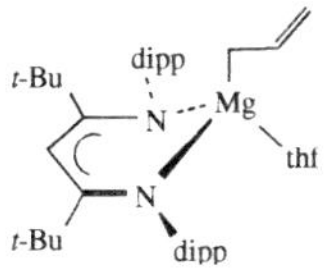 t-Bu, dipp, N, Mg(allyl)(thf), N, t-Bu, dipp	20
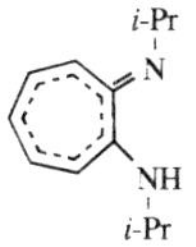 i-Pr, N, NH, i-Pr (aminotroponiminate)	Me$_2$Mg	Toluene	17	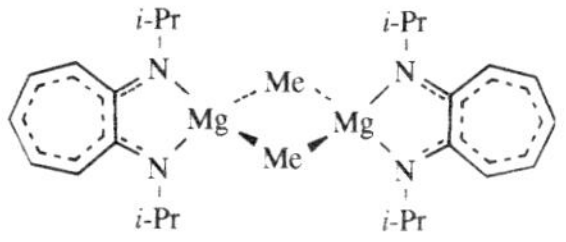 Dimeric [Mg(Me)(L)]$_2$ with bridging Me groups	15
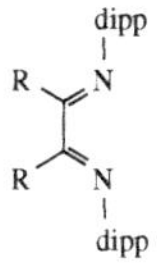 dipp, R, N, R, N, dipp (α-diimine)	Me$_2$Mg	Et$_2$O	18a,b	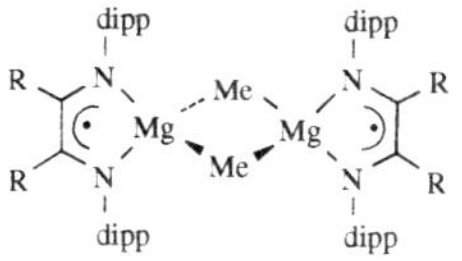 Dinuclear Mg–Me bridged diimine complex; R = 1,8-naphthdiyl; R = Me	21, 22
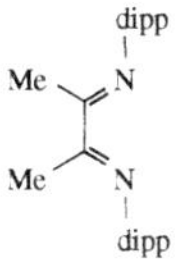 dipp, Me, N, Me, N, dipp	Me$_2$Mg	Toluene/THF	19	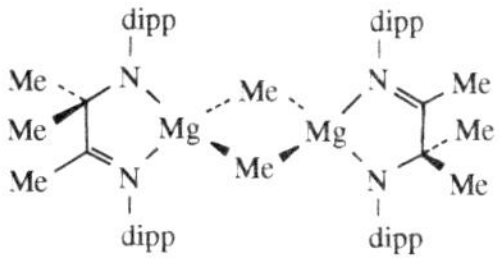 Dinuclear Mg–Me bridged complex	21, 22
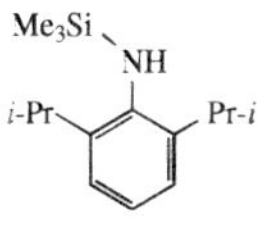 Me$_3$Si, NH, i-Pr, Pr-i	Bu$_2$Mg	Hexane/THF	20	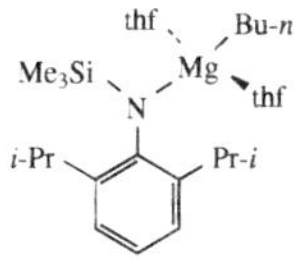 Me$_3$Si, N, Mg(Bu-n)(thf)$_2$, i-Pr, Pr-i	23
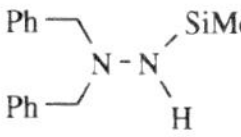 Ph, N–N, SiMe$_3$, Ph, H	Bu$_2$Mg	Hexane	21	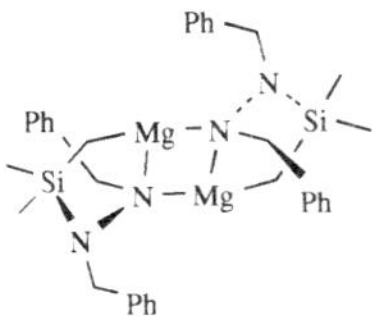 Mg$_2$Si$_2$N$_4$ cage with Ph and CH$_2$Ph groups	24
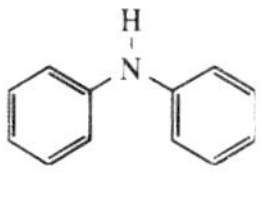 Ph, H, N, Ph (diphenylamine)	R$_2$Mg, R = Et, i-Pr	THF	22a,b	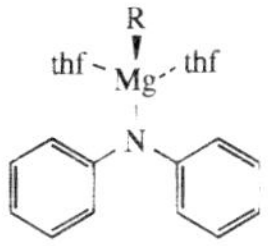 thf, R, Mg, thf, N(Ph)$_2$	25

(continued overleaf)

TABLE 1. (*continued*)

Amine/Amide	R_2Mg	Solvent	Product number	Product	Reference
PhCN	Cp_2Mg	Et_2O	23	(Ph-substituted bis-imino cyclopentadienyl Mg complex with NCPh)	26
NHRR^1 R=R^1=Ph R=H, R^1=CH(i-Pr)$_2$ R=H, R^1=2,6-i-Pr$_2$-C$_6$H$_3$ R=i-Pr, R^1=CH$_2$Ph	[CpMgMe(OEt$_2$)]$_2$	Et_2O	24a–d	(Cp$_2$Mg$_2$ dimer with bridging NRR^1 amides)	27
(2,5-bis(dimethylaminomethyl)pyrrole)	[CpMgMe(OEt$_2$)]$_2$	Et_2O	25	(pyrrolyl Mg Cp complex with Et$_2$O)	27
(N,N,N'-trimethylethylenediamine)	[CpMgMe(OEt$_2$)]$_2$	Et_2O	26	(Cp$_2$Mg$_2$ dimer with diamide ligand)	27
dipp–N(t-Bu)–N(H)–dipp (guanidine)	[CpMgMe(OEt$_2$)]$_2$	Et_2O	27	(t-Bu guanidinate Mg Cp)	28
dipp–N(t-Bu)–N(H)–dipp (guanidine)	[CpMgMe(OEt$_2$)]$_2$	THF	28	(t-Bu guanidinate Mg Cp·thf)	28
β-diketiminate (Bu-t, Bu-t) H	[CpMgMe(OEt$_2$)]$_2$	Et_2O	29	(β-diketiminate Mg Cp, Bu-t substituents)	29
β-diketiminate (Pr-i, Pr-i) H	[CpMgMe(OEt$_2$)]$_2$	(4-t-Bu-Py) Et_2O	30	(β-diketiminate Mg Cp·4-t-Bu-py, Pr-i substituents)	29

10. Organomagnesium-group 15- and 16-bonded complexes

TABLE 1. (continued)

Amine/Amide	R$_2$Mg	Solvent	Product number	Product	Reference
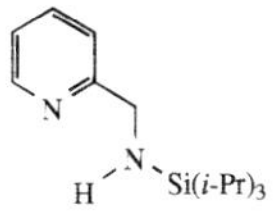 [2-pyridylmethylamine with Si(i-Pr)$_3$]	Me$_2$Mg	Toluene/THF	**31**	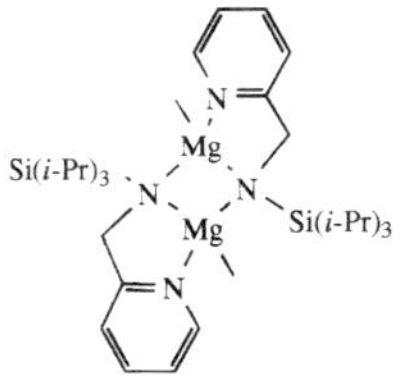 [dimeric Mg complex with Si(i-Pr)$_3$ groups]	30
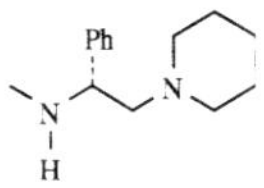 [PhCH(NHMe)CH$_2$-piperidine]	n-Bu$_2$Mg	Et$_2$O	**32**	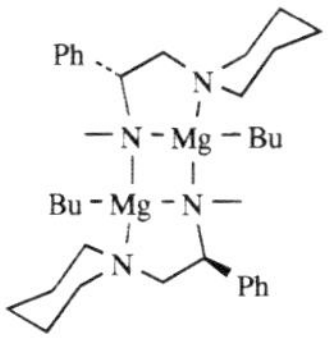 [dimeric Mg–Bu complex]	31
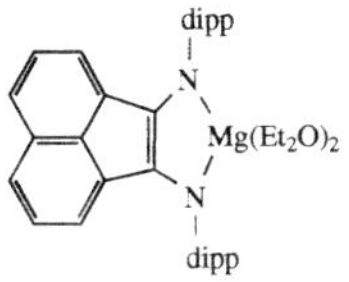 [acenaphthene-diimine with dipp, Mg(Et$_2$O)$_2$]	i-Pr$_2$Mg	Toluene	**33**	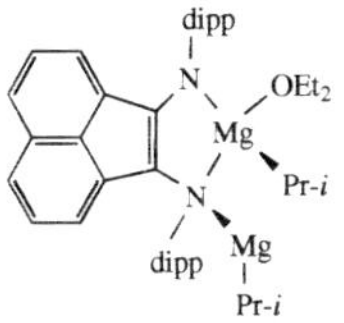 [dimeric Mg complex with OEt$_2$, Pr-i]	32
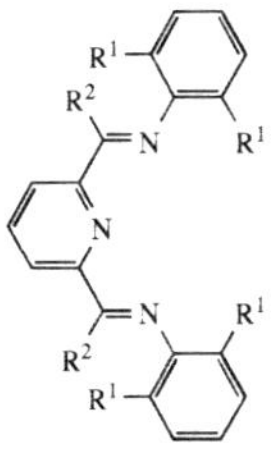 [2,6-bis(arylimino)pyridine]	R$_2^3$Mg	Toluene/Et$_2$O	**34a–d**	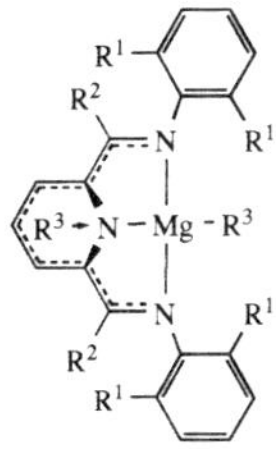 [Mg–R^3 pyridine diimine complex] $R^1 = R^2 = Me, R^3 = Et$; $R^1 = R^2 = Me, R^3 = i\text{-Pr}$; $R^1 = Et, R^2 = Me, R^3 = i\text{-Pr}$; $R^1 = i\text{-Pr}, R^1 = R^2 = Me$	33
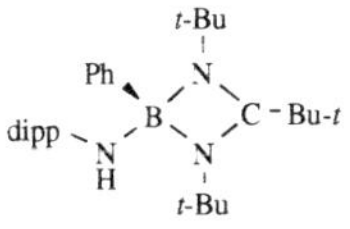 [boron-containing amidinate with Ph, dipp, t-Bu]	Bu$_2$Mg	Hexane/Et$_2$O	**35**	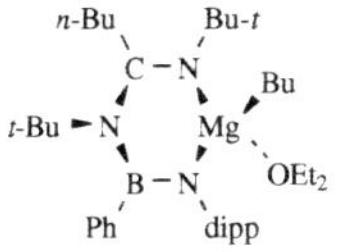 [Mg complex with n-Bu, t-Bu, OEt$_2$, Ph, dipp]	34
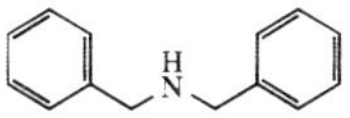 [dibenzylamine]	Bu$_2$Mg, t-BuLi	Heptane	**36**	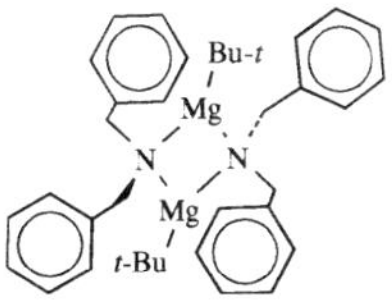 [dimeric Mg benzylamide with t-Bu]	35

(*continued overleaf*)

TABLE 1. (continued)

Amine/Amide	R₂Mg	Solvent	Product number	Product	Reference
(i-Pr)₂NH	Bu₂Mg, t-BuLi	Heptane	37	[(i-Pr)N−Mg(t-Bu)−N(i-Pr)−Mg(t-Bu)] dimer	35
2,6-(i-Pr)₂C₆H₃NH₂	Bu₂Mg	Heptane (TMEDA)	38	ArNH−Mg(n-Bu)(TMEDA)	35
pyrrole-2-CH=N(i-Pr)	t-Bu₂Mg	Toluene	39	bis(pyrrolyl-imine)Mg₂(t-Bu)₂	36
i-Pr₃Si−NH₂	Me₂Mg	THF	40	[(i-Pr₃SiNH)Mg(Me)(thf)]₂	37
MeO(CH₂)₂N(H)(CH₂)₂OMe	R₂Mg, R=Me, Et, Np	Et₂O	41	dimeric Mg amide with OMe chelation	38
aza-crown HN	R₂Mg, R=Me, Et, Np	THF/Et₂O	42	dimeric Mg amide with crown ether	38

of this route. For example, if the dialkylamide is small, the reaction may not cease at the alkyl(amido) stage but proceed to the dialkyl and bis(amido) species (Scheme 2)[39]. In such instances it is likely that the formation of insoluble polymeric products drives the reaction. In some cases mixed alkyl/aryl (methyl/cyclopentadienyl) R₂Mg bases have been employed. For example, compounds **24–30** were synthesized using [CpMgMe(OEt₂)]₂ as a convenient reagent (Table 1)[27–29]. In each case the alkyl group acts as the base, producing methane gas, whereas the cyclopentadienyl ring remains π-coordinated to the

$Me_2NH + Et_2Mg \longrightarrow Me_2NMgEt \longrightarrow 1/2\,(Me_2N)_2Mg + 1/2\,MgEt_2$

SCHEME 2

magnesium center. Several extensions and exceptions to these alkane elimination reactions have also been utilized, and are outlined below.

Ligand redistribution between magnesium bis(amides) and diorganomagnesium may take place. For instance, simply mixing the two metal reagents together results in the preparation of the heteroleptic organomagnesium amide complex **33**. Mixed-metal reagents have also been employed, including a mixed magnesium/lanthanum complex for the synthesis of β-diketiminate **14**[18]. A mixed-metal route was also used to prepare compounds **36** and **37**[35]. Specifically, heptane solutions of n-Bu(s-Bu)Mg were reacted with dibenzylamine and diisopropylamine to produce mixtures of the respective n- or s-BuMgNR$_2$ complexes. These mixtures proved difficult to separate but subsequent treatment with t-BuLi gave the t-BuMgNR$_2$ complexes, which were crystallized as pure solids from solution[35]. Another series of reactions using a second metal center are those involving the preparation of the {η^3-tris(pyrazolyl)borato} derivatives **3–6**. These compounds are conveniently synthesized by reaction of diorganomagnesium reagents with either the thallium or potassium precursors rather than protonated ligand[6–11]. Thus, these reactions proceed by metathesis rather than by deprotonation.

Addition reactions between diorganomagnesium compounds and organic nitriles may be used to produce organomagnesium imides[40]. The 1,2-cyclopentadienyl diimine complex **23** was synthesized from the reaction between Cp$_2$Mg and benzonitrile[26]. This is an unusual reaction in the organomagnesium amide series as one of the cyclopentadienyl rings remains π-bound to the magnesium center while two benzonitriles add sequentially to the second ring (Scheme 3)[26]. Protons from the cyclopentadienyl ring are transferred to the benzonitriles, reducing the nitrile group to a carbon–nitrogen double bond. Another example of an addition reaction comes from the preparation of compound **34**, where N-alkylation occurs on reaction of the base with the bis(imino)pyridine ligand[33].

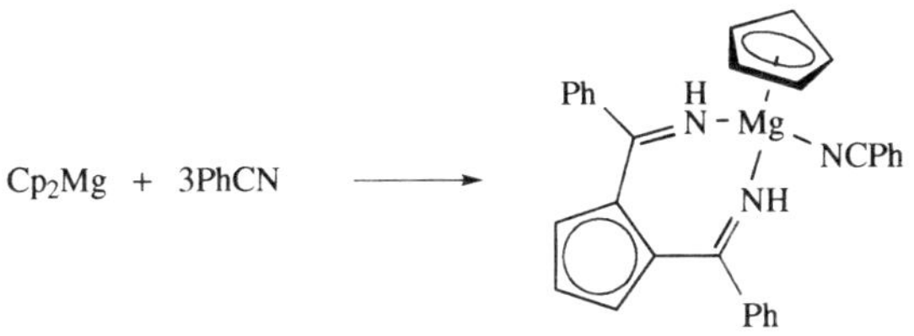

$Cp_2Mg + 3PhCN \longrightarrow$ (**23**)

SCHEME 3

Compound **21** is formed upon reaction of Bu$_2$Mg with a silyl hydrazine (Scheme 4)[24]. In this instance the base removes a proton from the hydrazine followed by an unexpected migration of the benzyl group, and subsequent deprotonation of the trimethylsilyl group[24].

The reactions involving diorganomagnesium reagents may be carried out in solvents ranging from saturated hydrocarbons to polar etheral solvents. The ability of these reactions to be conducted in hydrocarbon solvents gives them a distinct advantage over the alternative Grignard route. The Grignard route involves the reaction of a metal amide (typically an alkali metal amide) with a classic Grignard reagent of the form RMgX (X = halide). An alkali metal halide is eliminated upon formation of the organomagnesium

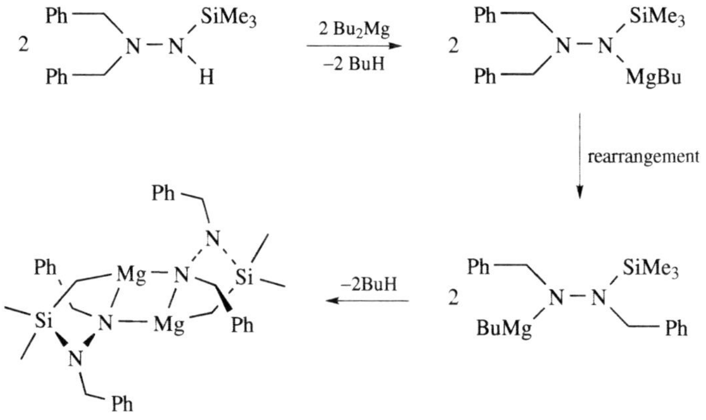

SCHEME 4

amide complex. Table 2 contains a complete list of organomagnesium amides produced from Grignard reagents which have been characterized in the solid state. These reactions are typically carried out in mixed solvent systems. The Grignard reagent, which is generally insoluble in hydrocarbons, is prepared as a solution in polar solvents such as THF or Et_2O, then combined with a hydrocarbon/arene solution of the metal amide. Although this method provides a simple and straightforward means to organomagnesium amides, it is limited in its application for subsequent organic syntheses because of these solvent restrictions. Furthermore, the presence of Lewis base donor solvent is a drawback in some instances as it may cause disproportionation of organomagnesium complexes, resulting in the formation of bis(amide) and dialkylmagnesium species[41, 47]. This possibility will be discussed in more detail in Section II.A.3.

A final set of miscellaneous organomagnesium amides that have been structurally characterized is outlined in Table 3. These were synthesized by neither R_2Mg bases nor by Grignard reagents. Compounds **11a, 55** and **56** were generated by heating existing compounds **10b, 16** and **47**, respectively, under vacuum[15, 20, 44]. The coordinated solvent in each of the precursor compounds was eliminated, yielding desolvated dimeric, hexameric and monomeric species, respectively[15, 20, 44]. Compounds **57–59** were synthesized by reacting a magnesium bis(amide) base with an acetylene[25, 48]. The terminal carbon was deprotonated, yielding the corresponding alkynylmagnesium amide species. Finally, reaction of $MgBr_2$ with a potassium amide was used to prepare compound **60**[49].

2. Structural characterization

By the end of 2006 over seventy organomagnesium amides were present in the Cambridge Structural Database. Rather surprisingly, although Magnuson and Stucky reported the first crystal structure of this class of compound in 1969, it has only been in the last few years that the majority of work has appeared[4]. The complex $[Me_2NCH_2CH_2N(Me)MgMe]_2$, **1**, is shown in Figure 1, and consists of two metal centers bridged by two amido nitrogen centers[4]. The tetracoordinate coordination sphere of the metals is completed by binding to a methyl unit and a chelating dimethylamido function. This early structural analysis possesses several features that have proved to be typical for these compounds. In general the metals tend to be tetracoordinate, either through chelation or

TABLE 2. Structurally characterized organomagnesium amides synthesized from the reaction of metal amides with Grignard reagents

Amide	Grignard	Solvent	Product number	Product	Reference
Ph-CH₂-N(M)-CH₂-CH₂-N(Me)-Me, M = Li or Na	n-BuMgCl	Hexane/THF	43	dimeric Mg complex with Ph-CH₂-N bridges and n-Bu groups	41
t-Bu, Bu-n triazine with Ph substituents and Li(thf)₃	MeMgCl	Toluene/THF	44	t-Bu, Bu-n triazine with Ph substituents, Mg(thf)₂	42
β-diketiminate (dipp) Li	MeMgCl	THF	10b	β-diketiminate (dipp) Mg-thf	15
β-diketiminate (dipp) Li—OEt₂	i-PrMgCl	Et₂O	45	β-diketiminate (dipp) Mg(i-Pr)(OEt₂)	43
β-diketiminate (dipp) Li	i-PrMgCl	Toluene/Et₂O	46	β-diketiminate (dipp) Mg—Pr-i	44
β-diketiminate (dipp) Li	PhMgBr	Toluene/Et₂O	47	β-diketiminate (dipp) Mg(Ph)(OEt₂)	44
β-diketiminate (dipp) Li	MeMgBr	Toluene/Et₂O	11a	bis(β-diketiminate) dimeric Mg₂Me₂	44

(*continued overleaf*)

TABLE 2. (*continued*)

Amide	Grignard	Solvent	Product number	Product	Reference
(bis-cyclohexyl diimino cyclopentadienyl Li complex)	MeMgBr	Toluene/THF	48	(corresponding Mg-Me·thf complex)	45
2,2,6,6-tetramethylpiperidide Na	BuMgCl	Hexane/Et$_2$O	49	(bis-TMP Mg$_2$(n-Bu)$_2$ dimer)	46
(acenaphthylene-bis(dipp)diamide Na(Et$_2$O))	i-PrMgCl	Hexane/Et$_2$O	50	(acenaphthylene-bis(dipp)diamide Mg(Pr-i)(OEt$_2$))	32
(n-Bu, Bu-t, t-Bu, Ph, dipp boraamidinate Li)	RMgX	Toluene/Et$_2$O	51	(corresponding Mg–R·L complex) R = Me, X = Br, L = OEt$_2$ R = t-Bu, X = Cl, L = none R = Mes, X = Br, L = none	34
(n-Bu, Bu-t, t-Bu, Ph, dipp boraamidinate Li)	RMgCl R = Ph, i-Pr	Toluene/THF	52	(corresponding Mg–R·thf complex)	34
(i-Pr)$_2$N–Na	t-BuMgCl	Hexane/Et$_2$O	37	(bis(diisopropylamide) Mg$_2$(t-Bu)$_2$ dimer)	35
(Me$_3$Si)$_2$N–Na	t-BuMgCl	Hexane/Et$_2$O	53	(bis(bis(trimethylsilyl)amide) Mg$_2$(t-Bu)$_2$ dimer)	35
2,2,6,6-tetramethylpiperidide Na	t-BuMgCl	Hexane/Et$_2$O	54	(bis-TMP Mg$_2$(t-Bu)$_2$ dimer)	35

TABLE 3. Structurally characterized organomagnesium amides synthesized by miscellaneous methods

Mg precursor	Precursor 2	Solvent	Product Number	Product	Reference
	—	150 °C vacuum	**55**		20
	—	150 °C vacuum	**56**		44
	—	150 °C vacuum	**11a**		15

(continued overleaf)

TABLE 3. (continued)

Mg precursor	Precursor 2	Solvent	Product Number	Product	Reference
$(i\text{-}Pr_2N)_2Mg$	$R-C\equiv C-H$ $R=Ph, SiMe_3$	THF	**57**		25
$(i\text{-}Pr_2N)_2Mg$	$t\text{-}Bu-C\equiv C-H$	THF	**58**		25
(acenaphthylene-diamide)Mg(thf)$_3$ with dipp-N groups	$Ph-C\equiv C-H$	Toluene/THF	**59**		48
$MgBr_2$	β-diketiminate K salt with t-Bu, SiMe$_3$, N-SiMe$_3$	pentane	**60**		49

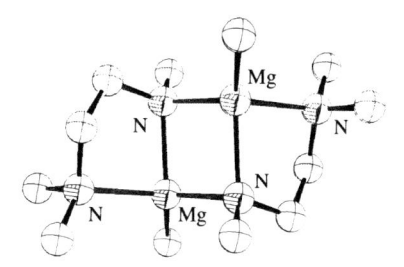

FIGURE 1. Dimeric molecular structure of **1** with hydrogens omitted

through interactions with donor solvents. A small number of three-coordinate compounds are known, where sterically bulky groups are bound to the metal center. The dimeric arrangement of **1**, with bridging nitrogen groups, is also commonplace[4]. Preferential bridging by the alkyl units is limited to instances where the nitrogen centers are part of a large ligand set such as some β-diketiminates.

Monomeric complexes typically only arise when the materials are crystallized in the presence of donor solvents, producing solvated solid-state compounds. A few exceptions to the solvation of monomers can be seen. One notable example is compound **12**, which was prepared by the reaction of t-Bu$_2$Mg with the corresponding protic amine (Figure 2)[14]. The related β-diketiminate **46** (Table 2) was also obtained as an unsolvated monomeric compound and was prepared from a reaction conducted in toluene with only small amounts of diethyl ether present[44]. The η^3-tris(pyrazolyl)borate complexes **3–5** (Table 1) also typically crystallize solvent-free, as the metal achieves tetracoordination through binding to the tridentate ligand and the terminal organic fragment[6–10].

There are only two examples of structurally characterized R$_2$NMgR compounds which have aggregation states larger than dimers. The remarkable dodecameric complex [DippN(H)MgEt]$_{12}$, **9** (Table 1), forms a ring structure composed of twelve interconnected MgNMgC rings (Figure 3)[13]. (Dipp = 2,6-diisopropylphenyl). It is noteworthy that both the amine and the ethyl groups bridge between the magnesium atoms. The large ring is slightly bowed, deviating at most ca 0.24 Å from the average plane of the magnesium atoms. In turn, the 2,6-diisopropylphenylamido groups point out from the magnesium atoms and away from the ring, while all of the smaller ethyl groups project towards the center of the ring. The second highly aggregated structure is allylmagnesium β-diketiminate **55** (Table 3), which is obtained upon sublimation of the monomeric THF

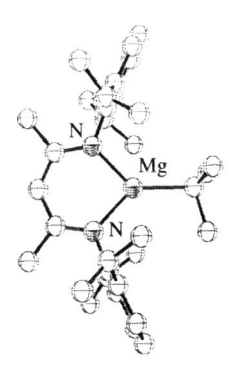

FIGURE 2. Monomeric three-coordinate structure of **12** with hydrogens omitted

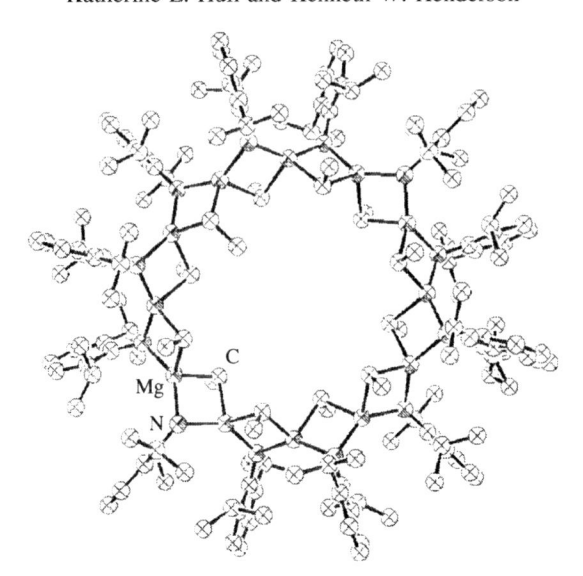

FIGURE 3. Ring hexameric structure of **9** with hydrogens omitted

derivative[20]. This hexameric ring is similar to **9**, although the deviations from the mean plane of the ring are slightly more significant. Each of the magnesium atoms is bridged by allyl groups while the bulky amides project outward from the ring.

The η^3-tris(pyrazolyl)borato-based compounds **3–6** (Table 1) are a unique subset of the organomagnesium amides[6–11]. The sterically-demanding environment of the η^3-tris(pyrazolyl)borato ligand strongly controls the overall structure of these compounds. Three nitrogens from each of three five-membered pyrazolyl rings coordinate to the magnesium center and provide an overall -1 charge to the complex. Two of the pyrazolyl rings are coplanar with each other and the magnesium, while the third ring sits perpendicular to this plane above the magnesium center. This magnesium atom, which is highly protected within this amido pocket, is then available to bond to a variety of organic fragments to complete its coordination sphere. The compounds within this series are exclusively monomeric and, though sometimes synthesized in ethereal solvent, compound **6** is the only solvated complex[11]. The nature of the organic group within this set of complexes appears to have little influence upon the metrical parameters of the overall structures.

Many organomagnesium complexes containing β-diketiminate ligands have been structurally characterized. Typically, the metal center lies in the NCCCN plane and is equivalently coordinated to both nitrogen atoms and is further bonded to an alkyl fragment. The nature of the alkyl group on the magnesium and the presence of polar solvent both influence the aggregation state of the resulting solid-state structure. For example, methyl- and butyl-substituted compounds **11a** and **11b** (Table 1) form dimeric aggregates in the solid state when generated in non-polar solvent[14–16]. However, the analogous compound **12** (Table 1), which is t-butyl-substituted, is monomeric when crystallized from non-polar solvent[14]. The steric bulk of the t-butyl ligand blocks the magnesium's coordination sphere and prevents dimerization. Furthermore, when solvated monomeric **10a** and **10b** are compared with unsolvated dimeric **11a**, it becomes evident that the presence of donor solvent reduces the aggregation state of the complex through solvation of the magnesium center[14–16].

The cyclopentadienyl-containing compounds **23–30** (Table 1) adopt another structure type for organomagnesium amides[26–29]. In all cases, the structures consist of one η^5-cyclopentadienyl ligand that is bound to the magnesium centers with the metal–cyclopentadienyl centroid distances lying in a narrow range between 2.0 and 2.1 Å. Both monomeric and dimeric aggregates are observed, which again is primarily related to the steric bulk present on the amide group.

3. Reactivity studies

a. Metalation reactions. The most commonly studied reaction of organomagnesium amides is their use in metalation reactions. It is generally assumed that the organic group is the most reactive unit of the reagent, and is consequently involved in these reactions. It should, however, be noted that the involvement of the alkyl or the amide unit has been a subject of some debate in the reactions of the mixed-metal reagents[50,51]. A number of general reactivity patterns for organomagnesium amides has been demonstrated using the η^3-tris(pyrazolyl)-hydroborate framework. As shown in Scheme 5, the complex {η^3-HB(3-*t*-Bupz)$_3$}MgMe, **3a**, has been employed in a wide range of reactions[6,52]. In all of these cases the methyl unit is used as a base to perform a series of deprotonation reactions. The use of the bulky η^3-tris(pyrazolyl)-hydroborate ligand prevents complications due to oligomerization of the products or ligand rearrangement.

Similarly, the unsolvated β-diketiminate complex {HC(C(Me)N-2,6-(*i*-Pr)$_2$C$_6$H$_3$)$_2$}MgBu, which is generated *in situ* by reacting Bu$_2$Mg with HC(C(Me)N-2,6-(*i*-Pr)$_2$C$_6$H$_3$)$_2$, has been shown to be reactive towards alcohols, amines and carboxylic acids to form the corresponding amidomagnesium alkoxides, amides and carboxylates[44]. Another noteworthy example of a reaction involving a β-diketiminate complex is shown in Scheme 6. Reaction of **11a** with Me$_3$SnF under mild conditions yields a fluorine-bridged dimer[16]. This is a rare example of a molecular magnesium fluoride complex and it is presumably stabilized towards disproportionation to MgF$_2$ by the bulk of the amide ligand[16].

Organomagnesium amides have been utilized as alternatives to standard Grignard reagents[53–55]. These reagents are believed to be less nucleophilic than classic Grignard reagents. The slight reduction in their reactivity allows their use in reactions where a mild base is desired. A significant potential advantage of using alkylmagnesium amides over either lithium amides or magnesium bis(amides) in the deprotonation of relatively weak acids is that the reactions are driven to completion due to the irreversible loss of alkane. This is particularly useful in instances when the pKa of the carbon acid and the amine are similar. The reactions of commercially available butylmagnesium diisopropylamide, BuMgDA, with cyclopropane carboxamides are good examples of this reactivity[56]. As shown in Scheme 7, BuMgDA reacts with cyclopropane carboxamides to give the β-magnesiated species, which readily undergoes a variety of substitution reactions. Another useful variation in this reactivity pattern was found by altering the stoichiometry of amide and alkyl present in the magnesium base reagent. Specifically, whereas BuMgDA reacts with the cyclopropanes to give the β-magnesiated species, mixing Bu$_2$Mg and diisopropylamine in a 1:0.5 molar ratio produces a system that gives predominantly the α-metallated product. It was assumed that this is due to the kinetic selectivity of this reagent mixture. In any event, the intermediate may be trapped to produce α-carboxy, α-iodo or α-alkyl products that are difficult to prepare by other means. Another demonstration of the utility of this reagent has been the β-metalation of amide-activated cyclobutanes[57]. This reaction is notable since equivalent lithium amide systems are unreactive with such weakly acidic substrates.

The magnesium bis(amide) Mg(TMP)$_2$ (TMP = 2,2,6,6-tetramethylpiperidide) has been shown to be a useful base in the selective deprotonation of arenes to produce arylmagnesium amide intermediates[54]. For example, reaction of Mg(TMP)$_2$ with methyl benzoate

SCHEME 5

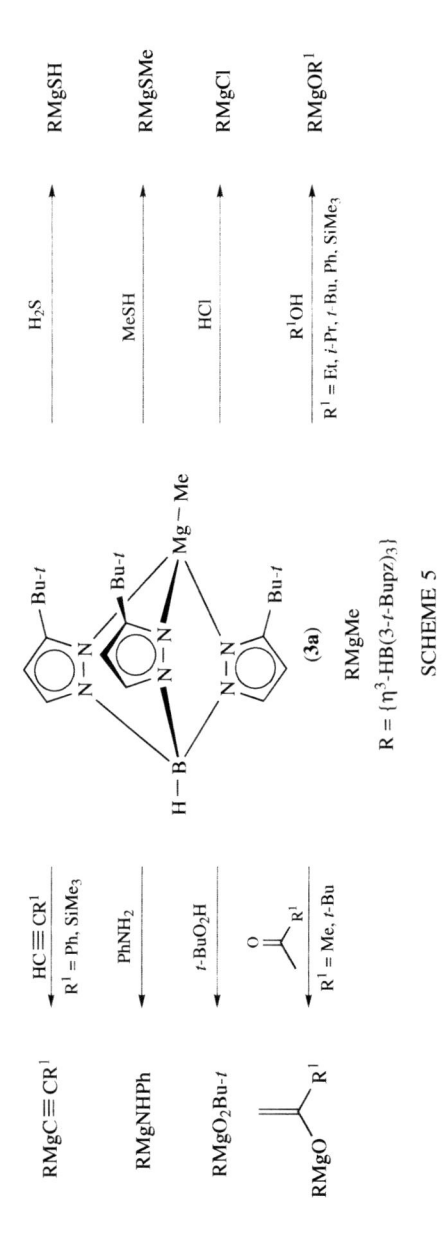

(3a) RMgMe
R = {η³-HB(3-t-Bupz)₃}

Reactions shown:
- H₂S → RMgSH
- MeSH → RMgSMe
- HCl → RMgCl
- R¹OH → RMgOR¹ (R¹ = Et, i-Pr, t-Bu, Ph, SiMe₃)
- HC≡CR¹ (R¹ = Ph, SiMe₃) → RMgC≡CR¹
- PhNH₂ → RMgNHPh
- t-BuO₂H → RMgO₂Bu-t
- R¹C(O)Me (R¹ = Me, t-Bu) → RMgOC(R¹)=CH₂

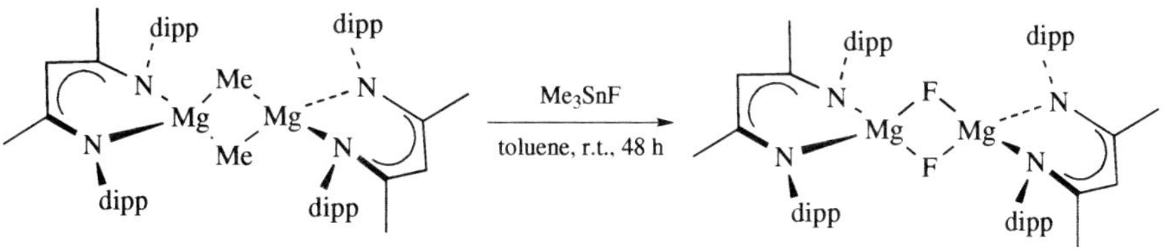

SCHEME 6

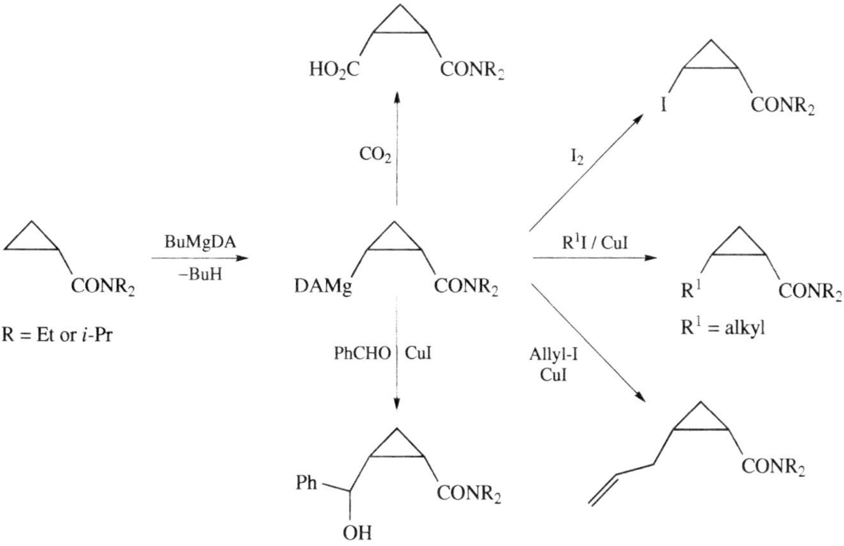

SCHEME 7

followed by quenching with carbon dioxide gives dimethyl *ortho*-phthalate in over 80% yield. In comparison, reaction with amido Grignard reagents, R_2NMgX, results in condensation with the ester group. Similarly, arylmagnesium amides are proposed to be key intermediates in the directed metalation of benzamides, cyclopropanes and carbocubanes.

Organomagnesium amide complexes have also been studied for use as reagents in the halogen–magnesium exchange reactions of halogenated arenes and indoles[58]. The reagent i-PrMgN(i-Pr)$_2$ proved to be useful for magnesiation of iodophenoxyalcohols. However, poor yields were obtained using iodophenols and iodoindoles. Also, related bromine–magnesium exchange reactions using i-PrMgN(i-Pr)$_2$ with bromophenoxyalcohols, phenols and indoles were unsuccessful, requiring application of the mixed-metal reagents of the type i-PrMgBu$_2$Li. Organomagnesium amides have been applied to the carbomagnesation of olefins, although the yields of each of the addition products are substantially lower than when using the dialkylmagnesium analogues[53]. Also, a small number of these complexes have been used as catalysts for the polymerization of *rac*-lactide and ε-caprolactone[18,34].

b. Disproportionation reactions. Many organomagnesium amide complexes are sensitive to the presence of coordinating solvents. Addition of polar solvents to arene or hydrocarbon solutions of alkyl(amido)magnesium species may result in disproportionation, yielding the bis(amido) and dialkylmagnesium complexes[25, 41, 47]. At least in some instances the driving force for the disproportionation is the increase in coordination number at the metal center. As shown in Scheme 8, chelation of two (2-pyridyl)amido units on the metal center allows coordination by addition donor solvent, increasing the coordination number at the metal from four to six[47]. Studies have shown that modest variations of the organic unit on the (2-pyridyl)amido substituent does not effect this reaction. Another important factor in such disproportionation reactions appears to be the relative strength of the donor solvent. The alkyl magnesium amides [Ph$_2$NMgR(THF)$_2$] (R=Et or *i*-Pr) are readily crystallized from THF solutions upon reaction of MgR$_2$ with HNPh$_2$[25]. However, addition of the strong donor solvent HMPA, (Me$_2$N)$_3$PO, results in exclusive isolation of the bis(amide) [(Ph$_2$N)$_2$Mg(HMPA)$_2$]. Therefore, the nature of the equilibrium between the hetero- and homoleptic magnesium species is similar to the Schlenk equilibrium in Grignard reagents and related complexes[59].

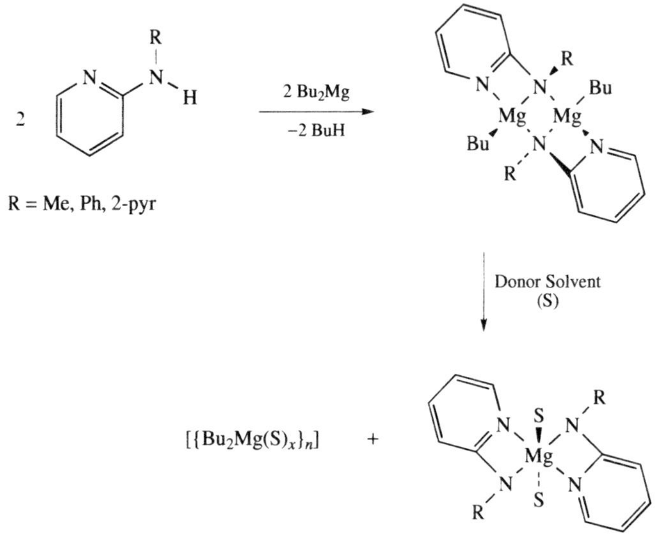

SCHEME 8

Many monomeric organomagnesium amide solvates may be transformed on heating under vacuum. In some instances this leads to simple desolvation of the complexes and in turn gives rise to dimerization or even further aggregation[15, 20, 28, 29, 44, 60]. Another outcome is disproportionation[15, 28, 29, 60]. For example, the disproportionation of an ether-solvated β-diketiminate complex upon sublimation is shown in Scheme 9[29].

c. Reactions with oxygen. Organomagnesium amides are air- and moisture-sensitive, and several studies have been carried out demonstrating their reactivity towards O$_2$[8, 9, 15, 17]. The most common outcome of this reaction is insertion of oxygen into the metal–carbon bond to form either alkoxide or alkylperoxide species. Reaction of the solvated magnesium β-diketiminate complex, [MeMg{η^2-(*i*-Pr$_2$)ATI}(THF)], where ATI = aminotroponiminate, with O$_2$ produced the methoxy-bridged dimer [MeOMg{η^2-(*i*-Pr)$_2$)ATI}(THF)]$_2$[15].

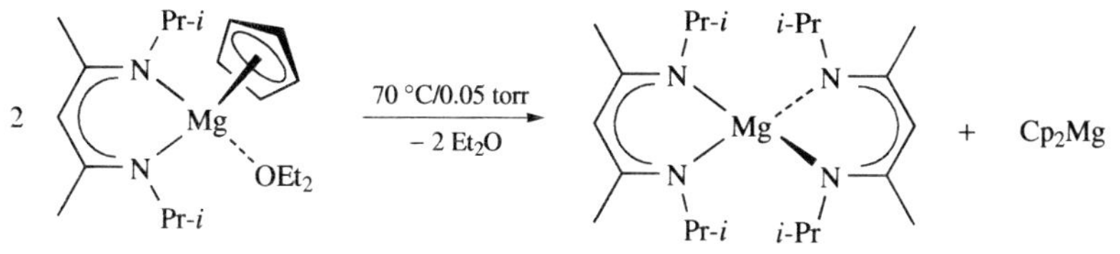

SCHEME 9

In situ ^{1}H NMR monitoring of this reaction in THF-d_8 showed the loss of the methyl signal and the concomitant appearance of the methoxy signal, confirming the insertion of O_2 into the magnesium–carbon bond. The solvated β-diketiminate complex, **13**, has also been observed to undergo O_2 insertion[17]. The ^{13}C{^{1}H} NMR spectrum obtained upon addition of dry O_2 gas to a benzene-d_6 solution of **13** revealed the presence of two species in a 2:1 ratio. Crystallization yielded both the benzyloxo and benzylperoxo products, which are shown in Scheme 10. An unusual reaction was observed on addition of O_2 to {η^3-HB(3-t-Bupz)$_3$}MgCH$_2$SiMe$_3$, **4**. In this case the siloxide {η^3-HB(3-t-Bupz)$_3$}MgOSiMe$_3$ was produced due to cleavage of the Si–C bond[8,9]. It was proposed that this reaction involves a radical process whereby the initially prepared organoperoxide rearranges to form the thermodynamically stable Si–O bond and formaldehyde.

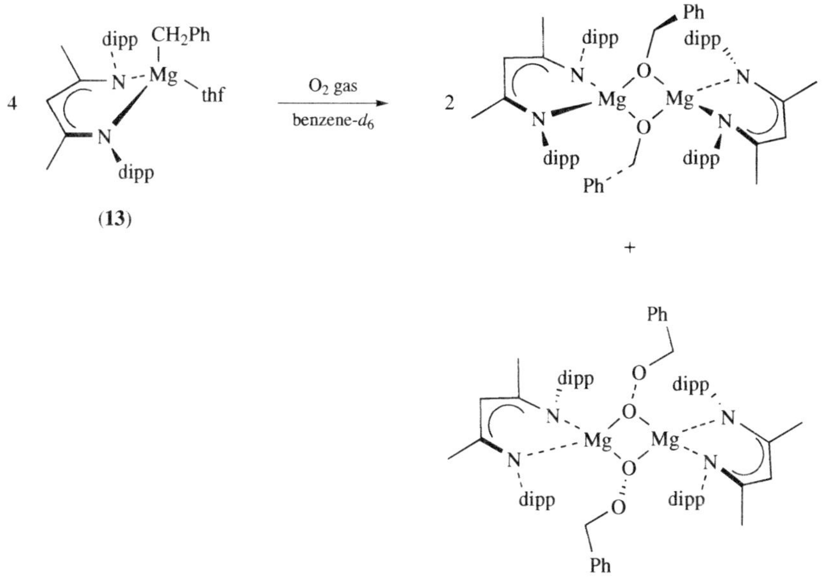

SCHEME 10

d. Reactions with aldehydes and ketones. The most common outcomes of the reaction of organomagnesium amides with aldehydes or ketones are reduction, enolization and addition, as shown generically in Scheme 11. The specific reaction occurring (or competitive reactions) is determined by the interplay of the sterics and electronic effects of the reagent and substrate carbonyl compound.

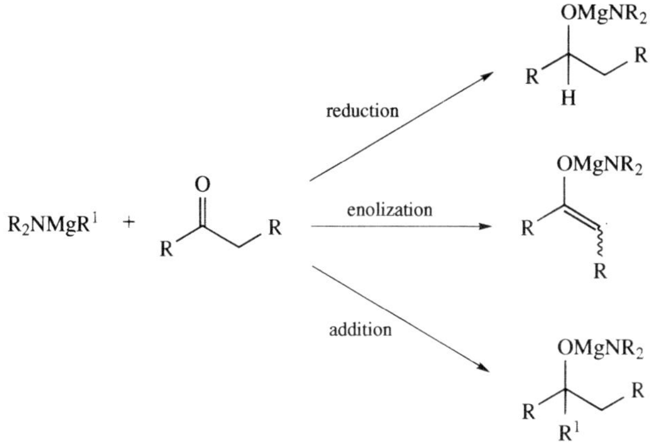

SCHEME 11

Generally, enolization reactions will occur when the organo group on the reagent is relatively large and the ketone contains an acidic α-proton. This is the preferred pathway for the reaction of β-diketiminate complex **46** with 2′,4′,6′-trimethylacetophenone (Scheme 12)[44]. Enolization is also favored in this case as the ketone is sterically protected toward attack by nucleophilic addition[61,62]. Another feature of this reaction is that the structure of the products is dependent upon the solvent media present. In THF, the amido(enolate) is a solvated monomer whereas in toluene, an unusual dimer is produced which utilizes both the carbon and the oxygen centers of the enolate group to bridge the metal centers. However, addition of THF to the toluene solution containing the dimer produces the same monomeric solvate generated directly in THF solution.

An interesting case is the reaction of $\{\eta^3\text{-HB}(3\text{-}t\text{-Bupz})_3\}$MgMe, **3a**, with acetone and t-butyl methyl ketone[10,16]. As outlined in Scheme 5, despite carrying a small methyl unit the reagent acts as a base rather than a nucleophile on reaction with unhindered ketones to produce enolates. In this instance it appears that the steric bulk of the η^3-tris(pyrazolyl)-hydroborate ligand dominates the reaction pathway. Enantioselective deprotonation reactions of conformationally-locked ketones have also been mediated by organomagnesium amides, which carry chiral amide groups (Scheme 13)[63]. These reagents show similar selectivities to their bis(amido) counterparts but have the advantage of requiring only half the amount of chiral starting material[64–67]. It is also worth noting that the heteroleptic reagents react chemoselectively with the ketones under study to produce only the enolate products. In comparison, reaction of the dialkylmagnesium starting material with the substituted cyclohexanones results in substantial quantities of both secondary and tertiary alcohol after workup due to participation of competitive reduction and alkylation reactions.

Addition reactions dominate when the organic group is relatively small and can act as a good nucleophile for attack of unhindered ketones or aldehydes. For example, reaction of the methylmagnesium amides, R_2NMgMe ($NR_2 = N(\text{Pr-}i)_2$, NPh_2 and $c\text{-}NC_5H_8Me_2$) with either 4-t-butylcyclohexanone or the more sterically encumbered 2,2,6,6-tetramethyl-4-t-butylcyclohexanone have been reported to display good stereoselectivity for alkylating ketones[68]. A combination of the steric bulk of the amide and the ketone, as well as the nature of the solvent media present was found to effect the selectivity obtained. Asymmetric alkylation reactions have also been completed using these reagents. Optically active aldehydes have been shown to react with alkylmagnesium amides to produce chiral

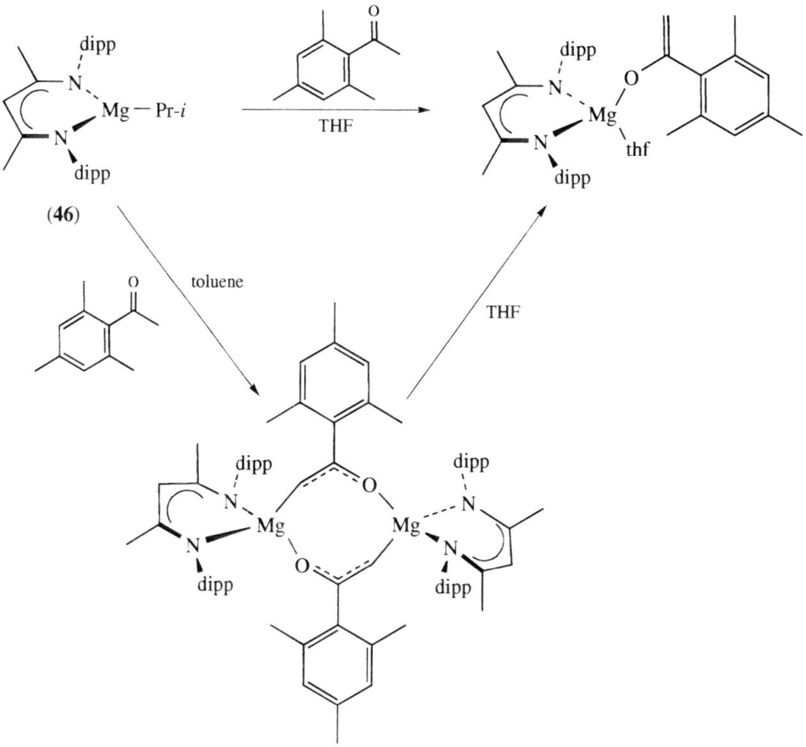

SCHEME 12

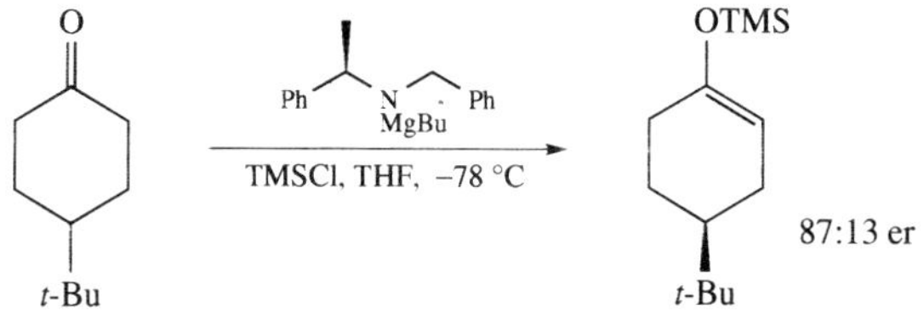

SCHEME 13

secondary alcohols with essentially complete Cram selectivity[69]. Furthermore, incorporation of a chiral amide unit into the reagent allows the possibility of heteromolecular asymmetric induction reactions with unsaturated groups. This approach has been demonstrated to be highly successful using potentially chelating chiral amides, including the structurally characterized complex **32**[31]. These very simply-prepared reagents display selectivities up to 91:9 er using a variety of alkyl and aryl nucleophiles, and also for a wide range of aldehydes (Scheme 14).

Reduction of ketones may occur if the alkyl group contains a β-hydrogen that is available for abstraction. For example, addition of benzophenone to the *in situ* prepared complex BuMgN(SiMe$_3$)$_2$ results in β-hydride transfer from the butyl group leading to the formation of the reduction product[70]. This type of reaction has also been conducted in an

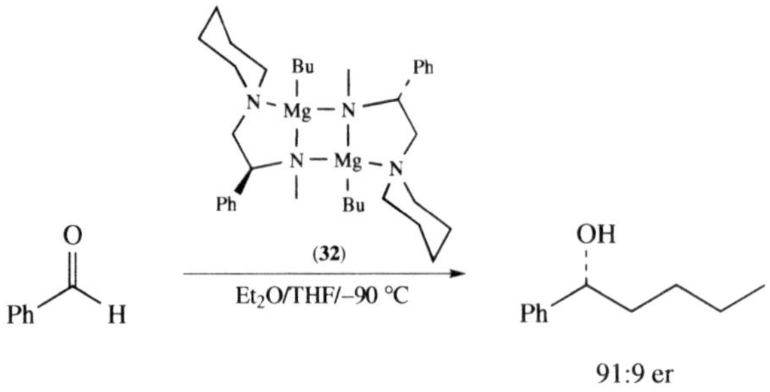

SCHEME 14

asymmetric manner through application of chiral amides as described previously for the alkylation reactions[31]. Specifically, reaction of *in situ* prepared chiral organomagnesium amides with a number of aldehydes yield secondary alcohols in excellent yields and selectivities (typically >95% and >85% respectively).

Another interesting example of an insertion reaction is found through the addition of benzophenone to complex **59** (Scheme 15)[48]. In this case hydrogen is abstracted from an amine group with addition of an alkyne unit across the carbonyl to produce a radical anion.

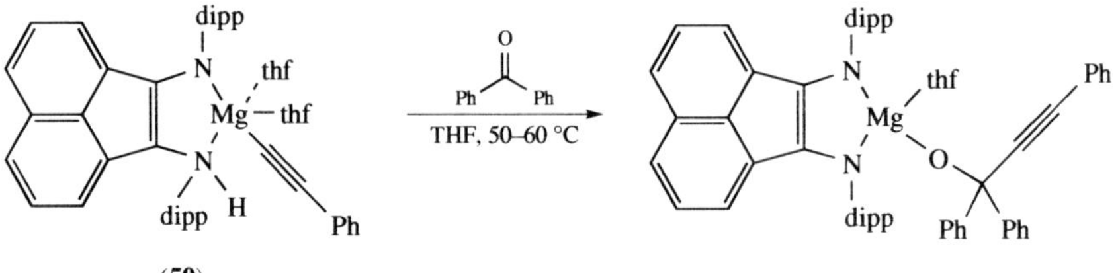

SCHEME 15

A useful application of organomagnesium amides is in the enantioselective conjugate addition to enamidomalonate to prepare β-amino acid derivatives (Scheme 16)[71]. The alkylmagnesium amide complexes provided both high yields and high selectivity in the organic transformation.

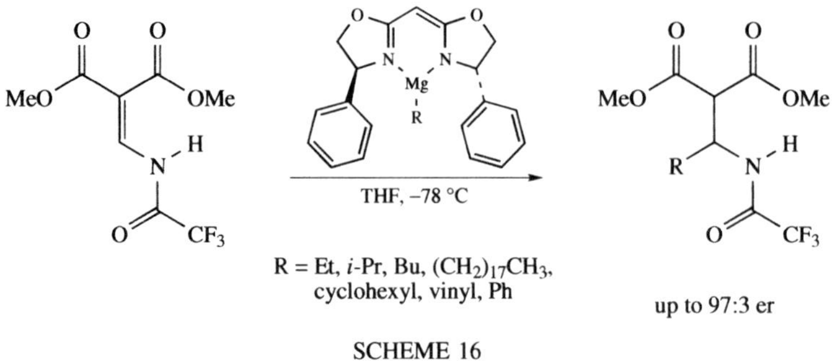

R = Et, *i*-Pr, Bu, (CH$_2$)$_{17}$CH$_3$, cyclohexyl, vinyl, Ph

SCHEME 16

B. Organomagnesium Heavy Pnictogenides

Organomagnesium complexes of heavy group 15 elements are much more rare than their amido analogues. In fact, only three examples of structurally authenticated complexes have been reported in the literature, all containing magnesium–phosphorus bonds[72,73]. The limited number of these compounds is at least in part a consequence of the lability of the bonds between magnesium and the heavy group 15 elements. However, this is an area that has received generally little attention and certainly merits further study.

The first example of a structurally characterized organomagnesium phosphanide only appeared in 1998 with the synthesis of complex **61**[72]. As shown in Scheme 17, this complex is produced upon the addition reaction between magnesium bis[(bis(trimethylsilyl) phosphanide] and 1,4-diphenylbutadiyne[72]. Compound **61** is dimeric in the solid state, forming a central Mg_2P_2 ring with magnesium–phosphorus bond lengths of 2.559(2)/ 2.569(2) Å. The second phosphorus atom of the ligand then forms a dative interaction to each magnesium center, with a Mg–P distance of 2.708(3) Å. Replacing magnesium for barium in this reaction results in a more reactive intermediate that immediately undergoes further reaction with butadiene present in solution to produce a phosphacyclopentadienide[72].

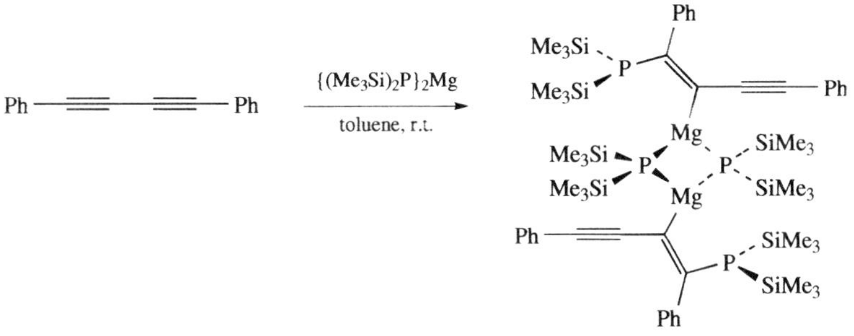

SCHEME 17

The organomagnesium phosphanide complexes **62** and **63** shown in Scheme 18 were directly prepared by metalation of the appropriate secondary phosphanes[73]. Reaction of two equivalents of the phosphanes with Bu_2Mg again only produced **62** and **63** rather than the expected bis(phosphanide) derivatives. It was speculated that this may be a consequence of steric hindrance caused by chelation in the heteroleptic complexes. Both complexes again form dimers with tetracoordinated metals and central Mg_2P_2 rings, with magnesium–phosphorus distances of 2.5760(8)/2.5978(8) Å for **62** and 2.5765(17)/2.5730(16) and 2.6138(16)/2.6105(17) Å for **63**. These distances are comparable to the magnesium–phosphorus bonds of the four-membered ring in **61**. They are also similar to the magnesium–phosphorus distances in bisphosphanides[74,75]. These compounds are found to rapidly decompose in THF-d_8 solution and are believed to undergo ligand degradation.

III. ORGANOMAGNESIUM-GROUP 16-BONDED COMPLEXES

Organomagnesium complexes of the group 16 elements have been even less studied than their group 15 analogues. A summary of the known and relevant chemistry of these species is given below.

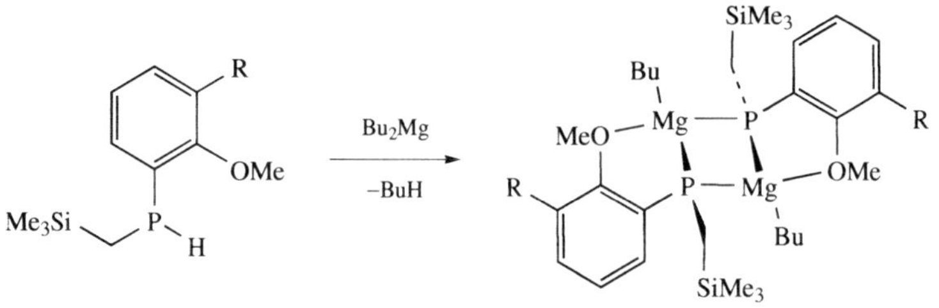

[Scheme 18 diagram]

R = H (**62**), Me (**63**)

SCHEME 18

A. Organomagnesium Alkoxides and Aryloxides

1. Synthesis

Organomagnesium alkoxides and aryloxides are typically synthesized by methods which are comparable to the synthesis of organomagnesium amides. The two most common routes again are the deprotonation of alcohols by R_2Mg bases or the reaction of a Grignard reagent with a metal alkoxide. Table 4 gives a summary of the structurally characterized organomagnesium alkoxides and aryloxides, and details of their methods of preparation. Complexes **64–67** were generated by the alkane elimination method[76–78]. Similarly, a series of methyl- and cyclopentadienylmagnesium alkoxides has also been prepared in this manner, although they have not been structurally characterized[83,84]. The Grignard transmetalation procedure was used in the synthesis of **68–70**[77,79,80], and also to prepare a series of phenyl- and butylmagnesium alkoxides[85]. Phenylmagnesium carboxylates have also been synthesized by reacting sodium salts of carboxylic acids with phenylmagnesium bromide[86]. Alternative methods, however, were used in the synthesis of compounds **71** and **72**[81,82]. Complex **71** was unexpectedly formed via cleavage of 2,1,1-cryptand upon addition of dineopentylmagnesium[81]. The mixed alkyl, amide, alkoxide complex **72** was first formed as a low yield product by reacting n-BuMgCl with NaN(H)Dipp in ether. It was then rationally prepared by combining stoichiometric quantities of Bu_2Mg, $Mg[N(H)Dipp]_2$ and n-BuOH in heptane[82].

2. Structural characterization

Monomeric, dimeric and tetrameric aggregation states of organomagnesium alkoxides and aryloxides have all been observed. Monomeric structures **66** and **67** consist of magnesium centers that are coordinated by the sterically encumbering donor ligands 18-crown-6-ether and TMEDA, preventing further aggregation[78]. Complex **68** is the sole tetrameric cubane structure for this class of compounds that has been characterized in the solid state thus far (Figure 4)[79]. This complex was prepared by transmetalation in a mixture of toluene and THF (5:1) followed by sublimation. Nevertheless, the solution chemistry of a variety of alkylmagnesium alkoxides has been studied in detail and found to form numerous oligomers[83,87]. As expected, the type of aggregate formed is determined by the extent of the branching of the alkyl and alkoxy groups and the strength of the donor solvent present. In contrast with organomagnesium amides, none of the structures consists of organo-bridged magnesium centers.

TABLE 4. Structurally characterized organomagnesium alkoxides and aryloxides, showing their methods of preparation

Mg precursor	Precursor 2	Solvent	Product number	Product	Reference
Bu$_2$Mg	2,6-di-t-Bu-phenol	Hexane/Toluene	**64**	[dimeric Mg aryloxide structure]	76
Bu$_2$Mg	Mes$_2$B–OH	THF	**65**	[Mg/B/O cluster with thf and Bu ligands]	77
i-Bu$_2$Mg	2,6-di-t-Bu-phenol	Et$_2$O (crystallization from benzene after addition of 18-crown-6-ether)	**66**	[i-Bu-Mg aryloxide with 18-crown-6]	78
Et$_2$Mg	2,6-di-t-Bu-phenol	Et$_2$O (crystallization from benzene after addition of TMEDA)	**67**	[Et-Mg aryloxide with TMEDA]	78

(*continued overleaf*)

TABLE 4. (*continued*)

Mg precursor	Precursor 2	Solvent	Product number	Product	Reference
MeMgBr	*t*-BuOK	Toluene	**68**	Mg/O/Bu-*t* cubane cluster	79
MeMgCl	Mes₂B–OLi	Et₂O	**69**	Me/Mg/O/B(Mes)₂ ring complex	77
t-BuMgCl	2,2,6,6-tetramethylpiperidide Na	Hexane/THF/O₂	**70**	*t*-Bu/Mg/O/thf complex	80

| (Me₃CH₂)₂Mg | 71 | 2,1,1-cryptand | Benzene/Et₂O | 81 |
| Bu₂Mg | 72 | Heptane (BuOH) | | 82 |

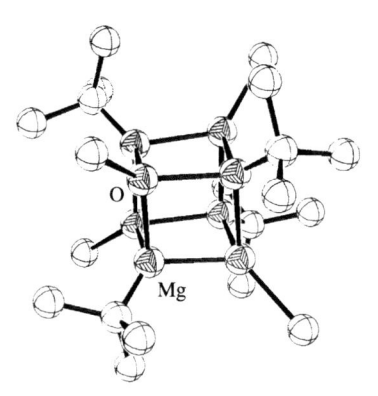

FIGURE 4. Tetrameric cubane structure of **68** with hydrogens omitted

3. Reactivity studies

Organomagnesium alkoxides and aryloxides have been utilized in only a few applications. Methylmagnesium t-butoxide **68** has been used in the chemical vapor deposition of MgO films onto silicon substrates[79]. MgO films with good crystallinity were grown at 800 °C on Si(111) surfaces, whereas polycrystalline films were formed at 400 °C. Intermediate temperatures produced multiple crystallite orientations. Similar results were obtained for deposition onto Si(100) surfaces over this range of temperatures.

Tri- and tetra-substituted alkenenitriles can be generated by the addition of Grignard reagents to γ-hydroxyalkynenitriles[88]. It was proposed that deprotonation of the hydroxyl group by t-butylmagnesium chloride followed by addition of a second Grignard reagent, R^2MgX, results in the formation of a chelated organomagnesium alkoxide intermediate (Scheme 19). Subsequent addition of t-butyllithium to this intermediate followed by alkylation with an electrophile yields tetra-substituted nitriles. Alternatively, the cyclic magnesium chelate can be protonated to yield tri-substituted nitriles.

SCHEME 19

B. Organomagnesium Heavy Chalcogenides

Analogous to heavy group 15 organomagnesium complexes, there are very few organomagnesium heavy group 16 complexes which have been synthesized or structurally characterized. Indeed, only three examples of structurally characterized organomagnesium sulfides have appeared, and no heavier chalcogenides are known. A series of cyclopentadienyl-based thiol complexes has been prepared by treating Cp_2Mg with three different alkanethiols (Scheme 20)[89].

Each of the three complexes was obtained as a crystalline solid, but only the *t*-butyl derivative **73** has been structurally characterized. X-ray crystallography reveals a tetrameric cubane structure composed of four magnesium centers each coordinated to three sulfur atoms. Each magnesium center is additionally bonded to the π-face of a cyclopentadienyl ring. When THF or 4-*t*-butylpyridine is added to a solution of dichloromethane, two new complexes **74** and **75** are formed. Both compounds were found to be dimeric with central Mg_2S_2 rings, as expected upon solvation of the cubane complex. The two dimeric aggregates have slightly different geometries. The cyclopentadienyl rings of the thf-coordinated dimer **74** are oriented in *cis* fashion whereas they are *trans* in the 4-*t*-butylpyridine-coordinated dimer **75**.

Limited reactivity studies of organomagnesium sulfides have also been conducted. These complexes have recently been employed as modified Grignard reagents in the cross-coupling of benzonitriles[90]. The advantage of these complexes over Grignard reagents is that nucleophilic addition across the nitrile is inhibited. The alkylmagnesium sulfide complexes shown in Scheme 21 were prepared *in situ* by transmetalation, then reacted with the appropriate benzonitrile species. The aryl alkanes were produced in good yields upon heating the THF solutions at reflux overnight.

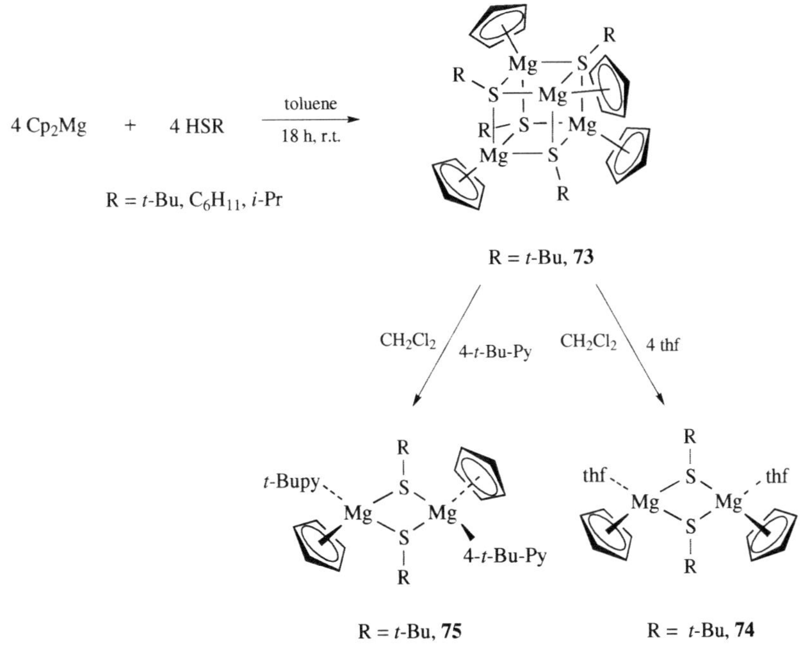

SCHEME 20

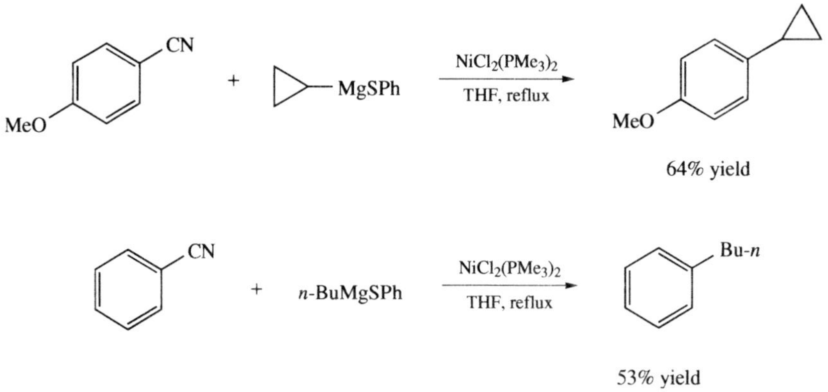

SCHEME 21

IV. REFERENCES

1. R. E. Mulvey, *Organometallics*, **25**, 1060 (2006).
2. R. E. Mulvey, F. Mongin, M. Uchiyama and Y. Kondo, *Angew. Chem., Int. Ed.*, **46**, 3802 (2007).
3. L. Menunier, *Compt. Rend.*, **136C**, 758 (1903).
4. V. R. Magnuson and G. D. Stucky, *Inorg. Chem.*, **8**, 1427 (1969).
5. L. M. Engelhardt, B. S. Jolly, P. C. Junk, C. L. Raston, B. W. Skelton and A. H. White, *Aust. J. Chem.*, **39**, 1337 (1986).
6. R. Han, A. Looney and G. Parkin, *J. Am. Chem. Soc.*, **111**, 7276 (1989).
7. R. Han and G. Parkin, *J. Am. Chem. Soc.*, **112**, 3662 (1990).
8. R. Han and G. Parkin, *Organometallics*, **10**, 1010 (1991).
9. R. Han and G. Parkin, *Polyhedron*, **9**, 2655 (1990).
10. J. L. Kisko, T. Fillebeen, T. Hascall and G. Parkin, *J. Organomet. Chem.*, **596**, 22 (2000).
11. M. H. Chisholm, N. W. Eilerts, J. C. Huffman, S. S. Iyer, M. Pacold and K. Phomphrai, *J. Am. Chem. Soc.*, **122**, 11845 (2000).
12. N. Kuhn, M. Schulten, R. Boese and D. Bläser, *J. Organomet. Chem.*, **421**, 1 (1991).
13. M. M. Olmstead, W. J. Grigsby, D. R. Chacon, T. Hascall and P. P. Power, *Inorg. Chim. Acta*, **251**, 273 (1996).
14. V. C. Gibson, J. A. Segal, A. J. P. White and D. J. Williams, *J. Am. Chem. Soc.*, **122**, 7120 (2000).
15. P. J. Bailey, C. M. Dick, S. Fabre and S. Parsons, *J. Chem. Soc., Dalton Trans.*, 1655 (2000).
16. H. Hao, H. W. Roesky, Y. Ding, C. Cui, M Schormann, H.-G. Schmidt, M. Noltemeyer and B. Žemva, *J. Fluorine Chem.*, **115**, 143 (2002).
17. P. J. Bailey, R. A. Coxall, C. M. Dick, S. Fabre, L. C. Henderson, C. Herber, S. T. Liddle, D. Loroño-González, A. Parkin and S. Parsons, *Chem. Eur. J.*, **9**, 4820 (2003).
18. L. F. Sánchez-Barba, D. L. Hughes, S. M. Humphrey and M. Bochmann, *Organometallics*, **25**, 1012 (2006).
19. P. J. Bailey, R. A. Coxall, C. M. Dick, S. Fabre and S. Parsons, *Organometallics*, **20**, 798 (2001).
20. P. J. Bailey, S. T. Liddle, C. A. Morrison and S. Parsons, *Angew. Chem., Int. Ed.*, **40**, 4463 (2001).
21. P. J. Bailey, R. A. Coxall, C. M. Dick, S. Fabre, S. Parsons and L. J. Yellowlees, *Chem. Commun.*, 4563 (2005).
22. P. J. Bailey, C. M. Dick, S. Fabre, S. Parsons and L. J. Yellowlees, *Dalton Trans.*, 1602 (2006).
23. W. Vargas, U. Englich and K. Ruhlandt-Senge, *Inorg. Chem.*, **41**, 5602 (2002).
24. H. Sachdev and C. Preis, *Eur. J. Inorg. Chem.*, 1495 (2002).

25. K.-C. Yang, C.-C. Chang, J.-Y. Huang, C.-C. Lin, G.-H. Lee, Y. Wang and M. Y. Chiang, *J. Organomet. Chem.*, **648**, 176 (2002).
26. N. Etkin, C. M. Ong and D. W. Stephan, *Organometallics*, **17**, 3656 (1998).
27. A. Xia, M. J. Heeg and C. H. Winter, *Organometallics*, **21**, 4718 (2002).
28. A. Xia, H. M. El-Kaderi, M. J. Heeg and C. H. Winter, *J. Organomet. Chem.*, **682**, 224 (2003).
29. H. M. El-Kaderi, A. Xia, M. J. Heeg and C. H. Winter, *Organometallics*, **23**, 3488 (2004).
30. M. Westerhausen, T. Bollwein, N. Makropoulos, S. Schneiderbauer, M. Suter, H. Nöth, P. Mayer, H. Piotrowski, K. Polborn and A. Pfitzner, *Eur. J. Inorg. Chem.*, 389 (2002).
31. K. H. Yong and J. M. Chong, *Org. Lett.*, **4**, 4139 (2002).
32. I. L. Fedushkin, A. A. Skatova, M. Hummert and H. Schumann, *Eur. J. Inorg. Chem.*, 1601 (2005).
33. I. J. Blackmore, V. C. Gibson, P. B. Hitchcock, C. W. Rees, D. J. Williams and A. J. P. White, *J. Am. Chem. Soc.*, **127**, 6012 (2005).
34. T. Chivers, C. Fedorchuk and M. Parvez, *Organometallics*, **24**, 580 (2005).
35. B. Conway, E. Hevia, A. R. Kennedy, R. E. Mulvey and S. Weatherstone, *Dalton Trans.*, 1532 (2005).
36. J. Lewiński, M. Dranka, I. Kraszewska, W. Sliwiński and I. Justyniak, *Chem. Commun.*, 4935 (2005).
37. M. Westerhausen, T. Bollwein, N. Makropoulos and H. Piotrowski, *Inorg. Chem.*, **44**, 6439 (2005).
38. E. P. Squiller, A. D. Pajerski, R. R. Whittle, and H. G. Richey, Jr., *Organometallics*, **25**, 2465 (2006).
39. G. E. Coates and D. Ridley, *J. Chem. Soc. A*, 56 (1967).
40. E. C. Ashby, L. C. Chao and H. M. Neumann, *J. Am. Chem. Soc.*, **95**, 5186 (1973).
41. K. W. Henderson, R. E. Mulvey, W. Clegg and P. A. O'Neil, *J. Organomet. Chem.*, **439**, 237 (1992).
42. D. R. Armstrong, K. W. Henderson, M. MacGregor, R. E. Mulvey, M. J. Ross, W. Clegg and P. A. O'Neil, *J. Organomet. Chem.*, **486**, 79 (1995).
43. J. Prust, K. Most, I. Müller, E. Alexopoulos, A. Stasch, I. Usón and H. W. Roesky, *Z. Anorg. Allg. Chem.*, **627**, 2032 (2001).
44. A. P. Dove, V. C. Gibson, P. Hormnirum, E. L. Marshall, J. A. Segal, A. J. P. White and D. J. Williams, *Dalton Trans.*, 3088 (2003).
45. P. J. Bailey, D. Loroño-González and S. Parsons, *Chem. Commun.*, 1426 (2003).
46. E. Hevia, A. R. Kennedy, R. E. Mulvey and S. Weatherstone, *Angew. Chem., Int. Ed.*, **43**, 1709 (2004).
47. K. W. Henderson, R. E. Mulvey and A. E. Dorigo, *J. Organomet. Chem.*, **518**, 139 (1996).
48. I. L. Fedushkin, N. M. Khvoinova, A. A. Skatova, and G. K. Fukin, *Angew. Chem., Int. Ed.*, **42**, 5223 (2003).
49. C. F. Caro, P. B. Hitchcock, and M. F. Lappert, *Chem. Commun.*, 1433 (1999).
50. P. C. Andrikopoulos, D. R. Armstrong, H. R. L. Barley, W. Clegg, S. H. Dale, E. Hevia, G. W. Honeyman, A. R. Kennedy and R. E. Mulvey, *J. Am. Chem. Soc.*, **127**, 6184 (2005).
51. M. Uchiyama, Y. Matsumoto, D. Nobuto, T. Furuyama, K. Yamaguchi and K. Morokuma, *J. Am. Chem. Soc.*, **128**, 8748 (2006).
52. R. Han and G. Parkin, *J. Am. Chem. Soc.*, **114**, 748 (1992).
53. U. M. Dzhemilev and O. S. Vostrikova, *J. Organomet. Chem.*, **285**, 43 (1985).
54. P. E. Eaton, C.-H. Lee and Y. Xiong, *J. Am. Chem. Soc.*, **111**, 8016 (1989).
55. H. Böhland, F. R. Hofmann, W. Hanay and H. J. Berner, *Z. Anorg. Allg. Chem.*, **577**, 53 (1989).
56. M.-X. Zhang and P. E. Eaton, *Angew. Chem., Int. Ed.*, **41**, 2169 (2002).
57. P. E. Eaton, M. X. Zhang, N. Komiya, C. G. Yang, I. Steele and R. Gilardi, *Synlett*, 1275, (2003).
58. J. Xu, N. Jain and Z. Sui, *Tetrahedron Lett.*, **45**, 6399 (2004).
59. J. F. Allan, W. Clegg, K. W. Henderson, L. Horsburgh and A. R. Kennedy, *J. Organomet. Chem.*, **559**, 173 (1998).
60. R. Han and G. Parkin, *J. Organomet. Chem.*, **393**, C43 (1990).
61. J. F. Allan, K. W. Henderson, A. R. Kennedy and S. J. Teat, *Chem. Commun.*, 1059 (2000).
62. Z. S. Sales, R. Nassar, J. J. Morris and K. W. Henderson, *J. Organomet. Chem.*, **690**, 3474 (2005).
63. E. L. Carswell, D. Hayes, K. W. Henderson, W. J. Kerr and C. J. Russell, *Synlett*, 1017 (2003).

64. M. J. Bassindale, J. J. Crawford, K. W. Henderson and W. J. Kerr, *Tetrahedron Lett.*, **45**, 4175 (2004).
65. K. W. Henderson, W. J. Kerr and J. H. Moir, *Tetrahedron*, **58**, 4573 (2002).
66. J. D. Anderson, P. García García, D. Hayes, K. W. Henderson, W. J. Kerr, J. H. Moir and K. Pai Fondekar, *Tetrahedron Lett.*, **42**, 7111 (2001).
67. K. W. Henderson, W. J. Kerr and J. H. Moir, *Chem. Commun.*, 479 (2000).
68. E. C. Ashby and G. F. Willard, *J. Org. Chem.*, **43**, 4094 (1978).
69. M. F. Reetz, N. Harmat and R. Mahrwald *Angew. Chem., Int. Ed. Engl.*, **31**, 342 (1992).
70. K. W. Henderson, J. R. Allan and A. R. Kennedy, *J. Chem. Soc., Chem. Commun.*, 1149 (1997).
71. M. P. Sibi and Y. Asano, *J. Am. Chem. Soc.*, **123**, 9708 (2001).
72. M. Westerhausen, M. H. Digeser, H. Nöth, T. Seifert and A. Pfitzner, *J. Am. Chem. Soc.*, **120**, 6722 (1998).
73. S. Blair, K. Izod, W. Clegg and R. W. Harrington, *Eur. J. Inorg. Chem.*, 3319 (2003).
74. E. Hey, L. M. Engelhardt, C. L. Raston and A. H. White, *Angew. Chem., Int. Ed. Engl.*, **26**, 81 (1987).
75. M. Westerhausen and A. Pfitzner, *J. Organomet. Chem.*, **479**, 141 (1994).
76. K. W. Henderson, G. W. Honeymoon, A. R. Kennedy, R. E. Mulvey, J. A. Parkinson and D. C. Sherrington, *Dalton Trans.*, 1365 (2003).
77. S. C. Cole, M. P. Coles and P. B. Hitchcock, *Organometallics*, **23**, 5159 (2004).
78. A. D. Pajerski, E. P. Squiller, M. Parvez, R. R. Whittle and H. G. Richey, Jr., *Organometallics*, **24**, 809 (2005).
79. M. M. Sung, C. G. Kim, J. Kim and Y. Kim, *Chem. Mater.*, **14**, 826 (2002).
80. B. Conway, E. Hevia, A. R. Kennedy, R. E. Mulvey and S. Weatherstone, *Dalton Trans.*, 1532 (2005).
81. E. P. Squiller, R. R. Whittle and H. G. Richey, Jr. *Organometallics*, **4**, 1154 (1985).
82. E. Hevia, A. R. Kennedy, R. E. Mulvey and S. Weatherstone, *Angew. Chem., Int. Ed.*, **43**, 1709 (2004).
83. E. C. Ashby, J. Nackashi and G. E. Paris, *J. Am. Chem. Soc.*, **97**, 3162 (1975).
84. O. N. D. Mackey and C. P. Morley, *J. Organomet. Chem.*, **426**, 279 (1992).
85. S. Gupta, S. Sharma and A. K. Narula, *J. Organomet. Chem.*, **452**, 1 (1993).
86. P. N. Kapoor, A. K. Bhagi, H. K. Sharma and R. N. Kapoor, *J. Organomet. Chem.*, **369**, 281 (1989).
87. G. E. Coates, J. A. Heslop, M. E. Redwood and D. Ridley, *J. Chem. Soc. A*, 1118 (1968).
88. F. F. Fleming, V. Gudipati and O. W. Steward, *Tetrahedron*, **59**, 5585 (2003).
89. A. Xia, M. J. Heeg and C. H. Winter, *J. Organomet. Chem.*, **669**, 37 (2003).
90. J. A. Miller and J. W. Dankwardt, *Tetrahedron Lett.*, **44**, 1907 (2003).